Oceans, Rivers and Lakes: Energy and Substance Transfers at Interfaces

Developments in Hydrobiology 131

Series editor
H. J. Dumont

Oceans, Rivers and Lakes: Energy and Substance Transfers at Interfaces

Proceedings of the Third International Joint Conference on Limnology and Oceanography held in Nantes, France, October 1996

Edited by

J.-C. Amiard, B. Le Rouzic, B. Berthet & G. Bertru

Reprinted from Hydrobiologia, volumes 373/374 (1998)

Springer-Science+Business Media, B.V.

Library of Congress Cataloging-in-Publication Data

A C.I.P. Catalogue record for this book is available from the Library of Congress.

ISBN 978-94-010-6216-9 ISBN 978-94-011-5266-2 (eBook)
DOI 10.1007/978-94-011-5266-2

Printed on acid-free paper

Originally published by Kluwer Academic Publishers in 1998

Hydrobiologia **373/374**, v–vii, 1998.
J.-C. Amiard, B. Le Rouzic, B. Berthet & G. Bertru (eds),
Oceans, Rivers and Lakes: Energy and Substance Transfers at Interfaces.

Contents

Session 5: Fluxes at Interfaces

Hydrobiologia **373/374:** ix, 1998.
J.-C. Amiard, B. Le Rouzic, B. Berthet & G. Bertru (eds),
Oceans, Rivers and Lakes: Energy and Substance Transfers at Interfaces.

Preface

Since the First International Joint Conference on Limnology and Oceanography in Marseille (June 1989) was highly successful, it was decided to continue in the same vein. Thus, a Second Congress was held in Evian, France in May 1993. The Third International Congress on Limnology and Oceanography was held in Nantes, France, October 7–9, 1996, hosted by the Cité des Congrès.

The Congress was organized round the main theme 'Interfaces – Oceans, Rivers and Lakes: Energy and Substance Transfers at Interfaces', split into five sessions:

- Methodologies and modelling.
- Interface between medium and living beings.
- Interface between nutrients and physical medium.
- Interfaces between pollutants, meduium and living beings.
- Fluxes at the interfaces.

One hundred and seven oral and poster presentations were proposed to 250 attendees originating from 18 countries. Probably as a consequence of the proposed topic, presentations in the field of Oceanography mainly concerned estuarine and coastal areas, with a special interest in salt marshes and empoldered zones. In the part devoted to continental waters, the attention was focussed on hydrodynamics, biogeochemistry and ecology of wetlands and alluvial zones in hydrosystems of the main French rivers (Loire, Rhone, Seine, etc.). For both salt and fresh waters, numerous presentations took into account sedimentary particles, either deposited or suspended. The sediment–water interface and fluxes between these compartments is a common concern for both limnology and oceanography: sorption–desorption of heavy metals, degradation and allocation of organic matter, fluxes of nutrients, particularly phosphorus. In most of the presentations, the biological compartment was approached indirectly and often at the species level only, although environmental stress and disturbances concern the biological communities as a whole.

The congress organizers and editors hope that discussions about these different topics may be a source of future cooperation between European marine and freshwater scientists.

Manuscripts selected by the Scientific Committee and finally proposed for publication were reviewed by the editors and several anonymous reviewers, to whom the editors offer their grateful thanks.

The organizing committee would like to thank all public agencies, offices, and private companies that contributed to the conference and supported the publication of this volume.

On behalf of all participants, I thank Henri J. Dumont, Editor-in-chief of *Hydrobiologia* and René Mijs, Kluwer Academic Publishers, for their interest in the production of this special volume.

JEAN-CLAUDE AMIARD
CNRS/Université de Nantes
1, rue Gaston Veil
F-44035 Nantes Cedex
France

Hydrobiologia **373/374:** xi–xiv, 1998.
J.-C. Amiard, B. Le Rouzic, B. Berthet & G. Bertru (eds),
Oceans, Rivers and Lakes: Energy and Substances Transfers at Interfaces.

List of participants

ALGERIA
ARAB Lamia, *Alger*
MENASRIA Réda, *Alger*

BELGIUM
BILLEN Gilles, *Bruxelles*
BRION Natacha, *Bruxelles*
SERVAIS Pierre, *Bruxelles*

BRAZIL
OLIVEIRA A. H., *Belo Horizonte*

CANADA
BOURGET Edwin, *Quebec*
COSSA Daniel, *Montreal*
GAGNE Jean-Pierre, *Rimouski*
HUDIER Eric, *Rimouski*
PELLERIN-MASSICOTTE Jocelyne, *Rimouski*
PHAM Thanh-Thao, *Montreal*
PINEL-ALLOUL Bernadette, *Montreal*
PLANAS Dolors, *Montreal*
VINCENT Bruno, *Rimouski*

IVORY COAST
CECCHI Philippe, *Bonabe 01*

FRANCE
ADAM Christelle, *St Paul-lez-Durance*
ALEYA Lofti, *Besancon*
AMBLARD Christian, *Aubiere*
AMIARD Jean-Claude, *Nantes*
AMIARD-TRIQUET Claude, *Nantes*
ARAUJO Moacyr, *Toulouse*
ARRIGNON Jacques, *Compiegne*
AWERBUCH Nathan, *Paris*
BARTHELEMY Dominique, *Toulouse*
BAUDET Joseph, *Nantes*
BAUDIN Jean-Pierre, *St Paul-lez-Durance*
BELAMIE René, *Lyon*
BERARD Annette, *Thonon*
BERTHO Marie-Laure, *Wimereux*
BERTRU Georges, *Rennes*
BESSAT Frédéric, *Perpignan*
BLANC Paul, *Thonon*
BODINEAU Laurent, *Villeneuve d'Ascq*
BOGE Gérard, *La Garde*
BOISSON Jean-Claude, *Vaulx en Velin*
BONIN Daniel, *Marseille*
BOUCHARD Virginie, *Rennes*
BOURDIER Gilles, *Aubiere*
BOURNET Emmanuel, *Noisy-le-Grand*
BOURRAIN Xavier, *Orleans*
BRIENT Luc, *Rennes*
BRIGAULT Sandrine, *Bouguenais*
BROSSE Sébastien, *Rennes*
CANN Charles, *Rennes*
CAPBLANCQ, *Toulouse*
CARMOUZE Jean-Pierre, *Montpellier*
CARRIAS Jean-François, *Aubiere*
CAZAUBON Arlette, *Marseille*
CHARPIN Marie, *Aubiere*
CHARRIER Christophe, *Vaulx-en-Velin*
CHAUVELON Philippe, *Arles*
CHAUVET Eric, *Toulouse*
CLARET Cécile, *Le Bourget-du-Lac*
CLAVIER Jacques, *Plouzane*
CLEMENT Bernard, *Rennes*
COIFFARD Samuel, *Rennes*
COSTIL Katherine, *Rennes*
CREACH Véronique, *Rennes*
D'ELBEE Jean, *Biarritz*
DALIGAULT Patrick, *St Laurent-de-T.*
DANIEL Hervé, *Rennes*
DARTUS Denis, *Toulouse*
DAUTA Alain, *Toulouse*
DAUVIN Jean-Claude, *Paris*
DAVOULT Dominique, *Wimereux*
DEBOUDT Karine, *Wimereux*
DEROUX J. M., *Ales*
DESCHAUX Pierre, *Limoges*
DESLOUS-PAOLI Jean-Marc, *Sete*
DEVAUX Jean, *Aubiere*
DION Patrick, *Pleubian*
DOMAIZON Isabelle, *Aubiere*

DORIOZ J. Marcel, *Thonon*
DUCLOS Yves, *Paris*
EVEN Stéphanie, *Fontainebleau*
FAUVET Guillaume, *Le Bourget du Lac*
FAYOLLE Stéphanie, *Marseille*
FEUILLADE Mauricette, *Thonon les Bains*
FEUILLADE Jacques, *Thonon les Bains*
FLAMENT Pascal, *Wimereux*
FONTVIEILLE Dominique, *Le Bourget-du-Lac*
FOSSATI Odile, *Montpellier*
FOULQUIER Luc, *St Paul lez Durance*
FRONTIER Serge, *Villeurbanne d'Asq*
FUSTEC Eliane, *Paris*
GAGNEUR James, *Toulouse*
GARNIER Josette, *Paris*
GAUDY Raymond, *Marseille*
GAYTE Xavier, *Le Bourget-du-Lac*
GERET Florence, *Nantes*
GILBERT Franck, *Marseille*
GILBERT Daniel, *Aubiere*
GOLASZEWSKI Geneviève, *Paris*
GOLTERMAN Han, *Arles*
GOMEZ-HERNANDEZ Elena, *Montpellier*
GONZALEZ Jean-Louis, *La Seyne/Mer*
GOULEAU Dominique, *l'Houmeau*
GRARE Sonia, *Rennes*
GREGOIRE Alain, *Le Bourget*
GREINER Isabelle, *Paris*
GREMARE Antoine, *Banyuls/Mer*
GRIMAUD Daniel, *Tours*
GUERLESQUIN Micheline, *Angers*
HAURY Jacques, *Rennes*
HORN Michel, *Saint-Germain La Blanche Herbe*
IDLAFKIH Zabid, *Paris*
IRACE Sandrine, *Paris*
JAIRY Aïcha, *Paris*
JAMET Jean-Louis, *La Garde*
JIGOREL Alain, *Rennes*
JOURNEL Blandine, *Villefranche-sur-Mer*
KHALIFA Abder, *Saint-Nazaire*
LAFFAILLE Pascal, *Rennes*
LAGADIC Laurent, *Rennes*
LAIR Nicole, *Aubiere*
LASCOMBE Claude, *Pierre-Bénite*
LE BEC Claude, *Sete*
LE GALL Jean-Yves, *Rennes*
LE NORMAND Catherine, *Chatou*
LE ROUZIC Bertrand, *Rennes*
LECAT Camille, *St Leu la Forêt*
LECLERCQ Stéphanie, *Villeneuve d'Asq*
LEGLISE Lionel, *Metz*
LEITAO Maria, *Angers*
LEROY Jean, *Dijon*
LOQUET Nicolas, *St Valery:Somme*
LUCAS Yves, *La Garde*
LUCAS Françoise, *Rennes*
MARMONIER Pierre, *Le Bourget-du-Lac*
MARTIN Jean-François, *Orleans*
MARTIN Eric, *Villeneuve d'Asq*
MARTIN Christian, *Besancon*
MARTIN Delphine, *Douai*
MASSERET Estelle, *Aubiere*
MASSON Nathalie, *Metz*
MAUREL Philippe, *Toulouse*
MAUREL Alain, *St Paul-lez-Durance*
MENNERAT Marc, *Jouac*
MERLE Christophe, *Villeurbanne*
MEYBECK Michel, *Paris*
MEZIANE Tarik, *Dinard*
MICHALET Anne, *Villeneuve d'Asq*
MIGON Christophe, *Villefranche-sur-Mer*
MOREAU Christophe, *Nantes*
MOREAU Samuel, *Rennes*
MOTAMED Bahram, *Caen*
MOUNY Pascal, *Paris*
MOUTIN Thierry, *Marseille*
OUBELKASSE Mohamed, *Tours*
PALES-ESPINOSA Emmanuelle, *Nantes*
PATTEE Eric, *Villeurbanne*
PAUC Henri, *Perpignan*
PAVILLON Jean-François, *Paris*
PELTE Thomas, *Thonon*
PETIT Daniel, *Villeneuve d'Asq*
PETIT Muriel, *Toulouse*
PICOT Bernadette, *Montpellier*
PIERRE Jean-François, *Vandoeuvre*
PINTE G., *Gif-sur-Yvette*
PITOIS Frédéric, *Rennes*
PLANTE-CUNY Marie-Reine, *Marseille*
POGGI Robert, *Nantes*
POULIN Michel, *Fontainebleau*
PRYGIEL Jean, *Douai*
RABETTE Chantal, *Aubiere*
RAINGLET Florence, *Nantes*
RAVAIL-LEGRAND Brigitte, *La Rochelle*
RESTITUITO Françoise, *Orcet*
REVEL Gilles, *Gif-sur-Yvette*
REYES-MARCHAND Patricia, *Jozerand*
ROCHE Bernard, *Bastia*
ROMANEIX Christian, *Periers*
ROSTAN Jean-Claude, *Villeurbanne*
RUBAN Véronique, *Bouguenais*

SANCHEZ Nathalie, *Paris*
SAURIAU Pierre-Guy, *L'Houmeau*
SCHANEN Olga, *Paris*
SIME-NANGO Télesphore, *Aubiere*
SOGON Stéphane, *Meudon*
STOESSEL Jean, *Metz*
STORA Georges, *Marseille*
TALEB Rabah, *Orleans*
TASSIN Bruno, *Noisy-le-Grand*
TCHEUKAM TOKO Denis, *Le Havre*
TEXIER Hervé, *Caen*
THOMAS Pierre, *Saint-Nazaire*
TOUTAIN Joseph, *Mt-St-Aignan*
TROCCAS Olivier, *Rennes*
VERAN Marie-Pierre, *St Paul-lez-Durance*
VERNEAUX Valérie, *Besancon*
VEZIE Chantal, *Rennes*
VINÇON-LEITE Brigitte, *Noisy-le-Grand*
VIVIER Jean-Paul, *Rennes*
WAHYUDI Imam, *St Nazaire*

GERMANY
LIEBEZEIT Gerd, *Wilhelmshaven*
MOLL Andreas, *Hamburg*

IRELAND
WILSON James, *Dublin 2*

ITALY
LA HURA Virginia, *Napoli*
RAVERA Oscar, *Venezia*
RICCARDI Nicoletta, *Venezia*
SPEZIE Giancarlo, *Napoli*
MARKOS Ruzena, *Ispra*

MOROCCO
AII ABDELAALI Naïma, *Marrakech*
ALAOUI-MHAMDI Mohammed, *Fes*
ALIFRIQUI Bouchra, *Marrakech*
BADRI Abdel-Jalil, *Marrakech*
BOUALLAM Samira, *Marrakech*
BOUHADDIOU Abderrahim, *Fes*
CHEGGOUR Mohamed, *Marrakech*
ID-HALLA M., *Agadir*
MAAMRI Abdellatif, *Oujda*
MOUHRI Khadija, *Marrakech*
SAADI EL HASSANI Amina, *Marrakech*
SEDKI Azzedine, *Marrakech*
TIFNOUTI Abdelaziz, *Marrakech*

MONACO
JAUBERT Jean-M., *Monaco*

PORTUGAL
FALCAO Manuela, *Olhao*
FERNANDEZ Paulo, *Evora*
PESSAO Fernanda, *Monte de Caparica*

SPAIN
ABOAL Marina, *Murcia*
MARTINEZ-TABERNER Antoni, *Palma de Mallorca*
MOYA Gabriel, *Palma de Mallorca*
RAMON Guillermo, *Palma de Mallorca*

SWEDEN
PETTERSON Kurt, *Norrtaje*
STROMBECK Niklas, *Norrtalje*

SWITZERLAND
MAZOUNIE Nabila, *Lausanne*
VERNET Jean-Pierre, *Versoix*

TUNISIA
CHAFFAI Amel, *Sfax*

UKRAINE
KOVALCHUK Andrij A., *Kiev-3*

U.S.A.
SIMENSTAD Charles, *Seattle*

List of sponsors

Ministère de l'Education Nationale, de l'Enseignement Supérieur et de la Recherche
Ministère de l'Environnement
Ministère des Affaires Etrangères

Conseil Général de Loire-Atlantique Ville de Nantes

Agence de l'Eau Loire-Bretagne
CEMAGREF
Intitut Français de Recherche pour le Développement en Coopération (ORSTOM)

Comité de Pilotage Environnement de l' Université de Nantes
Université de Nantes

Crédit Mutuel

Hydrobiologia **373/374:** 1–20, 1998.
J.-C. Amiard, B. Le Rouzic, B. Berthet & G. Bertru (eds),
Oceans, Rivers and Lakes: Energy and Substance Transfers at Interfaces.

Man and river interface: multiple impacts on water and particulates chemistry illustrated in the Seine river basin

Michel Meybeck
Laboratoire de Géologie Appliquée, Université de Paris VI, 5 Place Jussieu, F-75252 Paris Cedex 05, France

Key words: river, water quality, major ions, nutrients, Co, Ni, Hg, Zn, Seine, Paris

Abstract

We analyse average water and particulate chemistry (nutrients, major ions, heavy metals) in the Seine basin at 10 key positions, from stream order one to river mouth (order 8, 67 000 km^2), and for a population density gradient from one to 20 000 people km^{-2}. Particulates are studied on stream deposited sediments and on recent riverflood deposits collected over two years. The impact of Paris megalopolis (10 M people for 2 300 km^2) is considered both on the main river course and on periurban and urban streams. Average concentrations at each position are normalized for all variables to pristine levels, mostly determined on a set of forested headwaters, in order to define the 'change ratio'. In the main river course these ratios vary from less than 0.5 for dissolved SiO_2 to more than 10 for Na^+, K^+, particulate Hg and Zn, and exceeds 50 for NH_4^+, PO_4^{-3} and NO_2^-. They reach over two order of magnitude for NH_4^+ and PO_4^{-3} in urban streams now covered and used as sewers, the ultimate anthropogenic impact. Few variables, such as Mg^{++} and particulate Co, are not affected by human activities, and Ca^+ and HCO_3^- are regulated by calcite precipitation linked to river eutrophication. The change ratios can be used to describe the spatial structure of impacts. For each variable, the maximum impact position depends on the pollution mode and origin: NO_3^- maximum is already noted in small agricultural streams but PO_4^{-3} maximum occurs at the most downstream stations. The maximum impact of the Paris megalopolis is observed more than 75 km downstream of the city centre (proximal impact) but the river water quality is still affected 200 km downstream (distal impact). In addition to this classical longitudinal impact mode, the megalopolis also creates radial impacts, and 'inverse' impacts due to flow regulations in upstream river reaches.

Introduction

Since the early developments of agriculture, and of related deforestation, man has continuously modified – some say altered – the natural conditions of continental waters. The consequences of human activities are a change of the quality or quantity of surface water characteristics. In turn many activities which require a specific water quality may be impeded or limited by many means (Meybeck et al., 1989; WHO/UNEP, 1992; Gleick, 1993; Meybeck, 1996). In most cases these changes result from the addition of chemical substances through diffuse and point sources, in a few others they may also be caused by physical changes in river basins such as channelization and water withdrawal, (Naiman & Decamps, 1990; Petts & Calow, 1996a–c). Examples of water quality modifications are now known for all continents particularly through the compilation of the GEMS-Water programme launched in 1978 by UNEP and driven by WHO (Meybeck et al., 1989, Fraser et al., 1995; Meybeck & Ragu, 1996). Generalized damming, which will be partly considered here, is another major change of river basins on all continents (Dynesius & Nilsson, 1994; Vörösmarty et al., 1997) with profound impacts on ecological functioning (Petts & Calow, 1996b).

Few river basins combine all water quality issues resulting from high population density, intensive agriculture, major industries, mining activities, major channelization of river course, etc. They are mostly encountered in Western Europe: Rhine (Van der Weijden & Middelburg, 1989; Van Dijk & Marteijn, 1993) Maas, Scheldt, Elbe, Visla, Rhone (Santiago et al., 1994), Po (WRI, 1991); in Eastern Europe: Dniepr, Don (Tsirkunov et al., 1998), Volga, and in North America the Mississippi (Meade, 1995) etc. The multiple impacts of man on water quality have generally been studied for small stream orders (Webb & Walling, 1996). On large river basins examples of multiple chemical impacts are known only for a few main river courses (Van der Weijden & Middleburg, 1989; Santiago et al., 1994; Meade, 1995). Impacts of dam cascades are generally better documented (Dodge, 1989; Petts & Calow, 1996a–c).

The Seine river basin is a typical example of a river system exposed to multiple and heavy human pressure over a hundred year span. It features most water quality issues except those resulting from mining activities. These include thermal pollution and radioactive contamination which will not be discussed here. Paris was the fourth largest city in the world with a population of 550,000 people in 1700; it is now a 10 million person megalopolis with numerous industrial activities. Despite the well-recognized action of all the Seine basin authorities which collect and spend $ 1.3 billion annually to fight the deterioration of basin quality, with half of this sum funded by the specific Agence de l'Eau Seine Normandie (AESN) created some 30 y ago, the river Seine is still in a critical condition, particularly in its reach downstream of Paris. In 1989, an interdisciplinary programme named PIREN-Seine was launched by the CNRS (National Scientific Research Center) and other key French scientific institutions, and was funded by all bodies concerned with water uses in the Seine basin (Agence de l'Eau, City of Paris, water supply companies, local and national administrations etc.). The PIREN-Seine has worked ever since on various topics concerning the basin such as eutrophication (Billen et al., 1994, Garnier et al., 1995), wetland issues, micropollutants (Garban et al., 1996; Idlafkih et al., 1997) impacts of urban sewers (Servais & Garnier, 1993; Mouchel et al., 1994; Boët et al., 1994), agricultural land-use impacts (Penven & Muxart, 1995), pollutants flux estimates (Idlafkih & Meybeck, 1996), and fish distribution (Belliard et al., 1996). In 1993 a workshop summarized the first part of this programme (Fustec & de Marsily, 1993).

This paper is based on some of the PIREN-Seine studies, but it presents mostly the personal views of the author concerning the spatial structure of the basin chemistry from headwaters to river mouth, excluding the estuary, and particularly under the Paris megalopolis constraint. Most of the water quality data used here were obtained by the author and his students (Thibert, Idlafkih, Biger) within the PIREN-Seine, including analyses of sediment and water from small monolithologic basins, and of recent flood deposits, performed by A. Horowitz (U.S. Geol. Survey, Atlanta). Some of the important studies and surveys, performed outside the PIREN-Seine by various water administrations are also considered. The objectives here are the following: (i) establish the natural background composition of the Seine water and particulates for major ions, nutrients and selected metals (Co, Ni, Hg, Pb, Zn), (ii) estimate their trends over one hundred years where possible, (iii) study the spatial distribution of these elements from small streams to the river mouth, (iv) study the human impact on water and particulate chemistry over an extended gradient of population density and of basin size, as appreciated by stream order.

The Seine river basin

The Seine river basin drains an area of 67 500 km^2 at the Poses station, last lock on the river before the estuarine section (Figure 1). It is mostly located on sedimentary rocks (97.5% of the basin, including 78.2% of various carbonate rocks as chalk and limestones, Thibert, 1994) and is characterized by a gentle relief with a maximum altitude of 902 m in the upper reach of the Yonne in the Morvan. The major Seine tributaries are the Aube (4 750 km^2), the Yonne (11250 km^2), the Marne (13 160 km^2) joining the river in Paris city, and the Oise (16 900 km^2) joining the river downstream from Paris at Conflans.

The Strahler stream orders have been used here to describe the basin scales from streams to the whole basin. This approach, used in watershed ecology (Naiman, 1982) and in river eutrophication (Billen et al., 1994) is tested here for water quality. Stream orders were defined using 1/50 000 maps (Garnier et al., 1993). The smallest streams have an order 1, and the confluence of two orders 1 makes an order 2. Confluence of two orders 2 makes an order 3, etc. The stream order is generally linked to the drainage area but sometimes, similar basins may have different orders as the Marne and the Oise at their confluence with the Seine (respectively 6 and 7). The Seine upstream of its confluence with the Marne is also referred here to as Upper Seine (32 200 km^2). Within Paris city the Seine has a stream order seven which increases to order eight downstream the confluence with the Oise.

In addition to its geological and geomorphological homogeneity, the Seine basin is also characterized by a single water regime which is very constant over the

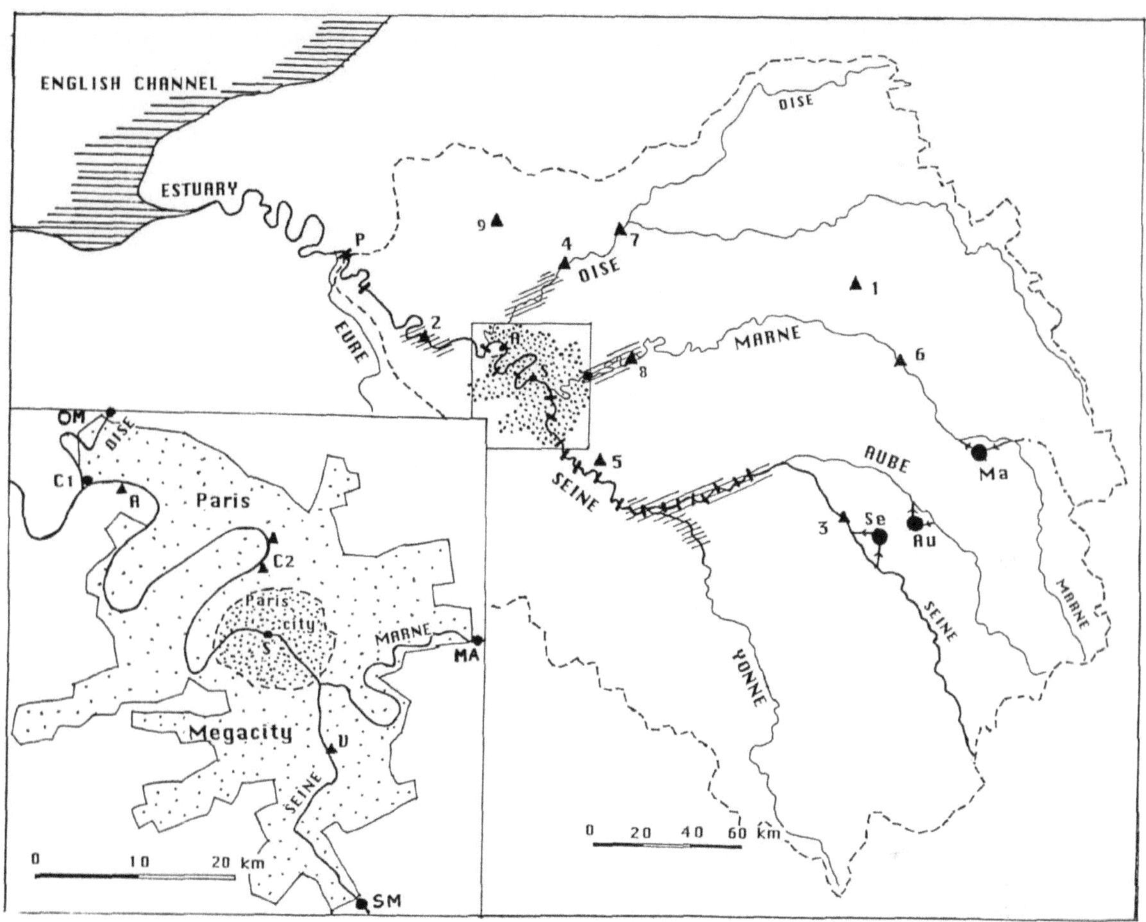

Figure 1. Principal features of the Seine River basin. Dotted area: Paris megalopolis. Major other cities exceeding 50 000 people (1: Reims, 2:Mantes, 3: Troyes, 4: Creil, 5: Melun, 6: Chalons en Champagne, 7: Compiègne, 8: Meaux, 9: Beauvais). Hatched area: major sand pit fields. Reservoirs on diverted rivercourse (Au = Aube, , Ma = Marne, Se = Seine). Principal sampling locations: P = Poses station (river mouth), A = Achères treatment plant, S = Solférino bridge. C_1 = Conflans, C_2 = Clichy and La Briche combined sewer overflows, V = Valenton treatment plant. — limit of Seine basin at Poses; MA: Marne at Annet, SM: Seine at Morsang, OM: Oise at Mery. -/- major locks on the Seine mainstem.

whole basin, i.e. pluvial oceanic, with a high water stage from December to February (9.5 l s^{-1} km^{-2}) and a low water stage in summer (2.2 l s^{-1} km^{-2}). The average discharge at Poses is 450 m^3 s^{-1} or 6.65 l s^{-1} km^{-2} (215 mm y^{-1}). An excellent description of the Seine basin ranging from hydrology to water uses in the early 1970's has been published by the Agence de l'Eau (A.E.S.N. 1973–1978) in 12 volumes. Historical reviews of the Seine and Paris relationship since the gallo-roman times are presented by Guillerme (1990) and by Maneglier (1992).

Upstream of the Paris megalopolis the land-use is mostly agricultural with some of the highest crop yields in Europe resulting in massive use of fertilizers and pesticides. There are nine cities of limited sizes (from 50 000 to 200 000 people) (AESN 1973–1978) (Figure 1) and many have important industries (food, textiles, chemicals etc.). However, most of the heavy industries of the basin, including chemicals, car factories, and oil refineries, are located in the western Paris suburbs and between the Seine–Oise confluence and the estuary. The whole Seine basin, the estuary included, contains about 30% of the total France population and 40% of its industrial activity, 50%, 67%, 35% of wheat, sugar beet, and barley crops, respectively (A.E.S.N., 1991). When considering the whole basin above Poses, 23.2% of the population lived in 1968 in rural settlements and small towns (< 10,000 people) and 64.4% in Paris megalopolis. Paris' share has now increased to 74%.

Paris city proper is 10 km in diameter and the average population density reaches 20 000 people. km^{-2}. The megalopolis is here defined as a much larger coalescing urban area including Versailles, Trappes, Pontoise, Evry, Marne la Vallée, etc. It has an average diameter of 50 km and a total population of ca 10 million (in 1990) for 2,300 km^2, with an average density of about 2 000 people km^{-2} in the suburbs. Most of the megalopolis population is connected to sewer systems and water treatment (primary and secondary treatments), such as the Valenton plant (300 000 equivalent habitant), and the enormous Achères plant started in the 1940's which now has a total connected population of 8.1 million people (Figure 1). Sewage treated at Achères is discharged four km upstream of the Seine-Oise confluence i.e. some 75 km downstream of Paris centre with an average discharge of 25 $m^3\ s^{-1}$ in dry weather conditions. Since Paris sewers are combined ones, during rainstorm events the combined sewers overflow directly to the Seine, particularly through the two major collectors of Clichy and La Briche injected 20 km downstream of the Eiffel tower (Figure 1). Their total collected volume reaches 1 million m^3 during typical rainstorms with a peak discharge of 50 $m^3\ s^{-1}$(Estèbe, 1996). The pollution load of the Paris megalopolis for the whole basin, estuarine zone included, accounts for 85% domestic wastes and 44% of industrial toxic wastes (calculated from AESN, 1991).

The physical characteristics of the river course have been much changed in its middle and lower courses as the result of navigation, water discharge regulation, and sand extraction. This resulted in significant lateral connections, cut off from phreatic aquifers. Most of the Oise, the lower part of the Marne, and the Lower and Middle Seine have been used for navigation for more than 150 y resulting in a total of 17 major locks (exceeding 180 m in length) in the Seine and Marne (Guillerme, 1990; Allardi et al., 1993) (Figure 1). The total navigated, and periodically dredged, river length is 900 km plus 1200 km for secondary navigated canals. The Seine basin accounts for about 60% of the whole fluvial traffic in France (AESN, 1973–1978). Major sand and gravel extractions from alluvial deposits for the supply of Paris buildings and related infrastructures (highway network, new railroad tracks, thermal and nuclear power plants) have resulted in hundred of sand-pits in the flood plain, sometimes as far as 150 km from the centre of Paris along the river course (Figure 1).

Since the 1960's a set of important regulation reservoirs has been built on partially diverted courses of the Aube (1989–91), upper Seine (1966–67), and Marne (1974–75) (Figure 1), with respective volumes of 170, 205 and 350 M m^3. These reservoirs were mostly built to raise the low-water levels in summer and, to some extent, to limit winter flooding of Paris (Miquel & Levassor, 1993). They will not prevent the catastrophic millenial flood, as in 1910 when the centre of Paris was flooded for several weeks, but may reduce the peaks levels by at least 50 cm, an important issue when considering the flooding of the river bank motorways. Actually, without these additionnal summer discharges of high water quality, more than half of the Seine river discharge downstream of the Achères treatment plant would originate from secondary treated sewage, since the 10 y drought flow is estimated at only 51 $m^3\ s^{-1}$ in Paris (Miquel & Levassor, 1993).

Long term trends of Seine water quality compared to other rivers

The first chemical analyses of the Seine river in Paris date back to the founding of chemistry by Lavoisier and his followers as Sainte Claire Deville in 1848 (quoted by Livingstone, 1963) (Table 1, # 1). The same chemists also analyzed the Geneva lake water. Because the composition of this lake was still unchanged in 1930, Sainte Claire Deville analysis can be validated, except for Mg^{++} and SiO_2 which should be questioned. In 1848 the Seine river was no longer in pristine condition, since at this time industrial effluents already affected the water quality (Guillerme, 1990), but the sewage system was not yet installed. Since 1886 few water quality determinants as Cl^-, NO_3^-, NH_4^+ and some indicators of faecal pollution were regularly monitored by the City of Paris Water Quality Control Laboratory (now CRECEP) at Ivry a few km upstream from the Seine-Marne confluence, (Naves et al., 1990; Mangerel, 1982) (Figure 2). These records are probably some of the longest for a river together with the Thames record and the Cl^- records on the Rhine (Ackerman et al., 1970; Meybeck et al., 1989). For other water quality determinants much shorter series are now available since the beginning of the French regular water quality survey in 1971, the Reseau National de Basin (RNB). The key station is the Poses station located just upstream the estuarine section and the Seine-Eure confluence (Figure 1, Table 1, # 10).

Another way to assess the water quality evolution is to compare the actual concentrations measured at

Table 1. Average major ions and nutrients (mg l^{-1}) of stream, tributaries and main river course in the Seine basin.

		Ca^{++}	Mg^{++}	Na^{+}	K^{+}	Cl^{-}	$SO_4^{=}$	HCO_3^{-}	$N\text{-}NO_3^{-}$	$P\text{-}PO_4^{-3}$	$N\text{-}NH_4^{+}$	SiO_2
Seine 1848	(1)	74	(1.6)	7.3	2.2	7.5	21.8	202				(24.4)
Country	(2)	96	10.3	6	1.25	9	25	295	0.5	0.013	0.025	12
streams	(3)	97	7.7	7	1.4	13.5	32	250	5.0	0.014	0.018	14
1990's	(4)	105	6.4	8.5	2.1	20	25	270	4.7	0.035	0.03	89
Tributaries	(5)	90	10	8.0	2.0	17	22	275	3.8	0.030	0.060	4
and main	(6)	100	8	8.0	2.2	18	25	275	4.5	0.19	0.25	5
river	(7)	98	4.5	9.8	2.4	20.8	27	251	6.0	/	/	9.5
course	(8)	(100)	/	8.8	2.7	24.5	25.4	/	4.5	0.23	0.32	3.4
1990's	(9)	100.7	/	12.7	3.5	28.1	31.5	/	4.28	0.41	1.64	3.8
	(10)	104.5	9.4	20.2	4.8	38.4	56.9	252	4.4	0.70	2.41	6.3
Urbanized	(11)	105	11.0	36	6.4	55	88	270	4.6	1.14	3.12	/
streams	(12)	110.6	11.8	77.1	17.7	87.2	143.5	404	0.063	4.52	21.2	11.7
1990's	(13)	113.6	11.7	81.4	17.9	91.5	139	435	0.343	3.63	24.7	11.6

( ): doubtfull values, (1): Sainte Claire Deville quoted by Livingstone, 1963, (2): Median of unpolluted forested streams on the Seine basin (see text), (3): Median of representative monolithologic streams draining agricultural fields (see text), (4): Median of representative monolithologic streams draining fields and villages (see text), (5): Average for stream orders 4 and 5, Marne basin (Thibert, 1994), (6): Average Marne, Seine and Oise upstream confluences, (RNB, Thibert, 1994; and CGE-SEDIF, 1996). (7): Average distributed water at Paris (Thibert, 1994). (8): Average at Solférino bridge, Paris, 1978–1984 (Person, 1990). Adjusted for NO_3-trend, (9): Theoretical Conflans station (see text), (10): Poses, river mouth, 1990–92 average (Cossa et al., 1994), (11): Average of 7 periurban streams (orders 3 and 4). (12): Achères untreated sewage average of 10 weekly composite samples 1991–92 (Thibert, 1994). (13): Achères treated sewage (d.o).

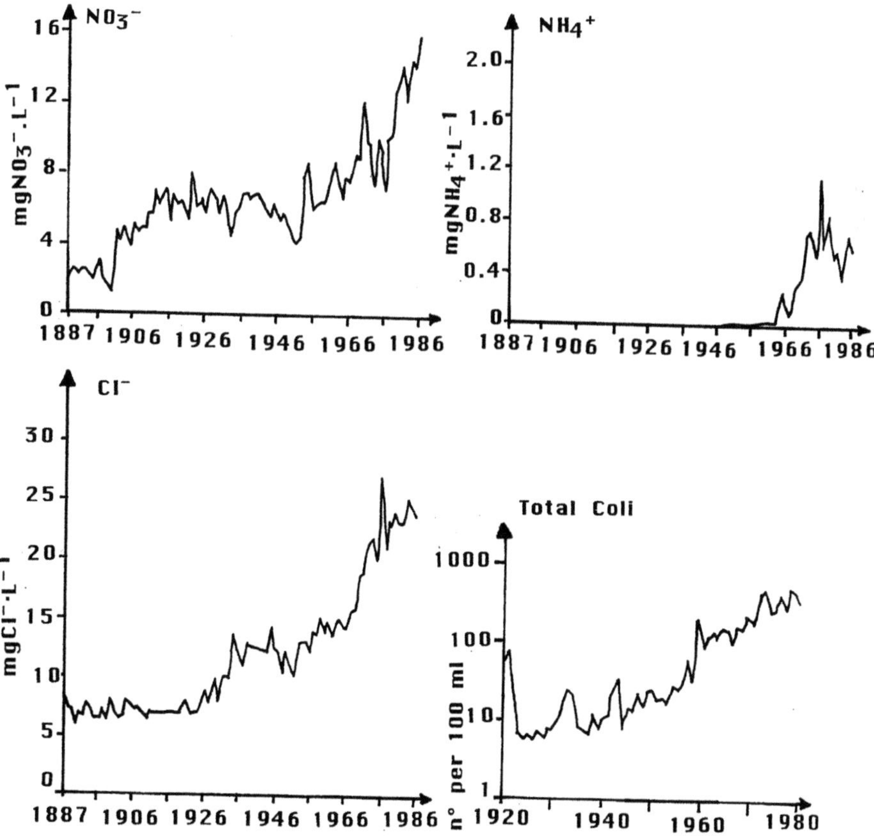

Figure 2. Schematized long-term trends of Seine River quality at Ivry; Cl^{-}, NO_3^{-}, NH_4^{+} (Naves et al., 1990), total coliforms (Mangerel, 1982).

the river mouth with estimated natural background values. For this, a set of small pristine forested streams located on various rocks types has been considered (Table 1, # 2). These streams were carefully selected to be free of any anthropogenic impact by Thibert (1994), according to a methodology used for the whole French territory (Meybeck, 1986).

Based on these various water quality data sets, several trend patterns of major ions from 1848 to 1990 can be found in the Seine at the mouth (Table 1, Figure 2).

Stable concentrations: Ca^{++}, Mg^{++}; HCO_3^-, DOC

These major ions are still found presently at levels similar to the 1848 analysis and/or to average forested pristine stream. Domestic wastes can be considered to partly explain this stability. The analysis of untreated sewage at Achères treatment plant (Table 1, # 12) compared to the average tap water of Paris (Table 1, # 7) shows that urban wastes barely affect Mg^{++} but do affect Ca^{++}. On the basis of these two analyses annual *per capita* release of elements (in kg cap^{-1} y^{-1}) can be computed combining the excess ionic content in sewage to the sewage flow (25 m^3 s^{-1} for Achères) and to the connected population (Thibert, 1994). Such computation was done for example in Montreal (Caillé et al., 1973), and Brussels (Verbanck et al., 1989). The *per capita* inputs of Mg^{++} are lower compared to those of other major ions: 0.7 for Mg^{++} in Paris and Montreal, while these figures are much higher in these three cities for Na^+(6.4 to 6.5), for K^+ (1.0 to 1.6), and for Ca^{++} (1.2 to 3.2).

For Ca^{++} and HCO_3^- there is a regulation mechanism in the river which is close to the calcite saturation limit. Any added Ca^{++}, from agriculture, industries or cities is precipitated at pH exceeding 8.3. In summer the average pH exceeds this levels and nychthemeral pH cycles range between 8.0 to 8.6 (CGE-SEDIF, 1996). Such high pH values are caused by the generalized eutrophication of the Seine for Strahler stream orders 6 to 8 in its mid and lower courses and of major tributaries (Garnier et al., 1993; Billen et al., 1994). In other eutrophicated river basins where carbonate rocks are dominating, as the River Loire, the autochthonous calcite may represent up to 30% of the summer suspended matter (Manickam et al., 1985; Meybeck et al., 1988).

According to the SNS survey at the river mouth the dissolved organic carbon (DOC) has also been stable or slightly decreasing (<10%) over the last 25 y (Ficht, 1995).

Moderate increase: Na^+, Cl^-, K^+, $SO_4^=$, TDS

These ions have multiple sources: agriculture (fertilizers), industries, domestic wastes, de-icing salts. They have regularly increased since 100 y as Cl^- (Table 1, # 1, 2 and 10; Figure 2). As a result of these trends the Total Dissolved Solids (TDS) have regularly increased over the last 25 y at the river mouth at a rate of + 1.6 mg l^{-1} y^{-1}.

Marked increase: NO_3^-, NH_4^+ PO_4^{-3}, Total coliforms

Nitrate increase started well before World War II but accelerated in the 1950's (Figure 2) when the use of N fertilizer grew exponentially. Such a trend is similar to the one observed for British rivers (Roberts & Marsh, 1987). The NO_3^- maximum in the early 1900's is still questioned and could be due to errors in reporting units or to analytical problems. From 1965 to 1981 nitrate increased at a rate of + 1.1 mg NO_3^- l^{-1} (Ficht, 1995). Since 1981 this rate is lower, +0,25 mg NO_3^- l^{-1}. Nitrate is now close to the recommended limit of drinking water (25 mg NO_3^- l^{-1}) athough still below the critical WHO limit (50 mg NO_3^- l^{-1}). At the Ivry water intake (Figure 2) near the Seine-Marne confluence (Figure 1), ammonia started to increase in the early 1960's when sewage collection in S.E. Paris suburbs was performed. Ammonia peaked in the 1970's then stabilized, probably as the result of sewage treatment. The Total coliforms pattern at Ivry (Figure 2) presented a marked increase in the early 60's then stabilized in the 1980's (Mangerel, 1982). Phosphate has only been documented for the last 25 y (Ficht, 1995); a marked long term increase is likely when the forest stream level (13 μg P l^{-1}) is compared to the present day level at Poses (700 μg P l^{-1}). The flux is now relatively stable around 50 t PO_4^{-3} y^{-1} from 1974 to 1994. A similar stabilization of phosphate level is also noted for the last five years at three major Paris megalopolis water intakes not affected by Achères outputs, on the Oise (Mery), the Seine (Choisy), and the Marne (Neuilly) (CGE-SEDIF, 1996).

Marked decrease: dissolved SiO_2

Although SiO_2 has only been regularly monitored since the mid 1980's, a marked decrease of silica is most likely. Analyses performed on small streams (Table 1, # 2 and 3), show median values around 12 and 14 mg SiO_2 l^{-1} while the yearly average at Poses is only 6.3 mg l^{-1} and summer averages even less. This is related to spring and summer uptake of SiO_2 by the aquatic vegetation as diatoms (Thibert, 1994; Garnier et al., 1995). Such silica uptake has also been reported for the Loire river (Meybeck et al., 1988) where an inverse correlation between SiO_2 and total pigments has been found. Silica trends are rarely documented. In the Mississippi a SiO_2 decrease in the 1960's and 1970's has been reported by Turner & Rabalais (1991), and also attributed to diatom development. Another decreasing trend was noted in Latvian rivers (Tsirkunov et al., 1991).

The Seine compared to other rivers

These Seine river trends are probably typical of many rivers in Europe and in populated parts of North America. However, exceptions can be observed in some few river systems. For instance in the Rhine, NH_4^+ and PO_4^{-3} have markedly decreased (Van Dijk & Marteijn, 1993) since the mid 1970's. In this basin industrial and domestic wastes treatment is probably more efficient than in the Seine: phosphorus is now commonly treated in sewage works in Germany and Switzerland and these countries have banned or severely restricted the use of phosphorus in detergents for more than 10 y. A phosphate detergent ban is still being debated by the French Ministry of the Environment. Is it related to the fact than one of the major european phosphorus product manufacturer is French?

In terms of NO_3^-, NH_4^+, and PO_4^{-3} the Seine river ranks now in the top seven among a set of 200 world major rivers, exceeding 10,000 km^2 or 300 m^3 s^{-1}, ranked in decreasing concentrations (Meybeck & Ragu, 1996 & 1997). It is only surpassed by few other European rivers (Trent, Thames, Scheldt, Tevere, Arno).

The salinity status of the Seine (TDS = 486 mg l^{-1}, Cl^- = 38 mg l^{-1}), even greatly affected by human activities, is still much better than the one of the Weser (TDS = 2463 mg l^{-1}, Cl^- = 1233 mg l^{-1}), Elbe (TDS = 726 mg l^{-1}; Cl^- = 174 mg l^{-1}), Wisla (TDS = 583 mg l^{-1}; Cl^- = 147 mg l^{-1}) and Rhine (TDS = 599 mg l^{-1}; Cl^- = 173 mg l^{-1}). All these rivers have been impacted by mining acitivities and are characterized by high levels of Na^+, Cl^- and/or $SO_4^=$ compared to all other temperate rivers (Meybeck & Ragu, 1996). The Cl^- trend in the Rhine is particularly well documented: it has gradually increased about 20 times from the early 1900's to the 1960's. It is now maintained close to the critical level of 200 mg l^{-1}. This Cl^- excess is mostly related to Alsace potash mines (MDPA) and Lorraine salt mines. In the Rhine over the last 30 y, Ca^{++} has slightly increased together with $SO_4^=$ while bicarbonates and Mg^{++} are stable (Van der Weijden & Middelburg, 1989).

Another important trend often found in world rivers is the decline of Total Suspended Solids (TSS) resulting from reservoir trapping, either downstream of reservoir cascades as in the Colorado, Rio Grande, Missouri, Don, Dniepr, Volga, etc. or downstream of a single major dam as for the Nile and the Columbia (Milliman & Meade, 1983; Meade & Parker, 1986). This process is now a major change in river systems at the global scale (Vörösmarty et al., 1997). In the Seine, although there is no appropriate long-term TSS monitoring, the reservoir trapping is thought to be limited. The total quantity of detrital organic and inorganic material trapped in the three major reservoirs to which water from the Aube, Marne, and Upper Seine is derivated, (Figure 1) is estimated thought to be much less than 10% of the TSS load at the river mouth, i.e. similar to the proportion of intercepted basin area. The annual average TSS load is estimated between 650,000 and 700,000 t/year at Poses (Ficht, 1995; Idlafkih, pers. com.). In the Seine basin only a limited part of the higher stream orders five to eight are affected by such sediment retention. Another trapping process is observed mostly during dry periods. It occurs between locks in the navigated reaches of the Seine, Marne and Oise, particularly within the Paris megacity (Estèbe, 1996).

Characterization of present basin water chemistry

The spatial variations are studied here for the early 1990's at different basin scales from few 10 km^2 (first and second stream orders) to the lower main course (67 500km^2, order 8). Data used here are only averages of time or space distributions: the impacts of individual point sources is not our objective excepted of course for Paris megalopolis. Spatial averages for streams and tributaries upstream from Paris (orders 1 to 6) are justified by the relative homogeneity of the Seine basin in terms of climate, lithology, land use, and population distribution. Since population density of stream basins does not follow hydrological order, the Paris megalopolis impact is considered at two levels: (i) along the Seine main river course over more than 100 km, (ii) on periurban and urban streams. The average water chemistry of all streams and river reaches are presented for each key position in Table 1, and their general characteristics in Table 2.

Table 2. Mean characteristics of schematized riverine sections on the Seine with regard to impact types.

Impact position	Strahler order	Area (1) (km^2)	K P (2) (km)	Pop. density (3) ($p\ km^{-2}$)	Pop. stress ($p\ l^{-1}\ s$) (4)
Pristine PR	1–3	4–40	200–700	< 1	< 0.1
Agricultural AG	1–3	4–40	200–750	1–10	0.15–1.5
Rural RU	1–3	10–100	200–750	10–40	1.5–6
Mid stream orders MO	4–5	650–2500	150–300	40–80	6–12
Periurban main course PUM	6–7	10 000–30 000	215	80–150	12–22
Urban main course URM	7	44 000	140–207	120	18
Proximal downstream PD	8	61 000	130	280	42
Distal downstream DD	8	67 000	0	240	36
Periurban streams PUS	1–4	10–500	140–300	500–2000	75–300
Urban streams URS	1–4	10–500	180–220	2000–20 000	300–3000

(1) Typical basin area or area at station, (2) Distance from mouth along river course, (3) Average population density in the whole watershed (people km^{-2}), (4) 'Population stress' for average flow: people l^{-1} s.

Pristine state and rural environment

For the smallest streams our data set is only based on our studies performed within the PIREN-Seine since there is practically no regular water quality monitoring for streams below order four within the RNB. Three types of land-use have been distinguished for these streams: the 'pristine' level PR (Table 1, # 2), 'agricultural' impact AG (Table 1, # 3), and 'rural' state RU, i.e. the impact of agriculture plus small villages (Table 1, # 4). These basins are located, respectively in forested areas (population density < 1 people km^{-2}) in basins draining only agricultural fields ($1 < d < 10$ people km^{-2}), and basins draining fields and small villages ($10 < d < 40$ people. km^{-2}) (Table 2). Both major ions and nutrients where analyzed (n = 100 to 140) at 56 sites (Meybeck, 1986; Thibert, 1994, Biger, 1996). The spatial coverage was favoured with regard to the temporal variations although it is known that during rainstorm events the stream nutrients concentrations as $NO_3{}^-$ may be quite variable (Muxart et al., 1993). The median concentration of stream quality have been considered here for each set. Since the major ions were one target of this study only small monolithologic basins were selected covering most kinds of sedimentary rocks such as chalk, limestones, siliceous sandstone, interbedded clay and carbonate terrains, and a few hard rocks basins (only 2.5% of the whole basin). In many cases the human impact has been checked directly on twinned basins, located at the same sites but with different land use, such as forested vs agricultural fields.

Table 3. Heavy metal levels in river particulates in the Seine compared to other temperate rivers ($\mu g\ g^{-1}$).

	Cd	Cr	Cu	Hg	Pb	Zn
Seine						
Pristine (1)	0.35	45	20	0.05	19	60
Pristine (2)	0.34	/	5	0.034	27	34
1976–82 (2)	24	/	330	3.5	195	1100
1990–92 (3)	4.95	/	174	1.22	184	611
1974–94 (4)	11.6	183	231	5.2	294	1187
Rhine						
1971–75 (5A)	50	1700	720	10	600	2900
1977–81 (5B)	17	724	325	/	596	1702
1985–88 (5A)	3	180	70	1.0	120	1100
Rhône (6)	1.5	136	131	1.2	97	220

(1): this work, based on 5000 y old sediments and forested watersheds sediments (< 100 μm). (2): Avoine et al. (1986) based on ancient river deposits, (3): average contents in suspended matter at Poses, Cossa et al. (1995), Idlafkih et al. (1995). (4) average in suspended matter at Poses: Ficht A. (1995). (5A): U.N.W.C. (1992). (5B): Van der Weijden & Middleburg (1989). (6): Martin et al. (1989) and, Guieu et al. (1991) for Cd, Pb, Zn, quoted in Nolting et al. (1996), Cossa & Martin (1991) for Hg, Santiago et al. (1994) for Cr and Cu.

Nutrient levels found in forested streams of the Seine basin are similar to those previously observed in a similar set of 250 unpolluted French streams of various geological background, also characterized by low median values of N-$NO_3{}^-$ (0.14 mg l^{-1}), P-$PO_4{}^{-3}$ (0.006 mg l^{-1}), and K^+ (0.7 mg l^{-1}), (Meybeck, 1986). When passing from forests to agricultural fields there is in the Seine basin streams, a ten-fold increase of $NO_3{}^-$ while $PO_4{}^{-3}$ is on the average not affected. This nitrate increase is generally attributed to excessive fertilization. Ammonia is also not much changed and may even be higher in some forest streams than in agricultural watersheds. It must be noted that PO_4^{-3} and NH_4^+ patterns are specific to the Seine basin: in other French regions such as Brittany, cattle raising can be a major source of both NH_4^+ and PO_4^{-3} which reach much higher levels in agricultural basins. In the Seine the marked increase of PO_4^{-3}, NO_2^- and NH_4^+, is only noted when the collection of domestic wastes is performed ($10 < d < 40$ peoples km^{-2}) while NO_3^- levels remain as high as in the agricultural streams. The silica pattern is opposite: as soon as the forest cover of streams disappears, dissolved silica drops (Thibert, 1994), as a result of uptake by benthic diatoms and macrophytes development for the smaller orders (2 to 4), then by planktonic diatoms at higher orders (Garnier et al., 1995).

Small tributaries and Seine main stem outside Paris

For 'mid Strahler stream orders' 4 and 5 MO (Table 1, # 5, Table 2,) upstream of Paris megacity we have used the average values determined in the Marne basin in 1990/91 by Thibert (1994). The 'periurban main course' PU_M (orders 6 and 7) (Table 1, # 6, Table 2) is characterized by the Oise, Marne, Aisne, Yonne and Upper Seine at their confluences for major ions (French national network, RNB). For the average nutrients we used the average 1991 to 1995 levels established at the three water intakes on the Marne, Upper Seine and Oise (CGE-SEDIF, 1996). The average population density of river basins at these sites increases from 80 to 150 people. km^{-2} when approaching the outer part of the megalopolis of Paris. In these middle reaches many water quality determinants, such as PO_4^{-3}, NH_4^+, NO_2^-, Cl^-, Na^+, present an important increase while average SiO_2 is still decreasing. Some other determinants are not affected- or are within the natural range observed in the basin, as for Ca^{++}, Mg^{++}, $SO_4^=$ and HCO_3^-. The influence of small cities (10 000 to 100 000 inhabitants typically), and various industrial activities are responsible for this gradual change of the river quality.

In this part of the river system some major hydrological changes also occur: water diversion, sand pits excavations, channelization. The three major reservoirs built on the upper Seine, the Marne and the Aube (Figure 1) have various local and regional influences, positive or negative, on water quality. A local negative influence is noted on the few 10 km long diverted reaches, (Barillier et al., 1993), particularly in winter, since the reservoirs are mostly filled in during the high water stage and the water mostly released in summer, thus completely inverting the natural water regime. A major positive impact is the important denitrification process in these reservoirs: about 1.0 10^9 g N y^{-1} in the Marne reservoir for 48 km^2 (Garnier et al., 1998). If this rate is extrapolated to the other major reservoirs it would correspond to a proportion of 4 to 8% of the total N-NO_3^- flux at the river mouth. Another important positive impact of these reservoirs is noted in summer, when the Paris sewers dilution capacity of the river is significantly increased by the released reservoir water.

The Seine maincourse through the Paris megalopolis

The 'urban main course' UR_M extends (Table 1, # 8, Table 2) from the Seine–Marne confluence i.e. from the city of Paris, to the Seine–Oise confluence at Conflans. We have used the average water quality observed daily by the Laboratoire d'Hygiène de la Ville de Paris (LHVP) from 1978 to 1984 at the Solférino Bridge in the centre of Paris (Figure 1) In order to take into account the increase of NO_3^- from 1978–84 to the early 1990's a 10% increase in 10 y has been assumed, shifting N-NO_3^- concentrations from 4.0 to 4.5 mg l^{-1}. Indeed, the water quality at Solférino Bridge is surprisingly similar to the previous one (Table 1, # 6) for all major ions, and most nutrients. The quality of tap water distributed in Paris (Table 1, # 7), is also similar, except for Mg^{++}. This relatively moderate impact on the Seine within Paris can be explained as such: most of the waste waters of Paris city and suburbs collected during dry weather period are treated, then discharged in the Seine at the Achères plant, some 75 km downstream the Eiffel tower. The Valenton treated sewage waters and few other sewers, upstream or within the centre of Paris, are relatively minor point sources of pollution in the Paris megalopolis during dry weather periods.

During storm events (on average there are five events per year exceeding a total volume discharged of 0.5 M m^3, Estèbe, 1996) the combined sewer overflows (CSO), of Clichy and La Briche mixing both domestic sewage and urban runoff, (Figure 1) may reach several 10 m^3 s^{-1} at peak flows. In summer when the Seine water discharge is 100 m^3 s^{-1} and less, the dilution of these high BOD and COD levels, respectively 88 and 315 mg l^{-1} (Mouchel, 1993), is only partially realized. This may result in major anoxic events and NH_4^+ peaks in the river, lasting 8 to 24 h (Chocat et al., 1994; Chebbo et al., 1995), and in massive fish kills (Boët et al., 1994). The contaminated particles injected into the Seine river during these events rapidly settle within 10 to 20 km where they are stored in the river bed (Estèbe, 1996) until they are flushed downstream during winter floods, from December to March, if the river discharge exceeds $10\,l\,s^{-1}\,km^{-2}$ (Idlafkih et al., 1997). The impact of the urban storm runoff is therefore maximum during summer low flows some 30 km downstream of the Eiffel tower where it is combined with the impact of the western Paris suburbs industries. Evidence of such acute industrial and/or CSO pollutions are found in surficial sediment (Garban et al., 1996; Estèbe, 1996) and in cores taken upstream of locks as in Chatou (Estèbe, 1996).

Lower course of the Seine

The lower course is here defined as starting upstream Achères at the western margin of the Paris suburbs (Figure 1) and extending some 150 km downstream Paris city centre. The full mixing between Achères treated sewer (Table 1, # 13), injected on the left bank of the river, the Seine and the Oise which meet at Conflans is, at low flow, only achieved near the city of Mantes some 40 km downstream. The impact of Achères treated wastes is actually a complex one that has drawn the attention of the first part of the PIREN–Seine programme (Chesterikoff et al., 1991, 1992; Servais & Garnier, 1993). However, the study of water river quality is very difficult at this site since the Achères effluents are injected on the left bank of the river and are not fully mixed when the Seine meets the Oise at Conflans a few km downstream (Chesterikoff et al., 1991, 1992). The Mantes RNB station, (# 2, Figure 1) is also impacted by additional local industrial inputs between Conflans and Mantes.

An average theoretical quality corresponding to a 'proximal downstream' PD position downstream Conflans has been calculated for the mixing of the Seine, the Oise and the Achères effluent (Table 1, # 9; Table 2). This is made as such: an average discharge of $420\ m^3\ s^{-1}$ for the Oise and Seine combined with an average quality as observed at the Solferino Bridge, and $25\ m^3\ s^{-1}$ of Achères treated sewer with an average quality as measured by Thibert (1993) (Table 1, # 13). It is here assumed that the river quality of the Oise at Conflans (about 1/4 of the water discharge) is close to that observed for the Seine at the Solférino bridge. This theoretical average quality does not take into account the internal biogeochemical processes affecting the nitrogen and phosphorus species downstream between Achères: most of these processes require long distances to fully develop – up to 100 km – as for the nitrification (Chesterikoff et al., 1992).

This theoretical Conflans station (Table 1, # 9) shows marked water quality deterioration for PO_4^{-3}, NH_4^+, NO_2^-, Na^+, Cl^-, K^+ and $SO_4^=$. Chesterikoff et al. (1991) have shown that at low stage, the Seine river between Conflans and Mantes is also affected by the daily variations of Achères inputs. Dial cycles of electrical conductivity, Cl^-, NH_4^+ were noted and a PO_4^{-3} cycle is likely.

River mouth station

The Poses station corresponding to the 'distal downstream' DD position of the river basin is one of the best monitored water quality station in France (Table 1, # 10; Table 2). It is operated by the Service de la Navigation de la Seine (SNS, Rouen) since the 1960's (Ficht, 1995) and within the French RNB network since 1971. We also used here the results of a 1900–1992 pilot study performed by the French ocean research institute (Ifremer), the SNS, and some other institutions, to optimize the Seine monitoring for pollutant flux assessment (Cossa et al., 1994; Idlafkih et al., 1995, 1997). The last major tributary, the Eure River, reaches the Seine a few km downstream Poses.

The influence of the Paris megalopolis does not decrease at this station. An increase of NO_2^- and NH_4^+ between averages analyses # 9 and # 10, (Table 2) may originate from sediment pore water diffusion and/or to partial ammonification of organic N (Chesterikoff et al., 1992). Additional sources of dissolved salts (Na^+, K^+, Cl^-, $SO_4^=$...) in this river reach are from industrial origins. At Poses PO_4^{-3} and NH_4^+, still present a marked dilution pattern of Achères inputs with river discharge (Chesterikoff et al., 1992; Cossa et al., 1994; Ficht, 1995).

Small streams within Paris megalopolis

The 'periurban streams' PU_S (Table 1, # 11) are those encountered at the outskirts of the megalopolis. Their population density is between 500 and 2 000 people km^{-2}, and their stream order is up to four. Their water quality is based on four streams South of Paris monitored within the RNB: the Meauldre (at Mareil), the Orge (Athis Mons), the Yerres (Villeneuve St Georges), the Yvette (Epinay sur Orge). Their five-year average (1991–1995), courtesy of S. Romon (ASSI) has been mixed with three other streams sampled by the author, the Ru de Gally (West of Paris), the Croult (North of Paris) and the Mort Bras (East of Paris), to propose a first estimate of the average quality of periurban streams (orders 1 to 4). These streams present a marked contamination and may reach aver-

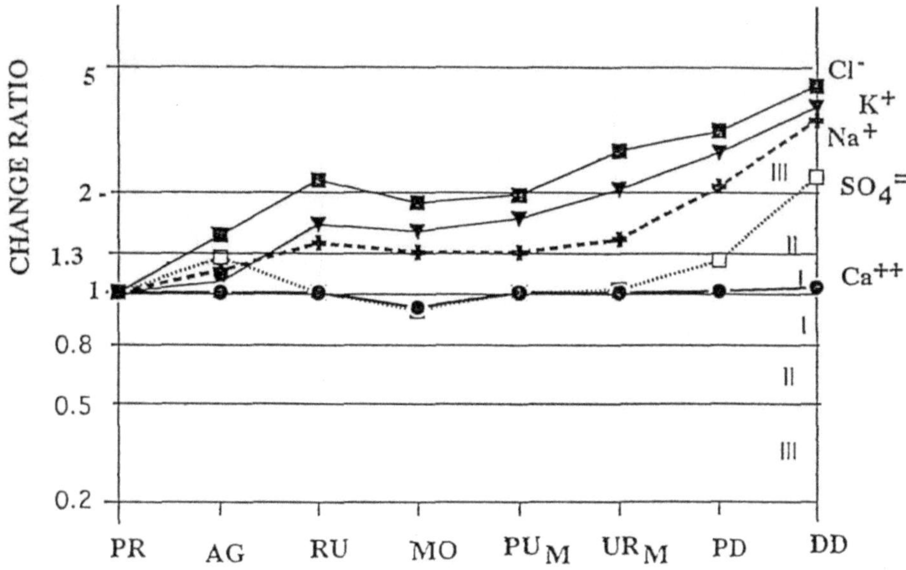

Figure 3. Schematized longitudinal variations of major ions along the Seine River from pristine headwaters (PR) to river mouth (DD). Change ratio: mean analyses for the early 1990's normalized to pristine values. I: unsignificant change from pristine value, II: minor change, III: important change. Position of sites see Table 2.

age contents as high as 0.5 mg N-NO_2^- l^{-1}, 8.6 mg N-NH_4^+ l^{-1}, 2.5 mg P-PO_4^{-3} l^{-1}, 12.4 mg K^+ l^{-1}, 79 mg Cl^- l^{-1}, 101 mg $SO_4^=$ l^{-1}, etc. These figures are generally well above those observed for the Seine main course at any site.

Urban streams

'Urban streams' UR_S (Table 1, # 12) are understood here as the ultimate impact of the megalopolis on surface waters. They are generally fully covered and transformed into domestic and/or rainstorm sewers. Examples of such 'streams' are found in the centre of the Paris megalopolis but their headwaters may start in rural or periurban conditions as for the Croult (North of Paris), and the Bièvre (South of Paris). Such basins have population densities from 2 000 to 20 000 km^{-2}, (Table 2) i.e 2 to 4 orders of magnitude higher than those found for agricultural streams. They have been characterized here by the average chemistry of Achères untreated sewers. Ammonia and phosphate are so high in these waters that they should be considered as major ions. Nitrate levels are near zero: due to anoxic conditions inorganic nitrogen is mostly present as NH_4^+. Total Dissolved Solids are well above those measured in periurban streams, due to the addition of Na^+, K^+, Cl^-, $SO_4^=$ from street runoff, domestic wastes, small industries. The solubility index of calcite is therefore higher and observed HCO_3^- are 50% higher than at other sites.

The change ratio, a common tool for water quality

The comparison of chemical analyses at various sites characterized by concentrations ranging over four orders of magnitude is difficult. Therefore all analyses were normalized for each element to the average pristine reference values for the whole basin (Table 1, # 2; Table 2, PR). These ratio are referred to as the 'change ratio'.

This normalisation allows for an easy comparison of all the water quality determinants. In order to visualize the various longitudinal gradients of the change ratio ranging from less than 30% to nearly two orders of magnitude, a logarithmic scale of these normalized concentrations has been preferred (Figures 3 to 7). In the range of 0.8 to 1.3 change ratio, concentrations are not significantly different from natural background values and it is assumed there is no impact. Change ratio of 1.3 to 2.0, and 0.5 to 0.8, define a 'minor change', probably without much consequences for most water quality determinants. Change ratio of 2 to 5 and 0.5 to 0.2 are characteristic of an 'important change' and between 5 to 10 and 0.1 to 0.2 they qualify as 'major change'. Change ratio of more than 10 and less than 0.1 indicate a 'severe modification' for the determinants investigated. Such approach has long

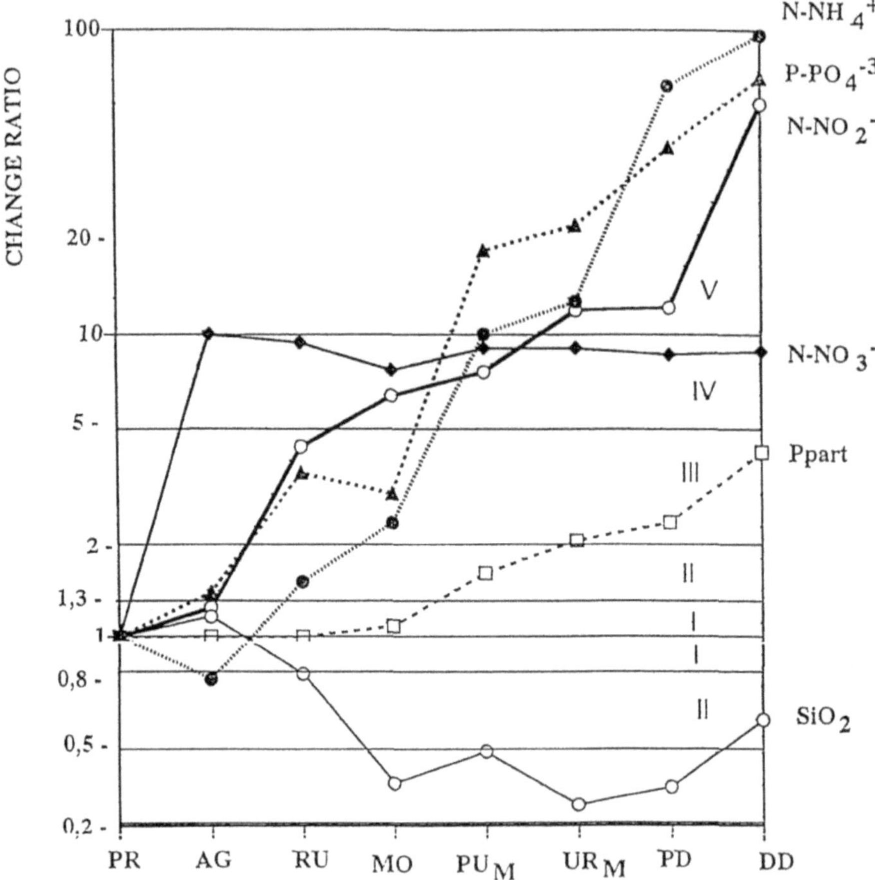

Figure 4. Schematized longitudinal variations of nutrients along the Seine River from pristine headwaters (PR) to river mouth (DD). Change ratio: mean analyses for the early 1990's normalized to pristine values. I: insignificant change from pristine value, II: minor change, III: important change, IV: major change, V: severe change. P part = particulate phosphorus from flood deposits. Position of sites see Table 2.

been used for particulate metals specially for which pristine levels could be determined by core analyses ('environmental archives'). It is extended here to dissolved components for which a average pristine reference has been set up from a representative set of small unpolluted basins.

The ten major water quality situations observed within the Seine basin can be investigated on the basis of the change ratio (Tables 1 and 2) and ordered along two major gradients: an upstream–downstream evolution as expressed by their stream order and an increasing population density.

Water chemistry and the stream order gradient

The gross longitudinal variations of water chemistry from pristine forested basins (orders 1 to 3) to the river mouth (order 8) is presented on Figure 3 (major ions) and Figure 4 (nutrients).

For major ions the impact of human activities is still moderate in the Seine basin. There is no significant change for Ca^{++} and Mg^{++}. This confirms the limited trend observed for Ca^{++} between 1848 and today. The anthropogenic impact for $SO_4^{=}$ is only significant downstream Paris (Figure 3, PD and DD) partly because natural $SO_4^{=}$ level is very high (25 mg l^{-1}) due to numerous gypsum deposits leached by groundwater. Sodium, potassium and chloride are much more sensitive to human impacts. There is first a minor increase due to diffuse agricultural and domestic sources, but de-icing salts on country roads cannot be ruled out, then a stabilization with little additional impacts of small towns. Then, the Paris megalopolis, and its industries, are responsible for an 'important change'.

Dissolved nutrients present three major patterns (Figure 4). Agricultural practices are responsible for

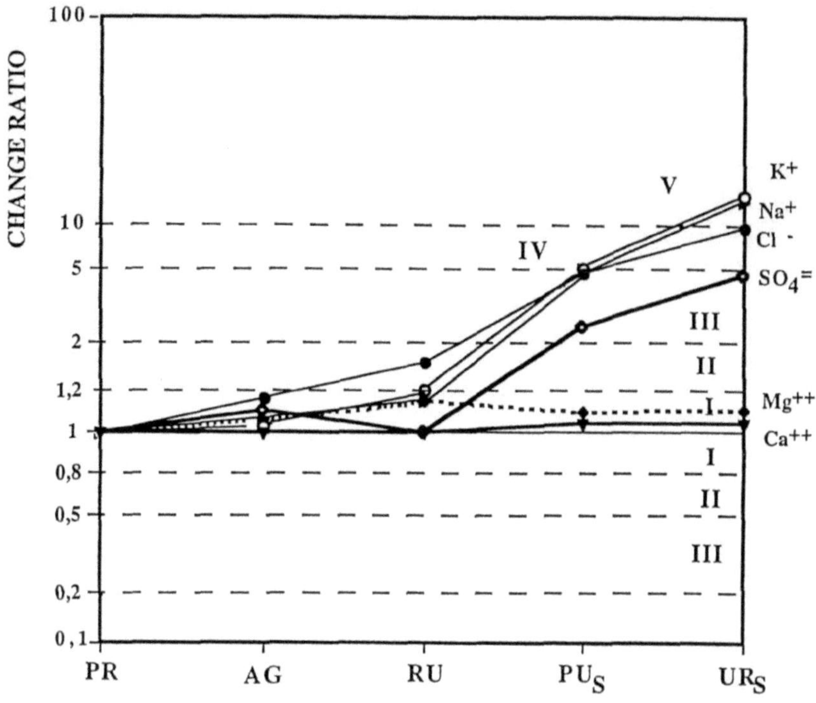

Figure 5. Schematized evolution of major ions on small streams (order 1 to 4), PR: pristine forested, AG: agricultural, RU: rural, PUs: periurban streams, URs urban 'streams' (sewers). Change ratio: mean analyses for the early 1990's normalized to pristine values. I: insignificant change from pristine value, II: minor change, III: important change, IV: major change, V: severe change. Position of sites see Table 2.

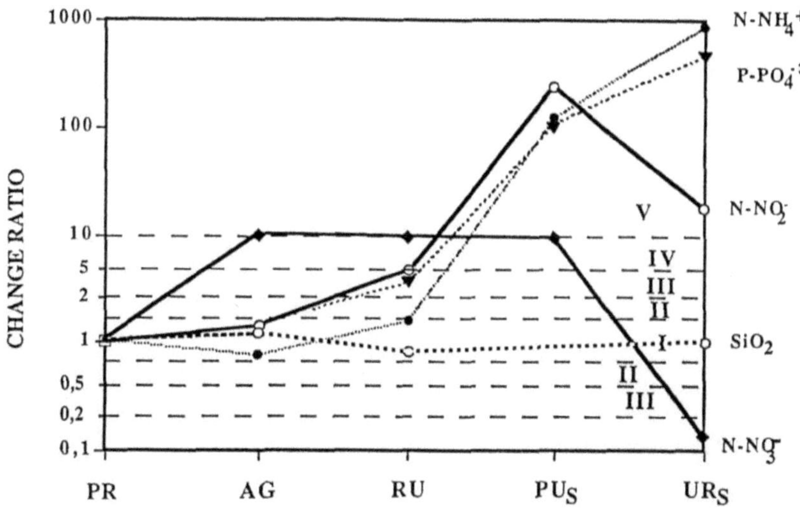

Figure 6. Schematized evolution of nutrients on small streams (order 1 to 4), PR: pristine forested, AG: agricultural, RU: rural, PU_S: periurban streams, UR_S urban 'streams' (sewers). Change ratio: mean analyses for the early 1990's normalized to pristine values. I: unsignificant change from pristine value, II: minor change, III: important change, IV: major change, V: severe change. Position of sites see Table 2.

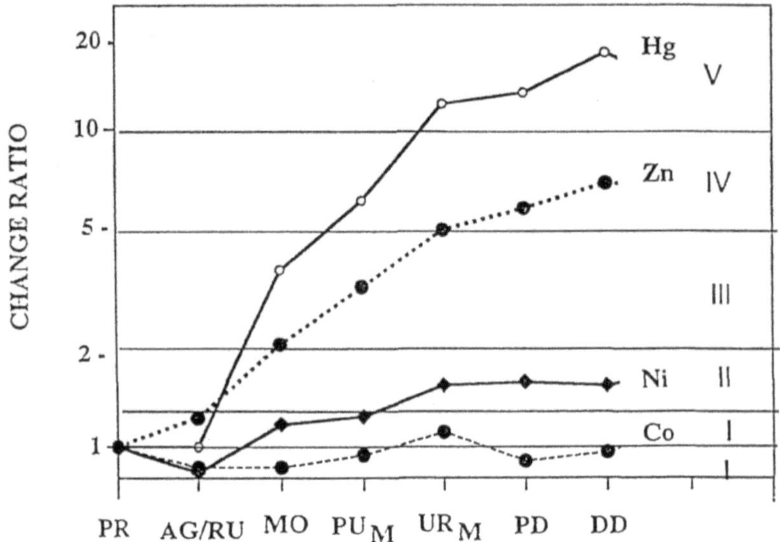

Figure 7. Schematized longitudinal variations of selected metals along the Seine River from pristine headwaters (PR) to river mouth (DD). Change ratio: mean analyses for flood deposits (January 1994 and 1995) normalized to pristine values. I: insignificant change from pristine value, II: minor change, III: important change, IV: major change, V: severe change. (Horowitz, Meybeck, Idlafkih, & Biger, in prep.). Position of sites see Table 2.

a dramatic increase of NO_3^- by a factor 10 outside forested basins (from PR to AG). This level is then very much constant throughout the Seine basin even across Paris. Domestic waste influence starts in rural conditions (RU) – particularly for PO_4^{-3} and NO_2^-, less for NH_4^+ – then regularly increases to the river mouth station (DD), where the maximum impact of Paris is observed for NO_2^- and NH_4^+.

The silica pattern is completely different (Figure 4). Natural levels are relatively high in the Seine basin, (12 mg SiO_2 l^{-1}) due to the presence of amorphous minerals in sedimentary deposits such as chalk (Meybeck, 1986). Along the fluvial continum silica is gradually taken up by aquatic vegetation, and average annual levels are less than 50% of natural values in the mid-order streams (MO to UR_M). In summer, concentrations can be much lower.

Water chemistry and the population density gradient

The evolution of major ions and nutrients in streams (10 to 100 km^2 typically), is presented on Figures 5, and 6, along a population density gradient from less than 1 km^{-2} (PR) to more than 10 000 km^{-2} (UR_S).

The pattern of major ions evolution is highly variable. Ca^{++} and Mg^{++} are not affected at all; Na^+, K^+, Cl^- and $SO_4^=$ present a major increase from rural settlements to 'urban streams' (sewers) where they are now five to 15 times higher than their expected pristine levels, 2000 y ago when Lutetia was founded.

Again the nutrient pattern is complex. In periurban conditions (PU_S) NO_3^- levels are still similar to the ones found in rural area (RU), but NH_4^+, NO_2^- and PO_4^3- increase proportionally to the population density i.e. by a factor 50. In urban streams (UR_S) NO_3^- is nearly completely reduced, NO_2^- is low, NH_4^+ and PO_4^{-3} reach their maximum values. Along this population gradient, silica is practically not affected.

Human impacts on river particulate metals from recent flood deposits

There is presently no regular monitoring of suspended particles chemistry in the Seine river as for the great majority of monitored rivers in the world (Meybeck et al., 1989). However, it is now well recognized that many micropollutants as heavy metals are preferentially transported in river systems in association with the particulates (Horowitz, 1995) which are now more and more considered in river monitoring. Therefore their spatial distribution from natural background to maximum human impacts should be studied on the particulate matter. Two types of samples are classically considered: suspended matter and deposited sediments (Thomas & Meybeck, 1992). Suspended particles study requires a specific sampling in order to take into account time variations, particularly during floods when particles fluxes are maximum (Idlafkih et al., 1997). The composition of deposited sediments is more stable in time but in contaminated sections, their spatial variability can be high and it is difficult to obtain representative samples, like downstream of Paris (Garban et al., 1996; Estèbe, 1996). A third approach, based on flood deposits has been developed here for the Middle and Lower Seine. For the upstream reference levels the deposited sediments from forest streams basins have been considered.

The average reference for four selected particulate metals (Co, Ni, Zn and Hg) and phosphorus (Horowitz, Meybeck, Biger, Idlafkih in prep) is based on fine bed sediments (less than 100 μm) at 28 pristine forested (PR) and 34 agricultural (AG) or rural (RU) sites from the monolithologic basins already considered for water analyses. They were analysed by ICP-MS after HNO_3-HF acid attack at the G.B.E. laboratory in Montpellier. For the pristine levels the averages of 15 analyses in forested limestone basin and 13 analyses in marl basins were compared and combined with two analyses of ancient flood plain deposit dating from 5000 BP obtained from an archaeological excavation at Bercy, few km downstream the Seine/Marne confluence. The reference values are 0.05 $\mu g.g^{-1}$ for Hg, 60 for Zn, 7 for Co and 14.4 for Ni. The two other sets of stream bed sediments from agricultural ($n = 18$) and rural ($n = 16$) sites have been included. For all the other types situations previously defined for water chemistry the flood deposits analyses were considered.

Flood deposits are relatively easy to obtain in amount > 100 g at sites of decreasing current velocity, on river banks, against bridges, steps, etc. When carefully collected, they integrate flood events over some days to more than one week and they are thought to be more time-representative than a few samples of suspended load during the flood, and much cheaper than samples obtained from a 6h continuous flow-through centrifugation. They were sampled by hand by the author shortly after the peak stage of the January 1994 and 1995 major floods, both exceeding the decennal floods with runoff reaching nearly 30 l s^{-1} km^{-2}. Most samples were taken under 30 cm of water, some were scooped soon after emmersion (1 to 3 d). All samples were analysed at the U.S. Geological Survey laboratory in Atlanta by A. Horowitz according to Horowitz, 1991. In the Seine basin the silt fraction (2–50 μm) is generally dominating in such deposits. Flood deposits grain size is similar to the one of suspended load collected by river sediment traps and their compositions is close to the TSS-weighted average contents in particulates transported by the Seine as checked at few stations on suspended particles. However, it must be noted that these exceptionnal flood deposits are usually less contaminated than suspended particles sampled at low flows and minor floods (Cossa et al., 1994; Idlafkih et al., 1997; Garban et al., 1996; Estèbe, 1996). Their contamination level should be considered as minimum since they mix up the polluted particles with much less polluted ones resulting from soil erosion.

The mid stream orders four to seven (MO), upstream of the Paris megalopolis, are defined by the median of 10 samples of flood deposits collected on the Seine upstream Montereau, the Yonne, Grand Morin, Aisne, Oise upstream Compiègne, and along the medium course of the Marne. The influence of Paris suburbs (PU_M) is appreciated by the mean of five samples on the Seine and Marne at their confluence, and on the Oise at Conflans. The lower reach of the Oise is here considered to be as impacted as that of the Seine and Marne, resulting from the inputs of Compiègne, Beauvais and Creil and of many industries along the Oise river (Figure 1). The Seine main course within Paris (UR_M) is defined on four samples taken near the Alma bridge and the Ile de la Jatte. The proximal impact (PD) is defined on three samples at Maison Lafitte, Conflans, and Mantes i.e downstream of the inputs of the Paris combined sewers overflows or of Achères sewers. Finally the distal influence of the megacity (DD) and of local pollution sources downstream Mantes is defined by three deposits taken at Poses in 1994 and 1995. The particulate chemistry of periurban and urban streams, not yet completed, is not discussed here.

The interpretation of particulate chemistry is also based on the change ratio: all average analyses are normalized to the reference levels. Change ratio are presented along the river course from the pristine levels (PR) to the distal downstream impact of Paris (DD) for particulate phosphorus (Figure 4) and for Co, Hg, Ni, and Zn (Figure 7). As for dissolved constituents, different patterns of elemental evolution in river particulates along the longitudinal gradient can be described. Samples in agricultural and rural sites have identical median values and were combined. They are not significantly different ($\pm 30\%$) from those of the forested streams. Downstream the rural sites to the river mouth (DD) cobalt stays stable across the whole stream order gradient. Nickel starts to significantly increase in the urban section of the main river course (UR_M), then becomes stable. Zinc and mercury present the maximum sensitivity to human impacts. They increase continously from the rural sites to the river mouth where mercury levels during floods stages are nearly 20 times higher than for the pristine stream. Particulate phosphorus is stable from forest streams to the mid-order streams (change ratio < 1.3) then significantly increases in the suburbs (PU_M) and continues to grow until the river mouth station, where the change ratio is > 4, an indication of a 'major impact'.

Typology of anthropogenic impacts in the Seine River

The Seine River illustrates the two well-described impacts, the diffuse sources impacts and the point sources impacts. Both affect the river downstream of pollution sources as the Paris city impacts at proximal and distal positions. Two other types, so far less studied and linked to the megacity, can be presented: the 'radial impacts' and the 'inverse impacts'.

Pollutants originating from diffuse sources of pollution, as nitrate and fertilizer contamination, affect all stream orders in the Seine basin from order one or two to the river mouth. Dispersed or semi-diffuse sources of pollution characterize the dissolved and particulate phosphorus inputs which occur as soon as domestic wastes are collected and released in the river network without appropriate treatment. Both very high nutrient inputs are responsible for a marked eutrophication process starting at Strahler order six for the phytoplankton, closely controlled by riverine morphology and hydrology (Garnier et al., 1995).

The 'downstream impacts' of point sources are classical but megalopolis impactsd are here peculiar for two reasons: (i) the ratio of total population over river discharge, or 'population stress' (Moody & Battaglin, 1995) is high (Table 2). For the lower Seine it is on the average, 35 people l^{-1} s and reaches, 200 people per l^{-1} s at low flows. This figure is probably one of the highest in the world, only surpassed by a few other megalopolis like Dehli and Caracas. For instance the yearly average population stress is only between 4 and 10 along the Mississippi river course (Moody & Battaglin, 1995). (ii) the centre of gravity of impacts on the main river course during the dry periods is not in middle of the megalopolis but shifted 75 km downstream due to the collection and partial treatment of most sewage waters at Achères (see PD situation on Figures 3, 4 and 7). This impact structure is modified during rainstorm events over Paris when the maximum impact of urban runoff is observed only 30 to 40 km downstream of the Eiffel tower. It must be noted that, if the Achères treatment plant did not exist during the summer droughts, the main river course would be anoxic then highly hypoxic over more than 100 km. This phenomenon has been observed in the past for instance during strike periods of the treatment plant or before its completion. During these periods the denitrification rate was so high that NO_3^- contents were less than 0.1 mg N-NO_3^- l^{-1} downstream of Achères (Chesterikoff et al., 1992).

'Radial impacts' describe the megalopolis influence on small local streams (10 to 100 km^2) from the center to the periphery. The highest impacts are of course noted for the ultimate alteration stage when streams are fully covered and converted into sewers networks. This type of impact defining 'urban streams' is noted for Paris at 10 to 15 km around the Eiffel tower, where more than 80% of basin area are urbanized. For 'periurban streams' there is a gradual decrease of the megalopolis impact from 15 to 50 km off the city center, where the urbanization rate drops from 80% to 10%.

Other radiating impacts of the megalopolis can be observed, such as: (i) physical modification of the river course related to wetland filling, river channelization of the riverbank, sand pits excavations in alluvial plains. (ii) land-fills contamination with urban wastes, and field fertilization with treated sewage sludges still rich in many micropollutants – as far as 100 km from Achères; and (iii) polluted atmospheric fallout. Concerning the latter, the impact of Paris and its industries is proven by a major plume of excess Cl^- (3 to 4 times) and $SO_4^{=}$ (10 to 20 times) in rainfall with regards to ocean-derived salt based on the Na^+ content of rain. This plume extends from the Rouen–Paris corridor and spreads eastward of Paris for at least 100 km (Thibert, 1994). The urban atmospheric fallout is also an important source of heavy metals and organochlorine compounds (Granier et al., 1992). Yet, impact on the Pb or Zn contents of forested soils is not proven (Biger, 1996).

'Inverse impacts' are those caused in the upstream river course. They are mostly linked to hydrological changes related to megalopolis water management. The demand for sustained water discharge during low flows for safe Paris drinking water supply and for navigation requirements, has resulted in the creation of multi-purpose reservoirs sometimes far upstream of the megalopolis (more than 150 km for the Seine, Aube and Marne reservoirs). They mostly concern the major tributaries (here orders 5 and 6). Their influence on the river system is multiple and can be seen as beneficial for some of the water quality management issue as nitrates. However, such reservoirs also affect the river channel morphology downstream of impoundments or on diverted reaches (Brookes, 1996), and they invert the water regime along the diverted reaches. Their impact on the aquatic biota (e.g. the reproduction of pike in floodplains; Boët pers. com.) is under investigation. In other basins the inverse impacts resulting from dam construction are often related to the energy demand, or to flood control, of the megacity.

Perspectives

(1) Chemical impacts on water and particulates are highly variable within a given river basin. Chemical determinants can be stable as, (here Mg^{++} or particulate Co), chemically regulated (Ca^{++} and HCO_3^-), moderately modified (Na^+, K^+, Cl^-, $SO_4^=$, particulate P and Ni), or severly modified (NO_3^-, PO_4^{-3}, particulate Hg and Zn). Some variables may even be found at levels lower than in pristine conditions, as dissolved SiO_2 and TSS below dams. The longitudinal impact pattern is not the same for all determinants and is directly linked with the pollution mode and location. For instance NO_3^- can already be at their maximum at stream order one while PO_4^{-3} is maximum in the most downstream reach of the Seine. The 'Change ratio' with regard to pristine reference values is a good working tool to compare the time and space evolution of water quality within basins and between basins, provided that a pristine reference can be established.

(2) Water quality can be appreciated from several view points (Meybeck, 1996). The first one is the water quality management perspective comparing the actual water chemistry, with regard to various existing or potential uses. It is generally based on multipurpose regulatory water quality thresholds. A second one is the environmental assessment in a conservation perspective: the evolution of the water and particulates chemistry is considered with regards to their initial or pristine levels without any reference to specific water uses. This purely descriptive viewpoint was developed here through the change ratio. A third view point takes into account the potential toxicity of chemicals on the aquatic biota. For this, the change ratio is sometimes not operational: for instance the aquatic biota may well tolerate a 10 times increase of Cl^- (e.g. from 5 to 50 mg l^{-1}) while a doubling of particulate phosphorus content or of some dissolved metals may have serious consequences. A sound environmental management should be based on all these different approaches, thus using several water quality scales. Actually most of the existing management tools refer to a single combined water quality scale poorly adapted to the various management objectives.

(3) Documenting natural or pristine conditions is essential to set up environmental management. But they are difficult to be established in populated, industrialized, and agricultural river basins. For the Seine basin we postulated that such conditions are met in forested stream basins area where minimum nutrients, Na^+, K^+, Cl^-, and particulate heavy metals levels were actually found. This approach was validated for major ions using past water analyses, and for particulate metals with analyses of pre-historic fluvial deposits: in both cases the present average forest streams levels were very close to the other estimates. However, such small forested basins are usually not surveyed by most water authorities, since little use of the water is made there. For many water authorities in Western Europe the 'non polluted' reference levels are still based on stream orders three and four, which may be already quite modified by agricultural and domestic impacts.

(4) Point sources impact, as sewage outfalls, have long been the major water quality issue considered. In the 1970's the acidification of stream ecosystem by diffuse atmospheric sources began to be studied, then nutrient spiralling (Newbold, 1996), and the role of ecotones (Naiman & Decamps, 1990). Other developping approaches take into account the whole river basin and its hydrological structure based on stream order to understand the origins and fate of organic carbon in pristine river systems (Wetzel & Ward, 1996) as for the Moisie in Quèbec (Naiman, 1982). Recent eutrophication, models developped for the Seine basin (Billen et al., 1994) take into account nutrients inputs, channel morphology, hydrological variations and light penetration along the Seine from stream order one to eight. They simulates the average phytoplancton and bacterial biomasses, the nutrients levels, and the average 0_2 level for each stream order. When considering the impact of a megalopolis such as Paris a specific water quality and hydrological models are coupled to the Strahler eutrophication models taking into account sewers inputs in dry and wet conditions.

(5) The 'population stress' of the aquatic environment also looks promising for environmental assessments and can be tested on nutrients, major ions and some micropolluants in other river basins, where

it may explain much of their variability in time and space. Similar indicators of other human pressures on aquatic ecosystems should be developed to account for other impacts from pollutant sources such as urban surface runoff, industries, and mines, for the impacts of dams, for channel artificialization etc.

Acknowledgements

The author is grateful to students and colleagues who provided basic data used here: E. Biger, Z. Idlafkih & S. Thibert (Géologie Appliquée; Université de Paris VI); D. Cossa (Ifremer, Nantes), A. Ficht (SNS, Rouen) & R. Person (LHVP, Paris), S. Romon (ASSI). Communication of unpublished metal analyses by A. Horowitz (US Geol. Survey, Atlanta) is warmly appreciated as well as the support of M. Monnin (CNRS, Montpellier) in sediment analyses of pristine basins and corrections by V. Lemaire-Drinkwater. This work greatly benefited from many discussions with several colleagues of the PIREN-Seine programme managed by G. de Marsily & E. Fustec (Géologie Appliquée, Université de Paris VI) and from the scrupulous corrections of one reviewer. All institutions financing the PIREN-Seine are acknowledged.

References

A.E.S.N. (1973–1978). Les bassins de la Seine et des cours d'eau normands. Agence de l'Eau Seine-Normandie, Nanterre, 12 vols + Atlas.

A.E.S.N., 1991. Memento statistique, Agence de l'Eau Seine-Normandie, Nanterre, 126 pp.

Ackerman, W. C., R. H. Harmeson & R. A. Sinclair, 1970. Some long-term trends in water quality of rivers and lakes. Eos, Trans. Am. Geoph. Union, 51: 516–522.

Allardi, J., P. Boet, J. Belliard & L. Boughida, 1993. Influences des actions anthropiques sur le peuplement ichtyologique de la Seine. In Fustec, E. & G. De Marsily (eds) 1993. La Seine et son Bassin: de la Recherche à la Gestion. Colloque, CNRS, Paris: 222–237.

Antweiler, R. C., D. A. Goolsby & H. E. Taylor, 1995. Nutrients in the Mississippi River. In Meade, R. H. (ed.), Contaminants in the Mississippi River, 1987–1992. US Geol. Survey Circ. 1133: 73–86.

Barillier, A., J. Garnier & M. Coste, 1993. Experimental water release: impact of the water quality on a river 60 km downstream (upper Seine river, France). Wat. Res. 27: 635–643.

Belliard, J., P. Boet & E. Tales, 1996. Regional and longitudinal patterns of fish community structure of the Seine river basin, France. Envir. Biol. Fish. 50: 133–147.

Biger, E., 1996. Mesure du bruit de fond géochimique naturel et mise en évidence de l'influence anthropique en zone rurale dans le bassin de la Seine. Diplome d'Etudes Approfondies, Lab. Géologie Appliquée, Univ. Paris VI, 37 pp.

Billen, G., J. Garnier & P. Hanset, 1994. Modelling phytoplankton development in whole drainage networks: the Riverstrahler Model applied to the Seine river system. Hydrobiologia 289: 119–137.

Boet, P., B. Duvoux, J. Allardi & J. Belliard, 1994. Incidence des orages estivaux sur le peuplement piscicole de la Seine à l'aval de l'agglomération parisienne (bief Andresy-Mericourt). La Houille Blanche, 1–2: 141–147.

Brookes, A. 1996. River Channel Change. In Petts, G. E. & P. Calow (eds), River Flows and Channel Forms, Blackwell Science London: 221–241.

Caille, A., P. Campbell, M. Meybeck & J. L. Sasseville, 1973. Etude du fleuve St Laurent: effluent urbains de l'agglomération de Montréal. Rapport Interne, INRS Eau, Université du Québec, Québec, 203 pp.

CGE-SEDIF (1996). Qualité des rivières aux prises d'eau du Syndicat des eaux d'Ile de France, bilan de l'année 1995, Compagnie Générale des Eaux, Paris, 124 pp.

Chebbo, G., J. M. Mouchel, A. Saget & M. Gousailles, 1995. La pollution des rejets urbains par temps de pluie: flux, nature et impact. Techn. Sci. Munic. L'Eau, 1995: 796–806.

Chesterikoff, A., B. Garban, G. Billen & M. Poulin, 1992. Inorganic nitrogen dynamics in the river Seine downstream from Paris (France). Biogeochem. 17: 147–164.

Chesterikoff, A., B. Garban & D. Ollivon, 1991. Daily rhythms in the river Seine: relative impacts of natural and anthropogenic factors. Wat. Res. 25: 1523–1528.

Chocat, B., M. Cathelain, A. Mares & J. M. Mouchel, 1994. La pollution due aux rejets urbains par temps de pluie: impacts sur les milieux récepteurs. La Houille Blanche, 1/2: 97–105.

Cossa, D., M. Meybeck, Z. Idlafkih & B. Bombled, 1994. Etude pilote des apports en contaminants par la Seine. Report IFREMER, Nantes, R. Int. Del. 94. 13, 151 pp.

Dessery, S., C. Dulac, J. M. Laurenceau & M. Meybeck, 1984. Evolution du carbone organique particulaire ≪algal≫ et ≪détritique≫ dans trois rivières du Bassin Parisien. Arch. Hydrobiol. 100: 235–260.

Dodge, D. P. (ed.), 1989. International Large River Symposium (LARS). Can. Spec. Publ. Fish. aquat. Sci., 106, 629 pp.

Dynesius, M., & C. Nilsson, 1994. Fragmentation and flow regulation of river systems in the Northern Third of the world. Science 266: 753–762.

Estebe, A., 1996. Impact de l'agglomération parisienne et de ses rejets de temps de pluie sur les concentrations en métaux des matières en suspension et des sédiments en Seine en période estivale. Thèse de doctorat de l'Université de Paris XII – Val de Marne, 237 pp.

Ficht, A., 1995. Etudes de pollutions en Seine à l'aval de Poses. Bilan de l'année 1994. (Yearly reports on the quality of the Seine at Poses and in its estuary). Service de la Navigation de la Seine, Rouen.

Fraser, A. S., M. Meybeck & E. D. Ongley 1995. Water Quality of World River Basins, UNEP, Nairobi, Earthwatch Environment Library 14, 39 pp.

Fustec, E. & G. De Marsily (eds), 1993. La Seine et son Bassin: de la Recherche à la Gestion. Colloque CNRS, available at: Lab. Géologie Appliquée, Université de Paris 6, 75252, Paris, 383 pp.

Garban, B., D. Ollivon, A. M. Carru & A. Chesterikoff, 1996. Origin, retention and release of trace metals from sediments of the river Seine. Wat. Air Soil Pollut. 87: 363–381.

Garnier, J., G. Billen & M. Coste, 1995. Seasonal succession of diatoms and chlorophycae in the drainage network of the river Seine: observations and modelling. Limnol. Oceanogr., 40: 750–765.

Garnier, J., G. Billen, P. Hanset, P. Testard & M. Coste, 1993. Developpement algal et eutrophisation dans le réseau hydrographique de la Seine. In Fustec, E. & G. De Marsily (eds), 1993. La Seine et son Bassin de la Recherche à la Gestion. Colloque, CNRS, Paris: 289–310.

Granier, L., M. Chevreuil, A. M. Carru & A. Chesterikoff, 1992. Atmospheric fallout of organochlorines and heavy metals on the Paris area (France). Sci. Tot. Envir. 126: 165–172.

Gleick, P.H. (ed.), 1993. Water in Crisis: a Guide to the World's Fresh Water Resources, Oxford Univ. Press, 730 pp.

Guillerme, A., 1990. Le testament de la Seine. Rev. Geogr. Lyon, 65: 240–250.

Horowitz, A. J., 1991. A Primer on Sediment-Trace Element Chemistry, 2nd edn. Lewis Publ. Cy., Mich.

Horowitz, A. J., 1995. The use of suspended sediment and associated trace elements in water quality studies. Int. Ass. Hydrol. Sci. Spec. Publ. 4, 58 pp.

Idlafkih, Z., M. Meybeck, J. F. Chiffoleau, D. Cossa & A. Ficht, 1997. Comportement des métaux particulaires (Al, Fe, Mn, Cd, Cu, Hg, Pb et Zn) dans la Seine à Poses en période de hautes eaux (1990/95). Int. Ass. Hydrol. Sci. Publ. 243: 45–58.

Idlafkih, Z., D. Cossa & M. Meybeck, 1995. Comportement des contaminants en trace dissous et particulaires (As, Cd, Cu, Hg, Pb, Zn) dans la Seine. Hydroécol. Appl. 7: 127–150.

Idlafkih, Z. & M. Meybeck, 1997. Rivers flux monitoring optimisation: two case studies on the Seine and Rhine rivers. In Ottens, J. J., F. A. M. Claessen, P. G. Stoks, J. G. Timmernan & R. G. Ward (eds), Monitoring Taylor Made II, int. workshop Nunspeet, Sept 1996, Nederland: 331–338.

Livingstone, D. A., 1963. Chemical composition of rivers and lakes. Data of geochemistry, Chapter G. U.S. Geol. Surv. Prof. Paper 440G: G1–G64.

Maneglier, H., 1992. Une grande ville et un bassin. Paris et la Seine. Rev. Géogr. Lyon 67: 299–303.

Mangerel, P., 1982. La Seine et ses affluents autour de Paris: teneurs en coliformes de 1920 à nos jours. In. Assoc. Française de Limnol., Annual Meeting, SRAE, Orléans: 200–205.

Manickam, S. L. Barbaroux & Ottmann, 1985. Composition and mineralogy of suspended sediment in the fluvio-estuarine zone of the Loire river, France. Sedimentology 32: 721–741.

Meade R. H. & R. S. Parker, 1985. Sediments in rivers of the United States. In National Water Summary 1984, US. Geol. Survey Water-Supply Paper 2275: 49–60.

Meade, R. H. (ed.), 1995. Contaminants in the Mississippi river, 1987–1992. US Geol. Survey Circ. 1133: 13–30.

Meybeck, M., 1986. Composition naturelle des ruisseaux non pollués en France. Sci. Géologique Bull. Strasbourg, 39: 3–77.

Meybeck, M., 1996. River water quality, global ranges, time and space variabilities, proposal for some redefinitions. Verh. Int. Ver. Limnol. 26: 81–96.

Meybeck, M., G. Cauwet, S. Dessery, M. Somville & D. Gouleau, 1988. Nutrient (organic C, P, N, Si) in the eutrophic river Loire (France) and its estuary. Estuarine, Coastal and Shelf Science 27: 595–624.

Meybeck, M., D. Chapman & R. Helmer, 1989. Global Freswater Quality, a First Assessment. Blackwell, London, 306 pp.

Meybeck, M., & Helmer, R., 1989. The quality of rivers: from pristine stage to global pollution. – Global Planet. Change 1: 283–309.

Meybeck, M. & A. Ragu, 1996. River discharge to the oceans: an assessment of suspended solids, major ions, and nutrients. UNEP/WHO, Environment of Information and Assessment Div., UNEP, Nairobi, 245 pp.

Meybeck, M. & A. Ragu, 1997. Presenting Gems Glori, a compendium of world river discharges to the oceans. Int. Ass. Hydrol. Sci. Publ. 243: 3–14.

Milliman, J. D. & R. H. Meade, 1983. World-wide delivery of river sediment to the oceans. Journal of Geology: 1–19.

Miquel, J. & A. Levassor, 1993. Les barrages réservoirs dans le contexte du bassin de la Seine. In Fustec, E. & G. De Marsily (eds), 1993. La Seine et son Bassin de la Recherche à la Gestion. Colloque CNRS, Paris: 23–48.

Mouchel, J. M., 1993. Impacts des rejets urbains des temps de pluie de l'agglomération parisienne sur les concentrations d'ammonium et d'oxygène à Chatou. In Fustec, E. & G. De Marsily (eds), La Seine et son bassin, de la recherche à la gestion, CNRS, Paris: 187–200.

Mouchel, J. M., L. Simon & M. A. Maldiney, 1994. Impacts en Seine des rejets urbains de temps de pluie sur les concentrations d'oxygène dissous. La Houille Blanche 1/2: 135–141.

Mouchel, J. M. & M. Seidl, 1993. Impacts des rejets urbains en temps de pluie sur les concentrations d'ammoniaque et d'oxygène à Chatou. In Impacts des surverses d'orage sur la qualité des eaux de la Seine. Rapport de Synthèse IV 1989–92, Piren-Seine, Université de Paris VI.

Muxart, T., M. J. Penven, N. Fauchon & M. Garmouna, 1993. Les mesures de la qualité des eaux des rivières du Vannetin, de l'Aubertin, du Grand Morin et de la Marne. In Fustec, E. & G. De Marsily (eds), 1993. La Seine et son Bassin: de la Recherche à la Gestion. Colloque, CNRS, Paris: 127–147.

Moody, J. A. & W. A. Battaglin, 1995. Chemical character of the Mississippi River. In Meade R. H. (ed.), Contaminants in the Mississippi river, 1987–1992. US Geol. Survey Circ. 1133: 31–39.

Naiman, R. J., 1982. Characteristics of sediment and organic carbon export from pristine boreal forest watersheds. Can. J. Fish. aquat. Sci. 39: 1699–1718.

Naiman, R. J. & H. Decamps (eds), 1990. The Ecology and Management of Aquatic-Terrestrial Ecotones. Man and Biosphere Séries, vol. 4, Unesco-Parthenon Pub.

Naves, J., G. Bousque, P. Leroy, P. Hubert & R. Vilagines, 1990. Evolution de la qualité de l'eau de la Seine à Ivry-Sur-Seine (France) de 1887 à 1986. In Int. Symp. Application des modèles mathématiques à l'évolution des modifications de la qualité de l'eau, ENIT, Tunis, 7–12 Mai 1990: 35–44.

Newbold, J. D., 1996. Cycles and spirals of nutrients. In Petts, G. E. & P. Calow (eds), River Flows and Channel Forms, Blackwell Science London: 130–159.

Ollivon, D., B. Garban & A. Chesterikoff 1995. Analysis of the distribution of some polycyclic aromatic hydrocarbons in sediments and suspended matter in the river Seine (France). Water, Air, Soil Pollution, 81: 135–152.

Pelletier, J., 1990. Sur les relations de la ville et des cours d'eau. Rev. Geogr. Lyon 65: 233–239.

Penven, M. J. & T. Muxart, 1995. Le drainage agricole: un rôle fondamental dans les transferts d'eau et de matière, l'exemple du Plateau Briard. Annales de Géographie 581: 582: 88–104.

Petts, G. E. 1989. Perspectives for ecological management of regulated rivers. In Gore, J. A. & G. E. Petts (eds), Alternatives in Regulated River Management, CRC Press: 3–24.

Petts, G. E. & P. Calow (eds), 1996a. River Restoration, Blackwell Science London, 231 pp.

Petts, G. E. & P. Calow (eds), 1996b. River Biota, Blackwell Science London, 257 pp.

Petts, G. E. & P. Calow (eds), 1996c. River Flows and Channel Forms. Blackwell Science London, 262 pp.

Reynolds, C. S. 1996. Algae. In Petts, G.E. & P. Calow (eds), River Flows and Channel Forms. Blackwell: 6–26.

RNB: Reseau National de Bassin. Regular National Water Quality Survey; 1971–1995. Data available from the Agence de l'Eau Seine Normandie, 92000, Nanterre, France.

Roberts, G. & T. Marsh, 1987. The effects of agricultural pratices on the nitrate concentrations in the surface water domestic supply sources of western Europe. Int. Ass. Hydrol. Sci. Publ. 164, 365–380.

Santiago, S., R. L. Thomas, G. Larbaigt, C. Corvi, D. Rossel, J. Taradellas, D. J. Gregor, L. McCarthy & J. P. Vernet, 1994. Nutrient heavy metal and organic pollutant composition of suspended and bed sediments in the Rhone river. Aquat. Sci. 56: 220–242.

Servais, P. & J. Garnier, 1993. Contribution of heterotrophic bacterial production to the carbon budget of the river Seine (France). Microbiol. Ecol. 25: 19–33.

Thibert, S. 1994. Exportations naturelles et anthropiques des ions majeurs et des éléments nutritifs dans le bassin de la Seine. Thèse de doctorat de l'Univ. de Paris VI, 204 pp.

Thomas, R. L. & M. Meybeck, 1992. The use of particulate matter. In Chapman, D. (ed.), Assessment of the Quality of the Aquatic Environment through Water, Biota and Sediment. Chapman and Hall, London: 317–369.

Tsirkunov, V. V., I. K. Akuz & A. A. Zenin, 1998. The lower Don Basin. In Kimstach, V., M. Meybeck and E. Barsudy (eds), From Dnieper to Baikal, an Assessment of Water Quality Issues in the Former Soviet Union, UNEP/WHO, E & FN Spon, London, in press.

Tsirkunov, V. V., I. K. Akuz, & A. A. Zenin, 1998. The lower Don Basin. In Kimstach, V. & M. Meybeck (eds), From Dnieper to Baikal, an Assessment of Water Quality Issues in the Former Soviet Union, UNEP/WHO.

Tsirkunov, V. V., A. M. Nikanorov, M. M. Laznik & Z. Dongwei, 1991. Analysis of long-term and seasonal river water quality changes in the Latvian S.S.R., Soviet Union. Wat. Res. 26: 1203–1216.

Turner, R. E. & N. N. Rabalais, 1991. Changes in Mississippi river water quality this century and implication for coastal food webs. Bioscience: 140–147.

Van der Weijden, C. H., J. J. Middelburg, 1989. Hydrogeochemistry of the River Rhine: long term and seasonal variability, elemental budgets, base levels and pollution. Wat. Res. 23: 1247–1266.

Van Dijk, G. M. & E. C. L. Marteijn (eds), 1993. Ecological rehabilitation of the River Rhine. Ecological Rehabilitation of the rivers Rhine and Meuse, Rpt 50, RIZA/RIVM, Nederland, 62 pp.

Verbanck, M., J. P. Vanderborght & R. Wollast, 1989. Major ions content of urban waste water: assessment of per capita loading. Res. J. Wat. Poll. Contr. Fed. 61: 1722–1728.

Vörösmarty, C. J., M. Meybeck, B. Fekete & K. Sharma, 1997. The potential impact of neo-castorization on sediment transport by the global network of rivers. Int. Ass. Hydrol. Sci. Publ. 245: 261–282.

Webb, B. W. & D. E. Walling, 1996., Water Quality. II Chemical characteristic. In Petts, G. E. & P. Calow (eds), River Flows and Channel Forms, Blackwell Science London: 102–129.

Wetzel, R. G. & A. K. Ward, 1996. Primary production. In Petts, G. E. & P. Calow (eds), River Flows and Channel Forms, Blackwell Science London: 168–183.

WHO/UNEP, 1992. Water Quality. Progress in the implementation of the Mar del Plata Action Plan and a Strategy for the 1990s, WHO, Geneva, 79 pp.

WRI, 1991. Special report on the water quality of the Po river basin, Water Research Institute (CNR), Roma, Quad. Ist. Ric. Acque, 92, 400 pp.

Hydrobiologia **373/374:** 21–25, 1998.
J.-C. Amiard, B. Le Rouzic, B. Berthet & G. Bertru (eds),
Oceans, Rivers and Lakes: Energy and Substance Transfers at Interfaces.

Mechanisms for internal loading of phosphorus in lakes

Kurt Pettersson
Erken Laboratory, Institute of Limnology, Uppsala university, Norr Malma 4200, S-761 73 Norrtälje, Sweden

Key words: internal loading, phosphorus, fractionation

Abstract

Lake sediments play an important role in the phosphorus metabolism in lakes. The impact depends on the tendencies to retain and to release phosphorus.The internal loading will often determine the eutrophication status of the lake and the time lag for recovery after reduction of the external loading. Internal loading is most important during the summer. The potential source of phosphorus in the surficial sediments is very large in comparison to the pools in the water column. This means that even if only a very small amount is released, it will have significant impact on the phosphorus concentration in the lake water.

The distribution of phosphorus forms in sediments have been investigated since the 1950s. Generally, vertical profiles of sediment phosphorus content, expressed on dry weight basis, show an increasing total content towards the sediment surface. The vertical profile continues up into the water when looking at the particulate matter.

A number of environmental factors are important in the mobilisation processes. Most studies indicate that sediment bacteria have a significant role in uptake, storage and release of phosphorus including anaerobic release of iron-bound phosphorus. Several phytoplankton species have resting stages overwintering on the sediment. When growth is induced, they leave their habitat in order to shift their life-form to a pelagic one. In Lake Erken the cyanobacteria *Gloeotrichia echinulata* has been shown to contribute significantly to the internal loading of phosphorus during the summer.

Introduction

Lake sediments play an important role in the phosphorus metabolism in lakes. The impact depends on, firstly, the tendency to retain phosphorus and thereby decrease the availability in the water and, secondly, on the ability to release phosphorus when environmental conditions change. The composition, prehistory and binding capacity of the sediment govern its reactions. The general view until the 1970s was that sediments acted as sinks for phosphorus with the exception of periods of anoxia when iron-bound phosphate is released to the hypolimnion of stratified lakes (Einsele, 1936; Mortimer, 1941, 1942). During the last decades more knowledge has been gathered (Boström et al., 1982) and a much more complicated picture has developed emphasizing the important role of internal loading for lake water quality and especially summer blooms of phytoplankton in lakes.

Inorganic and organic particles are almost continuously deposited on lake bottoms. Allochtonous particles are to a significant amount minerogenic and more or less inert. The retention in the sediment is high. Autochtonous plankton and detritus deliver organic material with a high phosphorus content to the sediment. Precipitation events in the lake water will also contribute to inorganic compounds in the sediment as do the shells of diatoms. Due to various mechanical, physical, chemical and biological processes the phosphorus from these particles can be released into the water again. This release is generally called internal loading to be separated from external loading. In hypertrophic lakes, the role of the sediments come into focus after a reduction of the external loading. The internal loading will then determine the eutrophication status of the lake and the time lag for recovery. The process is most important during the summer when the external loading generally is at its minimum in temperate lakes. In the summer it is not unusual for the phosphorus concentration in the water of eutrophic lakes to be several times higher than in the inflowing water.

The potential phosphorus source in the surficial sediments is very large in comparison to the pools in the water column (Boström et al., 1982). This means that even if only a very small amount is released, it will have significant effect on the phosphorus concentration in the lake water. Phosphorus can leave the sediment in a dissolved state mainly as a phosphate or as particles. In the latter case, there are two major ways: as resuspended sediment particles or by migration of resting stages of phytoplankton. To leave the sediment in the form of a dissolved state, phosphorus must be mobilised by physical, chemical or biological processes and then transported to the lake water.

Material and methods

Results are presented from two lakes in south-eastern Sweden, Lake Erken and Lake Limmaren. Lake Erken is a moderately eutrophic, dimictic lake with an area of 24 km^2 and a drainage area of 144 km^2. The average depth is 9.0 m and the maximum depth 21 m. The theoretical water residence time is 7 years. Lake Limmaren is a non-stratified eutrophic lake with a surface area of 5.5 km^2 and a watershed of 22.5 km^2, a mean depth of 4.6 m and a maximum depth of 7.8 m. The water residence time has been estimated to 3 years. Both watersheds are dominated (more than 60%) by forests.

Fractionation of sediment phosphorus was performed on sediments as well as suspended and settling matter in Lake Erken following the scheme of Hieltjes and Lijklema (1980). In addition total phosphorus was determined by persulfate digestion in the sodium hydroxide extract. Suspended particles were collected on Whatman GF/F filters before analysis. Settling particles were collected fortnightly in sedimentation traps (no preservatives) with 0.05 m diameter and 0.3 m length. Pore water concentrations of phosphate in the sediments of Lake Erken and Lake Limmaren were determined using the method of Murphy & Riley (1962) This was determined after a five day incubation (darkness) period at the ambient temperature of the lake water in the laboratory. Plastic rods covered with a dialysis membrane (Löfgren & Enell, 1988) were placed in undisturbed sediment cores.

Fractional composition of sediment phosphorus

The distribution of phosphorus forms in sediments have been investigated since the 1950s when Chang & Jackson (1957) presented an extraction scheme for soils that was adapted to investigations of lake sediments (Williams et al., 1971, 1976). With this method as a base, a number of modifications have been presented. In later years the method of Hieltjes & Lijklema (1980) has been widely used to separate inorganic phosphorus forms. In this scheme labile, iron- and aluminium-bound and calcium-bound phosphorus is distinguished. The extraction agents are ammonium chloride, sodium hydroxide and hydrochloric acid, respectively. Furthermore, organic phosphorus extracted with sodium hydroxide can be determined using a total phosphorus analysis from which the amount of phosphate is substracted. Total organic phosphorus in sediments is usually determined as the difference between total sediment phosphorus and the sum of inorganic fractions.

Generally, vertical profiles of sediment phosphorus content expressed on dry weight basis, show an increasing total content towards the sediment surface (Boström et al., 1982). Organic as well as iron- and aluminium-bound phosphorus are the two fractions explaining this fact. In Lake Erken, the fractional composition of phosphorus in suspended matter, sedimenting particles and surficial sediment was determined during summer (Figure 1). There were drastic changes from a very high amount of labile phosphorus in the suspended matter to only a few percent in the surficial sediment. The other inorganic forms increased in proportion when going from suspended particles to settling ones and lastly to the sediment. Organic phosphorus extracted with sodium hydroxide decreased drastically from more than 50% in the suspended matter to less than 25% in the sediment. Furthermore, the total phosphorus content decreased from 5.7 mg g^{-1} dw in suspended matter to 2.2 mg g^{-1} dw in the settling particles and less than 2 mg g^{-1} dw in the sediment. Thus the vertical profile continues up in the water when looking at the particulate matter. This clearly indicates that phosphorus rich material in the water releases phosphorus on its way down to the sediment and that a further release takes place from the sediment. In the case when iron-rich material is transported to the sediment, the release to lake water can be replaced by a binding to iron and an accumulation in the sediment. Potentially mobile phosphorus in sediments is poorly defined. Phosphorus adsorbed

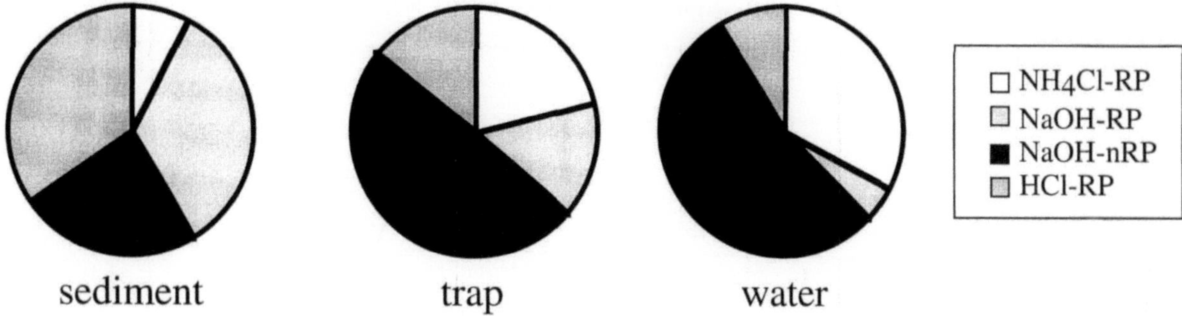

Figure 1. Phosphorus forms in sediment, settling particles (trap) and suspended matter (water) in Lake Erken. The shares are given as percent of the sum of the four fractions. The fractionation was performed according to the scheme of Hieltjes & Lijklema (1980) with the determination of total phosphorus in the sodium hydroxide extract. NaOH-nRP is the difference between total phosphorus concentration and that of reactive phosphorus in the extract.

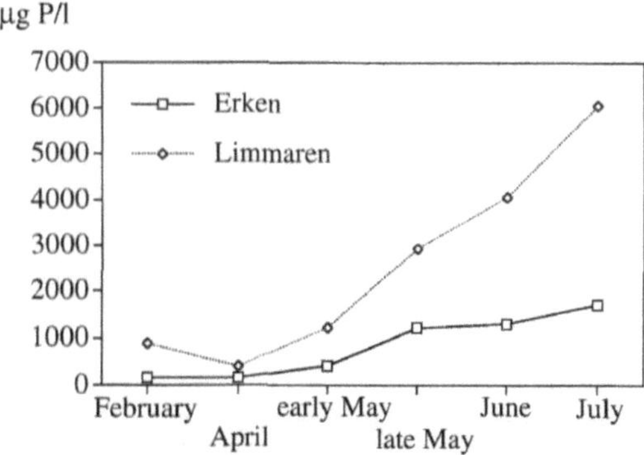

Figure 2. The seasonal development of pore water concentration of phosphate in the surficial (0–4 cm) sediments of Lake Erken and Lake Limmaren. The sediments were studied by Å. Eckerrot in 1992.

to iron surfaces is highly available during anaerobic conditions. Easily degradable organic phosphorus is potentially mobile.

Mobilisation and transport

Phosphorus can leave the particulate state in several ways. The splitting mechanisms include desorption, dissolution of precipitates and complexes, ligand exchange mechanisms and enzymatic hydrolysis of organic ester bonds (Boström et al., 1982). In the two latter cases, bacteria can play a significant role by production of organic acids and hydrolytic enzymes, respectively. A number of environmental factors are important in the mobilisation processes. The redox-potential affects reduction of iron(III) to iron(II) and a subsequent release of phosphate. At higher pH levels hydroxide ions replace phosphate causing ligand exchange. A higher temperature increases the bacterial activity giving lower redox potential and a higher production of enzymes. The effects are indirect, but are very important as shown in Figure 2, describing the increase in phosphate concentration found in the pore waters of Lake Erken and Lake Limmaren during a seasonal study. The effects could also be seen in the lake water with elevated phosphate concentrations during summer. With sediment from Lake Esrom Kamp-Nielsen (1975) could also show that the release of phosphorus to aerobic water was as high as that to anaerobic water when the temperature was higher than 18 C. Different kinds of chelating agents can also replace phosphate and cause a release. The significance of bacteria in the release of phosphate has been discussed by several authors. Most studies indicate that sediment bacteria have a significant role in uptake, storage and release of phosphorus including anaerobic release of iron-bound phosphorus (Gächter and Meyer, 1993). Polyphosphate metabolism together with anaerobic manganese and iron respiration are the mechanisms that give sediment bacteria a unique ecological niche (Boström et al., 1988; Davelaar, 1993).

The transport of phosphorus to the water can be separated into phosphate release and particulate transport. Generally, there is a sharp diffusion gradient from the sediment pore water to the overlying lake water. The gradient can be kept sharp by currents and turbulence induced by wind or gas ebullition. Furthermore, these processes can also cause transport of particles from the sediment to the lake water, i.e. resuspension. During resuspension, the particles meet with new environmental conditions that might induce further release of phosphate to the water. Another sig-

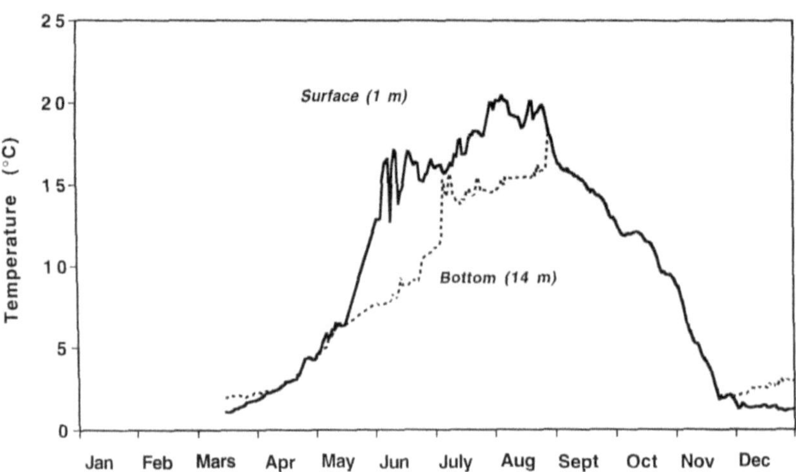

Figure 3. Temperature stratification in Lake Erken during summer as exemplified by the year 1995.

nificant transport process is the migration of resting stages from the sediment to the water. Several phytoplankton species have resting stages overwintering on the sediment. When growth is induced, they leave their habitat in order to shift their life-form to a pelagic one. In Lake Erken, the cyanobacteria *Gloeotrichia echinulata* has been shown to contribute significantly to the internal loading of phosphorus during the summer (Pettersson et al., 1993). Lake Erken is stratified during the summer (Figure 3) and phosphate usually accumulates in the hypolimnion. The concentration varies from year to year and seems to depend on the water temperature in the hypolimnion that is affected by the onset of the stratification. Thus weather conditions during the spring and the early summer affect the phosphate concentration level in the hypolimnion of Lake Erken. The stability of the stratification and the magnitude of storm events and internal seiches affect the transport of phosphate through the metalimnion. In Lake Erken phosphorus reaches the epilimnion in summer through two major pathways, phosphate transport through the metalimnion and migration of Gloeotrichia colonies from the sediment. Additionally, the hatching of *Ceratium hirundinella* cysts can contribute to the phosphorus transport some years. In the two lake examples given here, it is important to note that neither of them has experienced high external loading. Nevertheless, they both have intensive waterblooms of cyanobacteria during the summer that can only be explained by internal processes such as phosphate release from the sediments and/or migration of resting stages to the pelagic system. In these two lakes with very long water residence time, internal processes govern the water quality during the summer.

Acknowledgements

I express my gratitude to Åsa Eckerrot, Anu Toom and Helena Enderskog for assistance in sampling and analyses as well as to Ulf Lindqvist for preparing the graphs. I thank two anonymous referees for valuable comments.

References

Boström, B., M. Jansson & C. Forsberg, 1982. Phosphorus release from lake sediments. Arch. Hydrobiol. Beih. 18: 5–59.

Boström, B., J. M. Andersen, S. Fleischer & M. Jansson, 1988. Exchange of phosphorus across the sediment–water interface. Hydrobiologia 170: 133–155.

Chang, S. C. & M. L. Jackson, 1957. Fractionation of soil phosphorus. Soil Sci. 84: 133–144.

Davelaar, D., 1993. Ecological significance of bacterial polyphosphate metabolism in sediments. Hydrobiologia 253: 179–192.

Einsele, W., 1936. Uber die Beziehungen des Eisenkreislaufs zum Phosphatkreislauf in eutrophen See. Arch. Hydrobiol. 29: 664–686.

Löfgren, S. & M. Enell, 1988. Phosphorus in interstitial water: methods and dynamics. Hydrobiologia 170:103–132.

Gächter, R. & J. S. Meyer, 1993. The role of microorganisms on mobilization and fixation od phosphorus in lake sediments. Hydrobiologia 253: 103–121.

Hieltjes, A. H. M. & L. Lijklema, 1980. Fractionation of inorganic phosphates in calcareous sediments. J. envir. Qual. 9: 405–407.

Kamp-Nielsen, L., 1975. A kinetic approach to the aerobic sediment-water exchange of phosphorus in Lake Esrom. Ecol. Modelling 1: 153–160.

Mortimer, C. H., 1941. The exchange of dissolved substances between mud and water in lakes I. J. Ecol. 29: 280–329.

Mortimer, C. H., 1942. The exchange of dissolved substances between mud and water in lakes II. J. Ecol. 30: 147–201.

Murphy, J. & J. P. Riley, 1962. A modified single solution method for the determination of phosphate in natural waters. Anal. Chim. Acta 27: 31–36.

Pettersson, K., E. Herlitz & V. Istvánovics, 1993. The role of Gloeotrichia echinulata in the transfer of phosphorus from sediments to water in Lake Erken. Hydrobiologia 253: 123–129.

Williams, J. D. H., J. K. Syers, D. E. Armstrong & F. Harris, 1971. Fractionation of inorganic phosphate in calcareous lake sediments. Soil. Sci. Soc. am. Proc. 35: 736–739.

Williams, J. D. H., J.-M. Jaquet & R. L. Thomas, 1976. Forms of phosphorus in surficial sediments of Lake Eire. J. Fish. Res. Bd Can. 33: 413–429.

Hydrobiologia **373/374:** 27–45, 1998.
J.-C. Amiard, B. Le Rouzic, B. Berthet & G. Bertru (eds),
Oceans, Rivers and Lakes: Energy and Substance Transfers at Interfaces.

River ecosystem modelling: application of the PROSE model to the Seine river (France)

Stéphanie Even[1], Michel Poulin[1], Josette Garnier[2], Gilles Billen[3], Pierre Servais[3], André Chesterikoff[4] & [5]Michel Coste

[1]*Centre d'Informatique Géologique, École des Mines de Paris, 35, Rue Saint Honoré, F-77305 Fontainebleau cedex, France (E-mail: even@cig.ensmp.fr, poulin@cig.ensmp.fr)*

[2]*Laboratoire de Géologie Appliquée, URA 1367 CNRS, Université Pierre et Marie Curie (Paris VI), 4, Place de Jussieu, BP 123, F-75252 Paris Cedex 05, France (E-mail: jgarnier@biogeodis.jussieu.fr)*

[3]*Groupe de Microbiologie des Milieux Aquatiques, Université Libre de Bruxelles, Campus de la Plaine CP 221, Boulevard du Triomphe, B-1050 Bruxelles, Belgium (E-mail: gbillen@ulb.ac.be, pservais@ulb.ac.be)*

[4]*Institut d'Hydrologie et de Climatologie. Laboratoire de Chimie Analytique, Université Pierre et Marie Curie (Paris VI), 4, Place de Jussieu, BP 123, F-75252 Paris Cédex 05, France*

[5]*Centre d'Étude du Machinisme Agricole et du Génie Rural des Eaux et Forêts, 50, rue de Verdun Gazinet, BP3, F-33610 Cestas, France*

Key words: Seine river, ecological modelling, validation, anthropic impacts

Abstract

The Seine river crosses the most densely populated and industrialized area of France, Paris (16 million inhabitants), surrounded by fertile land with intensive agriculture.

In the framework of a CNRS (Scientific Research National Center) research project, computer programs have been designed to tackle problems related to eutrophication, non-point pollution and the impact of sewage during dry or wet periods (urban runof and sewage network overflow). The PROSE software has been specially designed to simulate the behaviour of the most disturbed stretches of the Seine ecosystem on the last 300 kilometers of the river, upstream of the estuarine area.

The 1-D hydraulic sub-model of PROSE is based on a finite difference solution of Saint-Venant equations solved with the Preissman scheme. It simulates steady state situations as well as highly transient situations such as fast changes in river discharge during rainy periods or dam motions. The biological sub-model is based on the RIVE model, describing the major processes in a river ecosystem: primary production, heterotrophic bacterial activity and organic matter decomposition, major nutrients species (nitrogen, phosphorus), nitrifying activity and oxygen balance. Water column and sediment variables are simulated. Most of the parameters have been estimated during laboratory experiments or field studies.

Different situations observed between 1989 and 1991 allowed a detailed validation of the model. The model was then used to explore the reaction of the ecosystem (particularly its oxygen status) to changes in physical constrains (discharge, reoxygenation at dams) or in biological processes (release of microorganisms accompanying waste water discharge).

Introduction

The Seine River is strongly affected by human pressure. Besides intense agricultural activity in its upstream basin, it flows through one of the largest town in Europe, Paris, with about 10 millions of inhabitants (including the suburbs). As a consequence of diffuse and point source pollution, eutrophication, chronic or accidental desoxygenation and micropollutant pollution constitute major features of the River. A comprehensive interdisciplinary programme was started in

1989, initiated by the CNRS (National Center of Scientific Research) with the aim of understanding the ecological functioning of the river and to establish mathematical models for water quality management.

Regarding eutrophication, the RIVERSTRAHLER model (Billen et al., 1994; Garnier et al., 1995), established at the scale of the Seine River Basin, allowed to understand the development of phytoplankton from headwaters to downstream sectors and to test scenarios leading to a reduction in algal blooms. This model couples a simple hydrological model based on the concept of stream order (Strahler, 1957), with a detailed biological model (Billen et al., 1994; Garnier et al., 1995) taking into account the main biological processes of the ecosystem. The model can simulate at a seasonal and geographical scale, for a period of ten days (assuming permanent conditions), the variations of the 26 state variables taken into account. Such a model is not adapted to deal with the problem of chronic desoxygenation due to urban runoff and sewage network overflow, nor to describe as finely as sometimes required, the detailed ecological behaviour of the Seine.

The model PROSE (Even, 1995), presented in this paper, has been specially designed for studying the behaviour of the fluvial part of the system where urban impact prevails. The PROSE software considers a fine description of the physical system on a course of 300 km of river, from upstream Paris to the mouth of the estuary. The hydraulic model is coupled with the same biological model (RIVE) as the RIVERSTRAHLER model so that the latter can feed the former with suitable boundary conditions.

A main challenge in the paper is to validate the coupling of the two physical and biological complex submodels by comparing the simulations by PROSE with observations gathered from 1989 to 1992 in the downstream sector of the Seine River. Whereas several aspects of ecosystem functioning have already been interpreted in detail in several other papers (Garnier et al., 1992a; Garnier et al., 1995; Chesterikoff et al., 1992; Garban et al., 1993; Brion & Billen, 1997), we here give an overview of the ecosystem functioning, focusing on oxygenation. We will explore the specificity of the PROSE model in dealing with transient situations, searching whether a release by the upstream reservoirs, i.e. an increasing discharge, can improve oxygen downstreams of Paris. A comparison of the behaviour of the PROSE model with classical water quality models, such as the QUAL2 model developed at US-EPA, or the KALITO Model developed by the Water Authority on the water Seine, is also given. The models implemented during the PIREN Seine are also qualified to tackle complex ecosystem evolution, providing a fairly good understanding and representation of the dominant processes, while other validations remain more questionable.

Study area

The Seine catchment is 78 900 km^2 in area and represents 15% of the French territory. Intensive agriculture is developed in the whole basin, in the Beauce, Gâtinais, Champagne and Vexin, with 40% of the French industrial activity concentrated between Paris and the estuary. The Seine is a highly modified river, navigable along two thirds of its course. Upstream reservoirs on the Seine and main tributaries (Marne, Aube, Yonne) contribute to a minimal flow of 70 to 100 m^3 s^{-1} during summer.

The 16 millions of inhabitants of Île de France represent 28% of the total population of France.

Paris and its suburbs constitute the major anthropic pressure within the basin. On the one hand, the rivers Seine and Marne (its main tributary) provide the major part of drinking water requirements (approximately 25 m^3 s^{-1}) for the 10 millions inhabitants of Paris and its agglomeration. On the other hand an important sewer network brings wastewater to seven treatment plants along the course of the two rivers. The Achères plant, located 50 km downstreams of Paris (see Figure 1), evacuates 27 m^3 s^{-1} and is the second largest waste water treatment plant in the world. It is responsible for chronic desoxygenation. Urban runoff or combined sewer overflow during rainfall events are responsible for accidental desoxygenation during low flow periods. Fish mortality is then sometimes observed.

Diffuse and point source pollution contributes to the eutrophication of the system. Every year, high values of phytoplankton biomass are observed, from spring to autumn.

The PROSE model has been designed to simulate and interpret the behaviour of the most disturbed part of the Seine ecosystem (Figure 1), from Paris and its suburbs to the estuary, representing 200 km. We also simulate a reference reach of 100 km upstream from Paris, beginning at the junction with the Yonne river.

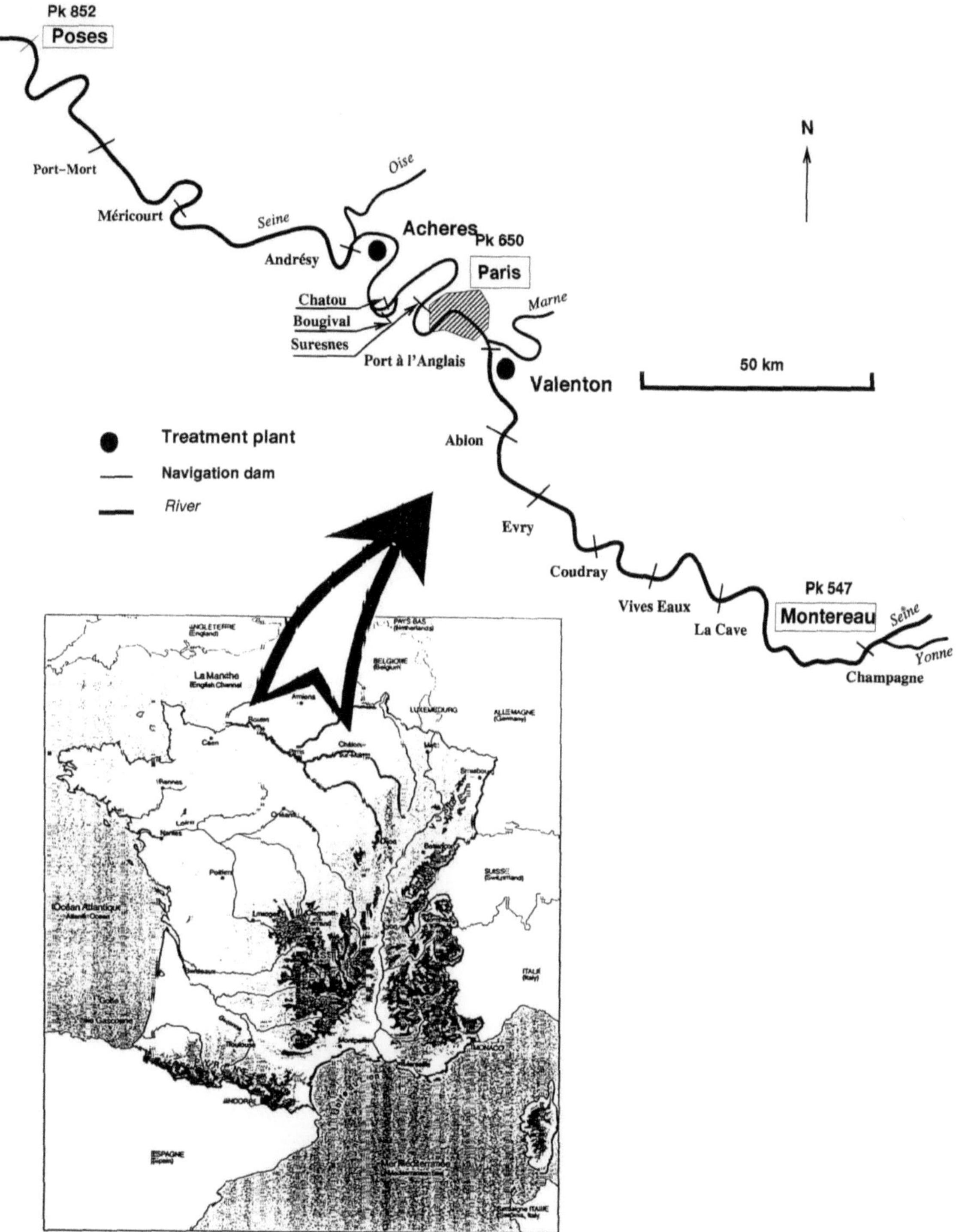

Figure 1. Seine river between Montereau, just after the junction with the river Yonne, 100 km upstream of Paris, to Poses, at the entry to the estuary, 200 km downstream to Paris.

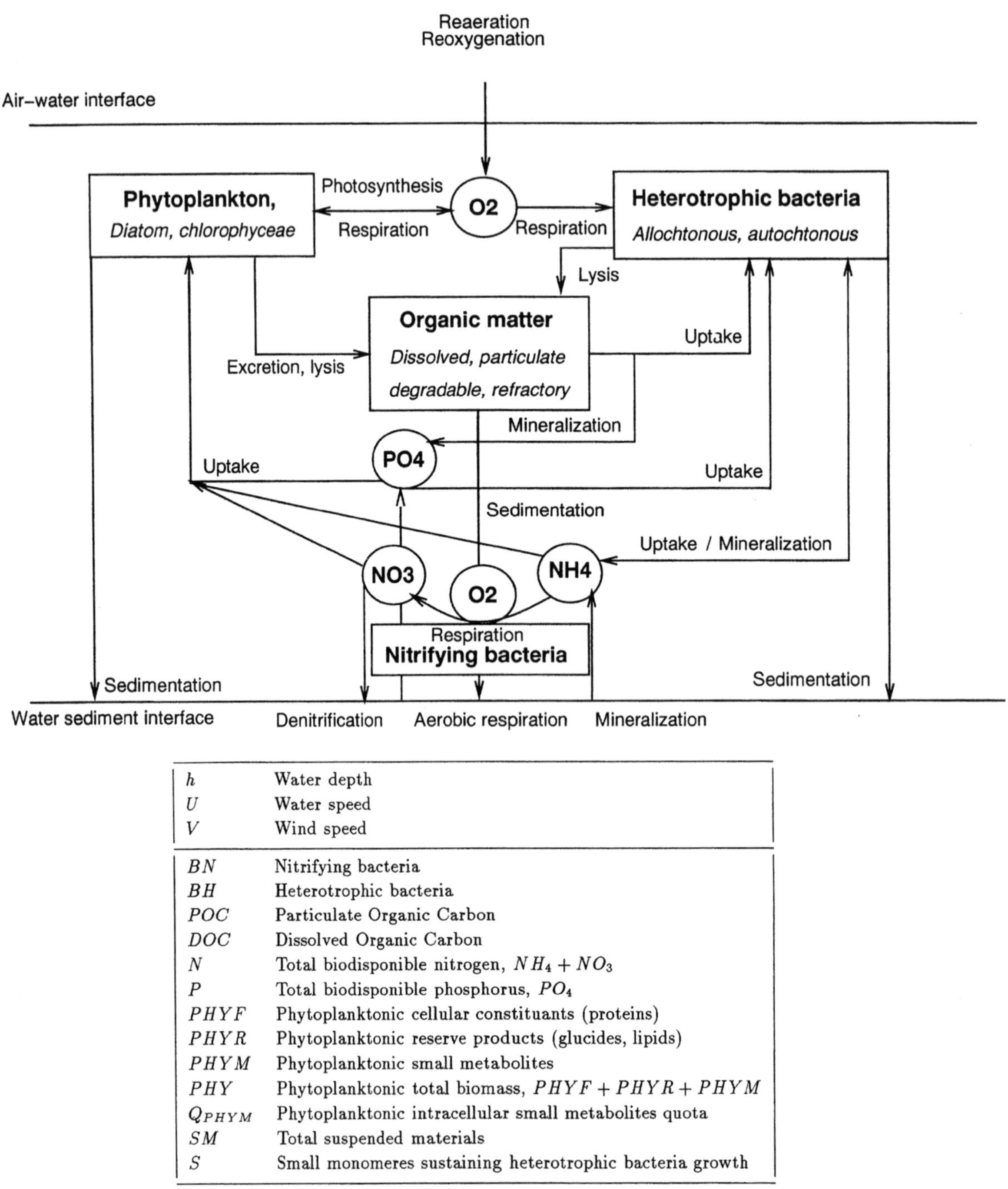

h	Water depth
U	Water speed
V	Wind speed
BN	Nitrifying bacteria
BH	Heterotrophic bacteria
POC	Particulate Organic Carbon
DOC	Dissolved Organic Carbon
N	Total biodisponible nitrogen, $NH_4 + NO_3$
P	Total biodisponible phosphorus, PO_4
$PHYF$	Phytoplanktonic cellular constituants (proteins)
$PHYR$	Phytoplanktonic reserve products (glucides, lipids)
$PHYM$	Phytoplanktonic small metabolites
PHY	Phytoplanktonic total biomass, $PHYF + PHYR + PHYM$
Q_{PHYM}	Phytoplanktonic intracellular small metabolites quota
SM	Total suspended materials
S	Small monomeres sustaining heterotrophic bacteria growth

Figure 2. Conceptual biological scheme taking into account organic matter degradation through heterotrophic bacteria activity, phytoplanktonic production, nitrification activity, nutrients mineralization and uptake, benthic fluxes.

Table 1.

Process	Expression	Parameters
Photosynthesis	$\underbrace{P_{moy}}_{\frac{\sum_0^h P(z)dz}{h}} [PHYF]$ $P(z)=P_{max}(1-e^{(\frac{\alpha I(z)}{P_{max}})})e^{(-\frac{\beta I(z)}{P_{max}})}$ $I(z)=I_0e^{-\eta z}, \eta=\eta_0+\eta_1[SM]+\eta_2[chla]$	$P^T_{max} = 0.15^D - 0.3^C\ h^{-1}$ $\alpha = 0.0014\ h(\mu E/m^2/s)^{-1}$ $\beta = 0\ h(\mu E/m^2/s)^{-1}$
Phyto. respiration	$\underbrace{R_{phy}}_{R_{cst}+R_\mu \mu} [PHYF]$	$R^T_{cst} = 0.0014\ h^{-1}, R_\mu = 0.5$
Phyto. excretion	$\underbrace{E}_{E_{cst}+E_{phot}.Phot.} [PHYF]$	$E_{cst} = 0.001\ h^{-1}, E_{phot} = 0.0006$
Growth	$\underbrace{\mu_{phy}}_{\mu_{max}.(\frac{[Q_{PHYM}]}{[Q_{PHYM}]+K_M}.\frac{[N]}{[N]+K_N}.\frac{[P]}{[P]+K_P})} [PHYF]$	$\mu^T_{max,phy} = 0.05^D - 0.1^C\ h^{-1}$ $K_M = 0.08, K_N = 0.014\ mgN/l$ $K_P = 0.015^D - 0.05^C\ mgP/l$
	$\underbrace{\mu_{BH}}_{\mu_{max}.\frac{[S]}{[S]+K_S}} [BH]$, $\underbrace{\mu_{BN}}_{\mu_{max}.(\frac{[NH_4]}{[NH_4]+K_{NH_4}}.\frac{[O_2]}{[O_2]+K_{O_2}})} [BN]$	$\mu^T_{max,BH} = 0.05^{NB} - 0.15^{AB}\ h^{-1}$ $K_S = 0.1_mgC/l$ $\mu^T_{max,BN} = 0.0036\ h^{-1}$ $K_{NH_4} = 1.12\ mgN/l, K_{O_2} = 0.6\ mg/l$
Lysis	$M_{phy}.[PHY], M_{BH}.[BH], M_{BN}.[BN]$	$M^T_{phy} = 0.006^D - 0.02^C\ h^{-1}$ $M^T_{BH} = 0.02\ h^{-1}, M^T_{BN} = 0.02\ h^{-1}$
Sedimentation	$\frac{W_{phy}}{h}.[PHY], \frac{W_{BH}}{h}.[BH], \frac{W_{POC}}{h}.[POC]$	$W^T_{phy} = 0.005\ h^{-1}, W^T_{POC} = 0.02\ h^{-1}$ $W^T_{BH} = 0.0^{NB} - 0.04^{AB}\ h^{-1}$
POC Hydrolysis	$k_C.[POC]$	$k^T_C = 0.008^L - 0.00025^{ML} - 0^R\ h^{-1}$
DOC Hydrolysis	$\underbrace{H}_{H_{max}\frac{[DOC]}{[DOC]+K_H}} [BH]$	$H^T_{max} = 0.75^L - 0.25^{ML} - 0^R\ h^{-1}$
DOC uptake	$\frac{\mu_{BH}}{Y}[BH]$	$Y_{BH} = 0.25$
Nitrification	$\frac{\mu_{BN}}{Y}.[BN]$	$Y_{BN} = 0.08\ molC/molN$
Bacterial N recycling	$\mu_{BH}.(\frac{1}{Y}(\frac{N}{C})_{CO} - (\frac{N}{C})_{BH}).[BH]$	$\frac{C}{N} = 7\ mgC/mgN$
Bacterial P recycling	$(\frac{P}{C})_{CO}H - (\frac{P}{C})_{BH}\mu_{BH}[BH]$	$\frac{C}{P} = 40\ mgC/mgP$
Phyto. N uptake	$(\frac{N}{C})_{phy}.\mu_{phy}.[PHYF]$	
Phyto. P uptake	$(\frac{P}{C})\mu_{phy}.[PHYF]$	
Denitrification	if $O_2 + \Delta O_2 < 0, NO_3$uptake $= \frac{4}{5}(O_2 + \Delta O_2)$	
O2 phyto. production	$dbo.P_{moy}.[PHYF]$	$dbo = 1.25\ molO_2/molC$
O2 respiration	$dbo.(R_{tot}[PHYF] + (1-Y).\frac{\mu_{BH}}{Y}[BH] + 2\frac{\mu_{BN}}{Y}[BN])$	
Air-water reaeration	$\underbrace{K_{rea}}_{K^D_{rea}+K^V_{rea}} ([O_2]^{sat}(T) - [O_2])$ $K^D_{rea}=\frac{(DU)^{\frac{1}{2}}}{h^{\frac{3}{2}}}, K^V_{rea}=A\frac{D^{\frac{2}{3}}V^{2,23}}{h}$	$D = 1.8\ cm^2/s$

^{L} : Labil, ^{ML} : Moderatly labil, ^{R} : Refractory

^{C} : Chlorophyceaes, ^{D} : Diatoms

^{NB} : Native bacteria, ^{AB} : Allochtonous bacteria

$f(T) = f(T_{opt}).e^{(\frac{(T-T_{opt})^2}{\sigma^2})}, T_{opt} = 32^C, 21^D, 24^{NB}, 24^{AB}oC$

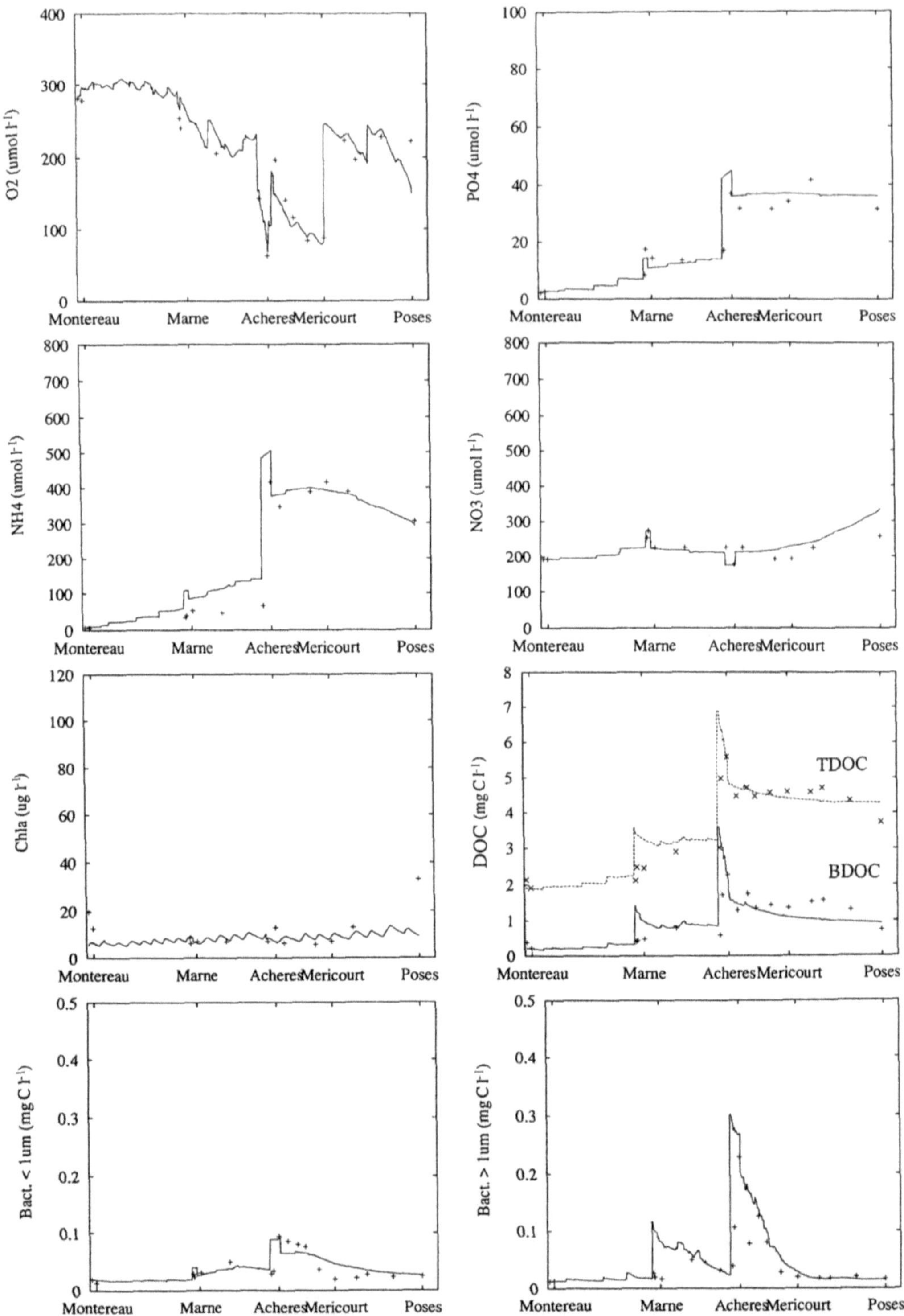

Figure 3. Results of simulations by the PROSE model for a same following water mass in October 1989 for major variables. Observations (+) are shown in comparison.

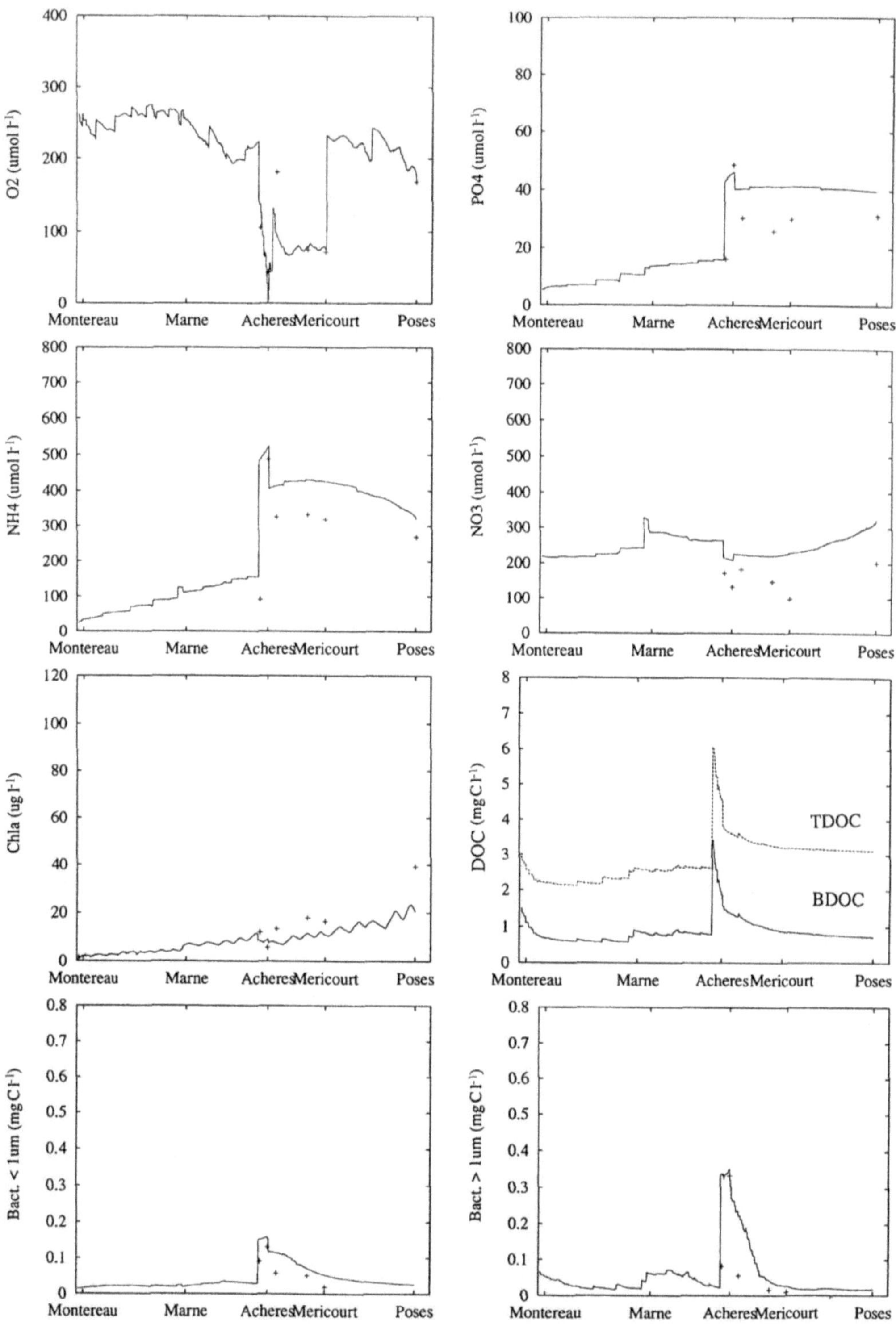

Figure 4. Results of simulations by the PROSE model for a same following water mass in June 1990 for major variables. Observations (+) are shown in comparison.

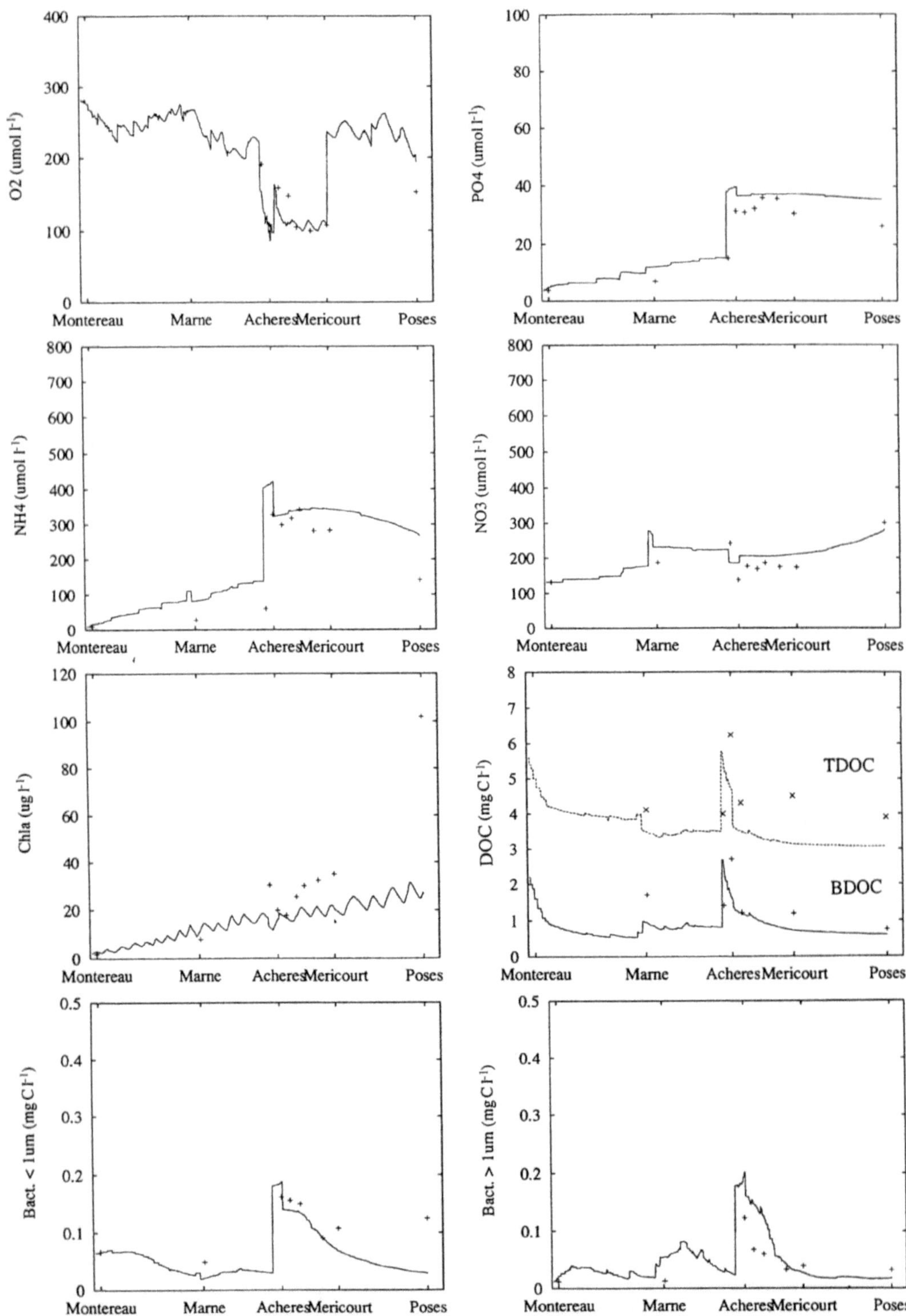

Figure 5. Results of simulations by the PROSE model for a same following water mass in September 1990 for major variables. Observations (+) are shown in comparison.

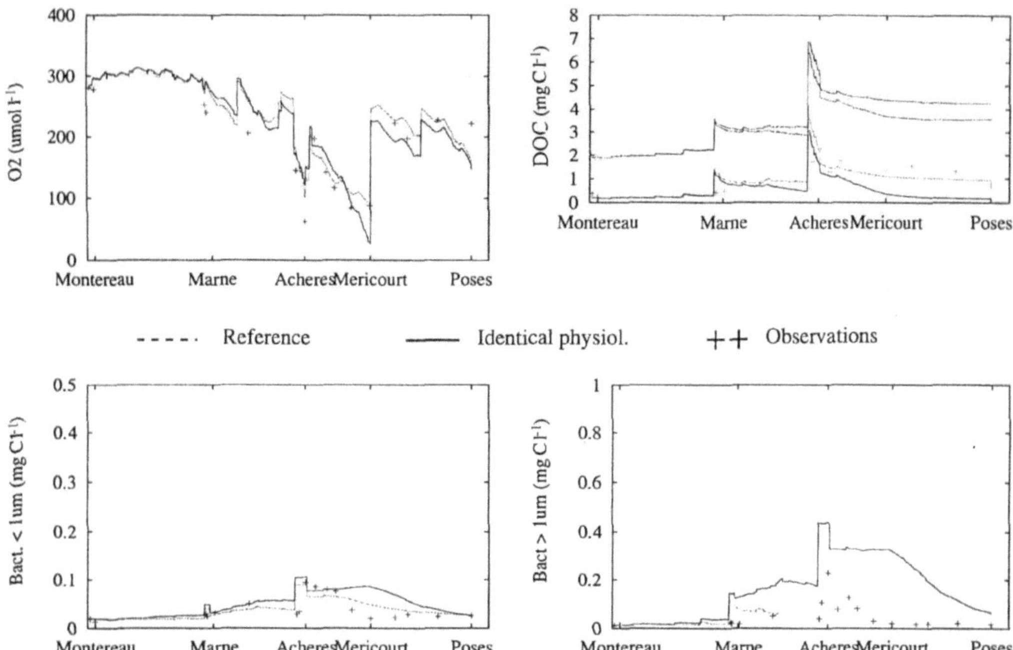

Figure 6. Sensitivity of the model PROSE to heterotrophic bacteria physiological characteristics. First reference simulation considers different physiological characteristic (higher growth and sedimentation rates for bacteria > 1 μm). Second simulation considers allochtonous bacteria physiological characteristics for both bacteria types.

Model description

Description of the PROSE model

PROSE (Even, 1995) describes physical and biological variables in a fluvial ecosystem. The hydraulic sub-model in based on the Saint-Venant equations (Hug, 1975),

$$\frac{\partial S}{\partial t} + \frac{\partial Q}{\partial x} = q, \tag{1}$$

$$\frac{\partial Q}{\partial t} + \frac{\partial}{\partial x}(\frac{Q^2}{S}) + gS\frac{\partial z}{\partial x} = -gSJ, \tag{2}$$

where S is the wet section, Q the flow rate, q the linear flow rate of effluents, $\boldsymbol{g}$ the gravity acceleration, J the energy slope. Energy slope is expressed using the Manning-Strickler formula

$$J = \frac{Q^2}{K^2 R_H^{\frac{4}{3}} S^2},$$

K designing the Strickler coefficient and R_H the hydraulic radius. Strickler coefficients have been determined for every reach of the river, using upstream and downstream water level (three per day) and daily flow rate measurements on four years. Strickler coefficients vary as a function of flow rate.

The system of equations 1 and 2 is solved with the semi-implicit finite-difference scheme of Preissmann (Cunge et al., 1980), whose stability and convergence properties have been well studied (Samuels & Skeels, 1990; Thiriot & Benayada, 1993; Ponce & Simons, 1978). Highly transient situations such as fast changes in river discharge as well as navigation dam operations are represented. Roughness coefficients have been estimated as a function of the river discharge (Even, 1995). The hydraulic model representing dam sill positions, calculates part of the discharge that overflows (Q_{over}) or underflows (Q_{under}). Calculated reoxygenation at dams uses this information, with two types of formula depending on the nature of the flow. Formulation of Holler (Holler, 1971) for the overflow case or the Avery & Novack (1978) expression for underflow are used. Reaeration at the air-water interface is a function of diffusivity and wind (Table 2) (Thibodeaux et al., 1994).

The conservative equation for every constituent of the model is

$$\frac{\partial M}{\partial t} = -\frac{\partial F}{\partial x} + R, \tag{3}$$

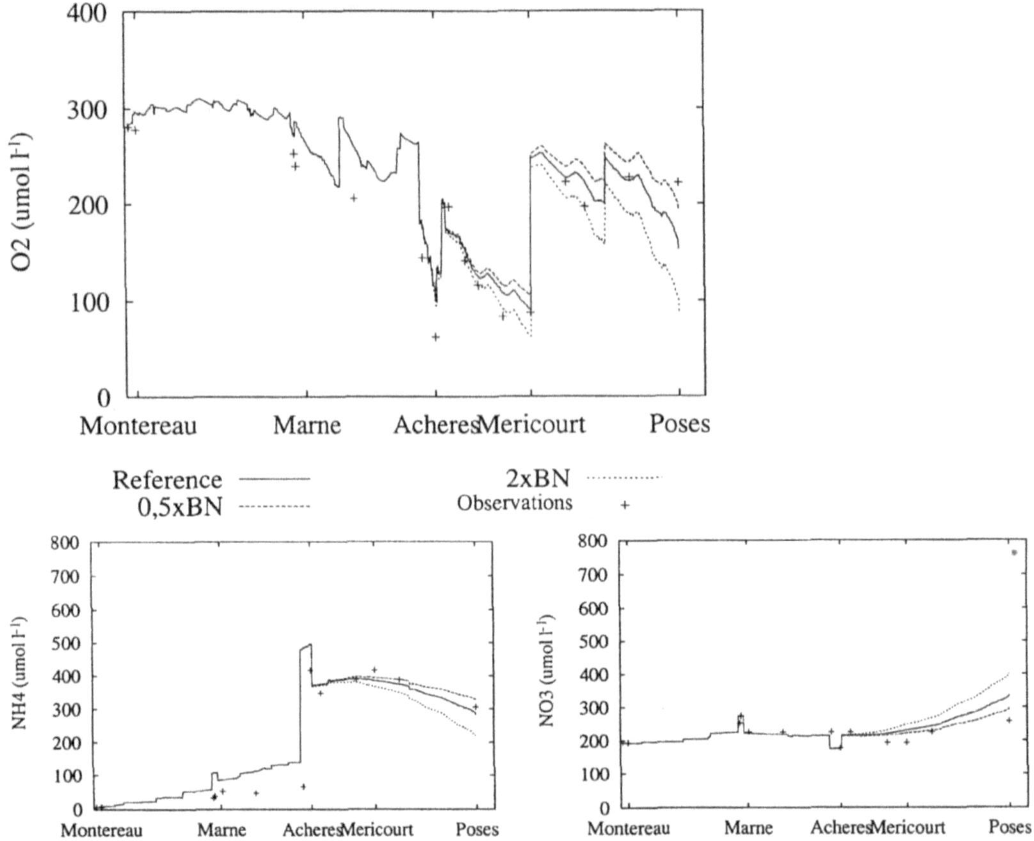

Figure 7. Sensitivity of the model PROSE to nitrifying bacteria contribution from Acheres treatment plant. Contribution multiplied and divided by 2 with regard to the reference situation.

where M is the total mass of a constituent considered in an elementary volume, F is the total flux of constituent passing the boundary of the volume in the longitudinal direction and R is the balance of the production and consumption *via* biological and chemical reactions with other constituents in the volume. The advective and the dispersive fluxes represent the total fluxes exchanged at the boundary:

$$F = QC - D\frac{\partial SC}{\partial x}, \tag{4}$$

C designing the concentration of the constituent, D the longitudinal dispersive coefficient, varying with flow rate, S and Q corresponding to the previous definition.

The transport sub-model solves the 1-D advection-dispersion equation, independently of biological and chemical reactions. It was also possible to introduce numerous numerical schemes to solve the advective equation, which simulation results are sensible to numerical precision. Precise numerical techniques were also used like the finite-difference scheme of Bott (1988) or characteristic techniques of Holly & Preissmann (Holly & Usseglio-Polatera, 1984; Holly & Preissmann, 1977) or Rash & Williamson (1990). The choice of the numerical scheme is determined by precision/calculation time consideration, principally depending on the simulated context.

The biological sub-model is based on the RIVE sub-model (Billen et al., 1994; Garnier et al., 1995; Even, 1995) and simulates the major biological processes in the ecosystem (Figure 2 and Table 1), with time steps from minutes to hours. The kinetics of phytoplankton activities is based on the AQUAPHY model (Lancelot et al., 1991) which distinguishes the processes of photosynthesis dependent on light from the process of growth controlled by the availability of intracellular carbon and nutrients (see Table 1). Platt's (1980) formula is used to describe photosynthesis (see Table 1). Dominant phytoplankton groups are taken into account (Garnier et al., 1995). Organic matter, considered in three classes of biodegradability, is de-

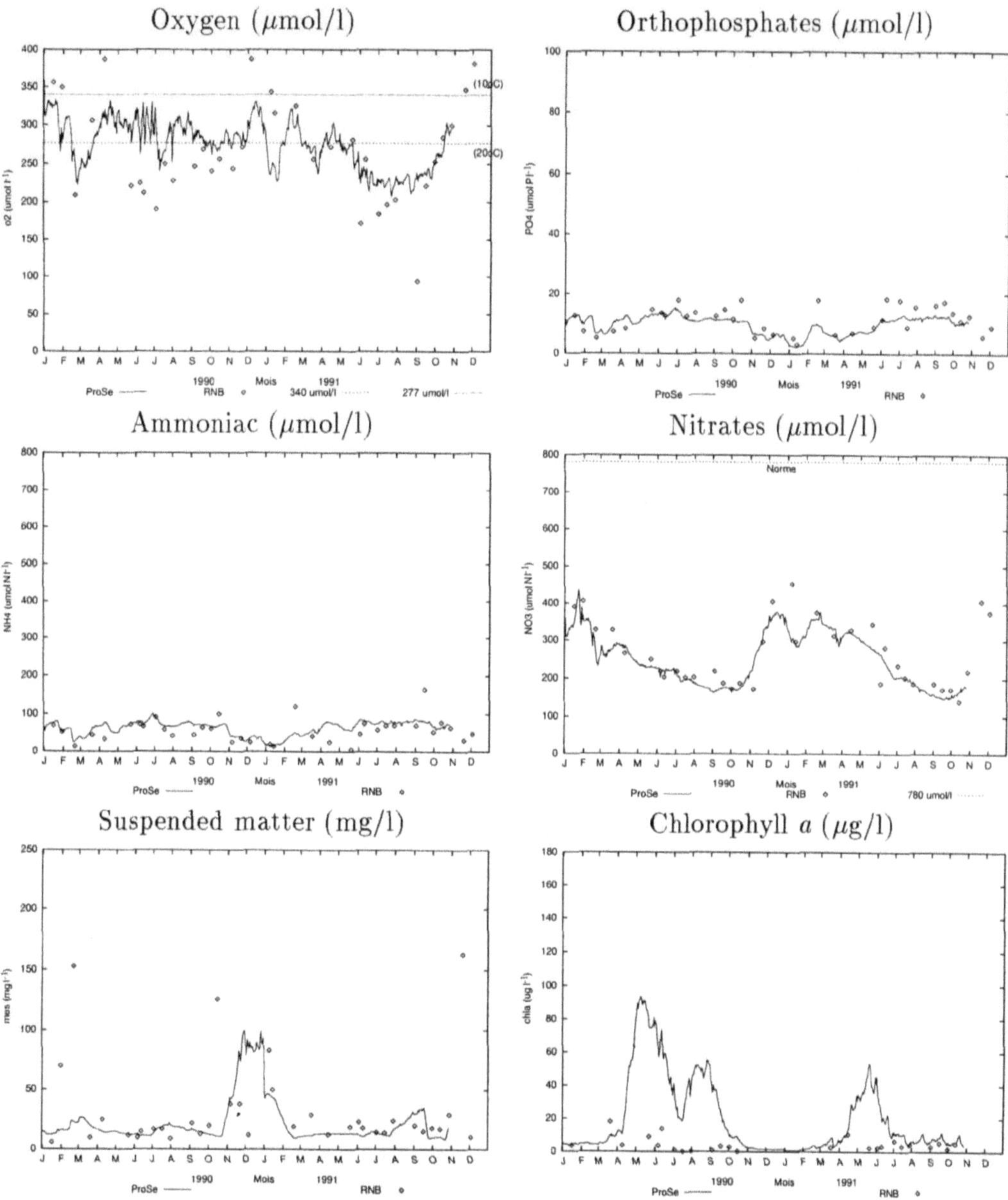

Figure 8. Seasonal survey in Tolbiac (Pk 653.75) in 1990 and 1991. Comparison between simulations and National Basin Network measurements.

graded by two size classes of heterotrophic bacteria (Billen & Servais, 1989; Garnier et al., 1992a). Benthic organic matter degradation and nutrient recycling are calculated according to algorithms developed by Lancelot & Billen (1985) and Billen et al. (1989). These algorithms express the mass balance of organic carbon, oxygen, inorganic forms of nitrogen, phosphorus and silica in the sedimentary column. Considering a diagenetic steady state, these equations allow to calculate the different fluxes across the sediment water-interface as a response to sedimented organic material. Nitrification and reversible orthophosphate adsorption are also included. Many of the parameters and kinetics of elementary processes were determined during the research program.

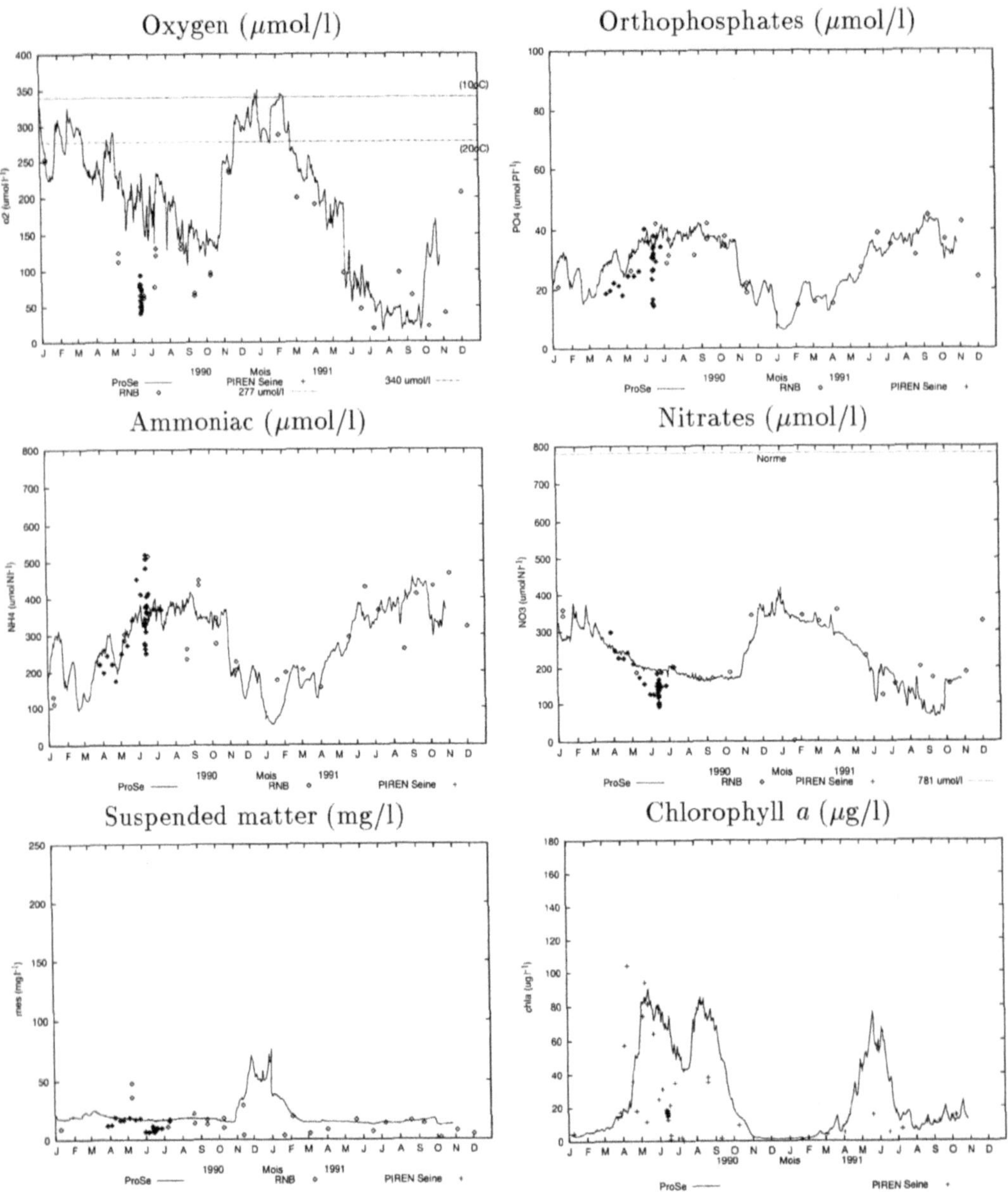

Figure 9. Seasonal survey in Méricourt (Pk 778.0) in 1990 and 1991. Comparison between simulations and National Basin Network measurement

Data base

Besides experimental investigations set up to determine the kinetics of the processes and/or the parameters of the kinetics in the biological model, several campaigns for validation were carried out, so that most of the state variable of the model have been measured (nutrients, suspended matter, dissolved and particulate organic matter in total and biodegradable fractions, bacteria within two sizes, phytoplankton biomass of the dominant groups. Three longitudinal profiles were sampled in October 1989, June and September 1990 at 15 to 18 stations between Montereau and Poses (Figure 1). These three campaigns were typical of low summer discharge when intense desoxygenation is often observed.

In addition, the data base comprises classical variables of water quality (nutrients, chlorophyll, suspended matter, ...) originated by the Seine river Water

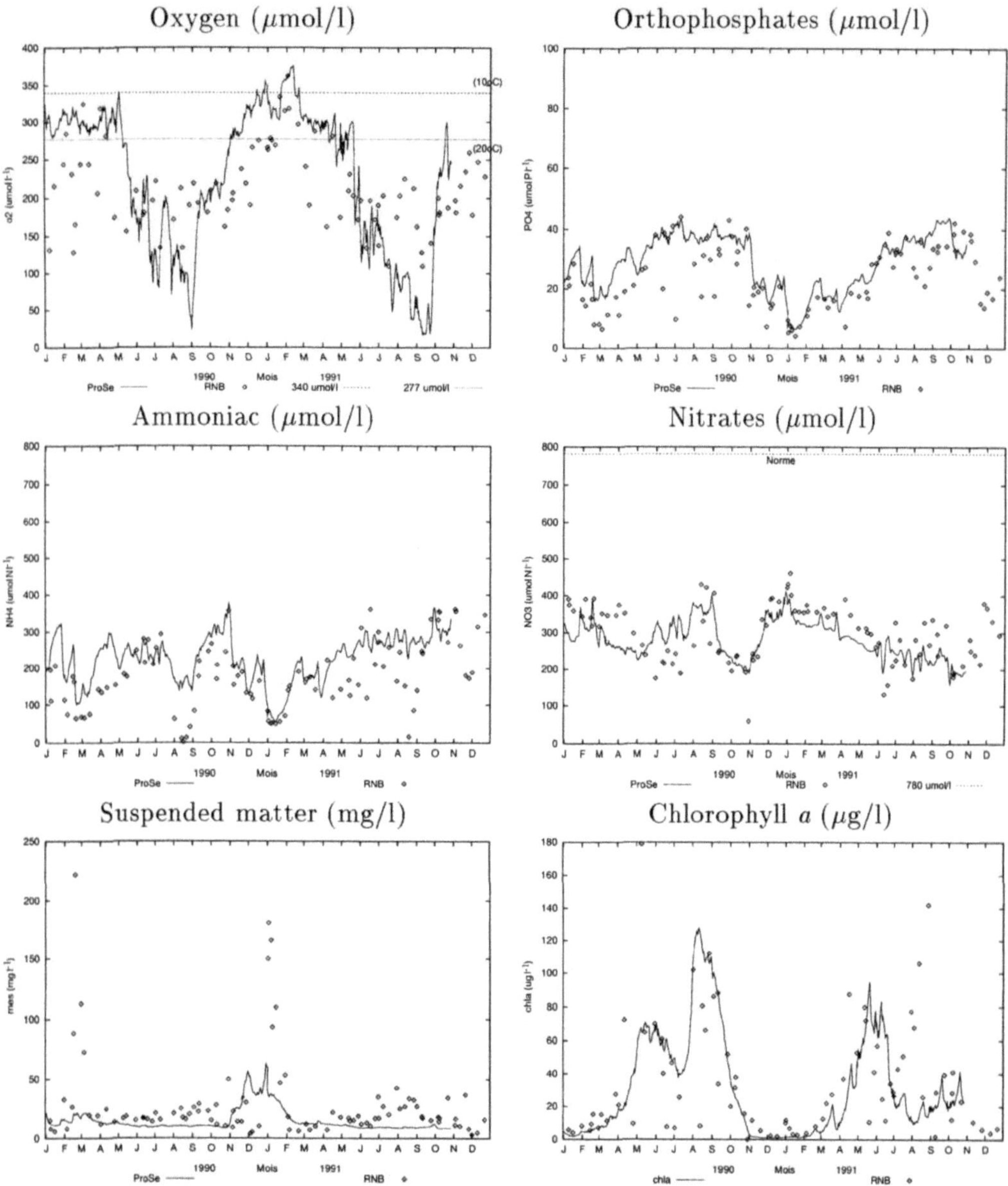

Figure 10. Seasonal survey in Poses (Pk 852.0) in 1990 and 1991. Comparison between simulations and National Basin Network measurements.

Authority (Agence de l'Eau Seine Normandie). Light intensity, wind and temperature were made available by the National Meteorologic Service and Seine river Water Authority. Dam positions, river flow and flow of the main tributaries were registered by the Navigation Service. Information on treatment plants (flow and quality criteria like BOD and suspended matter) are given by the Seine Basin Water Authority. Pollution loads were converted into C, N and P using factors determined during the research program (Servais et al., 1997).

Boundary conditions were extrapolated from available measurements.

Results

The course of the river is expressed in km (pK), Honfleur, downstream the estuary, being set at pK 1000. The pK increases from upstream to downstream. Poses, at the mouth of the estuary, is at pK 852, Paris

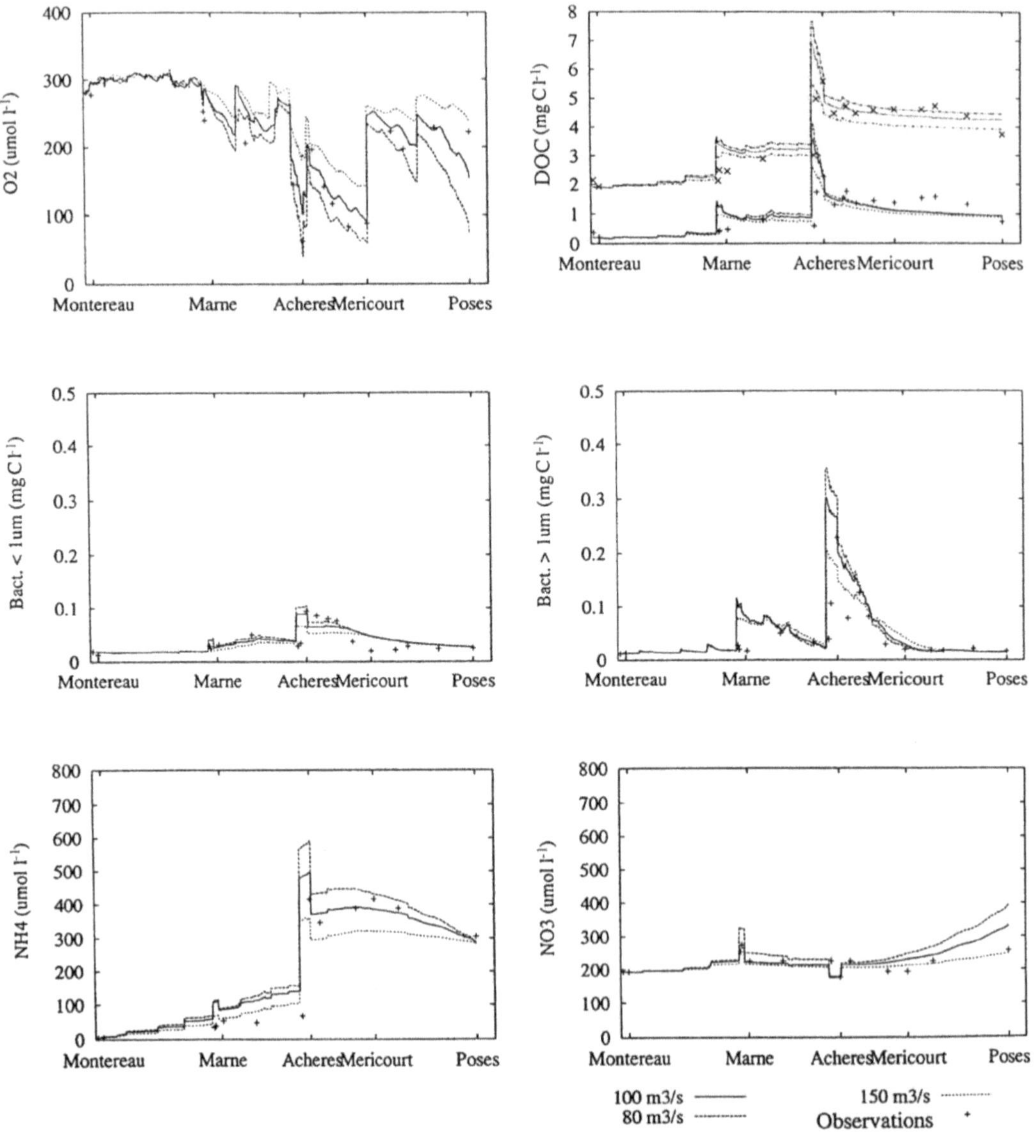

Figure 11. Sensitivity of the PROSE model to flowrate. Reference situation (100 m^3 s^{-1}) compared to lower (80 m^3 s^{-1}) and higher (150 m^3 s^{-1}) flowrate.

is at pK 650, and Montereau, the upstream boundary limit for the PROSE model, is at pK 547.

Typical features of river functioning in the sector downstream of Paris

Phytoplankton is generally dominated by centric diatoms (90% of total biomass) during spring bloom. Maximum biomass values higher than 100 μgchla l^{-1} are commonly measured at that time. Phytoplankton can therefore constitute an organic load that may consume high amounts of oxygen. On the other hand, up to 20 μgchla l^{-1} (frequently observed in Paris) phytoplankton contributes significantly to reoxygenation by photosynthesis, but this may be reduced during cloudy periods or by any variation of turbidity. As phytoplankton activity is in many ways responsible for the fragilization of the system, and can, in few days during the development or breakdown of a bloom, produce supersaturation or consume high amont of oxygen, the understanding of phytoplankton development is essential to approach desoxygenation problems. During summer, a mixed community of diatoms and

Chlorophycae is observed. In 1990, a second diatom bloom was observed during autumn. The phytoplankton biomass within the different taxonomic groups is converted into chlorophyll a according to Garnier et al. (1995), taking into account the respective contributions of the taxonomic groups to total biomass. Seasonal succession of phytoplankton development at the scale of the drainage network is discussed in Garnier et al. (1995).

Among the two types of heterotrophic bacteria considered by the model, the autochtonous one coming from upstream is characterized by small size (< 1 μm) and lower disappearance rates, whereas the allochtonous population coming from the effluents is mostly composed of larger bacteria (> 1 μm) with higher growth and disappearance rates (Garnier et al., 1992a). As shown by Garnier et al. (1992b), the physiological characteristics of the two populations significantly influence the degradation of organic matter and oxygenation.

Longitudinal profiles

A good agreement is found between the simulations and the observations of the major variables of the model (algal biomass, oxygen, ammonium and nitrates, inorganic dissolved phosphorus, inorganic suspended matter, bacterial biomass and activities, organic particulate and dissolved carbon, see Figures 3 to 8).

Longitudinal variations show similar trends for the three periods investigated (Figures 3, 4, 5).

From Montereau (PK 547) to the entrance of Paris, oxygen approaches saturation. Bacteria and phytoplanktonic biomass are relatively low. Ammonium and phosphorus concentrations are increasing slowly from the upstream limit to the entrance of Paris. In all cases, the major trend and the level of concentration of the variables are adequately simulated (Figures 3, 4 ,5).

From the entrance of Paris and after the main waste water treatment plant (Achères), an important decrease in oxygen level is observed. The Oise river (PK 718) has an important dilution role.

Heterotrophic bacteria biomass is high, dominated by large-sized bacteria and heterotrophic activity is the dominant process within the sixty kilometers between Achères and the Méricourt dam (PK 770). Despite their high growth rate, allochtonous bacteria have a limited impact (not more than 30 km downstreams of Achères), as they disappear rapidly from the water column.

To estimate the impact of variable physiological characteristics between autochtonous and allochtonous bacteria, the same physiological characteristics have been considered for both bacteria types. The test, presented on Figure 6, consisted in introducing growth and sedimentation rates characteristics of the autochtonous bacteria for both types of bacteria. Impact on oxygen downstream the effluent would also be more pronounced and extended. Competition between small and large sized bacteria has been discussed in Garnier et al. (1992b).

At the Méricourt dam (a chop of 4 to 5 meters) reoxygenation was important and is maintained to the mouth of the estuary. Intense nitrification was observed in the last fifty kilometers, before the river Seine enters the estuarine zone. Transformation of ammonium to nitrate was well simulated by the model in October 1989 and September 1990. As nitrifying bacteria have been shown to originate from the Achères plant (Brion & Billen, 1997), simulations considering nitrifying bacteria contribution from the Achères plant multiplied or divided by two have also been performed. The results show that bacteria in the effluents are the determining factor for nitrification (Figure 7). Due to their low growth rate, nitrifying bacteria had a significant impact in the downstream section of the River, depending significantly on its the flow rate in the river (Figure 7). Transformation of ammonium to nitrate was, however, not so well reproduced in June 1990.

In the three cases investigated, phytoplankton developed along the course of the river and reaches its maximum level just upstream of the estuary (Figures 3, 4, 5). Phytoplankton increase was badly simulated in September 1990.

In fact, to facilitate the field sampling, the duration of the campaigns compared to the real residence time of the water was always shorter. This approximation is acceptable as far as the river system is more or less at steady state. But this cannot always be the case for phytoplanktonic activity, with its sharp blooms or breakdowns within only a few days, as well as weekly variations in biomass that can vary from 5 to 20 μgchla l^{-1} during summer. The model also simulates the development of phytoplankton biomass sampled at the begining of the campaign, at the speed of the water. In this case the residence time was about three weeks. On the other hand the duration of the campaigns was around one week to ten days. This in itself can explain the deviation of simulations from actual measurements.

Finally, in the last 80 km, oxygen concentrations result from a fragile balance between its production by photosynthetic activity and respiration by nitrifying bacteria. As simulated phytoplankton biomass was always underestimated in this reach, simulated oxygen concentrations may be underestimated, as is the case in October 1989. In the two other cases, only one point had been sampled downstream of Méricourt, making difficult any refined interpretation. It seems however that the underestimation of nitrification in September 1991 lead to an underestimation of oxygen concentration.

On the other hand, phytoplankton nutrient uptake in this area was negligeable for ammonia and orthophosphate.

Seasonal evolution

Simulations have been performed for the years 1990 and 1991 and were compared with data available from the water Authority of the Seine Basin. Only major constituents, such as oxygen, nitrate, ammonium, inorganic dissolved phosphorus, chl *a* and suspended matter are presented. Critical periods are thus placed in a general context. Three stations were analyzed: Tolbiac (PK 654) in Paris *intra muros*, Méricourt (PK 778) just before the dam and Poses (PK 852) at the mouth of the estuary.

Seasonal variations in oxygen concentration were characterized by a severe decrease in summer (from May to October) downstream of Achères treatment plant. Whereas oxygen concentration remain near the saturation in Paris (pK 653), the effluents were responsible for anoxic conditions until Méricourt (pK 778), a restoration occurring at Poses (pK 852) (Figures 8, 9, 10). Oxygen concentrations in Paris and downstream of Achères are well simulated. The model underestimate the summer concentrations at Poses (Figure 10). Ammonium and inorganic phosphorus concentrations were low all the year in Paris. Significant seasonal fluctuations were observed downstream from Achères due to a dilution effect of the winter flows (Figures 8, 9, 10). Due to soil leaching in the absence of vegetation cover, nitrate variations were opposite, with high concentrations in winter (Chesterikoff et al., 1992). The comparison between calculated and observed concentrations of nitrate and ammonium during early spring shows an underestimation of ammonium and overestimation of nitrate. Ammonium was also underestimated in summer at Poses.

Discussion

Impact of domestic effluents on the oxygenation of the river Seine downstream from Paris

Discharge of treated or untreated wastewater of domestic or industrial origin is a major factor affecting the ecological functioning of most large rivers. It has recently been stressed that the direct impact of waste water effluents cannot be understood by only considering their release of pollutants. Large biomasses of microorganisms have been shown to accompany organic matter and nutrients in the effluents (Garnier et al., 1992a; Garnier et al., 1992b; Brion & Billen, 1997; Servais et al., 1997) and this input of microorganisms determines the processes in the system (nitrification, heterotrophic activity), as shown in this paper. Moreover biodegradable characteristics of organic matter as well as differentiating between dissolved and particulate stages consitute other major factors. Some work is actually needed to evaluate the respective contributions between dissolved and particulate organic matter to oxygen deficits, with the first species responsable of direct and the second delayed impacts. It was shown that oxygen depletion downstream from effluents could not be understood without benthic oxygen consumption. Organic matter sedimentation and erosion characteristic are some of the main processes, that should be studied.

In addition, the information on wastewater discharge, often available as BOD_5 or inhabitant-equivalent has been converted into the proper format required by the biological model RIVE. A number of relationships have been established for the case of the Seine River and given in Servais *et al.* (in prep.). These converting factors have been used to determine the effluents contribution in organic matter and heterotrophic and nitrifying bacteria. The model simulations are close to observations: they constitute a backward validation of these parameters.

PROSE have also been implemented to simulate combined sewer overflows after storm events on Paris suburb (Even et al., 1996). Overflows and resulting navigation dam management lead to highly transient regimes during few hours. It was shown that oxygen deficit and minimum levels in the river were partly influenced by previous oxygenation and oxygen concentrations. Phytoplankton is therefore one of the major state variables. Since transient evolution is at the scale of hours to one day, precise representation of photosynthetic activity and respiration is required.

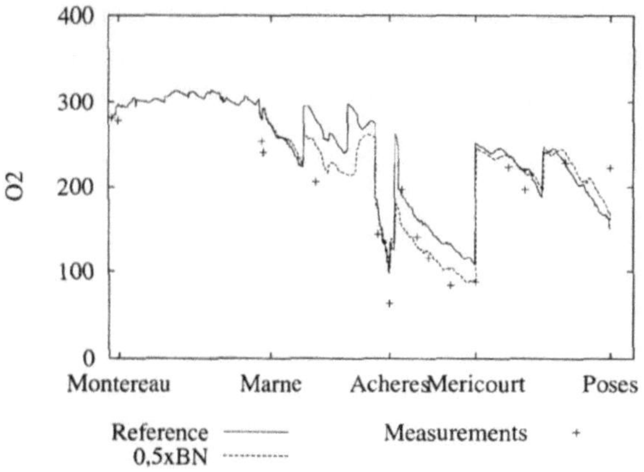

Figure 12. Sensitivity on oxygen of dam sill position through Paris.

As far as phytoplankton activity is concerned, photosynthetis and growth are indeed distinguished in the model. Observed dark growth and delay of maximum growth rate relative to maximum light intensity are simulated if we consider reserve compounds and small metabolites.

Impact of hydraulic managements

Besides pollution, the Seine is influenced by hydraulic management. Residence time in the system is long, due to low natural flow on one hand, but also depending on upstream reservoir releases, and on high water levels maintained by navigation dams on the other hand. These hydrological and morphological constrains are of major importance for biological processes and water quality.

As low flow largely depends on upstream reservoir release, the range of observed summer flow was explored with the PROSE model, as shown in the Figure 11, between 80 $m^3\ s^{-1}$ corresponding to the end of June situation without reservoir releases up to 100 and 150 $m^3\ s^{-1}$, characteristic for summer flow. Effects of flow sustainment by reservoir releases lead to an improvement of the oxygenation in the downstream part of the river.

The hydraulic submodel allows to test different stategies about dam management. Water can run partly above the sill (overflow), partly under the dam (underflow). Reoxygenation is much more important when water is overflowing. Figure 12 illustrates two scenarios of reoxygenation at dams, one considering the current configuration, that is bottom opening from 10 to 50 cm in Paris and downstream from Achères, the second one considering only overflow. The model tends to demonstrate that if underflow discharges at dams were suppressed, an improvement of 1 mg l^{-1} in oxygen concentration would be obtained.

Comparison of PROSE with other river models

Frequently used water quality models are based on simple differential equations (i.e. first order kinetics), for nitrification or biological oxygen demand (BOD) for example, following the model proposed by Streeter & Phelps (1925). The models QUALII and WASP4 of the US-EPA (Brown & Barnwell, 1987; Ambrose et al., 1988), the KALITO model developed at the Seine Basin Authority (Lesouef & André, 1982), the MIKE 21 water quality and eutrophication model of the Danish Hydraulic Institute, the DELFT Hydraulics water quality model, belong to this category, even if some of them include many processes such as phytoplankton, zooplankton or benthic activities. Equations are simple functions of one key variable, such as ammonia for nitrification or biological oxygen for demand oxygen deficit. Most of the parameters introduced have no direct biological meaning and are calibrated on the basis of a comparison between simulation results and *in situ* concentrations measurements.

A different approach is used in the PROSE-RIVE model, which is based on a complete description of the kinetics of the basic processes involved. The formulation of these kinetics as well as the value of the corresponding parameters is derived from direct experimental measurements. No calibration is therefore carried out and the comparison of the simulation with surveys is a true measure of our understanding of system behaviour.

The QUAL2, KALITO and ProSe models have been implemented on the same reach of the river Seine, along the 300 km between Montereau and Poses.

With QUAL2 and KALITO models, parameters for nitrification, BOD decrease or reaeration have to be calibrated for three main reaches (Lesouef & André, 1982; Aristaghes & Wagner, 1989; Even, 1990). Also, these parameters must be changed for past and present situations (Billen et al., 1995) because the number of waste water treatment plants and the quality of the treatment have been improved. In adition, the significance of the variables taken into account to decribe the processes is questionable. For example BOD_5 measurement cannot distinguish bacterial from phytoplanktonic respiration and is not suitable to simulate the impact of effluent input in eutrophication phenomena. Similarly, using a first order function of ammonia concentration does not take into account the time lag necessary for the development of a consequent bacteria population. As a consequence, to simulate properly nitrification downstream of Achères, at least two parameters are necessary in the KALITO model, one being quasi zero in the first ten kilometers, a second one more important after Méricourt. Moreover, since the distance of intense nitrification is a function of flow rate, as it is controlled by nitrifying bacteria development, the reach of application and the amplitude of the parameters should vary in a complex way.

On the contrary, no adjustment of parameters is required for the model PROSE implemented in identical conditions (Billen et al., 1995) and simulations of the impact of the sewage network are consistent with the measurements.

Conclusion

The river Seine is a highly perturbed and managed river, with moderate summer flow rate. Domestic effluents are responsable for severe dysfunction of the system in the urbanized part. As a result of human occupation in the basin, eutrophication is another major problem. Mathematical models developed in the PIREN Seine research program have been used to synthetize and analyze ecosystem functioning. They are designed to be implemented as management tools.

A concerted approach have been adopted between the models, each being designed to tackle different time and space scale problems. The RIVER-STRAHLER model has been used on the entire basin scale to study seasonal eutrophication problems. The PROSE model is more refined with respect to the physical description and numerical resolution. It has been specifically designed to study impacts of urban and industrial activities on the river in a global context. Both models are based on the RIVE biological and chemical model which considers major processes of an aquatic ecosystem and represents elementary reactions.

PROSE-RIVE has been validated in a broad range of real situations, which showed the relevance of the values of the process description. PROSE-RIVE model represents therefore a powerful tool for scientific and management purposes.

Many past and present situations, including longitudinal changes over a 300 km reach of river, and natural to very polluted areas, have been simulated with the same set of parameters. Classical water quality models such as the US-EPA QUAL2 or KALITO model, based on first order kinetics, have been implemented in the same reach of the Seine. In that case calibrated parameters were variable, depending on the reach of the river and on the period of time simulated.

It appears that precise knowledge of the chemical and biological composition of the effluents, including bacteriological aspects, is necessary to realistically simulate polluted waters. In addition, as bacterial and phytoplankton activities have a major impact, the hydraulic context must also be precisely described. As a matter of fact, dilution of effluents and residence times of dissolved and particulate species have great impact on river ecological behaviour.

Finally the modelling approach developed in the framework of the PIREN Seine provides a new point of view on model conceptualization, validation and water quality management.

Acknowledgements

This work has been undertaken within the framework of the Research Project of the CNRS on the Seine basin, with the implication of numerous institutions concerned with water management like Agence de l'Eau Seine-Normandie, Service de la Navigation de la Seine, Météo-France, Syndicat Intercommunal pour l'Assainissement de l'Agglomération Parisienne, Lyonnaise des Eaux and Compagnie Générale des Eaux. We are grateful to our colleagues involved in this scientific program from the following institutions: CEMAGREF (Paris and Bordeaux), Laboratoire de Géologie Appliquée and the Institut d'Hydrologie et de Climatologie (Paris VI University), Groupe de Microbiologie des Milieux Aquatiques (Bruxelles University). The development of the RIVE model has been supported by the CE program BIBOCULARS (DGXII).

References

Ambrose, R. B., T. A. Wool, J. P. Wool & R. W. Schanz, 1988. WASP 4, A hydrodynamic and water quality model. Model theory, user's manual guide and programmer's guide. Department of civil Engineering, Tufts University, Medford, MA 02155 – Environmental research laboratory office of research and development – U.S. epa, Environmental Research Laboratory – Office of Research and Development – U.S. Environmental Protection Agency – Athens, Georgia 30613.

Aristaghes, M. & A. Wagner, 1989. Utilisation du modèle QUAL2 pour la gestion de la qualité des rivières: application à la Seine entre Montereau et Poses. Technical report, Société Hydrotechnique de France.

Billen, G., P. Dégardin, S. Even & W. Thomas, 1995. Intercomparaison des modèles KALITO, MONET et ProSe. Technical report, PIREN Seine.

Billen, G., S. Dessery, C. Lancelot & M. Meybeck, 1989. Biogeochemistry, volume 1, chapter Seasonal and inter-annual variations of nitrogen diagenesis in the sediments of a recently impounded basin, pages 73–100. Kluwer Academic Publishers.

Billen, G., J. Garnier & P. Hanset, 1994. Modelling phytoplankton development in whole drainage networks: The riverstrahler model applied to the Seine river system. Hydrobiologia: 119–137.

Billen, G. & P. Servais, 1989. Modélisation des processus de dégradation bactérienne de la matière organique en milieu aquatique. In Micro-organismes dans les écosystèmes océaniques: 219–245. Masson.

Bott, A., 1988. A positive advection scheme obtained by nonlinear renormalization of the advective fluxes. Monthly Weather Rev. 117: 1006–1015.

Brion, N. & G. Billen, 1997. Wastewaters as a source of nitrifying bacteria in river systems: the case of the river Seine downstream from Paris. Biogeochem., in press.

Brown, L. C. & T. O. Barnwell, 1987. Enhanced stream water quality models, QUAL2E and QUAL2E UNCAS - Documentation and user's TechReport. Department of civil Engineering, Tufts University, Medford, MA 02155 – Environmental research laboratory office of research and development - U.S. Environmental Protection Agency, Environmental Research Laboratory – Athens, GA Office of research and Development – U.S. Environmental Protection Agency – Athens, Georgia 30613, Rapport EPA/600/3-87/007:

Chesterikoff, A., B. Garban, G. Billen & M. Poulin, 1992. Inorganic nitrogen dynamics in the river Seine downstream from Paris (France). Biogeochemistry 17: 147–164.

Cunge, J., F. Holly & A. Vervey, 1980. Practical aspects of computational river hydraulics. Editions Pitman Advanced Publishing Program.

Even, S., 1990. Modèle de qualité d'eau de rivière: étude bibliographique, application de QUAL2E au cas de la Seine, proposition d'un modèle Seine. Master, ENPC, ENGREF, Paris XII.

Even, S., 1995. Modélisation d'un écosystème fluvial: la Seine. Le modèle ProSe. Phdthesis, cole des Mines de Paris, Centre de Géologie Informatique.

Even, S., M. Poulin, J.-M. Mouchel & G. Billen, 1996. Simulating the impact of cso's from greater paris on the seine river using the model prose. In Seventh International Congress on Urban Drainage Storm Water. IAWQ, EWPCA, IAHR and ATV.

Garban, B., D. Ollivon, M. Poulin & A. Chesterikoff, 1993. Phénomènes d'échanges à l'interface eau-sédiment en seine à l'aval de Paris. tentative de bilan. Technical report, PIREN Seine.

Garnier, J., G. Billen & M. Coste, 1995. Seasonal succession of diatoms and chlorophyceau in the drainage network of the river Seine: Observations and modelling. Limnol. Oceanogr. 40: 750–765.

Garnier, J., G. Billen & P. Servais, 1992a. Physiological characteristics and ecological role of small and large sized bacteria in a polluted river (Seine river, France). Arch. Hydrobiol. Beih. 37: 83–94.

Garnier, J., P. Servais & G. Billen, 1992b. Bacterioplankton in the Seine river (France): impact of the parisian urban effluent. Can. J. Microbiol. 38: 56–64.

Holler, A., 1971. The mechanism describing oxygen transfer from the atmospher to discharge through hydraulic structure. Proceeding XIV Congress A.I.R.H. Document A45.

Holly, F. & A. Preissmann, 1977. Accurate calculation of transport in two dimensions. J. Hyd. Eng. 103: 259–1277.

Holly, F. & J. Usseglio-Polatera, 1984. Dispersion simulation in two-dimensional tidal flow. J. Hyd. Eng. 110: 905–926.

Hug, M., 1975. Mécanique des fluides appliquée. Editions Eyrolles.

Lancelot, C. & G. Billen, 1985. Carbon-nitrogen relationships in nutrient metabolism of coastal ecosystems. In Advances in Aquatic microbiology, pages 263–321. Academic Press.

Lancelot, C., C. Veth & S. Mathot, 1991. Modelling ice-edge phytoplankton bloom in the Scotia-Weddell sea sector of the southern ocean during spring 1988. J. mar. System 2: 333–346.

Lesouef, A. & A. André, 1982. Mise au point d'un modèle de qualité de la seine de montereau à poses. Technical report, Société Hydrotechnique de France. XVIIe journées de l'Hydraulique (Nantes).

Platt, T., C. Gallegos & W. Harrisson, 1980. Photoinhibition of photosynthesis in natural assemblages of marine phytoplancton. J. mar. Res. 38: 687–701.

Ponce, V. & D. B. Simons, 1978. Convergence of four-point implicit water wave models. J. Hyd. Div. 104: 947–958.

Rash, P. J. & D. L. Williamson, 1990. On shape-preserving interpolation and semi-lagrangian transport. Siam Journal Science Stat. Comput. 11: 656–687.

Samuels, P. & C. Skeels, 1990. Stability limits for preissmann's scheme. J. Hydr. Eng. 116: 997–1012.

Servais, P., J. Garnier, N. Demarteau, N. Brion & G. Billen, 1997. Supply of organic matter and bacteria to the aquatic ecosystems through waste water effluents., in press.

Strahler, A., 1957. Quantitative analysis of watershed geomorphology. Géophys. Union Trans.: 913–920.

Streeter, H. & E. B. Phelps, 1925. A study of the pollution and natural purification of the Ohio river. Technical report, U.S. Public Health Service, Treasury Department, Washington DC. Public Health Bulletin.

Thibodeaux, L., M. Poulin & S. Even, 1994. A model for enhanced aeration of streams by motor vessels with application to the River Seine. J. Hazardous Materials: 459–473.

Thiriot, C. & L. Benayada, 1993. Essai critique sur la diffusion numérique dans un algorithme d'écoulement transitoire à surface libre. La Houille Blanche: 527–536.

Hydrobiologia **373/374:** 47–60, 1998.
J.-C. Amiard, B. Le Rouzic, B. Berthet & G. Bertru (eds),
Oceans, Rivers and Lakes: Energy and Substance Transfers at Interfaces.

Forecasting of water quality in lakes: a predictive use of a one-dimensional model. Application to lake Bourget (Savoie, France)

Bruno Tassin & Brigitte Vinçon Leite[1]
[1] *Cergrene: ENPC-ENGREF – 6–8 rue Blaise Pascal, Cité Descartes, F-77455 Marne-la-Vallée cedex 2, France*

Key words: lake, predictive modelling, water quality forecasting, temperature, phosphorus, dissolved oxygen

Abstract

In order to assess the efficiency of eutrophication control programs, predictive models are necessary. We propose a methodology for implementing such a model, based on the coupling of a biogeochemical model of a lake and the use of long term time series of meteorological data. This methodology is applied to lake Bourget (Savoie, France). It allows to obtain both mean and standard deviation (first and second order moments) of the state variables of the model on a 1 y period. The sensitivity of the model to the various forcing variables, as well as to the initial conditions is analyzed as well as the linear or non-linear behavior of the model. Finally, the propagation of the uncertainties (standard deviations) in time and space, for both water temperature and dissolved oxygen are assessed.

Introduction

Since the beginning of the seventies and the increase of the use of computers, many lake water quality modeling attempts have been emphasized. Various modeling directions have been explored ranging from simple input-output models (OCDE, 1982) to complicated 3-dimensional models including several species at various trophic levels (Simons, 1979). As pointed out by several authors (Jørgensen, 1989), the practical application of eutrophication spatially discretized models has shown that they can be used as a tool in environmental management.

Moreover modeling is often a useful tool for improving the knowledge of lake ecosystems, highlighting the differences between a conceptual representation, included in the model, of the behavior of the system and on-site observations (Vinçon-Leite et al., 1995).

Nevertheless, one of the major objective of modeling, which is to predict the evolution of a lake, remains rarely studied compared to other mathematical modeling studies. The success of Vollenweider's approaches (Vollenweider, 1968; OCDE, 1971; OCDE, 1982) is probably based on the acute need of effective solutions to the problem of pollution. Thus, predictive limnology (Peter, 1986) aims at helping in both eutrophication control and lake management. The first attempts used important amounts of data (Chapra & Reckhow, 1983) and produced empirical statistical models (Dillon & Rigler, 1975; OCDE, 1982). Despite the success of these approaches, several problems arose. Between others we mention:

- The models are mostly steady-state models and they only produce a description of the final state of the lake. They are not able to produce information concerning the state of the lake between the initial state and the final state;
- As a consequence of the previous point, they provide little understanding of the behavior of the lake;
- the uncertainty of the prediction is not sufficiently precise.

More recently, dynamic phosphorus budget models have been developed (Havis & Ostendorf, 1989), describing the evolution of the lake between the initial and the final state, and as a consequence trying to suppress the first shortcoming mentioned above.

All of these models give only a global response of a lake considered as an homogeneous basin, as they are mostly interested in the prediction of the mean total phosphorus concentration.

Table 1. Lake Bourget morphometric and hydrologic characteristics.

Lake surface area	4210^6 m^2
Volume	310^9 m^3
Maximal depth	145 m
Mean depth	80 m
Mean depth of the thermocline	15 m
Length	18 km
Maximum width	3 km
Catchment area	560 km^2
Mean discharges	16 m^3 s^{-1}
Mean residence time of water	8 y

Thus it seems necessary to combine the spatially discretized modeling approach and the predictive modeling approach, in order to produce prediction of the lake behavior, for different state variables and for the most important layers, at least the epi- and the hypo- limnion of the lake.

In this paper we propose a predictive modeling approach of a lake : Lake Bourget (Savoie, France), based on a vertical one-dimensional water quality model (Vinçon-Leite, 1991). After a brief description of lake Bourget, we will successively focus on the modeling methodology used and on the results obtained for a one year prediction. This period is obviously too short for providing valuable prediction for management purposes. Nevertheless this first step allows to check the propagation of uncertainties due to variable meteorological forcing in the water quality model (Hondzo & Stefan, 1992). On the management side, this one year predictive model helps:

- in the planning of the field surveys
- in the understanding of the impact on the lake ecosystem of the specific meteorological characteristics of one year.
- in the determination of initial conditions at the beginning of the year which lead to a specific state later, e.g. hypolimnic anoxia.

Description of Lake Bourget

Lake Bourget is one of the largest lakes in France. Located in the vicinity of two important cities (Chambéry and Aix les Bains) – 400 000 inhabitant-equivalents – the lake has suffered from eutrophication. The diversion of the main waste water discharges at the beginning of the eighties, reduced phosphorus and nitrogen loading by 50% and 70%, respectively. Despite the fact that some water quality indicators exhibit a real improvement (total phosphorus concentration), others like primary production, chlorophyll or algae species, remain unsatisfactory, especially during spring (Vinçon-Leite, 1991). The Secchi Depth varies from 12 m in winter to 2.5–3 m in summer. The main morphometric characteristics of the lake are listed in Table 1. Two physical characteristics are of primary importance, the maximum depth of the lake which does not allow a yearly complete overturn and the mean residence time, which causes only small year to year changes in the ecosystem characteristics, outside of the effect of the perturbations caused by the yearly meteorological specificities.

Lake Bourget was recently intensively surveyed during two periods (1987–1989 and 1994–1996). Concurrently a lake water quality model has been implemented and validated over the 1981–1988 period. A meteorological station depending on Meteo France is located at the south end of the lake (Voglans). Thus meteorological data for the main variables are available at a daily time step over the 1974–1990 period.

Methodology

Lake dynamic predictive models, either on physical or biogeochemical aspects, are quite scarce. For example a 1-D temperature predictive mode has been developed and applied to Lake Calhoun in order to check the propagation of uncertainty in lake temperature models (Hondzo & Stefan, 1992). The same authors and others developed recently long term temperature and oxygen simulations associated with fish habitats (Stefan et al., 1996). Other authors (Canale & Effler, 1989) developed a predictive stochastic phosphorus model, based on a well-known total phosphorus budget model.

Predictive modelling is always the coupling of two models:

1. a model of the forcing variables which aims at describing their stochastic behaviour – in our case it is a model of the meteorological variables.
2. a model of the state variables, which is fed by the results of the previous model.

Various methods are available for taking into account this coupling as:

- Monte-Carlo simulations: in this technique, statistical distributions of the model state variables are obtained through numerous simulations. For

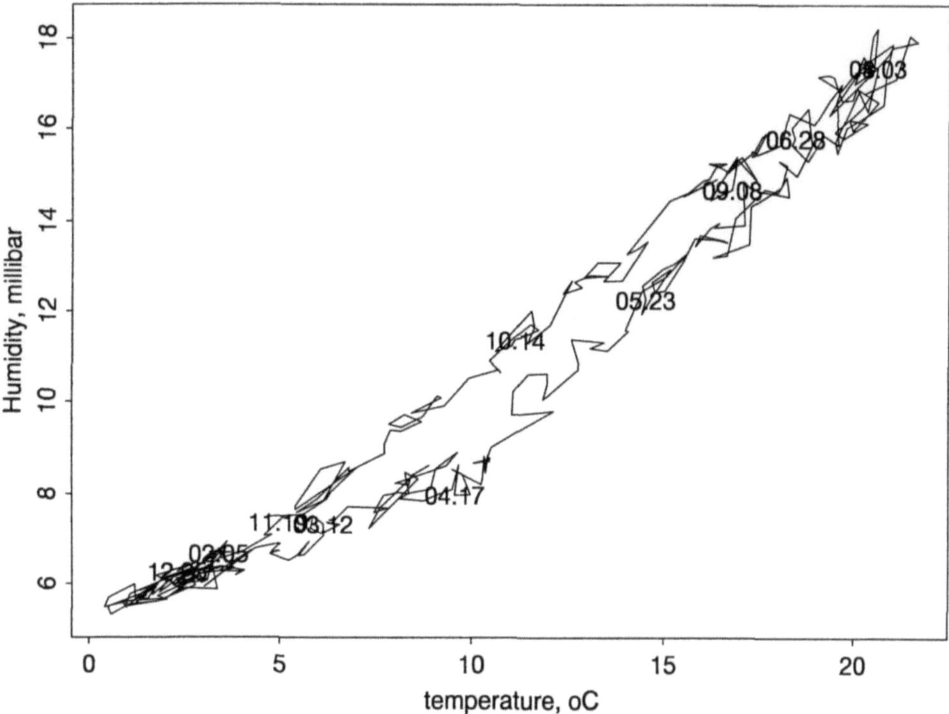

Figure 1. Daily temperature (C) and humidity (mbar) at Voglans Météo-France station. Each point represents the average over the 1974–1990 period of the daily temperature and humidity measurements. 12 points are identified, showing the month and the day in the month.

each simulation the forcing variables are randomly changing according to their statistical specificities, giving a unique value for each state variable. Then these values are gathered in order to calculate their statistical distribution. This method needs an a priori analysis of the time series of the forcing variables – i.e the forcing variables model, – and is computer time consuming, but quite easy to undertake.

- Perturbation methods: in this technique the model of the state variables is linearized, giving access to a first-order estimate of the mean and covariance of the state variables. In this approach the time series of the forcing variables are reduced to the residual series, by removing periodic means and standard deviations (Richardson, 1981).

In our case this two-step methodology has been applied as follows.

The model of the forcing variables

Reduction of the meteorological variable number

As mentioned above, six meteorological variables used as forcing variables in the model. For simplicity, it is interesting to minimize this number, looking for strong correlations between the variables.

As pointed out by other authors (Hondzo & Stefan, 1992; Bacsko et al., 1991) air temperature and air humidity are clearly correlated (Figure 1). This biunivoque relation is easily modeled and humidity is then described as a temperature function.

Short wave insolation has been calculated using theoretical calculations of the length of the day along the year and on the basis of cloudiness measurements using empirical formulas (Henderson-Sellers, 1986).

Moreover, in our case atmospheric pressure exhibits only small variations along the year, and has only a weak importance in the calculations of the fluxes at the air winter interface. For this reason we consider that atmospheric pressure is constant across the year. Under these hypotheses, only three forcing variables have to be studied: temperature, cloudiness and wind velocity.

Temperature

Daily temperatures averaged over the 1974–1990 period show a sinusoidal pattern. For each day the standard deviation falls between 1.5 and 2.5 C, with

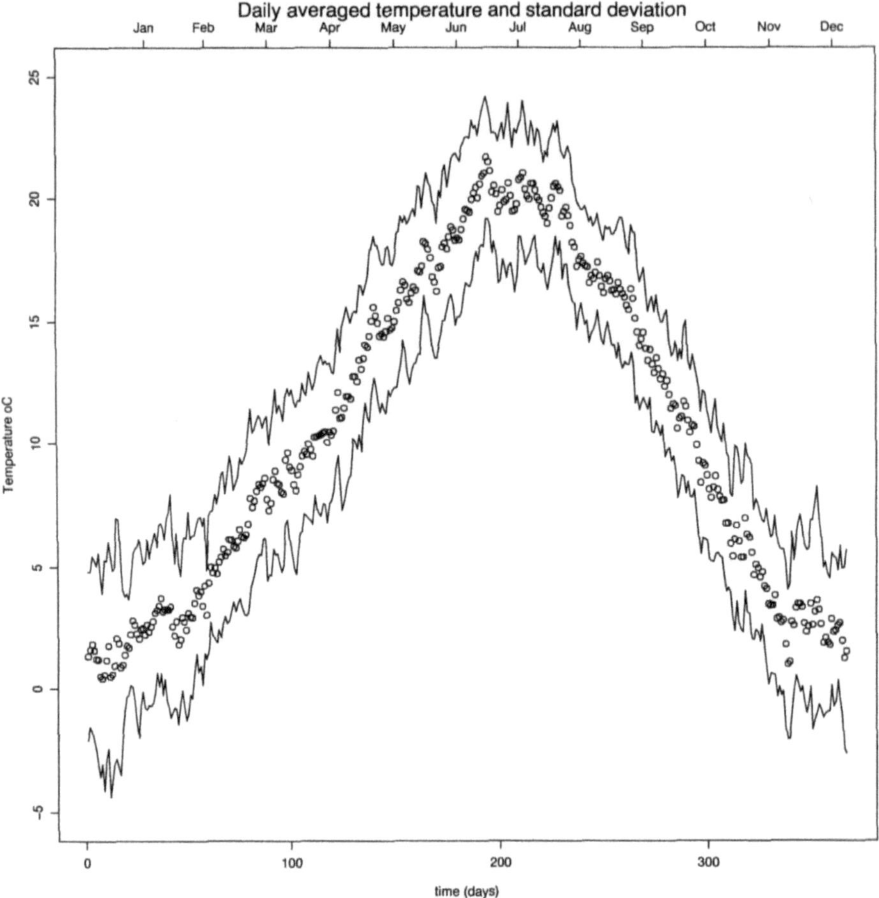

Figure 2. Mean daily temperature and standard deviation at Voglans Station over the 1974–1990 period.

a maximum value in winter (Figure 2). An analysis of the autocorrelation factors of the daily deseasonalized time series shows that meteorological events separated by more than 10 d are independent.

Cloudiness

As for temperature, averaged daily cloudiness over the 1974–1990 period exhibit a seasonal pattern (Figure 3). The auto correlation frequency corroborates this seasonal behavior, showing a 1 y period.

Wind velocity

Wind velocity is a rapidly varying variable, and thus the raw time series did not exhibit a seasonal pattern. Looking at the monthly distribution of the wind velocity, it appears that high wind events occur mostly during winter and spring. In summer, wind velocities lie always under four m s^{-1}. Wind velocities cannot be described using a classical time series model. The best modeling approach is to generate wind velocities randomly, using the observed monthly distribution.

Simulation of the forcing variables

Temperature and cloudiness have been described using classical time series models (Box & Jenkins, 1977). Nevertheless we tried unsuccessfully to generate automatically forcing temperature and cloudiness time series. This failure is partly due to the structure of the models used, which sometimes caused unrealistic excursions of the generated variables. Thus the model of the forcing variables was composed of the 17 real meteorological years at our disposal from 1974 to 1990.

The model of the state variables

The model used in this study is a deterministic, one dimensional water quality model. It is based on the coupling of a thermal and a biogeochemical submodel.

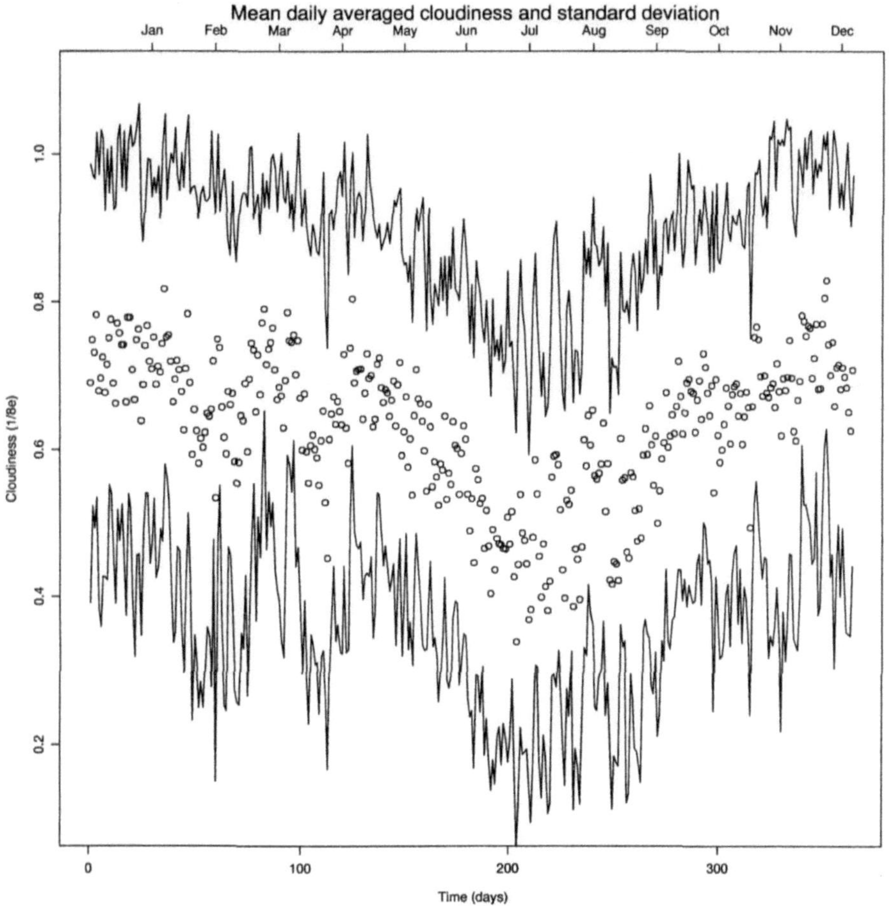

Figure 3. Mean daily averaged cloudiness and standard deviation at Voglans station over the 1974–1990 period.

The model (Tassin, 1987) has been successfully applied on various systems for the thermal submodel (Tassin, 1987; Vinçon-Leite, 1991; Tusseau et al., 1996) and various meteorological conditions, over long time periods (Tassin, 1986; Vinçon-Leitte et al., 1995). It is based on the advective – dispersive equation of heat, integrated on the horizontal dimension (equation 1). It is solved using a semi-implicit finite differences scheme with 1 m thick layers and a 3-h time step.

$$\begin{aligned} \frac{\partial V(z,t)}{\partial t} = & \frac{1}{A(z)} \frac{\partial}{\partial z} \left[A(z)\, K(z,t) \frac{\partial V(z,t)}{\partial z} \right] && \text{Dispersion} \\ & - \frac{1}{A(z)} \frac{\partial}{\partial z} [A(z)\, W(z,t)\, V(z,t)] && \text{Advection} \\ & + SS(z,t) && \text{Sources and sinks} \end{aligned} \quad (1)$$

where $V(z, t)$ is the variable considered at depth z and time t, $A(z)$ is the cross-sectional lake area at depth z, $K(z, t)$ is the eddy diffusion coefficient and $W(z, t)$ is the vertical velocity of water.

$SS(z, t)$ is the term of sources and sinks corresponding for each variable to all the fluxes in the system.

Thermal submodel

The thermal submodel is characterized by:

- the surface heat budget as well as the inlake heat sources ($SS(z, t)$) due to short wave radiations absorption. The surface heat budget is described using classical formulation of the radiative fluxes (daily insolation time, atmospheric longwave radiation, back radiation from the water) and of the convective fluxes (sensible and latent heat fluxes) (Henderson-Sellers, 1986). These fluxes are based on meteorological forcing variables (air temperature, humidity, direct solar radiation, cloudiness, wind velocity, air pressure).

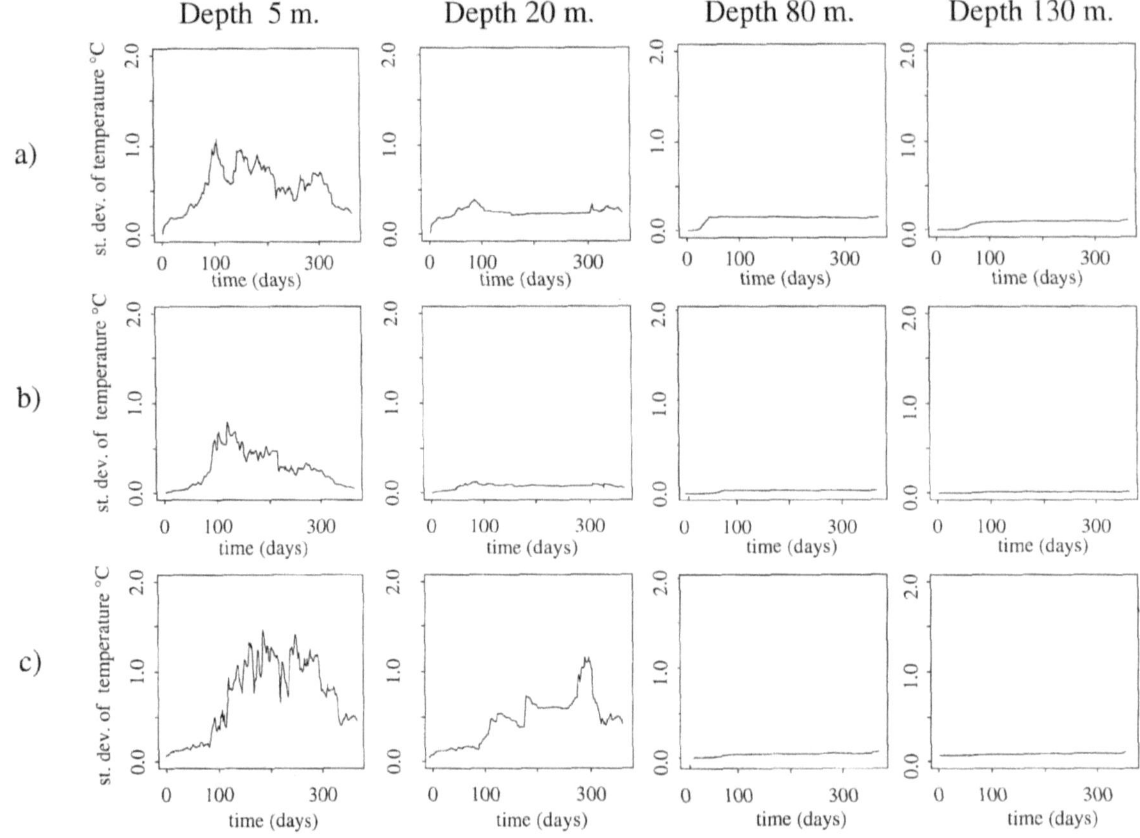

Figure 4. Standard deviation of the simulated distribution of temperature at 5, 20, 80 and 130 m. *a* Only air temperature has a year to year variation. *b.* Only cloudiness has a year to year variation. *c.* Only wind velocity has a year to year variation.

- the vertical dispersion coefficient ($K(z, t)$). It has a different formulation in the epilimnion and in the hypolimnion. The epilimnic dispersion coefficient is based on the exponential decay of the horizontal velocity of the water, due to the wind under the water surface (Simons, 1989). The hypolimnic dispersion coefficient is based on a stability function depending on the Brunt-Vaïsala frequency (Tassin, 1987).

Biogeochemical submodel

Four state variables are included in the model :

- Dissolved Oxygen
- Soluble Reactive Phosphorus
- Algal Phosphorus
- Zooplanktonic Phosphorus.

The most important processes which are taken into account are:

- photosynthesis with light extinction, nutrients and temperature limitations
- mineralization of organic matter
- predation of algae by herbivorous zooplancton
- sedimentation of particulate matter
- phosphorus release from the sediment
- benthic oxygen demand

More details on the equations which come into play within this model can be in found elsewhere (Vinçon, 1991). However, it is assumed that initial conditions for the state variables are available and consistent with the space scale of the model. As a consequence there is no need for a spinup period for the simulation of the water quality of the lake.

The external sources which are taken into account are the oxygen exchange at the air–water interface and the input of phosphorus considered as available phosphorus from the tributaries.

The coupling of the two models

Each predictive simulation of the state variables was based on 17 one-year simulations of these variables, with the same initial conditions and with the forcing

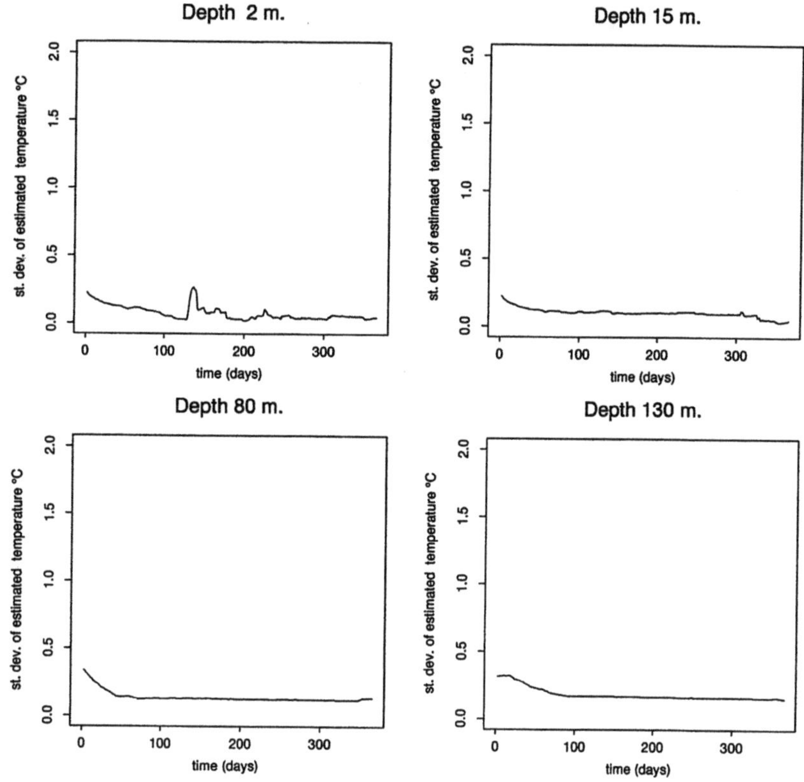

Figure 5. Standard deviation of the simulated distribution of temperature at 5, 20, 80 and 130 m. Initial conditions are varying according to the measurements in January (first decade) over the period (1988–1995).

variables corresponding to the 17 meteorological years at our disposal.

Then the results of each one-year simulation are gathered and statistically analyzed in terms of mean and standard deviation. This method is partly similar to the method used for predictive temperature simulation in the Calhoun lake (Hondzo & Stefan, 1992). It can be considered as a Monte Carlo method, despite the fact that the forcing variables are not generated but are real observed time series.

Then the predictive model has been used in various ways for:

1. Assessing the role of the main forcing variables on the behavior of the lake. For this purpose a sensitivity analysis of the model to the forcing variables has been conducted. Due to calculation limitations, no synergy effect between the forcing variables is considered .
2. Assessing the role of the initial conditions on the lake response along the year. As Lake Bourget is a meromictic lake, initial winter conditions are varying from year to year and the behavior of the lake is the consequence of both these initial conditions and of the forcing variables time series. In order to assess the specific role of the initial conditions, a sensitivity analysis to these conditions has been conducted on the basis of the measured vertical profiles of the state variables during the last ten years.
3. Assessing the linear behavior of the lake. Linear models are often more easy to use that non-linear models, thus testing the linear behavior of the lake has very practical consequences for modeling considerations.
4. assessing the uncertainties in the results of the simulation. Taking into account in the simulation the interannual variability of the forcing variables leads to results which are no more a unique deterministic answer but a probability distribution. We propose to analyze the second order moment of this distribution as a way for checking the uncertainties of the model results.

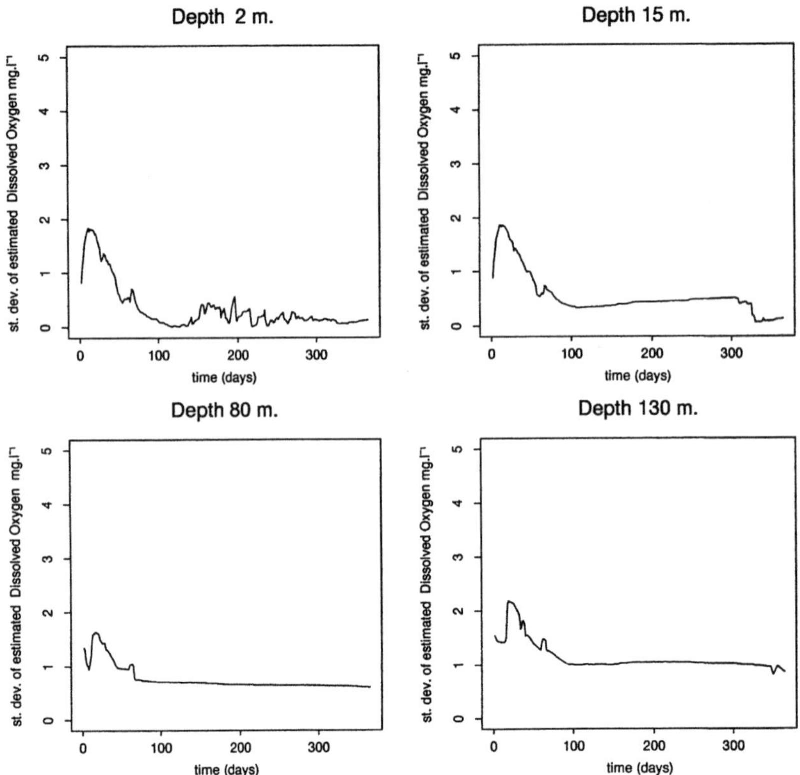

Figure 6. Standard deviation of the simulated distribution of dissolved oxygen at 5, 20, 80 and 130 m. Initial conditions are varying according to the measurements in January (first decade) over the period (1988–1995).

Results and discussion

Sensitivity analysis

Concerning the forcing variables

The sensitivity analysis has been conducted for each of the three basic forcing variables: air temperature, cloudiness and wind velocity. It is based on 17 simulations with the same yearly distribution of the meteorological variables, excepted for the forcing variable under study, for which we used the observed values over the 1974–1990 period. January 1988 profiles were used as initial conditions. They are typical of beginning of winter profiles in the lake. At this date, the first 60 meters of the lake are completely mixed. A 0.5 C gradient remains in the deeper layer between 60 m and the bottom of the lake. The simulation is conducted over a one year period.

As the water temperature is itself a forcing variable of the water quality model state variables, we here only present results concerning water temperature. Figure 4 shows the standard deviation of the water temperature distribution at 5 m (upper layer), 20 m (metalimnion), 80 m (deep layers), 130 m (near bottom layers). As a consequence of the simulation duration, results at the beginning of the year (first month) are affected by the initial conditions and are not significant in terms of sensitivity.

Concerning temperature, most important effects are located in the upper layers. Immediately under the surface, from April to November, the calculated standard deviation (CSD) lies between 0.5 and 1 C. This seasonal trend of the CSD is surprising, showing a rapid increase during the end of winter and spring and a slow decrease from the beginning of summer to the end of the year. This shows that the system is very sensitive to temperature variation at the beginning of the year and that the second part of the year partly corrects and reduces the spring perturbation, progressing towards a robust mean value. From the thermocline to the deep layers, the CSD exhibits the same pattern, with an increase in the first term of the year followed by a constant value from April to December. This pattern confirms firstly the importance of the air temperature at the beginning of the year, which allows or

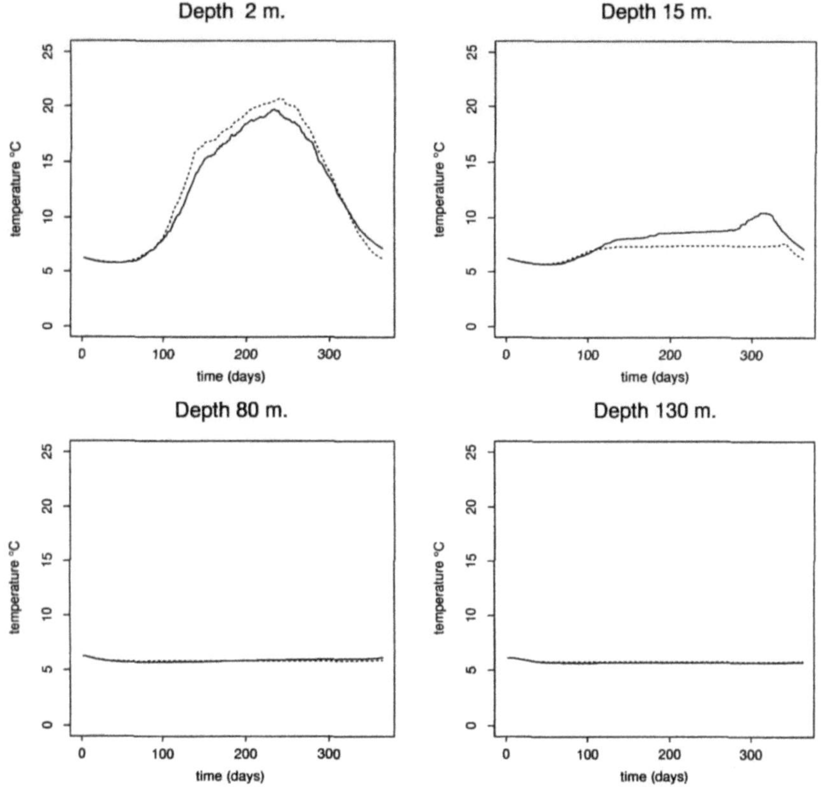

Figure 7. Comparison of the estimated temperature using the averaged meteorological data and the mean of the results of the predictive model.

not a complete mixing of the lake. Secondly, it shows the memory effect of the deep layers; when the lake is stratified, it is no possible to modify the deep layers' temperature.

The sensitivity to cloudiness is less important than sensitivity to air temperature. For this variable CSD does rarely exceed 0.5 C, excepted in the upper layers during the onset of the thermocline. As for temperature, the CSD decreases from the middle to the end of the year. Effects under the thermocline are quite null.

Surprisingly, the effect of the wind velocity is very important. This observation is not in accordance with other similar studies reported in the literature (Hondzo & Stefan, 1992). The CSD exceeds 1 C in the upper layers of the lake, from the end of spring to the middle of fall. It must be pointed out that the CSD yearly evolution concerning wind velocity has not the same pattern than the CSD concerning air temperature. Air-temperature CSD rises up at the beginning of spring and wind-velocity CSD is delayed to the beginning of summer, when the lake is strongly stratified. Thus air-temperature and wind velocity interannual variabilities have cumulative effects in the upper layers, the first one acting mostly during spring and the second one during summer and fall. The role of the thermocline on the wind velocity CSD is also clearly shown on the CSD yearly profile at 20 m. A peak value is observed in fall, during the beginning of the destratification. No effect of sensitivity due to yearly wind variability is observed in the deep layers.

In conclusion, on lake Bourget, wind velocity is of primary importance on the yearly water temperature structure. Moreover, only air-temperature has the ability to modify the deep layers' water temperatures. These observations strengthen the chosen methodology which is based on the use of observed meteorological time-series.

Concerning the initial conditions

Another sensitivity analysis of the complete lake model has been conducted for checking the impact of the initial conditions on the seasonal evolution of the lake. Using the averaged daily meteorological variables, we simulated the behavior of the model state variables, on the basis of the measured profiles, between January 1st and January 10th over the period

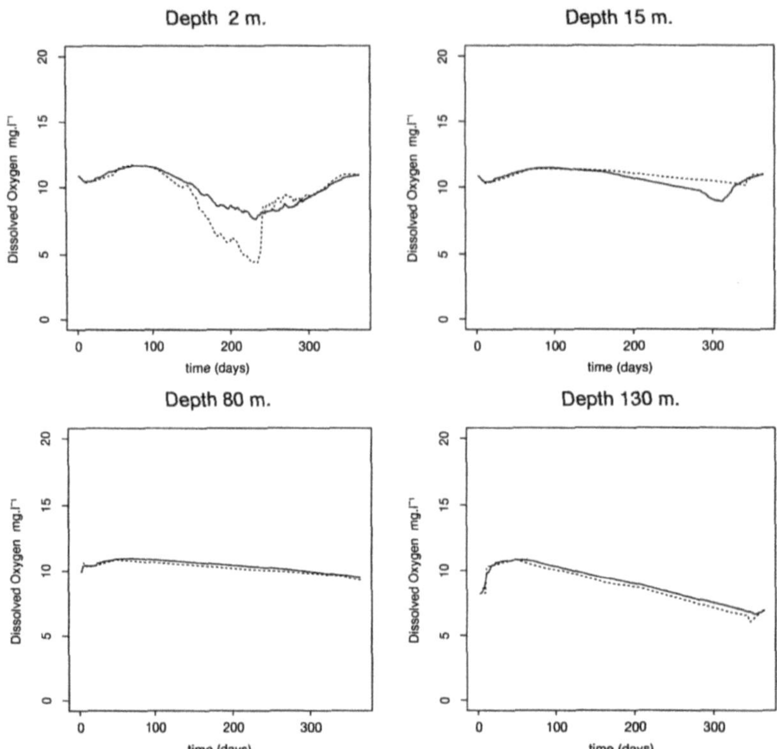

Figure 8. Comparison of the estimated dissolved oxygen concentrations using the averaged meteorological data and the mean of the results of the predictive model.

1988–1995. We present here the standard deviation of the water temperature (Figure 5) and of dissolved oxygen (Figure 6) at 2, 15, 80, and 130 m.

Concerning water temperature, initial conditions have a weak impact (CSD values remain under 0.3 C) but they concern all the water column. Thus, depending on the initial water temperature profile, a complete overturn will occur or not, with the well known consequences on the biogeochemical behavior of the lake. In the upper layers a small increase of CSD is observed during the onset of the thermocline and rapidly disappears. This result corroborates with winter predictive simulation conducted on Lake Geneva (Tassin, 1986).

Concerning dissolved oxygen, the results are quite different. Depending on the initial conditions and on the occurrence or not of a complete mixing, oxygen concentrations may greatly vary and the CSD values in the bottom layers lie between 1 and 2 $mgO_2\ l^{-1}$. On the contrary, near the surface, the CSD decreases from the beginning of the year to the beginning of spring from 1 to 0 $mgO_2\ l^{-1}$. Small variations, without any general consequences of the CSD are also observed during summer, depending mostly on the algal growth fluctuations.

Comparison between the predictive model and the deterministic model using the mean meteorological year

As mentioned previously this is a way for checking if a linear approximation of the model is possible.

The period under consideration is the 17 y long period from 1974 to 1990 where meteorological data are at our disposal.

We have to compare the results of two modeling alternatives:

1. Proceed to several yearly simulations with the same initial conditions and different forcing variables data. Considering a specific day of the year,

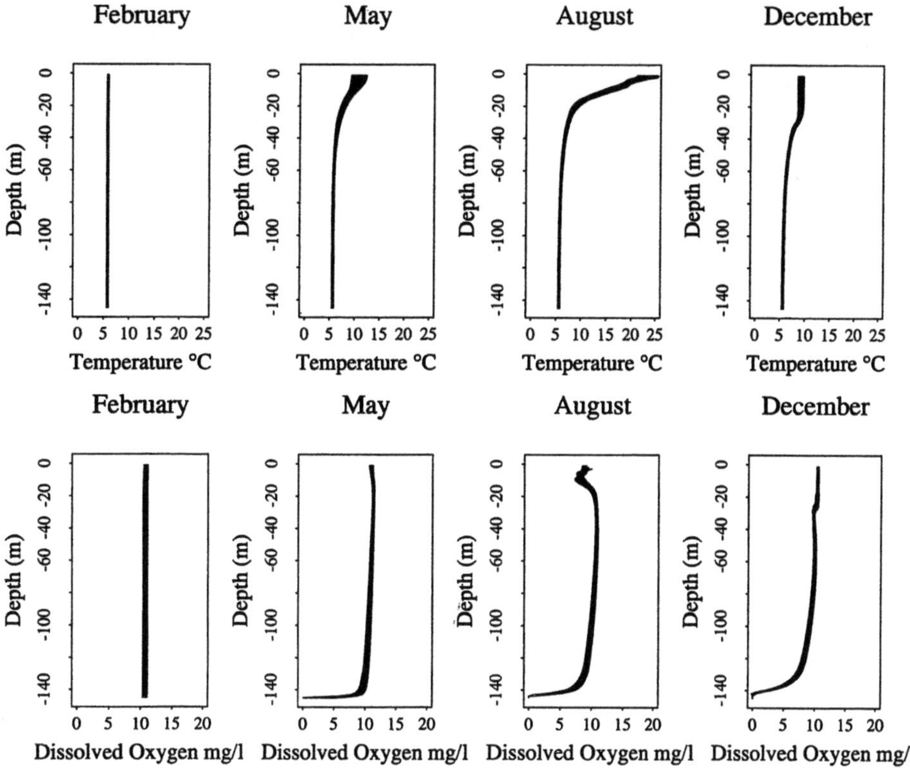

Figure 9. Vertical profiles of temperature and dissolved oxygen (first day of each month) using data of January 1995 as initial conditions. The shaded area represents the + 1 and − 1 standard deviation from both sides of the mean.

the results for this day are obtained by averaging the results for the same day for each simulation.

2. Proceed to one yearly simulation, with the same initial conditions than previously, but the forcing variables for a specific day are obtained by averaging the forcing variables for the same day over the 1974–1990 period.

We present, simultaneously for both alternatives, the evolution of mean temperature (Figure 7) and mean dissolved oxygen (Figure 8) at 2, 15, 80 and 130 m.

In the analysis of the results we have to separate the hypolimnion from the epilimnion. In the hypolimnion, the results are quite similar for the two compared methods and for both temperature and dissolved oxygen. On the contrary, in the upper layers significant differences occur for both temperature and dissolved oxygen. Concerning temperature, from the onset to the decline of the thermocline, differences rise up to over than 2 C (10%), immediately under the surface and at 15 m, in the metalimnion, the difference exceeds 3 C during the decline of the stratification ($\approx$25%).

This is mostly due to a difference in the determination of the location of the thermocline in both modeling approaches. Concerning dissolved oxygen, important differences also appear in the surface layer, during the intensive algal production period (from the middle of spring to September). The results obtained with the second alternative are unrealistic. This is a direct consequence of the averaging of the forcing variables and especially the wind. Averaging the daily wind values over 17 y leads to significantly lower values of wind velocity and suppresses the high wind periods. As a consequence, no wind mixing occurs in the lake and surface reaeration is also reduced. Moreover, during this period, light intensity is high and primary production takes place at a depth of 10 m, but in the absence of mixing the photosynthetic oxygen remains at its production depth and does not reach the upper layers.

Thus, the linear approximation of the behavior of the model for deep layers is undoubtedly valid. Concerning the upper layers, non linearities in the model cause significant differences between the two modeling approaches, not only for water temperature but also for dissolved oxygen and other state variables. Thus, the use of averaged daily meteorological data

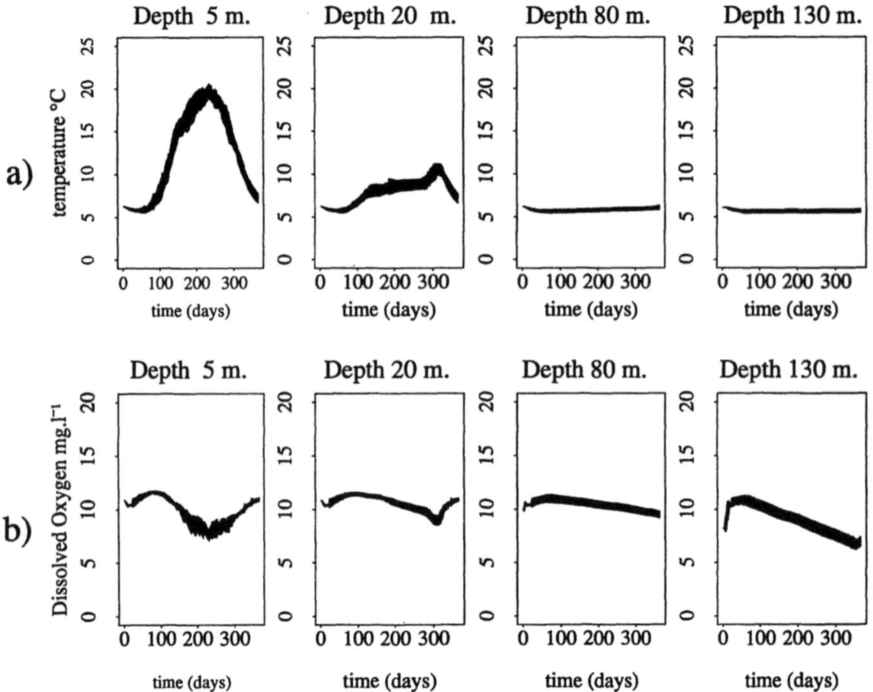

Figure 10. Yearly evolution of temperature (a) and Dissolved Oxygen (b) using data of January 1995 as initial conditions. The shaded area represents the + 1 and − 1 standard deviation from both sides of the mean.

for assessing the mean behavior of the lake is not strictly valid. Other averaging processes of the results (averaging over the epilimnion, over the hypolimnion) will probably decrease the differences observed in the results of the two modeling approaches (Hondzo & Stefan, 1992). Nevertheless the main interest of a one dimensional model is to produce a fine description of the vertical gradients; other simpler modeling approaches are available for producing a two-layer description of a lake behavior.

Second order moment analysis

The second moment analysis provides an assessment of the uncertainty of the results of the model and is one of the most useful points of the proposed methodology. Using the January 1995 initial conditions, we present the results for temperature and dissolved oxygen, both for the vertical profiles (Figure 9), and for yearly evolution (Figure 10).

The vertical profile of temperature (mean + or - standard deviations) indicates that uncertainty propagates downwards to the deep layers of the lake during winter. As previously mentioned this is caused by the inter-annual air temperature variability and by the meromictic behavior of the lake. Under the thermocline, uncertainty does not decrease with depth and remains constant in time. This observation agrees with calculations of dispersion coefficients in the hypolimnion of deep lakes, and is a direct consequence of the formulation of the dispersive coefficient in the model. During the stratification the uncertainty concerning temperature in the upper layer is important, especially in spring and fall (onset and deepening of the thermocline).

Concerning dissolved oxygen, as well as water temperature, uncertainty propagates downwards during winter, then it remains constant. This is surprising as dissolved oxygen concentrations are a consequence of all the biogeochemical processes and integrate all their uncertainties. Specifically, organic particulate material production and its consecutive sinking from the epilimnion to the bottom layer of the lake varies from year to year due to meteorological forcing (cloudiness and air-temperature). This yearly variable sinking had to lead to a variable oxygen consumption in the deeper layers of the lake and an increase of the uncertainty, which is not observed in the results of the model. Thus, despite the fact that high uncertainties for water temperature are observed in the upper layers, they are filtered by the biogeochemical processes, and do not have any consequence on the dissolved oxygen profile, except during the more or less complete overturn period.

Conclusions

We proposed and tested on lake Bourget a simple methodology for predictive lake water quality modeling. The main results are :

1. Because of the sensitivity of the model to the wind velocity, and because realistic time series of this variable are difficult to model, working with real long term forcing variables time series, seems to be a good solution.
2. Initial profiles at the beginning of the year are of primary importance for meromictic lakes. All the behavior of the ecosystem during the year is conducted by the initial profiles.
3. For deep layers the model acts linearly. This is not the case for the upper layers, especially during spring and summer, where non-linearities seem to have significant consequences on both physical and biogeochemical aspects.
4. Uncertainties propagate from the upper to the deep layers only in winter in connection with mixing processes. With this method, it is possible to compare a particular year with the averaged computed one. Concerning biogeochemical variables like dissolved oxygen, the uncertainties remain constant all the year after the winter mixing event.

Of course one of the most important aims of predictive modeling is to assess the efficiency of eutrophication control processes. For progressing in this direction it is necessary to work on longer predictive periods. For Lake Bourget, eight years, which is the water retention time seems to be an appropriate simulation time. This objective raises important difficulties:

- on the modeling aspects: on longer periods, it becomes necessary to take into account the structural changes of the lake. This means that the lake is not in steady state, and that the ecosystem dynamics will change according to modifications of forcing variables like nutrient loadings, which were neglected in the 1-y approach. It is well known (Lehman, 1988; Pelletier et al., 1990) that during restoration, structural changes occur. Unfortunately, despite interesting modeling attempts (Jørgensen, 1988; Jørgensen, 1992), important weaknesses remain in the quantitative description of these processes. For the particular case of Lake Bourget, working in a predictive way needs to improve the relationship between particulate material and especially algae and light extinction.
- On the computing aspects: the simple extension of the proposed methodology to longer periods will lead to an exponential increase of computation time. Improvements have to be made in order to escape from this unrealistic situation. We have seen that the use of averaged meteorological data provides quite satisfactory results concerning temperature and dissolved oxygen in the hypolimnion. Attention must be drawn to the real impact of the discrepancies observed in the upper layers on longer simulations.

Acknowledgements

This work was conducted at CERGRENE in cooperation with GRETI (Université de Savoie) and the 'Cellule Technique du Lac du Bourget' and was supported by SILB and SIAC (lakeside intercommunal councils) and by the Agence de l'Eau Rhône Méditerranée Corse.

We are grateful to all of them as well as to the anonymous reviewers for their valuable comments.

References

Box, G. E. P. & G. M. Jenkins, 1976. Time series analysis: forecasting and control. Holden Day.

Canale, R. P. & S. W. Effler, 1989. Stochastic phosphorus model for Onondaga lake. Water Res. 23: 1009–1016.

Chapra, S. C. & K. H. Rickhow, 1983. Engineering limnology, volume Vol1: Data analysis and empirical modelling. Butterworth Publishers.

Dillon, P. J. & F. H. Rigler, 1975. The phosphorus chlorophyll relationship in lakes. Limnol. Oceanogr. 19: 767–773.

Havis, R. N. & D. W. Ostendorf, 1989. Approximate dynamic lake phosphorus budget model. J. Envir. Eng. 115: 809–821.

Henderson-Sellers, B., 1986. Calculating the surface energy balance for lake and reservoir modelling: A review. Rev. Geophys. 23: 625–649.

Hondzo, M. & H. G. Stefan, 1992. Propagation of uncertainty due to variable meteorological forcing in lake temperature models. Water Resour. Res. 28: 2629–2638.

Jørgensen, S. E., 1988. Use of models as experimental tool to show that structural changes are accompanied by increased energy. Ecol. Modell. 41: 117–126.

Jørgensen, S. E., 1989. Eutrophication models: the state of the art. In Salánki, J. & S. Herodek (eds), Conservation and Management of Lakes, volume 38. Symp. Biol. Hung., Budapest, Adememiai Kiado, 609–621.

Jørgensen, S. E., 1992. Parameter, ecological constraints and exergy. Ecol. Model. 62: 163–170.

Lehman, J. T., 1988. Hypolemnetic metabolism in Lake Washington: relative effects of nutrient load and food web structure on lake productivity. Limnol. Oceanogr. 33: 1334–1347.

OCDE, 1971. Les bases scientifiques de l'eutrophisation des lacs et eaux courantes sous l'aspect particulier du phosphore et de l'azote comme facteurs d'eutrophisation. OCDE.

OCDE, 1982. Eutrophication of waters. OCDE.

Pelletier, J., G. Balvay & J. C. Druart, 1990. Evolution du plancton du Léman (Campagne, 1989). Rapport technique, CIPEL – Rapports sur les études et recherches entreprises dan le bassin lémanique, Laussanne, Suisse.

Peter, R. H., 1986. The role of prediction in limnology. Limnol. Oceanogr. 31: 1143–1159.

Racsko, P., L. Szeidl & M. Semenov, 1991. A serial approach to local stochastic weather models. Ecol. Modell. 57: 27–41.

Richardson, C. W., 1981. Stochastic simulation of daily precipitation temperature and solar radiation. Water Res. Res. 28: 2629–2638.

Simons, T. J. 1979. Assessment of water quality simulation capability for lake Ontario. Scientific Series 111, CCIW.

Simons, T. J., 1989. The seasonal climate of the upper ocean. Data analysis and model development. Rapport technique, National Water Research Institute.

Stefan, H. G., M. Hondso, X. Frang, J. G. Eaton & J. H. McCormick, 1996. Simulated long-term temperature and dissolved oxygen characteristics of lakes in the north-central United States and associated fish habitats limits. Limnol. Oceanogr. 41: 1124–1135.

Tassin, B., 1986. Contribution à la modélisation écologique du lac Léman: modèles physiques et biogéochimiques. Thèse de Doctorat, ENPC.

Tassin, B., 1987. Un modèle de l'évolution saisonnière du phosphore dans le lac Léman. Sci. de l'eau 6: 67–95.

Tusseau, M. H., C. Lancelot, J. M. Martin & B. Tassin, 1996. 1-d couped physical-biological model of the northwestern mediterranean sea. Deep. sea res. (II) 44: 851–880.

Vinçon-Leite, B., 1991. Contribution de la modélisation mathématique à l'étude de la qualité de l'eau dans les lacs sub-alphins: le lac du Bourget. Thèse de Doctorat, ENPC.

Vinçon-Leite, B., B. Tassin & J. M. Jaquet, 1995. Contribution of mathematical modeling to lake ecosystem understanding: Lake Bourget (Savoy France). Hydrobiologia 300/301: 433–442.

Vollenweider, R. A., 1986. Water management research. Number 63.26 in OECD/DAS/SCI.OCDE.

Hydrobiologia **373/374:** 61–73, 1998.
J.-C. Amiard, B. Le Rouzic, B. Berthet & G. Bertru (eds),
Oceans, Rivers and Lakes: Energy and Substance Transfers at Interfaces.

Comparison of Neutron Activation Analysis and ICP-MS used for river water pollution control

G. Pinte[1], M. A. R. Veado[2], A. H. Oliveira[2], M. Khalis[3], S. Ayrault[1] & G. Revel[1]
[1] *Pierre Süe Laboratory, CEA-CNRS, C.E./Saclay, F-91191 Gif sur Yvette Cedex, France*
[2] *Department of Nuclear Energy, UFMG, Av. do Contorno 824, CEP 30110-060 Belo Horizonte, Brazil*
[3] *M[ly] Ismail University, Faculty of Sciences, BP 4010, Bni Mhamed, Meknes, Morocco*

Key words: INAA, ICP-MS, pollution, river water, sediments

Abstract

This paper compares the performance and limitations of two analytical methods for samples from the environment: Instrumental Neutron Activation Analysis (INAA) and Inductively Coupled Plasma with Mass Spectrometry (ICP-MS). A Brazilian river, the Das Velhas, was chosen as the test-site. This river runs through a Fe, Mn and Au mining area. The two methods were initially tested on three international reference materials: SLRS-3 (riverine water, Research Council Canada), GS-N (granite, ANRT, France) and SOIL-7 (soil, International Atomic Energy Agency, Vienna, Austria). Actual river samples (water and sediments) were analysed next. The INAA and ICP-MS results were generally in good agreement. However, ICP-MS analysis occasionally suffered from random flask contamination (e.g. Zn). The concentrations obtained by ICP-MS for certain elements (e.g. Zr) were lower than the certified values and results from INAA, probably because the dissolution of their natural forms (oxides) was not complete. For INAA, only one irradiation run was used. Some elements with short half-life radioisotopes (e.g. Mn, Mg or V) and usually analysable by NAA were not detected.

The results obtained for the Das Velhas river showed a strong relationship between mining activity and pollutant concentrations. Increased concentrations of certain metals corresponded to passage of the river through the industrial mining area.

The use of both INAA and ICP-MS enabled accurate results for water and sediment samples to be obtained for approximately 50 elements. For water analysis, ICP-MS was clearly the most convenient method, as INAA was handicapped by the effect of radiolysis which prevented irradiation of liquid samples. For sediment analysis, sample preparation time for ICP-MS was time consuming. INAA avoided these problems and could be used as a reference method. Routinely, a particular method was selected for each element.

Introduction

The need to monitor the elemental composition of a river ecosystem affected by a source of pollution is clear. As some metallic pollutants (Cr, Cd, Co) can accumulate in sediments, aquatic plants and animals, their concentration, even at trace level in water, must be checked. Neutron Activation Analysis (NAA) has been widely employed for the determination of multielemental concentrations in water (Batley, 1989; Stoeppler & Merian, 1991). It is a sensitive and reliable technique for environmental monitoring (Lamotte & Revel, 1991). Its main drawback is the need for a neutron source (nuclear reactor or neutron accelerator). Inductively Coupled Plasma – Mass Spectrometry (ICP-MS) is a newer technique for trace analysis of liquids (Jarvis et al., 1992). Interest in this method has been increasing for the last ten years and equipment for ICP-MS is becoming more and more common in analytical laboratories. Less expensive than NAA and not requiring nuclear equipment, ICP-MS can be used for more than fifty analyses per day, including standardisation and control procedures. Some elements not analysable by NAA (e.g. Pb) and of great interest for environmental management can be determined by ICP-MS.

The Pierre Süe Laboratory has 30-years of experience with NAA and was equipped with ICP-MS three years ago. Because of the increasing interest in environmental studies, it was decided to compare the performance and limitations of these two analytical methods for river pollution studies. Three rivers in Brazil and one in Morocco, each presenting different pollution problems and geological environments, were chosen as test-sites. The conclusions for the Moroccan river were published in Khalis (1996) and Khalis et al. (1997). In this paper, we present and discuss the results obtained for one of the Brazilian rivers, the Das Velhas. Prior to analysis of actual river samples (water and sediments), the two respective methods were tested on reference materials.

Description of sites studied

The Das Velhas river in the Minas Gerais State of Brazil, runs through a region rich in Fe, Au and Mn ores. The state itself is the world's second largest producer of minerals. Waste water from about 24 legal industries and several illegal ones (e.g. mercury gold extraction) are discharged into the Das Velhas basin. In spite of environmental controls carried out by the most important industries, the mines release significant quantities of mineral matter, the accumulation and dissolution of which causes serious pollution. This pollution is of highest priority in the Minas Gerais State. Studies of fresh water ecosystems in Minas Gerais are just beginning and ecological problems have been identified in rivers, lakes and reservoirs (Pinto-Coelho et al., 1994). Figure 1 shows the location of the Das Velhas river in Brazil, and the location of the samples sites, designed S1, S2, S4 and S5 for those on the river, and S3 and S6 for the sites sampled on the two tributaries of the river.

Materials and methods

Ultrapure acids, all Normaton grade, were from Prolabo, France. Ultrapure water was obtained by a Milli-Q System (i.e. 18 mΩ cm) (Millipore S. A., St Quentin, France).

We conducted the analysis of international reference materials to monitor elemental losses and contamination during analysis. The reference materials used were: SLRS-3 (riverine water, National Research Council of Canada), GS-N (granite, ANRT, France) and SOIL-7 (soil, International Atomic Energy Agency, Vienna, Austria).

Sampling

Water samples were collected in 125 ml acid-cleaned Nalgene polyethylene bottles at a depth of 15 cm near the river bank and stored at a temperature of 4 °C in the dark. At each sampling site, three types of water samples were collected: (i) filtered and acidified with ultrapure nitric acid to pH = 2; (ii) non filtered but acidified; (iii) non filtered, non acidified. We also investigated particle separation by centrifugation (4300 rpm for 10 minutes). Sediment samples were collected at the sampling site in Sterilin polystyrene tubes, stored at 4 °C in the dark and heated for 24 hours at 80 °C before analysis.

Neutron Activation Analysis (NAA)

There are extensive descriptions of this method (De Soete et al., 1972; Kruger, 1971; Alfassi, 1994), its application to biological materials (Guinn & Hoste, 1980) and its place amongst other analytical methods (Revel, 1987). For the present study, we used NAA without any chemical separation (Instrumental Neutron Activation Analysis or INAA), and with only one irradiation run.

Due to the effect of radiolysis, water samples must be prepared in a solid form prior to irradiation for INAA. 10 ml sample were evaporated in ultrapure quartz vials (Suprasil grades provided by Heraeus, Les Ulis, France) at 80 °C for 24 hours. Prior to use, the vials were cleaned with fluorhydric acid, rinsed with ultrapure water and dried. Vials containing samples and an empty vial were then heat-sealed. They were irradiated simultaneously with standards in the experimental nuclear reactor OSIRIS (*Commissariat à l'Energie Atomique, Saclay, France*) at a neutron flux of 1.4×10^{14} n cm^{-2} s^{-1} for 17 hours. After a decay time of 10 and 20 days, the subsequent gamma radioactivity was measured for 8 hours with a 100 cm^3 coaxial ultrapure germanium crystal coupled to a 4096 pulse height analyser. The computational k_0 method (Piccot et al., 1997) was used to determine the concentration of the elements of interest.

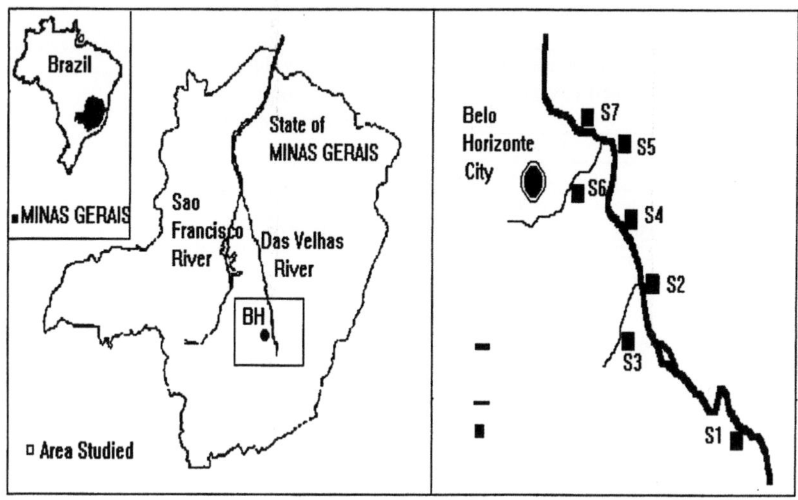

Figure 1. Location of the area studied and sample sites on the Das Velhas river and its two tributaries.

The sediment samples (150 mg in pure aluminium foils) were irradiated for 3 hours in ORPHEE, the second experimental nuclear reactor at Saclay, using a neutron flux of 2×10^{13} n cm^{-2} s^{-1}. The decay time was 10 days for the measurement of the radioactivity due to isotopes with half life of several days (Sm, La, Ca, Rb, Nd, U, etc) and 20 days for long-lived isotopes (Co, Fe, Cr, Sc, Hf, Zr, Th, Ce, etc).

Inductively Coupled Plasma – Mass Spectrometry (ICP-MS)

ICP-MS combines the ease of sample introduction characteristic of ICP with the high detecting power of mass spectrometry. Samples are introduced to the ICP apparatus by the direct nebulisation of aqueous solutions. Total acquisition times of a few minutes are typical; therefore, high sample throughput is possible. Current methodology has been described in Jarvis et al. (1992).

All measurements were carried out on a PlasmaQuad PQ2+ inductively coupled argon plasma mass spectrometer (Fisons, VG Elemental, Winsford, Cheshire, UK). Instrumental settings and operating conditions were as reported in Table 1.

To control any drift of the apparatus response during an analytical run, internal standards at a concentration of 10 μg l^{-1} were added to the sample and standard solutions. Be, In and Re were chosen because they are distributed through the mass range, their natural concentrations are very low and they were not analysed. Standard solutions were prepared by dilution of SPEX (Instrumental S.A., Longjumeau, France) certified stock solutions (10 μg ml^{-1}).

For the analysis of water, all non-filtered samples were centrifuged, diluted twenty times using ultrapure water for the determination of major elements elements (concentrations greater than 50 μg l^{-1}), and acidified with HNO_3 to pH = 2 (no dilution for trace analysis).

Table 1. ICP-MS operating conditions

Plasma–	
R.f. power	
Forward	1.3 kW
Reflected	< 5 W
Gas control	
Intermediate	0.85 l min^{-1}
Outer	14 l min^{-1}
Aerosol	0.75 l min^{-1}
Nebulizer	Meinhard type pumped at 1 ml min^{-1}
Spray chamber	Scott-type water cooled
Vacuum	
Expansion	2.10^2 Pa
Intermediate	< 10 Pa
Analyzer	0.1 Pa
Data acquisition–	
Dwell time	200 ms
Acquisition time	60 s

Table 2. Results (in $\mu g\ l^{-1}$) obtained by INAA and ICP-MS for SLRS-3 water reference (NRCC)

Element	Concentrations measured ± standard deviation		Certified values ± conf. limits (significance level 0.05)
	INAA (6)	ICP-MS (9)	
Li		0.54±0.02	
B		4.3±0.4	
Na	2450±250		**2300±200**
Mg		1430±100	
Al		29±1	**31±3**
Ca	5700±600	5900±400	**6000±400**
Sc	0.0060±0.0001	<0.36	
V		0.30±0.03	**0.30±0.02**
Cr	0.38±0.04	0.38±0.04	**0.30±0.04**
Fe	110±10	103±20	**100±2**
Co	0.042±0.004	0.04±0.03	**0.027±0.003**
Ni		1.03±0.06	**0.83±0.08**
Cu		1.5±0.1	**1.35±0.07**
Zn	2.7±0.3	0.26±0.08	**1.04±0.09**
As	0.67±0.07	0.67±0.06	**0.72±0.05**
Rb	1.75±0.18	1.7±0.1	
Sr	32±3	31.7±0.5	**28.1***
Y		0.13±0.02	
Mo	<0.2	0.23±0.03	**0.19±0.01**
Cd		<0.02	**0.013±0.002**
Sb	0.17±0.02	0.14±0.02	**0.12±0.01**
Cs	0.010±0.003	<0.01	
Ba	14±2	12.6±0.2	**13.4±0.6**
Pb		0.08±0.03	**0.068±0.007**
U	<0.04	0.04±0.02	**0.045***
Th	0.017±0.002	<0.02	
Rare earths			
La	0.26±0.03	0.26±0.02	
Ce	0.33±0.03	0.30±0.03	
Sm	0.045±0.004	<0.04	
Eu	0.0060±0.0006	<0.01	
Tb	0.0040±0.0004	<0.01	

* Information values only
( ) Number of analyses

For the ICP-MS analysis of sediments, a mineralization step was necessary. A sample of 250 mg obtained by random sampling was placed in a teflon bomb. Three millilitres of HNO_3, 1 ml of HCl and 1 ml of HF (concentrated acids) were added. The bomb was placed in a closed stainless steel vessel. The apparatus was held in an oven at 110 °C over 17 hours. After cooling, the tube was opened and the teflon container placed on a sand bath until evaporation to almost dryness. One millilitre of HF plus 4 ml of ultrapure water were added and the evaporation was continued over 1 hour to eliminate the silica (SiF_4 is volatile). To eliminate fluoride ions, especially CaF_2, 2 ml of $HClO_4$ were added and an evaporation to dryness ($\approx$ 1 hour) was performed. To eliminate chloride ions, 3 ml of HNO_3 were added and the sample placed on the sand bath for 1 hour. This operation was repeated to ensure the total elimination of chloride ions which are undesirable for ICP-MS analysis. After cooling, the acid solution was diluted with 5 ml of ultrapure water and reboiled. After cooling, the solution was made up to 100 ml with ultrapure water so that the final concentration of the solid was 2.5 mg per ml of solution. The resultant solution was limpid. Preservation of these

Table 3. Comparison with the certified concentrations ($\mu g\ g^{-1}$) of the concentrations obtained for the GS-N reference material (ANRT) using INAA and ICP-MS

Elements	Concentration measured ± standard deviation		Certified values ± confidence limits
	INAA (7)	ICP-MS (2)	
Li		54 ± 1	55.0 ± 4.5
Mg		14160 ± 790	13900 ± 302
Al		77450 ± 4900	77700 ± 477
Ca	18000 ± 950	20780 ± 1000	17860 ± 360
Sc	7.3 ± 0.6	7.4 ± 0.1	7.3 ± 0.4
Ti		3451 ± 80	4050 ± 180
V		61 ± 3	65 ± 8
Cr	55 ± 5	50 ± 2	55 ± 4
Fe	24830 ± 1200	28760 ± 5320	26230 ± 280
Co	65 ± 4	63 ± 2	65 ± 4
Ni		35 ± 3	34 ± 4
Cu		20 ±1	20.0 ± 0.7
Zn	48 ± 12	48 ± 3	48.0 ± 3.3
Ga		15 ± 5	22 ± 3
As	1.1 ± 0.1	1.3 ± 0.1	1.6 ± 0.3
Rb	180 ± 10	184 ± 1	185 ± 5
Sr		572 ± 44	570 ± 19
Y		15.6 ± 0.1	16.0 ± 0.6
Zr	250 ± 20	63 ± 1	235 ± 15
Mo		1.6 ± 0.3	1.2 ± 0.3
Cd		0.4 ± 0.1	0.04*
Sn		3.1 ± 0.1	3.0 ± 0.2
Sb	0.69 ± 0.05	1.26 ± 0.01	0.70 ± 0.17
Cs	5.4 ± 0.4	5.6 ± 0.3	5.40 ± 0.26
Ba	1380 ± 200	1460 ± 190	1400 ± 44
Hf	6.1 ± 0.6		6.20 ± 0.33
Ta	2.8 ± 0.1		2.6 ± 0.2
W	450 ± 150	419 ± 17	450 ± 63
Pb		52 ± 5	53 ± 4
Th	41 ± 4	43 ± 4	41.0 ± 3.5
U	7.7 ± 0.8	7.1 ± 0.1	7.5 ± 0.8
Rare earths			
La	70 ± 6	72 ± 1	75.0 ± 2.7
Ce	140 ± 8	138 ± 1	135 ± 7
Pr		14.6 ± 0.5	14.5 ± 0.6
Nd	47 ± 3	47 ± 1	49.0 ± 1.5
Sm	8 ± 1	7.2 ± 0.7	7.5 ± 0.22
Eu	1.6 ± 0.3	1.74 ± 0.04	1.70 ± 0.06
Gd		6.0 ± 0.3	5.2 ± 0.3
Tb	0.54 ± 0.04	0.72 ± 0.02	0.60 ± 0.04
Dy		3.2 ± 0.1	3.1 ± 0.3
Ho		0.54 ± 0.03	0.60 ± 0.16
Tm		0.2 ± 0.1	0.22 ± 0.03
Er		1.7 ± 0.1	1.50 ± 0.15
Yb	1.4 ± 0.1	1.4 ± 0.2	1.40 ± 0.15
Lu		0.22 ± 0.02	0.22 ± 0.03

*** Information values only**
( ) Number of analyses

Table 4. Comparison with the certified concentrations (μg g^{-1}) of the concentrations obtained for the SOIL-7 reference material (IAEA) using INAA and ICP-MS

Element	Concentrations measured ± standard deviation		Certified concentrations (conf. limits range)
	INAA (8)	ICP-MS (4)	
Li		30±1	**31*** (15-42)
Mg		11900±500	**11300*** (11000-11800)
Al		66550±2000	**47000*** (44000-51000)
Ca	164000±16400	145000±15000	**163000*** (157000-174000)
Sc	8.3±0.8	8.7±0.4	**8.3** (6.9-9.0)
Ti		2100±200	**3000*** (2600-3700)
V		54±5	**66** (59-73)
Cr	63±6	63±2	**60** (49-74)
Fe	25700±2000	25000±1000	**25700*** (25200-26300)
Co	8.6±0.8	10.5±1.0	**8.9** (8.4-10.1)
Ni		38±8	**26*** (21-37)
Cu		13±2	**11** (9-13)
Zn	105±50	110±30	**104** (101-113)
Ga		11±2	**10*** (9-13)
As	13.3±1.3	14.8±1.2	**13.4** (12.5-14.2)
Rb	49±5	53±1	**51** (47-56)
Sr		108±1	**108** (103-114)
Y		16±2	**21** (15-27)
Nb		7±2	**12*** (7-17)
Mo		1.2±0.2	**2.5*** (0.9-5.1)
Cd		1.6±0.2	**1.3*** (1.1-2.7)
Sb	1.8±0.2	1.6±0.1	**1.7** (1.4-1.8)
Cs	5.5±0.5	5.7±0.1	**5.4** (4.9-6.4)
Ba	141±20	138±2	**159*** (131-196)
Hf	4.9±0.5		**5.1** (4.8-5.5)
Ta	0.7±0.1		**0.8** (0.6-1.0)
Pb		59±2	**60** (55-71)
Th	8.1±0.8	8.0±0.2	**8.2** (6.5-8.7)
U	2.5±0.3	2.0±0.1	**2.6** (2.2-3.3)
Rare earths			
La	28±3	28±1	**28** (27-29)
Ce	58±6	55±1	**61** (50-63)
Pr		6.6±0.2	
Nd	31±3	24±2	**30** (22-34)
Sm	4.8±0.5	5.0±0.2	**5.1** (4.8-5.5)
Eu	0.80±0.08	1.1±0.1	**1.0** (0.9-1.3)
Gd		5.8±0.2	
Tb	0.60±0.06	0.8±0.1	**0.6** (0.5-0.9)
Dy		3.9±0.1	**3.9** (3.2-5.3)
Ho		0.5±0.1	**1.1*** (0.8-1.5)
Er		1.7±0.1	
Tm		0.20±0.01	
Yb	2.6±0.4	1.2±0.2	**2.4** (1.9-2.6)
Lu		0.20±0.01	**0.3*** (0.1-0.4)

*** Information values only**
( ) Number of analyses

Table 5. Results obtained in natural water from Das Velhas river by INAA and ICP-MS ($\mu g\, l^{-1}$)

Site S2			Site S4			Site S7		
Element	AAN (1)	ICP-MS (6)	Element	AAN (1)	ICP-MS (6)	Element	AAN (1)	ICP-MS (6)
Li-2)		1.8±0.4	Li		7.5±1.1	Li		14.7±3.7
B-2)		2.3±1.0	B		11±2	B		6.5±0.8
Mg-2)		1660±160	Mg		1580±160	Mg		5100±450
Al-2)		360±140	Al		2900±300	Al		12100±1500
Ca-2)	2230±670	3000±300	Ca	3670±1000	4060±500	Ca	19400±2000	17200±2000
Sc-1)	2.2±0.2	2.4±0.7	Sc	1.2±0.1	4.0±0.3	Sc	3.5±0.4	6.5±3.0
Ti-2)		1.4±1.2	Ti		4.5±1.8	Ti		4.1±1.2
V-2)		0.7±0.2	V		0.8±0.1	V		5.5±1.5
Cr-1)	2.1±0.4	1.4±0.5	Cr	8.7±0.8	8.5±2.7	Cr	15±2	16±2
Fe-1)	640±60	700±80	Fe	3600±400	3870±400	Fe	5900±600	7400±800
Co-1)	1.0±0.1	1.6±0.5	Co	30±3	39±2	Co	63±7	77±10
Ni-2)	<5	2.8±1.2	Ni	-	30±4	Ni	35±10	57±16
Cu-2)		1.4±0.3	Cu		26±3	Cu		47±6
Zn-1)	<10	<12	Zn	40±20	<13	Zn	80±40	40±20
As-1)	0.30±0.03	0.4±0.2	As	<0.38	<0.4	As	2.3±0.2	2.5±0.3
Rb-2)	2.3±0.30	2.8±0.6	Rb	-	5.7±1.6	Rb	-	7.0±1.2
Sr-2)		13.6±1.5	Sr	<23	25±3	Sr	60±20	80±7
Y-2)		1.0±0.3	Y		19±4	Y		32±4
Zr-1)	<7	<0.04	Zr	<7	<0.06	Zr	<7	0.7±0.2
Nb-2)		<0.01	Nb		<0.01	Nb		<0.01
Mo-2)	<0.4	<0.06	Mo	<0.4	0.12±0.06	Mo	<0.4	0.15±0.06
Ag-2)	<0.09	<0.03	Ag	<0.04	<0.03	Ag	<0.4	<0.03
Cd-2)	<0.12	<0.08	Cd		0.20±0.04	Cd	<0.1	0.34±0.06
Sn-2)		<0.05	Sn		0.23±0.04	Sn		<0.05
Sb-1)	0.05±0.02	0.03±0.01	Sb	0.050±0.005	0.06±0.02	Sb	0.06±0.02	0.04±0.01
Cs-1)	0.06±0.01	0.09±0.03	Cs	0.16±0.05	0.38±0.10	Cs	0.30±0.03	0.30±0.03
Ba-2)	8±3	8.3±2.0	Ba	170±20	220±20	Ba	280±30	320±40
Hf-1)	<0.02	<0.03	Hf	<0.05	<0.03	Hf	<0.02	<0.03
Ta-1)	<0.006	<0.02	Ta	<0.006	<0.02	Ta	<0.006	<0.01
W-2)	<0.2	<0.07	W	<1	<0.07	W	<8	<0.07
Au-1)	<0.01	<0.03	Au	<0.01	<0.04	Au	0.05±0.01	0.06±0.02
Tl-2)		<0.06	Tl		0.27±0.02	Tl		0.37±0.04
Pb-2)		1.3±0.3	Pb		22±4	Pb		42±7
Th-2)	0.08±0.01	0.03±0.02	Th	-	0.29±0.10	Th	1.2±0.2	1.4±0.3
U-2)	0.13±0.02	0.16±0.03	U	1.7±0.2	2.7±0.3	U	5.3±0.5	5.6±0.4
Rare earths								
La-1)	1.0±0.1	1.1±0.3	La	36±4	30±2	La	58±6	62±6
Ce-1)	2.1±0.2	2.9±1.3	Ce	46±5	57±7	Ce	100±10	110±10
Pr-2)		0.28±0.07	Pr		7.8±0.8	Pr		16±4
Nd-2)	<1.3	0.9±0.2	Nd	25±3	29±3	Nd	52±5	60±4
Sm-1)	0.30±0.03	0.23±0.05	Sm	6.5±0.6	5.8±0.7	Sm	15±2	12±2
Eu-1)	0.040±0.004	0.05±0.01	Eu	0.80±0.08	1.3±0.2	Eu	1.7±0.2	2.7±0.3
Gd-2)		0.21±0.03	Gd		5.2±0.7	Gd		10.6±0.8
Tb-1)	0.03±0.01	0.03±0.01	Tb	0.46±0.04	0.7±0.1	Tb	<2	1.4±0.1
Dy-2)		0.17±0.04	Dy		3.9±0.3	Dy		7.2±0.5
Ho-2)		0.03±0.01	Ho		0.7±0.1	Ho		1.20±0.13
Er-2)		0.09±0.01	Er		2.0±0.2	Er		3.5±0.4
Tm-2)		<0.01	Tm		0.28±0.03	Tm		0.45±0.04
Yb-1)	0.08±0.01	0.09±0.03	Yb	1.5±0.2	1.6±0.2	Yb	2.7±0.3	2.8±0.3
Lu-1)	0.012±0.001	0.013±0.006	Lu	0.20±0.02	0.23±0.04	Lu	0.40±0.04	0.40±0.04

Recommended method : 1) INAA 2) ICP-MS

( ) Numbers of analyses

Table 6. Results obtained in natural sediments from Das Velhas river by INAA and ICP-MS ($\mu g\ g^{-1}$)

Element	AAN (2)	ICP-MS (4)	Element	AAN (2)	ICP-MS (4)	Element	AAN (2)	ICP-MS (4)
Li-2)		32±3	Li		30±3	Li		32±3
B-2)		-	B		-	B		-
Mg-2)		2900±300	Mg		2800±280	Mg		740±70
Al-2)		52000±5200	Al		74000±7400	Al		70000±7000
Ca-2)	-	33100±3310	Ca	-	33900±3400	Ca	-	5500±550
Sc-1)	15±2	20±5	Sc	16±2	22±3	Sc	4.4±0.4	4.6±0.4
Ti-2)		2650±260	Ti		2700±270	Ti		3700±370
V-2)		83±8	V		110±10	V		36±4
Cr-1)	140±15	145±15	Cr	170±15	175±20	Cr	260±30	200±20
Fe-1)	55000±5000	47000±5000	Fe	115000±12000	85000±10000	Fe	470000±50000	320000±60000
Co-1)	23±2	26±3	Co	18±2	19±2	Co	14±2	23±8
Ni-2)	66±30	92±9	Ni	-	72±7	Ni	-	56±6
Cu-2)		47±5	Cu		53±5	Cu		45±5
Zn-1)	270±130	115±50	Zn	260±70	110±30	Zn	<20	50±25
As-1)	56±6	63±6	As	<14	19±2	As	560±200	280±100
Rb-2)	92±9	103±10	Rb	74±8	90±9	Rb	<70	7.3±0.7
Sr-2)	<60	46±5	Sr	<200	36±4	Sr	<300	11±1
Y-2)		25±3	Y		13±1	Y		5.0±0.5
Zr-1)	225±30	109±10	Zr	190±70	110±10	Zr	<400	50±5
Nb-2)		12±1	Nb		6.7±0.7	Nb		5.2±0.5
Mo-2)	<20	1.9±0.2	Mo	<17	1.2±0.1	Mo	<20	1.8±0.2
Ag-2)	<0.8	2.0±0.2	Ag	<2	0.6±0.1	Ag	<3	1.1±0.1
Cd-2)	-	0.45±0.05	Cd		0.14±0.01	Cd		0.15±0.02
Sn-2)		6.0±0.6	Sn		3.6±0.4	Sn		0.9±0.1
Sb-1)	2.6±0.8	1.8±0.5	Sb	7.0±0.7	4.0±0.4	Sb	6.0±0.6	8.0±2.0
Cs-1)	5.7±0.6	7.0±0.7	Cs	3.8±0.4	5.0±0.5	Cs	<1	0.35±0.04
Ba-2)	344±30	330±30	Ba	<400	360±40	Ba	<500	74±7
Hf-1)	8.2±0.2		Hf	5.6±0.6	<0.2	Hf	3.3±0.3	1.7±0.8
Ta-1)	1.6±0.2		Ta	1.2±0.2		Ta	<1	
W-2)	<5	4.0±0.4	W	<40	1.3±0.1	W		1.6±0.2
Au-1)	<0.01	<0.03	Au	<0.02		Au	0.38±0.04	0.22±0.10
Tl-2)		0.90±0.09	Tl		0.58±0.06	Tl		0.26±0.03
Pb-2)		29±3	Pb		23±2	Pb		17±2
Th-2)	16±2	16±2	Th	10±1	9.3±1.0	Th	3.6±0.5	3.3±0.3
U-2)	5.5±0.5	6.7±0.6	U	<5	3.4±0.3	U	<7	4.0±0.4
Rare earths								
La-1)	45±5	48±5	La	23±2	24±2	La	11±1	12±2
Ce-1)	88±9	95±10	Ce	60±6	60±6	Ce	21±2	27±5
Pr-2)		10±1	Pr		6.0±0.6	Pr		2.9±0.3
Nd-2)	37±4	34±3	Nd	<60	21±2	Nd		10±1
Sm-1)	6.8±0.7	6.6±0.7	Sm	4.0±0.4	3.4±0.3	Sm	1.5±0.2	1.6±0.2
Eu-1)	1.0±0.1	2.0±0.8	Eu	0.72±0.07	0.60±0.06	Eu	0.30±0.03	0.20±0.03
Gd-2)		10±1	Gd		5.8±0.6	Gd		2.9±0.3
Tb-1)	0.8±0.3	2.6±0.3	Tb	0.50±0.05	1.3±0.1	Tb	<1	0.8±0.1
Dy-2)		12±1	Dy		6.0±0.6	Dy		3.5±0.3
Ho-2)		0.80±0.08	Ho		0.43±0.04	Ho		0.22±0.02
Er-2)		4.5±0.5	Er		3.0±0.3	Er		1.4±0.1
Tm-2)		0.30±0.03	Tm		0.24±0.02	Tm		0.10±0.01
Yb-1)	3.2±0.3	2.4±0.3	Yb	2.2±0.2	1.40±0.01	Yb	0.80±0.08	0.70±0.07
Lu-1)	0.6±0.1	0.3±0.1	Lu	0.40±0.04	0.24±0.02	Lu	0.18±0.02	0.11±0.02

Recommended method : 1) INAA 2) ICP-MS ; ( ) Numbers of analyses

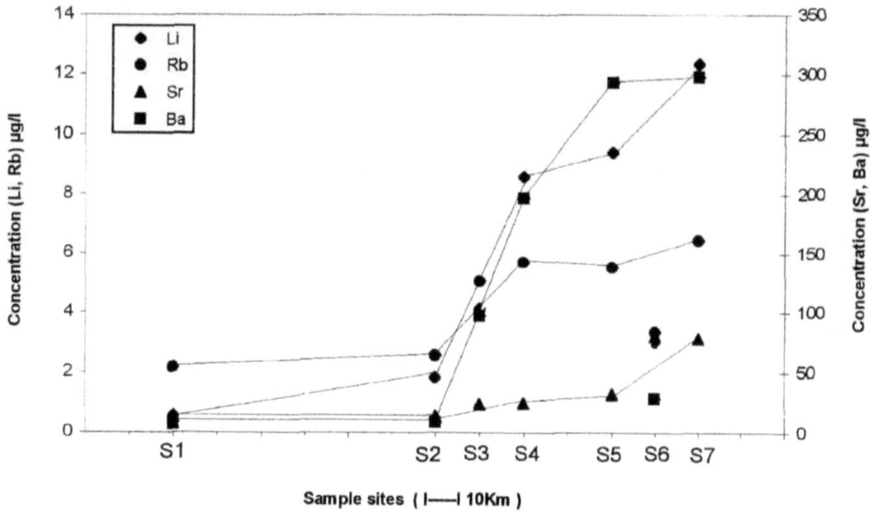

Figure 2. Concentration of alkaline and alkaline earth elements in the water of the Das Velhas river.

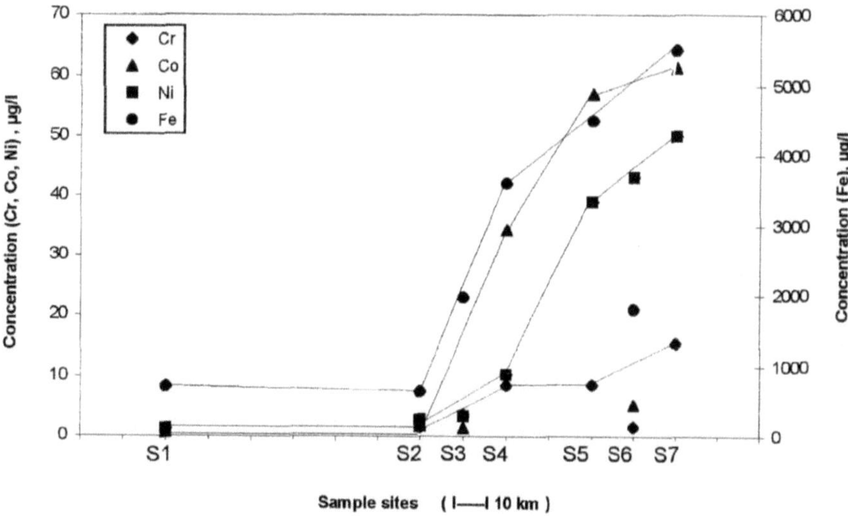

Figure 3. Concentration of some transition elements in the water of the Das Velhas river.

solutions was good. For analyses, these solutions were diluted by a factor of 10.

Results

Preliminary tests showed significant contamination of ultrapure water by filtration using 0.2 μm Millipore cellulose nitrate filters rinsed with nitric acid. Successive washing steps with pure water did not ever prove conclusive (Veado et al., 1997). The most accurate results were obtained using the Millipore polycarbonate 0.45 μm filters rinsed with the water to be sampled before use. These samples were also centrifuged prior to analysis because the presence of particles smaller than 0.45 μm was suspected.

The first water analysis was conducted one or two weeks after collection and analyses were repeated each month for 9 months. No change was observed in the case of the acidified samples, but concentrations fell in the samples not acidified, particularly for metals and rare earth elements, while the concentrations of alkaline elements remain stable.

The INAA and ICP-MS results for the international reference materials SLRS-3, GS-N and SOIL-7 compared to the certified values are shown in Tables 2, 3 and 4 for each respective material.

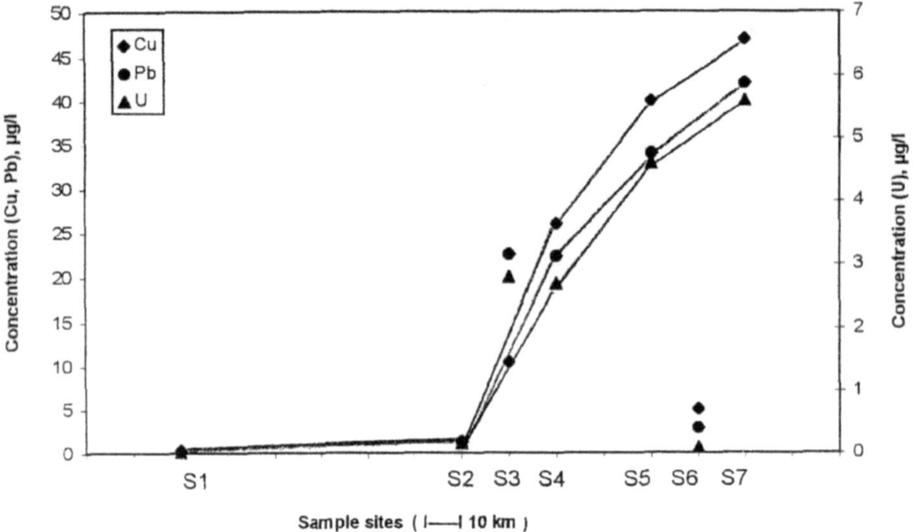

Figure 4. Concentration of Cu, Pb and U in the water of the Das Velhas river.

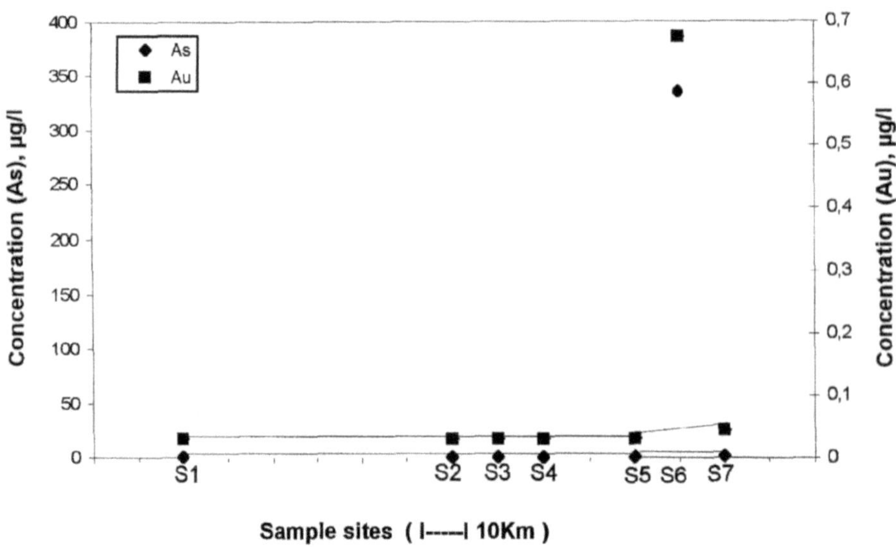

Figure 5. Concentration of Au and As in the water of the Das Velhas river.

Several thousand determinations were performed along the course of four different rivers under investigation in our laboratory. The results for river samples presented in this paper are those obtained from the Das Velhas in Brazil. Figures 2 to 6 show the variation in the concentrations of some elements in water along the course of this river. These concentrations were measured by ICP-MS on non filtered and acidified samples. Tables 5 and 6 give the concentration measured on the Das Velhas river water and sediments samples, both by INAA and by ICP-MS.

Discussion

Acidification of water samples was necessary to avoid losses by adsorption of elements onto walls of containers. This acidification enabled preservation of samples for several months, without significant variation of concentrations, as verified by the analysis of two samples each month over a 9-month period. However, acidification dissolved a proportion of elements retained on suspended particulate matter. In these circumstances, filtration before acidification appeared to provide more accurate analyses. This operation always has the risk of contamination. Furthermore, to fully benefit from all the information provided by the filtration, it would be necessary to use several filters of decreasing size down to 0.02 μm to retain colloids. The waters sampled were cloudy (dry extract: 2 g l^{-1} at S1 to 5 g l^{-1} at S7). Successive filtrations increase contamination risk and would be impractical with water with a lot of particulate matter rapidly clogging 0.45 μm filters and necessitating preliminary decantation. Centrifugation appears to cause less contamination, but it is difficult to conduct on site. For these preliminary studies aimed at controlling pollution levels and not at studying the transport mechanisms, we chose to retain the samples unfiltered and acidified. Analysis of samples provided an assessment of the total quantities of elements carried by the river water (as dissolved ions and suspended matter).

NAA and ICP-MS are two multi-elemental, selective and sensitive analytical methods, well adapted to trace analyses. However, their respective principles, implementation and application are very different.

Activation analysis is a comparatively old method, born in 1936 with the paper of von Hevesy & Levi. NAA requires very heavy means (nuclear reactor for this study). The time taken for the analyses depends on the half-life of the radioisotopes used and is usually long: in the order of several days to several weeks. On the other hand, this method by principle is particularly accurate and can be used as a reference method (Revel, 1987). The detection limits obtained in this study were poor in comparison to the limits obtained for a pure material such as silicon (Revel et al., 1984). Due to the effect of radiolysis, it is not possible to irradiate liquid samples and the detection limits of NAA for water analysis depend on the blank value introduced by the evaporation support (impurities and weight). For sediment analysis, we chose an instrumental method (without any chemical separation after irradiation) to eliminate dissolution problems and to reduce the time for the analysis. Detection limits could be greatly improved by radiochemical separation after irradiation to remove the major elements (e.g. Fe) or those elements providing high radioactivity (e.g. Na). The irradiation conditions could be also multiplied to increase the sensitivity for some elements and to determine some other elements analysable by NAA, in particular those giving radioisotopes having short or medium half-life (e.g. Al, Mg, Mn, Ti or V). The method previously used for the analysis of plants and soils (Pham Thi Huynh et al., 1997) could be used, but this method requires three different irradiations and the total time necessary for the three analyses is not practical when a large number of samples is under investigation.

The ICP-MS method is considerably faster than NAA, except if mineralization is required, and we can perform about one hundred water analyses each day. However, care must be taken when examining results because the risk of error is significant. These risks are mainly due to the isobaric interference by compounds of the argon-light element and to contamination. Consequently, it is necessary to control the method repeatedly both with standard solutions and high purity water. We in fact conducted these determinations several times a day; thus the number of analysed samples fell. It was also necessary to completely dissolve the sediments to obtain very dilute solutions ($<$ 1 g l^{-1}). This dissolution is often difficult, due to the presence of refractory metal oxides. This decreases the sensitivity (reagents bring contamination) and the accuracy (some elements can be lost). Except for these disavantages, ICP-MS is a very powerful method for mineral pollution studies, especially for liquid samples.

Results of the analyses of SLRS-3 (Table 2) were in good agreement for the two methods as well as with the certified values, except for Zn. Analysis of this element occasionally suffers from random flask contamination and the results obtained by ICP-MS probably arose due to excessive correction. For the solid reference materials, granite GS-N (Table 3), and sediment SOIL-7 (Table 4), the trace elements were determined by INAA after only one irradiation for 3 hours at 2 $\times$ 10^{13} n cm^{-2} s^{-1}. Because of the radioactivity induced by certain radioisotopes, particularly the ^{24}Na ($T_{1/2}$ = 15.0 h), the first measurements were done only 8 days after the irradiation and it was then not possible to detect several elements usually analysable by NAA, for example: ^{28}Al ($T_{1/2}$ = 2.25 min), ^{64}Cu ($T_{1/2}$ = 12.7 h), ^{56}Mn ($T_{1/2}$ = 2.58 h) and ^{52}V ($T_{1/2}$ = 3.75 min). It was not possible to analyse Li and Pb by NAA, nor Dy and Pr by INAA (Figueiredo et al., 1987). The two latter determinations require chemical separations. Very dilute solutions are strictly necessary for ICP-MS analysis because higher concentrations lead to plasma instability and signal attenuation. Thus, an element present at a concentration in solid of 1 μg g^{-1} is determined at a concentration of 0.25 ng ml^{-1} in our analysed solution. For analysis of Soil-7, the values for As and Yb were outside the confidence interval, probably because of their low concentrations. Iron was determined using the ^{57}Fe isotope (abundance 2.2%), because the measurement of ^{56}Fe (abundance 91.7%) is confounded by ArO^+. Hf and Ta were not determined by ICP-MS because of their poor stability in nitric solutions. In addition, the analysis of Ga was affected by Ba^2 interference.

For GS-N, the concentration obtained for Zr using ICP-MS was lower than the certified value, probably because the natural form of Zr (ZrO_2), is not easily dissolved. We chose INAA for the determination of Zr in our studies, but it was necessary to take into account interference by Uranium (fission products Zr^{95}) (Revel, 1984).

Taking into account the observations made for the reference materials, we examined the concentrations measured on the actual river samples. A part of the results obtained both by INAA and ICP-MS are listed in Tables 5 and 6. The INAA and ICP-MS results are generally in good agreement. However, we routinely select a particular method for each element. We have indicated in the tables which method is best adapted for analysis of the concentration of each element.

Al, Mg, Ti, Cu, Mn and V were analysed by ICP-MS, although Al and Ti oxides are difficult to dissolve, because they were not measured in the INAA conditions used. Hf, Ta and Zr were analysed by INAA which appeared more reliable. Fe was determined by INAA. Some others (Li, Y, Nb, Pb, Pr, Gd, Dy, Ho, Er and Tm) were analysed by ICP-MS because their determination was not possible by NAA (no convenient radio-isotope), or because their determination by INAA is comparitively insensitive without chemical separation.

However, a number of elements are well analysed by both methods. Our choice was thus based on the characteritics of our samples and on the conditions of irradiation we used. Ca was analysed by ICP-MS because of the presence of high concentrations of Al and Na in sediments, and the isotope used for NAA (^{46}Ca) is present as a very low proportion of the element (0.003%). Mo, Sr and Ba, usually well determined by NAA, were also analysed by ICP-MS. The NAA determination of Ni and U, presenting low sensitivity by (n,γ) nuclear reactions, demands the use of fast neutrons irradiation, thus we chose ICP-MS. To avoid losses during the mineralization step, we used INAA for As determination in sediments, although the detection limit is high without chemical separation. We would rather retain the INAA method for Sc (in ICP-MS, its rays are often masked by those of neighbor elements, Ca and Ti, naturally far more abundant in natural samples), Cr, Co and Zn (an important blank is often introduced by the container in ICP-MS). These elements present high cross section and well differentiated gamma rays. Rb, Cs, Sb and Th are well analysed by the two methods, we made a quite arbitrary choice.

Except for Pr, Nd, Gd, Dy, Ho, Er and Tm which need chemical separation to be analysed by NAA, rare earths are well determined by the two methods. ICP-MS appears to be very convenient for the analysis of all the rare earth elements in geological materials if all interference is well controlled (Valladon et al., 1995). We thus chose ICP-MS for the rare earth elements mentioned earlier, and INAA for La, Ce, Eu and Tb to confirm the ICP-MS results.

Even if a method showed an obvious advantage, its results were checked with those obtained by the other method.

Figures 2 to 4 show a strong relationship between the mining activity and pollutant concentrations. The increase of the concentrations in metals between sites 2 to 4 corresponded to the passage of the river through the industrial mining area (Figure 2, for alkaline and alkaline earth elements, and Figure 3 for some transition elements). Cu, Pb and U contamination was partly causes by the tributary S3 (Figure 4). The very high Au concentration in the tributary S6 (Figure 5) was correlated with the intensive exploitation of gold, often illegal, on this tributary. The surprisingly high concentration of As in S6 could be explained by the presence of an old arsenic mine, abandoned and situated a few kilometres upstream from S6.

Among heavy metal and toxic elements, only Hg was not analysed. Analysis of this element is possible by both INAA and ICP-MS. However, in natural samples, Hg is often present as very volatile organic compounds and its conservation in our samples was uncertain. A special sampling process dedicated to the analysis of Hg must be used for the accurate determination of this element.

Conclusion

Combined use of INAA and ICP-MS is a powerful procedure for the study of river pollution by metallic elements. This combination provided numerous accurate results for waters and sediments samples, and enabled the analysis of about fifty elements.

For water analysis, ICP-MS is indisputably the most convenient method, provided that the different sources of error: isobaric interference, ions bi-charged, contamination, etc, are correctly controlled. INAA is handicapped by the effect of radiolyse which prevent the irradiation of liquid samples. It is best used to control the determination of some elements more particularly subject to error in ICP-MS method (e.g. Zn).

For sediment samples, ICP-MS could also be applied, provided that the dissolution of the numerous elements in oxide form (e.g. Al, Ti, Zr) is complete. Methods of dissolution have to be well adapted both to the analytical equipment and to the nature of the sediment. Such methods are often time consuming to develop and apply. Furthermore, they are not always absolutely reliable. INAA avoids these problems.

When the use of ICP-MS requires chemical separation or mineralization processes, INAA is a usefull reference method. In this work, INAA was more reliable, but its turnover was smaller than ICP-MS. The precision of INAA is sometime inferior to the precision of ICP-MS. In our study, INAA detection limits for a number of elements (e.g. As, Cu, Ba) were high, but a simple chemical separation would allow significant reduction of these limits (Alaux-Negrel, 1991). Our preference for one method was strongly related to our objectives: to analyse, in a minimum delay with the maximum accuracy, a large number of elements in numerous samples.

References

Alfassi, Z. E., 1994. Chemical Analysis by Nuclear Methods, John Wiley & Sons, UK, 556 pp.

Allaux-Negrel, G., 1991. Etude de l'évolution des eaux profondes en milieux granitiques et assimilés – Comportement des éléments traces. PhD Thesis, Paris, France, 216 pp.

Batley, G. E., 1989. Trace Element Speciation: Analytical Methods and Problems. CRC PRESS, Florida, 350 pp.

De Soete, D., R. Gijbels & J. Hoste, 1972. Neutron Activation Analysis (Chemical Analysis Series, Vol. 34), Wiley-Interscience, NewYork, USA, 836 pp.

Figueiredo, A. M. G., S. May & G. Pinte, 1987. Détermination de la teneur des éléments dysprosium, europium, gadolinium, praséodyme, samarium et lanthane dans la roche étalon granite GS-N par activation neutronique. Analusis 15: 179–182.

Guinn, V. P. & J. Hoste, 1980. Elemental Analysis of Biological Materials, Technical Reports Series n° 197, IAEA, Vienna: 105–140.

Jarvis, K. E., A. L. Gray & R. S. Houk, 1992. Handbook of ICP-MS edited by Blackie, Chapman and Hall, New York, 380 pp.

Khalis, M., 1996. Etude de l'influence du milieu environnant sur la composition minérale d'une rivière marocaine. PhD Thesis, Meknes, Morocco, 156 pp.

Khalis, M., A. Sekkaki, J. L. Irigaray, F. Carrot & G. Revel, 1997. Environmental control of Boufekrane river water basins by NAA and ICP-MS, Wat. Res. 31: 2930–2934.

Kruger, P., 1971. Principles of Activation Analysis, John Wiley & Sons, UK, 522 pp.

Lamotte, A. & G. Revel, 1984. L'analyse élémentaire dans le domaine des traces et des ultratraces: quelques problèmes posés à l'analyste. Analusis 12: 423–431.

Pham Thi Huynh, M., F. Carrot, S. Chu Pham Ngoc, M. Dang Vu & G. Revel, 1997. Determination of rare earth elements in rice by INAA and ICP-MS, J. Radioanal. Nucl. Chem. 217: 95–99.

Piccot, D., C. Deschamps, R. Delmas & G. Revel, 1997. K0-Labsue: a quasi non-dependent data format package for K0-quantification. J. Radioanal. Nucl. Chem. 215: 263–269.

Pinto-Coelho, R. M. & A. Giani, 1994. E. V. Sperling (eds), Ecology and Human Impact on Lakes and Reservoirs in Minas Gerais. SEGRAC, Belo Horizonte, 193 pp.

Revel, G., 1984. L'analyse élémentaire dans le domaine des traces et des ultratraces: quelques risques d'erreur en analyse par activation. Analusis 12: 506–513.

Revel, G., 1987. Euroanalysis VI, Reviews on Analytical Chemistry. In E. Roth (ed.), Les Editions de Physique, Les Ulis, France, 168–181.

Revel, G., N. Deschamps, C. Dardenne, J. C. Pastol, B. Hania & M. Nguyen Dinh, 1984. Characterization of solar grade silicon by NAA. J. Radioanal. Nucl. Chem. 85: 135–150.

Stoeppler, M. & E. Merian (ed.), 1991. Analytical Chemistry of Metals and Metal Compounds – Metals and Their Compounds in the Environment. VCH, Weinheim, 1438 pp.

Valladon, M., B. Dupré & M. Polve, 1995. A new method for interferences and shifts corrections. Terra Nova 7: 347–362.

Veado, M. A. R., 1997. Appliquation de l'analyse par activation neutronique et de l'ICP-MS pour l'identification et le dosage des éléments métalliques apportés par les rejets miniers dans la rivière Das Velhas, Brésil. PhD Thesis, Orsay, France, 154 pp.

Veado, M. A. R., G. Pinte, A. H. Oliveira & G. Revel, 1997. Application of INAA and ICP-MS by studying river pollution in the state of Minas Gerais. J. Radioanal. Nucl. Chem. 217: 101–106.

von Hevesy, G. & H. Levi, 1936. The action of neutrons on the rare earth elements. Det Kgl. Danske Videnskab. Selskab. Math-fys. Medd. K. Dans K. Acad. 14: 1–34.

Hydrobiologia **373/374**: 75–87, 1998.
J.-C. Amiard, B. Le Rouzic, B. Berthet & G. Bertru (eds),
Oceans, Rivers and Lakes: Energy and Substance Transfers at Interfaces.

Fate of plant detritus in a European salt marsh dominated by *Atriplex portulacoides* (L.) Aellen

V. Bouchard[1], V. Creach[1], J. C. Lefeuvre[1], G. Bertru[1] & A. Mariotti[2]
[1] *Laboratoire d'Evolution des Systèmes Naturels et Modifiés, UMR 6553 'ECOBIO', Université de Rennes 1, Avenue du Général Leclerc, 35042 Rennes Cedex, France*
[2] *Laboratoire de Biogéochimie Isotopique, Université Pierre et Marie Curie -INRA-CNRS, UMR 1756, case courrier 120, 4 place Jussieu, 75252 Paris Cedex 05, France*

Key words: salt marshes, primary production, decomposition, sediment, carbon isotopes

Abstract

This paper deals with the organic matter dynamics of a dwarf shrub species, *Atriplex portulacoides*, which characterizes the majority of ungrazed European salt marshes. These dynamics were investigated by the simultaneous estimation of primary production, litter production, decomposition processes and accumulation of organic matter in sediment. We used harvested biomass techniques and the litter-bag method, as well as more recent techniques such as determination of natural isotopic compositions. The results of this study have been compared with those of other studies of salt marshes of the Eastern Coast of America dominated by *Spartina alterniflora.*

Studies of the fate of halophytic organic matter in salt marshes have shown that processes such as consumption, decomposition, or flooding rapidly could distribute organic material to different compartments of the marsh system. Aboveground productivity of *A. portulacoides* was 1700 g m^{-2} yr^{-1}. Approximately 80% of this production was transformed rapidly in necromass. A large part of this dead material (86%) fell as litter directly onto the sediment at the sampling site. The remainder was washed away by tides, either out of the marsh, or into the upper marsh where it accumulated in drift lines. Decomposition of *A. portulacoides* on the sediment was followed using litter-bags. Decay constants, based on the first order decay function, ranged between 0.0404 d^{-1} and 0.0113 d^{-1} and appeared high when compare with data obtained in other American and European salt marshes. Nitrogen immobilisation was not detectable in this decomposition experiment. During decay, nitrogen concentrations increased, whereas absolute amount of nitrogen decreased. After microbial decomposition, plant detritus was rapidly incorporated into sediment. The carbon isotopic composition of organic matter in sediment showed that organic matter in the middle marsh came from both aboveground production of *A. portulacoides* and microalgae. The patterns for these two kinds of organic matter in different granulometric fractions depended on tidal submersion frequency. During the summer, when the salt marsh was not disturbed by tides, the clay fraction got richer in plant organic matter, whereas, during the other periods, this fraction was characterized by microalgal organic matter.

This study provides new data about the dynamic of organic matter. It remains still difficult to establish comparisons between North American and European salt marsh functioning according to the variability of primary production, microbial decomposition and tidal range in these ecosystems.

Introduction

In the marsh-estuarine system, it is generally accepted that halophytic vegetation, covering the high intertidal zone, includes some of the most productive communities (Long & Mason, 1983). This organic matter produced in salt marshes supports a considerable and complex detrital food web in the ecosystem (Teal, 1962; Odum & de la Cruz, 1967; Odum & Heald, 1975) which leads to fluxes of detrital material towards coastal sea waters according to the outwelling hypothesis (Odum, 1968; Haines, 1977; Nixon, 1980). The relatively small herbivore consumption in salt marshes means that most of the annual production becomes detritus: detritus can be either washed away by tides or become litter which decomposes on sediment or in wrack in salt marshes (Teal, 1962; Mann, 1972).

This dominance of the detrital path of organic matter over the herbivore path makes the salt marsh a suitable environment in which to study the fate of primary production, from decomposition to its incorporation in sediment (Marinucci, 1982; Tenore, 1984; White & Howes, 1994).

Most salt marsh studies have been conducted on the East coast of North America. Because of the dominance of *Spartina alterniflora* Loisel there, most work has focused on this species (Marinucci, 1982; Dame, 1989). It is generally accepted that about 5 to 10% of the productivity of *S. alterniflora* is grazed directly by invertebrates (Mann, 1972). Moreover, less than 1% of the production leaches out of the plant as dissolved organic matter during tidal flooding (Gallagher et al., 1976). In fact, most of the production (up to 80% of the aboveground production and 100% of the belowground production according to Valiela et al., 1982) decomposes *in situ* through microbial activity. However, it is still unclear whether generalisations about productivity and decomposition can be made from *S. alterniflora* salt marshes. Both differences in floristic composition (Chapman, 1974) and in environmental factors, particularly flooding regime (Lefeuvre & Dame, 1994), make it hard to compare American and European marshes.

Studies of European salt marshes have focused principally on the North West coast (Wolf et al., 1979; Dankers et al., 1984; De Leeuw et al., 1990). Given such scarcity of data, recent studies have been initiated (Boorman et al., 1994; Lefeuvre et al., 1994). These studies have adopted two complementary approaches: (1) the evaluation of primary production and decomposition to assess organic matter turn-over and fluxes and (2) the evaluation of halophytic organic matter transfer through the marsh-estuarine ecosystem using molecular markers such as stable isotopes. Natural isotopic compositions have been already used with success to determine the importance of macrophytes in estuarine food webs in the Sapelo Island or in the Mississippi estuary (Haines, 1977; Haines & Montague, 1979; Hackney & Haines, 1980). These biological markers provide clues to both the origin and transformation mode of organic matter (Fry & Sherr, 1984). Contrary to North American salt marshes, C_3 plants cover most of European sites. Their $\delta^{13}C$ values vary between –26 to –23‰ (Créach, 1995) and differ from C_4 plants and microalgae whose carbon isotope ratios are around –13‰ and –17‰, respectively (O'Leary, 1981; Currin et al., 1995).

The aim of this study was to determine the production and fate of organic matter produced by one of the most abundant species of European salt marshes, *Atriplex portulacoides* (L.) Aellen. We took into account the primary production and litter production of this species in order to provide a rough estimate of the detritus fluxes under tide action. The litter decomposition was studied in the field with litter-bag experiments to assess the pattern of organic carbon loss and the evolution of total nitrogen. Finally, the incorporation of organic matter to sediments was determined by the measurement of organic carbon percentages and natural isotopic composition in different granulometric fractions.

Description of the site studied

The study area is located in France, in the 'baie du Mont Saint-Michel' (Figure 1). This bay is situated in the South of the Normandy gulf, between Brittany and the Cotentin peninsula. In this area, tides are considerable and can reach 16 m during exceptional spring tides. The tides are semi-diurnal. In this macrotidal bay, mudflats cover about 19 000 ha while the high tidal zone is occupied by 4 000 ha of salt marshes.

The ungrazed salt marsh studied was approximately 1.8 km wide, with a well-developed creek system and clear zonation of plant communities. The vegetated zone is located in a 1.8 m elevation range (Levasseur, 1993), and plant communities are distributed in space in relation to the frequency and duration of flooding, sediment budget and soil properties, all of which are directly linked to marsh topography (Guillon, 1984).

Depending on flooding frequency and vegetation zonation, the salt marsh can be divided in four areas: the pioneer area (10 ha), the lower marsh (20 ha), the middle marsh (37 ha) and the upper marsh (47 ha) (Levasseur, 1993; Bouchard, 1996). On the mudflats, the pioneer area is dominated by *Salicornia* spp. (L.) and *Spartina anglica* (Curt.). The lower marsh is dominated by *Suaeda maritima* (Dum.), *Puccinellia maritima* (Huds.), *Salicornia* spp. and *Aster tripolium* (L.). The middle marsh is a monospecific area, covered by a woody perennial shrub *Atriplex portulacoides* (L.). This vegetation is considered to represent the climax in the salt marsh (Gehu & Gehu-Franck, 1982). *Elytrigia aetherica* (Hack.) grows in the upper marsh and extends along the creeks, while the marsh along the sea wall is occupied by an area dominated by grasses

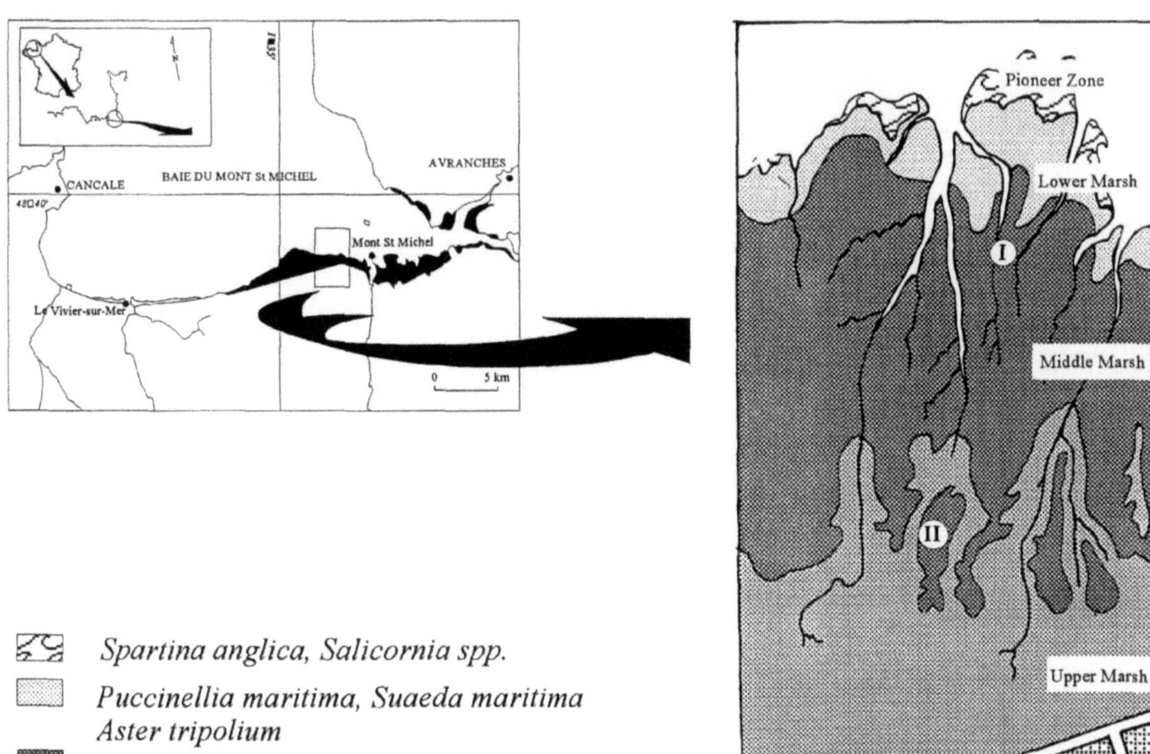

Figure 1. Map of salt marsh in bay of Mont Saint-Michel with the locations of Sites I and II. Insert shows the North Brittany location.

– *Elytrigia aetherica, Festuca rubra* (Meyer) – mowed once a year in mid July.

Methods

Primary production

The *A. portulacoides* community was sampled during one year (1994) at 9 sites chosen randomly (Bouchard & Lefeuvre, 1996). Vegetation was cut monthly for a year (from February 1994 to February 1995), just above the marsh surface at each sampling site in a 30 × 30 cm quadrat. We used a three-dimensional cuboid frame with removable sides to avoid practical problems especially with the sampling of tall vegetation. The harvested material was sorted to separate living and dead material. Dry weights (D.W.) were obtained by drying in an oven at 60 °C to constant weight. Net aerial primary production (NAPP) was calculated from these biomass data using Smalley's method (Smalley, 1959). This method was chosen for the sake of comparison since it has been widely used in studies on primary production in European salt marshes (Groenendijk, 1984; Boorman et al., 1994). The annual net aerial primary production (aNAPP) was calculated as the sum of each monthly production (mNAPP).

The turn-over of *A. portulacoides* was calculated as the ratio between a NAPP and peak total biomass. This ratio made it possible to estimate total necromass production for one year (Groenendijk & Vink-Lievaart, 1987).

To conduct isotopic analyses, the aerial structures of *A. portulacoides*, plants were randomly sampled in the middle march each month. The samples were washed with deionised water, dried at 50 °C and ground (2 mm) prior to immediate analyse.

Litter production

Litter production of the *A. portulacoides* community was estimated for one year. In each site previously selected for measurement of primary production, 10 litter traps (20 cm × 20 cm) were placed on the sediment surface beneath the vegetation. Traps made out of wooden frames and nylon mesh were located randomly at each site. Litter was collected frequently (every 5 or 7 days) for one year to limit decomposition and subsequent weight loss. Collected litter was taken to the laboratory in plastic bags and sorted into plant components, i.e., leaves, flowers and stems. This material was dried at 60 °C to constant weight.

Litter decomposition

Decay rates of *A. portulacoides* leaves were measured on the sediment under the vegetation using the litter-bag method. Bags were placed at two sites: in the lower part (Site I) and in the upper part (Site II) of the *A. portulacoides* community. Litter-bags were prepared by collecting *A. portulacoides* in fresh wrack, just after high tide. Fresh litter (ca. 2 g) was placed in pre-weighed, dried, nylon mesh litter-bags. Initial dry masses were calculated from the mean of six litter samples which were oven-dried (60 °C) to constant mass. We used the ratio between initial dry mass and initial fresh mass to assess the dry mass of litter placed in each bag at time zero. Two kinds of mesh were used: large mesh (2 mm) and fine mesh (0.3 mm). Litter-bags were placed in the field in mid-November with subsequent collections timed to detail the important phases of decay (5, 10, 25, 50, 100 and 200 days).

Litter-bags were collected at low tide and sealed into individual pre-weighed plastic bags. Litter was then washed gently with freshwater over a 0.03 mm mesh stainless steel screen until water appeared clear. Litter was dried at 60 °C to constant weight and milled. For each litter sample, carbon and nitrogen content were determined using a Perkin Elmer 2400 elemental analyser. Means and standard deviations of carbon and nitrogen concentrations of reference samples were 70.89 ± 0.17 (n = 13) and 10.48 ± 0.08 (n = 13), respectively.

All decomposition studies of *A. portulacoides* leaves showed significant fit to the negative exponential decay model (Buth & Wolf, 1985; Hemminga & Buth, 1991). Decay rates (k) were therefore determined through this first-order decay function, $X_t = X_0 e^{-kt}$. Half-lives were calculated using the equation $t_{half} = 0.693/k$. Decay rates were reported as % organic C loss per day.

Sediment sampling and analyses

Five sediment cores were extracted in the middle marsh and in the lower part of the *A.portulacoides* community (Site I) during two periods: (1) in March (A) when the salt marsh is regularly submerged. (2) in August (B) when it is rarely submerged. A 1 cm thick sample was removed from each core, one at the surface and at three other at depths of 5–6 cm, 10–11 cm, 20–21 cm. Samples from the same level were pooled and frozen. Before analysis, samples were thawed, freed of living roots, and treated with diluted hydrochloric acid (pH = 2.0) to remove carbonate contamination (40%–50% in the sediment). The treated samples were washed with deionized water, centrifuged, dried and sieved (2 mm).

Carbon isotope ratio in samples was measured with an ANA-1500-CHN analyser (Carlo-Erba) coupled with an isotopic mass spectrometer (Sira-10, VG). The isotopic analysis was reported in δ values, relative to the PDB (Craig, 1957):

$$\delta^{13}C = [(R_{sample}/R_{standard}) - 1] \times 10^3 \quad R = {}^{13}C/{}^{12}C$$

Internal standards were combusted with each group of samples. Mean and standard deviation of reference samples (cyclohexanone) were −26.02±0.05‰ (n = 12) for carbon isotope ratio and 0.5% for organic carbon content.

Decarbonated samples were measured for granulometry. A Robinson pipette was used for 2 μm, 5 μm, 10 μm and 20 μm fractions, using a series of AFNOR (French Association for Standardization) sieves for the fractions equal to or greater than 50 μm. Sediments were classified with a system frequently used in France (Np18-304): clay (< 2 μm), silt (2–20 μm), fine sand (20–200 μm) and coarse sand (0.2–2 mm).

Results

Organic matter production

Living material of *A. portulacoides* was present throughout the year and its amount was always significantly superior to that of dead material (Figure 2), the mean ratio (live biomass/dead biomass) being to 5.08 ± 1.62 (n = 13). Live biomass increased from April to September, and then decreased gradually until late winter. Peaks of biomass observed in winter (March 1994 and February 1995) were not significant (t-test, $p > 0.05$). Two peaks in the weight of dead material were noted in April and October respectively,

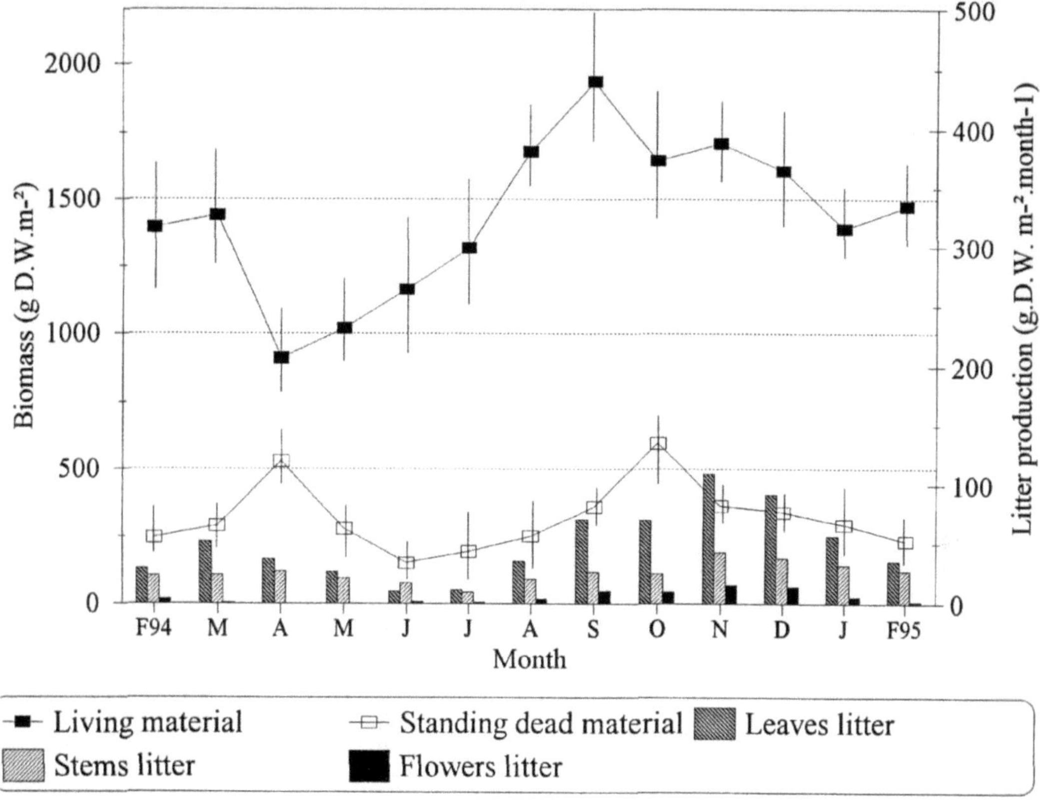

Figure 2. Evolution of *A. portulacoides* biomass (living and standing dead material) with standart errors and of litter production on the sediment during one year.

while in the other months, the quantity of total dead material remained stable and was low all through the year.

Using Smalley's method the primary production of *A. portulacoides* was estimated as 1770 g m^{-2} yr^{-1} and the turn-over of the community as 0.77 yr^{-1}.

Macro-detritus dynamics

The annual litter production of *A. portulacoides* was estimated to be 1170 ± 189 g m^{-2} yr^{-1}. Leaf litter comprised the largest proportion (62%) of the total annual litter, of which flowers represented only 7%. Four-weekly totals for the litterfall indicated a marked seasonality in leaf and flower litterfall in the marsh (Figure 2). The period of high litterfall was during August-January. A second and smaller peak of leaf litterfall was observed in March and April. Stem litterfall followed a similar pattern but with a smaller peak in winter and early spring.

These estimations of primary production, turn-over and litter production made it possible to assess the proportion of *A. portulacoides* necromass which was not deposed directly at the production site (Figure 3). It appeared that approximately 86% (about 1170 g D.W. m^{-2} yr^{-1}) of the necromass produced during one year fell directly under the vegetation, onto the sediment, whereas 14% (about 190 g D.W m^{-2} yr^{-1}) was exported from this production site.

Litter decomposition

Mass and organic carbon losses from litter-bags were correlated. Therefore, only carbon loss, the more direct measurement of decay, was used to estimate exponential decay rate (k) and half-life of litter (Table 1). Analysis of variance indicated that decomposition was faster in litter-bags of large mesh than in litter-bags of small mesh ($F_{2,88} = 9.45$ and $p < 0.01$). In large mesh bags, 50% of organic carbon was lost after 17.3 days and 24.9 days of decay at Site I and Site II. At the same sites, this loss took 58.2 and 66.6 days in small mesh bags. Moreover, for the same kind of litter-bags, decomposition was faster at Site I than at Site II

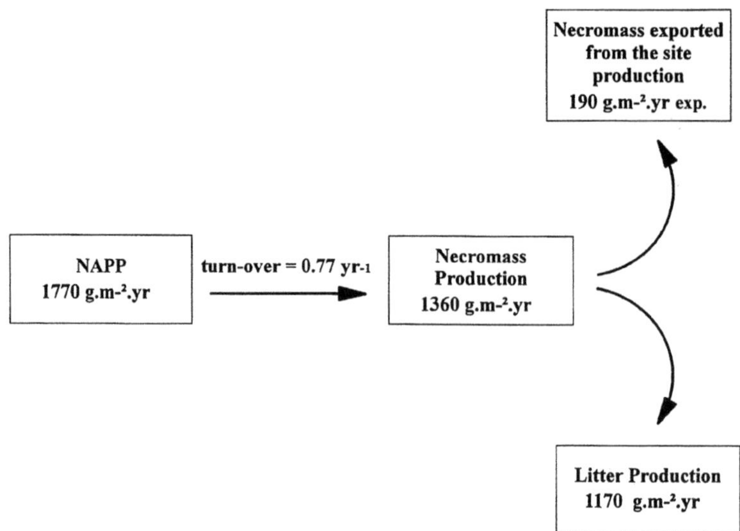

Figure 3. Dynamic of *A. portulacoides* macro-detritus through necromass production, litter production and litter exportation.

Table 1. Exponential decay rate after 100 days of decay (k), regression coefficients for decay rate constants (r^2), half life and carbon remaining (CR) at the end of the experiment according to each litter-bag experiment

		Site I		Site II	
		lm	fm	lm	fm
November 1993					
k (days^{-1})	(r^2)	0.0404 (0.98)	0.0119 (0.97)	0.0278 (0.96)	0.0104 (0.97)
CR (%)		2.1 ± 0.9	8.4 ± 2.1	6.2 ± 1.3	11.1 ± 2.3
Half life (days)		17.3	58.2	24.9	66.6

lm: large mesh and fm: fine mesh

($F_{2,88} = 6.25$ and $p < 0.05$). In fact, in large mesh litter-bags, decay rate was estimated as 0.0404 d^{-1} at Site I and as 0.0278 d^{-1} at Site II.

The nitrogen content varied significantly over time ($F_{2,88} = 21.11$ and $p < 0.01$) in litter-bags but did not vary significantly within the site ($F_{2,88} = 4{,}12$ and $p > 0.05$) (Figure 4a). Fresh *A. portulacoides* litter of averaged 1.51% nitrogen. From the beginning of decay, there was a net increase in nitrogen content in all experiments. After the same period of incubation (100 days), nitrogen content was significantly higher in small mesh bag experiments (2.47 ± 0.03) than in large mesh bag experiments (2.25 ± 0.05) (t-test and $p < 0.05$).

However, there was a net decrease of the quantity of nitrogen contained in the litter, without significant site-related differences (Figure 4b). This decrease was faster in litter-bags with large-mesh whereas, after 100 days of incubation, 40% and 3% of the original nitrogen remained in litter placed in fine mesh litter-bags, against only 3% for litter placed in large mesh bags.

Carbon isotopic variations of plants and sediment

The carbon isotope ratios of aboveground plant structures varied between –26.6 to –23.6‰, over the year. The mean and standard deviation were –25.3 ± 0.8‰ ($n = 12$). During decomposition, the natural isotopic composition of *A. portulacoides* was depleted faster in litter-bags placed at Site I in the middle marsh than at Site II (Figure 5). But in the two cases, at the end of the decomposition experiment, the carbon isotope ratio of plant detritus in litter-bags attained the same value: –27.6‰.

The organic carbon in middle marsh sediment varied from 1.9 to 3.1%. It was correlated with the percentage of clay (0–2 μm) and silt (2–20 μm) in sediment ($r^2 = 0.57$, $p < 0.05$, $n = 8$). These were the two largest in sediment fractions and ranged from

A

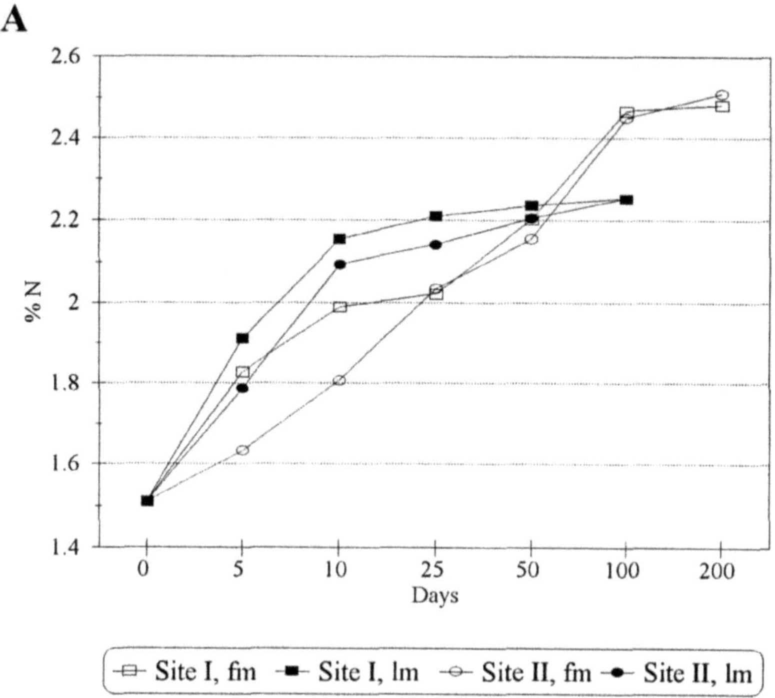

B

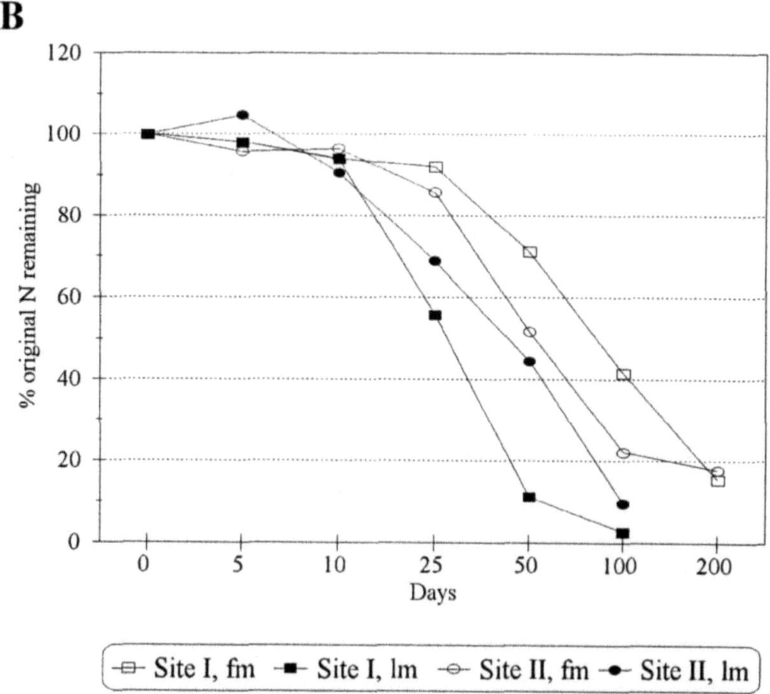

Figure 4. Dynamic of nitrogen in *A. portulacoides* leaves during decomposition: A- N concentrations; B- Absolute amount of N.

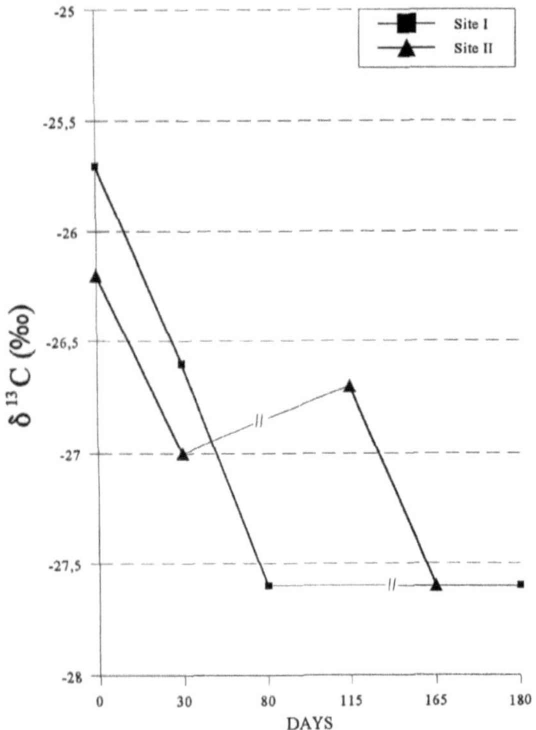

Figure 5. Variation of carbon isotope ratios of plant in litter-bags in Site I and Site II.

40.4 to 57.4% in March and from 47.6 to 66.5% in August. The percentage of organic carbon in the two granulometric fractions (clay: 0–2 μm and silt+sand: 2–2000 μm) was not very different, particularly in period B, where the values were similar at the surface and at 5 cm depth, compared with values found in period A (Figure 6). Organic carbon content was always highest in the clay fraction at 10 and 20 cm depths, independent of the period.

We determined the percentage of organic matter from *A. portulacoides* in the different fractions of sediment from the mixing equation:

$$\delta^{13}C_{SOM} = F_p\,\delta^{13}C_p + F_a\,\delta^{13}C_a \text{ where } F_p + F_a = 1$$

$\delta^{13}C_{SOM}$ is the isotope ratio of organic carbon in sediment, $\delta^{13}C_p$ is the isotope ratio mean for *A. portulacoides* (–27.6‰), $\delta^{13}C_a$ is the isotopic ratio of microalgae (–20.1‰) in salt marsh channels (Créach, 1995), and F_p and F_a are the proportions of plant and microalgae in the sedimentary organic matter. During the two periods, A and B, the natural isotopic composition was more enriched in the 0–2 μm fraction (–22.6‰ to –21.4‰) than in the 2–2000 μm fraction (–23.9‰ to –22.6‰). The percentage of organic matter derived from plant decomposition was highest in the 2–2000 μm fraction. It ranged from 34.7 to 50.7% in period A, and from 33.3 to 41.3% in period B (Figure 7) The percentage of organic carbon from plants in the 0–2 μm fraction increased in period B since it ranged from 17.3 to 25.3% in March, and rose 22.6 to 29.3% in August (Figure 7).

Discussion

Production of organic matter

The aboveground production of *A. portulacoides* reported in this study (1770 g m^{-2} yr^{-1}) was high compared to NAPP estimates reported in other tidal marsh communities of NW Europe (Wolff et al., 1979; De Leeuw et al., 1990). In Europe, higher production has been reported for *Spartina* spp. and *Elytrigia aetherica* (Groenendijk, 1984; Boorman et al., 1994). Moreover, other work on *A. portulacoides* has shown that this species is one of the most productive in European salt marshes: high productivity values have already been found by Boorman et al. (1994), Lefeuvre et al. (1994) and Bouchard & Lefeuvre (1996). According to these studies, *A.portulacoides* can produce up to 3.600 g D.W. m^{-2} yr^{-1}. The lower productivity recorded in this study can be explained by the invasion of a micro-lepidoptera which considerably reduces the productivity of this species (Bouchard & Lefeuvre, 1996). Nevertheless, in contrast to previous work which estimated that European salt marshes produced only 500 to 1 000 g m^{-2} yr^{-1} (Wolff et al., 1979; De Leeuw et al., 1990), these more recent studies (particularly on *A. portulacoides*) indicate that European salt marshes can produce as much as American salt marshes (up to 3.700 g m^{-2} yr^{-1} according to Gallagher et al., 1980).

Detritus dynamic and fluxes

The majority of the organic matter produced by *A. portulacoides* falls directly on the sediment. Only 14% of the necromass produced is displaced by the mechanical action of tides. In fact, populations located in the lower part of the community are completely flooded by 5% of tides, whereas the populations located in the upper part of the middle marsh are flooded by less than 0.2% of tides (Levasseur, 1993). On the East coast of North America, the largely dominant and productive stands of *S. alterniflora* occupy the space between mud flats and the below mid-tide level (Mc Kee &

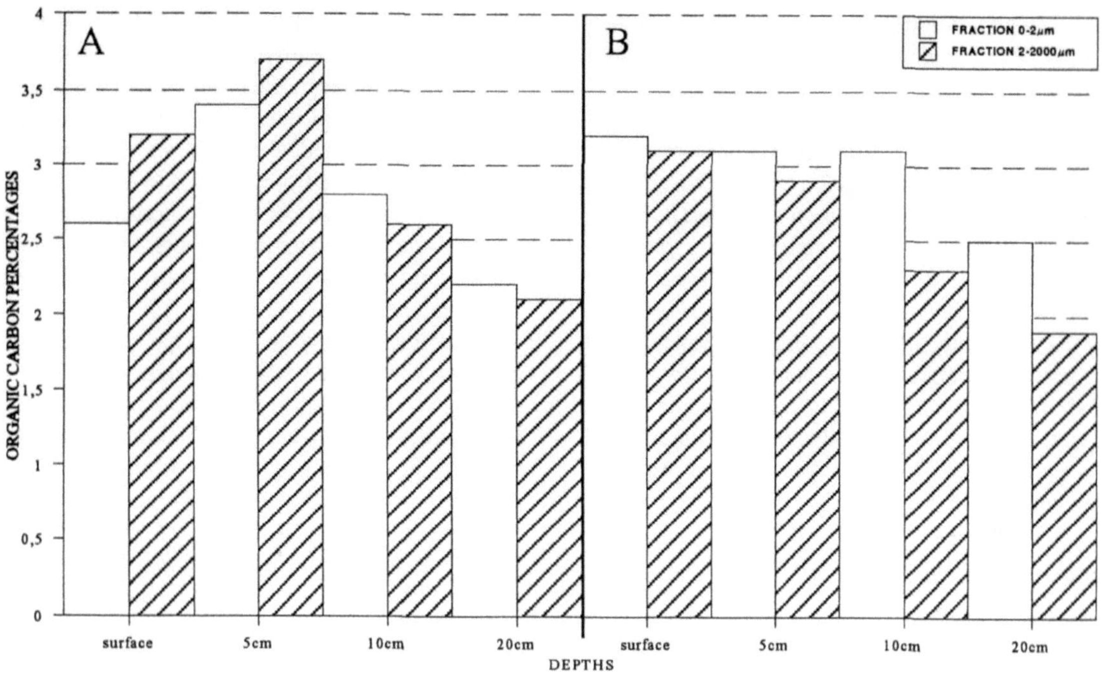

Figure 6. Percentages of organic carbon in clay (0–2 μm) and silt/sand (2–2000 μm) fractions of sediment beneath *A. portulacoides* during period regularly (A) and rarely (B) submerged by tides.

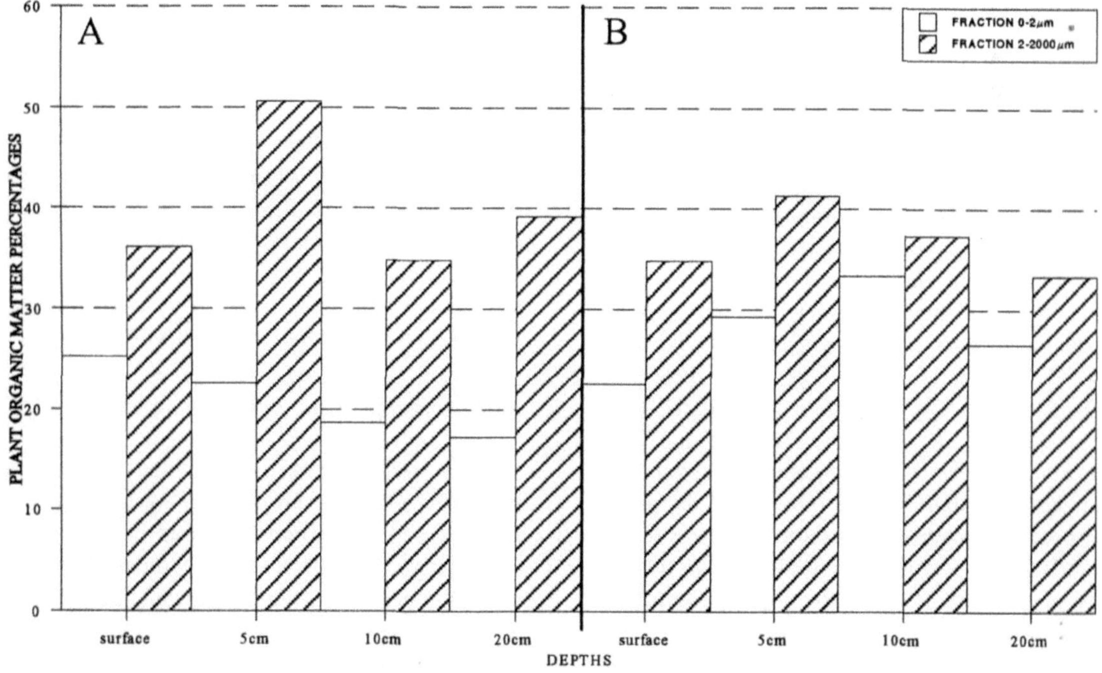

Figure 7. Percentages of organic matter from plants in clay (0–2 μm) and silt/sand (2–2000 μm) fractions of sediment beneath *A. portulacoides* during period regularly (A) and rarely (B) submerged by tides.

Patrick, 1988). This community is generally flooded twice a day by the tide. Indeed, in the macrotidal bay of Fundy, it seems that most of the production from *S. alterniflora* stands located in the lower marsh is exported to coastal waters around the year, whereas in the upper marsh 75% of the annual production can overwinter (Cranford et al., 1989; Gordon & Cranford, 1994). However, it has often been estimated that only 1% (Dame, 1982 and 1989; Dankers et al., 1984) or 10% (Wolff et al., 1979) of salt marsh primary production is washed away rapidly by the tide as litter. Most macro-detritus moved by tide accumulates into the marsh (Reidenbaugh & Banta, 1981; Dankers et al., 1984). Finally, a large part of European and American salt marsh primary production dies at the end of the growing season, and is decomposed primarily during the following winter and spring in the marsh.

Litter decomposition

As the greater part of the high production of *A. portulacoides* enters the detrital food chain *in situ*, litter decomposition processes represent the essential phase in the cycling of biogenic elements. Comparison with data from the literature shows that decomposition rates recorded in this study (from 0.0404 d^{-1} to 0.0119 d^{-1}) were 5 to 13 times higher: for *A. portulacoides*, decay rates already measured, have been estimated between 0.0031 d^{-1} and 0.0059 d^{-1} (Buth et al., 1982; Buth & Wolf, 1985; Buth & Voesenek, 1987) depending on litter-bag mesh-size and bag location in the marsh. Compared with decay rates measured for *S. alterniflora* leaves from 0.038 d^{-1} to 0.0019 d^{-1} (Marinucci, 1982), decomposition kinetics of *A. portulacoides* were similar or faster. The rate of litter decomposition appeared to be principally controlled by the physico-chemical nature of the plant material and by environmental factors (Valiela et al., 1984; Dame, 1989; Hemminga & Buth, 1991; Berg et al., 1993). Tidal flooding seems to be particularly important maintaining litter moisture (Hemminga & Buth, 1991; Halupa & Howes, 1995) and could explain the faster decomposition kinetics observed in the present study in the lower site (Site I), more frequently flooded by tide than the upper site (Site II). Moreover, the rapid decay of *A. portulacoides* litter in our marsh could have been due to higher winter temperature than in sites in Northern Europe (Christian, 1984; Hemminga & Buth, 1991). Fauna may also affect decomposition rates, directly by ingestion of small litter fragments, or indirectly by changing the abundance and activity of microbial decomposers (Rice & Tenore, 1981; Reice & Stiven, 1983; Kemp et al., 1990). This activity could partially explain the effect of litter-bag mesh-size on decay rates. However, the effect of detritus feeders in leaf breakdown will only become clearer with better knowledge of individual feeding mechanisms and activities (Boulton & Boon, 1981). In fact, analysis of carbon and nitrogen isotopic composition as well as gut analysis has shown that the diet of some deposit-feeders under *A. portulacoides* consisted of more than 50% plant detritus (Créach et al., 1997).

After decomposition, the organic matter can be exported towards coastal sea waters or can be used directly *in situ* by salt marsh food chains. The quality of salt marsh detritus as food source for consumers depends on a significant degree of nitrogen availability for assimilation and growth (Tenore, 1984). Many types of detritus – and mostly refractory ones – become enriched in nitrogen as they decay. Different field and laboratory experiments have shown increases in both concentration and absolute amounts of nitrogen in decaying salt marsh plants (Marinucci, 1982; Valiela et al., 1985; Newell et al., 1989). However, degrees of nitrogen immobilization are due to the slow rate at which litter decomposes (Shaver & Melillo, 1984; White & Howes, 1994). In the present study, *A. portulacoides* litter appears to decompose so quickly that this disappearance prevents nitrogen enrichment. In fact, estimates, using a variety of methods, have suggested that a maximum of only 20–25% of net nitrogen increase in litter is due to microbial biomass (Lee et al., 1980; Hicks et al., 1991). More recently, White & Howes (1994) found that, by decomposition midpoint, 50 to 65% of the total detrital nitrogen pool was from external sources. The present study suggests that the immobilization of nitrogen in *A. portulacoides* litter is not detectable. It seems that nitrogen is not held in litter but is rapidly distributed to the system.

Fate of organic matter in the sediment

Several studies have demonstrated that microbial processes are the main pathway by which carbon is recycled within sedimentary food chains. However other processes described by Nyman et al. (1993) can influence the partition of organic matter, such as above and belowground primary production and flooding frequency. In the middle marsh, where cores were extracted, percentages of organic matter in sediment were higher (2.6 to 2.9%) than in the lower marsh (0.2 to 0.9%) as described in Créach (1995). This organic matter content was low but similar to that found by Ember et al. (1987) in a salt marsh of South Carolina covered by tall and medium *S. alterniflora*. Sedimentary organic compounds are represented principally by various carbohydrates, amino acids and lipids which are more or less easy to identify depending on degree of microbial degradation (Meyers, 1994). Hence the determination of natural isotopic composition offers a valuable tool for identification of the origin of organic matter (Fry & Sherr, 1984).

The carbon isotope ratio is not the same for all primary producers. It depends mainly on the photosynthetic pathways of vascular plants (Bender, 1971; Smith & Epstein, 1970). *Atriplex portulacoides*, the dominant salt marsh plant is a C_3 plant. Its mean carbon isotope ratio was –25.3‰. This value was close to the $\delta^{13}C$ values of other C_3 plants present in the salt marsh such as *Salicornia* spp. (–26.2‰), *Suaeda maritima* (–25.9‰), *Puccinellia maritima* (–26.4‰) and of those reported in the literature (Haines & Montague, 1979; Peterson & Howarth, 1987). The variations observed in the natural isotopic composition of this perennial plant probably reflect environmental conditions (Tiezsen, 1991).

During decomposition, the isotopic fingerprint of *A. portulacoides* is slightly modified. The $\delta^{13}C$ value of plant detritus was –27.6‰ and was the same whatever the location of litter-bags in the field. This depletion has not been observed in microcosms (Créach, 1995), where the $\delta^{13}C$ values of the C_3 plants do not change, in contrast to those of C_4 plants such as *S. anglica*, where natural isotopic composition was depleted from 3 to 4‰ during microbial decomposition (Benner et al., 1984; Fogel et al., 1989). This variation can be explained by the preservation of lignin components depleted (4 to 7‰) in ^{13}C relative to polysaccharides which comprise the bulk of plant tissues (Benner et al., 1987).

The carbon isotope ratios measured in the sediment of the middle marsh indicate a dual origin of organic matter, namely plants and microalgae. Organic matter was largely preserved in the clay and silt fractions of the sediment. Depending on tidal frequency, the organic matter derived from plants was distributed in different ways in these two fractions. When the salt marsh was regularly submerged, the coarsest fraction was the richest in plant detritus, principally from the litter deposed on the marsh because *A. portulacoides* does not present a well developed belowground structure. During summer, the processus of decomposition was not disturbed by tides, and detritus was broken down into the smallest size particles by mechanical means or by colonisation of the substratum by bacteria, fungi and protozoa as observed by Vernberg (1993). This process involves an enrichment of macrophytic organic matter in the finest fraction (0–2 μm) for which the isotopic value was close to that of the coarsest fraction. Thus, the origin of the organic matter depends on tidal range, aboveground production of *A. portulacoides*, and microbial decomposition. Variability on these factors could account for the difference between the results of our study and that of Haines (1977) in which 90–95% of sedimentary organic matter inferior to 57 μm was found.

Conclusion

The interest of this study was to investigate simultaneous different ecological processes which govern the organic matter distribution in order to assess element (like carbon and nitrogen) cascades in such ecosystems. In regard to important studies on the major species (*S. alterniflora*) growing in American salt marshes, we focused on *A. portulacoides,* the most abundant species in ungrazed European salt marshes. Our study shows that primary production in European salt marshes can be as high as in North American salt marshes. In both sytems, the major part of this organic matter is decomposed in place. However, decay rates are 5 to 13 times higher in our study area. These results support the idea that salt marsh functioning is largely dominated by the detrital pathway, particularly in the middle marsh covered by *A. portulacoides*, in comparison with more frequently flooded marshes. Since rates of accumulation of organic matter in sediments and regeneration of nutrient through mineralization depend on decay rates, the knowledge of how microbial activity control takes place, principally via the chemistry of litter and environmental factors, seems to be a key to important ecological features on which the next studies should focus.

Acknowledgements

We thank all people who made this project possible, particularly Valérie Adamandidis, Luc Brient, Micheline Grably, Cyril Girardin and Gérard Barnoux for technical assistance in the field and in the laboratory. This research was supported by EEC (DG XII), contracts n° EV4V-0172-F(EDB) and EV5V/CT92/0098.

References

Bender, M. M., 1971. Variations in the $^{13}C/^{12}C$ ratios of plants in relation to the pathway of photosynthetic carbon dioxide fixation. Phytochemistry 10: 1239–1244.

Benner, R., A. E. Maccubin & R. E. Hodson, 1984. Anaerobiodegradation of the lignin and polysaccharide components of lignocellulose and synthetic lignin by sediment microflora. Appl. envir. Microbiol. 47: 998–004.

Benner, R., M. L. Fogel, E. K. Sprague & R. E. Hodson, 1987. Depletion of ^{13}C in lignin and its implications for stable carbon isotope studies. Nature 329: 708–710.

Berg, B., M. P. Berg, P. Bottner, E. Box, A. Breymeyer, R. Calco de Anta, M. Coûteaux, A. Escudero, R. Gallardo, W. Kratz, M. Madeira, E. Mälkönen, C. Mc Claugherty, V. Meentemeyer, F. Munoz, P. Piussi, J. Remacle & A. Virzo de Santo, 1993. Litter mass loss rates in pine forests of Europe and Eastern United States: some relationships with climate and litter quality. Biogeochemistry 20: 127–159.

Boorman, L. A., J. Hazelden, P. J. Loveland, J. G. Wells & J. E. Levasseur, 1994. Comparative relationship between primary production and organic and nutrient fluxes in four European salt marshes. In W. J. Mitsch (ed.), Global Wetlands: Old World and New. Elsevier Science B.V.: 181–200.

Bouchard, V., 1996. Production et devenir de la matière organique des halophytes dans un marais salé européen en système macrotidal. Thesis Rennes University, 209 pp.

Bouchard, V. & J. C. Lefeuvre, 1996. Hétérogénéité de la productivité d'*Atriplex portulacoides* (L.) Aellen dans un marais salé macrotidal. C. R. Acad. Sci. 319: 1027–1034.

Boulton, A. J. & P. I. Boon, 1991. A review of methodology used to measure leaf litter decomposition in lotic environments: time to turn over an old leaf? Aust. J. mar. Freshwat. Res. 42: 1–43.

Buth, G. J. C., P. F. M. Verdunschot & L. de Wolf, 1982. Decomposition of three halophyts in different habitats of an Eastern Scheld salt marsh. Hydrobiol. Bull. 16: 103–112.

Buth, G. C. C. & L. Wolf, 1985. Decomposition of *Spartina anglica*, *Elytrigia pungens* and *Halimione portulacoides* in a Dutch salt marsh in association with fauna and habitat influences. Vegetatio 62: 337–355.

Buth, G. J. C. & L. A. C. J. Voesenek, 1987. Decomposition of standing and fallen litter of halophytes in a Dutch salt marsh. In A. H. L. Huiskes, C. W. P. M. Blom & J. Rozema (eds), Vegetation between land and sea. Dr. W. J. Pub.: 146–162.

Chapman, V. J., 1974. Salt Marshes and Salt Deserts of the World. 2nd ed., J. Cramer Verlag, Bremerhaven.

Christian, R. R., 1984. A life table approach to decomposition studies. Ecology 65: 1693–1697.

Craig, H., 1957. Isotopic standards for carbon and oxygen and correction factors for mass spectrometric analysis of dioxide. Geoch. Cosmoch. Acta 12: 133–149.

Cranford, P. J., D. C. Jr Gordon & C. M. Jarvis, 1989. Measurement of cordgrass, *Spartina alterniflora*, production in a macrotidal estuary. Estuaries 12: 27–14.

Créach, V., 1995. Origines et transferts de la matiere organique dans un marais littoral: Utilisation des compositions isotopiques naturelles du carbone et de l'azote. Thesis Rennes University, 134 pp.

Créach, V., M. T. Schricke, G. Bertru & A. Mariotti, 1997. Stable isotopes and gut analyses to determine feeding relationships in salt marsh macroconsumers. Estuar. coast. Shelf. Sci. 320: 339–347.

Currin, C. A., S. Y. Newell & H. W. Paerl, 1995. The role of standing dead Spartina alterniflora and benthic microalgae in salt marsh food webs: considerations based on multiple stable isotope analysis. Mar. Ecol. Prog. Ser. 121: 99–116.

Dame, R. F., 1982. The flux of floating macrodetritus in the North Inlet estuarine ecosystem. Mar. Ecol. Prog. Ser. 16: 161–171.

Dame, R. F., 1989. The importance of *Spartina alterniflora* to Atlantic coast estuaries. Aquat. Sci. 1: 639–660.

Dankers, N., M. Binsbergen, K. Zegers, R. Laane & M. Rutgers, 1984. Transportation of water, particulate and dissolved organic and inorganic matter between a salt marsh and Ems-Dollard Estuary, The Netherlands. Estuar. coast. Shelf. Sci. 19: 143–165.

De Leeuw, J., H. Olff & J. P. Baker, 1990. Year-to-year variation in peak above-ground biomass of six salt-marsh angiosperm communities as related to rainfall deficit and inundation frequency. Aquat. Bot. 36: 139–151.

Ember, L. M., D. F. Williams & J. T. Morris, 1987. Processes that influence carbon isotope variations in salt marsh sediments. Mar. Ecol. Prog. Ser. 36: 33–42.

Fogel, M. L., E. K. Sprague, A. P. Gize & R. W. Frey, 1989. Diagenesis of organic matter in Georgia Salt Marsh. Estuar. coast. Shelf. Sci. 28: 211–230.

Fry, B. & E. B. Sherr, 1984. ^{13}C measurements as indicators of carbon flow in marine and freshwater ecosystems. Contrib. Mar. Sci. 27: 49–63.

Gallagher, J. L., W. J. Pfeiffer & L. R. Pomeroy, 1976. Leaching and microbial utilization of dissolved organic carbon from leaves of Spartina alterniflora. Estuar. coast. Shelf. Sci. 4: 467–471.

Gallagher, J. L., R. J. Reimold, R. A. Linthurst & W. J. Pfeiffer, 1980. Aerial production, mortality, and mineral accumulation-export dynamics in *Spartina alterniflora* and *Juncus roemerianus* plant stands in a Georgia salt marsh. Ecology 61: 303–312.

Gehu, J. M. & J. Gehu-Franck, 1982. Etude phytosociologique analytique et globale de l'ensemble des vases salés et des prés salés et saumâtres de la façade atlantique française. Bull. Ecol. 13: 357–386.

Guillon, L. M., 1984. Les schorres de la baie du Mont Saint-Michel. Unités de végétation et facteurs de milieu. Rapport inédit. CEE/Ministère de l'environnement. Lab. Evol. Syst. Nat. Mod., Université de Rennes 1 et Museum National d'Histoire Naturelle.

Groenendijk, A. M., 1984. Primary production of four dominant salt marsh angiosperms in the SW Netherlands. Vegetatio 57: 143–152.

Groenendijk, A. M. & M. A. Vink-Lievaart, 1987. Primary production and biomass on a Dutch salt marsh: emphasis on the below-ground component. Vegetatio 70: 21–27.

Gordon, D. C. Jr. & P. J. Cranford, 1994. Export of organic matter from macrotidal salt marshes in the upper bay of Fundy, Canada. In W. J. Mitsch (ed.), Global Wetlands: Old World and New. Elsevier Science B.V.: 257–264.

Hackney, C. T. & E. B. Haines, 1980. Stable carbon isotope composition of fauna and organic matter collected in a Mississipi Estuary. Estuar. coast. Shelf. Sci. 10: 703–708.

Haines, E. B., 1977. The origin of detritus in Georgia salt marshes estuaries. Oikos 29: 254–260.

Haines, E. B. & C. L. Montague, 1979. Food sources of estuarine invertebrates analysed using $^{13}C/^{12}C$ ratios. Ecology 60: 48–56.

Halupa, P. J. & B. L. Howes, 1995. Effects of tidally mediated litter moisture content on decomposition of *Spartina alterniflora* and *S. patens*. Mar. Sci. 123: 379–391.

Hemminga, M. A. & G. J. C. Buth, 1991. Decomposition in salt marsh ecosystems of the S.W. Netherlands: the effects of biotic and abiotic factors. Vegetatio 92: 73–83.

Hicks, R. E., C. Lee & A. C. Marinucci, 1991. Loss and recycling of amino acids and protein from smooth cord grass (*Spartina alterniflora*) litter. Estuaries 14: 430–439.

Kemp, P. F., S. Y. Newell & C. S. Hopkinson, 1990. Importance of grazing on the salt marsh grass *Spartina alterniflora* to nitrogen turnover in a macrofaunal consumer, *Littorina irrorata*, and to decomposition of standing-dead *Spartina*. Mar. Biol. 104: 311–319.

Lefeuvre, J. C. & R. F. Dame, 1994. Comparative studies of salt marsh processes in the New and Old Worlds: an introduction. In W. J. Mitsch (ed.), Global Wetlands: Old World and New. Elsevier Science B.V.: 169–180.

Lefeuvre, J. C., G. Bertru, F. Burel, L. Brient, V. Créach, Y. Gueuné, J. Levasseur, A. Mariotti, A. Radureau, C. Retière, B. Savouré & O. Troccaz, 1994. Comparative Studies on Salt Marsh Processes: Mont Saint-Michel bay, a multi-disciplinary study. In W. J. Mitsch (ed.), Global Wetlands: Old World and New. Elsevier Science B.V.: 215–234.

Levasseur, J. E., 1993. Physiographie, végétation et production primaire. In J. C. Lefeuvre (ed.), Comparative studies on salt marsh processes. Commission of the European Community. EEC contract n° E5V-0098: 33–54.

Long, S. P. & C. F. Mason, 1983. Saltmarsh Ecology. Glasgow, Blackie.

Mann, K. H., 1972. Macrophyte production and detritus food chains in coastal waters. Mem. Ist. ital. Idrobiol. 29 Suppl.: 353–383.

Marinucci, A. C., 1982. Trophic importance of *Spartina alterniflora* production and decomposition to the marsh-estuarine ecosytem. Biol. Conserv. 22: 35–58.

Meyers, P. A., 1994. Preservation of elemental and isotopic source identification of sedimentary organic matter. Anal. Chem. 56: 1859–1861.

Mc Kee, K. L. & W. H. Patrick, 1988. The relationship of smooth cordgrass (*Spartina alterniflora*) to tidal datums: a review. Estuaries 1: 143–151.

Newell, S. Y., R. D. Fallon & J. D. Miller, 1989. Decomposition and microbial dynamics for standing, naturally positioned leaves of the salt-marsh grass *Spartina alterniflora*. Mar. Biol. 101: 471–481.

Nyman, J. A., R. D. Delaune, H. H. Roberts & W. H. Patrick Jr., 1993. Relationship between vegetation and soil formation in a rapidly submerging coastal marsh. Mar. Ecol. Prog. Ser. 96: 269–279.

Nixon, S. W., 1980. Between coastal marshes and coastal waters – a review of twenty years of speculation and research on the role of salt marshes in estuarine productivity and water chemistry. In P. Hamilton & K. B. Macdonald (eds), Estuarine and Wetland Processes with Emphasis on Modelling. Plenum Press, New York: 437–525.

Odum, E. P. & A. A. de la Cruz, 1967. Particulate organic matter in a Georgia salt marsh-estuarine ecosystem. In G. H. Lauff (ed.), Estuaries. AAAS, Washington.

Odum, E. P., 1969. The strategy of ecosystem decelopment. Science 164: 262–270.

Odum, E. P. & E. J. Heald, 1975. The detritus based food web of an estuarine mangrove community. In L. E. Cronin (ed.), Estuarine Research, Vol. 1., Academic Press, New York: 265–286.

O'Leary, M. H., 1981. Carbon isotope fractionation in plants. Phytochemistry 20: 553–567.

Reice, S. R. & A. E. Stiven, 1983. Environmental patchiness, litter decomposition and associated faunal patterns in a *Spartina alterniflora* marsh. Estuar. Coast. Shelf. Sci. 16: 559–571.

Reidenbaugh, T. G. & W. C. Banta, 1980. Origin and effects of *Spartina* wrack in a Virginia salt marsh. Gulf Research Report 6: 393–401.

Rice, D. L. & K. R. Tenore, 1981. Dynamics of carbon and nitrogen during the decomposition of detritus derived from estuarine macrophytes. Estuar. Coast. Shelf. Sci. 13: 681–690.

Shaver, G. R. & J. M. Melillo, 1984. Nutrient budget of marsh plants: efficiency concepts and relation to availability. Ecology 65: 1491–1510.

Smalley, A. E., 1959. The growth cycle of *Spartina* and its relation to the insect populations in the marsh. In R. A. Ragotzkie, J. M. Teal, L. R. Pomeroy & D. C. Scott (eds), Proc. Salt Marsh Conf., Athens, University of Georgia: 96–100.

Smith, B. N. & S. Epstein. 1970. Biogeochemistry of the stable isotopes of hydrogen and carbon in salt marsh biota. Plant Phys. 46: 738–742.

Teal, J. M., 1962. Energy flow in the salt marsh ecosystem of Georgia. Ecology 43: 614–624.

Tenore, K. R., 1984. Nitrogen in benthic food chains. In T. H. Blackburn & J. Sorensen (eds), Nitrogen Cycling in Coastal Marine Environments. Willey: 191–206.

Tieszen, L. L., 1991. Natural variations in the carbon isotope values of plants: implications for archaeology, ecology and paleoecology. J. Arch. Sci. 18: 227–248.

Valiela, I., B. Howes, R. Howarth, A. Giblin, K. Foreman, J. M. Teal & J. E. Hobbie. 1982. The regulation of primary production and decomposition in a salt marsh ecosystem. In B. Gopal, R. E. Turner, R. G. Wetzel & D. E. Whigham (eds), Wetlands: Ecology and Management. National Institute of Ecology, Jaipur, India: 151–168.

Valiela, I., J. Wilson, R. Buchsbaum, C. Riestsam, D. Bryant, K. Foreman & J. Teal, 1984. Importance of chemical composition of salt marsh litter on decay rates and feeding by detritivores. Bull. mar. Sci. 35: 261–269.

Valiela, I., J. M. Teal, S. D. Allen, R. van Etten, D. Goehringer & S. Volkmann, 1985. Decomposition in salt marsh ecosystems: the phases and major factors affecting disappearance of aboveground organic matter. J. exp. mar. Biol. Ecol. 89: 29–54.

Vernberg, F. J., 1993. Salt marsh processes: a review. Envir. Toxic. Chem. 12: 2167–2195.

White, D. S. & B. L. Howes, 1994. Nitrogen incorporation into decomposing litter of *Spartina alterniflora*. Limnol. Oceanogr. 39: 133–140.

Wolff, W. J., M. N. van Eeden & E. Lammens, 1979. Primary production and import of particulate organic matter on a salt marsh in the Netherlands. Neth. J. Sea. Res. 13: 242–255.

Hydrobiologia **373/374:** 89–100, 1998.
J. C. Amiard, B. Le Rouzic, B. Berthet & G. Bertru (eds),
Oceans, Rivers and Lakes: Energy and Substance Transfers at Interfaces.

Benthic influence on the metabolism of a shallow tropical lagoon (Lagoa da Barra, Brazil)

Jean-Pierre Carmouze[1], Bias de Farias[2], Marcelo Corrêa Bernardes[3] & Kátia Naomi Kuroshima[4]
[1] *ORSTOM, B.P. 165, F-97323 Cayenne Cedex, France*
[2] *Laboratorio de Hidrobiologia, Universidade Federal do Rio de Janeiro, Ilha do Fundao, Rio de Janeiro, Brazil*
[3] *Departamento de Geoquimica, Universidade Federal Fluminense, Morro do Valonguinho, 24030 – Niteroí, RJ, Brazil*
[4] *Universidade do Vale do Itajaí, Facimar, rua Uruguai, 458 – Caixa Postal 360. 88 302 Ita jaí, SC, Brazil*

Key words: tropical lagoon, autotrophic activity, heterotrophic activity, pelagic carbon metabolism, benthic carbon metabolism

Abstract

Pelagic and benthic carbon metabolism in the shallow tropical Barra Lagoon was studied weekly between 1990 and 1993, in periods characterized by temporary nitrogen limitation, fish kills and transitory colonization of the bottom by macro-algae. Diurnal and nocturnal metabolic rates were measured by monitoring changes in CO_2 content in entire water columns (free water and waters incubated in plexiglass tubes with and without inclusion of sediment). Total community carbon metabolism exhibited a marked autotrophism and high short term changes, dampening seasonal trends. Autotrophism and heterotrophism of the lagoon were controlled by phytoplankton biomass variations. In the presence of sediment, diurnal net benthic production was 11 to 64% of total diurnal net production, and nocturnal net benthic mineralization was 21 to 52% of total nocturnal net mineralization. The imbrication between the pelagic and benthic organic carbon cycles regulated carbon metabolism of the total community. A severe temporary nitrogen limitation, a fish kill and a temporary shift in primary producers did not substantially change metabolic rates because appropriate shifts occurred from the pelagic carbon cycle to the benthic carbon cycle, and *vice-versa*.

Introduction

A series of coastal lagoons in the State of Rio de Janeiro, Brazil (23° S, 42–43° W), has been altered during the past decades by agro-industrial and domestic pollution. Some lagoons are fairly well preserved, others have become profoundly modified. Three lagoons have been intensively investigated, Guarapina, Saquarema and Barra, to study the steps of the complex process of eutrophication with regard to carbon metabolism of total communities. Most biological activity studies have been conducted using *in situ* dissolved oxygen and carbon incubation techniques (Alem & Samaan, 1969; Tundisi et al., 1973; Pages et al., 1981; Vaulot & Frisoni, 1986; Randall & Day, 1987; Flores-Verdugo et al., 1988 ; Sandoval-Rojo et al., 1988; Moreira & Knoppers, 1990). For Saquarema and Barra, we chose to assess the basic rates of organic carbon production and mineralization using diurnal curve 'free water' measurements of carbon dioxide, as recommended by Odum & Hoskin (1958). As few other authors (Park et al., 1958; Qasim, 1979; Sournia et al., 1981; Reyes & Merino, 1991; Carmouze et al., 1991), we also applied *in situ* incubation techniques with free water measurements in order to collect more information on total pelagic and benthic carbon metabolism and their mutual relations. These aspects are of particular interest to the discussion on whether coastal systems are autotrophic or heterotrophic (Smith & MacKensie, 1987; Smith 1988; Reyes & Merino, 1991).

In this paper, we report measurements of the pelagic and benthic C-metabolism of Barra Lagoon, carried out from October 90 to August 93; we also analyse how they interact and how they control total community C-metabolism.

Description of the site, Materials and methods

Barra Lagoon (28° S and 43° W) has a surface area of 7 km^2, a mean depth of 1.10 m and salinities from 3 to 12‰. It belongs to the Marica system which is composed of three sublagoons (Figure 1). With a mean tidal range < 0.01 m, it may be defined as a 'choked' system in the classification of Kjerfve (1986). It is dominated by phytoplankton production, with *Synechocystis* f. *salina aquatilis*, as the dominant species (Domingos et al., 1994).

One hundred and twelve weekly surveys of 24–36 h duration were carried out from October 90 to April 91 and from October 91 to August 93 at a single, central station of the lagoon. Forty of them included *in situ* incubations: during the hot season (from October 90 to April 91 and from January 93 to March 93) and the cold season (from mid-June 93 to mid-September 93) and only one day in September 92, describing a transitory sediment colonization by macro-algae.

Each survey consisted of 5–6 daily measurements of temperature, salinity, pH, alkalinity and dissolved oxygen in free water, and, if such should be the case, in incubated water (18 h, 6 h, 10 h, 12 h, 15 h 30, 18 h). These variables were measured at 0.10 m from the surface, at mid-depth, and above the bottom. Hydrological and meteorological data were simultaneously monitored: air temperature, intensity and direction of wind, lagoon turbulence, and Secchi disk depth. Insolation, cloudiness and precipitation were supplied by the meteorological station of Maricá. Water samples for analysis of nutrients and chlorophyll were collected at 18 h. Temperature and salinity were measured by a YSI salinometer (Model NCT) and pH with a 0.001 high resolution unit (Metrohm pHmeter 654) and a high precision pH electrode assembly (Metrohm N° 65102). Alkalinity was determined by sulfuric acid titration using Gran's function (Gran, 1952). Samples for nutrients were filtered and frozen at – 18 °C, before analysis by Standard Methods (Strickland & Parsons, 1972). Particulates for chlorophyll analysis were retained on Whatman GFC filters and held at – 18 °C until analysis in the laboratory (Lorenzen, 1967). Incubations were done in two plexiglass tubes isolating the entire water column (length = 2 m, diameter = 18 cm), with and without inclusion of superficial sediment. Total inorganic dissolved carbonic species, TCO_2, were assessed by the pH-alkalinity method, using the ionic association model of Garrels & Thomson (1962), modified by Carmouze 1984. The stagnant film model (Kanwisher, 1960) was used to assess CO_2 exchanges at the water–air interface. As recommended by Stumm & Morgan (1981), TCO_2 flux was considered, instead of CO_2 flux, with a view to correct chemical reactions of the $[CO_2\text{-}H_2O]$ system occurring during the diffusion. The K_{CO_2} value given by (Weiler, 1975) was selected and multiplied by the ratio of molecular diffusion coefficient of HCO_3^- and CO_2 (Lerman, 1979) to give the mass transfer coefficient of TCO_2, K_{TCO_2} (as far as HCO_3^- is the predominant carbonate species at the pH range 7–9). Then, $K_{TCO_2} = 0.008 \text{ m} \times \text{h}^{-1}$ for V (wind speed) $< 5 \text{ m} \times \text{s}^{-1}$ and $K_{TCO_2} = 9.5 \times 10^{-4} \times V2$ m h^{-1} for $V > 5 \text{ m s}^{-1}$. Finaly, TCO_2 flux, F_{TCO_2}, was estimated from: $F_{TCO_2} = K_{TCO2}$ ($[TCO_2]_{eq} - [TCO_2]_s$) with $[TCO_2]$eq = TCO_2 concentration related to CO_2 saturation value and $[TCO_2]_s$ = ambient surface TCO_2 concentration ($\approx$ 0.10 m). Positive F_{TCO2} values indicate that the TCO_2 flux is directed toward the water column.

In this mostly phytoplankton-based lagoon, total community metabolism was described by (1) daily averages of TCO_2 saturation % in the column water, %TCO_2, (concentration ratio of the measured TCO_2 and the calculated TCO_2 in equilibrium with atmospheric pCO_2), which characterize either dominant heterotrophic activity (supersaturated waters) or dominant autotrophic activity (sub-saturated waters) (2) weekly biological activity balance given by changes in TCO_2 content in the water column calculated from two 24-h measurement cycles separated by 7 d, after correction for CO_2 exchanges at the air-water interface (3) diurnal net production and nocturnal net mineralization of C_{org}, calculated from changes in CO_2 content in the water column, between 6 h and 18 h and between 18 h and 6 h and simultaneous CO_2 exchanges at the air-water interface. The same procedure was performed for incubated tubes.

The experimental measurements supply (Figure 2): (1) diurnal net C_{org}-production in the free water + sediment system (tube in contact with the sediment): $P_{\text{net tot}} = P_{FW-d} + P_{FW\,rec.-d} + P_{sed-d} - M_{FW-d} - M_{sed-d}$, with P_{FW-d} = diurnal pelagic

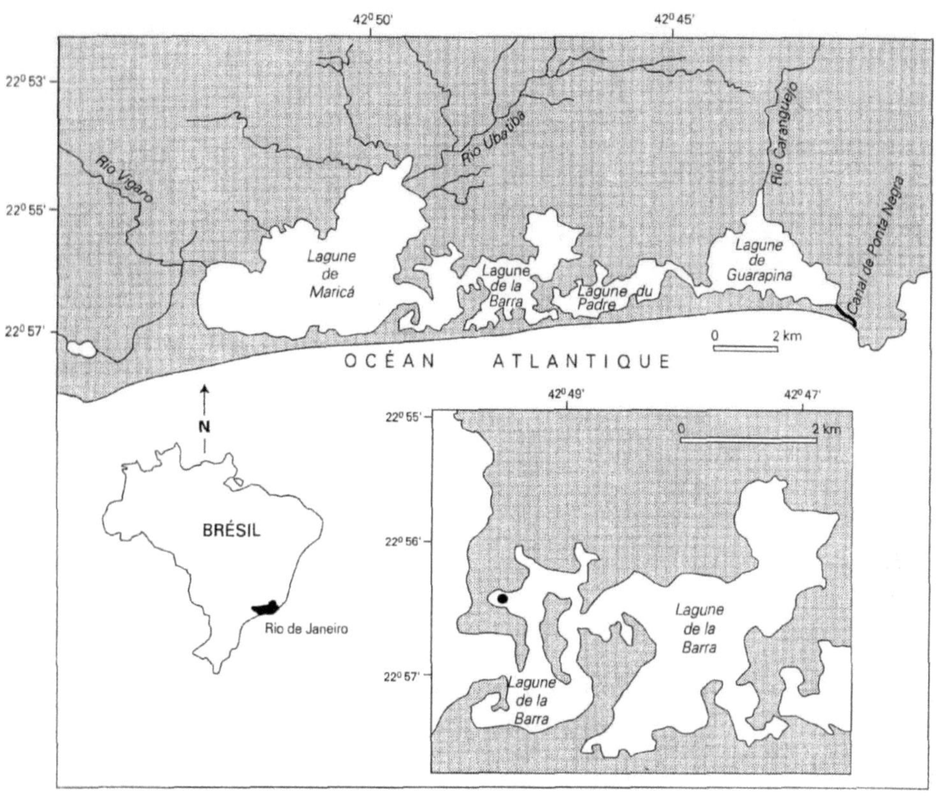

Figure 1. The lagoonal system of Maricá-Guarapina, Brazil.

Table 1. Diurnal and nocturnal rates of pelagic and benthic C_{org}-metabolism. $P_{net\,tot}$ = diurnal net production of C_{org} in free water with sediment inclusion; P_{netFW} = diurnal net production of C_{org} in free water without sediment inclusion; $P_{net\,sed}$ = diurnal net production of C_{org} in the sediment and in free water, originated by nutrients recycled in the sediment; $M_{net\,tot}$ = nocturnal net mineralization of C_{org} in free water with sediment inclusion; M_{netFW} = nocturnal net mineralization of C_{org} in free water without sediment inclusion; $M_{net\,sed}$ = nocturnal net mineralization of C_{org} in the sediment.

	$P_{net\,tot}$	P_{netFW}		$P_{net\,sed}$		$M_{net\,tot}$	$M_{net\,EL}$		$M_{net\,sed}$	
	mmoles m^{-2}	mmoles m^{-2}	% $P_{net\,tot}$	mmoles m^{-2}	% $P_{net\,tot}$	mmoles m^{-2}	mmoles m^{-2}	% $M_{net\,tot}$	mmoles m^{-2}	% $M_{net\,tot}$
From 20/10/90 to 26/03/91 ($n=24$)	219 ± 64	194 ± 88	78.5	25 ± 55	11.5	− 159 ± 39	− 125 ± 40	78.5	− 34 ± 25	21.5
23/09/92	191	70	36.5	121	63.5	− 225	− 111	49.5	− 114	50,5
From 12/01/93 to 23/03/93 ($n=11$)	173 ± 41	122 ± 34	70.5	51 ± 44	29.5	− 173 ± 27	− 83 ± 30	48	− 90 ± 33	52
From 15/06/93 to 10/9/93 ($n=5$)	159 ± 24	108 ± 41	68	51 ± 50	32	− 155 ± 25	− 99 ± 45	64	− 56 ± 39	36

C_{org}-production due to biogenic elements recycled in the water column (photosynthesis, biosynthesis), $P_{FW\,rec.-d}$ = diurnal pelagic C_{org}-production due to biogenic elements recycled at the sediment-water interface (photosynthesis, biosynthesis), P_{sed-d} = diurnal benthic C_{org}-production (photosynthesis, biosynthesis), M_{Fw-d} = diurnal pelagic C_{org}-mineralization (respiration) and M_{sed-d} = diurnal benthic C_{org}-mineralization (respiration, fermentation), (2) diurnal net C_{org}-production in the free water (tube without contact with sediment): $P_{net\,FW} = P_{FW-d} - M_{FW-d}$, (3) nocturnal net C_{org}-mineralization in the water + sediment system (tube in contact with the sediment): $M_{net\,tot} = M_{FW-n} + M_{sed-n} - P_{FW-n} - P_{sed-n}$, with $M_{FW-n} =$

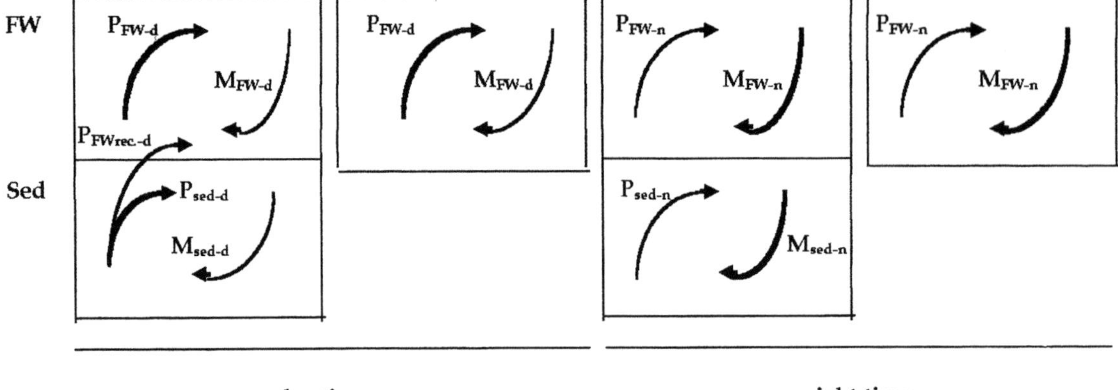

Figure 2. Diurnal and nocturnal pelagic and benthic *C*-metabolisms: P_{FW-d} = diurnal C_{org}-production due to biogenic elements recycled in the water column; $P_{FWrec.-d}$ = diurnal C_{org}-production due to biogenic elements recycled in sediment–water interface; M_{Fw-d} = diurnal pelagic C_{org}-mineralization; P_{sed-d} = diurnal C_{org}-production, M_{sed-d} = diurnal benthic C_{org}-mineralization; P_{FW-n} = nocturnal C_{org}-production in the water column; M_{FW-n} = nocturnal pelagic C_{org}-mineralization; P_{sed-n} = nocturnal C_{org}-production; M_{sed-n} = nocturnal benthic C_{org}-mineralization.

nocturnal pelagic C_{org}-mineralization (respiration), M_{sed-n} = nocturnal benthic C_{org}-mineralization (respirations, fermentations), P_{FW-n} = nocturnal pelagic C_{org}-production (biosynthesis) and P_{FW-n} = nocturnal pelagic C_{org}-production (biosynthesis), (4) nocturnal net mineralization of C_{org} in free water (tube without inclusion of sediment): $M_{net\,FW} = M_{FW-n} - P_{FW-n}$.

The diurnal activity difference between the two tubes represents net production of C_{org} in the water column originated from benthic nutrient recycling, plus net C_{org}-production in the sediment, minus diurnal benthic C_{org}-mineralization: $P_{net-sed} = P_{FW\,rec.-d} + P_{sed-d} - M_{sed-d}$. Exceptionally, when $M_{sed-d} > P_{FW\,rec.-d} + P_{sed-d}$, the diurnal activity difference represents a net mineralization. The nocturnal activity difference between the two tubes represents nocturnal benthic mineralization: $M_{net\,sed} = M_{sed-n} - P_{sed-n}$.

Results and discussions

The carbon metabolism of the total community

Barra Lagoon exhibited strong autotrophy, as autotrophic mostly exceeded heterotrophic activity: over the entire period, the resulting biological activity, calculated from weekly balances of C_{org} showed a net average production of 30 mmoles m^{-2} d^{-1}, while the quasi permanent sub-saturation of water column in TCO_2 (mean value = 89.7%) produced an input of 29.5 mmoles of atmospheric $CO_2/m^{-2}/d^{-1}$. This counter-balanced the net amount of CO_2 assimilated (Figure 3a and b). This autotrophic feature has been reported from similar lagoon systems: the southern part of the Laguna Madre, USA (Odum & Wilson, 1962; cited in Reyes & Morino, 1991) and Bojorguez Lagoon, Mexico (Reyes & Merino, op. cit.). But, net heterotrophy was also detected in Bissel Pond, USA (Nixon et al., 1976) and in the El Verde Lagoon, Mexico (Flores-Verdugo et al., 1988) as well as balanced metabolisms in The Terminos Lagoon (Day et al., 1988) and in Saquarema Lagoon, Brazil (Carmouze et al., 1991, Carmouze & Vasconcelos, 1992). This last lagoon, adjacent to Barra Lagoon, exhibited net autotrophy in summer and net heterotrophy in winter during two consecutive annual cycles. So, the authors supposed that photosynthetic activity was more sensitive to environmentally associated factors than heterotrophic activity, i.e. more stimulated in summer and less in winter. This seasonal auto/heterotrophic shift was not confirmed in Barra Lagoon, where a weak heterotrophy was noted in summer 1992–93 ($P_{net\,tot}$ = 178 ± 41 mmoles m^{-2}; $M_{net\,tot}$ = 189 ± 27 mmoles m^{-2}, n = 12), and a marked autotrophy in winter 92 ($P_{net\,tot}$ = 203 ± 64 mmoles m^{-2}; $M_{net\,tot}$ = 166 ± 54 mmoles m^{-2}, n = 11). In fact, the auto-heterotrophic feature seems to be directly associated with algal biomass changes, because growing phytoplankton phases always coincided with increasing net autotrophy (December 91, May–June 91, November 92) and degenerescence phytoplankton phases with decreas-

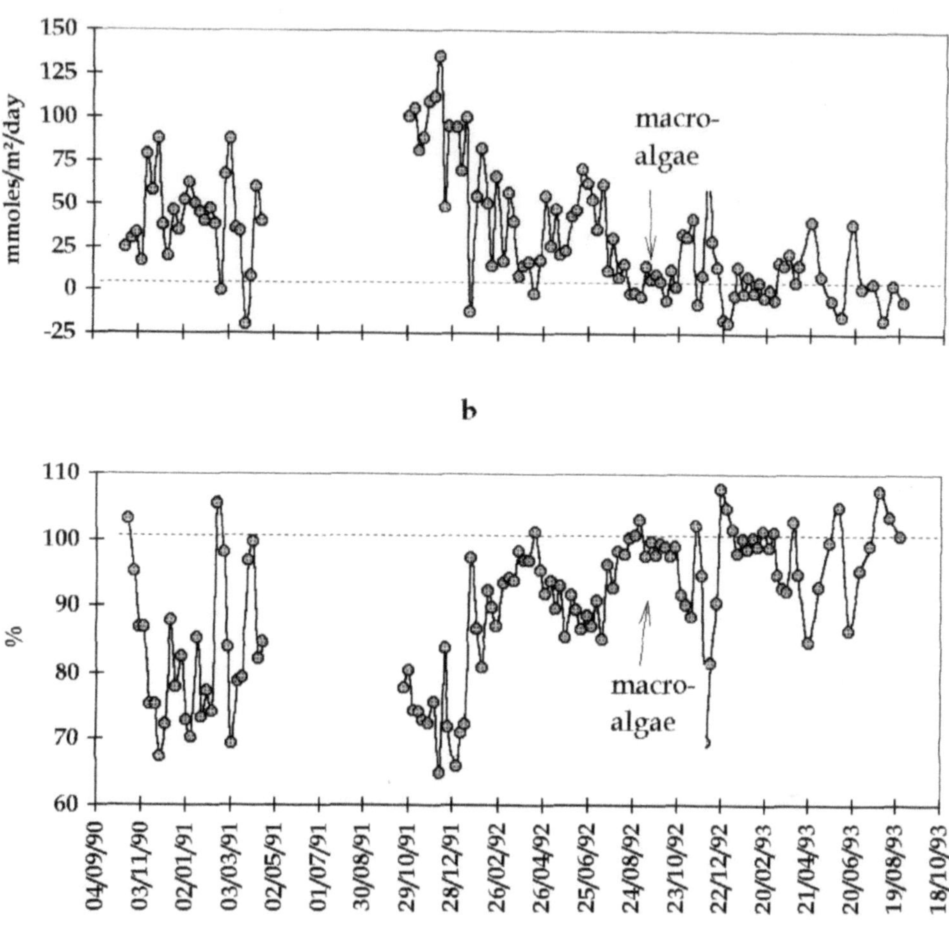

Figure 3. (a) Weekly biological activity balance of organic carbon, based on changes of T_{CO_2} contents in the water column calculated from two 24-h measurement cycles separated by 7 d, after correction of CO_2 exchanges at the air–water interface, during the period studied. (b) Daily averages of T_{CO_2} saturation % in the water column (concentration ratio of the measured T_{CO_2} in the column and the calculated T_{CO_2} in equilibrium with atmospheric P_{CO_2}) during the period studied.

ing net autotrophy (February–March 92; September 92 and January–February 93) (Figure 3a and b, Figure 4a). However, results of Barra and Saquarema are not conflicting as far as, in Saquarema Lagoon, highest and lowest phytoplankton biomass coincided with summer and winter respectively.

C-metabolic rates monitored in Barra Lagoon were high, compared to literature data. Average annual values of $P_{\text{net tot}}$ and $M_{\text{net tot}}$, were respectively 215 ± 87 mmoles m^{-2} and 179 ± 57 mmoles m^{-2} ($n = 112$). Knoppers (1994) mentions that, in most coastal lagoons, primary production range between 45 and 115 mmoles m^{-2} d^{-1}. Metabolic rates of the adjacent lagoons of Saquarema and Guarapina are lower: in Saquarema Lagoon, $P_{\text{net tot}} = 110 \pm 68$ mmoles m^{-2} and $M_{\text{net tot}} = 105 \pm 49$ mmoles m^{-2} ($n = 40$); in Guarapina Lagoon, diurnal production (evaluated by ^{14}C-method, i.e. a value between gross and net production) was 120 ± 45 mmoles m^{-2} ($n = 22$). Knoppers et al. (1991), analyzing the differences between these three lagoons established that trophic state was related to water turn-over time, with additional anthropogenic nutrient loading affecting the relationship.

The standard-deviation of metabolic rates, recorded in Barra Lagoon, as well as in Guarapina Lagoon and Saquarema Lagoon, demonstrated that biological activities were variable at the daily scale. In Barra, standard-deviations of $P_{\text{net tot}}$ and $M_{\text{net tot}}$ were equal

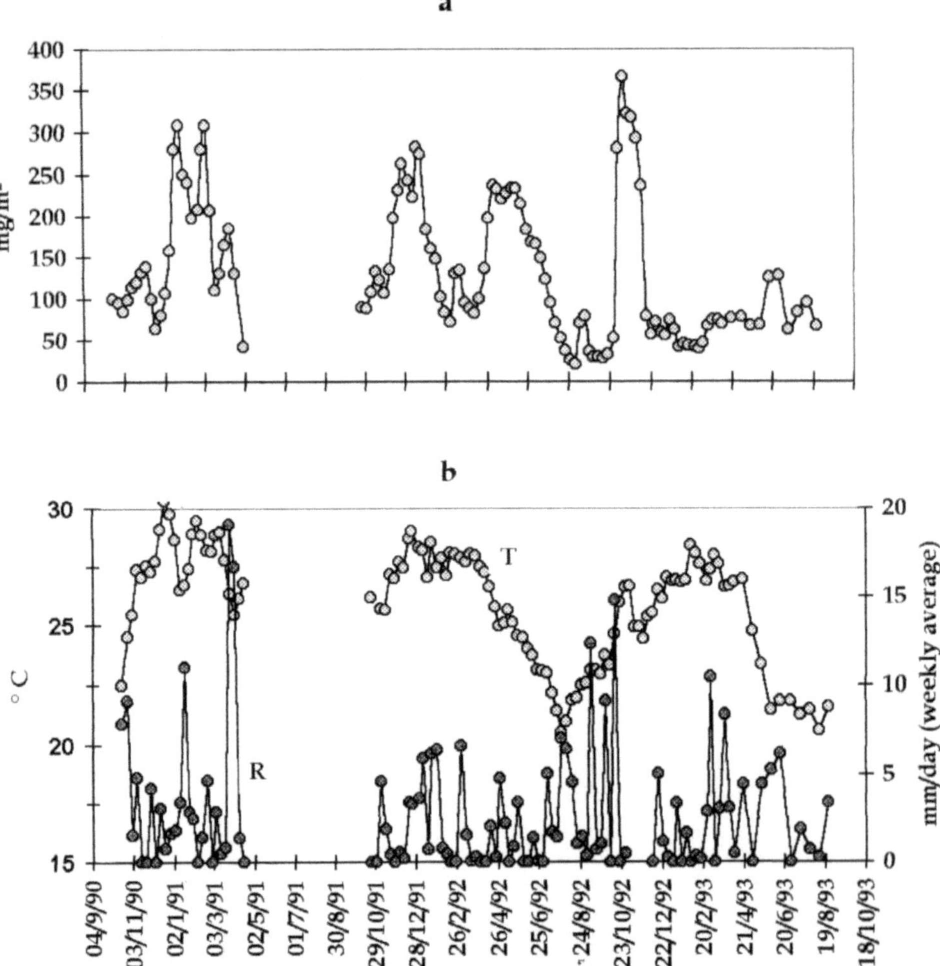

Figure 4. (a) Sum of Chlorophyll a and pheophytin concentrations during the period studied. (b) Daily averages of temperature, T, and weekly averages of rain fall, R, during the period studied.

to 40% and 32% of their mean values (the wider dispersion of $P_{\text{net tot}}$ is due to its insolation dependance) respectively. This variability is partly due to frequent polar fronts, which cause marked short-term meteorological changes such as temperature decrease, rainfall, river flooding and marine intrusions to the lagoon (Bernardes, 1952). Typical dry-cold and wet-hot seasons, as encountered in tropical and sub-tropical climates, were less apparent (Figure 4b).

Seasonal patterns of C_{org}-metabolism were severely affected by yearly differences in the passage of such fronts which occur with irregular frequency and variable intensity, as shown by the temporal evolution of diurnal $P_{\text{net tot}}$ and nocturnal $M_{\text{net tot}}$, in Figure 5. Highest $P_{\text{net tot}}$ rates were monitored at the beginning of the hot seasons. $M_{\text{net tot}}$ rates similarly changed during the two first summers, increasing in late January and progressively decreasing in April, but during the third summer, they decreased from November to March. In winter (May to August), both $P_{\text{net tot}}$ and $M_{\text{net tot}}$ decreased, but unequaly: in 92, $P_{\text{net tot}} > M_{\text{net tot}}$, while in 93, $P_{\text{net tot}} < M_{\text{net tot}}$. $P_{\text{net tot}}$ and $M_{\text{net tot}}$ were poorly correlated, although heterotrophic activity would be linked to photosynthetic activity, as far as it is regulated by the rate at which utilizable matter becomes available. The coefficient of correlation was 0.52 for the entire period, with a maximum of 0.65 in summer 93 and a minimum of 0.12 in winter 93. In Saquarema Lagoon, this coefficient ranged from 0.30 to 0.75 (Carmouze et al., 1991).

Thus, short term meteorological may deeply alter C_{org}-metabolism and mask seasonal trends. The well-

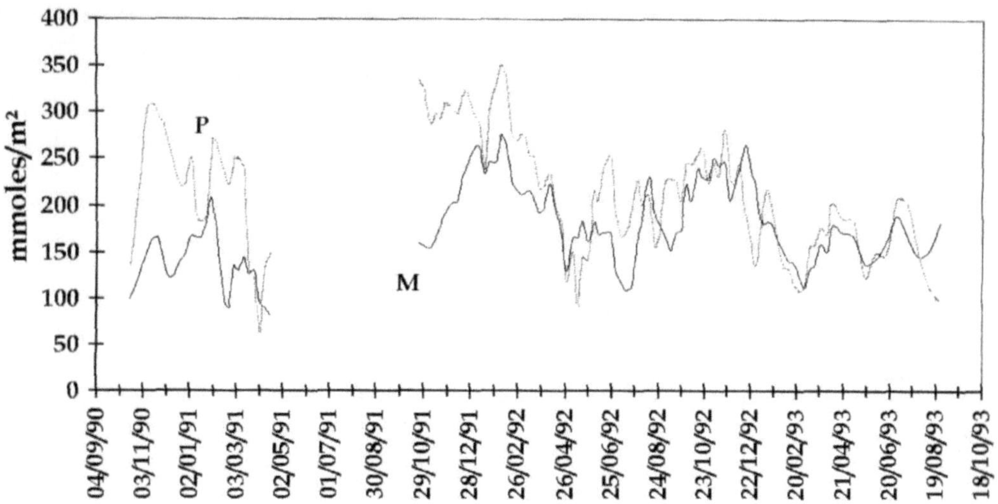

Figure 5. Diurnal net production, $P_{\text{net tot}}$ (P in the graph), and nocturnal net mineralization, $M_{\text{net tot}}$ (M in the graph), of C_{org}, calculated from changes of CO_2 contents in the water column, respectively between 6 AM and 6 PM and between 6 PM and 6 AM, after correction of CO_2 exchanges at the air–water interface, during the period studied.

established seasonal influence of temperature on metabolic rates (Harris & Pucini, 1977; Tison & Pope, 1980; Carmouze & Caumette, 1985; Carmouze et al., 1991) is not entirely verified here: from late July to early September 92, $P_{\text{net tot}} = 211 \pm 76$ mmoles m^{-2} and $M_{\text{net tot}} = 176 \pm 56$ mmoles m^{-2} ($n = 8$), with $t < 22.5$ °C, while, from mid-January to mid-March 93, $P_{\text{net tot}} = 145 \pm 42$ mmoles m^{-2} and $M_{\text{net tot}} = 144 \pm 30$ mmoles m^{-2} ($n = 9$), with $t > 27.5$ °C.

Besides climatic control, two non cyclic events occurred during the study period, contributed to dampen seasonal trends of carbon metabolism still more: a severe nitrogen limitation in late spring and summer 1990–91 (which indirectly caused a fish kill in February 91) and macro-algal colonizing of the bottom (*Cladophora vagabunda* and *Enteromorpha*, sp) from mid-August to mid-October 93. The nitrogen deficiency (molar ratio of seston N/C = 0.04 in that period, against 0.10 the rest of the period studied), probably due to intense denitrification at the beginning of the hot season, promoted cyanophycean growth (*Synechocystis*). Two months later, this species, now degenerescent, produced toxins which killed the filter fish *Brevoortia tyranus* (Azevedo & Carmouze, 1994). This fish kill did not durably alter the carbon metabolism: $P_{\text{net tot}}$ and $M_{\text{net tot}}$ were 224 ± 82 and 156 ± 43 mmoles m^{-2} (with $n = 17$) before mortality and to 205 ± 38 and 165 ± 29 mmoles m^{-2} (with $n = 7$) after mortality. Nor did partial substitution of phytoplankton by macro-algae in September 93 significantly modify $P_{\text{net tot}}$ and $M_{\text{net tot}}$ values (Figure 5). Indeed these two events did not seriouly affect the C-cycle, but they probably modified benthic and pelagic C-cycles and their mutual dependence, by the organic matter sedimented after the fish mortality, and the presence of benthic macro-algae. *In situ* incubations, during the first and last year of the study, will give complementary data to better describe these internal changes and their consequence on C_{org}-metabolism.

The carbon metabolism of pelagic and benthic communities

Daily balances of benthic and pelagic C_{org}-metabolism induced by benthic activity (differences in C_{org}-metabolism monitored in the two tubes) exhibited net but unequal C_{org}-mineralization rates: an average of 9 mmoles m^{-2} 24 h^{-1} from mid-October 90 to the end of March 91; 5 mmoles m^{-2} 24 h^{-1} from mid-January 93 to late March 93, and 39 mmoles m^{-2} from mid-June 93 to mid-August 93. These differences are consequences of time variable benthic participation in nocturnal mineralization and diurnal production of C_{org}.

Highest rates of nocturnal benthic mineralization, M_{sed-n}, occurred in summer 1992, representing 52% of the total mineralization against 21.5 and 36%, in summer 1991 and winter 93. M_{sed-n} seems related to stocks of NH_4^+ in the interstitial water of the first twenty centimeters of sediment and also phytoplankton biomass. The lowest values of M_{sed-n} rates in summer 1990–91 (34 ± 21 mmoles m^{-2}, $n = 26$) coincided with lowest NH_4^+ stocks (9.5 ± 1.8 mmoles m^{-2}, $n = 26$) and highest Chla + pheophytin (147 ± 73 mg m^{-2}, $n = 12$). Period of highest M_{sed-n} (90 ± 33 mmoles m^{-2}, $n = 11$) overlaid that of higher stocks of NH_4^+ (69 ± 12.5 mmoles m^{-2}, $n = 11$) and lowest pigment sum (57 ± 16.5 mg m^{-2}, $n = 13$). Benthic nocturnal mineralization, M_{sed-n}, decreased and pelagic nitrogen recycling increased with increasing phytoplankton and vice-versa.

Nocturnal benthic activities in Barra Lagoon (M_{sed-n} rates and $M_{sed-n}/M_{net\,tot}$ ratios) were well within the range of estuarine and coastal systems (Nixon, 1981; Florek & Rowe, 1983, 1985; Balzer et al., 1986; Machado & Knoppers, 1988; Carmouze et al. 1991). In Saquarema Lagoon, M_{sed-n} represented an average of 37% of $M_{net\,tot}$.

Day-time incubations mostly demonstrated increasing net C_{org}-production in the tube with sediment, compared to the other. This increase, $P_{net-sed}$, sum of $P_{FW\,rec.-d} + P_{sed-d} - M_{sed-d}$ was unequal all along the study period. It was only 11.5% of $P_{net\,tot}$ in summer 1990–91 (25 ± 50 mmoles m^{-2}, $n = 24$), while it rose to 29.5% in summer 93 (51 ± 44 mmoles m^{-2}, $n = 11$) and 32% (51 ± 50 mmoles m^{-2}, $n = 5$) in winter 93. Similar incubations carried out in Saquarema Lagoon indicated that the elimination of nutrients liberated from the bottom may reduce net diurnal production even more (Pereira, 1991).

Day as well as night-time incubations pointed out that, in summer 1990–91, the benthic loop of C-metabolism was less important than in summer and winter 1993 (Figure 6 and 7). Perhaps nitrogen limitation, which marked the first period, forced the system to adopt a shorter cycle, chiefly located in the pelagic zone. On the contrary, in 1993, benthic metabolism was enhanced, at the expense of the pelagic metabolism. The greater nocturnal nutrient recycling in the sediment contributed to increased photosynthesis of pelagic phytoplankton, and probably of benthic algae, because this last community gained better light conditions (in summer and winter 93, Secchi disk was 75% and 90% of corresponding bottom depths). Although, mean $P_{net-sed}$ values were equal for these two periods, gross production in the sediment may have been higher in summer because mean $M_{net-sed}$ values were lower in winter (56 ± 39 mmoles m^{-2}, $n = 5$, against 90 ± 33 mmoles m^{-2}, $n = 11$).

In fact, the seasonal trends we pointed out masked a more complex daily scale reality. For example, in summer 1990–91, diurnal balance of benthic activity, $P_{FWrec.-d} + P_{sed-d} - M_{sed-d}$, was not always equal to net production of C_{org}. Some days, M_{sed-d} counterbalanced or was higher than $P_{FWhrec.-d} + P_{sed-d}$. Diurnal benthic heterotrophy occurred for two weeks after the fish kill (Figure 6a). Lack of recycled nutrients was not a limiting factor for photosynthesis in the tube isolated from the bottom because of the presence of high phosphate and ammonium concentrations in the water. The probable limiting factor was a reduction of light in the tube in contact with the sediment because of particulate matter resuspension, partly from the fish kill. On the other hand, M_{sed-d} may have increased in the presence of this new organic matter, but, surprisingly, nocturnal benthic mineralization did not significantly change (Figure 6b).

In early summer 1993, M_{sed-d} was again higher than $P_{FWhrec.-d} + P_{sed-d}$ (Figure 7a). This period coincided with an important reduction in phytoplankton biomass. Photosynthesis did not depend on benthic recycled nutrients, while at the same time, M_{sed-d} increased because of dead phytoplankton accumulated on the sediment. A new, opposite sequence started, with a new phase of phytoplankton growth. Benthic influence tended once more to decrease at the end of the study period. This tendency was corroborated by the stock of NH_4^+ in interstitial waters, which returned to the lowest values exhibited during the first year.

Bottom colonization by macro-algae in August-September 1993, as we have seen, did not significantly change global C_{org}-metabolism (Figure 5), but the benthic loop increased as the pelagic loop decreased: the single 24 h incubation, performed in that period, indicated that 51% of $M_{net\,tot}$ and 63.5% of $P_{net\,tot}$ were carried out in the bottom.

According to this experiment, the global 24-h balance in the tube in contact with the sediment showed a net C_{org}-mineralization. In fact, during this period, $M_{net\,tot}$ counterbalanced $P_{net\,tot}$ as shown by TCO_2 concentrations, which oscillated around a saturation value (Figure 3).

Conclusions

Like most shallow lagoons, Barra lagoon is sensible to short-term meteorological changes. In that tropical zone, passages of polar fronts, of unequal intensity and irregular frequency, induce high day to day changes

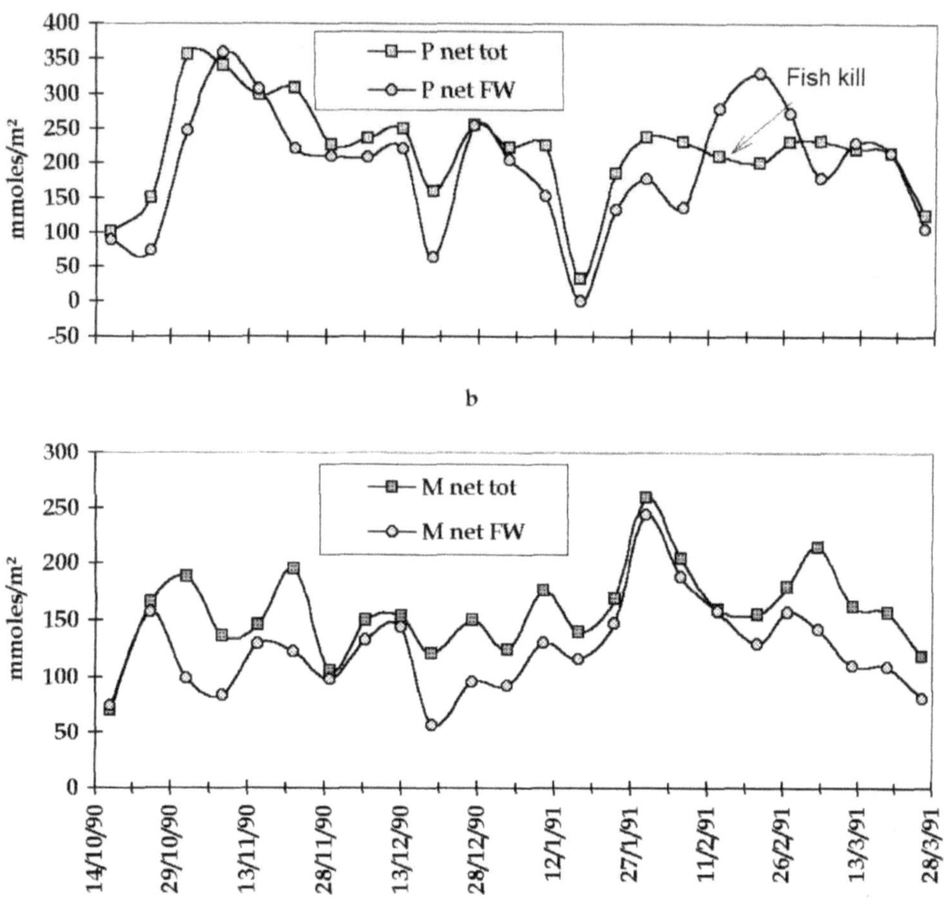

Figure 6. (a) Diurnal net organic carbon production in plexiglass tubes with and without inclusion of sediment, respectively $P_{\text{net tot}}$ and $P_{\text{net FW}}$ from October 90 to March 91. (b) Nocturnal net organic carbon production in plexiglass tubes with and without inclusion of sediment, respectively $M_{\text{net tot}}$ and $M_{\text{net FW}}$ from October 90 to March 91.

in C_{org}-metabolism, causing deep modifications and severely dampening seasonal trends. The chaotic character of the carbon metabolism is thus enhanced, preventing any serious predictions on the future evolution of the system.

However, global C-metabolism is not greatly altered when consistent structural changes occur. It is dampened by shifts from pelagic to benthic C-metabolism and vice-versa. For example, severe nitrogen limitation in January 91 was compensated by an increased pelagic C-metabolism at the expense of benthic C-metabolism. Diurnal C_{org}-production and nocturnal C_{org}-mineralization in September 93 did not significantly change when macro-algae substituted phytoplankton. On that occasion, biological activities moved from the pelagic to the benthic zone. The fish kill event was dampened by a rapid pelagic → benthic → pelagic succession concerning C-metabolism dominance.

The tendency of the ecosystem to be more autotrophic or more heterotrophic, which depends on the living organisms inhabiting the system and on their activities, is not easy to identify. The C-metabolism balance must be established at least on an annual basis. In spite of that, it may severely change from year to year: the August 92–August 93 annual cycle exhibited an equilibrated carbon balance (a net production of 3 mmoles m^{-2} d^{-1}), while the October 91–October 92 period was markedly autotrophic (a net production of 47 mmoles m^{-2} d^{-1}). Besides, in view of the wide array of methodological approaches in use, the divera

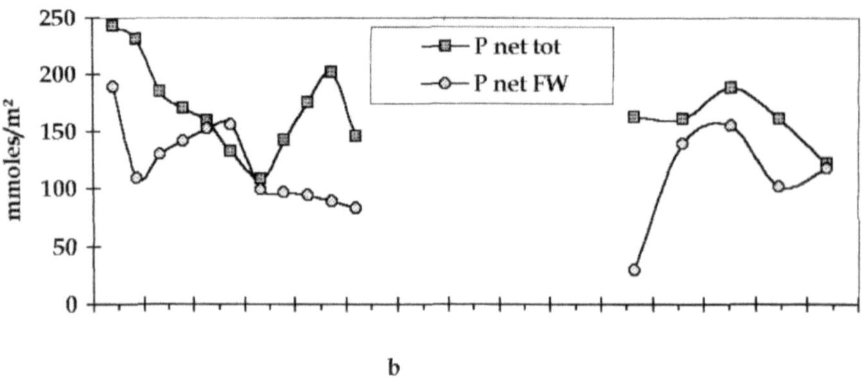

b

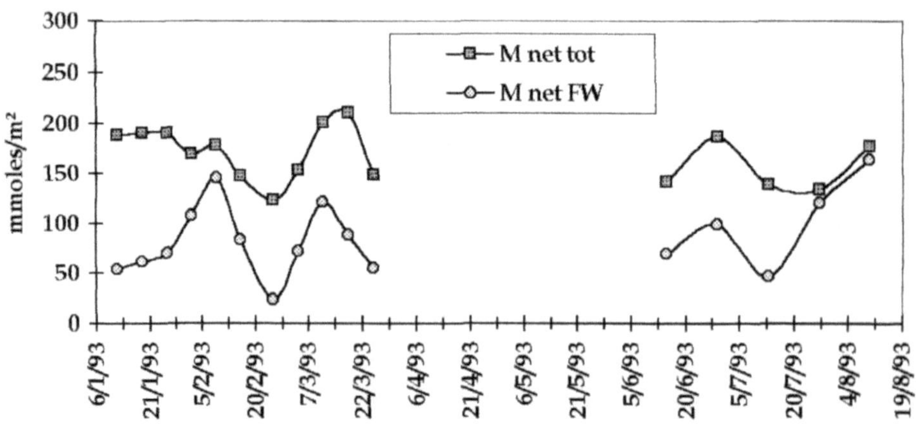

Figure 7. (a) Diurnal net organic carbon production in plexiglass tubes with and without inclusion of sediment, respectively $P_{\text{net tot}}$ and $P_{\text{net FW}}$ from January 93 to August 93. (b) Nocturnal net organic carbon production in plexiglass tubes with and without inclusion of sediment, respectively $M_{\text{net tot}}$ and $M_{\text{net FW}}$ from January 93 to August 93.

sity of plant communities involved, and the scarcity of studies on lagoon metabolism, a common pattern is still difficult to attribute to these habitats. The Smith & Mackensie proposal (1987), which states that, with some exceptions, shallow water systems tend toward net heterotrophy must be reconsidered. Seasonal shift between autotrophy in summer and heterotrophy in winter, proposed by Carmouze et al. (1991) and considered as a common feature of phytoplankton based lagoons by Knoppers (1994), cannot be generalized. Autotrophic and heterotrophic alternances seem to be strongly linked to growing and degenerescence phases of phytoplankton, which do not always coincide with hot and cold seasons.

The daily benthic C-metabolism (defined here as the sum of the benthic C-metabolism and the pelagic C-metabolism fraction induced by benthic nutrient recycling), almost always showed a net mineralization. Its relative importance, compared to that of the pelagic C-metabolism, partly controlled the auto-heterotrophism of the system. The autotrophy of the system decreased when the benthic loop increased at the expense of the pelagic loop. That was the case in summer 93, with the increase of benthic mineralization rate due to accumulation of dead macro-algae and phytoplankton in the sediment and with the decrease of primary production because of low abondance of phytoplankton. It was also true in late winter–early spring 92, when macro-algae partly substituted phytoplankton. On the other hand, in summer 91, when the phytoplankton-based cycle predominated, the system became more autotrophic.

References

Azevedo, S. & J. P. Carmouze, 1994. Une mortalité de poissons dans une lagune tropicale (Brésil) durant une période de dominance de Cyanophycées. Coïncidence ou conséquence? Rev. Hydrobiol. trop. 17: 265–272.

Alem, A. A. & A. A. Samaan, 1969. Productivity of Lake Mariut, Egypt. Part II. Primary production. Int. Rev. ges. Hydrobiol. 54: 491–527.

Balzer, W., F. Pollehene & H. Erlenkeuser, 1986. Cycling of organic carbon in a coastal marine system. In Sly, P. G. (ed.), Sediments and Water Interactions. Springer Verlag, Berlin: 325–330.

Bernardes, L. M. C., 1952. Tipos de Clima do Estado do Rio de Janeiro. Rev. Bras. Geog. 14: 57–80.

Carmouze, J. P. 1984. Généralisation d'une méthode de détermination du carbone minéral total par pH- métrie dans les eaux. Rev. Hydrobiol. trop. 17: 175–189.

Carmouze, J. P., B. Knoppers & P. Vasconcelos, 1991. The eutrophication of the lagoon of Saquarema. Biogeochemistry 14: 129–148.

Carmouze, J. P. & P. Vasconcelos, 1992. The eutrophication of the lagoon of Saquarema, Brazil. Science of the total Environment, Supp. 1992, Elsevier Science Publishers B.V., Amsterdam.

Carmouze, J. P., 1994. O metabolismo dos ecossistemas aquáticos. Edgar Blusher, São Paulo, 254 pp.

Carmouze, J. P. & P. Caumette, 1985. Les effets de la pollution organique sur les biomasses et activités du phytoplancton et des bactéries hétérotrophes dans la lagune Ebrié (Côte d'Ivoire). Rev. Hydrobiol. trop. 18: 175–246.

Day, J. W., Jr., C. J. Madden, F. Ley-Lou, R. L. Wetzel & A. M. Navarro, 1988. Aquatic primary productivity in Términos lagoon. In Yañez-Arancibia, A. & J. W. Day, Jr. (eds), Ecology of Coastal Ecosystems in the Southern Gulf of Mexico The Términos lagoon region Universidad Nacional Autónoma de México, México: 221–235.

Domingos, P., V. L. Huszar & J. P. Carmouze, 1994. Évolution de la communauté phytoplanctonique d'une lagune tropicale, au cours d'une période marquée par une mortalité de poissons. Rev. Hydrobiol. trop.

Florek, R. J., & G. T. Rowe, 1983. Oxygen consumption and dissolved inorganic nutrient production in marine coastal and shelf sediments of middle Atlantic Bight. Int. Rev. Ges. Hydrobiol. 68: 73–112.

Flores-Verdugo, F. J., J. W. Day Jr., L. Mee & R. Briseño-Dueñas, 1988. Phytoplankton production and seasonal biomass variation of Seagrasses, *Ruppia maritima* L., in a tropical Mexican lagoon with an ephemeral inlet. Estuaries 11: 51–56.

Garrels, R. M. & Thompson, 1962. A chemical model for sea water at 25 °C and one atmosphere total pressure. Amer. J. Sci. 260: 57–66.

Gran, G., 1952. Determination of equivalent point in potentiometric titration. Analyst 77: 661–671.

Harris, G. P. & B. B. Pucini, 1977. Photosynthesis by natural phytoplankton populations. Arch. Hydrobiol. 80: 405–457.

Kanwisher, J., 1960. P_{CO_2} in sea water and its effects on movement of CO_2 in nature. Tellus 12: 207–209.

Kjerfve, B., 1986. Comparative oceanography of coastal lagoons. In Wolfe D. A. (ed.), Estuarine Variability, New York, Academic Press: 63-81.

Knoppers, B., 1994. Aquatic Primary Production in Coastal Lagoons. Coastal Lagoon Processes edited by B. Kjerfve (Elsevier Oceanography Series, 60) Elsevier Science Publishers: 243–286.

Knoppers, B., B. Kjierfve & J. P. Carmouze, 1991. Trophic state and water turn-over time in six choked coastal lagoons in Brazil. Biogeochemistry 14: 149–166.

Lerman, A., 1979. Geochemical Processes. Wiley New York: 74–120.

Lorenzen, C. J. 1967. Determination of Chlorophyll and phaeopigments: spectrophotometric equations. Limnol. Oceanogr. 12: 343–346.

Machado, E. C. & B. Knoppers, 1988. Sediment oxigen consumption in an organic rich sub-tropical lagoon, Brazil. Sci. Tot. Envir. 75: 341–349.

Moreira, P. F. & B. Knoppers, 1990. Um ciclo anual da produção primária e nutrientes na Lagoa de Guarapina, R.J. Acta Limnol. Bras. 3.

Nixon, S. W., 1981. Remineralization and nutrient cycling in coastal marine ecosystems. In Neilson, B. & L. E. Cronin (eds), Estuaries and Nutrients Humana, Press Clifton, New Jersey: 111–138.

Nixon, S. W., C. A. Oviatt & S. S, Hale, 1976. Nitrogen regeneration and the metabolism of coastal marina bottom communities. In Anderson, J. M. & A. Macfaydan (eds), The Role of Terrestrial and Aquatic Organisms in Decomposition Processes. Blackwell Scientific Publications, Oxford: 269–283.

Odum, H. T. & C. M. Hoskin, 1958 Comparative studies on the metabolism of marine waters. Publ. Inst. mar. Sci., University of Texas 5: 16–46.

Pages, J., L. Lemasson & P. Dufour, 1981. Primary production measurements in a brackish tropical lagoon. Effect of light, as studied by the C-14 method. Rev. Hydrobiol. Trop. 14: 3–15.

Park, K., D. W. Hood & H. T. Odum, 1958. Diurnal pH variation in Texas bays, and its application to primary production estimation. Publ. Inst. mar. Sci., University of Texas 9: 404–453.

Pereira, C. E., 1991. – Estudo experimental da influência do sedimento sobre as atividades metabólicas na laguna de Saquarema. Tese de Mestrado. Universidade Federal Fluminense, Niterói, Brasil, 86 pp.

Qasim, S. Z., 1979. Primary production in some tropical environments. In Dunbar, M. J. (ed.), Marine Production Mechanisms. International Biological Program-IBP 20, Cambridge University Press, Cambridge: 31–69.

Randall, J. & J. W. Day, Jr., 1987. Effects of river discharge and vertical circulation on aquatic primary production in a turbid Louisiana (USA) estuary. Neth. J. Sea Res. 21: 231–242.

Reyes, M. & M. Merino, 1991. Diel dissolved oxygen dynamics and eutrophication in a shallow well-mixed tropical lagoon (Cancun, Mexiso). Estuaries 14: 372–381.

Sandoval-Rojo, L. C., F. J. Flores-Verdugo, U. Z. Araújo, J. W., Jr. Day, & A. E. Mercado, 1988. Phytoplankton productivity in Barra de Navidad coastal lagoon on the Pacific coast of México. Rev. Hydrobiol. Trop. 21: 101–108.

Smith, S. V., 1988. Mass balance in coral reef-dominated areas. In Jansson, B. O. (ed.), Coastal-Offshore Ecosystem Interactions. Lecture Notes on Coastal and Estuarine Studies 22, Springer Verlag, Berlin: 209–226.

Smith, S. V. & F. T. Mackenzie, 1987. The ocean as a net heterotrophic system: implications from the carbon biogeochemical cycle. Global Biogeochemical Cycles (American Geophysical Union) 1: 187–198.

Sournia, A., B. Delesalle & M. Ricard, 1981. Premiers bilans de production organique et de calcification d'un récif-barrière de la Polynésie Française. Oceanol. Acta 4: 423–431.

Strickland, J. D. H. & T. R. Parsons, 1972. A practical handbook of seawater analysis, 2nd edn. Bull. Fish. Res. BD Canada, 167, 319 pp.

Stumm, W. & J. Morgan, 1981. Aquatic Chemistry. An introduction emphazing chemical equilibria in natural waters. Wiley-Interscience Publication John Wiley & Sons. New York, 779 pp.

Tison, D. L. & D. H. Pope, 1980. Effects of temperature on mineralization by heterotrophic bacteria. Appl. Envir. Microbiol. 39: 584–587.

Tundisi, J., T. M. Tundisi & M. B. Kutner, 1973. Plankton studies in a mangrove environment. VII. Further investigations on primary production, standing stock of phyto- and zoo-plankton, and some environmental factors. Internationalen Revue der gesamten Hydrobiologie 58: 925–940.

Vaulot, D. & G.F. Frisoni, 1986. Phytoplanktonic productivity and nutrients in five Mediterranean lagoons. Oceanol. Acta 9: 57–63.

Weiler, R. R., 1975. Carbon dioxyde exchanges and productivity in lake Erie and lake Ontario. Verh. Int. Ver. Limnol. 19: 694–704.

Hydrobiologia **373/374:** 101–109, 1998.
J.-C. Amiard, B. Le Rouzic, B. Berthet & G. Bertru (eds),
Oceans, Rivers and Lakes: Energy and Substance Transfers at Interfaces.

Afforestation with *Eucalyptus globulus* and leaf litter decomposition in streams of northern Spain

J. Pozo[1], A. Basaguren[1], A. Elósegui[1], J. Molinero[1], E. Fabre[2] & E. Chauvet[2]
[1] *Lab. Ecología, Fac. Ciencias, Universidad del País Vasco/Euskal Herriko Unibertsitatea, Apdo. 644, 48080 Bilbao, Spain*
[2] *Centre d'Ecologie des Systèmes Aquatiques Continentaux (CNRS-UPS). 29, rue Jeanne Marvig, 31055 Toulouse Cedex 04, France*

Key words: Afforestation, eucalyptus, stream, litter breakdown, fungi, macroinvertebrates, nutrients

Abstract

To test the hypothesis that decomposition of leaf species in streams is influenced by afforestation with *Eucalyptus globulus*, we compared decay rates, nutrient levels, fungal biomass and macroinvertebrate assemblages on alder and eucalyptus leaf litter in three streams (two headwaters under different forests, and a mid reach) of the Agüera catchment (northern Spain).

Whatever the reach, alder always decomposed significantly faster than eucalyptus. Litter contents in nitrogen and phosphorus rose during breakdown at the mid reach, but not at the headwaters. No differences in fungal biomass were found between alder and eucalyptus leaves at the headwater reaches; however, at the mid reach, eucalyptus showed the highest values.

Alder litter, a high quality substratum, was readily colonized by shredders, and decayed rapidly at all sites. Eucalyptus, a low quality species, had lower nutrient contents and was less favoured by shredders. Under high nutrient levels (particularly phosphorus), however, it was readily colonized by fungi, thus shifting from medium to high breakdown rates. The potentially negative impact of afforestation with eucalyptus on streams can thus be reduced in situations of high concentrations of dissolved nutrients.

Introduction

Litter of terrestrial origin is the main trophic resource for the communities of many headwater streams (Petersen & Cummins, 1974; Vannote et al., 1980). Disturbance of the vegetation can affect the amount and quality of litter entering the stream (Golladay et al., 1989; Webster et al., 1990), and thus, may have substantial impacts on the functioning of stream systems.

There is an extensive literature on the effects of clearcutting on streams (Meyer & Jonhson, 1983; Griffith & Perry, 1991; Campbell et al., 1992; Tuchman & King, 1993; David et al, 1994), but the large-scale substitution of tree species can also have important effects (Griffith & Perry, 1991; Pozo et al., 1997), although these have not been thoroughly studied.

The large-scale plantation of eucalyptus outside its natural range has created some concern among limnologists (Graça, 1993; Bärlocher et al., 1995). Eucalyptus litter has frequently been quoted as of low quality and slow breakdown rate (Bunn, 1988a; Boulton, 1991), characteristics which are linked to its high content of phenolics and tannins, and to its waxy cuticles (Bunn, 1988a; Campbell et al., 1992; Bärlocher et al., 1995). However, the genus has hundreds of species growing in a wide variety of climatic and edaphic conditions, and showing broad differences in litter texture and composition (Mitchell, 1988).

Eucalyptus globulus Labill is one of the most widespread planted eucalyptus in the North of Spain (ICONA, 1980). Recent work by Abelho & Graça

(1996) in Portugal suggests that afforestation with this species can affect the breakdown rates of different leaf species (e.g., eucalyptus and chestnut). Some attempts to elucidate its impact on streams have addressed litter decomposition in the Agüera catchment (northern Spain). The first of these studies suggested that breakdown of low quality leaves such as eucalyptus was mainly influenced by concentrations of dissolved nitrogen and phosphorus (Pozo, 1993). In the headwaters, macroinvertebrates preferred alder over eucalyptus litter, suggesting that shredding activity could be responsible for the difference in decay rate between both species in nutrient-poor waters (Basaguren & Pozo, 1994). Later results (Molinero et al., 1996) confirmed that the loss rate of eucalyptus mass depended on the external nutrient supply (dissolved nutrients in stream water), and not on the type of riparian vegetation (deciduous forest *versus* eucalyptus plantation). Nutrient enrichment of materials during decomposition suggested that microbial activity is site-dependent and related to the nutrient concentration in water. Chauvet et al. (1997) showed that, whatever the stream, neither total conidial production nor number of hyphomycete species were different between eucalyptus and alder litter. Furthermore, at a site richer in phosphate, conidial production increased to the same extent in both types of leaves.

The present work addresses the breakdown of alder and eucalyptus litter in streams in relation to the riparian environment, nutrient availability, leaf quality, and colonization by fungi and macroinvertebrates.

Study sites

The Agüera is a third order stream in northern Spain, draining a catchment of 144 km^2. The uplands are covered by indigenous deciduous forests in the headwaters of the main channel, and by pine and eucalyptus plantations in the others; the lowlands are occupied by meadows and crops. The climate is temperate oceanic, with rainfall around 1500 mm yr^{-1}, and mean monthly temperatures ranging from 8 °C to 21 °C.

This work was carried out in three stream reaches: site A is located in the Salderrey stream, a first-order tributary surrounded by mature deciduous forest with oak, chestnut, alder and hazel; site C is in an unnamed first-order stream flowing through a plantation of *Eucalyptus globulus*; site B is in a third-order reach of the Agüera stream, surrounded by eucalyptus plantations but with riparian fringes of alder, sycamore and oak.

The water characteristics are similar in the two first-order reaches: low conductivity and phosphorus content (Table 1). Site B drains a more complex catchment, which includes some villages. As a result, the mineralization and the dissolved phosphate are higher, although the water quality is good (Elósegui et al., 1995).

Methods

On 15 November 1994, we filled a series of 5mm-mesh bags (4 replicates × 2 species × 3 sites × 6 dates) with 15 g of freshly fallen leaves of either *Alnus glutinosa* or *Eucalyptus globulus*. They were stored overnight in a refrigerator, and placed randomly the next day, attached to boulders in the study sites. In total, 144 bags were filled, 72 of each leaf type. After 1, 3, 8, 14, 20 and 26 weeks of immersion, four bags of each leaf species at each site were returned to the laboratory. Due to rapid breakdown, no alder material remained at site A after 14 weeks or at site C after 20 weeks. At site B, loss of bags forced termination of the experiment at week 8 (alder) or 14 (eucalyptus).

After collection, the contents of each bag were rinsed with distilled water on a 200 μm sieve, the macroinvertebrates were collected and preserved in 70% ethanol, and the leaves enclosed in plastic bags and frozen to –20 °C.

Thereafter, the leaf material was lyophilized, weighed, and sorted in subsamples for the determination of ash (combustion), nitrogen (Kjeltec Auto 1030), phosphorus (Allen et al., 1974) and ergosterol (Gessner et al., 1991). The macroinvertebrates were sorted under a binocular microscope, identified mostly to the species or genus level, and assigned to functional feeding groups (e.g., Cummins & Klug, 1979; Merritt & Cummins, 1996).

Leaf mass loss was expressed as percentage of initial AFDW remaining. We calculated breakdown rates according to the negative exponential model (Petersen & Cummins, 1974), by linear regression of ln-transformed data. Analyses of covariance, with time as covariant, were used to compare slopes, and significant differences were determined using Tukey's test (Zar, 1984). Two way ANOVAs (time *versus* species and time *versus* site) and Tukey's tests were used to test differences in nutrient and ergosterol contents and macroinvertebrate densities.

Table 1. Mean and range for selected physico-chemical characteristics of the water at the studied sites. (TIN = Total Inorganic Nitrogen)

	Site A	Site B	Site C
Temperature (°C)	8.8 (5.7–12.4)	10.9 (9.0–14.8)	10.5 (7.7–13.7)
Conductivity (μS cm^{-1})	99 (84–120)	206 (104–260)	64 (56–72)
pH	7.2 (7.0–7.8)	8.2 (7.8–8.8)	6.6 (6.3–7.2)
TIN (μg N l^{-1})	806 (473–1493)	642 (373–1132)	340 (127–824)
Orthophosphate (μg P l^{-1})	3.8 (2.7–7.5)	10.3 (3.0–25.2)	3.2 (2.1–5.0)

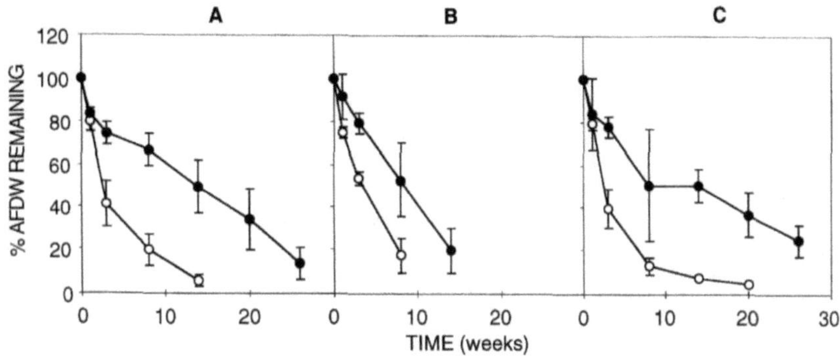

Figure 1. Breakdown dynamics (mean ± S.D.) of alder (○) and eucalyptus (●) litter at the three studied sites.

Results

Mass loss

There were large differences in breakdown dynamics between leaf types (Figure 1). Mass loss of alder during the first week was very similar in all sites, and was slightly greater than that of eucalyptus. After this period, differences between species became evident: alder decayed much more rapidly than eucalyptus, with less than 10% of the initial mass remaining after 14 weeks.

Breakdown was most rapid at site B and slowest at site C, although differences were significant only for eucalyptus (Table 2). Whatever the reach, alder always decayed at a significantly greater rate than eucalyptus (k from 0.0235 to 0.0321 d^{-1} for alder, from 0.0066 to 0.0169 d^{-1} for eucalyptus).

Nutrient content

There were significant differences in the initial composition of litter: nitrogen contents of 2.84% and 0.92%, and phosphorus contents of 0.080% and 0.035% were found in alder and eucalyptus respectively.

Figure 2 shows the changes in nitrogen and phosphorus contents of leaf litter over time. At site B, both species underwent nitrogen enrichment, but concentration was higher in alder, which resulted in increased differences between species. In headwater reaches, the N content in eucalyptus showed very small changes while in alder it increased sharply at site A during the first three weeks, and then fell to levels near the initial values; at site C, the N content in alder increased during the first week, then decreased, and after the eighth week recovered its initial value.

The difference in nitrogen content between species was significant in all sites (Table 3), the values always being higher in alder. For this species, there were significant differences between sites: nitrogen content was lowest at site C, medium at site A and highest at site B. In the case of eucalyptus, site B also showed significantly higher concentrations than the headwaters, but there was no difference between sites A and C.

The dynamics of phosphorus in the decaying litter differed greatly between headwater and downstream reaches (Figure 2). At sites A and C, litter showed a general downward trend, which was especially clear in alder. In contrast, at site B both leaf species exhibited an increase of phosphorus content, and eucalyptus litter reached values close to initial concentrations of alder leaves. ANOVAs stressed the difference between

Table 2. Breakdown rates (d^{-1}) and 95% confidence intervals of alder and eucalyptus in the studied reaches. Summary of multiple inter-site comparisons (Tukey's test) after analysis of covariance is also shown, underlining the cases where no significant differences were found

Site	Alder	Eucalyptus	Ancova
A	–0.0288 (±0.0050)	–0.0097 (±0.0023)	Euc. < Ald.
B	–0.0321 (±0.0094)	–0.0169 (±0.0053)	Euc. < Ald.
C	–0.0235 (±0.0055)	–0.0066 (±0.0023)	Euc. < Ald.
Inter-site comparisons	C A B	*CAB*	

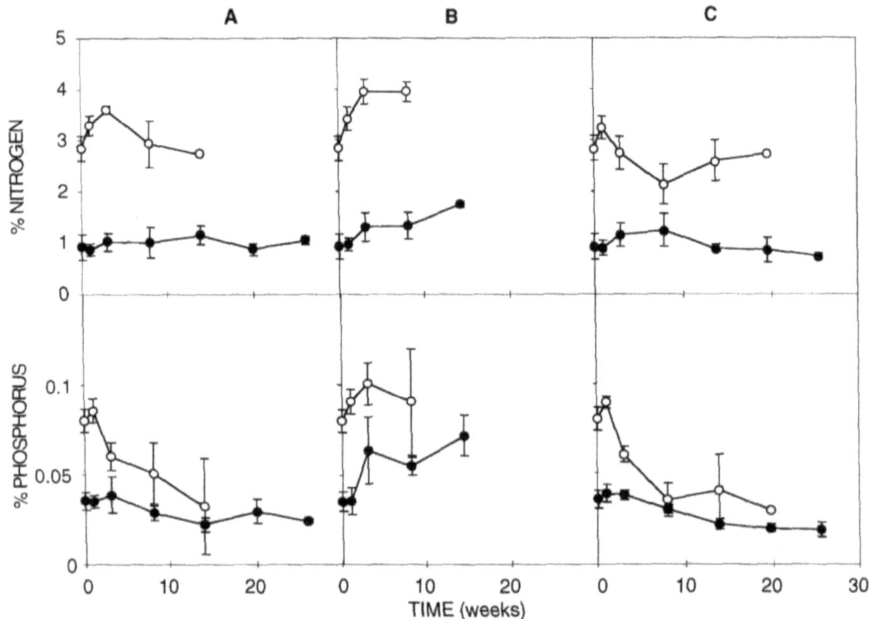

Figure 2. Nitrogen and phosphorus concentration (mean ± S.D.) in alder (○) and eucalyptus (●) litter.

species at all sites (Table 3), and the higher phosphorus content of litter downstream.

As a result of the variations of nitrogen and phosphorus content, the N:P ratio of litter, especially that of alder, increased in the headwaters and suffered small variations at site B (Figure 3). N:P ratio of alder was significantly higher than that of eucalyptus at all sites. Inter-site differences were only significant in the case of eucalyptus (Table 3).

Fungal biomass

Figure 4 shows the ergosterol content of decaying leaves. In headwaters, the general pattern of both species was very similar: although slightly different between site A and C, peaks of ergosterol occurred

Table 3. Results of Tukey's tests after two way ANOVAs (time × site and time × species) for nitrogen and phosphorus contents and N:P ratio. The cases where no significant differences were found appear underlined

Site	%N	%P	N:P
A	Euc. < Ald.	Euc. < Ald.	Euc. < Ald.
B	Euc. < Ald.	Euc. < Ald.	Euc. < Ald.
C	Euc. < Ald.	Euc. < Ald.	Euc. < Ald.
Species			
Alder	C < A < B	C A < B	B C A
Eucalyptus	A C < B	A C < B	B < C A

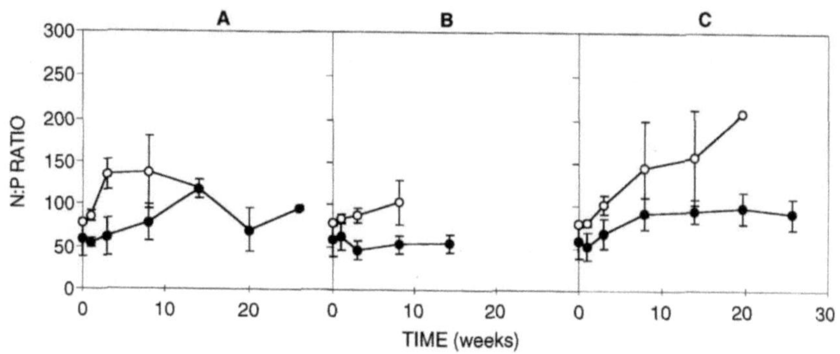

Figure 3. N/P atomic ratio (mean ± S.D.) of alder (○) and eucalyptus (●) litter.

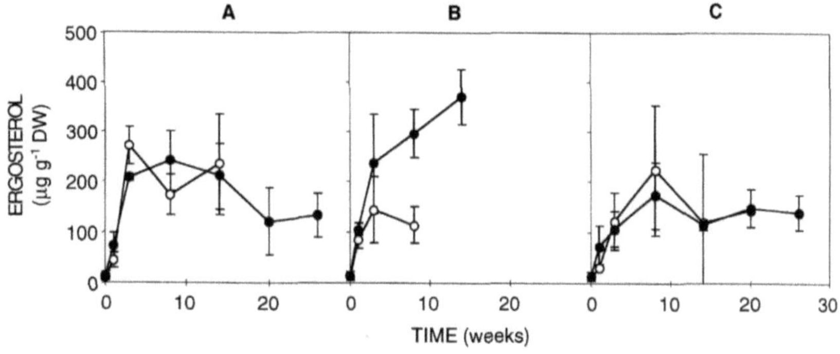

Figure 4. Ergosterol contents (mean ± S.D.) of alder (○) and eucalyptus (●) litter.

Table 4. Results of Tukey's tests after two way ANOVAs (time × site and time × species) for ergosterol contents. The cases where no significant differences were found appear underlined

Site	Ergosterol
A	Ald. Euc.
B	Ald. < Euc.
C	Ald. Euc.
Species	
Alder	B C A
Eucalyptus	A C < B

Table 5. Results of Tukey's tests after two way ANOVAs (time × site and time × species) for macroinvertebrate and shredder densities (ind g AFDW −1) in the litter bags. The cases where no significant differences were found appear underlined

Site	Invertebrates	Shredders
A	Euc. < Ald.	Euc. < Ald.
B	Euc. < Ald.	Euc. < Ald.
C	Euc. < Ald.	Euc. < Ald.
Species		
Alder	A B C	B < A C
Eucalyptus	A C < B	B < C A

during the first 8 weeks, then the levels decreased slowly, from 200–250 to around 150 μg g^{-1}. At site B an alder peak was reached within 3 weeks, and mean values were always lower than 150 μg g^{-1}. Concentrations in eucalyptus, on the other hand, continued to rise during the whole study period, reaching a maximum ergosterol concentration of 360 μg g^{-1} after 14 weeks. ANOVAs (Table 4) showed significant differences between species only at site B, where eucalyptus was significantly richer in ergosterol than any of the other site × leaf combinations.

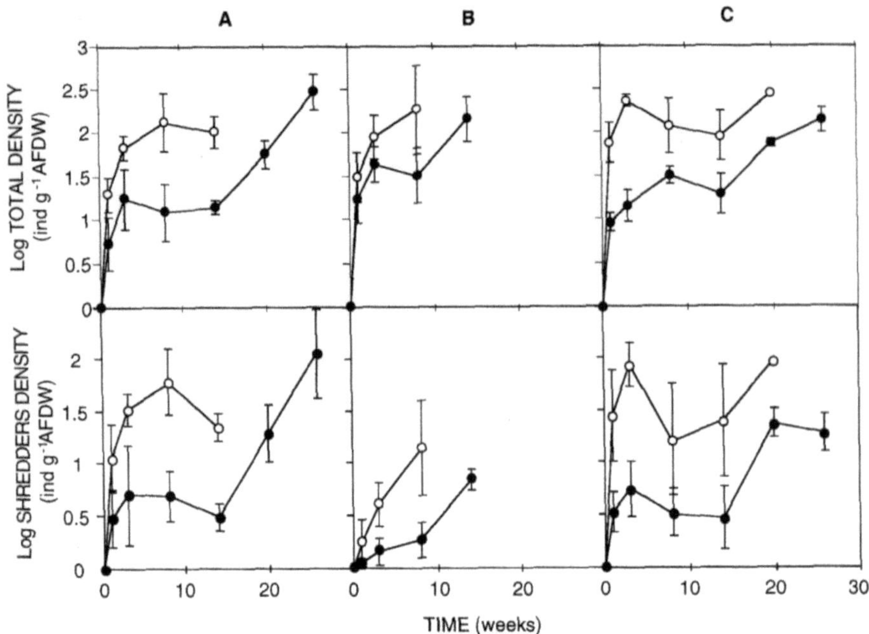

Figure 5. Log. of the number of macroinvertebrates and shredders g^{-1} AFDM (mean ± S.D.) in alder (○) and eucalyptus (●) litter bags.

Macroinvertebrates

The density of macroinvertebrates found in alder leaf bags rose during the first 3 or 8 weeks of colonization (depending on site), and then leveled off (Figure 5). The colonization of eucalyptus bags was slower, and the density of macroinvertebrates in this substrate continued to rise until the end of the experiment at all sites. The fauna associated with litter bags in headwaters was dominated by chironomids and shredders (e.g., *Protonemura* sp., *Echinogammarus tarraconensis*, *Nemoura* sp., *Capnioneura* sp. and *Leuctra* sp., at site A; *Capnioneura* sp., *Leuctra* sp. and *Protonemura* sp., at site C). In the mid reach, scrapers and collectors were dominant (e.g., *Potamopyrgus jenkinsi*, *Ephemerella ignita*, *Baetis* spp. and Simuliidae), shredders being less important (e.g., *Echinogammarus berilloni* and *Lepidostoma hirtum*). All the mentioned taxa were found in bags of both leaf species, although differences occurred in terms of density. The shredders followed dynamics quite similar to those of total macroinvertebrates, although at site B shredder density was much lower (Table 5) and the increase slower than at the headwaters (Figure 5). In all sites, the density of both macroinvertebrates and shredders on eucalyptus was significantly lower than on alder (Table 5). Although the density of macroinvertebrates on eucalyptus was significantly greater at site B than at the headwaters, shredder density was significantly lower.

Discussion

Litter breakdown is a complex process, which involves leaching, physical abrasion, microbial conditioning and processing by macroinvertebrates (Cummins et al., 1973; Cummins & Klug, 1979; Webster & Benfield 1986). The relative importance of these factors can change, resulting in highly variable species- and site-specific dynamics (Tuchman & King, 1993).

The present study has not shown an effect of the type of riparian vegetation (deciduous forest *versus* eucalyptus plantation) on the breakdown rates of alder or eucalyptus. In contrast, Abelho & Graça (1996) reported that, in Portugal, litter processing (eucalyptus included) was slower in streams under eucalyptus plantations than in those flowing through deciduous forests. However, these authors used fine mesh bags in their experiments compared to large mesh in the present study, which limits the comparison. Our results support previous observations from the Agüera stream (Pozo, 1993; Molinero et al., 1996), which suggested that, in this catchment, factors other than riparian vegetation influence the decomposition of several leaf species.

Pozo (1993) reported that alder decomposed more rapidly than eucalyptus in oligotrophic headwaters, while at mid and low reaches dissolved nutrient availability allowed enhancement of conditioning and palatability of eucalyptus, thus promoting decay rates as fast as those of alder. In the present study, although the quality of eucalyptus litter at site B was enhanced, alder always decayed faster. The breakdown rate for a given species varies from year to year within the same reach (Pozo et al., 1994). In addition, the present experiment was carried out with freshly fallen leaves, instead of the dry material used in previous studies. The alder leaves used here were richer in nutrients (2.8% N and 0.08% P) than the material (1.7% N and 0.04% P) used by Pozo (1993), while eucalyptus leaves showed only minor differences between both studies. Thus, the higher quality of alder in this study may have promoted a faster breakdown, and produced significant differences with respect to eucalyptus breakdown rates even at site B.

While nitrogen concentration often increases as litter decays (Suberkropp et al., 1976; Tate & Gurtz, 1986; Chauvet, 1987; Bunn, 1988a; Gessner, 1991), the pattern of change in phosphorus is less regular. Despite differences in the initial condition of leaves (fresh or dried), the nutrient dynamics of the present study are similar to those of Pozo (1993). The most interesting point is that nutrient enrichment of leaves is site-dependent. An increase of both N and P concentrations in leaves is clear only at site B, where dissolved phosphorus concentrations were always highest. Uptake of N and P at this site was observed by Molinero et al. (1996), while release of both elements predominated in nutrient-poor water reaches, such as sites A and C. The enhancement of microbial activity on leaf material by nutrients has also been shown in other studies (Jenkins & Suberkropp, 1995; Newell et al., 1995; Suberkropp & Chauvet, 1995). However, this enrichment of litter does not have the same effect on the decay rate of all species: only eucalyptus, the species of lowest quality, shows enhanced decomposition at the nutrient-rich site (Molinero et al., 1996).

The increase in N and P concentrations of alder leaves at site B changes the litter quality, but, given its rapid mass loss, it is difficult to prove an uptake of dissolved nutrients, because we did not find increases of absolute amounts in any of both elements. However, in eucalyptus litter, the absolute amounts increased by the third week in 44% and 93%, respectively. Molinero et al. (1996) have also pointed out that the relative immobilization of P at this site was higher than that of nitrogen. This fact was not seen neither at site A nor at C, where dissolved phosphorus was clearly lower (see Table 1). Phosphorus appears to be the limiting nutrient in stream water since inorganic N:P ratios were high at all sites (N:P > 100). The higher P concentrations at site B coinciding with higher P immobilization in leaves (Figure 2) which resulted in lower N:P ratios (Figure 3) are consistent with this hypothesis.

The dynamics of ergosterol on decaying leaves demonstrates the importance of fungal activity (e.g., Newell, 1992; Gessner & Chauvet, 1994; Maharning & Bärlocher, 1996). With the exception of site B, where ergosterol contents were higher in eucalyptus than in alder, fungal biomass showed similar dynamics in both species within a given site. Bärlocher et al. (1995) found that conidial production of fungi on *E. globulus* lagged 2 weeks behind alder, because of the slow leaching of phenols. The same lag was found in conidial production in our studies (Chauvet et al., 1997) but was not found for ergosterol. Peaks of ergosterol clearly occurred after those of conidial production, as has been shown for different leaf species (Maharning & Bärlocher, 1996). Although eucalyptus litter has usually been considered as a poor quality substrate for aquatic fungi, it appears to be comparable with alder, itself one of the most readily colonized substrates in European streams (Bärlocher, 1992). The sporulation rates and specific diversity of hyphomycetes measured in our samples were similar for alder and eucalyptus (Chauvet et al., 1997), which confirms the suitability of eucalyptus as a substrate for these fungi.

The identical ranking of sites by nutrient or ergosterol contents suggests that fungi are responsible for the enrichment of eucalyptus leaves. Furthermore, the correlation between ergosterol and nitrogen or phosphorus contents at site B is significant for eucalyptus (data not shown). Thus, increased loss rate, nutrient enrichment and high concentrations of ergosterol are characteristics that differentiate eucalyptus litter breakdown between the mid reach and the headwaters.

What role do macroinvertebrates play? Our results show that alder is colonized by significantly higher densities than eucalyptus, whatever the stream. The fast breakdown of alder has often been associated with its high nutrient contents, which would make it a prime substrate for shredders (Hassage & Harrel, 1986), even without prior microbial colonization (Gessner & Schwoerbel, 1989; Cortes et al., 1995). Our results point also in this direction, since nutrient contents, total macroinvertebrate and shredder densities were usually higher in alder than in eucalyptus. Only at the end of the experiment did eucalyptus reach faunal densities similar to those previously registered by alder. This suggests that eucalyptus leaves need a conditioning period before being used by macroinvertebrates, which agrees with other reported data (e.g., Bunn, 1988b; González et al., 1998).

There is some controversy on the relative importance of microbial *versus* macroinvertebrate activity on litter decay. For instance, Pearson et al. (1989) reported that the importance of invertebrates in the breakdown of different litter species is highly variable: in some cases 77%, but in others only 6% of the mass loss can be explained by invertebrates. Our results, and those from a previous experiment (Basaguren & Pozo, 1994), suggest that microbial activity is responsible for the rapid decay of eucalyptus in mid reaches, where shredders have lesser importance, and water chemistry makes the nutrient enrichment of litter easier. On the other hand, selection of macroinvertebrates against eucalyptus, due to its low nutrient enrichment, could explain the slow breakdown of this species in the headwaters, where microbial activity seems to be more limited, according to ergosterol levels or conidial production (Chauvet et al., 1997).

Thus, *Alnus glutinosa* is a species with high nitrogen and phosphorus contents, readily colonized by shredders, that breaks down rapidly in all sites. *Eucalyptus globulus*, on the other hand, has lower nitrogen and phosphorus contents, is less preferred by shredders, but under high nutrient levels (particularly phosphate) can be readily colonized by fungi, thus shifting from medium to rapid breakdown rates. Although afforestation with *E. globulus* can produce severe changes in the timing, amount and quality of litter inputs into streams (Pozo et al., 1997), in northern Spain it has no influence on the breakdown rates of leaf species. Moreover, the factors discussed above lead us to conclude that even material as low in quality as eucalyptus leaves can be readily used when an exogenous nutrient supply facilitates microbial activity, thus reducing the potential negative impact of such afforestation practices.

Acknowledgements

This work has been supported by the research project DGICYT PB 92-0459 (General Bureau of Scientific and Technic Research, Central Government, Madrid). Thanks to Professor David Harper (Leicester University) for improving the English style.

References

Abelho, M. & M. A. S. Graça, 1996. Effects of eucalyptus afforestation on leaf litter dynamics and macroinvertebrate community structure of streams in Central Portugal. Hydrobiologia 324: 195–204.

Allen, S. E., H. M. Grimshaw, J. A. Parkinson & C. Quarnby, 1974. Chemical analysis of ecological materials. Blackwell, Oxford.

Bärlocher, F., 1992. Research on aquatic hyphomycetes: historical background and overview. In F. Bärlocher (ed.), The Ecology of Aquatic Hyphomycetes. Springer Verlag, Berlin: 1–15.

Bärlocher, F., C. Canhoto & M. A. S. Graça, 1995. Fungal colonization of alder and eucalypt leaves in two streams in central Portugal. Arch. Hydrobiol. 133: 457–470.

Basaguren, A. & J. Pozo, 1994. Leaf litter processing of alder and eucalyptus in the Agüera stream system (Northern Spain). II. Macroinvertebrates associated. Arch. Hydrobiol. 132: 57–68.

Boulton, A. J., 1991. Eucalypt leaf decomposition in an intermittent stream in south-eastern Australia. Hydrobiologia 211: 123–136.

Bunn, S. E., 1988a. Processing of leaf litter in a northern jarrah forest stream, western Australia: I. Seasonal differences. Hydrobiologia 162: 201–210.

Bunn, S. E., 1988b. Processing of leaf litter in a northern jarrah forest stream, western Australia: II. The role of macroinvertebrates and the influence of soluble polyphenols and inorganic sediment. Hydrobiologia 162: 211–223.

Campbell, I. C., K. R. James, B. T. Hart & A. Devereaux, 1992. Allochtonous coarse particulate organic material in forest and pasture reaches of two south-eastern Australian streams. II. Litter processing. Freshwat. Biol. 27: 353–365.

Cortes, R. M., M. A. S. Graça, J. N. Vingada & S. Varandas de Oliveira, 1995. Stream typology and dynamics of leaf processing. Ann. Limnol. 31: 119–131.

Cummins, K. W., R. C. Petersen, F. O. Howard, J. C. Wuycheck & V. I. Holt, 1973. The utilization of leaf litter by stream detritivores. Ecology 54: 336–345.

Cummins, K. W. & M. J. Klug, 1979. Feeding ecology of stream invertebrates. Ann. Rev. Ecol. Syst. 10: 147–172.

Chauvet, E., 1987. Changes in the chemical composition of alder, poplar and willow leaves during decomposition in a river. Hydrobiologia 148: 35–44.

Chauvet, E., E. Fabre, A. Elósegui & J. Pozo, 1997. The impact of eucalypt on the leaf-associated aquatic hyphomycetes in Spanish streams. Can. J. Bot., 75: 880–887.

David, J. S., M. O. Henriques, T. S. David, J. Tomé & D. C. Ledger, 1994. Clearcutting effects on streamflow in coppiced *Eucalyptus globulus* stands in Portugal. J. Hydrol. 162: 143–154.

Elósegui, A., X. Arana, A. Basaguren & J. Pozo, 1995. Self-purification processes in a medium-sized stream. Envir. Mgmt 19: 931–939.

Gessner, M. O., 1991. Differences in processing dynamics of fresh and dried leaf litter in a stream ecosystem. Freshwat. Biol. 26: 387–398.

Gessner, M. O. & E. Chauvet, 1994. Importance of stream microfungi in controlling breakdown rates of leaf litter. Ecology 75: 1807–1817.

Gessner, M. O. & J. Schwoerbel, 1989. Leaching kinetics of leaf litter with implications for the current concept of leaf processing in streams. Arch. Hydrobiol. 115: 81–90.

Gessner, M. O., M. A. Bauchrowitz & M. Escautier, 1991. Extraction and quantification of ergosterol as a measure of fungal biomass in leaf litter. Microb. Ecol. 22: 285–291.

Golladay, S. W., J. R. Webster & E. F. Benfield, 1989. Changes in stream benthic organic matter following watershed disturbance. Holarctic Ecol. 12: 85–96.

González, J. M., A. Basaguren & J. Pozo, 1998. Mechanisms involved in leaf litter processing: influence of bag mesh size. In G. Bretschko & J. Helešic (eds), Advances in River Bottom Ecology. Backhuys Publishers, Leiden: 89–97.

Graça, M. A. S., 1993. Patterns and processes in detritus-based stream systems. Limnologica 23: 107–114.

Griffith, M. B. & S. A. Perry, 1991. Leaf pack processing in two Appalachian mountain streams draining catchments with different management histories. Hydrobiologia 220: 247–254.

Hassage, R. L. & R. C. Harrel, 1986. Allochthonous leaves as a substrate for macrobenthos. J. Freshwat. Ecol. 3: 453–465.

ICONA, 1980. Las frondosas en el primer inventario forestal nacional. Ministerio de Agricultura. Subdirección General de Protección de la Naturaleza. Madrid.

Jenkins, C. C. & K. Suberkropp, 1995. The influence of water chemistry on the enzymatic degradation of leaves in streams. Freshwat. Biol. 33: 245–253.

Maharning, A. R. & F. Bärlocher, 1996. Growth and reproduction in aquatic hyphomycetes. Mycologia 88: 80–88.

Merrit, R. W. & K. W. Cummins, 1996. An introduction to the aquatic insects of North America. 3rd. edn., Kendall/Hunt. Dubuque, Iowa.

Meyer, J. L. & C. Jonhson, 1983. The influence of elevated nitrate concentration on rate of leaf decomposition in a stream. Freshwat. Biol. 13: 177–183.

Mitchell, A., 1988. A field guide to the trees of Britain and Europe. 2nd. ed., Collins & Sons Ltd.

Molinero, J., J. Pozo & E. González, 1996. Litter breakdown in streams of the Agüera catchment: influence of dissolved nutrients and land use. Freshwat. Biol. 36: 745–756.

Newell, S. Y., 1992. Estimating fungal biomass and productivity in decomposing litter. In G. C. Carroll & D. T. Wicklow (eds), The Fungal Community. Marcel Dekker, New York.

Newell, S. Y., M. A. Moran, R. Wicks & E. Hodson, 1995. Productivities of microbial decomposers during early stages of decomposition of leaves of a freshwater sedge. Freshwat. Biol. 34: 135–148.

Pearson, R. G., R. K. Tobin, R. E. W. Smith & L. J. Benson, 1989. Standing crop and processing of rainforest litter in a tropical Australian stream. Arch. Hydrobiol. 115: 481–498.

Petersen, R. C. & K. W. Cummins, 1974. Leaf processing in a woodland stream. Freshwat. Biol. 4: 343–368.

Pozo, J., 1993. Leaf litter processing of alder and eucalyptus in the Agüera stream system (North Spain). I. Chemical changes. Arch. Hydrobiol. 127: 299–317.

Pozo, J., A. Elósegui & A. Basaguren, 1994. Aproximación sistémica al análisis de la cuenca del río Agüera. Limnetica 10: 83–91.

Pozo, J., E. González, J. R. Díez, J. Molinero & A. Elósegui, 1997. Inputs of particulate organic matter to streams with different riparian vegetation. J. N. am. benthol. Soc. 16: 602–611.

Suberkropp, K. & E. Chauvet, 1995. Regulation of leaf breakdown by fungi in streams: Inluence of water chemistry. Ecology 76: 1433–1445.

Suberkropp, K., G. L. Godshalk & M. J. Klug, 1976. Changes in the chemical composition of leaves during processing in a woodland stream. Ecology 57: 720–727.

Tate, C. M. & M. E. Gurtz, 1986. Comparisons of mass loss, nutrients, and invertebrates associated with elm leaf litter decomposition in perennial and intermittent reaches of tallgrass prairie streams. Southw. Nat. 31: 511–520.

Tuchman, N. C. & R. H. King, 1993. Changes in mechanisms of summer detritus processing between wooded and agricultural sites in a Michigan headwater stream. Hydrobiologia 268: 115–127.

Vannote, R. L., G. W. Minshall, K. W. Cummins, J. R. Sedell & C. E. Cushing, 1980. The river continuum concept. Can. J. Fish. aquat. Sci. 37: 130–137.

Webster, J. R. & E. F. Benfield, 1986. Vascular plant breakdown in freshwater ecosystems. Ann. Rev. Ecol. Syst. 17: 567–594.

Webster, J. R., S. W. Golladay, E. F. Benfield, D. J. D'Angelo & G. T. Peters, 1990. Effects of forest disturbance on particulate organic matter budgets of small streams. J. N. am. benthol. Soc. 9: 120–140.

Zar, J. H., 1984. Biostatistical Analysis Prentice-Hall, Englewood Cliffs, New Jersey.

Hydrobiologia **373/374:** 111–119, 1998.
J. C. Amiard, B. Le Rouzic, B. Berthet & G. Bertru (eds),
Oceans, Rivers and Lakes: Energy and Substance Transfers at Interfaces.

Short-term effect of nitrogen enrichment on the microbial communities of a peatland

D. Gilbert[1], C. Amblard[1], G. Bourdier[1] & A.-J. Francez[2]
[1] *Laboratoire de Biologie Comparée des Protistes, UPRES-A CNRS 6023, Université Blaise Pascal Clermont-Ferrand II, 63177 Aubière cedex, France*
[2] *Service d'Ecologie végétale, UMR-CNRS 65-53 'Ecobio', Université de Rennes I, Complexe scientifique de Beaulieu, 35042 Rennes cedex, France*

Key words: peatland, nitrogen, microbial communities, microbial loop

Abstract

The aim of this study was to assess the structure of the microbial loop in a *Sphagnum fallax – Carex rostrata* fen of the French Massif central, and the impact of nitrogen supplies on the different microbial communities.

Microalgae (46% of the total microbial biomass), Protozoa (26%) and heterotrophic Bacteria (17%) were the dominant microorganisms. Rotifera (5%), Cyanobacteria (3%), Fungi (2%) and Nematoda (< 1%) were also present. Testate Amoebae were well represented in *Sphagnum* peatland (14% of total microbial biomass). Thus, the structure of the protist community in the surface of fen peatlands is notably different from that recorded in other environments.

The input of nitrogen led to a steady increase in microbial biomasses, but only changed the structure of microbial communities significantly when the input was 50 kg ha^{-1}. At high inputs, nitrogen supply increased the relative importance of Cyanobacteria, Euglenophyceae, Bacillariophyceae and Ciliates, and decreased the relative importance of heterotrophic Bacteria, other microalgae and testate Amoebae. The increase in the values of photosynthetic assimilation and heterotrophic activities could also reflect changes in community functioning.

Introduction

Energy and material flows in aquatic ecosystems transit, in part, through the microbial food web (Dissolved Organic Matter → Bacteria → heterotrophic Protists (Ciliates, Flagellates, Amoebae) → Macrozoorganisms) (Azam et al., 1983). The relative importance of this pathway, a microbial loop, has been demonstrated in pelagic marine ecosystems (Porter et al., 1985; Gifford, 1991) and in lakes (Riemann & Christoffersen, 1993; Amblard et al., 1994), could be an early indicator of human disturbance on aquatic biological communities (Munawar & Weisse, 1989). Furthermore, it appears that Protozoa are less sensitive to the acidity of the medium than Metazoa (Simek & Straskrabova, 1992), which could thus have a dominant role in the trophic pathways in peatlands. For example, recent studies have shown that inputs of agricultural fertilizers to nutrient-poor, acid peatlands, can lead to considerable changes in the structure of the microbial loop (Gilbert et al., 1998).

Furthermore, N-based acid deposition increased in recent years as a result of fossil fuel combustion and use of N-fertilizers (Van Breemen, 1984). This increased deposition may affect community structure and nutrient cycling of peatlands, which receive a large proportion of their nutrients directly from precipitation, because N is often a limiting nutrient for plant growth (Austin & Wieder, 1987; Yesmin et al., 1995).

We tested the hypothesis that the microbial community structure and C-cycling in peatland ecosystems are quickly affected by disturbances such as inputs of nitrogen. Microorganisms may respond more rapidly than other components of the peatland ecosystem, because of their high turnover rate. Accordingly, the abundance and biomass of heterotrophic Bacteria, Cyanobacteria, Microalgae, Protozoa, Fungi, Rotifera and Nematoda and their auto- and heterotrophic activities were measured in a peatland in the French Massif Central subject to regular inputs of nitrogen.

Methods

The experiments were carried out in the Pradeaux peatland (Puy-de-Dôme, France, 3° 55′E, 45° 32′N), a drained fen with an area of 10 hectares, aged 9000 yr (Francez, 1992) and situated at an altitude of 1350 m.

Three 1 × 1 m quadrats were isolated in a *Sphagnum fallax – Carex rostrata* community. Four PVC cores 25 cm in diameter and 30 cm long were inserted vertically into each parcel. One core acted as control (control core); the others received 5, 15 or 50 kg ha^{-1} of nitrogen (cores 5 N, 15 N and 50 N). The nitrogen was supplied in 8 doses in the form of 250 ml of an aqueous solution of NH_4NO_3, at two week intervals starting 15 March 1995. These inputs were in addition to those derived from precipitation, which amounted to about 4.5 kg ha^{-1} of nitrogen in the first 6 months of 1995.

Samples were collected on 13 July 1995 (except for the measurements of photosynthetic assimilation, which were conducted on 30 June 1995), i.e. 4 months after the first nitrogen addition, by pressing the surface of the moss with a sieve (mesh size 1.5 mm) and sucking up the water with a syringe. This method, used by Grolière (1977) for ciliates and by Francez (1988) for rotifera, permits to extract most microorganisms Possibly, there is an under-estimation of the densities of microorganisms, but this does not invalidate comparisons between variables from different cores. By using three quadrats, triplicate measurements could be made (with the exception of measurements of potential heterotrophic assimilation, for which only a single measurement was made. Mean values for triplicates measurements are given with their standard deviation (M ± SD). During each site visit, temperature, pH, conductivity were measured *in situ* with a Checkmate M90 combined probe (T °C: ± 0.1 °C; O_2: ± 0.1 mg l^{-1}; pH: ± 0.01 unit; conductivity: ± 0.01 μS cm^{-1}).

Community counts and estimation of biomass

Samples were fixed in 2% glutaraldehyde and stored at 4 °C in the dark.

Heterotrophic bacteria

Subsamples (0.25 ml) were homogenized with 3 ml sterilized distilled water. Heterotrophic Bacteria were stained with DAPI (4,6 diamino-2 phenylindol) according to Porter & Feig (1980) and filtered onto 0.2 μm pore-size filters. Preparations were made within 24 h of sampling and stored at - 25 °C in the dark. Slides were examined at 1250 × magnification by epifluorescence microscopy in a dark room using an Olympus HBS microscope equipped with an epifluorescence illuminator HB2-RFL, a mercury lamp HBO-100W and a neofluar objective lens 100/1.25.

Other microorganisms

The biomass of Cyanobacteria, Fungi (mycelia and spores of actinomycetes and hyphomycetes), microalgae, Protozoa, Rotifera and Nematoda and their community composition were determined after Utermöhl (1958). Samples were stored at 4 °C in the dark. For < 100 μm micro-organisms, storage volumes of 1 ml were settled for at least 24 h in plankton chambers. Densities were estimated with an inverted Leitz microscope at 200 × and 600 ×, by scanning the whole chamber area to account for any heterogeneous distribution of the cells. For > 100 μm microorganisms, storage volumes of 10 ml were settled in plankton chambers. Densities were estimated with an inverted Leitz microscope at 100 ×.

Biovolumes and estimation of biomass

Biovolumes of each community were estimated by assuming geometric shapes and converted to biomass using the conversion factor: 1 $mm^3 = 10^{-6}$ mg of dry biomass.

Chlorophylls

Chlorophylls were determined spectrophotometrically from samples (10–50 ml) collected on Whatman GF/C glass fiber filters. Pigments were extracted with 90% acetone and concentrations were calculated from Scor-Unesco (1966) equations.

Auto- and heterotrophic activities

Photosynthetic and heterotrophic activities were measured from the assimilation of $NaH^{14}CO_3$ and ^3H amino-acids, using the double labeling technique. Tritium labeling was chosen because the specific activity of ^3H is up to 10 times greater than that of ^{14}C labeled compounds (Azam & Holm-Hansen, 1973), and because possible refixation by photosynthesis of inorganic products derived from bacterial heterotrophy (McKinley & Wetzel, 1979) or from algal respiration (Richardson & Fogg, 1982) is avoided or minimized. Activities were estimated using a protocol derived from Steemann-Nielsen's method (Steemann-Nielsen, 1952; 1977). For each core, three 125 ml glass stoppered bottles (two light and one dark bottles) were incubated after injection of $NaH^{14}CO_3$ (10 μCi/100 ml^{-1}, specific activity: 56 mCi/mmol^{-1}) and ^{3}H amino-acids (10 μCi/100 ml^{-1}, specific activity: 38 Ci/mmol^{-1}). The bottles were then incubated for three hours at solar midday. At the end of incubation, aliquots (5 ml) were taken from the bottles and filtered in parallel through Millipore membrane filters of 0.45 μm pore-size, at a vacuum pressure of less than 100 mm Hg. The radioactivity retained on the filters was counted using a liquid scintillation counter.

According to McKinley (1977) and McKinley & Wetzel (1979), the values for photoheterotrophic assimilation were calculated the difference between the activity measured in the light bottle and that measured in the dark bottle, on the basis that: (a) activity in the light bottle = potential photoheterotrophic activity + potential chemoheterotrophic activity + background and (b) activity in the dark bottle = potential chemoheterotrophic activity + background.

Results

Physico-chemical environment

The water temperature was between 17.7 and 21.2°C depending on cores.

The pH values were characteristic of an acid oligotrophic fen (from 4.2 ± 0.01 to 4.4 ± 0.2).

The conductivity values were low, between 19.9 and 55.1 μS cm^{-1}. The average values were 48.5 ± 6.9 μS cm^{-1} in the control cores, 24.5 ± 8.0 in 5N, 36.6 ± 0.1 in 15 N and 25.9 ± 2.7 in 50 N.

Bacterial abundance and biomass

The mean bacterial abundance was 6.2×10^{10}, 5.8×10^{10}, 7.5×10^{10} and 9.5×10^{10} cells l^{-1}, in the control, 5 N, 15 N and 50 N cores. Among the morphotypes observed, cocci largely dominated in all cores, accounting for about 79% of total abundance. The bacterial biomass was higher, on average, in the 50 N

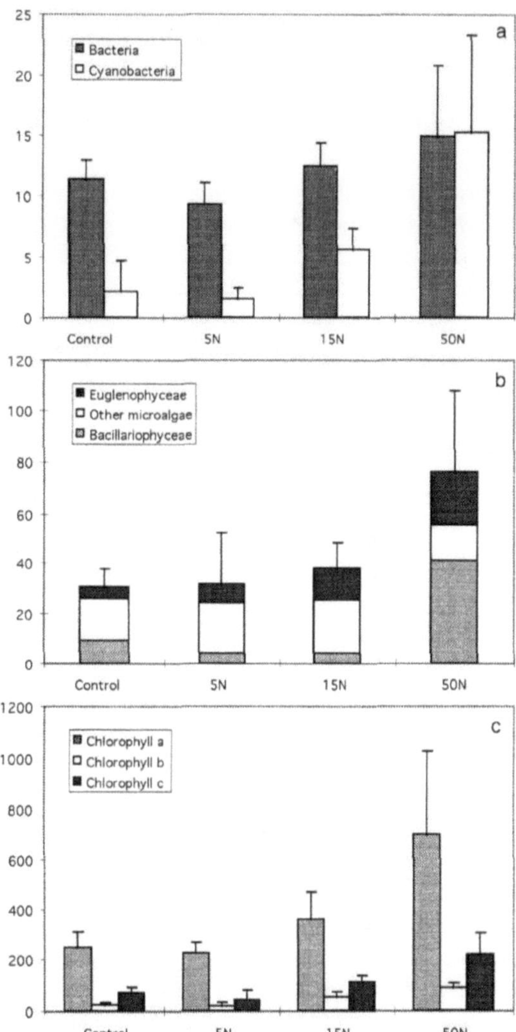

Figure 1. Spatial variations in dry biomasses (mg l^{-1}) of heterotrophic Bacteria and Cyanobacteria (a) and of Euglenophyceae, Bacillariophyceae and other microalgae (b) and spatial variations in chlorophyll *a*, *b* and *c* concentrations (μg l^{-1}) (c) in the control, 5 N, 15 N and 50 N cores.

(14.8 ± 6.0 mg l^{-1}) and in the 15 N (12.5 ± 1.8 mg l^{-1}) than in the control cores (11.3 ± 1.6 mg l^{-1}) and in the 5 N (9.4 ± 1.8 mg l^{-1}) (Figure 1a).

Abundance and biomass of cyanobacteria and microalgae and chlorophyll concentrations

The density of Cyanobacteria was 2.7×10^7 cells l^{-1} on average in the control cores, 2.1×10^7 cells l^{-1} in 5 N, 1.0×10^7 cells l^{-1} in 15 N and 0.7×10^7 cells l^{-1} in the 50 N. On the other hand, biomass was higher in the 50 N (15.2 ± 8.1 mg l^{-1}) and 15 N (5.7 ± 1.7 mg l^{-1}) than in the control (2.1 ± 2.6 mg l^{-1}) and 5 N (1.6 ± 0.9 mg l^{-1}) (Figure 1a). The two main species, *Anabaena cylindrica* and *Chroococcus turgicus*, were present in all samples, but cells of *A. cylindrica* were more abundant in the control and 5 N than cells of *C. turgicus*, whose cellular biovolume is greater than that of *A. cylindrica*, more abundant in 15 N and 50 N. The biomass of total Cyanobacteria increased considerably with inputs of nitrogen, but this increase was due to the development of the non nitrogen-fixing species *C. turgidus*, whereas the biomass of the nitrogen-fixing species *A. cylindrica* decreased with increasing nitrogen inputs.

On the whole, microalgal biomass was higher in the 50 N (75.6 ± 32.2 mg l^{-1}) than in the 15 N (37.7 ± 10.5 mg l^{-1}) or in the control and 15N (31.7 ± 20.8 and 30.4 ± 7.7 mg l^{-1}) (Figure 1b). Their abundance varied from 4.5×10^7 to 19.3×10^7 cells l^{-1}. The mean values were 8.9×10^7 cells l^{-1} in the control cores, 6.5×10^7 cells l^{-1} in 5 N, 8.2×10^7 cells l^{-1} in 15 N and 14.2×10^7 cells l^{-1} in 50 N.

Most of the species of Microalgae recorded were found in all samples. However, small-sized Bacillariophyceae (*Achnanthes* spp., *Navicula* spp., *Pinnularia subcapitata, Nitzchia* spp. etc.) were more abundant in the control, 5 N and 15 N, whereas species > 40 μm (*Pinnularia biceps, Eunotia pectinalis* and *Pinnularia viridis*) were more abundant in 50 N. The dominant Euglenophyceae belonged to *Rhabdomonas, Menoidium, Astasia, Trachelomonas* and *Phacus*, organisms which for the most part appear to be only slight pigmented or unpigmented. The Chlorophyceae present were of small size (< 20 μm) and belonged mainly to the Volvocales. Among the larger Chlorophyceae, only one species belonging to the genus *Mesotaenium* was abundant.

The chlorophyll *a* concentrations were higher in the 50 N (694 ± 332 μg l^{-1}) than in the 15 N (360 ± 109 μg l^{-1}) and in the control and 5 N (247 ± 69 μg l^{-1} and 227 ± 45 μg l^{-1}, respectively) (Figure 1c). The chlorophyll *b* and *c* concentrations followed the same pattern but were much lower than those of chlorophyll *a*. The chlorophyll *b* content varied from 89 ± 22 μg l^{-1} in 50 N to 17 ± 19 μg l^{-1} in 5 N. The chlorophyll *c* concentrations fluctuated from 216 ± 96 μg l^{-1} in 50 N to 42 ± 40 μg l^{-1} in 5 N (Figure 1c).

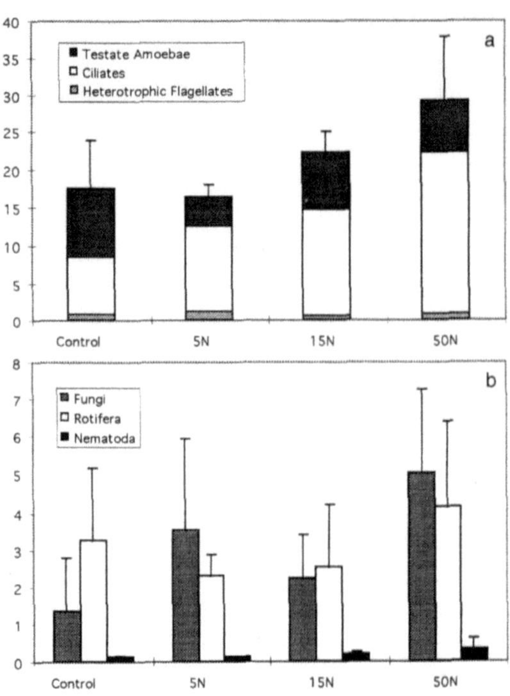

Figure 2. Spatial variations in dry biomasses (mg l^{-1}) of Testate Amoeba, Ciliates and heterotrophic Flagellates (a) and for Fungi, Rotifera and Nematoda (b) in the control, 5 N, 15 N and 50 N cores.

Abundance and biomass of Protozoa

On the whole, the protozoan biomass was higher in 50 N (29.2 ± 8.6 mg l^{-1}) and 15 N (22.2 ± 2.8 mg l^{-1}) than in the control and 5 N cores (17.4 ± 6.4 and 16.3 ± 1.8 mg l^{-1} respectively) (Figure 2a).

The density of heterotrophic Flagellates was rather uniform: it only varied between 1.1×10^7 cells l^{-1} (50 N cores) and 3.1×10^7 cells l^{-1} (5 N cores). Their biomass was low and on average amounted to 0.6 mg l^{-1} in 15 N and 1.2 mg l^{-1} in 5 N (Figure 2a).

The abundance of Ciliates varied between 3.5×10^6 cells l^{-1} in 5 N and 7.4×10^6 cells l^{-1} in 50 N. Their biomass increased with increasing nitrogen input (Figure 2a), being 7.6 ± 5.2 mg l^{-1} in the control cores, 11.2 ± 0.7 in 5 N, 13.8 ± 3.5 in 15N and 21.6 ± 6.3 in 50 N (Figure 2a). Several species of ciliates occur on the surface of peatlands. In addition to many small-sized hypotrichous species, various species belonging to *Colpoda, Trachelius, Spathidium and Spirostomum* were present. However, the highest biomass was accounted for by the development of large species of *Paramecium* (120 to 150 μm).

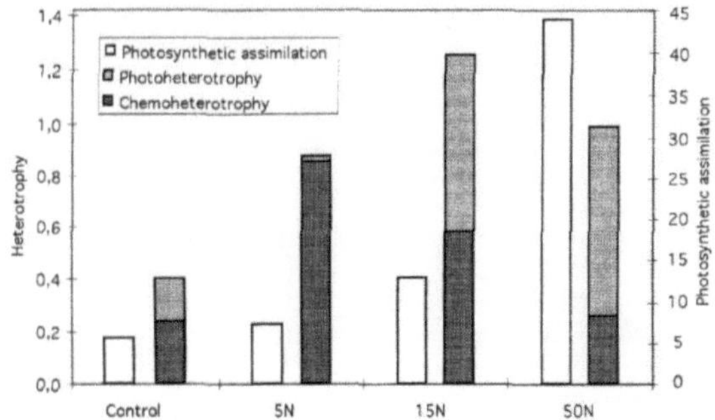

Figure 3. Spatial variations in photosynthetic assimilation (mgC m^{-3} h^{-1}) and in potential photoheterotrophic and chemoheterotrophic activities (μgC m^{-3} h^{-1}) in the control, 5 N, 15 N and 50 N cores.

The average density of testate Amoebae varied between 2.8×10^4 cells l^{-1} in 5 N and 8.1×10^4 cells l^{-1} in 50 N. Their biomass was 9.1 ± 2.8 mg l^{-1} in the control cores, 7.7 ± 5.5 in 15 N, 6.9 ± 2.2 in 50 N and 3.9 ± 1.8 in 5 N (Figure 2a). *Nebela collaris* and *Hyalosphaenia papilio* dominated in terms of biomass, but other species (*Nebela militaris, Nebela tincta, Phryganella* sp. *Euglypha strigosa, Placocysta spinosa, Heleoptera* sp. *Centropyxis aculeata* and *Arcella* sp.) occurred regularly.

Abundance and biomass of Fungi

The average biomass of Fungi was 1.3 ± 1.4 mg l^{-1} in the control cores, 3.5 ± 2.4 in 5 N, 2.2 ± 1.2 in 15 N and 5.0 ± 2.3 in 50 N (Figure 2b). This biomass, which did not include that of yeasts, consisted of 92 to 99% of mycelia and 1 to 8% of spores. Among the species encountered, *Helicoon pluriseptatum* was well represented. The average abundance of spores increased with increasing nitrogen input ($0.7 \pm 1.3 \times 10^5$ cells l^{-1} in the control cores, $2.2 \pm 2.2 \times 10^5$ cells l^{-1} in 5 N, $4.0 \pm 5.1 \times 10^5$ cells l^{-1} in 15 N and 12.8 ± 14.5 10^5 cells l^{-1} in 50 N).

Abundance and biomass of Rotifera

Average density varied between 4.0 and 8.8×10^4 ind l^{-1}. Mean biomass varied little between cores, 2.3 ± 0.6 mg l^{-1} in 5 N, 2.5 ± 1.7 in 15 N, 3.3 ± 1.9 in the control cores, and 4.1 ± 2.3 in 50 N (Figure 2b). Four main generals were identified: *Euchlanis, Trichocera, Notommata* and *Philodina*.

Abundance and biomass of Nematoda

Nematodes are frequent in peatlands. Their average density varied from 7.5×10^3 ind l^{-1} (control cores) to 25.3×10^3 ind l^{-1} (50 N cores). Their biomass was also higher in the 50 N (0.3 ± 0.3 mg l^{-1}) than in the 15 N, control and 5 N (0.2 ± 0.10, 0.1 ± 0.04 and 0.1 ± 0.02 mg l^{-1}, respectively) (Figure 2b). The most abundant nematodes measured between 150 and 350 μm.

Auto-and heterotrophic activities

Photosynthetic assimilation increased with nitrogen input. They were 5.6 mg m^{-3} h^{-1} C in the control cores, 7.2 in 5 N, 12.9 in 15 N and 44.0 in 50 N (Figure 3).

Generally, heterotrophic activity was higher in the fertilized cores than in the control cores (Figure 3). On average, the values of potential photoheterotrophic assimilation were lower in 5 N (0.02 μg m^{-3} h^{-1} C) and in the control cores (O.16 μg m^{-3} h^{-1} C) than in 15 N and 50 N (0.67 and 0.72 μg m^{-3} h^{-1} C, respectively) (Figure 3). On average, the values of potential chemoheterotrophic assimilation were higher in 5 N (0.85 μg m^{-3} h^{-1} C) and 15 N (0.58 μg m^{-3} h^{-1} C) than in 50 N and control cores (0.26 and 0.24 μg m^{-3} h^{-1} C) (Figure 3).

Discussion

Structure of the microbial communities in the Control cores

Three groups of micro-organisms dominated the biomass of this subaquatic peatland environment: Microalgae (46.1% of total microbial biomass), Protozoa (26.3%) and heterotrophic Bacteria (17.2%) (Figure 4). Among Microalgae, the other microalgae (25.1% of total microbial biomass), Bacillariophyceae (14.3%) and to a lesser extent Euglenophyceae were the main classes present (Figure 4). Protozoa mainly consisted of Ciliates (11.5% of total microbial biomass), testate Amoebae (13.7%) and heterotrophic Flagellates (1.1%) (Figure 4). Rotifera (4.9% of total microbial biomass), Cyanobacteria (3.3%), Fungi (2.0%) and Nematoda (0.2%) occurred regularly but only accounted for a low biomass (Figure 4).

A previous study conducted on less drained cores of the same peatland (Gilbert et al., 1998) and deal-

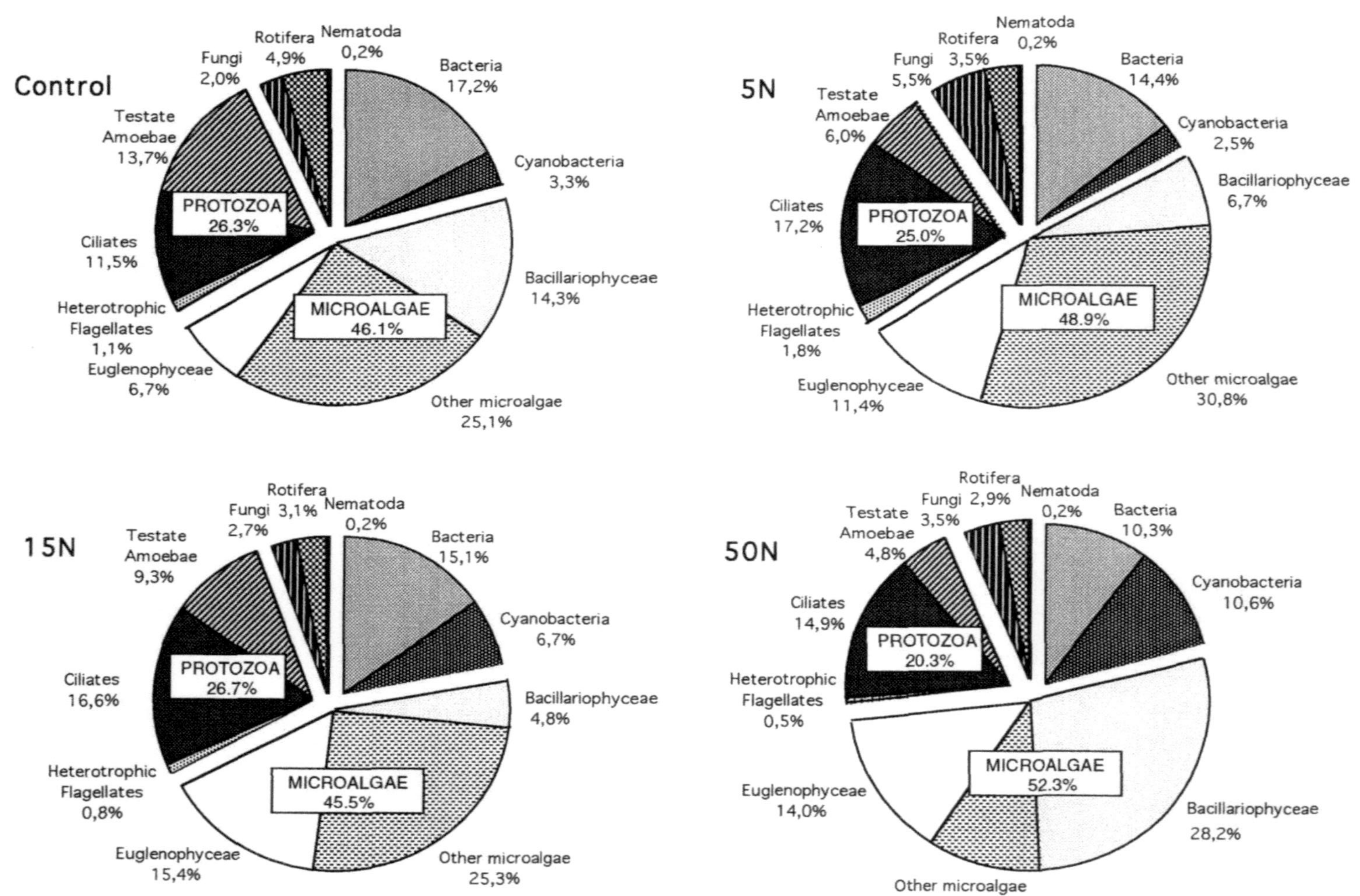

Figure 4. Relative proportions of the biomass of different microbial groups in the control, 5 N, 15 N and 50 N cores.

Table 1. Results of the Kruskal-Wallis test for measured biomasses and chlorophyll concentrations between the control, 5 N, 15 N and 50 N cores (* < 0.05).

	Con ≫ 5 N	Con ≫ 15 N	Con ≫ 50 N	5 N ≫ 15 N	5 N ≫ 50 N	15 N ≫ 50 N
Bacteria						
Cyanobacteria			*	*	*	
Fungi						
Bacillariophyceae					*	*
Euglenophyceae						
Other microalgae			*			
Total microalgae			*			
Heterotrophic Flagellates						
Ciliates			*		*	
Testate Amoebae						
Total Protozoa				*	*	*
Rotifera						
Nematoda						
Total Biomass			*			
Chlorophyll *a*			*	*	*	
Chlorophyll *b*			*	*	*	
Chlorophyll *c*			*	*	*	

ing with a more restricted range of micro-organisms (Heterotrophic Bacteria, Cyanobacteria, Microalgae and Protozoa), also found a dominance of microalgae, heterotrophic Bacteria and Protozoa, but in significantly different proportions. Heterotrophic Bacteria accounted for 31.9% of total microbial biomass, microalgae 35.2%, Protozoa 26.4% and Cyanobacteria 6.5%. These differences can be explained by the fact that the previous study was conducted on a habitat with a higher water content in the summer.

Overall, density and biomass values recorded during this study were in agreement with that previously found in these environments. Thus, the biomass for the entire micro-organism assemblage seems to be compatible with the values of biomass obtained by Williams and Sparling (1988), using an indirect method of fumigation-extraction. The abundance values of microalgae were higher than those obtained by Duthie (1965) on the surface of a upland peatbog, and those obtained by Tomaszewicz (1994) for desmids in peat excavations. The abundance of ciliates was close to that reported by Groliere (1977). The densities of testate Amoeba recorded were compatible with those reported by Warner (1987), who gave a value of about 10^4 cells l^{-1} in the top 5–10 cm of a *Sphagnum magellanicum* bog. The biomass values of fungal mycelia were 100 times lower than those reported by Collins (1978) obtained by direct counting in a peatland. We note that Czeczuga (1993) identified 35 species of Fungi in upland blanket bogs.The abundance of Rotifera in the present study was comparable with that reported by Francez (1987) in an oligotrophic fen characterized by a higher water content than in our study. The abundance of Nematoda was similar to that found by Kajak et al. (in Williams et al., 1996) in drained fens.

Two factors therefore seem to characterize the microbial communities in subaquatic peatland environments. Firstly, in contrast to what has been observed in humic lakes, where autotrophic micro-organisms account for more than 70% of the total microbial biomass (Amblard et al., 1995), heterotrophic micro-organisms (heterotrophic Bacteria, Protozoa, Fungi, Rotifera and Nematoda) account for nearly half of the microbial biomass in oligotrophic fens. However, it should be noted that Fungi, Rotifera and Nematoda were not taken into account in the study conducted on the lacustrine ecosystem. The potential chemo- and photoheterotrophic activities are also of the same order of magnitude in both ecosystems. Secondly, Testate amoebae are well represented in *Sphagnum* peatlands. They may be advantaged by their ability to circumvent climatic variations by encystment, and can use a wide range of prey (organic debris, Bacteria, Fungi, Microalgae, Protozoa, Rotifera and Nematoda) (Couteaux & Pussard, 1982). Thus, Testate Amoeba are likely to play a major role inside the microbial food web of *Sphagnum* peatlands. Ciliates also constitute a main part of microbial communities in these environments when heterotrophic flagellates only account for a low proportion of total microbial biomass. The relative great abundance of Ciliates is probably due to a high competitive capacity when bacteria develop in very numbers (Fenchel, 1987).

Impact of nitrogen inputs

The biomass and chlorophyll variables measured in the four study cores were compared using the Kruskal-Wallis test. In terms of biomass of the various microbial communities, only the input of 50 kg N ha^{-1} brought about a significant change in the population structure (Table 1). The input of 5 kg N ha^{-1}, which was reflected by a slight decrease in total microbial biomass, did not significantly change the structure of the microbial communities (Figure 4 and Table 1). The input of 15 kg N ha^{-1} led to a tendency to an increase in total biomass (Table 1), mainly caused by an increase in biomass of Cyanobacteria and Protozoa. Overall, the structure of the microbial communities in the control, 5 N and 15 N cores therefore remained similar. For example, the bacterial biomass accounted for between 14.4 and 17.2% of the total microbial biomass, the microalgae the biomass for 45.5 to 48.9% and that of Protozoa for 25.1 to 26.7% (Figure 4). On the other hand, the input of 50 kg N ha^{-1} led to a clear increase in biomass of Cyanobacteria, microalgae, Ciliates and also to an increase in chlorophyll *a*, *b* and *c* (Table 1). The proportion of Cyanobacteria increased in the 50 N cores, but this increase was due to the development of the non nitrogen-fixing *Chroococcus turgidus*, whereas the biomass of the nitrogen-fixing *Anabaena cylindrica* decreased with increasing nitrogen inputs.

The results obtained therefore show that the input of nitrogen led to a steady increase in total microbial biomass, but only changed the structure of microbial communities significantly when the input was 50 kg ha^{-1}. This contrasts with the general behaviour of *Sphagnum* in relation to the level of N inputs (Aerts et al., 1992; Williams et al., 1996). For example, experiments conducted by Austin and Wieder (1987) showed that the growth of *Sphagnum fallax* was significantly stimulated by low N inputs, while in the presence of high N inputs, the growth rate of this moss tended to decrease.

Acknowledgements

This work was supported by the European Community (Environmental programme, 3rd Framework, Contract No. EV5V-CT92-0099) and by the French Ministry of the Environment (EGPN subsidy, n° 95-093). We thank the CEREMCA Association, the mayor and the commune of Saint-Anthème and the Parc Naturel Régional Livradois-Forez for their collaboration.

References

Aerts, R., B. Wallén & N. Malmer, 1992. Growth-limiting nutrients in *Sphagnum*-dominated bogs subject to low and high atmospheric nitrogen supply. J. Ecol. 80: 131–140.

Amblard, C., G. Bourdier, T. Sime-Ngando, S. Rachiq & J. F. Carrias, 1994. Diel and vertical variations of the microbial stocks (Bacteria, heterotrophic Flagellates, Ciliates, phytoplankton) and their relative activities. Arch. Hydrobiol. Beih. 41: 125–144.

Amblard, C., J. F. Carrias, G. Bourdier & N. Maurin, 1995. The microbial loop in a humic lake: seasonal and vertical variations in the structure of the different communities. Hydrobiologia 300/301: 71–84.

Austin, K. A. & R. K. Wieder, 1987. Effects of elevated H^+, $SO_4{}^{2-}$, $NO_3{}^-$, and $NH_4{}^+$ in simulated acid precipitation on the growth and chlorophyll content of 3 North American *Sphagnum* species. The Bryologist 90: 221–229.

Azam, F. & O. Holm-Hansen, 1973. Use of tritiated substrates in the study of heterotrophic in sea water. Mar. Biol. 23: 191–196.

Azam, F., T. Fenchel, J. G. Field, J. S. Gray, L. A. Meyer & F. Thingstad, 1983. The ecological role of water column microbes in the sea. Mar. Ecol. Pro. Ser. 10: 257–263.

Collins, V. G., B. T. D'Sylva & P. M. Latter, 1978. Microbial population in peat. In Heal Ow. & D. F. Perkings (eds), Production Ecology of British Moors and Montane Grasslands, Springer-Verlag, Berlin: 94–112.

Couteaux, M. M. & M. Pussard M., 1982. Nature du régime alimentaire des protozoaires du sol. News Trends in Soil Biology. Proceedings of the VIII. Intl Colloquium of soil Zoology. Louvain-la-Neuve, Belgium: 179–195.

Czeczuga, B., 1993. Aquatic fungi of the Gorbacz and Ostrowki Peatbogs. Acta Mycologica 28: 69–75.

Duthie, H. C., 1965. A study of the distribution and periodicity of some algae in a bog pool. J. Ecol. 53: 343–359.

Francez, A. J., 1987. Successions écologiques dans les tourbières: le peuplement de rotifèresdu lac-tourbière de Chambedaze (Puy-de-Dôme, France). Bull. Ecol. 18: 31–38.

Francez, A. J., 1988. Le peuplement de rotifères libres de deux lacs-tourbières du Puy-de-Dôme (France). Vie et Milieu 38: 281–292.

Francez, A. J., 1992. Croissance et production primaire des sphaignes dans une tourbière des monts du Forez (Puy-de-Dôme, France). Vie Milieu 42: 21–34.

Fenchel, T., 1987. Ecology of protozoa. *Science* Tech. Publishers, Madison, Wisconsin/Springer Verlag, New-York, 197 pp.

Gifford, D. J., 1991. The Protozoan-Metazoan trophic link in pelagic ecosystems. J. Protozool. 38: 81–86.

Gilbert, D., C. Amblard, G. Bourdier & A. J. Francez, 1998. The microbial loop at the surface of a peatland: structure, functioning and impact of nutrients inputs. Microb. Ecol. 35: 83–93.

Grolière, C. A., 1977. Contribution à l'étude des ciliés des sphaignes: II-dynamique des populations. Protistologica 13: 335–352.

McKinley, K. R., 1977. Light-mediated uptake of 3H-glucose in a small hardwater lake. Ecology 58: 1356–1365.

McKinley, K. R. & R. G. Wetzel, 1979. Photolithotrophy, photoheterotrophy and chemoheterotrophy: patterns of resource utilisation on an annual and diurnal basis whitin a pelagic microbial community. Microb. Ecol. 5: 1–15.

Munawar, M. & T. Weisse, 1989. Is the 'Microbial loop' an early warning indicators of anthropogenic stress? in Munawar, M., G. Dixon, C. I. Mayfield, T. Reynoldson & M. H. Sadar (eds), Environmental Bioassay Techniques and their applications. Hydrobiologia 188/189: 163–174.

Porter, K. G. & Y. S. Feig, 1980. The use of DAPI for identifying and counting aquatic microflora. Limnol. Oceanogr. 25: 943–948.

Porter, K. G., E. B. Sherr, B. F. Sherr, M. Pace & R. W. Sanders, 1985. Protozoa in planktonic food webs. J. Protozool. 32: 409–415.

Richardson, K. A. & G. F. Fogg, 1982. The role of dissolved organic material in the nutrition and survival of marine dinoflagellates. Phycologia 21: 17–26.

Riemann, B. & K. Christoffersen, 1993. Microbial trophodynamics in temperate lakes. Marine Microb. Food webs 7: 69–100.

S.C.O.R.-U.N.E.S.C.O., 1966. Determination of photosynthethic pigments in sea water. S.C.O.R.-U.N.E.S.C.O., 69 pp.

Simej, K. & V. Straskrabova, 1992. Bacterioplankton production and protozoan bacterivory in a mesotrophic reservoir. J. Plankton Res. 14: 773–787.

Steemann-Nielsen, E., 1952. The use of radio-active carbon (^{14}C) for measuring organic production in the sea. J. Cons. Explor. Mer. 18: 117–140.

Steemann-Nielsen, E., 1977. The carbon-14 technique for measuring organic production by plankton algae. A report of the present knowledge. Folia Limnol. Scand. 17: 45.

Tomaszewicz, G. H., 1994. Abundance and composition of the desmid flora in a series of peat pits, in relation to pH and some other habitat parameters. Biologia, Bratislava 49/4: 519–524.

Utermölh, H., 1958. Zur vervollkommnung der quantative phytoplankton-methodik. Mitt. Int. Ver. Limnol. 9: 1–38.

Van Breemen, N., C. T. Driscoll & J. Mulder, 1984. Acidic deposition and internal proton sources in acidification of soils and waters. Nature 307: 599–604.

Warner, B., 1987. Abundance and diversity of testate amoeba (Rhizopoda, Testacea) in *Sphagnum* peatlands in southwestern Ontario, Canada. Arch. Protistenkd. 133: 173–180.

Williams, B. L. & G. P. Sparling, 1988. Microbial biomass carbon and readily mineralized nitrogen in peat and forest humus. Soil Biol Biochem 20: 579–581.

Williams, B. L., D. Silcock, A.-J. Francez, D. Gilbert, A. Buttler, P. Grosvernier., Y. Matthey, H. Vasander, J. Jauhiainen, A. Kajak, J. Petal & M. Ilomets, 1996. Impact of nitrogen deposition on the carbon balance in peatland ecosystems. EC Environment programme, 3rd Framework. Final report., 96 pp.

Yesmin, L., S. M. Gammack, L. J. Sanger & M. S. Cresser, 1995. Impact of atmospheric N deposition on inorganic- and organic-N outputs in water draining from peat. Sc. Total Envir. 166: 201–209.

Hydrobiologia **373/374:** 121–133, 1998.
J.-C. Amiard, B. Le Rouzic, B. Berthet & G. Bertru (eds),
Oceans, Rivers and Lakes: Energy and Substance Transfers at Interfaces.

Role of fish communities in particulate organic matter fluxes between salt marshes and coastal marine waters in the Mont Saint-Michel Bay

Pascal Laffaille[1], Sébastien Brosse[2], Eric Feunteun[1], Aurore Baisez[3] & Jean-Claude Lefeuvre[1]
[1] *U.M.R. ECOBIO Laboratoire d'Evolution des Systèmes Naturels et Modifiés, M.N.H.N., Université de Rennes I, Avenue du Général Leclerc, 35042 Rennes Cedex, France; e.mail: Pascal.Lafaille@univ-rennes1.fr*
[2] *Present address: CESAC, Laboratoire d'Hydrobiologie, Bat. 4R3, Université Paul Sabatier, 118 route de Narbonne, 31000 Toulouse, France*
[3] *Present address: CEMAGREF Bordeaux, Division Ressources Aquatiques Continentales, 50 avenue de Verdun, 33612 Cestas Cedex, France*

Key words: salt marsh, outwelling, organic matter fluxes, fish community, nursery, mullets

Abstract

Among the 90 fishspecies censused in the Mont Saint-Michel Bay (France), 23 colonise and forage in the salt marshes during flood. Therefore, this environment may play an important trophic and nursery role for these species. This community is largely dominated by mullets (81% of the biomass), *Liza ramada* and secondarily *L. aurata*. But gobies (mainly *Pomatoschistus minutus* and *P. lozanoï*) and sea bass (*Dicentrarchus labrax*) are also present; they represent respectively 11% and 4% of the biomass. During the tide cycles, mullets export from salt marshes about 7% of their body weight (FW) containing a mixture of sediment (43%), organic matter (24%) and water (33%). Gobies and sea bass mainly feed on dwelling macro-invertebrates, and they export respectively 4.5% and 10% of their body weight during a tide cycle. Thus, we estimated that 50 tonnes year^{-1} of particulate organic matter (dry weight POM) are exported from the 4000 ha of salt marshes to the marine coastal waters. These fish communities appear to be POM transporters and could play a significant role in the global energy budgets of coastal environments such as Mont Saint-Michel Bay. According to the seasons and the years, the energy exported by fish communities is assumed to range between 0 and 10% of the total POM output.

Introduction

Salt marshes are complex ecotones between terrestrial and marine systems, regularly flooded by tides. Fluxes of particulate organic matter (POM), dissolved organic matter (DOM) and inorganic nutrients occur between salt marshes and coastal environments. Research on these fluxes and the global functioning of these systems were undertaken by Odum (1968) in the development of outwelling hypothesis. According to this concept, up to 45% of organic matter produced in salt marshes may be exported to the coastal ocean that benefits from this fertilisation to support secondary production (Teal, 1962). However, several studies have shown that this concept should not be applied to all salt marshes, and that nutrient and organic matter fluxes between adjacent systems ought to be systematically measured (i.e. Pomeroy & Wiegert, 1981; Lefeuvre & Dame, 1994). Many authors thus have quantified these fluxes by abiotic vectors such the tidal currents and the drainage, between various salt marshes and adjacent marine coastal waters (e.i., Teal, loc. cit.; Valiela & Teal, 1979; Nixon, 1980; Dame et al., 1986; Chambers et al., 1992), and more specific to this study in the Mont Saint-Michel Bay (e.i., Lefeuvre et al., 1994; Troccaz et al., 1994; Bouchard, 1996; Troccaz, 1996; Troccaz & Giraud, 1996).

Functioning of salt marshes of Mont Saint-Michel Bay have been studied for 10 years (Lefeuvre et al., 1994). The salt marshes provide high primary productivity. Some vascular plants (e.g. *Atriplex portula-*

coides and *Elytrigia aetherica*) produce up to 3600 g DW m^2 y^{-1} (Lefeuvre et al., loc. cit.), among which less than 1% is exported to the sea as macrodetritus during spring tide (Bouchard, 1996). The remaining is decayed rapidly (Bouchard, loc. cit.) and exported as dissolved organic matter to the open sea by abiotic vectors (tidal currents and especially drainage) (Troccaz, 1996; Troccaz & Giraud, 1996). This contributes to sustain a high biodiversity in Mont Saint-Michel Bay, especially in fish and mollusks (Legendre, 1984; Feunteun & Laffaille, 1997). On the other hand, salt marshes globally trap sediment and particulate organic matter imported by marine waters (Troccaz, loc. cit.; Troccaz & Giraud, loc. cit.). Although significant fluxes of N and C occur between salt marshes, marine waters and atmosphere by rain and evapotranspiration (Hemond & Fifield, 1981), they had not been measured in Mont Saint-Michel Bay.

Beside these biogeochemical and physical processes, the role played by fauna, as biotic vector of organic matter fluxes, in the global functioning of estuarine systems has received scant attention (Bildstein et al., 1992). For instance, animals are able to export organic matter from salt marshes, where they forage, to other ecosystems where they rest or breed (i.e. 6% of phosphorus inputs were due to birds in a lake of western France (Marion et al., 1994; Feunteun & Marion, 1994)).

This role of animals in energy fluxes had not been studied in Mont Saint-Michel Bay either. However, the high primary productivity of Mont Saint-Michel Bay contributes to an exceptional invertebrate and vertebrate biodiversity. In particular, its fish community is very rich (Legendre, 1984; Feunteun & Laffaille, in press). Moreover, the Bay's mud flats represent one of the main nurseries of the French coast of the English Channel for many species such as sea bass, fishes and clupeids (Beillois et al., 1979; Legendre, loc. cit.). Many adult fishes also utilise the marshes and sustain a traditional fishery (Legendre, loc. cit.).

The objective of this study was thus assess the role of fishes in transferring particulate organic matter (POM) between salt marshes and marine waters in the Mont Saint-Michel Bay. Fish communities were studied in order to determine which species are most involved in these fluxes of organic matter, and to quantify and compare these fluxes with those due to abiotic vectors.

Table 1. Characteristics of the 12 tides sampled for fish in Mont Saint-Michel Bay, November 1995 to September 1996.

Dates	Tide amplitude (m)	Duration of submersions (min)	Sampling during flood tide	Sampling during ebb tide
07/11/95	11.60	100	Yes	Yes
23/11/95	12.45	155	Yes	Yes
07/03/96	12.00	125	No	Yes
18/03/96	12.25	150	Yes	Yes
05/04/96	12.30	130	No	Yes
17/04/96	12.20	120	No	Yes
03/05/96	12.15	125	No	Yes
17/05/96	11.80	100	Yes	Yes
02/06/96	12.05	120	No	Yes
01/07/96	12.55	140	No	Yes
31/07/96	12.65	160	No	Yes
01/09/96	12.60	155	No	Yes

Study site

The Mont Saint-Michel Bay (France) is a wide littoral zone situated in the Normano-Breton Gulf (lat. 48°40′ N, long 1°40′ W) and which extends over 500 km^2 (Figure 1). The tidal marshes cover 220 km^2, including 180 km^2 of mud flats and 40 km^2 of salt marshes. This Bay is characterised by strong tidal ranges which reach up to 16 m. Salt marshes are incised by a more or less dense creek network where sea water comes in during a number of tidal cycles. The study site is located in accreting salt marshes (30–50 ha per year) west of Mont Saint-Michel (Lefeuvre et al., 1994), dominated by *Halimione portulacoides* (Bouchard, 1996). It is situated at 2.5 km from the coastline on a 10 m wide creek which drains a 5.7 ha watershed. The water only reaches this canal in 43% of the tides when the water level is > 11.25 m, (Troccaz et al., 1994).

Materials and methods

Sampling fish community of the salt marsh

Fishes were caught in the creek between November 1995 and September 1996 during a total of 12 spring tides (Table 1). The samplings were made during ebb or ebb and flood tide in order to compare gut contents before and after residence in the salt marsh. Sampling selectivity was reduced by combining fyke net (4 mm

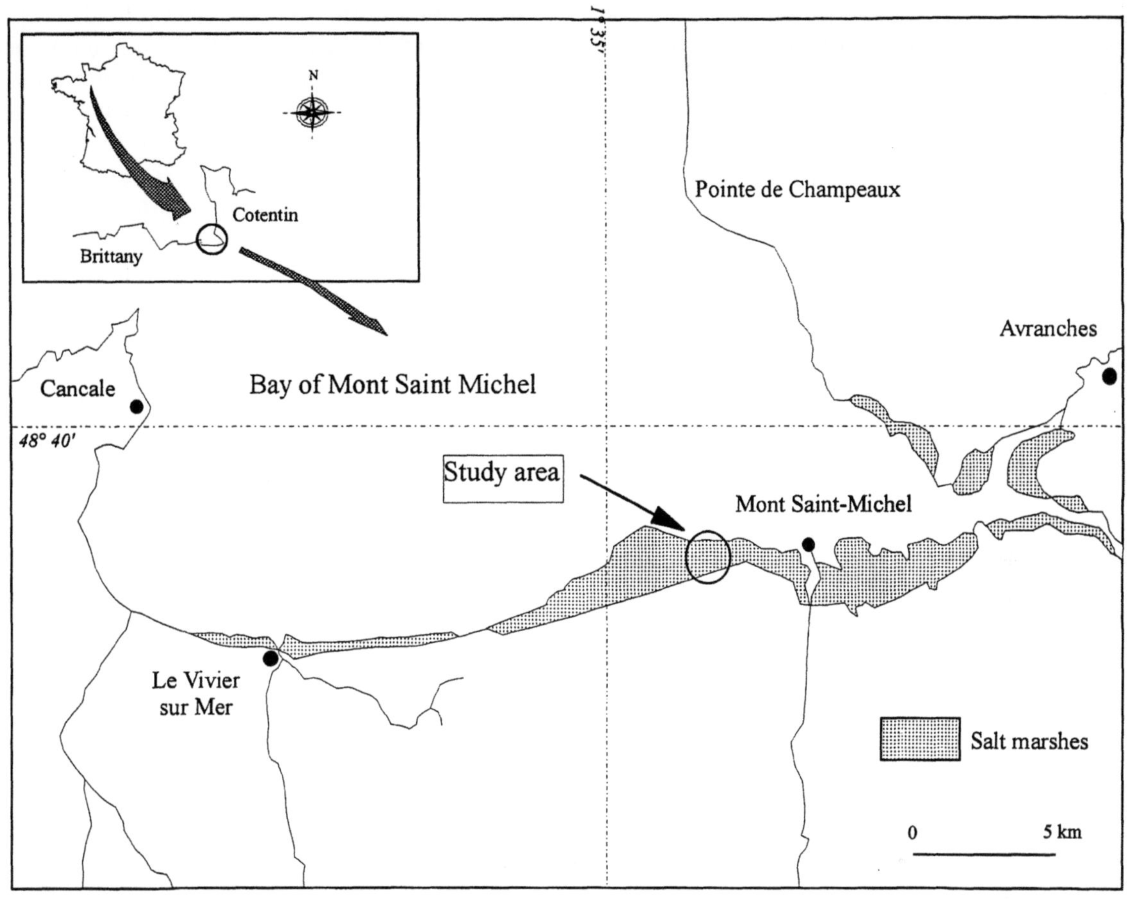

Figure 1. Location of the Mont Saint Michel Bay and study site.

mesh size, 5 m deep, 1.80 m high, 20 m long) to catch small fishes and trammel net (30 to 70 mm mesh size, 2 m high and 30 m long) to capture bigger fishes, especially mugilids, that could escape by avoiding the fyke net. Both devices were set in the canal to catch every fish that accessed the salt marsh during flood or left it during ebb. Sampling was conducted in order to quantify the fish which colonised the marsh and to analyse the variation of population and community structure during the tidal cycle. Every 15 min, a fyke and two trammel nets were set simultaneously. After five minutes, the fyke net was emptied. The trammel nets were taken up after 10 min. Thus, seven to 10 samples were collected depending upon the duration of the tidal cycle. Large fish were sampled during 66% of the time (10 min of fishing every 15 min) with trammel nets and fyke net. Small individuals such as larvae and juvenile, but also small species (i.e. gobies, gasterosteids, etc.), were caught 33% of the time (5 min of fishing every 15 min) with the fyke net.

All samples were stored in a refrigerated box and deep frozen at the laboratory (− 18 °C) until analysis. Later, they were identified to species using various guides (Wheeler, 1969; Russel, 1976; Farrugio, 1977; Reay & Cornell, 1988; Serventi et al., 1996). Individual fish was measured to the nearest 1 mm (fork length, FL) and weighed to the nearest 0.01 g (individual fresh body weight, BW). They were then weighed to the nearest 0.1 g for species (total fresh weight, TFW) to establish the biomass and numeric composition of the community during each sampling. The sampling design enabled us to assess the total numbers and biomass of fish that colonized the salt marsh at each tide studied by extrapolation with moving average of the samples: (i) relative number and biomass frequency of species at each tide and (ii) number and biomass of fish caught per min (CPUE). According to the sampling design, two methods were based on the calculation of estimating the composition (number and biomass abundance of species caught, %NA and %BA) of the fish community during this study: When the sampling was only conducted during ebb,

$$N_j = \frac{\left(\frac{N_i}{t_i}\right) + \left(\frac{N_i+1}{t_i+1}\right)}{2} x t j, \quad (1)$$

$$\text{CPUE} = {\sum N_j + \sum N_i}/{\sum t_j + \sum t_i}, \quad (2)$$

with:
t_i = sampling time (mn) for each sample and each device,
t_j = time between sample i and $i+1$;
N_i = quantity of fish (numbers or g) caught per sample and per species;
N_j = estimated quantity of fish (numbers or g) which entered in the canal between sample i and $i+1$.

When the sample was conducted during ebb and flood:

$$\text{CPUE} = \frac{\sum N_j + \sum N_i + \sum N_k}{\sum t_i + \sum t_j} \quad (3)$$

with:
N_k = Quantity of fish (numbers or g) caught per species and per sampling device during the flood.

Gut contents analysis

After thawing, measuring and weighing fishes, the entire content of the digestive tract was removed from every thin-lipped mullet (*Liza ramada*, FL > 200 mm), 213 sand gobies (*Pomatoschistus minutus*, FL > 29 mm) caught in March 1996 and every sea bass (*Dicentrarchus labrax*, FL > 59 mm) caught in November. The percentage of empty stomachs was noted (vacuity index, *V*). Then, the gut contents were weighed to the nearest 0.01 g (gut content fresh weight, FW). To assess the relationship between FW and DW, for each studied species, we also measured the dry weights of gut contents to nearest 0.01 g (gut content dry weight, DW) after desiccation at 50 °C in a drying oven.

In *L. ramada*, the percentage of particulate organic matter (POM) in the gut of 30 specimens caught during ebb was determined by loss on ignition at 480 °C. In order to remove mineral carbon, 10 samples of gut contents of thin-lipped mullets caught during the ebb tide were decarbonated by HCl (1 N, pH 2.0), rinsed with distilled water and lyophyllised. The proportion of total organic carbon (%Ctot) and nitrogen (%Ntot) and the molar C/N ratio was determined with a CHN analyser (Carlo-Erba 1500). Diatoms were also identified in the stomach contents of 15 *L. ramada* specimens. These were separated from sediment by centrifugation and density gradient in a 40% silica gel: Ludox HS solution (Blanchard et al., 1988). Diatom identifications were made according to Peragallo & Peragallo (1897), Hustedt (1927–1933; 1937 & 1959), Hendey (1964), Patrick & Relmer (1966), Germain (1981) and Ricard (1987). The frequency of occurrence (%FO) was estimated and the diatoms classified into three categories (I); accidental: FO < 0.10, secondary: 0.10 < FO < 0.50 and preferential: FO > 0.50 (Albertini-Berhault, 1973).

In sand gobies and sea bass, the food items of 100 and 26 stomach contents were identified to lowest possible taxonomic level and counted. The frequency of occurrence (%FO) and numerical frequency (%N) of macroinvertebrates were estimated according to Hureau (1970).

Estimation of POM exported by fish community

The amount of particulate organic matter exported by a fish community is mainly related to three factors: quantity of POM exported by species, weather conditions (seasonal fluctuation of composition of fish community) and duration of submersion at each tide. These variables were measured automatically. During each tide and for each species, we assessed the abundance and composition of fishes that grazed in the salt marshes, and, thus we estimated the POM they carried. For each species studied, we calculated the instantaneous ration (Ir), ratio between gut contents weight (FW) and body weight (BW). The difference of Ir between ebb and flood tide was assumed to represent the POM budget of each species during a tide cycle. Assuming that the quantity of fish which exploited the salt marsh is proportional to the total duration of the submersion, it is possible to assess the quantity of POM transferred per tide cycle and hence per month ($Q(\text{POM})_m$), where:

$$Q(\text{POM})_\text{m} = \text{CPUE}_\text{m} \times t_\text{m} \times \delta\,(\text{Ir}) \quad (4)$$

where: CPUE_m = average CPUE per species (g/mn^{-1}) during the month m, t_m = total duration (mn) of the submersion during the month m, and δ (Ir) = total dif-

Table 2. Composition of fish community caught in the canal between November 1995 and September 1996 during 12 tides. %Occ: percentage of samples in which the species was present,%BA: biomass abundance (%),%NA: numeric abundance (%), FL: fork lenght (mm), *n*: number of measured fish, ∗: no quantification of *A. anguilla*.

Families	Species	%Occ	%BA	%NA	FL min (mm)	FL max (mm)	FL mean (*n*)
Mugilidae	*Mugil* sp.	100	0.5	3.1	19	60	25 (454)
	Liza aurata (Risso, 1810)	33	1.7	0.1	95	200	136 (16)
	Liza ramada (Risso, 1826)	50	78.9	0.3	285	494	408 (50)
Gobiidae	*Aphia minuta* (Risso, 1810)	8	0.0	0.0	21	55	45 (9)
	Crystallogobius linearis (Von Düben, 1845)	8	0.0	0.0	33	38	36 (6)
	Pomatoschistus microps (Kroyer, 1838)	8	0.1	0.7	42	53	49 (7)
	Pomatoschistus minutus (Pallas, 1770)	83	8.1	60.3	19	58	40 (610)
	Pomatoschistus lozanoï (de Buen, 1923)	75	2.4	12.2	19	70	34 (117)
Serranidae	*Dicentrarchus labrax* (Linnaeus, 1758)	67	3.9	0.8	60	260	85 (200)
Atherinidae	*Atherina presbyter* (Cuvier, 1829)	17	1.8	1.4	61	85	72 (118)
Pleuronectidae	*Limanda limanda* (Linnaeus, 1758)	25	0.0	2.3	9	15	12 (150)
	Platichthys flesus (Linnaeus, 1758)	17	0.9	0.0	195	210	202 (4)
	P. flesus (larvae)	33	0.0	3.1	8	21	14 (200)
	Pleuronectes platessa (Linnaeus, 1758)	17	0.7	0.0	222	232	227 (2)
	P. platessa (larvae)	50	0.0	7.0	8	20	13 (200)
Clupeidae	*Clupea harengus* (Linnaeus, 1758)	25	0.2	1.3	23	110	37 (165)
	Sprattus sprattus (Linnaeus, 1758)	75	0.8	6.3	26	86	39 (515)
Gasterosteidae	*Gasterosteus aculeatus* (Linnaeus, 1758)	67	0.1	0.3	21	68	39 (45)
Ammodytidae	*Ammodytes tobianus* (Linnaeus, 1758)	8	0.0	0.4	14	20	17 (62)
Soleidae	*Solea vulgaris* (Quensel, 1806)	33	0.0	0.2	9	27	15 (30)
Syngnathidae	*Hippocampus hippocampus* (Linnaeus, 1758)	8	0.0	0.0	/	/	120 (1)
	Syngnathus abaster (Risso, 1826)	8	0.0	0.0	/	/	82 (1)
	Syngnathus rostellatus (Nilson, 1855)	25	0.0	0.1	65	99	81 (18)
Trachinidae	*Echiichtys vipera* (Cuvier, 1829)	8	0.0	0.0	/	/	11 (1)
Anguillidae	*Anguilla anguilla* (Linnaeus, 1758)	50	*	*	60	79	69 (35)

ference of Ir between flood and ebb during month m.

Results

Composition of the fish community

We caught 23 species belonging to 12 families (Table 2). Among these, eight occurred frequently and could be regarded as permanent species, including three species of mullets (*L. aurata*, *L. ramada* and *Chelon labrosus* (Risso, 1826)), two gobies (*P. minutus* and *P. lozanoï*), sea bass, sprat (*Sprattus. sprattus*) and three-spined stickleback (*Gasterosteus aculeatus*). Thus, the community's biomass was highly dominated by mullets (81.1%), especially by thin lipped grey mullet (78.9%) and secondarily by *Pomatoschistus* spp. (10.5%) and sea bass (3.9%). The community was numerically dominated by gobies (73.2%), especially by sand gobies (60.3%), flatfish larvae (12.6%) but only 3.5% are mugilids. Glass eels (*Anguilla anguilla*) and flatfish larvae were not quantified because the fyke net was not considered an efficient sampler of these small fishes. Except for gobies, mullets, sea bass and three-spined stickleback, the fishes were mainly found as larvae and juveniles and they only occurred for short period, in particular during early spring. During these early stages, most fishes feed on microplankton (Russel, 1976) which is transported the water during the tide; therefore, we assumed that they were not significantly involved in POM fluxes.

The salt marsh was colonised by fish every time the tide reached the creek. Nevertheless, abundances were variable: CPUE varied between 2.8 g min^{-1} (March 1996) and 400 g min^{-1} (September 1996) (Figure 2a). The first abundance peak occurred in November 1995;

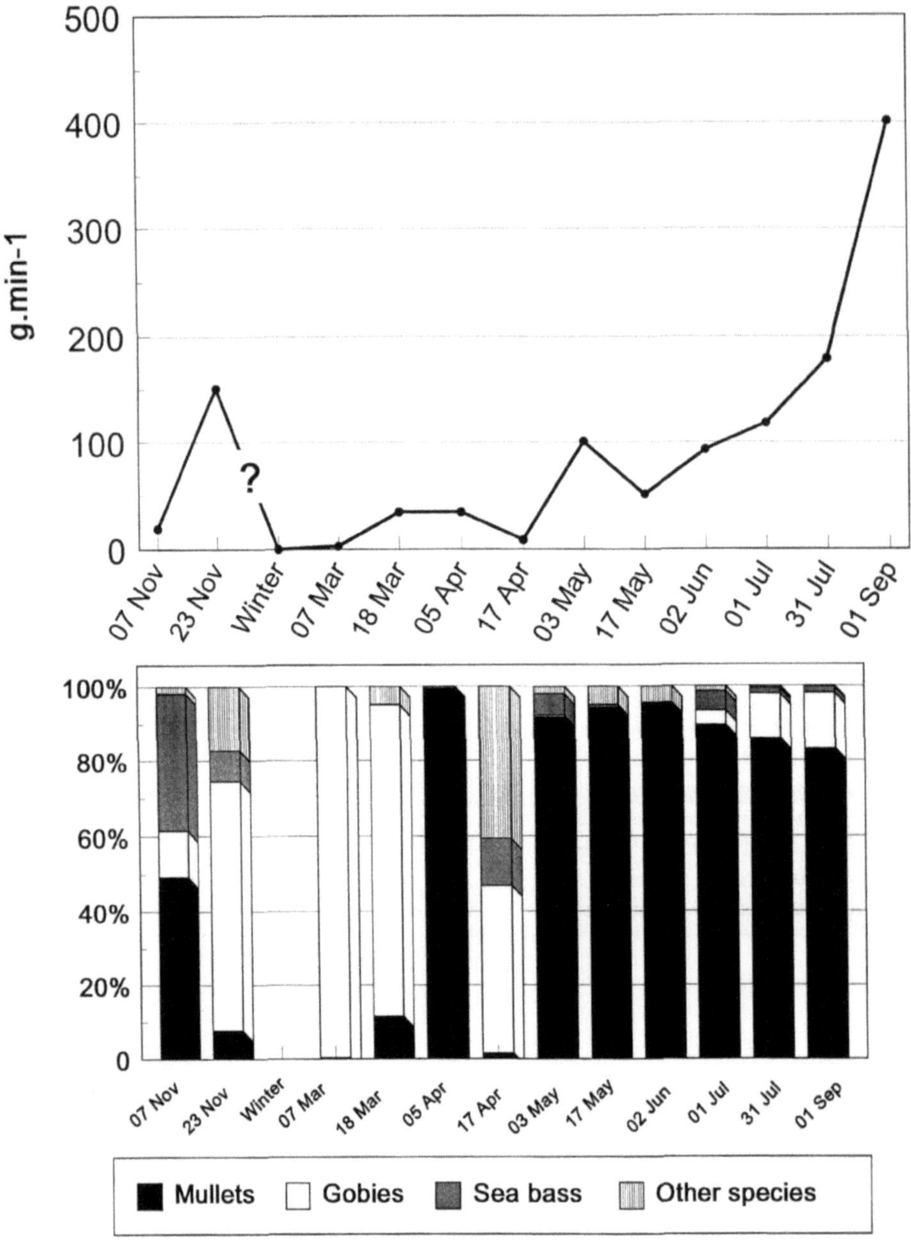

Figure 2. Seasonial fluctuations of the fish abundance. a. biomass (g) of fish caught per min (CPUE). b. biomass frequency (%).

it was mainly due to gobies (100 g.min^{-1}) (Figure 2b). No data was collected during winter but, considering the low water temperatures, we assume that abundances dropped and remained low in accordance with other sites (Feunteun, 1994) and in this creek in winter 1997 (Laffaille, unpublished data). The CPUE began to increase again at the beginning of March (2.8 g.min^{-1}) and in August (150 g.min^{-1}), when catches were generally dominated by mullets between April and September.

Thus, only five species were able to export significantly organic matter produced in the salt marshes or on the mud flats: two species of mullets (*Liza ramada* and *L. aurata*), two species of gobies (*Pomatoschistus minutus*, *P. lozanoï*) and the sea bass (*Dicentrarchus labrax*).

Table 3. Diatoms identified in the gut contents of *L. ramada* caught during ebb tide. %FO: frequency of occurrence, I: food items importance (P: preferential, S: secondary, A: accidental).

Food items	%FO	I	Benthic	Planktonic
Nitzschia sigma (Kützing) Wm. Smith	1.00	P	*	
Coscinodiscus excentricus Ehrenberg	1.00	P	*	
Scoliopleura tumida de Brébisson	1.00	P	*	
Podosira stelliger (Bailey) Mann	0.93	P	*	*
Surirella ovata Kützing	0.93	P	*	
Paralia sulcata (Ehrenberg) Kützing	0.93	P	*	
Actinoptychus undulatus (Bailey) Ralfs	0.87	P	*	*
Biddulphia rhombus (Ehrenberg) Wm. Smith	0.87	P	*	
Surirella gemma Ehrenberg	0.87	P	*	*
Raphoneis amphiceros Ehrenberg	0.87	P	*	
Navicula phyllepta Kützing	0.73	P	*	*
Auliscus sculptus (Wm. Smith) Ralfs	0.47	S	*	
Navicula palpebralis de Brélisson ex Wm. Smith	0.47	S	*	
Biddulphia mobiliensis Bailey	0.27	S	*	*
Navicula cincta (Ehrenberg) Van Heurck	0.27	S	*	
Melosira westii Wm. Smith	0.27	S	*	*
Gyrosigma obtusatum A. Boyer	0.20	S	*	
Navicula praetexta Ehrenberg	0.20	S	*	
Coscinodiscus rothii var. subsalsa (Juhl-Dannf) Hustedt	0.13	S	*	
Diploneis crabro Ehrenberg	0.13	S	*	
Navicula digito-radiata (Gregory) Ralfs	0.13	S	*	

Contribution to organic matter budget

Mullets

Mullets colonised the salt marshes at different life stages during the whole year. Large mullets (*L. ramada and L. aurata*) mainly occurred in November. Their abundances decreased in winter and started increasing again in April, reaching a peak in summer (Figure 2). This could be due to the fact that, in the northern hemisphere, mugilids graze more actively in spring and in summer (i.e, Hickling, 1970; Odum, 1970; De Silva & Witeryaratne, 1977; Albertini-Berhault, 1980). The scarceness of adult during winter could be related to the low temperatures, but also to a fasting period during gonad maturation between October and April (Cassifour, 1975).

Young mullets exclusively feed on zooplankton, but their diet focuses on phytobenthos with increasing age (i.e, Hickling, loc. cit.; Odum, loc. cit.; Blaber & Whitfield, 1977; Albertini-Berhault, loc. cit.). Therefore, we assume that only large individuals contribute significantly to organic matter fluxes. We found that the diet of large mullets was mainly composed of sediment, phytobenthos, zooplankton and various detritus. Most of the food items were diatoms. Among the 32 identified species, 11 dominant species were determined from the literature to be benthic forms (Table 3): *Nitzschia sigma*, *Coscinodiscus excentricus*, *Scoliopleura tumida*, *Podosira stelliger*, *Surirella ovata*, *Paralia sulcata*, *Actinoptychus undulatus*, *Biddulphia rhombus*, *Surirella gemma*, *Raphoneis amphiceros* and *Navicula phyllepta* (Peragallo & Peragallo, 1887; Hustdet, 1927–1933; 1937; 1959; Hendey, 1964; Patrick & Relmer, 1966; Germain,

Table 4. Comparison, between ebb tide and flow tide, of average instantaneous ration (%Ir) and vacuity index (%V) of *L. ramada*. sd: standard deviation, n: number of digestive contents analysed.

	% Ir		%V	*n*
	average	sd		
Flood tide	5.86	1.43	0	11
Ebb tide	12.71	3.15	0	39
Difference	*6,.5*			
T test	$p < 0{,}01\%$			

Table 5. Proportion of total organic carbon (%Ctot) and nitrogen (%Ntot) and the molar C/N ratio determined in mullet's digestive contents. sd: standard deviation, *n*: number of digestive contents analysed.

	%Ctot	%Ntot	C/N
Average	5.74	0.78	8.69
sd	1.34	0.21	0.89
n	10	10	10

1981; Ricard, 1987). At the same period, these species were also present in the creeks sediment (Brosse et al., 1996). This suggests that *L. ramada* grazed the superficial sediment layer and confirms several former studies (i.e. Hickling, loc. cit.; Odum, loc. cit.; Fagade & Olaniyan, 1973; Brusle, 1981; Almeida et al., 1993). We found a confirmation of these observations in the occurrence of grazing marks in the silt of the channels after the passage of mullets. None of the sampled mullets had empty guts ($V=0$) during ebb or during flood (Table 4). This shows that the grazing started on the mud flats adjacent to the marsh during flood. However, the average repletion index increased from 5.9%, when the mullets arrived from the sea with the flood, to 12.7% when they left the marsh at ebb. Thus, on average, mullets swallowed about 6.9% of their body weight while they grazed on the salt marsh at high tide.

A significant regression correlation ($r=0.96$) was found between the dry weight and fresh weight of the gut contents of 30 mullets. On average, DW is 66.9% of FW (sd = 7.44). However, no significant relation ($r=0.26$) was found between POM biomass and total gut contents (FW). The proportion of POM was comprised between 24.2% and 64.3% (average = 36.3%, sd = 13.1). Therefore, it was not possible to predict POM biomass from total ingested food biomass. This result was not surprising, because the proportion of organic matter depends on factors such as feeding behaviour, sediment composition, site, etc. For instance, Farrugio (1975) assessed that one third of the gut contents was composed of sediment, and Jacot (1920) showed that 40% of *Mugil* spp. stomach content were 'non digestible matter'. Tung (1971, *in* Brusle, 1981) showed that the proportion of organic matter varied from 65% of sediment when mullets fed on diatoms to only 35% when they fed on copepods. Moreover, the proportion of POM varies considerably according to seasons and to trophic areas (Almeida et al., 1993). The proportion of total organic carbon and nitrogen contained in the digestive tract of ten mullets (Table 5) averaged 5.7% (sd = 1.34) and 0.8% (sd = 0.21). These proportions were low, certainly because sediment was important in the gut contents. The average C/N ratio is 8.7% (sd = 0.89). Thus, inter-individual differences appeared to be relatively low.

Gobies

Five species of *Gobiidae* were sampled in the creek. But, because of their commonness, only *Pomatoschistus minutus* and *P. lozanoï* were significantly involved in organic matter fluxes. Gobies were present in the salt marsh during the whole study period except at the end of May (Figure 2 b). At the beginning of May, only 7.1% of males were present whereas 39.4% were mature females full of ovocytes. At the end of May, no gobies were caught and after this date high concentrations of larvae occurred (up to 125 gobies min^{-1}), but the biomass was low (3.6 g min^{-1}) which suggests that gobies did not participate significantly to organic matter flux at that period of year. A high abundance of adults occurred in autumn, when CPUE ranged between 60 and 100 g mn^{-1}.

Pomatoschistus minutus is considered as a generalist predator (Hamerlynck & Cattrijsse, 1994). But, with regard to the food items in the gut (Table 6), *Pomatoschistus* sp. can also be considered as an opportunist. They can either feed on microinvertebrates, but also on macroinvertebrates, and cases of cannibalism are occasionally observed. *Orchestia gammarellus* were the preferential food items in the canal (%OF = 0.600 and %N = 0.710). This amphipod is a resident species at Mont Saint-Michel salt marshes (Fouillet, 1986).

In *Pomatoschistus minutus*, V was low ($V=1.5\%$) during flood tides, but it dropped to 0% during ebb a few minutes later (Table 7). Meanwhile, Ir increased from 9.7% (flood tide) to 14.4% (ebb tide), indicating that these fishes fed actively while they colonised the salt marshes and that they exported on average 4.7% of their body weight during a tidal cycle. A significant correlation was found between dry weight and fresh weight of guts contents ($r=0.96$). On average, DW represents 59% (sd = 8.3) of gut contents.

Sea bass

Dicentrarchus labrax were present as juveniles stages, 0+ and 1+ (Table 2). They were absent in winter, but started arriving in the middle of spring (Figure 2). Un-

Table 6. Food items identified in the stomach contents of *D. labrax* ($n = 26$) and *P. minutus* ($n = 100$) caught during ebb tide. %FO: frequency of occurrence,%N: numeric frequency, *n*: number of stomach contents analysed.

Food items	Sea bass ($n = 26$)		Sand gobies ($n = 100$)	
	%FO	%N	%FO	%N
Orchestia gammarellus (Pallas)	0.923	0.666	0.600	0.710
Sphaeroma rugicauda (Leach)	0.538	0.244	0	0
Eurydice pulchra (Leach)	0.308	0.014	0.050	0.025
Corophium volutator (Pallas)	0.346	0.020	0	0
Bathyporeia sp.	0.192	0.028	0	0
Neris diversicolor (O.F. Müller)	0.269	0.011	0.180	0.076
Mysidacea n.i.	0	0	0.160	0.143
Crangon crangon (Linnaeus)	0.077	0.003	0	0
Carcinus maneas (Linnaeus)	0.115	0.003	0	0
Tellinidae sp.	0.038	0.001	0	0
Insect	0.115	0.005	0	0
Fish	0	0	0.050	0.025
Other	0.077	0.005	0.050	0.021

Table 7. Comparison, between ebb tide and flow tide, of average instantaneous ration (%Ir) and vacuity index (%V) of *P. minutus* caught the 18th March. sd: standard deviation, *n*: number of digestive contents analysed.

	% Ir		%V	*n*
	average	sd		
Flood tide	9.71	3.84	1.45	72
Ebb tide	14.4	4.61	0	141
Difference	*4.69*			
T Test	$p < 0{,}01\%$			

Table 8. Comparison, between ebb tide and flow tide, of average instantaneous ration (%Ir) and vacuity index (%V) of *D. labrax* caught in November 1995. sd: standard deviation, *n*: number of digestive contents analysed.

	% Ir		%V	*n*
	average	sd		
Flood tide	0.96	1.84	61.8	99
Ebb tide	11.04	2.24	0	68
Difference	*10.08*			
T Test	$p < 0{,}01\%$			

til the beginning of summer, their abundances were generally low (0 to 5.1 g min^{-1}), and peaked at the end of summer and especially in autumn (7 to 13 g min^{-1}).

The diet was exclusively composed of macroinvertebrates (Table 6), mainly the amphipod *Orchestia gammarellus*, which occurred in 92% of the stomach, and secondarily the isopod *Sphaeroma rugicauda*, found in 54% of the stomach contents. These preferential food items (%N = 0.910) are resident species in the Mont Saint-Michel salt marshes (Fouillet, 1986).

During flood, vacuity (Table 8) averaged 61.8% and dropped to 0% when the bass left the marsh at ebb. Ir (Table 8) increased from 0.96% at flood to 11.04% at ebb tide. The sea bass ingested about 10.08% of their body weight while they were in the salt marshes. This organic matter was then transferred to the marine coastal waters. There was a significant correlation between DW and FW of the gut contents ($r = 0.97$). Thus, it was possible to predict the DW from the FW as DW represented 62% (sd = 6.6) of the FW.

Discussion – Conclusion

At least 90 fish species are present in the Bay of Mont Saint-Michel (Beillois et al., 1979; Legendre, 1984; Feunteun & Laffaille, 1997). Of these, 23 species colonise the salt marsh mainly during their early life stages. Other species are also known to frequent salt marshes, for instance shad *Alosa alosa*

(Linnaeus, 1758) and salmon *Salmo salar* (Linnaeus, 1758) are occasionally caught in traditional fisheries (Lefeuvre, unpublished data), and juveniles of *Belone belone* (Linnaeus, 1758) and *Lepadogaster lepadogaster* (Bonnaterre, 1788) have also been collected during recent samplings (Laffaille, unpublished data). Therefore, the intense productivity of this environment, that is known to be exploited by birds (Schricke, 1983; Lefeuvre et al., 1994), is also actively foraged by fish communities. The Bay of Mont Saint-Michel (Beillois et al., loc. cit.) as with comparable shallow marine coastal waters such as estuaries and lagoons (i.e. Costa, 1988; Elie et al. 1990), are known to play a nursery role for many fish species. This function is usually described as restricted to the mud flats of the midlittoral zones in Mont Saint-Michel Bay. The present study shows that the salt marshes of the supralittoral zone also play an important nursery role (i.e. Boesch & Turner, 1984). Salt marshes are believed to serve as nurseries for many fishes as a result of the high production of vascular plant detritus and diatoms. The refuge role generally described in shallow bays may not be as significant in Mont Saint-Michel Bay; despite scarce predators (a few seals, piscivorous birds and large fish), salt marshes are submerged for too short period (max. 2 h) and the currents are too strong (megatidal system) to represent a signifiant refuge role for young fish.

We found that a minimum of five fish species feed actively in the salt marshes of Mont Saint-Michel Bay during tidal floods: two species of mullets, two species of gobies and juvenile sea bass. These five seem to colonise marshes to exploit their high productivity. Mullets exploit the primary productivity (mainly diatoms), whereas gobies and sea bass exploit the secondary productivity (especially *Orchestia gammarellus*). These resident species feed on diatoms and vascular plant detritus (Creach, 1995). Moreover, these fishes play a dominant role in the food web and in the fisheries of the Mont Saint-Michel Bay either as prey of other fishes and birds (gobies, and young mullets) or as valuable commercial species (sea bass) (Feunteun & Laffaille, 1997). Therefore, the primary productivity of salt marshes appears to play a fundamental role in the global functioning of Mont Saint-Michel Bay.

The magnitude of organic matter fluxes mainly depends on the biomass of fish that forage on the salt marsh during the tidal cycle. To summarise, during the tidal cycles, mullets export a mixture of sediment and organic matter from the salt marshes to the marine coastal waters of the Bay. During a given tide cycle, each fish exports about 6.9% of its fresh weight, comprising 42.6% of sediment, 24.3% of particulate organic matter and 33.1% of water. The dry gut contents POM averages 5.7% organic C and 0.8% organic N. Sand gobies and sea bass mainly feed and export respectively 4.5% and 10% of their body weight during a tide cycle. The difference of Ir takes into account excretion and gastric evacuation in the creek. Retention time of the ingested material is highly variable. For example, it ranges from 2 to 6 h for mullets (Brusle, 1981), which is much longer that the tidal submersion period of the creeks and the salt marsh. Therefore we assume that POM is low in the salt marsh and creeks during the tide cycle.

In order to assess the relative importance of POM fluxes from the salt marsh due to fish, we assume that respectively *Liza ramada* and *L. aurata*, and *Pomatoschitus minutus* and *P. lozanoï* export, on average an equal quantity and assumable quality of organic matter. Since sea bass and gobies' diets were largely dominated by *Orchestia gammarellus*, we used the proportion of Ctot and Ntot assessed by Creach (1995) in the Mont Saint-Michel Bay: Ctot $= 50.9 \pm 4.13$% ($n = 20$), Ntot $= 8.7 \pm 0.97$ ($n = 20$) and C/N $= 9.7 \pm 0.9$ ($n = 20$). On this basis, we estimated the organic matter fluxes due to these fish per tide cycle and per month (Figure 3). During the study, a total of 50 g ha^{-1} to 1700 g ha^{-1} of organic matter (DW) was exported monthly towards the marine coastal waters. On average, 67% of this organic matter is exported by mullets 23% by gobies and 10% by sea bass. But during late autumn, gobies export more than 70% of the organic matter transported by fish. The study site has a central location among the salt marshes of the Bay of Mont Saint-Michel. Therefore, assuming such a geographical representativity, the extrapolation of our data set permits to assess that the fish communities export about 50 tons DW of organic matter per year from the salt marshes towards the sea, comprising 10.3 tons of organic C and 1.7 tons of N. Of course, part of the POM exported by fish is decayed and can be imported later by tidal currents as POM, DOM or dissolved nutrients. However, we assume that the POM output by fish was very low compared to other sources during the study period. More, recent samplings have shown that the biomass of mullets present in the creek may be at least 15 times higher (unpublished data) which suggests that important interannual variations of fish contribution to organic matter fluxes between salt marshes and marine coastal waters are expected.

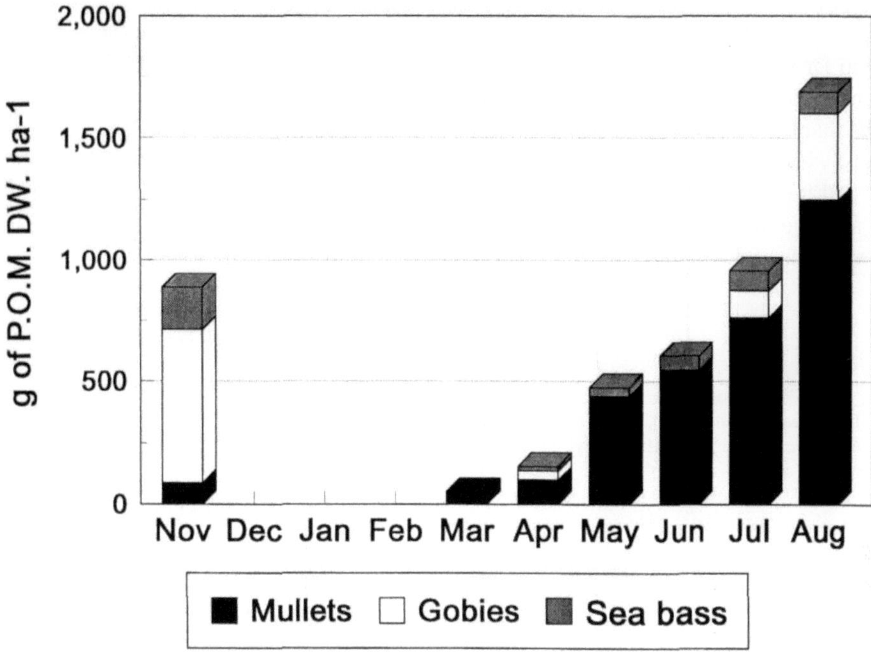

Figure 3. Monthly transfers of particulate organic matter by fishes between the studied salt marshes and marine coastal waters in the bay of Mont Saint Michel, between November 1995 and August 1996.

Even if the total input of OM to the Bay of Mont Saint-Michel has been estimated in the past years (Troccaz, 1996), we still do not know the proportion of this organic matter which is exported by fish. In order to improve the understanding of the coastal biogeochemical cycles, it is interesting to compare the POM budgets due to fish communities with those due to abiotic vectors (tidal currents, interstitial water, etc.). On a Bay-wide scale, the salt marsh traps POM during the tide cycles: for instance in 1991, the input was 2.9 kg m^{-2} (Troccaz, 1996; Troccaz & Giraud, 1996). In comparison, during the ebb tides between 1991 and 1994, Troccaz et al. (1994), Troccaz & Giraud, (loc. cit.) and Troccaz (loc. cit.) estimated that between 0.7 g m^{-2} y^{-1} and 5.8 g m^{-2} y^{-1} of total N, and 83.3 g m^{-2} y^{-1} to 114.2 g m^{-2} y^{-1} of total organic C was exported from these salt marshes. Of course, such comparison is difficult because of several uncertainties. Firstly, we noted an important temporal heterogeneity of the POM budgets due to fish transport and also to abiotic vectors. Therefore, a simultaneous survey of both POM fluxes vectors ought to be conducted a longer period. Secondly, the assumption that the study site is reprentative of the whole salt marsh will have to be verified because its productivity depends on several factors. For instance, in the parts of the marsh which are grazed by sheep, the original vegetation cover, dominated by *Atriplex portulacoides,* is replaced by *Puccinellia maritima*. Thirdly, the 4000 ha of salt marshes have variable ages, the oldest have had about 60 y of existence, whereas the youngest are only a few years old: 30 ha of pioneer mud flat zones are colonised every year, especially by *Spartina* spp., and *Salicorna* spp. (Levasseur et al., 1995). It has been shown that salt marshes have a tendancy to export more OM as they get older (Dame & Lefeuvre, 1994). Therefore, these variables add another variability source which has to be investigated in order to make accurate comparisons between the OM bugets due to fish communities and abiotic vectors. Nevertheless, as a preliminary approach, we assume that the fish community is responsible for part of the output of nitrogen and carbon from the salt marsh, which varies strongly according to years, and to characteristics of the salt marsh. We conclude that fish communities play an important role in the global functioning of the Bay and more generally of coastal environments, at least in terms of organic matter fluxes.

Acknowledgements

We are grateful to Dr M. T. Schricke and B. Savouré for identification of the macro-invertebrates and diatom species. We thank other members of Laboratoire d'Evolution des Systèmes Naturels et Modifiés for their laboratory and field assistance. We are indebted to V. Bouchard and two anonymous referees for their remarks and corrections of the manuscript.

References

Albertini-Berhault, J., 1973. Biologie des stades juvéniles de téleostéens mugilidae *Mugil aurata* Risso, 1810, *Mugil Capito* Cuvier, 1829 et *Mugil Saliens* Risso, 1810. 1. Régime alimentaire. Aquaculture 2: 251–266.

Albertini-Berhault, J., 1980. Biologie des stades juvéniles de *Mugilidae* dans la région marseillaise. Thesis, University of Aix-Marseille 2, 156 pp.

Almeida, P. R., F. Moreira, J. L. Costa, A. Assis & M. J. Costa, 1993. The feeding strategies of *Liza ramada* (Risso, 1826) in fresh and brackish water in the river Tagus, Portugal. J. Fish Biol. 42: 95–107.

Beillois, R., D. Desaunay, D. Dorel & M. Lemoine, 1979. Nurseries littorales en Baie du Mont Saint-Michel et du Cotentin Est. Rapport I.S.T.P.M.., 115 pp.

Bildstein, K. L., E. Blood & P. Frederick, 1992. The relative importance of biotic and abiotic vectors in nutrient transport. Estuaries 15: 147–157.

Blaber, S. J. M. & A. K. Whitfield, 1977. The feeding ecology of juvenile mullet (*Mugilidae*) in south-east African estuaries. Biol. J. linn. Soc. 9: 277–284.

Blanchard, G., M. J. Chretiennot-Linet, A. Dinet & J.-M. Robert, 1988. Méthode simplifiée pour l'extraction du microphytobenthos des sédiments marins par le gel de silice Ludox. C.R. Acad. Sci. Bio. Mar. Paris 307: 569–576.

Boesch, D. F. & R. E. Turner, 1984. Dependence of fishery species on salt marshes: the role of food and refuge. Estuaries 7: 460–468.

Bouchard, V., 1996. Production et devenir de la matière organique des halophytes dans un marais salé européen en système microtidal (Baie du Mont Saint-Michel). Thesis, University of Rennes 1: 218 pp.

Brosse, S., P. Laffaille, B. Le Rouzic & G. Pinay, 1996. Intersystem exchanges, role of epipelic algae in the retention and transfer of matter between a salt marsh and coastal sea water. International Congress of Limnology-Oceanography 'Ocean, rivers and lakes: energy and substance transfers at interfaces', Nantes (France), 7–9 October 1996.

Brusle, L., 1981. Food and feeding in grey mullet. In Orens O. H. (ed.), Aquaculture of the Grey Mullet. Cambridge University Press: 185–217.

Cassifour, P., 1975. Contribution à l'étude des Mugilidés des réservoirs à poissons du bassin d'Arcachon. Thesis, University of Bordeaux 1: 122 pp.

Chambers, R. M., J. W. Harvey & W. E. Odum, 1992. Ammonium and phosphate dynamics in a Virginia salt marsh. Estuaries 15: 349–359.

Costa, M. J., 1988. The Tagus and Mira estuaries (Portugal) and their role as spawning and nursery areas. J. Fish Biol. 33A: 249–250.

Creach, V., 1995. Origines et transferts de la matière organique dans un marais littoral: utilisation des compositions isotopiques naturelles du carbone et de l'azote. Thesis, University of Rennes 1: 134 pp.

Dame, R. F., T. Chrzanovski, K. Bildstein, B. Djerfue, H. Mc Kellar, D. Nelson, J. Spurrier, S. Stancyk, H. Stevenson, J. Vernberg & R. Zingmark, 1986. The outwelling hypothesis and North Inlet Estuary, South Carolina. Mar. Ecol., Prog. Ser. 33: 217–229.

Dame, N. & J. C. Lefeuvre, 1994. Tidal exchanges: import-export of nutrients and organic matter in New and old World salt marshes: conclusions. In Mitsch W. J. (ed.), Global Wetlands old World and New, Amsterdam: 181–201.

De Silva, S. S. & M. J. S. Witeryaratne, 1977. Studies on the biology of young grey mullet, *Mugil cephalus* L. Aquaculture 12: 157–167.

Elie, P., E. Feunteun & C. Rigaud, 1990. The inshore brackish water domain of the French Atlantic coast: ecological functions for the exploited species-impact of physical development. Bull. Ecol. 21: 33–38.

Fagade, S. O. & C. I. O. Olaniyan, 1973. The food and feeding interrelationship of the fishes in the Lagos lagoon. J. Fish Biol. 5: 205–225.

Farrugio, H., 1975. Les muges (Poissons Téléostéens) de Tunisie. Répartition et pêche. Contribution à leur étude systèmatique et biologique. Thesis, University of Montpellier, 201 pp.

Farrugio, H., 1977. Clés commentées pour la détermination des adultes et des alevins de *Mugilidae* de Tunisie. Cybium 3: 57–73.

Feunteun, E., 1994. Le peuplement piscicole du marais littoral endigué de Bourgneuf-Machecoul (France, Loire-Atlantique). Approche méthodologique pour une analyse quantitative de la distribution spatiale du peuplement piscicole et de la dynamique de certaines de ses populations. Thesis, University of Rennes 1: 240 pp.

Feunteun, E. & L. Marion, 1994. Assessment of Grey Heron predation on fish communities: the case of the largest European colony. Hydrobiologia 279/280: 327–344.

Feunteun, E. & P. Laffaille, 1997. Les peuplements piscicoles de la Baie du Mont Saint-Michel. Penn Ar Bed. 164: 50–56.

Fouillet, P., 1986. Evolution des peuplements d'Arthropodes des schorres de la Baie du Mont Saint-Michel. Thesis, University of Rennes 1, 330 pp.

Germain, H., 1981. Flore des diatomées. Diatomophycées. Eaux douces et eaux saumâtres du Massif Armoricain et des contrées voisines d'europe occidentale. Boubée (ed.), Paris, 444 pp.

Hamerlynck, O. & A. Cattrijsse, 1994. The food of *Pomatoschistus minutus* (Pisces, Gobiidae) in Belgian coastal waters, and a comparison with the food of its potential competitor *P. lozanoi*. J. Fish Biol. 44: 753–771.

Hendey, N. I., 1964. An introductory account of the smaller algae of British coastal waters. Part 5: Bacillariophyceae (Diatoms). Fisheries Invest. London ser. 4, 298 pp.

Hermond, H. F. & J. L. Fifield, 1981. Subsurface flow in salt marsh peat: a model and field study. Limnol. Oceanogr. 26: 795–800.

Hickling, C. F., 1970. A contribution a natural history of the English grey mullets (*Pisces*, *Mugilidae*). J. Mar. Biol. Ass. U.K. 50: 609–633.

Hureau, J., 1970. Biologie comparée de quelques poissons antarctiques (Nototheneiidae). Bull. Institute Oceanogr. Monaco 68: 1–244.

Hustedt, F., 1927–1933, 1937 & 1959. Die Kieselagen Deutschlands, Österreichs und der Schweiz mit Berücksichtigung der übrigen Länder Europas sowie der angrenzenden Meeresgebiete. In Rabenhorst Kryptogamen Flora. 7, Leitzig: 920 pp., 845 pp. & 816 pp.

Jacot, A. P., 1920. Age, growth and scale characters of *Mugil cephalus* and *Mugil curena*. Trans. am. Micros. Soc. 39: 199–230.

Lefeuvre, J. C., G. Bertru, F. Burel, L. Brient, V. Creach, Y. Guené, J. Levasseur, A. Mariotti, A. Radureau, C. Retière, B. Savouré & O. Troccaz, 1994. Comparative studies on salt marsh processes: Mont Saint-Michel Bay, a multi-disciplinary study. In Mitsh W. J. (ed.), Global Wetlands: Old World and New, Amsterdam: 215–234.

Lefeuvre, J. C. & R. F. Dame, 1994. Comparative studies of salt marsh processes in the New and the Old Worlds: an introduction. In Mitsh W. J. (ed.), Global Wetlands: Old World and New, Amsterdam: 169–179.

Legendre, C., 1984. La pêche artisanale sur le domaine intertidal de la Baie du Mont Saint-Michel. Rapport CEE Env. et Min. Env. Laboratoire d'Evolution des Systèmes Naturels et Modifiés, University of Rennes 1, 122 pp.

Levasseur, J., M. Ameline, L. Gresillon-Legea, M. C. Guenegou, M. J. Le Garrec, J. Le Rhun & G. Perez, 1995. Etude de la production primaire. In Lefeuvre J. C. (ed.), Transferts entre systèmes écologiques terrestre et marin en Baie du Mont Saint-Michel: rôle des marais salés. Laboratoire d'Evolution des Systèmes Naturels et Modifiés, University of Rennes 1: 7–44.

Marion, L., P. Clergeau, L. Brient & G. Bertru, 1994. The importance of avian-contributed nitrogen (N) and phosphorus (P) to Lake Grand-Lieu, France. Hydrobiologia 279/280: 133–147.

Nixon, S. W., 1980. Beetwen coastal marshes and coastal waters: A review of 20 y of speculation and research on the role of salt marshes in estuarine productivity and water chemistry. In Hamilton, P. & K. B. Mac Donald (eds), Estuarine and Wetland Processes. Plenum Publishing Co., New York: 437–525.

Odum, E. P., 1968. Energy flow in ecosystems: a historical review. Am. Zool. 8: 11–18.

Odum, W. E., 1970. Utilisation of the direct grazing and plant detritus food chains by the stripped mullet *Mugil cephalus*. In Steel J. H. (ed.), Marine Food Chain. Olivier & Boyd, Edimbourgh: 22–240.

Patrick, R. & C. W. Relmer, 1966. The diatoms of the United States. Vol. 1. Monographs of the Academy of Natural Sciences of Philadelphia, 687 pp.

Peragallo, H. & M. Peragallo, 1897. Diatomées marines de France et des districts maritimes voisins. M. J. Tempére (ed.), Grez-sur-Loing, Vol. 1: 540 pp.; Vol. 2: 137 pl.

Pomeroy, L. R. & R. G. Wiegert, 1981. The Ecology of a Salt Marsh. Ecological Studies 38, Springer-Verlag, New York, 271 pp.

Reay, P. J. & V. Cornell, 1988. Identification of grey mullets (Teleosteï, Mugilidae) juveniles from British waters. J. Fish Biol. 32: 95–99.

Ricard, M., 1987. Atlas du phytoplancton marin. Vol. 2. Les Diatomophycées. CNRS (ed.): 292 pp.

Russel, F. S., 1976. The Eggs and Planktonic Stages of British Marine Fishes. Academic Press, London, 524 pp.

Schricke, V., 1983. Distribution spatio-temporelle des populations d'Anatidés en transit et en hivernage en Baie du Mont Saint-Michel en relation avec les activité humaines. Thesis, University of Rennes 1, 299 pp.

Serventi, M., I. J. Harrison, P. Torricelli & G. Gandolfi, 1996. The use of pigmentation and morphological characters to identify Italian mullet fry. J. Fish Biol. 49: 1163–1173.

Teal, J. M., 1962. Energy flow in the salt march ecosystem of Georgia. Ecology 43: 614–634.

Troccaz, O., 1996. Evolution de la dynamique d'un marais salé: processus fonctinnels internes et relations avec le milieu côtier (la Baie du Mont Saint-Michel). Diplôme Doctoral de Recherches, University of Rennes 1, 105 pp.

Troccaz, O., F. Giraud, G. Bertru & J. C. Lefeuvre, 1994. Methodology for studying exchanges between salt marshes and coastal marine waters. Wetlands Ecol. Mgmt 3: 37–48.

Troccaz, O. & F. Giraud, 1996. A nutrient budget methodology in the Mont Saint-Michel Bay salt marsh, France. In INRA (ed.), Hydrologie dans les pays celtiques, Les Colloques n° 79, Paris: 369–376.

Valiela, I. & J. M. Teal, 1979. The nitrogen budget of a salt marsh ecosystem. Nature 280: 652–656.

Wheeler, A., 1969. The fish of the British Isles and North West Europe. Macmillan, London, 613 pp.

Hydrobiologia **373/374:** 135–148, 1998.
J. C. Amiard, B. Le Rouzic, B. Berthet & G. Bertru (eds),
Oceans, Rivers and Lakes: Energy and Substance Transfers at Interfaces.

Nitrate reductase activity of phytoplankton populations in eutrophic Lake Aydat and meso-oligotrophic Lake Pavin: a comparison

C. Mallet, M. F. Charpin & J. Devaux
Laboratoire de Biologie Comparée des Protistes, UPRESA CNRS 6023, Université Blaise Pascal Clermont II, 63177 Aubiere cedex, France

Key words: nitrate reductase activity, freshwater phytoplankton, meso-oligotrophic lake, eutrophic lake

Abstract

From April 1993 to May 1994, samples were taken monthly from the epilimnion, metalimnion and hypolimnion of two lakes: the eutrophic Lake Aydat, and the meso-oligotrophic Lake Pavin. Nitrate reductase (NR) activity was measured according to Hochman's (1986) modified procedure, to estimate potential nitrate assimilation by algal cells. Although the inorganic nitrogen (Ni) source was essentially in the oxidized form in both lakes ($77 \pm 24\%$ and $81 \pm 7\%$ respectively in Lake Aydat and in Lake Pavin), specific NR activity (sNR) was 10.5 times higher in Lake Pavin than in Lake Aydat. Moreover in this latter ecosystem the distribution of specific NR activity was sporadic. In Lake Pavin, NR activity was correlated with nitrate concentration in the medium, nitrogen nutrient ratios (NO_3/Ni, NH_4/Ni and NO_3/NH_4) and biomass of some diatoms and Chlorophyceae. However, in the eutrophic ecosystem the temporal changes in NR activity were more complex. The specific NR activity seemed to be independent of the available inorganic nitrogen substrate but linked to the species composition of the natural phytoplankton community. As in Lake Pavin, the distribution of NR activity seemed to be linked to the biomass of some diatoms. However, this was insufficient to explain the sporadic distribution of NR activity. As ammonification and NH_4 regeneration could occur rapidly and irregularly, NH_4 could be preferentially assimilated.

Introduction

In contrast to phosphorus, which is widely considered to limit algal growth, there have been few studies on the role of nitrogen in regulating phytoplankton populations in freshwater ecosystems. In such ecosystems, seasonal algal succession is largely conditioned by a combination of physical, chemical and biological changes (Reynolds, 1990). Although inputs are low during summer stratification in eutrophic lakes, this season is paradoxically typified by the development of large phytoplankton biomasses (Aleya & Devaux, 1992), suggesting enhanced nutrient uptake capacities or enzymatic adaptation to environmental constraints. In natural environments, nitrate is not the only source of nitrogen available to phytoplankton. But ammonium is the only inorganic nitrogen source to be taken up and metabolised directly by phytoplankton. Nitrate must be first reduced to ammonium by nitrate reductase and nitrite reductase before N-NO_3 can be incorporated. Measurements of nitrate reductase (NR) activity have been used as a qualitative indicator and quantitative measure of nitrate assimilation (i.e. incorporation into macromolecules) (Eppley et al., 1969; Collos & Slawyck, 1977).

To obtain a better understanding of the influence of different variables on NR activity in the field, a study was conducted on the spatial and temporal distribution of this enzyme activity in two lakes of different trophic status: eutrophic Lake Aydat (France) and meso-oligotrophic Lake Pavin (France). An attempt was made to compare the annual distribution of NR in these lakes of differing inorganic nitrogen nutrient availability, and to estimate impact of thermal structure on enzyme activity.

Description of the study site

Lake Aydat was formed when a lava flow dammed the small river Veyre. It is a dimictic lake (60 ha) with a maximum depth of 15.5 m (Millerioux, 1976). The lake becomes thermally stratified from spring onwards, the hypolimnion becoming oxygen-depleted. It could be defined as eutrophic because in the epilimnion, the summer inorganic nitrogen concentration was about 70 μmol l^{-1} N. According to Wetzel (1983), values between 35 and 107 μmol l^{-1} N are typical of eutrophic lakes. Nitrite never represented more than 2% of inorganic nitrogen. The soluble inorganic phosphorus concentration was between 10 and 40 μg l^{-1} P (Planche-Jamet, 1995). Our sampling station was in the centre of the lake at the deepest point.

The circular Lake Pavin of 44 ha with a maximum depth of 98 m (Lair, 1975), occupies a volcanic explosion crater, formed in the last glaciation. This lake is dimictic (with partial overturn) and meromictic. It is thermally stratified. The chemocline fluctuates between -50 and -60 m and the mixolimnion which accounts for 80% of the total water volume is the subject of this work. The anaerobic monimolimnion ($Z > -55$ m) rules out any life. This lake presents oligotrophic characters but during summer inorganic nitrogen concentrations in the epilimnion reach 42 μmol l^{-1} N. This value lies within the range of 14.3 to 46 μmol l^{-1} N typical of meso-oligotrophic ecosystems (Wetzel, 1983). Thus Lake Pavin could be described as a meso-oligotrophic lake. In this ecosystem, nitrite concentrations remained below the detection limit (1.5 nmol l^{-1}) throughout the sampling period. As in Lake Aydat, the soluble inorganic phosphorus concentration was between 10 and 40 μg l^{-1} P (Planche-Jamet, 1995). The sampling station was near the centre of the lake at the deepest point.

We here compare NR activities of the main phytoplankton communities encountered in these ecosystems in relation to different inorganic nitrogen fractions in both two lakes.

Material and methods

Natural phytoplankton populations were collected monthly from April 1993 to May 1994, in Aydat at 1 m (epilimnion), 5 m (metalimnion), 10 m (hypolimnion); and at 5 m (epilimnion), 10 m (metalimnion), 15 m (under the thermocline, where Dinophyceae developed abundantly), and 40 m (hypolimnion) in Pavin. Samples were collected using a Van Dorn bottle and were given a preliminary sieving though a 200 μm mesh to remove most zooplankton (Amblard, 1986). Then, they were transported refrigerated in dark PVC bottles to the laboratory where the samples were filtered and frozen (-20 °C) the same day.

Temperature and oxygen concentrations were measured with an YSI GRANT/3800 probe.

The same day (about 8 h after collection), dissolved inorganic nitrogen concentrations were determined spectrophotometrically on triplicate samples filtered on Whatman GF/F. External nitrate concentrations corrected for the presence of nitrite were measured using a Hach Nitraver® kit, based on the cadmium reduction method (limit of detection 2.0 μmol l^{-1} N). External ammonium concentrations were measured using indophenol blue (Solórzano, 1969) (limit of detection 0.1 μmol l^{-1} N). The total inorganic nitrogen concentration (Ni) was obtained by summing the concentrations of the nitrogen ions.

Algal cells were counted with an inverted microscope (Wild M40) according to Utermohl's (1958) method modified by Legendre & Watt (1971, 1972). The specific biomasses, evaluated from mean biovolume of each species, were converted to fresh weights by assuming a cell density of 1 g ml^{-1} (Lohmann's, 1908). The relative biomasses of different algal groups were calculated as a percent of total biomass. The different species presented in this study, represented the major species during the whole sampling cycle.

Photosynthetic pigments were harvested on GF/F glass-fibre filters with $MgCO_3$ to prevent degradation of chlorophyll *a* in the extract. The filters were stored at -20 °C for up to three weeks before they were assayed by spectrophotometry of acetone (90%) extracts. The chlorophyll *a* concentrations, which represented the chlorophyll *a* concentrations after correction of pheopigment concentrations, were calculated using the equations of Lorenzen (1967).

Potential nitrate reductase activity was measured according to a Hochman et al. (1986) simplified assay protocol. This protocol was modified: samples were filtered under low vacuum onto precombusted (500 °C, 12 h) GF/F (0.7 μm pore size) glass-fibre filters. They were immediately frozen in liquid nitrogen and transported to the laboratory to be kept at -20 °C. They were ground in a Teflon homogeniser in cold phosphate buffer (0.2 M pH 7.4) containing DL-dithiothreitol and polyvinylpyrrolidone (Hochman, 1982; Berges & Harrison, 1995a). The resulting enzyme extracts were incubated at 25 °C with a reaction substrate mixture consisting of NADH (6.5 μM) and KNO_3 (0.2 M), all in phosphate buffer (0.2 M pH 7.4). The nitrite formed by nitrate reduction was assayed using the diazotation procedure of Strickland & Parsons (1972).

During this study, Cyanobacteria accounted for 2.2 ± 3.8% of total biomass in Aydat and 6.6 ± 14.5% in Pavin (Charpin et al., 1997). As this percentage never exceeded 15%, only the NAD-dependent NR activity was measured. Given the anoxic Lake Aydat, the potential effect of nitrate reducing bacteria under anaerobic conditions was tested. The biomass of bacteria remained below 30%, 25% and 10% of the biomass collected onto GF/C filter respectively in the hypolimnion, the metalimnion and the epilimnion.

The resulting *in vitro* enzyme activities were corrected for *in situ* temperature with the modified equation of Arrhenius & Menten using an average activation energy of 15 kcal $mole^{-1}$ (Brandts, 1967 in Packard et al., 1971). Enzyme activity was expressed in nmoles of nitrogen formed l^{-1} h^{-1}. The specific nitrate reductase activity (sNR), normalised to the amount of chlorophyll *a* in the sample, was also calculated. It was expressed in nmoles N formed. $(\mu gchl)^{-1}$ h^{-1}.

Results

Lake Aydat

The patterns of oxygen, temperature, inorganic nitrogen distributions and phytoplankton succession have been described in detail by Charpin et al. (1996).

Between May and September 1993, the period of thermal stratification was characterised by an oxycline and a thermocline between 4 to 6 m. The overturn, starting in October 1993, resulted in a homothermal water column until April 1994. In February 1994, the lake was ice-covered. High oxygen concentrations linked to phytoplankton photosynthetic activity were recorded between 1 and 4 m during the whole sampling period. June and July 1993 were characterised by a deoxygenation from 5 to 14 m.

Nitrate concentrations, high in April 1993 (70.6 ± 12.6 μmol l^{-1} N, mean and standard deviation), exhibited seasonal fluctuations. During stratification, they decreased from 24.5 ± 13.0 μmol l^{-1} N in spring-summer 1993 to 5.0 ± 2.8 μmol l^{-1} N in September 1993 (Figure 1). Nitrate accounted for about 84% of dissolved inorganic nitrogen until August 1993. The overturn of October 1993 led to an increase in nitrate content, which remained at maximum levels from January to April 1994 (56.0 ± 4.0 μmol l^{-1} N). The mean concentration measured during this homothermy was 43.5 ± 15.5 μmol l^{-1} N. Ammonium concentrations in the epilimnion and the metalimnion remained constant until September 1993 (3.0 ± 1.0 μmol l^{-1} N at 1 m and 6.0 ± 2.0 μmol l^{-1} N at 5 m) however, high N-NH_4 were measured in July and August 1993 at 5 m (8.0 ± 1.0 μmol l^{-1} N) (Figure 2). This increase was concomitant with a decrease in oxygen concentration which led to a strong high increase in hypolimnic ammonium (72.0 μmol l^{-1} N in August 1993). Enhanced levels of ammonium content recorded during the autumnal mixing (9.0 ± 3.0 μmol l^{-1} N), fell to extremely low values from January 1994.

During the whole sampling period, the dominating algae were diatoms which contributed more than 75% of total biomass. During stratification, the phytoplankton higher in the 1–5 m layer, was dominated by small-sized Bacillariophyceae belonging to *Cyclotella* and *Stephanodiscus*, except in June 1993 at 1 and 5 m where Cryptophyceae (*Cryptomonas sp.*) represented 72% of total biomass (Figure 3). The phytoplankton biomass showed similar patterns of change in the epilimnion and metalimnion throughout summer. Chlorophyll *a* concentrations were usually higher in the epilimnion (9.9 ± 3.8 μg l^{-1}) than at other depths (5 and 10 m), although the maximum was reached in July at 5 m (Figure 4). The break of stratification homogenised phytoplankton biomass throughout the water column. It rose until April 1993. During autumn-winter 1993, the temporal distribution of chlorophyll *a* (8.6 ± 4.1 μg l^{-1}), was similar to that of biomass (Figure 4). In terms of species composition, September 1993 was characterised by a development of *Fragilaria crotonensis*. From November 1993, *Aulacoseira italica* developed abundantly and dominated until April 1994 (84.2% of total biomass). In May 1994, the phytoplankton was more diversified, as Chlorophyceae (*Ankistrodesmus sp.*) and Cryptophyceae developed (Figure 3).

NR activity showed a clear seasonal distribution. NR activities were detected at all depths only in July 1993 (1.3 nmol l^{-1} N h^{-1} at 1 m, 1.4 nmol l^{-1} N h^{-1} at 5 m and 0.8 nmol l^{-1} N h^{-1} at 10 m) (Figure 5). A peak was reached in August 1993 at 1 m (6.1 nmol l^{-1} N h^{-1}). Hypolimnic specific NR activity (normalised to chlorophyll *a*) was higher than in the epilimnion and the metalimnion (respectively 0.3, 0.1 and 0.1 nmol N$(\mu gchl)^{-1}$ h^{-1}) while ammonium concentrations were higher (Figure 2 and 5). At this time, the phytoplankton community was characterised by

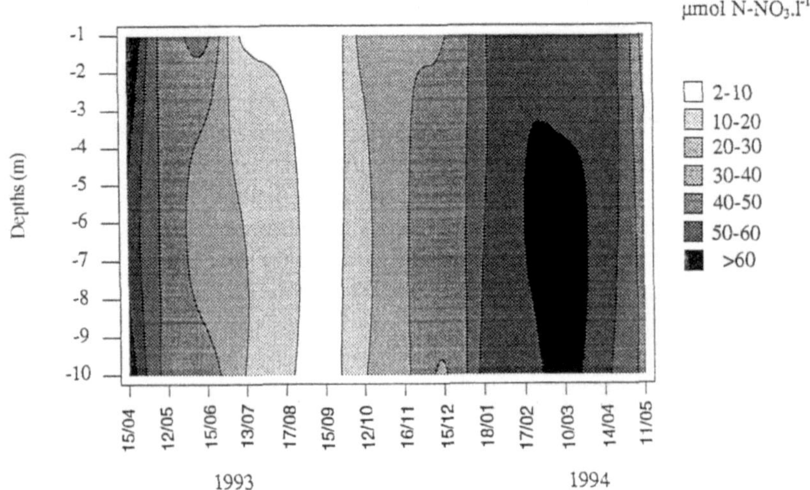

Figure 1. Annual distribution of external nitrate (N-NO_3) concentrations expressed as μmol l^{-1} N in the Lake Aydat.

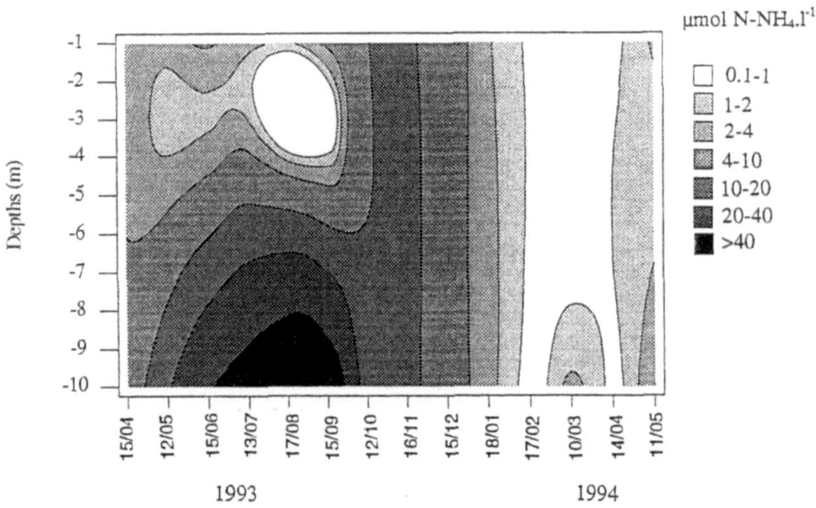

Figure 2. Annual distribution of external ammonium (N-NH_4) concentrations expressed as μmol l^{-1} N in the Lake Aydat.

diatoms at 1 and 5 m and Euglenophyceae at 10 m. During homothermy, NR activity (3.2 ± 3.7 nmol l^{-1} N at 1 m, 2.2 ± 2.1 nmol l^{-1} N at 5 m and 2.1 ± 3.2 nmol l^{-1} N at 10 m, i.e. a mean value of 2.5 ± 3.0 nmol l^{-1} N h^{-1}) was higher than during stratification (mean value 0.6 ± 2 nmol l^{-1} N h^{-1}) (Figure 5). No activity was detectable at 5 and 10 m in September and December 1993, and throughout the water column in January and February 1994. The highest values of specific NR activities were recorded in November and December 1993 (0.72 nmol N (μgchl)$^{-1}$ h^{-1} and 0.45 nmol N (μgchl)$^{-1}$ h^{-1}, respectively).

At the same time, high nitrate concentrations were found and a rise in chlorophyll *a* concentrations occurred, corresponding to the development of *Aulacoseira italica* population. Then from Mars to May 1994, NR activity increased concomitantly with the appearance of Chlorophyceae (*Ankistrodesmus* sp.) and the persistence of high nitrate concentrations.

Lake Pavin

The patterns of oxygen, temperature, inorganic nitrogen distributions and phytoplankton succession in Lake Pavin have been repported in Charpin et al. (1997).

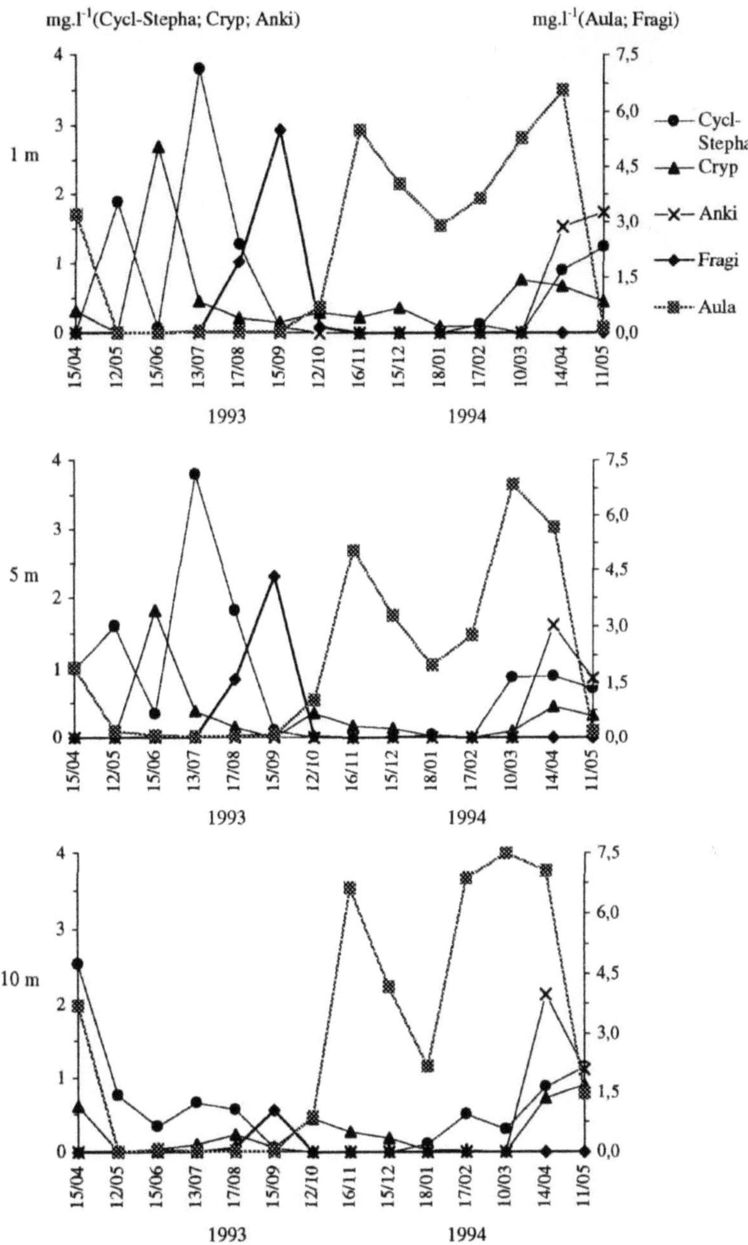

Figure 3. Annual distribution of the fresh weight phytoplankton biomass (mg l^{-1}) encountered during sampling (Cycl-Stepha = *Cyclotella-Stephanodiscus*; Cryp = *Cryptomonas sp.*; Anki = *Ankistrodesmus sp.*; Aula = *Aulacoseira italica*; Frag = *Fragilaria crotonensis*) in Lake Aydat.

The period of thermal stratification, recorded from May 1993 to October 1993, came with the formation of an oxycline and a thermocline located between 5 and 10 m, as in the case of Lake Aydat. From October 1993 to April 1994, the water column was homothermal. In January and February 1994, the lake was ice-covered, preventing sampling. High oxygen concentrations linked to phytoplankton photosynthesis were recorded between 5 and 15 m throughout the sampling period.

Contrary to Aydat, nitrate concentrations increased regularly from May 1993, to reach a peak in October

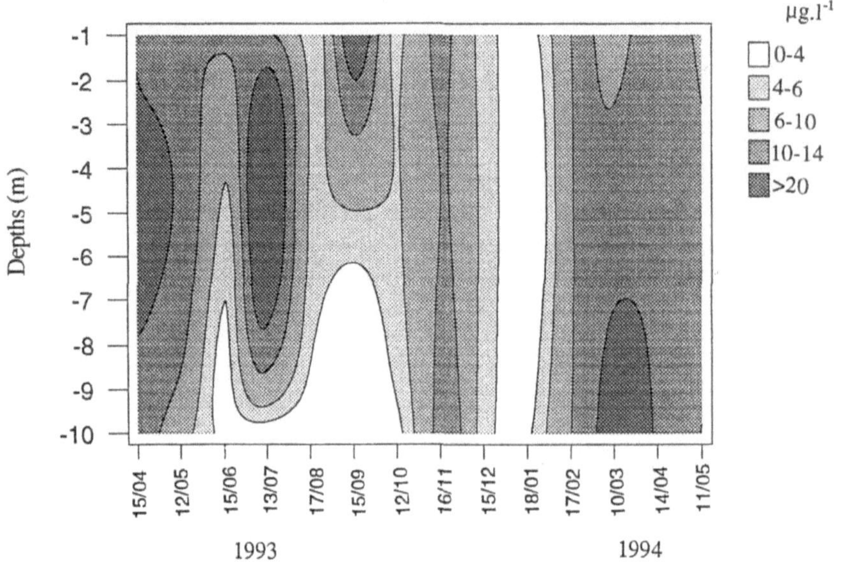

Figure 4. Annual distribution of chlorophyll *a* concentrations ($\mu g\ l^{-1}$) in the Lake Aydat.

1993 at 10 and 15 m (16.4 μmol l^{-1} N), in November 1993 at 5 m (15.0 μmol l^{-1} N) and in December 1993 at 40 m (21.4 μmol l^{-1} N) (Figure 6). However, these concentrations were lower than in Aydat. Moreover, in the hypolimnion, there was a decrease in nitrate concentration, as in the previous lake. All variables showed a uniform distribution throughout the water column from November 1993 onwards. The autumnal mixing was followed by a decrease in nitrate concentration, which remained low until April 1994 (4.3 μmol l^{-1} N in the upper 15 m and 6.4 μmol l^{-1} N at 40 m) (Figure 6). From spring 1993 to spring 1994, this form of nitrogen accounted for about 80% of total inorganic nitrogen, and was it most important source. Ammonium contents fluctuated between 0.7 and 2.1 μmol l^{-1} N during the whole sampling period with a mean concentration of 1.5 ± 0.5 μmol l^{-1} N (Figure 7).

The phytoplankton succession started with a proliferation of the dinophycean *Peridinium willei*, which made up more than 50% of total biomass until June 1993 at 5 and 10 m, August 1993 at 15 m and September 1993 at 40 m (Figure 8). But in contrary to summer succession in Aydat, the phytoplankton community was more diversified at 5 and 10 m, as the diatoms *Nitzschia sp.* and *Cyclotella comta*, and Chlorophyceae (*Oocystis lacustris*) developed (Figure 8). The highest biomasses (3.8 ± 5.4 mg l^{-1}) were recorded between 10 and 15 m close to the thermocline. In the epilimnion and the hypolimnion, mean phytoplankton biomasses were 0.8 ± 0.4 mg l^{-1} and 0.3 ± 0.2 mg l^{-1}. The chlorophyll *a* concentrations were roughly identical at 5, 10 and 40 m (respectively 0.6 ± 0.5, 0.9 ± 0.4 and 0.4 ± 0.3 $\mu g\ l^{-1}$) (Figure 9). At 15 m, the summer development of the *Peridinium willei* resulted in an increase in chlorophyll *a* (7.6 ± 11.1 $\mu g\ l^{-1}$) with a peak in July 1993 (23.3 ± 6.3 $\mu g\ l^{-1}$) (Figure 8 and 9). During homothermy, phytoplankton biomass was roughly homogeneous in the 5–15 m layer with a dominance of *Oocystis lacustis* from November to December (54.0 ± 25% of the total biomass). After the thaw, phytoplankton biomass consisted mainly of the Bacillariophyceae *Aulacoseira italica* (67.0 ± 25% of total biomass) and chlorophyll *a* concentrations rose to 3.4 ± 0.3 $\mu g\ l^{-1}$. Nevertheless, they remained lower than in Lake Aydat.

As in Aydat, NR activity exhibited seasonal and vertical variation. But in contrary to this lake, NR activity in Pavin was detected throughout the annual cycle, in the whole water column (Figure 10). Moreover, the annual mean of specific NR activity from all depths of Pavin (2.1 ± 2.6 nmol N.(μgchl)$^{-1}$ h^{-1}) was higher than in Aydat (0.2 ± 0.3 nmol N(μgchl)$^{-1}$ h^{-1}) while the mean of all other variates was lower. During stratification, specific NR activities were higher at 5 and 10 m than at 15 and 40 m (5.2, 2.1, 0.9 and 1.4 nmol N(μgchl)$^{-1}$ h^{-1}). The distribution of NR activity seemed to follow diatom and Chlorophyceae development in July and in August-September 1993

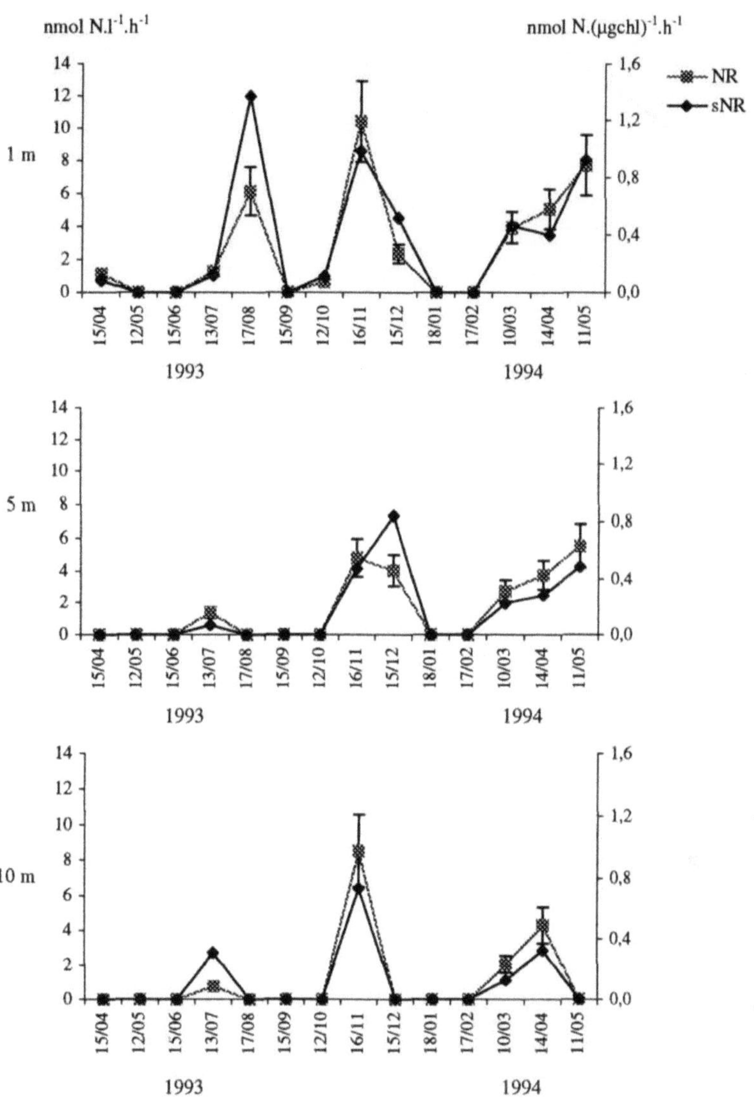

Figure 5. Annual distribution of the Nitrate Reductase (NR) activities expressed as nmol l^{-1} N h^{-1} and the specific Nitrate Reductase (sNR) (normalised to chlorophyll *a*) expressed as nmol N(μg chl)$^{-1}$ h^{-1} in Lake Aydat. Error bars represent standard errors of the mean of triplicate assays.

(Figure 8 and 10). At the same time, at 15 and 40 m, NR activity was lower. While the biomass of the dinophycean *Peridinium willei* increased abundantly at these depths, NR activity did not. Moreover, as at 15 m phytoplankton biomass was the highest recorded during this study, NR activity was the lowest (Figure 8 and 10). Then, in the whole water column, NR activity declined until the onset of homothermy. During homothermy, NR activity was higher at 15 and 40 m (3.4 ± 1.6 nmol l^{-1} N h^{-1} and 2.9 ± 1.7 nmol l^{-1} N h^{-1}) than at 5 and 10 m (2.4 ± 1.6 nmol l^{-1} N h^{-1} and 1.6 ± 0.7 nmol l^{-1} N h^{-1}) (Figure 10). The highest values were recorded in November–December 1993 (5.7 nmol l^{-1} N h^{-1} and 5.1 nmol l^{-1} N h^{-1} at 15 and 40 m) and coincided with a dominance of *Oocystis lacustris*. In spring 1994, contrary to Aydat, NR activities of Pavin were low, while chlorophyll *a* concentrations were high during the *Aulacoseira italica* bloom (Figures 8 and 10). Yet, at the same time, nitrate concentrations were low.

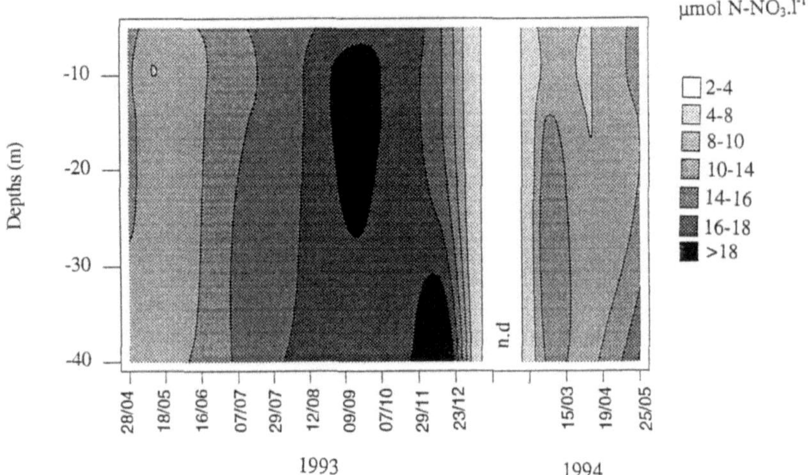

Figure 6. Annual distribution of external nitrate (N-NO_3) concentrations expressed as μmol l^{-1} N in the Lake Pavin (n.d.: no determination).

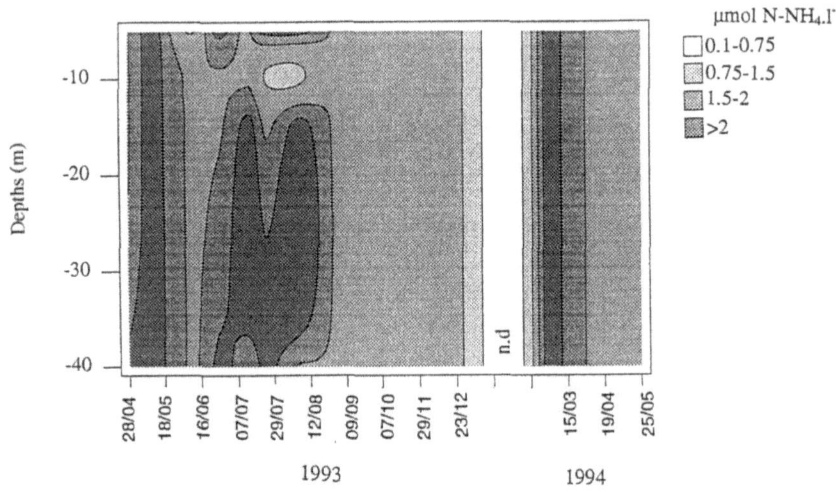

Figure 7. Annual distribution of external ammonium (N-NH_4) concentrations expressed as μmol l^{-1} N in the Lake Pavin (n.d.: no determination)

Discussion

The range of NR activities and specific NR activities differed between the two lakes. The mean value of NR activity over the entire water column was 1.8 ± 2.8 nmol l^{-1} N h^{-1} ($n=42$) in eutrophic Lake Aydat and 2.1 ± 1.5 nmol l^{-1} N h^{-1} ($n=52$) in the meso-oligotrophic Lake Pavin. Phytoplankton biomass in the eutrophic ecosystem was 2.6 times higher than in Lake Pavin. The means of the specific enzyme activities were 0.2 ± 0.3 nmol N(μgchl)$^{-1}$ h^{-1} in Lake Aydat and 2.1 ± 2.6 nmol N(μgchl)$^{-1}$ h^{-1} in Lake Pavin. In the latter ecosystem the specific NR activity, which was measurable throughout the annual cycle of sampling, was 10.5 times higher than in Aydat. In this eutrophic (lake) the distribution of specific NR activity remained sporadic all over the year. Paradoxically, the mean inorganic nitrogen concentration in Aydat was four times higher than in Pavin. Moreover, in both lakes the inorganic nitrogen source was essentially oxidized and accounted for 77 ± 24 % of inorganic nitrogen in Aydat, and 81 ± 7% in Pavin.

Many observations, both on phytoplankton cultures and in natural environments, support the idea that the presence of NO_3 or the decrease in NH_4 leads to NR synthesis and that ammonium ions inhibit it (Morris &

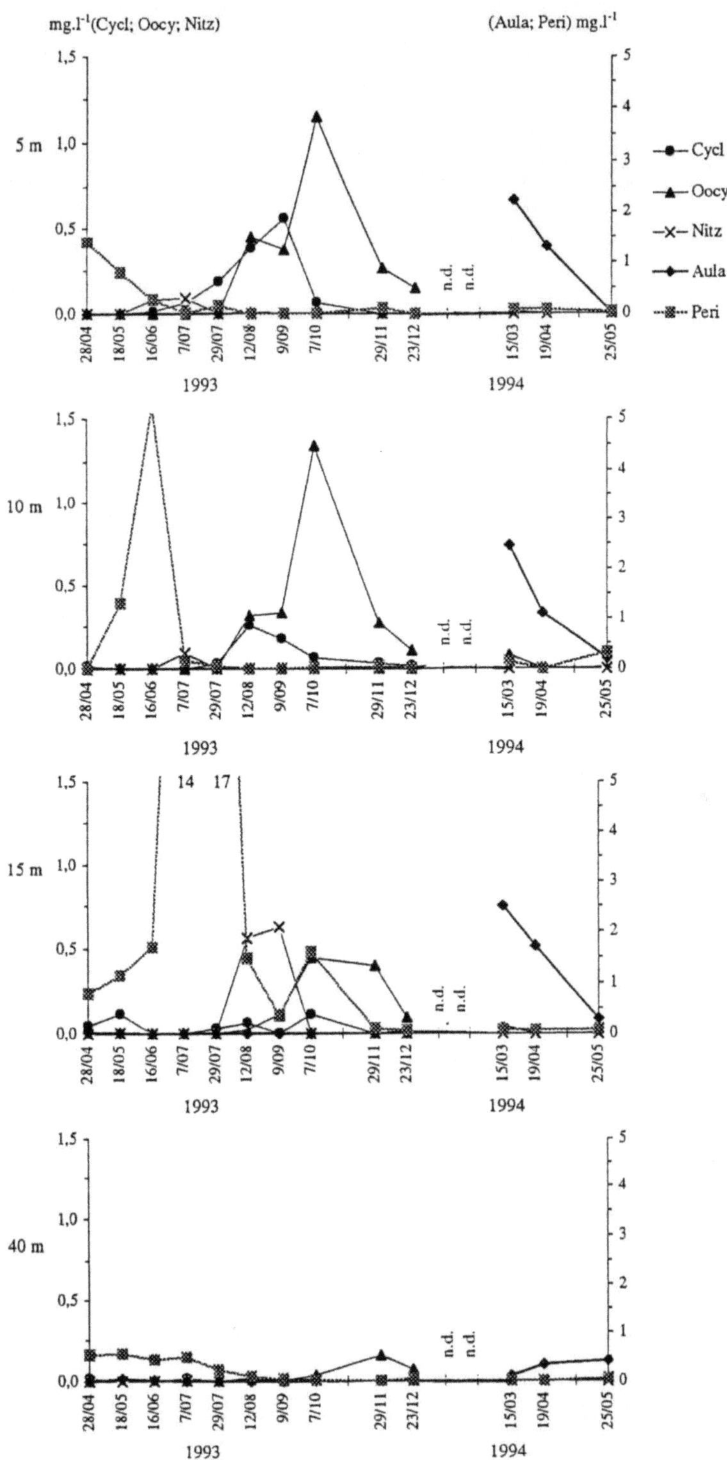

Figure 8. Annual distribution of the fresh weight phytoplankton biomass (mg l^{-1}) encountered during sampling (Cycl = *Cyclotella*; Oocy = *Oocystis lacustris*; Nitz = *Nitzschia sp.*; Aula = *Aulacoseira italica*; Peri = *Peridinium willei*) in Lake Pavin (n.d.: no determination).

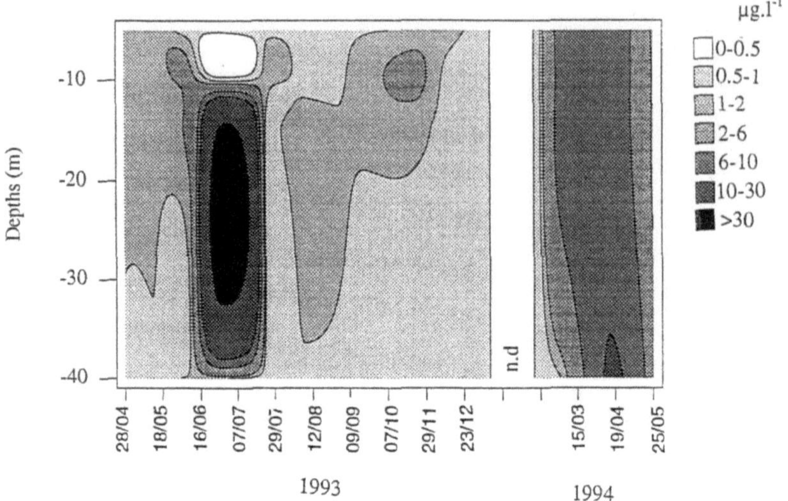

Figure 9. Annual distribution of chlorophyll *a* concentrations (μg l^{-1}) in the Lake Pavin (n.d.: no determination).

Table 1. Correlation coefficients (*r*), probabilities (*p*) and number of points (*n*) between specific NR activity (sNRA) and nitrate concentration, NO_3/Ni, NH_4/Ni, NO_3/NH_4 ratios and the biomass of main algal species in the Lake Pavin.

	Station	Period	Correlation (*r*)	*n*	*P*
NO_3	10, 15 m	All the annual cycle	0.479, 0.578	13, 13	< 0.05
NO_3/Ni	whole water column	Homothermy	0.629	20	< 0.05
NH_4/Ni	whole water column	Homothermy	– 0.623	20	< 0.05
NO_3/NH_4	whole water column	Homothermy	0.951	20	< 0.05
Cyclotella comta.	5–10–15 m	All thye annual cycle	0.318	39	< 0.05
Nitzchia sp.	5–10 m	Stratification	0.679	14	< 0.05
Peridinium willei	10, 15–40 m	Stratification	– 0.683, – 0.507	7, 14	< 0.05
Oocystis lacustris	15–40 m	All the annual cycle	0.445	26	< 0.05
Aulacosiera italica	15 m	Homothermy	–0.967	5	< 0.05

Syrett, 1965; Eppley et al., 1969; Pistorius et al., 1976; Funkhouser et al., 1980; Hipkin et al., 1980, 1983; Florencio & Vega, 1982; Blasco et al., 1984; Kristiansen, 1987; Tischner et al., 1989; Jimenez del Rio et al., 1994; Berges & Harrison, 1995 a). During summer 1993, specific NR activity was low in Aydat whereas it was higher and appeared throughout this period in Pavin. At this time of year, nitrate concentrations were low in these two lakes (10 ± 6 nmol l^{-1} N in Aydat and 10 ± 3 nmol l^{-1} N in Lake Pavin). In Aydat, specific NR activity first became measurable in July-August 1993 when nitrate concentration was lower than during spring 1993 (respectively 13 ± 5 nmol l^{-1} N and 47 ± 19 nmol l^{-1} N). During homothermy, nitrate concentration was higher in Aydat than in Pavin (respectively 44 ± 16 nmol l^{-1} N and 10 ± 5 nmol l^{-1} N), but specific NR activity in Aydat remained lower than in Pavin and still had a sporadic distribution. Moreover, as reported in numerous investigations (Eppley et al., 1970, Blasco & Packard, 1974; Blasco et al., 1984; Kristiansen, 1987) normalised NR activity in Lake Pavin was significantly correlated with nitrate concentration (Table 1). However, in eutrophic Lake Aydat, a similar direct relationship did not exist. So, our results were unable to confirm that a minimum threshold of nitrate was needed for the induction of NR activity.

A maximum threshold of ammonium, above which NR activity would be inhibited could, however, exist. Packard & Blasco (1974), found that a decrease in NR activity occurred when the ammonium concentration in samples increased above 0.2 μM. But in

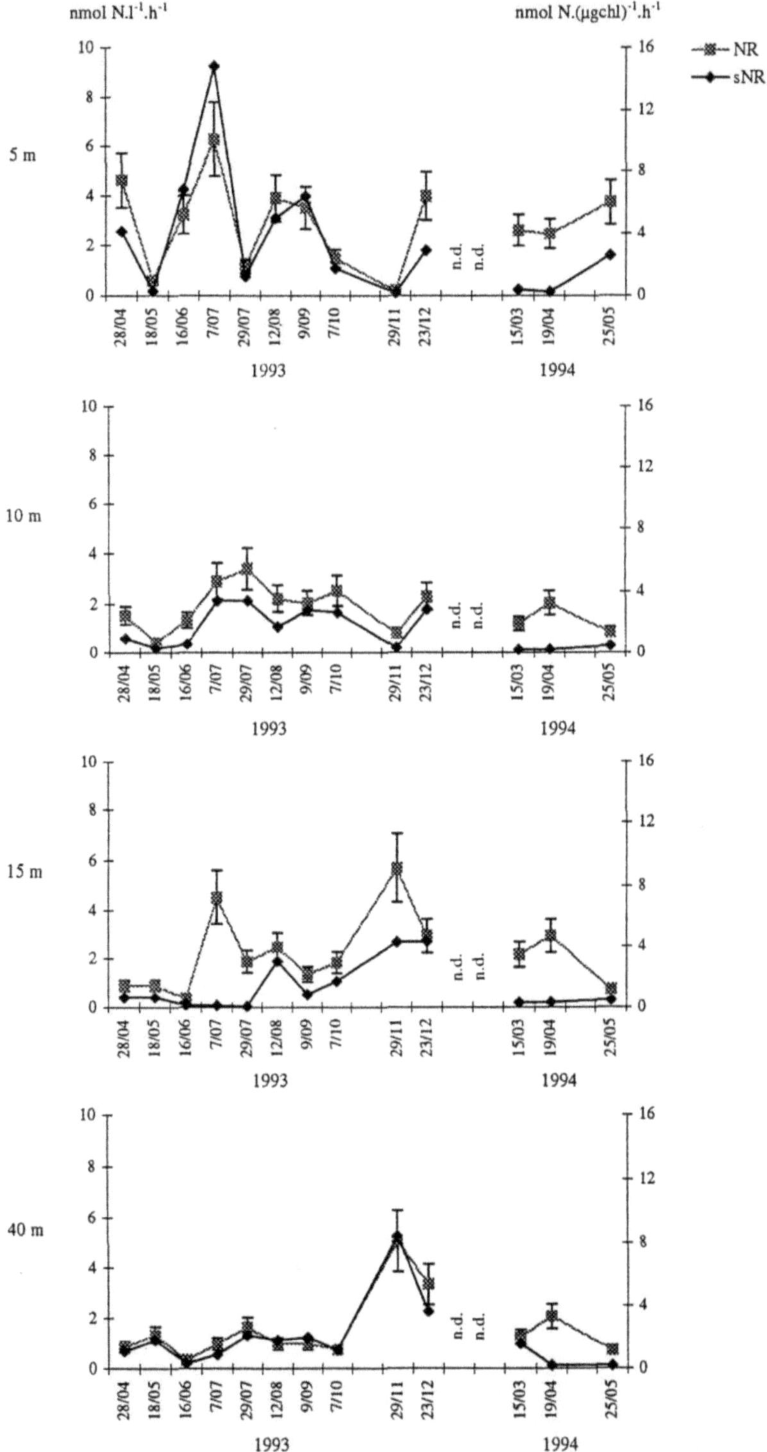

Figure 10. Annual distribution of the Nitrate Reductase (NR) activities expressed as nmol l^{-1} N h^{-1} and the specific Nitrate Reductase (sNR) (normalised to chlorophyll *a*) expressed as nmol N(μg chl)$^{-1}$ h^{-1} in Lake Pavin. Error bars represent standard errors of the mean of triplicate assays (n.d.: no determination).

Table 2. Coefficients of correlation, probabilities and number of points between specific NR activity (sNRA) and the biomass of main algal species in the Lake Aydat.

	Station	Period	Correlation (r)	n	P
Cyclotella and Stephanodiscus	1 m	Stratification	0.914	5	< 0.02
Cyclotella and Stephanodiscus	5 m	Stratification	0.911	5	< 0.02
Aulacosiera italica	whole water column	Homothermy	0.560	21	< 0.02
Ankistrodesmus sp	1–5 m	All the annual cycle	0.399	28	< 0.02

numerous investigations it has been shown that NH_4 did not directly inhibit NR activity (Blasco & Conway, 1982; Dortch & Conway, 1984). Moreover, according to Dortch (1990), ammonium inhibition could not be distinguished from a simple preference for ammonium (i.e. a direct interaction in which decreases in nitrate uptake and NR activity are independent of ammonium concentration). Furthermore, Berges et al. (1995 b), showed that ammonium 'inhibition' of NR activity in natural populations was apparently less complete and occurred over a much longer time scale than is generally appreciated. Our results were more in line with these studies, because in Aydat, during June and July 1993, NR activity was detected in the metalimnion and the hypolimnion at a time when ammonium concentration here was the highest recorded. Moreover, in Lake Pavin where mean ammonium concentration was 1.5 μg l^{-1} N, NR activity was also detected. Thus, we could not confirm the existence of an ammonium threshold. Also, the availability of inorganic nitrogen substrate was insufficient to produce the patterns of specific NR activity observed in these two lakes.

According to Packard & Blasco (1974), the enzyme activity of the NR seems more regulated by the relative proportions of various inorganic nitrogen species in the medium (NO_3/Ni or NH_4/Ni) than by actual concentrations of NH_4 or NO_3. In eutrophic Aydat, nitrate represented 58% ($\pm$ 22) of inorganic nitrogen during stratification and 86% ($\pm$ 14) during homothermy. Despite this discrepancy, no direct relationship was found between these ratios and specific NR activity. On the contrary, in meso-oligotrophic Pavin, nitrate represented 81% ($\pm$ 7) of inorganic nitrogen during homothermy and significant correlations were found between the NO_3/Ni, NH_4/Ni and NO_3/NH_4 ratios and the specific NR activity (Table 1). This latter result is in agreement with Kristiansen (1987), in the Oslo Fjord.

Nevertheless, factors other than NO_3 or NH_4 might affect the uptake of NO_3 (Dauchez, 1996; Wynne & Berman, 1990) and subsequent specific NR activity. Indeed, under identical culture conditions, different species of unicellular marine algae can exhibit differing NR activities. This was observed indirectly by Hipkin et al. (1983) and Berges & Harrison (1995a). According to Martinez et al. (1987), the species composition of natural phytoplankton has an influence on the spatial and temporal distribution of specific NR activity. In both Aydat and Pavin, the temporal changes in NR activity seemed linked with the diatom biomass. Indeed, in Lake Aydat, NR activity appeared simultaneously with the peak of *Cyclotella* and *Stephanodiscus* during summer. Moreover, in autumn and winter 1993–1994, when *Aulacoseira italica* dominated the phytoplankton, NR activity was high. Significant and positive correlations were obtained between these three diatoms and specific NR activity (Table 2). In Lake Pavin, during thermal stratification, the phytoplankton was dominated by the dinophycean *Peridinium willei*. Yet, at 5, 10 and 15 m, the biomass of *Peridinium willei* was negatively correlated with specific NR activity. An increase in NR activity occurred simultaneously with a shift from dinophycean to Bacillariophyceae dominance. This resulted in significant and negative correlations between specific NR and the biomass of *Peridinium willei*, and in positive relationships between enzyme activity and the biomass of the diatoms *Cyclotella comta* and *Nitzchia sp.* (Table 1). These results confirm those of Packard & Blasco (1974) and Serruya (1975), who demonstrated that *Gonyaulax peridinium* cells could have little or no NR activity when nitrate in the medium was high, implying that these cells were not actively utilizing this nitrogen source. McCarty et al. (1982) found that NH_4 was used preferentially to NO_3, Le Corre et al. (1993) that the *Gyrodinium* cf. *aureolum* bloom in the Hushant frontal region could be maintained almost exclusively by in situ remineralization of remnants of previous diatom blooms, and Packard & Blasco (1974) showed that diatoms seemed to have a lower half saturation constant and consequently a higher affinity for NO_3 than *Gonyaulax*. So, diatoms were enable to out- compete *Gonyaulax* at low NO_3 concentrations. As significant positive correlations were found between specific NR activity and the biomass of *Cyclotella comta* and *Nitzchia sp.*, this could imply that these probably accounted for most of the observed enzyme activity (Table 1).

However, during homothermy in Lake Pavin, no direct relationship was obtained with any phytoplankton species in the top ten meters when the Chorophyceae *Oocystis lacustris* and the diatom *Aulacoseira italica* dominated the phytoplankton. During this period, significant positive relationships were obtained between biomass of *Oocystis lacustris* and NR activity at 15 and 40 m (Table 1). These results could be explained by the changing physical structure which could promote transient accumulations of the epilimnic community in the meta- and hypolimnion, as reported in the meso-oligotrophic Lake Samish (Barbiero & McNair, 1996). However, only a negative correlation was obtained between the biomass of *Aulacoseira lacustris* and enzyme activity at 15 m (Table 1).

In Lake Pavin, the fact that NR activity was detected throughout the annual cycle suggests that this inducible enzyme was essential for the phytoplankton species present, which use nitrates as their main inorganic nitrogen source. In Lake Aydat, the appearance of NR activity was more complex. According to Wetzel (1983), eutrophic lakes are characterised by a decrease in nitrate concentration in summer caused by algal assimilation. In eutrophic Aydat, such a decrease occurred, but this was not due to algal assimilation, since NR activity remained low during summer. The decrease in nitrate could also be due to denitrification and ammonification which can occur rapidly in both anaerobic and aerobic conditions (Wetzel, 1983). Such phenomena could be suspected in the hypolimnion of Aydat, characterised by a simultaneous decrease in nitrate and a rise in ammonium during the summer deoxygenated period. Such a hypothesis could explain the sporadic specific NR activity in Aydat. Indeed, NH_4 is regenerated rapidly and might appear irregularly as a product of microbial action and animal excretion (McCarthy & Goldman, 1979). In most situations where both nitrogen sources are present, ammonium is preferentially assimilated by phytoplankton, and specific NR activity is consequently low (Eppley et al., 1969; Dortch, 1990; Berges et al., 1995 b). In such environmental conditions, available NH_4 is taken up in the vegetative cycle and the phytoplankton would assimilate NO_3 for only short periods of time.

Acknowledgements

We thank the Fondation Jean Moulin for allowing us to conduct field work on Lake Aydat and Dr C. Amblard and G. Bourdier for providing physical and chemical data and phytoplankton counts for Lake Pavin.

References

Amblard, C., 1986. Les nucléotides adényliques: intérêts pour l'étude de la biomasse, de l'activité métabolique et de la structure des peuplements phytoplanctoniques lacustres. Thèse Doct. Etat. Univ. Clermont II, 317 pp.

Aleya, L. & J. Devaux, 1992. The concept of seasonal succession theory applied to phytoplankton through the coupling use of diversity index and rang-frequencies diagrams in a eutrophic ecosystem. Int. Rev. Ges. Hydrobiol. 77: 579–591.

Barbiero, R. P. & C. M. McNair, 1996. The dynamics of vertical chlorophyll distribution in an oligomesotrophic lake. J. Plankton Res. 18: 225–237.

Berges, J. A. & P. J. Harrison, 1995a. Nitrate reductase activity quantitatively predicts the rate of nitrate incorporation under steady-state light limitation: a revised assay and characterization of the end in three species of marine phytoplankton. Limnol. Oceanogr. 40: 82–93.

Berges, J. A., W. P. Cochlan & P. J. Harrison, 1995b. Laboratory and field responses of algal nitrate reductase to diel periodicity in irradiance, nitrate exhaustion, and the presence of ammonium. Mar. Ecol. Prog. Ser. 124: 259–269.

Blasco, D. & T. T. Packard 1974. Nitrate reductase measurements in upwelling regions. 1. Significance of the distribution off Baja California and northwest Africa. Tethys 6: 239–246.

Blasco, D. & H. L. Conway, 1982. Effect of ammonium on the regulation of nitrate assimilation in natural phytoplankton populations. J. exp. mar. Biol. Ecol. 61: 157–168.

Blasco, D., J. J. MacIsaac, T. T. Packard & R. C. Dugdale, 1984. Relationship between nitrate reductase and nitrate uptake in phytoplankton in the Peru upwelling region. Limnol. Oceanogr. 29: 275–286.

Brandt, J. F., 1967. Heat effects on proteins and enzymes. In Rose A. H. (ed.), Thermobiology, Academic Press, New York: 25–72.

Charpin, M. F., L. Aleya & J. Devaux, 1996. GS activity in limnitic phytoplancton. Aquat. Sci. 58: 102–119.

Charpin, M. F., M. Richardot, N. Maurin & J. Devaux, 1997. Potential glutamine synthetase and glutamate dehydrogenase activities in the oligo-mesotrophic Lake Pavin: annual comparaison with an indirect estimation of inorganic nitrogen utilization. Archiv. Hydrobiol. (in press).

Collos, Y. & G. Slawyck, 1977. Nitrate reductase activity as function of *in situ* nitrate uptake and environmental factors of euphotic zone profiles. J. exp. mar. Biol. Ecol. 29: 119–130.

Dauchez, S., L. Legendre, L. Fortier & M. Levasseur, 1996. Nitrate uptake by size-fractionated phytoplankton on the Scotian Shelf (Northwest Atlantic): spatial and temporal variability. J. Plankton Res. 18: 577–595.

Dortch, Q. & H. L. Conway, 1984. Interactions between nitrate and ammonium uptake: variation with growth rate, nitrogen source and species. Mar. Biol. 79: 151–164.

Dortch, Q., 1990. The interaction between ammonium and nitrate uptake in phytoplankton. Mar. Ecol. Prog. Ser. 61: 183–201.

Eppley, R. W., J. L. Coatsworthy & L. Solórzano, 1969. Studies of nitrate reductase in marine phytoplankton. Limnol. Oceanogr. 14: 194–205.

Eppley, T., T. Packard & J. J. MacIsaac, 1970. Nitrate reductase in Peru current phytoplankton. Mar. Biol. 6: 195–199.

Florencio, F. J. & J. M. VEGA, 1982. Regulation of the assimilation of nitrate in *Chlamydomonas reinhardii*. Phytochemistry 21: 1195–1200.

Funkhouser, E. A., T. C. Shen & R. Ackermann, 1980. Synthesis of nitrate reductase in *Chlorella*. I. Evidence for an inactive protein precursor. Plant Physiol. 65: 939–943.

Funkhouser, E. A., T. C. Shen & R. Ackermann, 1980. Synthesis of nitrate reductase in *Chlorella*. 2. Evidence for synthesis in ammonia-grown cells. Plant Physiol. 65: 944–948.

Hipkin, C. R., B. A. AlBassam & J. P. Syrett, 1980. The roles of nitrate and ammonium in the regulaton of the developpement of nitrate reductase in *Chlamydomonas reinhardii*. Planta 150: 13–18.

Hipkin, C. R., R. J. Thomas & P. J. Syrett, 1983. Effects of nitrogen deficiency on nitrate reductase, nitrate assimilation and photosynthesis in unicellular marine algae. Mar. Biol. 77: 101–107.

Hochman, A., 1982. Studies of nitrate reductase in freshwater dinoflagellate *Peridinium cinctum*. Arch. Microbiol. 133: 62–65.

Hochman, A., Y. A. Nissan, D. Wynne, B. Kaplan & T. Berman, 1986. Nitrate reductase: an improved assay method for phytoplankton. J. Plankton Res. 8: 385–392.

Jimenez Del Río, M., Z. Ramazanov & G. García-Reina, 1994. Dark induction of nitrate reductase in halophilic alga Dunaliella salina. Planta 192: 40–45.

Kristiansen, S., 1987. Nitrate reductase activity in phytoplankton from the Oslo-fjord, Norway. J. Plankton Res. 9: 739-748.

Lair, N., 1975. Rôle du zooplancton dans les réseaux trophiques de deux lacs du Massif Central Français. Thèse Doc és Sci., Univ. Clermont II 321 pp.

Le Corre, P., S. L'Helguen & M. Wafar, 1993. Nitrogen source for uptake by *Gyrodinium* cf. *aureolum* in a tidal front. Limnol. Oceanogr. 38: 446–451.

Legendre, L. & W. D. Watt, 1971–1972. On a rapid technique for plankton enumeration. Inst. Oceanogr. Paris 58: 173–177.

Lohman, H., 1908. Untersuchungen zur Festtellung des vollstandigen Gehaltes des Meeres an Plankton. Wiss. Meereunters., Abt. Kiel N.F 10: 132–170.

Lorenzen, C. J., 1967. Determination of chlorophyll and pheopigments: spectrophotometric equations. Limnol. Oceanogr. 12: 343–346.

Martinez, R., T. T. Packard & D. Blasco, 1987. Light effects and diel variations of nitrate reductase activity in phytoplankton from the northwest Africa upwelling region. Deep Sea Res. 34: 741–753.

McCarthy, J. J. & J. C. Goldman, 1979. Nitrogenous nutrition of marine phytoplankton in nutrient-depleted water. Sciences (Wash., DC) 203: 670–672.

McCarthy, J. J., D. Wynn & T. Bermna, 1982. The uptake of dissolved nitrogenous nutrients in Lake Kinneret microplankton. Limnol. Oceanogr. 4: 673–680.

Millérioux, G., 1976. Contribution à l'étude du lac d'Aydat (Puy de Dôme). Thèse Doc. 3ème Cycle, Univ. Clermont II 92 pp.

Morris, I. & P. J. Syrett, 1965. The effect of nitrogen starvation on the activity of nitrate reductase and the other enzymes in *Chlorella*. J. Gen. Microbiol. 38: 21–28.

Packard, T. T., D. Blasco, J. J. MacIsaac & R. C. Dugdale, 1971. Variations of nitrate reductase activity in marine phytoplankton. Inv. Pesq. 35: 209–219.

Packard, T. T. & D. Blasco, 1974. Nitrate reductase activity in upwelling regions. 2. Ammonia and light dependence. Tethys 6: 269–280.

Planche-Jamet, D, 1995. Variations temporelles et verticales de l'activité phosphatase alcaline associée à différentes classes de taille du matériel particulaire au sein de trois lacs du Massif Central de degrés trophiques différents. (Lac pavin, lac d'Aydat, retenue de Villerest). Thèse Doc. 3ème cycle, Univ. Clermont II 180 pp.

Pistorius, E. K., H. S. Gewitz, H. Voss & B. Vennesland, 1976. Reversible inactivation of nitrate reductase in *Chlorella vulgaris in vivo*. Planta 128: 73–80.

Reynolds, C. S., 1990. Temporal scales of variability in pelagic environments and the response of phytoplankton. Freshwat. Biol. 23: 25–53.

Serruya, C., 1975. Nitrogen and phosphorus balances and load biomass relationships in lake Kinneret (Israel). Int. Ver. Theor. Angew. Limnol. Verh. 19: 1357–1369.

Solórzano, L., 1969. Determination of ammonia in natural waters by phenohypochlorite method. Limnol. Oceanogr. 14: 799–801.

Strickland, J. D. H. & T. R. Parsons, 1972. A practical handbook of seawater analysis. Bull. Fish. Res. Bd Can. 125.

Tischner, R., M. R. Ward & R. C. Huffaker, 1989. Evidence for a membrane-bound nitrate reductase involved in nitrate uptake of *Clorella sorokiniana*. Planta 178: 19–24.

Utermohl, 1958. Zur vervollkommung der quantitative Phytoplankton-Methodik. Mitt. Int. Ver. Limnol. 9: 1–38.

Wetzel, R. G., 1983. Limnology, 2nd edn. Saunders College Publishing, New York.

Wynne, D. & T. Berman, 1990. The influence of environmental factors on nitrate reductase activity in freshwater phytoplancton. I. Field studies. Hydrobiologia 194: 235–245.

Hydrobiologia **373/374:** 149–156, 1998.
J. C. Amiard, B. Le Rouzic, B. Berthet & G. Bertru (eds),
Oceans, Rivers and Lakes: Energy and Substance Transfers at Interfaces.

Influence of oyster culture on water column characteristics in a coastal lagoon (Thau, France)

Nabila Mazouni[1], Jean-Claude Gaertner[2] & Jean-Marc Deslous-Paoli[1]
[1] *IFREMER, Laboratoire DEL. Ecologie, B.P. 171, F-34203 Sète, France*
[2] *IFREMER, Laboratoire RH. Ecologie Halieutique, F-34203 Sète, France*

Key words: oyster, biofouling, fluxes, nutrients, oxygen, PCAIV

Abstract

The development of shellfish farming activities causes great changes in ecosystems functioning. In the Thau lagoon, oysters are reared on long constantly submerged lines, and these become fouled by several epifaunal species. The assemblage (oysters and epifauna) is defined as Oyster Culture Unit (OCU). The aim of our study was to estimate (i) how much the composition and the abundance of the epifaunal species can influence the nutrient and oxygen fluxes recorded at the shellfish–water interface and (ii) how these fluxes modify water column characteristics. We used Principal Component Analysis with Instrumental Variables (PCAIV). Two analyses were carried out, using sets of data on fluxes, the specific composition of the cultivated communities, and on oxygen, nutrient and chlorophyll a concentrations in the water column. The highest fluxes at the OCU–water interface were measured when epifaunal species richness was maximum. However, at our measurement scale (i.e. the oyster frame) no influence of this filter-feeders assemblage was observed on the chlorophyll a level. Conversely, we found a significant influence of oyster culture on the oxygen and dissolved nitrogen concentrations in the water column. The use of this recent factorial analysis was helpful to estimate the influence of the biofouling species composition on the fluxes at the OCU–water interface, and to estimate the potential impact of oyster cultures on the conditions prevailing in the water column.

Introduction

In semi-enclosed ecosystems, nutrient renewal in the water column is generally due to remineralization and turnover processes. Among the factors which control nutrient renewal, the influence of filter-feeder populations has been well described in the literature, either for natural populations (Cloern, 1982; Alpine & Cloern, 1992) or for cultivated species (Dame & Dankers,1988; Dame et al., 1989; Prins & Smaal, 1994). These populations of filter-feeding organisms also influence the phytoplankton primary production, as they release large amounts of dissolved nitrogen in the water column (Dame & libes, 1993; Prins & Smaal, 1994; Mazouni, 1995) and they constitute a phytoplankton sink (Jarry et al., 1990).

In the Thau lagoon (France), Japanese oysters (*Crassostrea gigas*, Thunberg) are reared on long lines constantly submerged, and these become fouled by several epifaunal species (David, 1970). This assemblage (oysters and epifaunal species) is called an Oyster Culture Unit (OCU).

We expected the development of this biofouling to have a great influence on the exchanges in the water-OCU interface. So, in this study, we use a Principal Component Analysis with Instrumental Variables (PCAIV) to test this hypothesis and to estimate (1) the influence of the oyster culture composition (species richness, biomass) on the fluxes at the OCU–water interface, and (2) the impact of the OCU on the conditions (i.e. nutrient concentrations, Chla, etc.) prevailing in the water column.

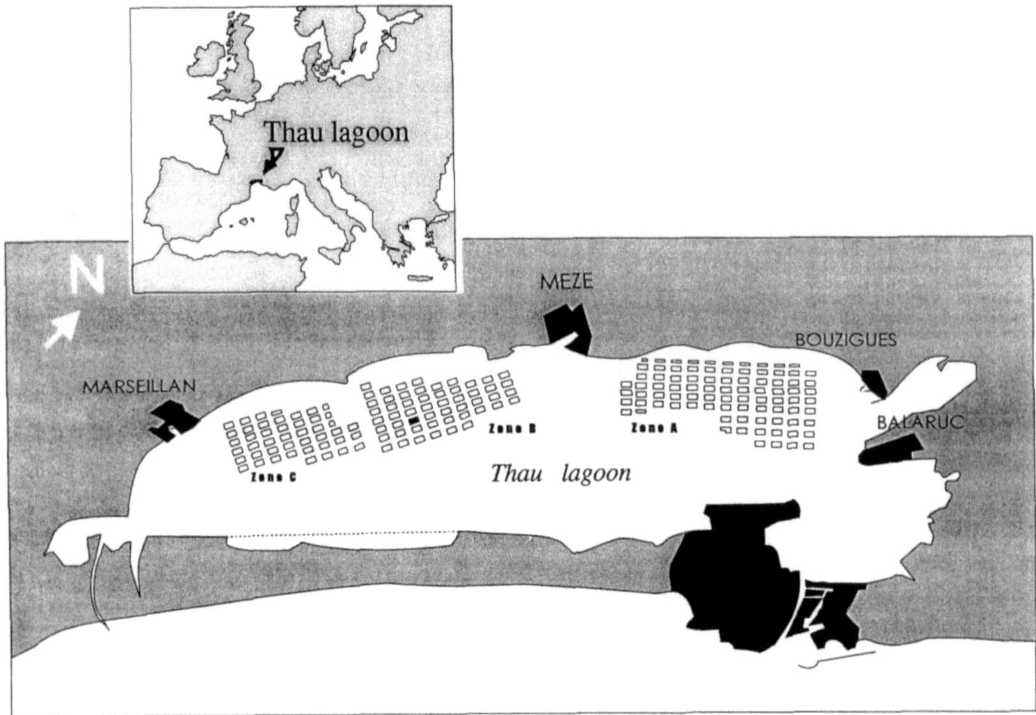

Figure 1. Geographical situation of the Thau lagoon.

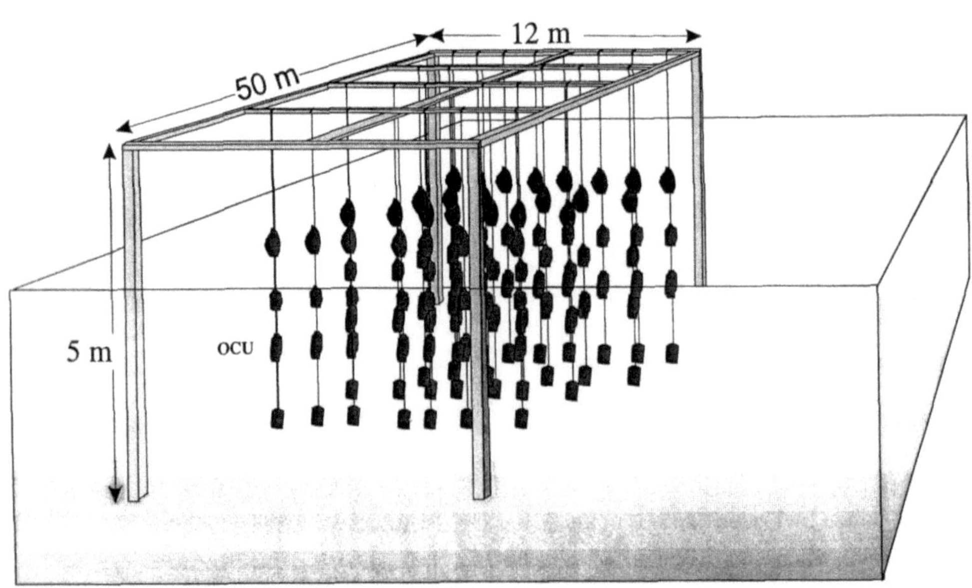

Figure 2. Presentation of the oyster farming technic used in the Thau lagoon, showing the structure of the oyster frames and the Oyster Culture Unit (OCU).

Materials and methods

The study site

Thau lagoon is a coastal Mediterranean lagoon of 7500 ha, with a depth of 4 m on average (Figure 1). This ecosystem is considered semi-enclosed, because the water renewal time is about three months (Millet, 1989). A semi-intensive oyster farming is developed in this basin, which provides 90% of the French Mediterranean oysters yield, and represents 10% of the French production. The oysters *Crassostrea gigas* (Thunberg) are reared on long lines suspended on metal frames (Figure 2). Each frame has a surface of 600 m^2 (50 × 12) and the immersed part of the lines are 4 m in length. We worked on long lines left constantly in immersion (no scraping away of the biofouling), because it was representative of the more frequent oyster farming technique in the lagoon (70%) and it does not disturb the biofouling life history. Several assemblages of oysters and their associated epibiota are distributed along the long lines. We considered each of these assemblages as an Oyster Culture Unit (OCU).

Table 1. Details of the codes of all the variables used in the Principal Component Analysis with Instrumental Variables.

Code	Variable	Unit
Group 1		
[CHLA]	Chlorophyll a concentration	$\mu g\ l^{-1}$
[O2]	Concentration of Oxygen	$ml\ l^{-1}$
[NH4]	Ammonium concentration	$\mu mol\ l^{-1}$
[NO]	Nitrate-nitrites concentration	$\mu mol\ l^{-1}$
[PO4]	Phosphate concentration	$\mu mol\ l^{-1}$
Group 2		
FIL	Filtration rate	$l\ h^{-1}$
F_{Oxygen}	Oxygen consumption	$ml\ h^{-1}$
$F_{Phosphate}$	Phosphate flux	$\mu mol\ h^{-1}$
$F_{Ammonium}$	Ammonium Production	$\mu mol\ h^{-1}$
$F_{Nitrate}$	Nitrate-nitrite Production	$\mu mol\ h^{1}$
Group 3		
Ascidians	Percentage	%
Polychaetes	Percentage	%
Oysters	Percentage	%
Bryozoans	Percentage	%
Macroalgae	Percentage	%

Data used

The variables used in this study were measured monthly from January 1992 to December 1992. We divided them into three groups. The codes used for all the variables are given in Table 1.

The *first group* contained the concentrations of oxygen (ml l^{-1}), ammonium (μmol l^{-1}), nitrate-nitrite (μmol l^{-1}), phosphate (μmol l^{-1}) and chlorophyll (μg l^{-1}).

The *second group*, was composed of variables representative of the fluxes at the OCU–water interface, measured with the Metabolic Enclosures System (MES, Mazouni, 1995). This system consisted of two transparent metacrylate hemispheres, fitted together by a plastic (PVC) ring. The total volume of the Metabolic Enclosure was about 35.7 l and each contained an Oyster Culture Unit. Three OCU were collected from different ropes but at the same depth (3 m), and studied in parallel. To estimate the filtration rate of the OCU communities, the MES was used in flow-trough conditions. The water circulation was performed with a peristaltic pump connected to each Metabolic Enclosure and two magnetic stirrers prevented stratification of the water. All the water 'entrances' were centralised to obtain homogeneous conditions for all the MES. We sampled the water at the entrance and exit of each MES to measure the influence of the OCU on the *in vivo* fluorescence. We made three replicates for each Metabolic Enclosure. We then stopped the pump and closed the system to use it in incubation to estimate nutrient and oxygen fluxes. Water (100 ml) was sampled every 30 min in each MES to estimate the evolution of nutrient concentrations during incubation. Oxygen concentrations were monitored with an oxygen probe, and recorded all along the experiment. More details on the experimental design and the variables used, can be found in Mazouni (1995).

The *third group*, was representative of the species composition of the OCU. It contained the percentages of the different species present in the assemblage.

Data measurements

We used a WTW LF 196 conductimeter to measure salinity and temperature. The concentration of dissolved oxygen was recorded with a polarographic oxymeter (YSI 58) calibrated on the air at 100% of saturation. The nutrient concentrations in the water column were measured using the method of Tréguer & Lecorre (1975) and Koroleff (1969). The filtration (FIL) of the OCU was calculated on the basis of changes in *in vivo* fluorescence (Lorensen, 1966) recorded on a fluorimeter (Turner 111).

The filtration rate of the OCU was expressed in l h^{-1} and calculated as: FIL = [(INIT -

FINAL)/INIT] · D, with INIT = initial *in vivo* fluorescence in the water column, FINAL = final *in vivo* fluorescence in the water after it was subjected to filtration by the OCU, D = flow rate in the Metabolic Enclosures ($l\,h^{-1}$).

Data analysis

Principal Component Analysis with Instrumental Variables (Rao, 1964; Sabatier et al., 1989) is a combination of classical multivariate ordination and multiple regression analysis. The first step of PCAIV concerns the elaboration of a model (linear regressions) to express each dependent variable (Y) as a linear combination of the whole set of explanatory variables (X). The second step of the analysis is a Principal Component Analysis (PCA) on the matrix obtained from the first step. In our study, normalized data were used, because of the heterogenity of variable units. This eigenvalue ordination is especially designed for direct analysis of the relationships between two multivariate ecological tables (Lebreton et al., 1991).

In this study, a first PCAIV was performed to quantitatively define the relationships between the composition of the Oyster Culture Unit (explanatory variables) and its activity (fluxes) (variables to explain). A second PCAIV was performed between the fluxes at the OCU-water interface (explanatory variables) and the water column characteristics (variables to explain). Among the constrained analysis methods, we used PCAIV rather than Canonical Correspondance Analysis (Ter, Braak, 1986), because of the quantitative nature of the dependent variables. The PCAIV and the permutation tests were performed with the ADE-4 software (Thioulouse et al., 1995), available in freeware distribution on internet (http: //pbil.univ-lyon1.fr/ADE-4.html).

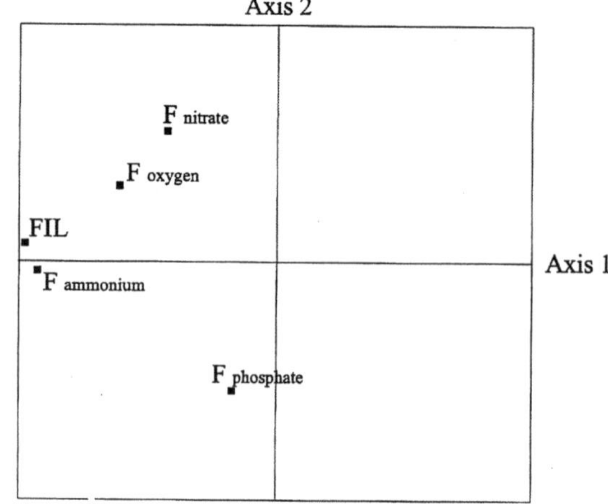

Figure 3. Projection of the dependent variables (fluxes at the OCU–water interface) on the first factorial plane of the Principal Component Analysis with Instrumental Variables.

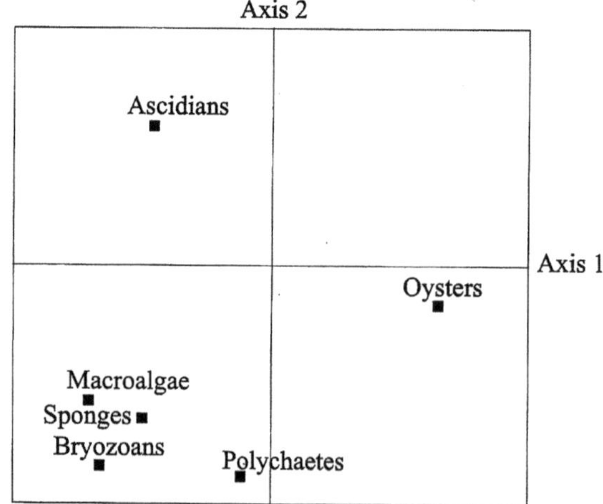

Figure 4. Correlation of the explanatory variables (species composition of the OCU) with the first two factorial axis of the Principal Component Analysis with Instrumental Variables.

Results

The species composition of the OCU explained 70% of the variability of nutrient and oxygen fluxes at the OCU–water interface and the permutation test was significant ($p < 0.05$). The first two factorial axes accounted for 89% of the total inertia of the model, with 67% for the first axis and 22% for the second. Thus we only present here the first factorial plane of the analysis.

The interpretation of the first PCAIV is performed through a comparative analysis of three graphs: the projection of dependent variables (Figure 3), the projection of explanatory variables (Figure 4) and the representation of the calculated model (Figure 5). The same approach is used to analyse the second PCAIV on respectively, Figures 6, 7 and 8.

The projection of the dependent variables (fluxes) on the first factorial plane (Figure 3) showed that the first axis of the analysis was characterized by maximum ammonium and oxygen fluxes, and a high filtration rate of the OCU in the summer. On the other hand, the phosphate fluxes and also the nitrate-nitrite fluxes explained the second axis, and thus seemed relatively independant of the temperature gradient found for the

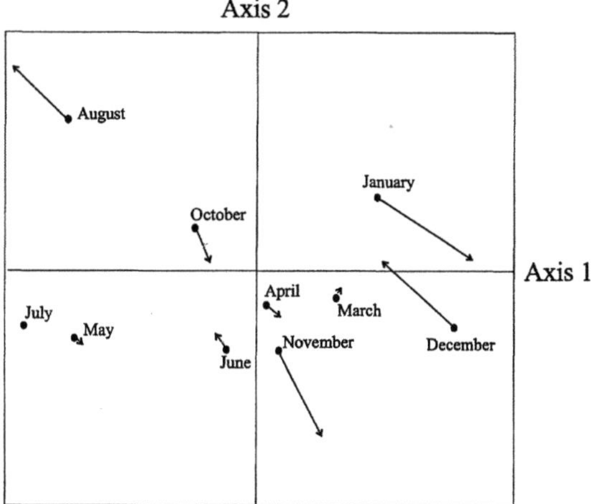

Figure 5. Adjustment of the first model obtained, on the first factorial axis of the Principal Component Analysis with Instrumental Variables. The head of the arrows illustrate the position of the samples (months) calculated by the model, whereas their origin indicate the observed position of the months on the first factorial plane.

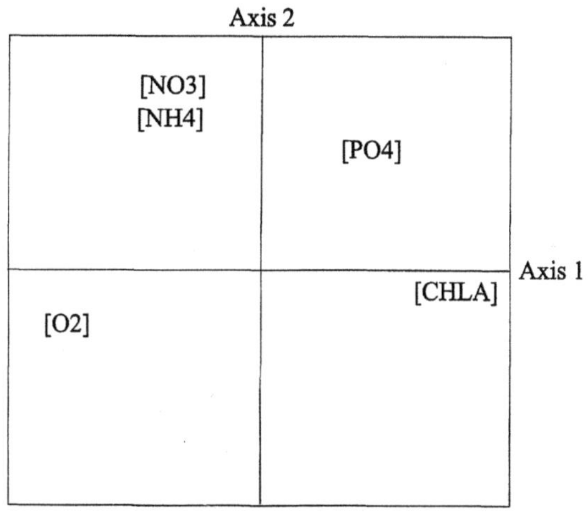

Figure 6. Projection of the dependent variables (water column characteristics) on the first factorial plane of the Principal Component Analysis with Instrumental Variables.

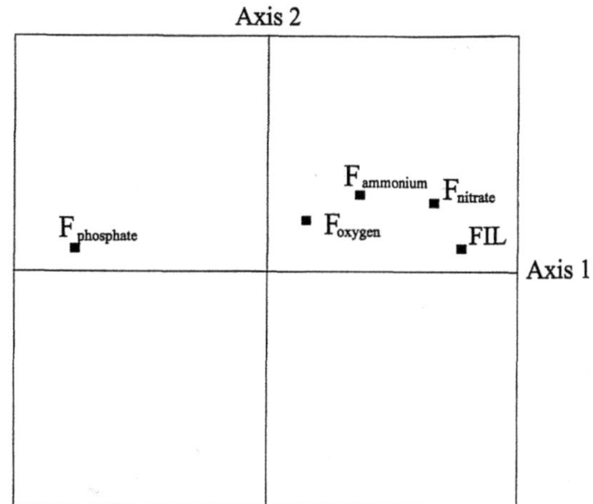

Figure 7. Correlation of the explanatory variables (fluxes at the OCU–water interface) with the first two factorial axis of the Principal Component Analysis with Instrumental Variables.

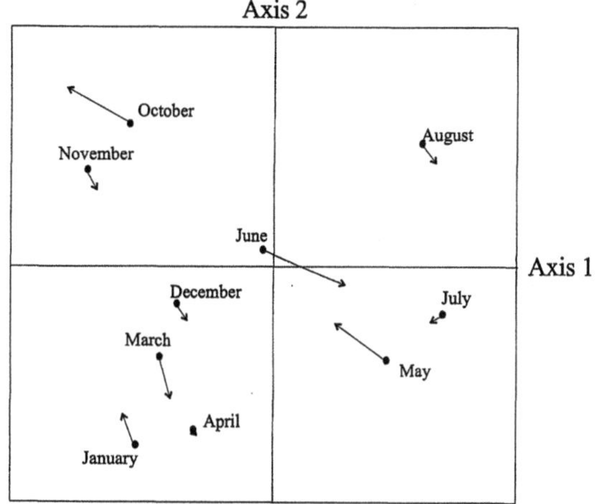

Figure 8. Adjustment of the second model obtained, on the first factorial axis of the Principal Component Analysis with Instrumental Variables. The head of the arrows illustrate the position of the samples (months) calculated by the model, whereas their origin indicate the observed position of the months on the first factorial plane.

axis 1, according to the distribution of the months (Figure 5).

The correlation of the explanatory variables (the species composition of the OCU) with the first two axes of the analysis (Figure 4), characterized an opposition between the percentages of sponges, macroalgae and bryozoa and the percentage of oysters. The species diversity of the assemblage seemed correlated to an important activity of this compartment (fluxes), especially in the spring. The percentage of polychaetes was independent of this relation and was opposed to the percentage of ascidians on axis 2. On the second axis, according to their position on the factorial plane, the abundance of polychaetes seemed associated to high phosphate fluxes, whereas the abundance of ascidians seemed associated to high nitrate–nitrite fluxes.

On Figure 5 (and Figure 8) the head of the arrows is representative of the positions of the samples (months) calculated by the model, whereas the origin of the arrows shows the observed position of the samples. We can see on Figure 5 that the model was well adjusted to our measurements, with the exception of January, August, November and December.

The second PCAIV was performed to determine whether the activity of the OCU had a significant influence on nutrient concentrations, chlorophyll level and oxygen content of the water column. In this second model, the activity of the oyster culture explained 68% of the total variability of the water column characteristics. Like in the first analysis, the permutation test was significant ($p < 0.05$). The first two axes accounted for 74% and 21% of the total variability of the model.

The projection of the dependent variables (water column characteristics) on the first axis of the factorial analysis (Figure 6), highlighted a strong opposition between the oxygen concentrations and the chlorophyll a level in the water column, and secondarily, between the nitrogen concentrations and the phosphate concentrations. On the second axis, the concentration of oxygen was opposed to that of phosphate and nitrogen.

On the first axis, the correlation of the explanatory variables (OCU activity) with the first two axes of the analysis (Figure 7) showed an opposition between, on one hand the fluxes of oxygen and nitrogen and the filtration and on the other hand, the fluxes of phosphate. The first were high in the summer period whereas the phosphate flux was weak. This latter was maximum during the autumn. Moreover, in winter when water temperature was less than 8 °C phosphate uptake was recorded at the OCU-water interface, although a phosphate release occurred the rest of the year. On the second axis no opposition between the variables was apparent. Thus, all the variables contributed to the formation of this second axis.

For this second PCAIV, the model was well adjusted, at the exception of October, May, June and March (Figure 8). Thus, the fluxes at the oyster culture-water interface seemed to have a significant influence on the water column, even if it occurred only in a period of the year.

Discussion

We showed that species composition of the OCU had a significant influence on the fluxes at the OCU–water interface. The higher fluxes were correlated with high species diversity and water temperature, for all elements considered, with the exception of phosphate. Indeed, no linear relation was observed between phosphate flux and water temperature. For this element, the maximum fluxes (peak) were associated to the presence of polychaetes and thus to the presence of a sediment compartment in the OCU. A same relation between phosphate fluxes and sediment compartment was found for a mussel bed (Dame et al., 1991).

The high filtration rates of the OCU were due to the presence of several filter-feeding organisms in the assemblage. Sponges are known to filter the water so efficiently that they remove almost the particles (Reiswig, 1972). On the other hand, ascidians are known to be continuous feeders that can regulate their clearance rate (Robbins, 1984) and particle retention efficiency (Osman et al., 1989; Lesser et al., 1992). Moreover, the retention efficiency of an assemblage is generally higher than that of its species (Branch, 1984; Stuart & Klumpp, 1984).

Although in literature filter-feeder populations are often considered as chlorophyll sinks (Cloern, 1982; Jarry et al., 1990), at our measurement scale (i.e. the oyster frame) no grazing influence was observed. Maximum filtration (as for chlorophyll level) was obtained during summer (July–August), when high water temperature and high nutrient content in the water column enhanced filter-feeder activity and phytoplankton biomass. Nevertheless, at this scale, the oyster cultures strongly influenced the phytoplankton population size spectrum (Mazouni et al., 1992). This result was in accordance with Smaal & Van Stralen (1990) who pointed out the limitation of the use of chlorophyll levels as a descriptor for trophic richness. Further improvements with flow cytometry are needed to test this hypothesis and to better understand the impact of the OCU on the primary producers in this coastal lagoon.

On the second axis of the first analysis, the opposition of the percentages of ascidians and polychaetes highlighted the difference of these epifaunal populations, and thus the complexity of the OCU structure. The ascidians as earlier colonists had a rapid development (Branch, 1984; Lamy, 1996), and an ability to quickly recolonize a substratum after perturbation. These earlier colonists, through their physiological activities and behaviour, tend to modify the environment to make it acceptable for later arrivals (Dayton, 1971; Sutherland, 1974, Dean & Hurd, 1980, Dean, 1981). On the OCU, they were responsible of an accumulation of organic matter and the formation of a sediment compartment suspended in the water column. On the assemblage, the polychaete population development was not only limited by their reproduction, but also by the nature of the substratum. Indeed, the polychaetes are generally considered as later arrivals, and their development needed the existence of a sediment compartment (Arakawa, 1990; Lamy, 1996). Thus, their development in the OCU seemed dependent on ascidian settlement and persistence.

The first model was in accordance with our measurements, at the exception of four months. During these months, two perturbations of the system occurred, an ecological one (anoxia, at the end of July–August) and a climatic one (extremely low water temperature, in January) described in Mazouni et al. (1996) which induced important mortalities of the epifaunal species. These results pointed out that a modification of the OCU stucture had a great influence on the exchanges at the OCU-water interface. Thus, in order to estimate the impact of the oyster cultures on an ecosystem, we must take into account the whole assemblage (oysters and epibiota).

In the second part of this study, an opposition between the oxygen concentration in the water column and the oxygen uptake by the OCU was obtained. This result pointed out the potential impact of the assemblage reared during bottom anoxia. The OCU assemblage could constitute an oxygen sink which contributes, in summer, to the spread of anoxic bottom conditions to the overlying water. This phenomenon, known locally as the Malaïgue (Mazouni et al., 1996), decimates the benthos and the reared species.

The oyster cultures, as in other shellfish ecosystems, had a significant influence on the availability of dissolved inorganic nitrogen (DIN) for the phytoplankton in the water column (Kaspar et al, 1985; Hammond et al., 1985; Dame & Libes, 1993). However, the maximun nitrogen fluxes were not correlated with maximum concentrations in the water column. So, in the Thau lagoon, other processes and compartments must be considered as they account significantly for the nutrient renewal in the water column: benthic remineralization (Kaspar et al, 1985; Hammond et al., 1985; Mazouni et al., 1996), protozoan (Caron & Goldman, 1990) and zooplankton (Lam Hoai, pers. com.) activities. The presence of macroalgae in the assemblage might also influence nutrient exchanges at the OCU–water interface. However, in our experiments their presence was too weak to be considered.

In conclusion, the use of this recent factorial analysis in ecology was helpful to define the influence of the biofouling communities on the fluxes at the OCU–water interface. Moreover, this method allowed us to quantitatively estimate the impact of an oyster culture on the water column. The approach could be applied to other data sets in order to improve the accuracy of some of the conclusions given by the present work, as for instance, the impact of the reared assemblage on the primary producers.

Acknowledgements

We thank Claude JUGE for the illustrations, and all the members of the IFREMER, DEL-ECO laboratory for their participation. This work was included in the PNOC-OXYTHAU and was financially supported by IFREMER.

References

Alpine, A. E. & J. E. Cloern, 1992. Trophic interactions and direct physical effects control phytoplankton biomass and production in an estuary. Limnol. Oceanogr. 37: 946–955.

Arakawa, K. H., 1990. Competitors and fouling organisms in the hanging culture of the Pacific oyster, Crassostrea gigas (Thunberg). Mar. Behav. Physiol. 17: 67–94.

Branch, G. M., 1984. Competition between marine organisms: ecological and evolutionary implications. Oceanogr. Mar. Biol. Ann. Rev. 22: 429-593.

Caron, D.A. & J.C Goldman, 1990. Protozoan nutrient regeneration. In Capriulo, G. M. (ed.), Ecology of Marine Protozoans, Oxford Univ. Press, New York: 283–306.

Cloern, J. E., 1982. Does the benthos control phytoplankton biomass in south San Francisco Bay? Mar. Ecol. Prog. Ser. 9: 191–202.

Dame, R., N. Dankers, T. Prins, H. Jongsma & A. Smaal, 1991. The influence of mussel beds on nutrients in the Western Sea and Eastern Scheldt estuaries. Estuaries 14: 130–138.

Dame, R. F. & N. Dankers, 1988. Uptake and release of materials by a wadden sea mussel bed. J. exp. mar. Biol. Ecol. 118: 207–216.

Dame, R. F. & S. Libes, 1993. Oyster reefs and nutrient retention in tidal creeks. J. exp. mar. Biol. Ecol. 17: 251–258.

Dame, R. F., J. D. Spurrier & T. G. Wolawer, 1989. Carbon, nitrogen and phosphorus processing by an oyster reef. Mar. Ecol. Prog. Ser. 54: 249–256.

David, A., 1970. Relations trophiques entre le plancton, les huîtres d'élevage et les ciones épibiontes (étang de Thau). Thèse Doct., Univ. Aix-Marseille II, 115 pp.

Dayton, P. K., 1971. Competition, disturbance, and community organization: the provision and subsequent utilization of space in a rocky intertidal community. Ecol. Monogr. 41: 351–389.

Dean, T. A, 1981. Structural aspects of sessile invertebrates as organizing forces in an estuarine fouling community. J. Exp. Mar. Biol. Ecol. 53: 163–180.

Dean, T.A. & L. E. Hurd, 1980. Development in an estuarine fouling community: the influence of early colonists on late arrivals. Oecologia 46: 295–301.

Hammond D. E., C. Fuller, D. Harmon, B. Hartan, M. Korosec, L. G. Miller, R. Rea, S. Warren, W. Berelson & S. W. Hager, 1985. Benthic fluxes in San Francisco Bay. Hydrobiologia 129: 69–90.

Jarry, V., M. Fiala, G.-F. Frisoni, G. Jacques, J. Neveux & M. Panouse, 1990. The spatial distribution of phytoplankton in a Mediterranean lagoon (Etang de Thau). Oceanol. acta. 13: 503–512.

Kaspar H. F., P. A. Gillespie, I. C. Boyer & A. L. MacKenzie, 1985. Effects of mussels aquaculture on the nitrogen cycle and benthic communities in Kenepuru Sound, Marlborough Sounds, New Zealand. Mar. Biol. 85: 127–136.

Koroleff F., 1969. Direct determination of ammonia in natural waters as indophenol blue. Int. Counc. Explor. Sea, C.M.C. 9: 1–6.

Lamy, N., 1996. Organisation, structure et dynamique des peuplements macrobenthiques d'une table conchylicole de l'étang de Thau (Hérault, France). Thèse de doc., Univ. Montpellier II, 292 p+ annexes.

Lebreton, J.D., R. Sabatier, G. Banco & A. M. Bacou, 1991. Principal component and correspondance analyses with respect to instrumental variables: an overview of their role in studies of structure – activity and species – environment relationships. In Devilliers, J & W. Karcher, (eds), Applied Multivariate Analysis in SAR and Environmental Studies. Brussels and luxembourg: 85–114.

Lesser, M. P., S. E. Shumway, T. Cucci & J. Smith, 1992. Impact of fouling organisms on mussel rope culture: interspecific competition for food among suspension-feeding invertebrates. J. Exp. Mar. Biol. Ecol., 165: 91–102.

Lorensen C. J., 1966. A method for the continuous measurement of *in vivo* chlorophyll concentration. Deep-Sea Res. 13: 223–227.

Mazouni N., J.-C. Gaertner, J.-M. Deslous-Paoli, S. M. Landrein & Geringer d'Oedenberg, 1996. Nutrient and oxygen exchanges at the water–sediment interface in a shellfish farming lagoon (Thau, France). J. Exp. Mar. Biol. Ecol. 205: 91–113.

Mazouni N., 1995. Etude *in situ* de l'influence des élevages ostréicoles sur le fonctionnement d'un ecosystème lagunaire méditerraneén. Thèse de doc., Univ. Aix-Marseille II, 256 pp.

Mazouni N., C. Courties, J.-M. Deslous-Paoli & A. Vaquer, 1992. Phytoplanktonic size spectrum modifications induced by shellfish cultures in Thau lagoon. Affiche et Communication orale, Congrès de cytométrie en flux et de cytométrie en image, Montpellier 63 pp.

Millet B., 1989. Fonctionnement hydrodynamique du bassin de Thau. Validation d'un modèle numérique de circulation (programme ECOTHAU). Oceanol. Acta 12: 37–46.

Osman, R. W., R. Withlatch & R. N. Zajac, 1989. Effects of resident species on recruitment into a community: larval settlement versus post-settlement mortality in the oyster Crassostrea virginica. Mar. Ecol. Prog. Ser. 54: 61–73.

Prins, T. C. & A. C. Smaal, 1994. The role of the blue mussel Mytilus edulis in the cycling of nutrients in the Oostershelde estuary (The Netherlands). Hydrobiologia 282/283: 413–429.

Rao, C. R., 1964. The use and interpretation of principal component analysis in applied research. Sankhya 26: 329–359.

Reiswig, H. M., 1972. Particle feeding in natural populations of three marine demosponges. Biol. Bull. mar. biol. Lab., Woods Hole. 141: 568–591.

Robbins, I. J., 1984. The regulation of ingestion rate, at higher suspended particulate concentrations, by some phleobranchiate ascidians. J. exp. mar. Biol. Ecol. 82: 1–10.

Sabatier, R., J.-D. Lebreton & D. Chessel, 1989. Principal component analysis with instrumental variables as a tool for modelling composition data. In Coppi, R. & S. Bolasco (eds), Multiway Data Analysis: 341–352.

Smaal, A. C. & M. R. Van Stralen, 1990. Average annual growth and condition of mussels as a function of food source. Hydrobiologia 195: 179–188.

Stuart, V. & D. W. Klumpp, 1984. Evidence for food-resource partitioning by kelpbed filter feeders. Mar. Ecol. Prog. Ser. 16: 27–37.

Sutherland, J. P., 1974. Multiple stable points in natural communities. Am. Nat. 108: 859–873.

Ter Braak, C. J. F., 1986. Canonical correspondance analysis : a new eigenvector technique for multivariate direct gradient analysis. Ecology 67: 1167–1179.

Thioulouse, J., S. Dolédec, D. Chessel & J. M. Oliver, 1995. ADE software: multivariate analysis and graphical display of environmental data. Proceeding of the 4th International Software Exhibition for Environmental Science and Engineering. Lyon: 57–62.

Tréguer, P. & P. Le Corre, 1975. Manuel d'analyses de sels nutritifs dans l'eau de mer. Utilisation de l'Autoanalyzer II Technicon. 2 ème édition, Laboratoire d'océanographie chimique, Univ. Bretagne occidentale, Brest.

Hydrobiologia **373/374:** 157–167, 1998.
J. C. Amiard, B. Le Rouzic, B. Berthet & G. Bertru (eds),
Oceans, Rivers and Lakes: Energy and Substance Transfers at Interfaces.

Laboratory experiments on trophic relationships and remote detection between two ciliates and *Cyclops vicinus vicinus*

Chantal Rabette[1], Antoine Thouvenot[1] & Nicole Lair[2]
[1] *Laboratoire de Biologie Comparée des Protistes, UPRES-A 6023 CNRS, Université Blaise PASCAL 63177 Aubière Cedex (France)*
[2] *Laboratoire de Géodynamique des milieux naturels et anthropisés, UPRES-A 6042 CNRS, Université Blaise PASCAL 63177 Aubière Cedex (France)*

Key words: Cyclops vicinus, ciliates, ingestion rate, remote detection, chemical communication

Abstract

Laboratory experiments were performed to study the feeding behavior of *Cyclops vicinus* (Copepoda, Cyclopoida), fed two ciliates (*Cyclidium* sp. and *Tetrahymena corlissi*) chosen for their different size and swimming behavior. All grazing experiments were conducted with predators starved for 24 hr. *Cyclidium* sp. was fixed with Glutaraldehyde and *Tetrahymena corlissi* with mercuric chlorid in order to reduce counting errors. The Incipient Limit Level (ILL) for *Tetrahymena corlissi* was reached at the initial concentration of 76 000 cell l^{-1} and corresponded to an ingestion rate of 340 cell ind^{-1} h^{-1}; that for *Cyclidium* sp. was reached at 12 000 cell l^{-1} with an ingestion rate of 54 cell ind^{-1} h^{-1}. The detection experiments were based on visual observations of *Cyclops* movements in experimental chambers with prey concentrations corresponding to the ILL. Both ciliates were used to evaluate the mechanisms involved in the detection of prey by *Cyclops vicinus*. Direct observations showed that *Tetrahymena corlissi* had a significant attractive effect on *C. vicinus*. Such behavior suggests that mechanoreception may be the overriding mechanism of remote detection. In contrast, *C. vicinus* does not react to the presence of *Cyclidium* sp. (at low concentration). These small prey, which swim actively in comparison with *T. corlissi*, are preyed on haphazardly. Ciliates can be a significant part of the diet of *Cyclops vicinus*, confirming their importance as mediators for energy transfer from the microbial loop to higher trophic levels; the mechanism of detection of such motile prey by *Cyclops vicinus* varies with prey swimming behavior.

Introduction

Ciliates may play an important role in nutrient cycling and energy transfer within aquatic food webs (Berk et al., 1977; Sherr et al., 1986; Stœcker & Capuzzo, 1990): by their repackaging action a fraction of the bacterial, pico- and nanoplanktonic production becomes available for the higher order consumers (Beaver & Crisman, 1989; Gifford, 1991). Early studies were concentrated on marine species (e.g. Stœcker & Capuzzo, 1990), nevertheless in freshwater ecosystems ciliated protozoans may have a high abundance, biomass and productivity (Finlay, 1977; Pace & Orcutt, 1981; Beaver & Crisman, 1982; Gates, 1984; Bienert et al., 1991), particularly in eutrophic waters.

Although the fact that ciliates feed on microorganisms, other protozoans and small metazoans has long been known, the extent of their consumption by larger metazoans is less well known partly because their soft bodies are seldom recognisable in gut contents (Archbold & Berger, 1985). There are many studies showing that calanoid copepods can graze a significant proportion of ciliate production, but little work has been carried out to determine whether cyclopoids can have a similar impact (reviewed in Sanders & Wickham, 1993). Calanoids feed by either filtering or actively capturing individual prey items, but cyclopoids feed raptorially (Kerfoot, 1978; Price et al., 1983). Thus, the effects of cyclopoid grazing on ciliates may not be the same as those of calanoids.

Cyclopoids are known to select their prey, with selection depending on such factors as prey hardness, shape and escape response as well as predator satiation (Brandl & Fernando, 1975; Kerfoot, 1978; Williamson, 1980; Stemberger, 1985; Gliwicz & Umana, 1994). There is now evidence that copepods use chemical cues to assess the quality of individual food items (Friedman, 1980), but experiments have focused on calanoid copepods and demonstrated their ability to discriminate between toxic and nontoxic (Fulton & Paerl, 1987; De Mott, 1989; De Mott & Moxter, 1991), between readily digested and undigestible algae (De Mott, 1989) and between live and dead individuals of the same species (Paffenhöffer & Van Sant, 1986; De Mott, 1988; Cowles et al., 1988). Cyclopoid are also known to detect some odoriferous molecules by chemoreception (Strickler, 1985; Atema, 1987; Price, 1988; De Mott, 1990). Chemosensorial structures allow contact selection (Vanderploeg & Paffenhöffer, 1985) as well as remote detection.

Another process in the prey detection is also the prey behavior. As reported by Uchima & Hirano (1986) and Uchima (1988) small marine cyclopoids prefer motile algae while freshwater ones prefer rotifers (Gilbert & Bogdan, 1981); but to our knowledge data about ciliates are poor. Hartmann et al., (1993) have observed that *Tetrahymena corlissi* did not visibly react when approaching calanoid copepods and was probably captured through such a reaching movement. By contrast *Stentor*'s contraction would reduce its chance of being detected by minimizing hydrodynamic signals to the predators (e.g. Jonsson & Tiselius, 1990; De Mott & Watson, 1991).

Recent studies have shown cyclopoids to be capable of ingesting algae (Adrian, 1987, 1991; Lair, 1992; Santer & van den Bosch, 1994; Santer, 1993,1994), but information on macrozooplankton predation on protozoans, and particularly on ciliates in freshwater systems remains limited. However ciliated protists are as small or smaller than most algae, the majority of planktonic ciliates ranging from 10 to 200 μm in size (Pace, 1982; Finlay et al., 1988). Ciliates, then, have the potential to be either small enough to be preferred cyclopoid prey, or too small to be detected or captured (Wickham, 1995). From at least one study, it would appear that ciliates may be easily captured, but are rarely ingested by cyclopoids (Williamson, 1980). However, it is not known whether this is a general phenomenon.

The two objectives of this study were therefore:

- To examine the ability of the adults of the raptorial-feeder *Cyclops vicinus* to feed in the laboratory on single species of differently sized ciliates over a wide range of prey concentrations.
- To determine whether those protists can be detected by *Cyclops vicinus*.

Materials and methods

The predator

Zooplankton was collected in the eutrophic Lake Aydat by vertical hauls with a 88 μm Juday mesh net and brought back to the laboratory. After a pre-filtration on a 200 μm mesh net, adults of *Cyclops vicinus* (1310.9 ± 198.6 μm) were sorted under a dissecting microscope (Wild M3Z) and quickly conditioned in 2 l flasks filled with lake water, filtered through 0.45 μm cellulose acetate filters (Sartorius membrane Type SM). In accordance with Vijverberg (1989), the maximum concentration in the rearing beakers was 50 individuals per litre. Prior to the experiments, the *Cyclops vicinus* were progressively adapted to a temperature of 15 ± 1 °C, in a culture room with a day-night cycle 16–8 h. Before running all the experiments, crustaceans were starved for 24 h.

The ciliates

The ciliates studied were chosen for their different size and swimming behavior; however it should be noted that while *Cyclidium* is fairly common among ciliates in lakes, *Tetrahymena* is not (even in Lake Aydat). The ciliated prey were cultured in 250 ml flasks at 20 °C in a culture medium consisting of ground brains and mineral water which provides enough bacteria for these bacterivorous ciliates.

- *Cyclidium* sp.: length 20.60 ± 1.55 μm; width: 13.30 ± 1.70 μm
- *Tetrahymena corlissi*: length 46.85 ± 6.10 μm; width: 21.50 ± 2.15 μm

In order to prevent contact of *Cyclops vicinus* with another potential food or chemical cues coming from the culture medium the ciliates have to be isolated. After a 12 h sedimentation in a test-tube, the largest particles in the culture medium were poured off and then two different treatments were applied, depending on the species. The fragile *Cyclidium* sp. were washed three times with filtered lake water by gentle filtration (pressure < 40 mm Hg), while the less delicate *Tetrahymena corlissi* were centrifuged (1000 to 1500 rpm for 30 sec to 1 min).

Ingestion experiments

(a) Preliminary experiments
Preliminary experiments were conducted with *Cyclidium* sp. for 3, 6, 12 and 24 hr with and without the predator, to determine the most appropriate experimental duration to obtain a significant decrease in the prey population. For all the experiments, the beakers contained adults of *Cyclops vicinus* starved for 24 hr, controls were without the copepods and each set of experiments (control and experiment) was repeated three times.

(b) Grazing experiments
Before the beginning of each experiment a feeding suspension was prepared with filtered lake water and ciliate concentrations ranging from 1 to 100 per ml to bracket the concentrations typically found in the field. Appropriate subsamples of washed ciliates were added to experimental and control beakers (time t_0). Inoculum cell concentrations were determined from counts under a microscope in a Sedgewick-Rafter cell. After adding the ciliate solution, 3 adult copepods without eggs were introduced and the volume was adjusted to 120 ml.

The beakers were incubated in the dark at 15± 1 °C. The experiments were ended by adding different fixatives depending on the ciliate species to prevent deformation and reduce counting errors according to Foissner (personal communication). *Cyclidium* sp. were fixed with 2.5 ml of glutaraldehyde and *T. corlissi* with a saturated $HgCl_2$ solution (69 g l^{-1}). Few drops of bromophenol blue solution (1%) were also added to stain the ciliates (Bereczky, 1985).

The total number of ciliates at time t_1 in each beaker was determined after counting under an inverted microscope (LEICA-DMIL, magnification 200).

Behavior experiments

These experiments were conducted in semi-cylindrical vessels (length: 40 cm; diameter: 8 cm) which appeared to be the most suitable material for such experiments. The experimental vessels were put in an illuminated culture room at 15 ± 1 °C, where in accor-

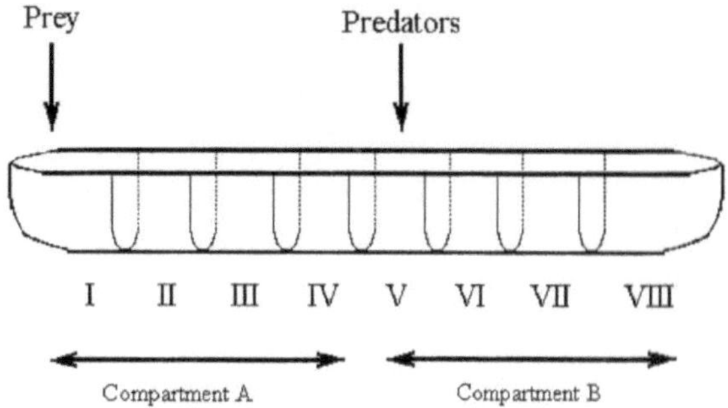

Figure 1. The experimental chamber.

dance with Hutchinson (1975) the light was less than 25 W m^{-2} to prevent phototropism. The displacement of the animals was visually observed each quarter of an hour. To simplify counting of the animals the cylinders were ≪virtually≫ divided into 5 cm segments. Segments I, II, III, IV correspond to compartment A (prey side), segments V, VI, VII, VIII correspond to compartment B. The prey were always introduced into segment I (Figure 1).

Before beginning the experiments, 40 adult copepods were introduced into the middle of each experimental chamber with filtered lake water; and the experiment was started only when the copepods were homogeneously distributed. The duration of the experiments was fixed at 1 h, the time to prevent prey diffusion in the experimental vessels and each set of experiment was performed in triplicates.

Behavioral experiments were run at concentrations similar or above the incipient limit level (ILL) as determined by the ingestion experiments: 10^4 cell $l^{-1}$for *Cyclidium* sp. and 4.10^5 cell $l^{-1}$for *T. corlissi.*

Data analysis

(a) Grazing experiments
The grazing rate (G) was determined from the initial ($C_{(t0)}$) and final ($C_{(e)}$) prey concentration in the experimental beakers; G was corrected by the prey growth coefficient (K) determined from prey counts in the control beakers at time t_0 and t_1. The formulas for G varied according to the position of $C_{(t0)}$ above or below the incipient limit level (ILL) for ciliate consumption by *C. vicinus* (Marin et al., 1986). For all the experiments $C_{(t0)} \leq$ ILL, i.e. the different feeding rates were calculated after determining the ILL through inspection of calculated clearance and ingestion rates.

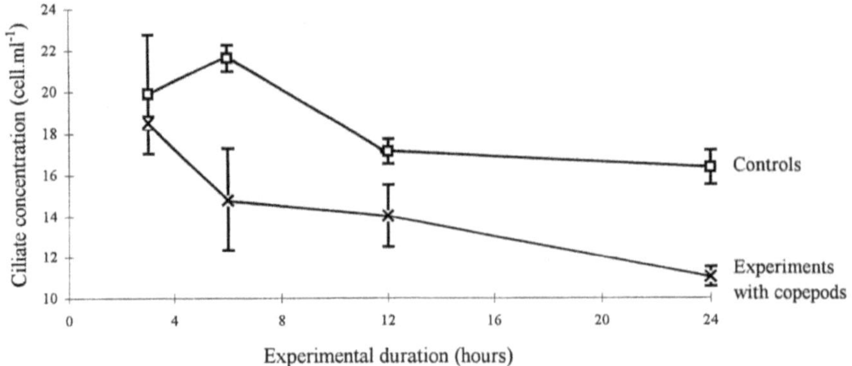

Figure 2. Changes of *Cyclidium* sp. concentration with or without the predators over 24 h.

$$\text{Growth coefficient: } K = \frac{\ln(N_{(t1)}/N_{(t0)})}{t}$$

$$\text{Grazing rate } G = \ln(N_{(t0)}/N_{(\text{ef})})\frac{1}{t} + K$$
$$\text{for } C_{(t0)} \leq \text{ILL}$$

$N_{(t0)}$: Initial Ciliate concentration
$N_{(t1)}$: Final Ciliate concentration in the control beakers
$N_{(e)}$: Final Ciliate concentration in the experimental beakers
t: Experimental duration

The clearance rate (*C*, millilitres per copepod per day) was determined from:

$$C = V \times G \times P^{-1}$$

where V = beaker volume and P = number of predators in the beaker. All changes in ciliate densities observed in the control beakers were assumed to have occurred independently of the presence or absence of the copepods.

Ingestion rates (*I*, cells per copepod per day) were calculated by assuming that cells encountered were captured, ingested or even destroyed and with reference to initial, rather than average prey concentration:

$$I = C \times N_{(\text{o})}$$

All rates were compared with the initial prey concentration (Marin et al., 1986). The data on feeding rates were compared by parametric statistical tests (*t*-test).

Holling type 2 or type 3 functional responses (Holling, 1959) were then fitted to the data using non-linear regression procedures.

Holling type 3 functional responses were fitted using the equation:

$$I = \frac{I_{\max} \times \text{cil}^2}{R_0^2 + \text{cil}^2}$$

Holling type 2 functional responses were fitted using the equation:

$$I = \frac{I_{\max} \times \text{cil}}{R_0 + \text{cil}}$$

where I_{max} = maximal ingestion rate, cil = prey density (ciliate ml^{-1}) and R_0 = ciliate density at which half the maximal ingestion rate occurs.

b) Behavior experiments

The patchiness index ($\overset{*}{x}/\bar{x}$) (Lloyd, 1967) was used to illustrate the copepod pattern of distribution in the experimental vessels at each time, where $\bar{x}$ is the average number of individuals observed in each segment of the experimental chamber at each time and S^2 the associated variance. The ratio of mean crowding ($\overset{*}{x}$) to mean ($\bar{x}$) at each time provides a suitable measure of dispersion.

The index is > 1 for a contagious distribution, $= 1$ for a random distribution, and < 1 for a regular distribution.

In the presence of prey, we postulated that a regular distribution involved an absence of copepod response, either because the prey was not detected, or because it did not induce a behavioral response. A contagious distribution in the A compartment indicated that the copepod is attracted by the prey, while a contagious distribution in the B compartment indicates a repulsion. This distribution pattern was considered as the expression of a behavioral response of the copepods

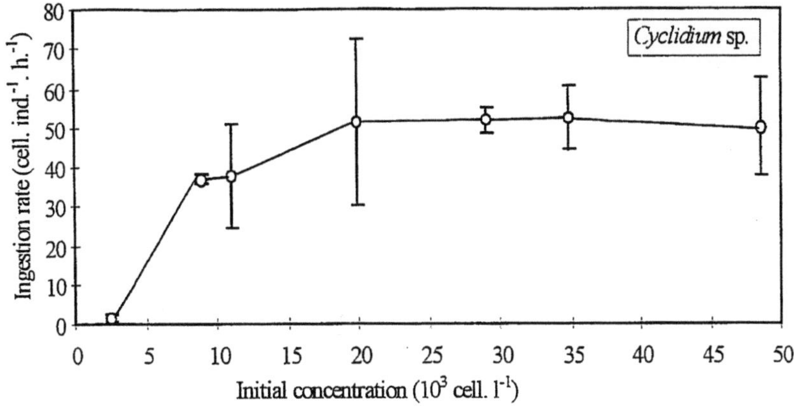

Figure 3. Ingestion rates of *Cyclops vicinus* adults feeding on different concentrations of *Cyclidium* sp. in laboratory experiments.

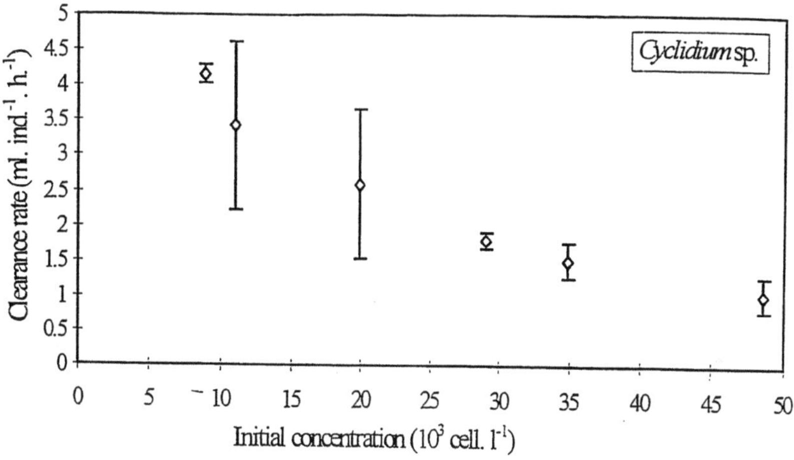

Figure 4. Clearance rates of *Cyclops vicinus* adults feeding on different concentrations of *Cyclidium* sp. in laboratory experiments.

to the prey. Wilcoxon's signed-rank test was used to compare experimental vessels with the controls.

Results

Feeding experiments

(a) Time duration determination

After the large variations observed during the first few hours of experiments, probably due to the switch of medium that the ciliates undergo, their density remained high in the controls, before a significant decrease between 6 and 12 h and a stabilization until 24 h (Figure 2). In presence of *Cyclops vicinus* the ciliates concentration decreased quickly after 6 h, then after the feeding activity continued until the end of the experiment in lower proportions.

The difference observed between controls and experiments was highly significant after 6 and 24 hr (t-test at $P = 0.05$; Table 1), so the experimental duration was fixed at 6 h.

(b) Feeding experiments

The raptorial-feeder *C. vicinus* adults significantly affected *Cyclidium* density. At the lower concentrations (from 9.10^3 to 11.10^3 cell l^{-1}) the mean ingestion rate fluctuated around 40 cell ind^{-1} h^{-1}. When the initial concentration was increased up to 12. 10^3 cell l^{-1} the ingestion rate remained stable at 53.87 ± 0.80 cell ind^{-1} h^{-1} (Figure 3). Over the range of concentrations used in this experiment the clearance rate for *Cyclidi-*

Table 1. Cyclidium sp. concentration, standard deviation and *t*-test values for the preliminary experiment.

Duration	Concentration (cell ml^{-1})		Standard deviation		*t*-test	
(h)	Controls	Experiments	Controls	Experiments	*t*-value	*P*
3	19.89	18.50	2.86	0.28	1	0.6141
6	21.60	14.81	0.64	2.47	5	0.0010
12	17.12	14.02	0.59	1.52	3	0.0500
24	16.36	11.02	0.82	0.48	4	0.0100

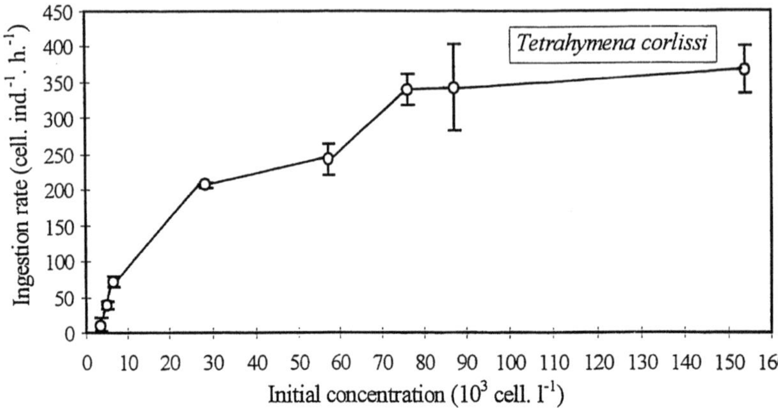

Figure 5. Ingestion rates of *Cyclops vicinus* adults feeding on different concentrations of *Tetrahymena corlissi* in laboratory experiments.

um sp. did not change significantly, it decreased from 4 to 1.5 ml ind^{-1} h^{-1} (Figure 4).

The ingestion rate of *C. vicinus* fed on *T. corlissi* increased with increasing concentration between 3.5 10^3 and 76. 10^3 cell l^{-1} and quickly achieved a plateau at 368.24 ± 33.88 cell ind^{-1} h^{-1}, in spite of the increase of available prey (Figure 5).

After 6 hr of feeding activity, at the lower concentrations (between 3.57 and 6.57. 10^3 cell l^{-1}) the clearance rate reached 10.94 ± 1.34 ml ind^{-1} h^{-1}; before decreasing to 2.4 ± 0.22 ml ind^{-1} h^{-1} at the higher concentrations (the highest being 154. 10^3 cell l^{-1}) (Figure 6).

When fed on single prey, the maximum ingestion of ciliates by *C. vicinus* varied greatly between prey type. However the results concerning the functional response for *Cyclops* preying on these two ciliates fitted the same Holling type functional response. As there was no obvious lower plateau of constant ingestion rate; the change in predation rates with density was fitted to a Holling type 2 response curve (Figure 7 and 8).

Behavior experiments

For the experiments with *Cyclidium* sp. either in the control, or at a prey concentration of 10^4 cells per litre, the *C. vicinus* distribution was not significantly different (at the 5% level), the copepod distribution being regular (Table 2).

For the experiments conducted with *T. corlissi* (Figure 9), the *C. vicinus* distribution was regular in the controls, at each run of the experiment. At the concentration of 4 10^5 cell l^{-1} the copepods distribution was contagious after 30 min ($\overset{*}{x}/\bar{x} = 1.13$) and 45 min ($\overset{*}{x}/\bar{x} = 1.05$) and the difference from the controls was significant ($P = 0.01$ see Table 2) which means that *Cyclops vicinus* was affected by *T. corlissi*.

Discussion

Microzooplankton have long been known to be supplementary food when marine copepods were feeding on low quality food particles or when phytoplankton were scarce (Berk et al., 1977). This suggestion of the exploitation of an alternative food source was con-

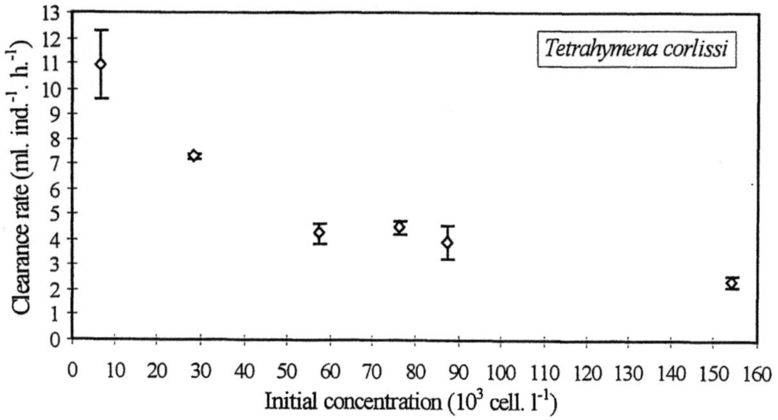

Figure 6. Clearance rates of *Cyclops vicinus* adults feeding on different concentrations of *Tetrahymena corlissi* in laboratory experiments.

Table 2. Results of patchiness index (R = Regular and C = Contagious) and Wilcoxon's signed – rank test (5% level).

	Concentration (cell l^{-1})	Time (min) 15	30	45	60	Comparison between controls and experiments	Spatial distribution of *Cyclops* in the experimental chamber	Attaction/ Repulsion
T. corlissi								
46.85 ± 6.10 μm								
Cl	1.00 E+05	R	C	C	R	$p = 0.0116$	Contagious distribution	Attraction
control	No prey	R	R	R	R	$p > 0.05$	Regular distribution	
Cyclidium sp.								
20.60 ± 1.55 μm								
Cl	1.00 E+04	R	R	R	R	$p > 0.05$	Regular distribution	
control	No prey	R	R	R	R	$p > 0.05$	Regular distribution	

firmed by the high clearance rate of protozoans relative to similar sized phytoplankton obtained by Atkinson (1996). The former study circumvented the size factor by comparing protozoans and diatom of similar dimensions.

Concerning freshwaters, our experimental results clearly indicate that the raptorial-feeder cyclopoid copepod *C. vicinus* preys efficiently on a wide range of protozoan concentrations in the absence of algae. Our results supplement those of Stœcker & Egloff (1987) who demonstrated that ciliates are a potentially important part of *Cyclops vicinus* food even when algae are present.

Most previous studies have concentrated on the impact of cladocerans or calanoids on ciliates and there are few estimates of cyclopoid ingestion rates with which to compare. Wickham (1995) calculated an ingestion rate of 200 ciliates individual^{-1} h^{-1} for *Cyclops abyssorum* fed on *Strobilidium velox*, whereas in the present study we recorded a higher maximal

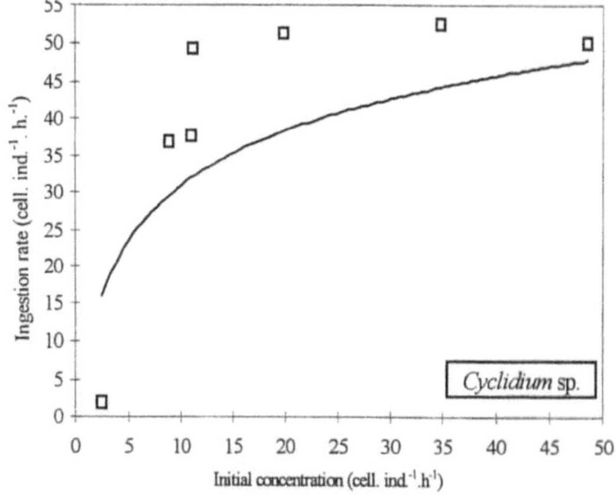

Figure 7. Holling type 2 functional response curves fitted to *Cyclops vicinus* fed on different initial concentrations of *Cyclidium* sp.

ingestion rate of 368 individual^{-1} h^{-1} for *C. vicinus* a smaller copepod fed on *Tetrahymena corlissi* (which

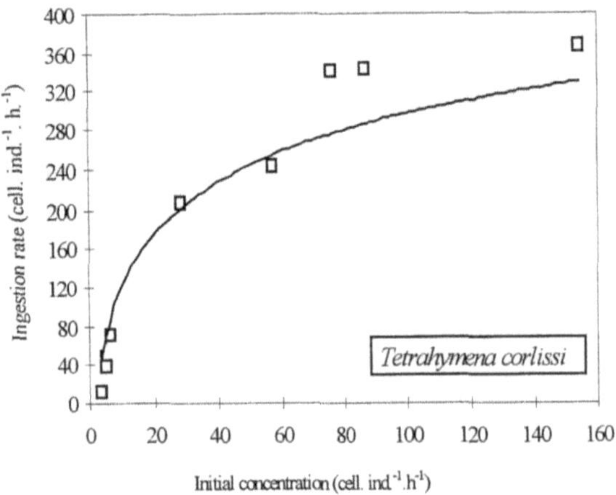

Figure 8. Holling type 2 functional response curves fitted to *Cyclops vicinus* preying on various densities of *Tetrahymena corlissi.*

is quite similar in size to *S. velox*). Moreover during this study we found a negative relationship between ciliate size and feeding rates of *C. vicinus* which could explain the great differences in the maximal ingestion rates recorded for the two different ciliate species.

For a better understanding of the copepods' response it must be kept in mind that prey size is not the only variable; predator behavior and prey response also play an important role. These experimental observations were thus consistent with other observations (e.g. Vanderploeg et al., 1988; Jonsson & Tiselius, 1990) indicating that behavioral models are essential for properly interpreting feeding preferences by copepods on large, motile prey. Analysis of prey detection, orientation, capture, handling, and ingestion, are required to completely understand the feeding process and preferences of such omnivorous copepods (e.g. Hartmann et al., 1993).

In the seventies, videocamera observations showed that cyclopoids actively orient and attack prey with considerable precision (Kerfoot, 1978). Now it is well known that they detect their prey with the help of mechanoreceptors on their first antennae. Prey that generate significant disturbance as they swim can be detected at distances of several millimetres (Williamson, 1991). Our results suggest that *T. corlissi*, a large slow-swimming ciliate, can be detected by mechanoreception. *Cyclidium* sp. is not detected or induces no behavioral response because this ciliates induced no changes in the predator distribution between controls and experiments. These results are consistent with those of De Mott & Watson (1991) who have shown that motile prey may favour mechanoreception. Another process in the prey detection is also prey behavior, lower feeding rates recorded for *Cyclidium* sp. could be due to their very fast swimming by jumps which can allow them to escape capture by *Cyclops*.

The maximal ingestion rate for *T. corlissi* occurred at higher densities than those commonly reported for natural ciliate communities. Although ciliate biomass increases with increasing trophy (Pace, 1986), total cell numbers rarely exceed 20 cells ml^{-1} (Pace & Orcutt, 1981; Hunt & Chein, 1983; Beaver et al., 1988; Berninger et al., 1993).

In considering the relevance of results from experimental feeding studies to natural trophic processes, it must also be kept in mind that protozoans and particularly ciliates, tend to be distributed very patchily in strongly stratified environments (Bark, 1985; Sime-Ngando & Hartmann, 1991) as well as not uniformly distributed in homogeneous water columns (Finlay et al., 1988; Sherr et al., 1991), and this may result in patches of high ciliate abundance, even when the average density is quite low. The same distribution is observed for zooplankton and probably result in much higher predator-prey encounter rates than those predicted from average prey densities (Williamson & Stoeckel, 1990; Folt et al., 1993).

Prey patchiness may expose copepod predators to periods of starvation. While it has been shown to induce variable feeding in copepods (Runge, 1980; Hasset & Landry, 1988), experimental studies rarely explain reasons for using either starved or satiated crustaceans in feeding studies (e.g. Stœcker & Capuzzo, 1990). We thus included hunger as an explicit parameter in our experiments. But it is clear that we thereby increased their capacity of prey ingestion while, according to Williamson (1987) with high concentration we favour the attack and capture probability.

These preliminary observations lay the groundwork for the development of definite behavioral capture models for ciliates by cyclopoid copepods, which must consider the importance of prey and predator behavior in the interpretation of feeding studies between motile prey and raptorial predators.

Acknowledgements

The authors want to thank Dr H. Taleb for helpful discussions, and two anonymous reviewers for their fruitful comments on the manuscript.

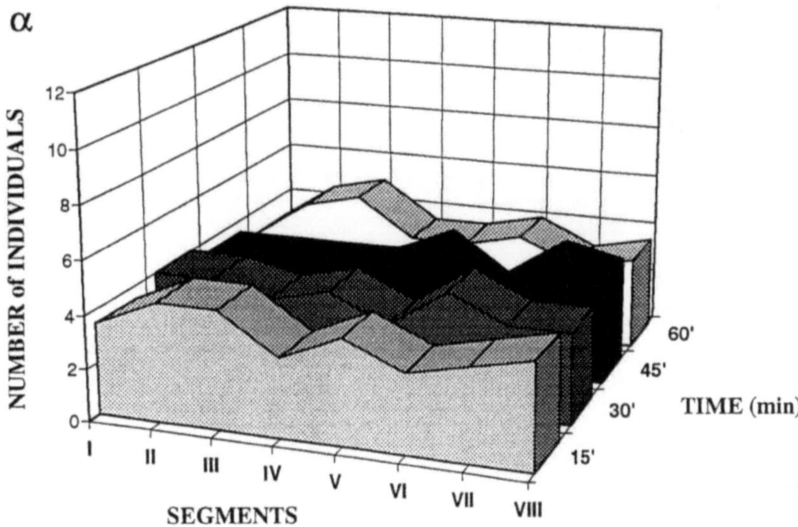

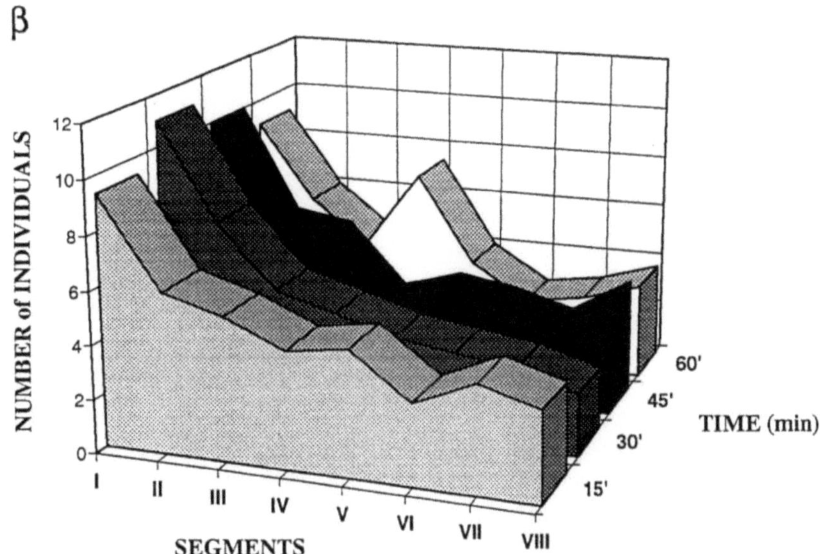

Figure 9. Distribution of *Cyclops vicinus* in the experimental chamber: α without prey (control). β in presence of *Tetrahymena corlissi* (4.10^5 cell l^{-1}) in segment I.

References

Adrian, R., 1987. Viability of phytoplankton in fecal pellets of two cyclopoid copepods. Arch. Hydrobiol. 110: 321–330.

Adrian, R., 1991 Filtering and feeding rates of cyclopoid copepods feeding on phytoplankton. Hydrobiologia 210: 217–223.

Archbold, H. G. & J. Berger, 1985. A qualitative assessment of some metazoan predators of *Halteria grandinella* a common freshwater ciliate. Hydrobiologia 124: 97–102.

Atema, J., 1987. Chemoreception in the sea: adaptations of chemoreceptors and behavior to aquatic stimulus conditions. In Atema, J., R. R. Ray, A. N. Popper & W. N. Tavolga (eds), Sensor Biology of Aquatic Animals. Springer, New York: 29–56.

Atkinson, A., 1996. Subantarctic copepods in an oceanic, low chlorophyll environment: ciliate predation, food selectivity and impact on prey population. Mar. Ecol. Prog. Ser. 130: 85–96.

Bark, A. W., 1985. Studies on ciliated protozoa in eutrophic lakes: 1. Seasonal distribution in relation to thermal stratification and hypolimnic anoxia. Hydrobiologia 124: 167–176.

Beaver, J. R. & T. L. Crisman, 1982. The trophic response of ciliated protozoans in freshwater lakes. Limnol. Oceanogr. 27: 246–253.

Beaver, J. R. & T. L. Crisman, 1989. The role of ciliated protozoa in pelagic freshwater ecosystems. Microb. Ecol. 17: 111–136.

Beaver, J. R., T. L. Crisman & J. R. W. Bienert, 1988. Distribution of planktonic ciliates in highly coloured subtropical lakes: comparison with clearwater ciliate communities and the contribution of mixotrophic taxa to total autotrophic biomass. Freshwat. Biol. 20: 51–60.

Bereczky, M. C., 1985. Fixations-und Färbungsschnellverfahren bei quantitativen ökologischen Untersuchungen von Protozoen in Binnengewässern. Arch. Protistenkd. 129: 187–190.

Berk, S. G., D. R. Brownlee, D. L. Heinle, H. J. King & R. R. Colwell, 1977. Ciliates as food source for marine planktonic copepods. Microbiol. Ecol. 4: 27–40.

Berninger, U. G., S. A. Wickham & B. J. Finlay, 1993. Trophic coupling within the microbial food web: a study with fine temporal resolution in a eutrophic freshwater ecosystem. Freshwat. Biol. 30: 419–432.

Bienert, R. W., J. R. Beaver & T. L. Crisman, 1991. The contribution of zooplankton biomass in an acidic, subtropical lake. J. Protozool. 38: 352–354.

Brandl, Z. & C.H. Fernando, 1975. Food consumption and utilization in two freshwater cyclopoid copepods (*Mesocyclops edax* and *Cyclops vicinus*). Int. Rev. Ges. Hydrobiol. 60: 471–494.

Cowles, T. J., R. J. Olson & S. W. Chishlom, 1988. Food selection by copepods: discrimination on the basis of food quality. Mar. Biol. 100: 41–49.

De Mott, W. R., 1988. Discrimination between algae and detritus by freshwater and marine zooplankton. Bull. Mar. Sci. 43: 486–499.

De Mott, W. R., 1989. Optimal foraging theory as predictor of chemically mediated food selection by suspension – feeding copepods. Limnol. Oceanogr. 34: 140–154.

De Mott, W. R., 1990. Retention efficiency, perceptual bias, and active choice as mechanism of food selection by suspension-feeding zooplankton. In Hughes, N.N. (ed.), Behavioral Mechanism of Food Selection. NATO Series G: Ecological Sciences Springer, Heidelberg, New York: 569–594.

De Mott, W. R. & F. Moxter, 1991. Foraging on cyanobacteria by copepods: responses to chemical defenses and resource abundance. Ecology 72: 1820–1834.

De Mott, W. R. & M. D. Watson, 1991. Remote detection of algae by copepods: responses to algal size, odors and motility. J. Plankton Res. 13: 1203–1222.

Finlay, B. J., 1977. The dependence of reproductive rate on cell size and temperature in freshwater ciliated protozoa. Oecologia 30: 75–81.

Finlay, B. J., K. J. Clarke, A. J. Cowling , R. A. Hindle, A. Rogerson & U. G. Berninger, 1988. On the abundance and distribution of protozoa and their food in a productive freshwater pond. Eur. J. Protistol. 23: 205–217.

Folt, C., P. C. Schulze & K. Baumgartner, 1993. Characterising a zooplankton neighbourhood: small scale patterns of association and abundance. Freshwat. Biol. 30: 289–300.

Friedman, M. M., 1980. Comparative morphology and functional significance of copepod receptors and oral structures. In Kerfoot, W. C. (ed.), Evolution and Ecology of Zooplankton Communities. The University Press of New England, Hanover (N.H.); Lond.: 185–197.

Fulton, R. S. & H. W. Paerl, 1987. Toxic and inhibitory effects of the blue algae *Microcystis aeruginosa* on herbivorous zooplankton. J. Plankton Res. 9: 837–855.

Gates, M. A., 1984. Quantitative importance in the planktonic biomass of lake ecosystems. Hydrobiologia 108: 233–238.

Gifford, D. J., 1991. The protozoan-metazoan trophic link in pelagic ecosystems. J. Protozool. 38: 81–86.

Gilbert, J. J. & K. G. Bogdan, 1981. Selectivity of *Polyarthra* and *Keratella* for flagellate and aflagellate cells. Verh. int. Ver. Theor. Angew. Limnol. 21: 1515–1521.

Gliwicz, Z. M. & G. Umana, 1994. Cladoceran body size and vulnerability to copepod predation. Limnol. Oceanogr. 29: 419–424.

Hartmann, H. J., H. Taleb, L. Aleya & N. Lair, 1993. Predation on ciliates by the suspension-feeding calanoid copepod *Acanthodiaptomus denticornis*. Can. J. Fish. aquat. Sci. 50: 1382–1393.

Hasset, R. P. & M. R. Landry, 1988. Short-term changes in feeding and digestion by the copepod *Calanus pacificus*. Mar. Biol. 99: 63–74.

Holling, C. S., 1959. The components of predation as revealed by a study of small-mammal predation of the European pine sawfly. Can. J. Ent. 91: 293–320.

Hunt, G. H. & S. M. Chein, 1983. Seasonal distribution, composition and abundance of the planktonic ciliata and testacea of Cayuga lake. Hydrobiologia 98: 257–266.

Hutchinson, G. E., 1975. A treatise on limnology, 3. Wiley & Sons, New York, 660 pp.

Jonsson, P. R. & P. Tiselius, 1990. Feeding behavior, prey detection and capture efficiency of the copepod *Acartia tonsa* feeding on planktonic ciliates. Mar. Ecol. Prog. Ser. 60: 35–44.

Kerfoot, W. C., 1978. Combat between predatory copepods and their prey: *Cyclops*, *Epischura*, and *Bosmina*. Limnol. Oceanogr. 23: 1089–102.

Lair, N., 1992. Daytime grazing and assimilation rates of planktonic copepods *Acanthodiaptomus denticornis* and *Cyclops vicinus vicinus*. Comparison of spatial and resources utilisation by rotifers and cladocerans communities in an eutrophic lake. Hydrobiologia 231: 107–117.

Lloyd, M., 1967. 'Mean-crowding'. J. Anim. Ecol. 36: 1–30.

Marin, V., M. E. Huntley & B. Frost, 1986. Measuring feeding rates of pelagic herbivores: analysis of experimental design and methods. Mar. Biol. 93: 49–58.

Pace, M. L., 1982. Planktonic ciliates: their distribution, abundance, and relationship to microbial resources in a monomictic lake. Can. J. Fish. aquat. Sci. 39: 1106–1116.

Pace, M. L., 1986. An empirical analysis of zooplankton community size structure across lake trophic gradients. Limnol. Oceanogr. 31: 45–55.

Pace M. L. & J. D. Orcutt, 1981. The relative importance of protozoans, rotifers and crustaceans in a freshwater zooplankton community. Limnol. Oceanogr. 26: 822–830.

Paffenhöfer, G. A. & K. B. Van Sant, 1986. The feeding response of a marine planktonic copepod to quantity and quality of particles. Mar. Ecol. Prog. Ser. 27: 55–65.

Price, H. J., 1988. Feeding mechanism in marine and freshwater zooplankton. Bull. mar. Sci. 43: 327–343.

Price, H. J., G. A. Paffenhöfer & J. R. Strickler, 1983. Modes of cell capture in calanoid copepods. Limnol. Oceanogr. 28: 116–123.

Runge, J. A., 1980. Effect of hunger and season on the feeding behavior of *Calanus pacificus*. Limnol. Oceanogr. 25: 134–145.

Sanders, R. W. & S. A. Wickham, 1993. Planktonic protozoa and metazoa: predation, food quality and population control. Mar. Microb. Food Webs 7: 197–223.

Santer, B., 1993. Potential importance of algae in the diet of adult *Cyclops vicinus*. Freshwat. Biol. 30: 269–278.

Santer, B., 1994. Influence of food type and concentration on the development of *Eudiaptomus gracilis* and implications for interactions between calanoid and cyclopoid copepods. Arch. Hydrobiol. 131: 141–159.

Santer, B. & F. van den Bosch, 1994. Herbivorous nutrition of *Cyclops vicinus*: the effect of a pure algal diet on feeding, development, reproduction and life cycle. J. Plankton Res. 16: 171–195.

Sherr, E. B., B. F. Sherr & G. A. Paffenhöfer, 1986. Phagotrophic protozoa as food for metazoans: a missing trophic link in marine pelagic food webs? Mar. Microb. Food Webs 1: 61–80.

Sherr, E. B., B. F. Sherr, T. Berman & O. Haddas, 1991. High abundance of picoplankton-ingesting ciliates during late fall in lake Kinneret, Israel. J. Plankton Res. 13: 789–799.

Sime-Ngando, T. & H. J. Hartmann, 1991. Short-term variations of the abundance and biomass of planktonic ciliates in an eutrophic lake. Eur. J. Protistol. 27: 249–263.

Stemberger, R. S., 1985. Prey selection by the copepod *Diacyclops thomasi*. Oecologia 65: 492–497.

Stœcker, D. K. & J. M. Capuzzo, 1990. Predation on protozoa: its importance to zooplankton. J. Plankton Res. 12: 891–908.

Stœcker, D. K. & D. A. Egloff, 1987. Predation by *Acartia tonsa* Dana on planktonic ciliates and rotifers. J. exp. mar. Biol. Ecol. 110: 53–68.

Strickler, J. R., 1985. Feeding currents in calanoid copepods: two new hypotheses. In Laverak, M. S. (ed.), Physiological Adaptations of Marine Animals. Soc. exp. Biol. 459–485.

Uchima, M., 1988. Gut content analysis of neritic copepods *Acartia omorii* and *Oĩthona davisae* by a new method. Mar. Ecol. Prog. Ser. 48: 93–97.

Uchima, M. & R. Hirano, 1986. Food of *Oĩthona davisae* (Copepoda: Cyclopoida) and the effect of food concentration at first feeding on the larval growth. Bull. Plankton Soc. Jap. 33: 21–28.

Vanderploeg, H. A. & G. A. Paffenhöfer, 1985. Modes of algal capture by the freshwater copepod *Diaptomus sicilis* and their relation to food-size selection. Limnol. Oceanogr. 30: 871–885.

Vanderploeg, H. A., G. A. Paffenhöfer & J. R. Liebig, 1988. *Diaptomus* versus net phytoplankton: effects of algal size and morphology on selectivity of a behaviorally flexible, omnivorous copepod. Bull. Mar. Sci. 43: 377–394.

Vijverberg, J., 1989. Culture techniques for studies on the growth, development and reproduction of copepods and cladocerans under laboratory and *in situ* conditions: a review. Freshwat. Biol. 21: 317–373.

Wickham, S. A, 1995. *Cyclops* predation on ciliates: species-specific differences and functional responses. J. Plankton Res. 17: 1633–1646.

Williamson, C. E., 1980. The predatory behavior of *Mesocyclops edax*: predator preferences, prey defenses, and starvation-induced changes. Limnol. Oceanogr. 25: 903–909.

Williamson, C. E., 1987. Predator-prey interactions between omnivorous diaptomid copepods and rotifers: the role of prey morphology and behavior. Limnol. Oceanogr. 32: 167–177.

Williamson, C. E., 1991. Copepoda: Ecology and classification of North American Freshwater Invertebrates 21: 787–822.

Williamson, C. E. & M. E. Stœckel, 1990. Estimating predation risk in zooplankton communities: the importance of vertical overlap. Hydrobiologia 198: 125–131.

Hydrobiologia **373/374:** 169–180, 1998.
J.-C. Amiard, B. Le Rouzic, B. Berthet & G. Bertru (eds),
Oceans, Rivers and Lakes: Energy and Substance Transfers at Interfaces.

Spatial and temporal distribution of chironomid larvae (Diptera: Nematocera) at the sediment–water interface in Lake Abbaye (Jura, France)

Valérie Verneaux & Lotfi Aleya
Laboratoire de Biologie des Organismes et Ecosystèmes, Université de Franche-Comté, Place Leclerc, 25030 Besançon Cedex, France

Key words: lakes, Chironomidae, bathymetric distribution, artificial substrate, biotypology

Abstract

A survey of the spatial and temporal distribution of the chironomid community was conducted in Lake Abbaye (France). Samples were taken monthly (from May to November) at five depths (– 1 m to – 20 m) using artificial substrates. The results showed that both depth and season contributed to variations in the chironomid community. However, correspondence analysis clearly showed that most of the variation was due to bathymetric gradients. These results indicate that information on the depth distribution of the chironomid species may be useful in lake classification. However, this classification requires multiple-season sampling, so as to include the different life-cycle patterns of species.

Introduction

Among the lacustrine macrofauna, the family Chironomidae has been without doubt the most studied. Chironomids are the major taxonomic group in standing water, often accounting for up to 50% of total macrobenthos density (Cranston, 1995). Since decades, the chironomid community has been reported to be a valuable tool for lake typologies (Thienemann, 1913). Brinkhurst (1974) gave a synthesis of these typologies from data recorded in several ecosystems. More recently, Saether (1975; 1979) and Wiederholm (1980) proposed a trophic typology of lakes based on the examination of the profundal chironomid assemblages. Although these schemes have been successfully tested by many authors (Aagaard, 1986; Lindegaard, 1992) they have also yielded some controversial results, sometimes making lake differentiations unclear (Forsyth, 1978; Timms, 1978 in Lindegaard, 1995). In addition, the profundal chironomid fauna is not always related to the trophic status of lakes (Real & Prat, 1992).

When we applied Saether's (1979) classification system to Lake Abbaye (Jura), we found little agreement between lake trophic status based on chlorophyll *a* concentration/mean depth, and the profundal chironomid community (Verneaux & Aleya, 1996). When examined in the context of Saether's (1979) and Brundin's (1949) classifications, the profundal Chironomids of nine other French lakes did not reflect differences between these lakes (Verneaux, 1996), even though they differed in other limnological variables (Chlorophyll *a*, organic matter, profundal oxygen content). The bathymetric distribution of macrobenthic communities identified at higher taxonomic levels seems to be more informative than a study of only profundal assemblages (Verneaux, et al., 1993; 1995; Verneaux et al., 1993). The quantitative and qualitative significance of chironomid larvae in these studies has caused us to focus our investigations on the bathymetric distribution of Chironomids based on species-level identification. Because this approach necessitates multiple-season samplings, the present study was undertaken to determine the degree of temporal stability

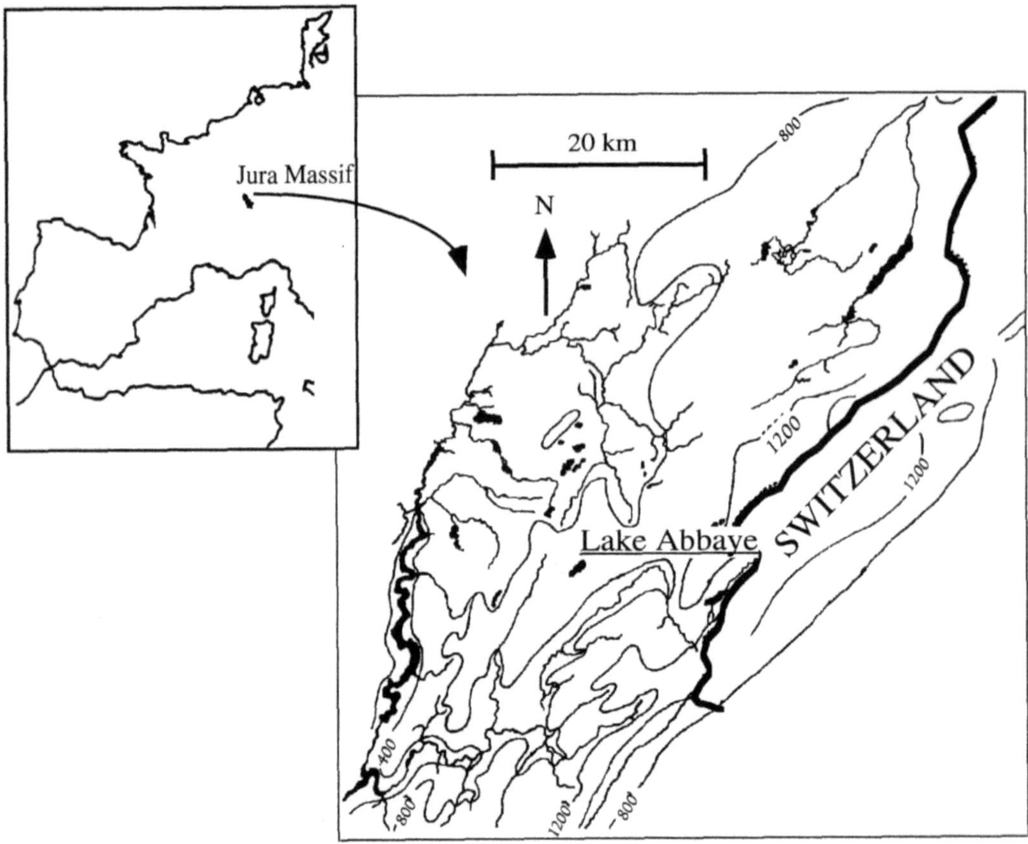

Figure 1. Geographical setting of Lake Abbaye.

in the bathymetric distribution of chironomid communities. This work was conducted in Lake Abbaye, which shows substantial seasonal variations in temperature and oxygen. According to Brundin (1949), these factors should have major effects on the bathymetric stucture of the larval chironomid community.

Study area and methods

Lake Abbaye is located 910 m a.s.l., in the Jura mountains in the east of France near Switzerland (Figure 1). It covers 80 ha with a maximum depth of – 20 m and a mean depth of – 7.3 m. The lake is asymmetric, with a stony bottom along the eastern shore but a gently sloping sedimentary zone extending from the western shore (Figure 2). Lake Abbaye has highly mineralized water (specific conductance 205–251 μS cm^{-1}) due to $CaCO_3$. The mean chlorophyll *a* concentration is 3.21 μg l^{-1}. Profundal sediment has a high level of organic matter (27.7% dry weight) of mainly terrestrial origin (Verneaux et al., 1991). The lake is holodimictic, and is ice-covered from late November to early April.

Chironomid larvae were collected monthly from May to November 1993 using artificial substrates, consisting of concrete plates (diameter: 20 cm; height: 3 cm) with gravel (diameter: 5 mm), crevices (diameter: 20 mm; depth: 1 cm) and synthetic brushes (height: 5 cm) embedded in the top. The colonizable area covered 250 cm^2. Each plate was placed in a basket surrounded by a 250 μm nylon mesh, which prevented the loss of organisms upon retrieval. We thus avoided the trad loss attributed to the use of artificial substrates (see Rosenberg & Resh, 1982). When the artificial substrate was set in the sediment-water interface for colonization, the mesh lay flat on the surrounding sediment, thus enabling species of all size to colonize.

Artificial substrates were placed at the sediment-water interface at five depths, in the sedimentary zone from the littoral to the maximum depth (at – 1, – 5,

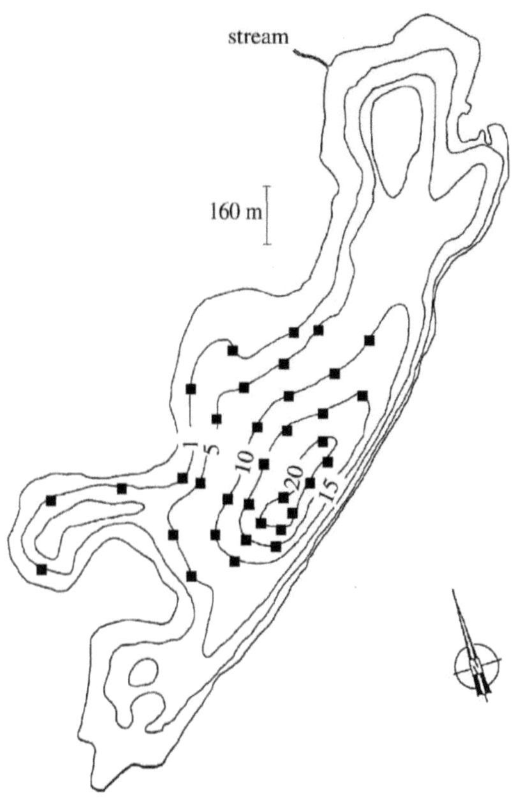

Figure 2. Bathymetric map of Lake Abbaye showing sampling sites (■). Depths are in meters.

– 10, – 15 and – 20 m). In April 1993, 28 substrates were submerged at each depth (see Figure 2 for the sampling sites). Four substrates were retrieved monthly from May to November from each depth. Two yielded the material to identify chironomid larvae at a generic or specific level. The other two were used for laboratory rearing of larvae in order to obtain specific identifications from adults and pupal exuviae. Each sample (one depth on one date) consisted of two artificial substrates.

For larval identification, substrates were washed through a 250 μm sieve and the collected larvae were preserved in formalin (5%). In the laboratory, they were cleared in cold KOH (10%) for 12 hours and mounted in Aquamount Gurr on microscope slides. For larval rearing, each artificial substrate was cleared of predatory fauna (Odonata, Sialidae, Dytiscidae), placed in an aquarium and covered with lake water. A 120 μm mesh net was fixed on top of the aquarium to catch emerging adults.

Temperature and oxygen concentration at the sediment–water interface were recorded monthly (from April to November) at each depth.

In order to determine the spatial and temporal distribution of chironomid taxa in Lake Abbaye, correspondence analyses (CA) were performed using the ADE program (Chessel et al., 1995).

Several months, and sometimes years (Evrard, 1995), depending on species, are needed to complete the life cycle of Diptera Chironomidae (from first larval instar to the emergence of the adult). As this study aimed at examining the impact of time on the bathymetric distribution of the larval chironomid communities and not on the temporal change in their species composition, we chose a mathematical transformation so that the data would not be weighted by the seasonal density variations of each taxon. The simple log $(x + 1)$ transformation does not allow such discrimination.

CA were performed on a matrix of 30 columns (the 30 most abundant larval taxa, mean density > 1 ind./artificial substrate) and 35 rows (seven dates, five depths). For each taxon, densities were transformed into 'seasonal bathymetric affinities'. It is simply a ratio of the taxon abundance at a given depth and date over its maximum abundance over the whole bathymetric profile (resulting in the range 0–1). The data transformation allows a rapid detection of possible temporal effects on larval bathymetric distribution. This transformation could not be referred to literature data since, to our knowledge, no reports have dealt with seasonal effect on the bathymetric distribution of chironomid larvae. Some authors have concentrated only on the temporal fluctuations on the composition of chironomid assemblages (Paterson & Fernando, 1970; Armitage, 1977; Aagard, 1982; Ketelaars et al., 1992). Others who took into account the bathymetric pattern did not include, temporal effects on the distribution of chironomid communities (Parrish & Wilhm, 1978; Lods-Crozet et al., 1985). And when this behaviour was considered, multivariate analyses were not performed (Heinis et al., 1989).

Results

Temperature and oxygen

The temperature at the sediment-water interface varied from 5.5 °C in the profundal zone in April, to a maximum of 21.6 °C in the littoral zone in August

Table 1. List of chironomid species collected using artificial substrates (a.s.) in Lake Abbaye. Density/a.s. (d/a.s.) and relative density [rd (%)] for larval taxa and reared species.

Taxa			Code	Larvae d/a.s.	Larvae rd(%)	Reared species d/a.s.	Reared species rd (%)
Chironomus anthracinus		Zetterstedt 1860	CHan	271	15.09	1	0.77
*Cladopelma viridula**		Linnee 1767	CLvi	44	2.42	7	4.60
Cryptochironomus spp.			CRyc	1	0.08		
*Dicrotendipes modestus**		Say 1823	DImo	25	1.37	6	3.83
Dicrotendipes nervosus		Staeger1839	DIne	1	0.08	10.77	
	E. carbonaria	Meigen 1804	EIca		74.50		
Einfeldia spp.			EInf	363	20.20		
	E. dissidens	Walker 1856	EIdi		241	5.80	
Endochironomus albipennis		Meigen 1830	ENal	125	6.98	8	5.17
Endochironomus tendens		Fabricius 1775	ENte	30.17	4	2.59	
Glyptotendipes gripekoveni		Kieffer 1913	GLgr	5	0.27		
Microtendipes chloris		Meigen 1818	MIch	311.72	6	4.21	
Pagastiella orophila		Edwards 1929	PAor	pr.	pr.		
	P. arcuatus	Goetg. 1919	PAar			1	0.57
	P. biannulatus	Staeger 1839	PAbi			pr.	pr.
Parachironomus spp.			PAch	10	0.54		
	*P. varus**	Goetg. 1921	PAva			pr.pr.	
	*P. vitiosus**	Goetg. 1921	PAvi			pr.pr.	
	P.tenuicaudatus	Malloch 1915	PAtc			pr.pr.	
*Paratendipes albimanus**		Meigen 1818	PAal	pr.	pr.	1	0.67
Pentapedilum sordens		V. D. Wulp 1874	PEso	2	0.10	pr.	pr.
Phaenopsectra flavipes		Meigen 1818	PHfl	pr.	pr.		
Polypedilum nubeculosum		Meigen 1804	POnu	30	1.37	4	2.87
Tripodura sp.			TRsp	1	0.08		
Pseudochironomus prasinatus		Staeger 1839	PSpr	13	0.72	pr.	pr.
	C. atridorsum	Kieffer 1924	CLat			16	10.44
	*C. iucundus**	Hirvenoja 1962	CLiu			3	1.82
Cladotanytarsus spp.			CLad	262	14.58		
	C. nigrovittatus	Goetg. 1922	CLni			85.36	
	C. pallidus	Kieffer 1922	CLpa			21.34	
Paratanytarsus tenuis		Meigen 1830	PAte	31	1.74	1	0.86
	T. debilis	Meigen 1830	TAde			11	7.38
Tanytarsus verrali gr.			TAve	33	1.84		
	*T. lactescens**	Edwards 1929	TAla			1	0.57
Tanytarsus miriforceps		Kieffer 1921	TAmi	29	1.61	1	0.48
Tanytarsus niger		Andersen 1937	TAni	134	7.43	pr.	pr.
	T. gregarius	Kieffer 1909	TAgr			5	3.64
	*T. mendax**	Kieffer 1925	TAme			11	7.57
Tanytarsus spp.			TAny	73	4.06		
	T. lestagei aggr.		TAle			95.94	
	T. usmaensis	Pagast 1931	TAus			1	0.38
Ablabesmyia longistyla		Fittkau 1962	ABlo	2	0.12	1	0.48
Ablabesmyia monilis		Fittkau 1962	ABmo	27	1.50	7	4.60
	*P. choreus**	Meigen 1804	PRch			1	0.86
	*P. crassinervis**	Zetterstedt 1838	PRcr			pr.	pr.

Table 1. Continued.

Taxa			Code	Larvae d/a.s.	Larvae rd(%)	Reared species d/a.s.	Reared species rd (%)
Procladius Holotanypus			PRoc	122	6.78		
	*P. signatus**	Zetterstedt 1850	PRsi			11	7.18
	P. flavifrons	Edwards 1929	PSfl			pr.	pr.
Procladius Psilotanypus			PSil	126	6.99		
	*P. rufovittatus**	V. D. Wulp 1874	PSru	36	23.85		
*Corynoneura scutellata**		Winnertz 1846COsc	pr.	pr.			
Cricotopus spp.			CRic	5	0.29	pr.	pr.
Nanocladius balticus		Palmen 1959	NAba	pr.	pr.		
*Nanocladius bicolor**		Zetterstedt 1838	NAbi	pr.	pr.		
Parakiefferiella bathophila		Kieffer 1912	PAba	18	0.99	pr.	pr.
*Psectrocladius limbatellus**		Holmgren 1869	PSli	6	0.33	1	0.86
Psectrocladius psilopterus		Kieffer 1906	PSps	10.04	pr.	pr.	
*Psectrocladius sordidellus**		Zetterstedt 1838	PSso	3	0.15	pr.	pr.
Total density/a.s.				1797	149		
Species richness				38		43	

* Species collected for the first time in the B8 zone; —— Species collected for the first time in France. pr. Presence; i.e. density < 1 ind./a.s.

(Figure 3A). After spring turnover, thermal stratification began in May. By November, autumn turnover resulted in a homogenous distribution of both temperature (Figure 3A) and oxygen concentrations throughout the water column (Figure 3B). Oxygen depletion (i.e., oxygen content below 3 mg l^{-1}, Stahl, 1986) lasted five months in the deepest zone (– 20 m), four months at – 15 m, and one month at – 10 m. Severe hypoxic conditions (oxygen content < 1 mg l^{-1}, presence of H_2S) prevailed in the two deepest zones through September and October.

Chironomid fauna

Thirty eight larval taxa and 43 reared species of Chironomids were identified (Table 1) for a total richness of 51 taxa; forty-nine were identified at the species level. The mean larval density collected was high, with 1,797 ind./artificial substrate, whereas larval rearing never produced more than 149 ind./artificial substrate. The tribe Chironomini was the major contributor to the larval stock, both qualitatively and quantitatively. This tribe accounted for 43% of total taxonomic richness and more than 50% of total chironomid density.

The most abundant taxon in the larval community was *Einfeldia* spp., mostly *E. dissidens* (Walk.), which comprised up to 20% of the total density. *Chironomus anthracinus* (Zett.) ranked second (15%), followed by *Tanytarsus* spp. (15%) and *Cladotanytarsus* spp. (14%). According to the results from larval rearing, the genus *Tanytarsus* was dominated by the species *T. mendax* (Kieff.) and *T. debilis* (G.), and the genus *Cladotanytarsus* by *C. atridorsum* (K.). The subfamily Orthocladiinae was the least abundant group.

Spatial and temporal patterns of Chironomids

Simple correspondence analysis (CA)

In Figure 4 the stations and the larval taxa are distributed on the plane of the first two axes of a simple CA. A great number of taxa had a maximal affinity for the littoral (– 1 m) and the sublittoral (– 5 m) zones whatever the sampling date. All littoral and a great number of the sublittoral stations are also grouped near the axis intersection. The distribution of stations along the first axis was a result of both temporal and spatial factors that were not easy to distinguish. Only a few taxa contribute heavily to the variance expressed on the first or second axis (> 25%). These discriminant taxa characterized some depths on particular dates. For example, *Chironomus anthracinus* characterized the deepest zones (– 15 m, – 20 m) in September and November, but these depths were characterized by the genus *Cricotopus* in June and August, respectively.

This first CA demonstrated a temporal effect on the bathymetric pattern of chironomid larval commu-

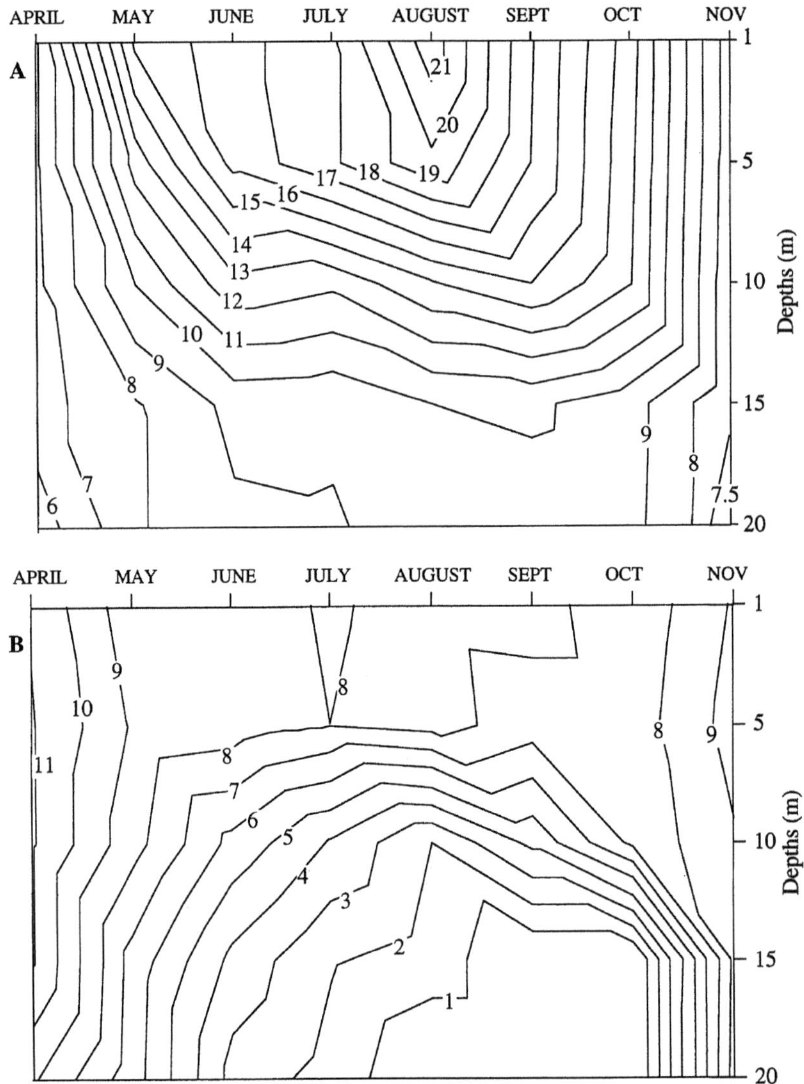

Figure 3. Annual pattern of (*A*) temperature (°C) and (*B*) oxygen concentrations (mg l^{-1}) based on measurements at five depths at the sediment-water interface.

nities. The observed seasonal effect was not a result of variations in taxon significance in the chironomid community, since the data transformation eliminated seasonal variations in density and relative abundance of taxa. Using bathymetric affinities only the position of taxa in the lake and their presence or absence were taken into account. Since all discriminant taxa had 71–100% occurrence in seasonal samples, it seems unlikely that phenological pattern can explain the temporal effects observed on the bathymetric pattern of the chironomid assemblage. Instead, the modifications of the depth distribution pattern resulted from seasonal variations in the depths where taxa were most abundant. The temporal effect was more marked in the deeper zone than in the shallower zone, as seasonal variability in the chironomid community increased at depths – 10 m, – 15 m, and – 20 m (Figure 5).

Is temporal variability more important than bathymetric variability? In other words, despite the seasonal effect, can the bathymetric structure of the larval community be considered to be constant?

In order to determine the relative importance of both factors in structuring the community, within- and between-groups CA were performed according

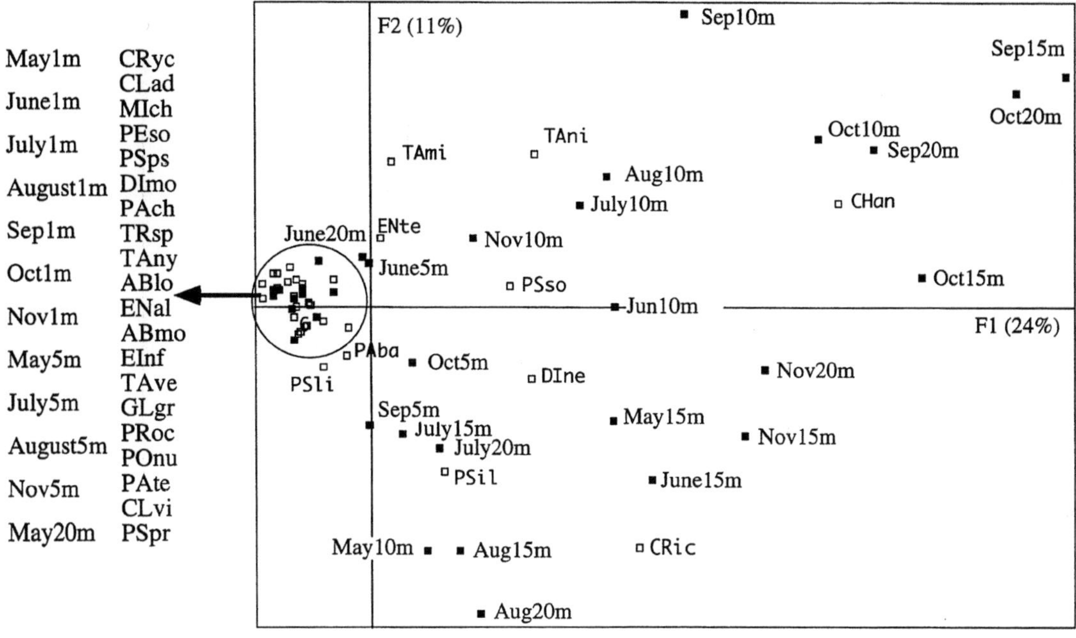

Figure 4. Distribution of taxa and stations on the F1∗F2 plane of a simple CA. 30 taxa, 35 stations (5 depths, 7 months).

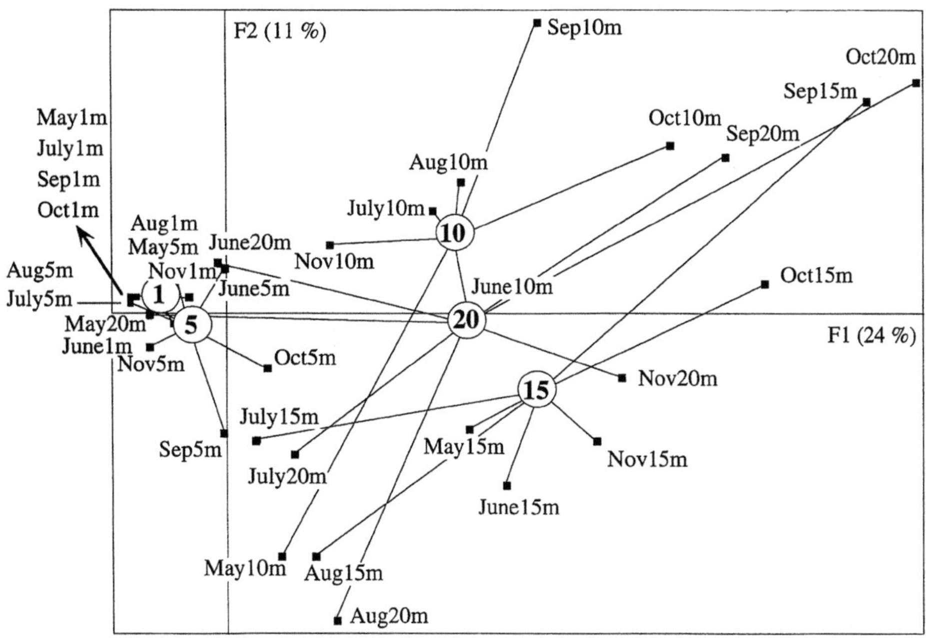

Figure 5. Distribution on the F1*F2 plane of a simple CA of the barycenters of bathymetric samples. This graph shows the seasonal variability of the data collected.

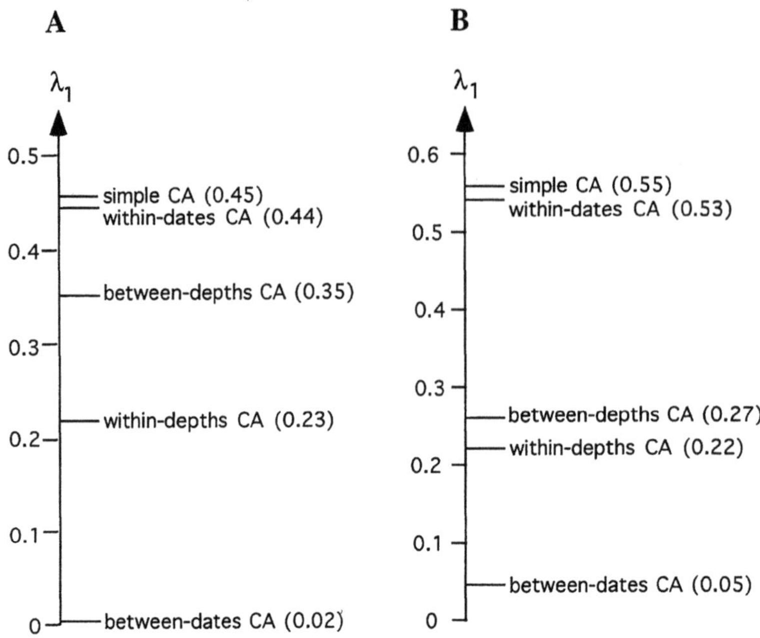

Figure 6. First eigenvalues (λ_1) of the within- and between-group CA applied to the larval (A) and reared (B) communities.

to Dolédec & Chessel (1989). These analyses permitted elimination of one or the other of these two effects (spatial or temporal) while illustrating the mean bathymetric or seasonal pattern.

Within- and between-groups correspondence analysis

The first eigenvalue (λ_1) of each CA expresses the proportion of variability explained by the first axis, and comparison of these first eigenvalues indicates how well each model explains the variance in the community data (Figure 6). This analysis was made on both larval taxa (Figure 6A) and reared species (Figure 6B), and in both cases, the λ_1 values of the within-dates CA (by which the temporal effect was eliminated) did not differ from those of the simple CA. This result clearly shows that spatial-temporal variance in the chironomid community distribution is chiefly due to the bathymetric variability. The variability between the bathymetric samples from each season (within-dates CA) was higher relative to that observed between the seasonal samples of a given depth (within-depths CA). These results demonstrate that, in Lake Abbaye, the bathymetric distribution of the chironomid fauna was stable thoughout the seasons.

Comparison between the larval average bathymetric distribution (Figure 7) and that for reared species (Figure 8), obtained by between-depths CA, emphasized the value of species-level data. At a specific identification level, a bathymetric succession of species appeared from *Paratendipes albimanus* (Mg.) to *Chironomus anthracinus* which did not appear when many taxa were combined at the generic level. The littoral and sublittoral zones were clearly differentiated along the second axis owing to the specific identification of individuals belonging to the genus *Cladotanytarsus* and *Einfeldia*. Thus, *C. atridorsum* showed a deeper bathymetric distribution than *C. iucundus* (Hirv.) and *C. nigrovittatus* (G.). The species *E. carbonaria* (Mg.) seemed to be more closely associated with the littoral zone than *E. dissidens* (Walk.) which showed a bathymetric preference for −5 m.

Although the species were distributed along a regular succession from the littoral to the sublittoral zone, several discontinuities were observed from −10 m (Figure 8). Examination of the average values of temperature and oxygen content recorded at each depth (Table 2) showed no significant variation between depths −1 m and −5 m. Significant decreases ($p < 0.05$) in the average temperature and oxygen content appear between −5 m and −10 m until −20 m. The profile that can be observed in the environmental variables that were measured, i.e., a progressive evolution from −1 m to −5 m, followed by a sudden decrease in oxygenation and temperature between −5 m and −10 m, could bring about a progressive distribution of the species between −1 m and −5 m and

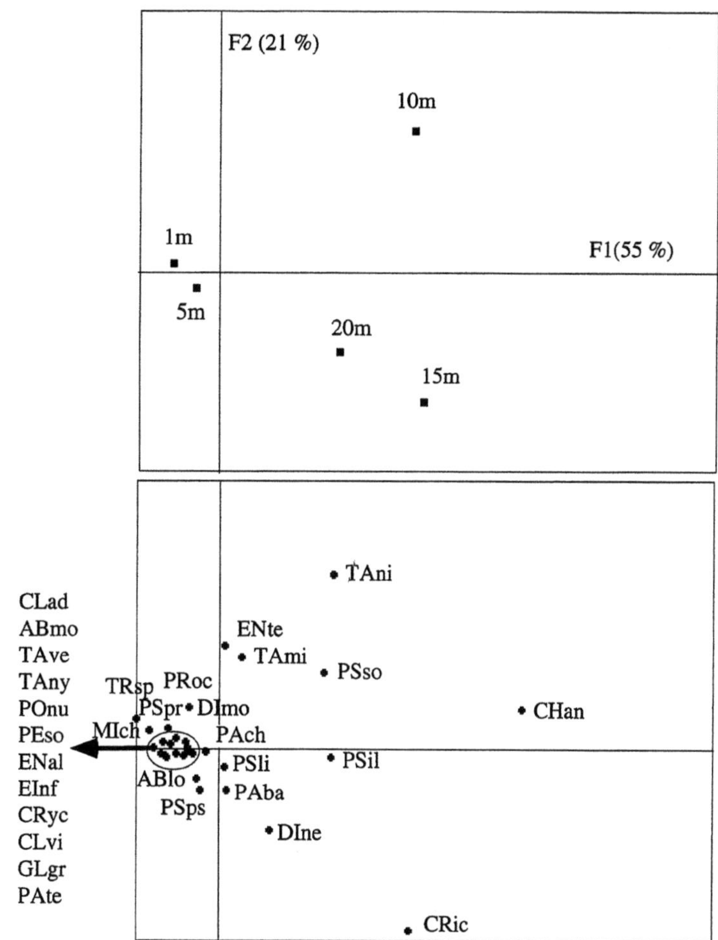

Figure 7. Mean bathymetric pattern of larval community obtained by a between-depths CA.

Table 2. Mean temperature (°C) and oxygen concentrations (mg l^{-1} and % saturation) recorded monthly from April to November 1993 ($n = 8$) at five depths and results of Student's t test.

Depths	Temperature (°C)		Oxygen (mg l^{-1})		Oxygen (% saturation)	
	Mean ± SD ($n = 8$)	t-test	Mean ± SD ($n = 8$)	t-test	Mean ± SD t-test	t-test
− 1 m	14.5 ± 5.1		8.6 ± 1.1		92.9 ± 9.5	
		1.737		− 0.591		− 0.772
− 5 m	13.9 ± 4.8		8.7 ± 1.2		93.9 ± 7.0	
		2.996*		2.724*		2.935*
− 10 m	11.4 ± 3.1		6.5 ± 3.0		63.9 ± 26.3	
		4.277**		2.647*		2.745*
− 15 m	8.7 ± 1.3		4.2 ± 4.1		39.4 ± 38.2	
		2.964*		2.324		2.672*
− 20 m	8.3 ± 1.4		3.5 ± 3.9		32.6 ± 35.4	

SD standard deviation; * Significant $p < 0.05$; ** Highly significant $p < 0.01$.

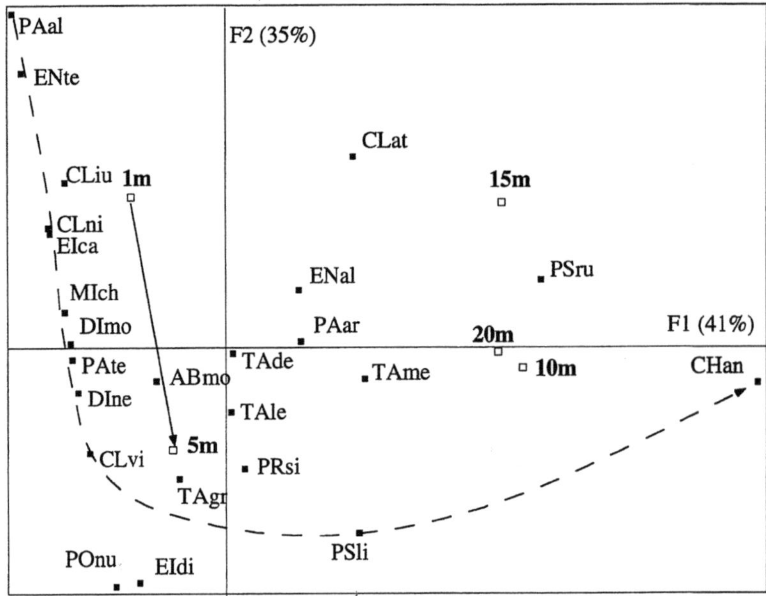

Figure 8. Mean bathymetric pattern of reared community obtained by a between-depths CA.

a change in the composition of the community from −10 m.

Discussion

The chironomid community

Among the 51 taxa collected in Lake Abbaye (Table 1), 16 species were recorded for the first time in the biogeographic zone B8 (north-east of France) defined by Serra-Tosio & Laville (1991). Two species were also recorded for the first time in France [*Cladotanytarsus nigrovittatus* and *C. pallidus* (Kieff.)]. Most of these species have been previously collected by Verneaux & Vergon (1974) in the north-east of France but at lower altitudes (< 500 m) in the B13 zone (Serra-Tosio & Laville, 1991). The large number of these first records suggests that the chironomid fauna has been poorly investigated in the B8 zone. This study also adds to the altitudinal distribution of many species.

In Lake Abbaye, the contribution of the tribe Chironomini to species richness (58%) was higher than that recorded in Lake Petichet (30%) (Serra-Tosio & Gay, 1978) and Lake Port-Biehl (33%) (Laville, 1971). This may be related to the high content of organic matter in sediment (OM max: 35% dry weight). Brundin (1949) showed an increase in Chironomini and a decrease in Orthocladiinae species richness as sediment organic matter increased in oligotrophic Swedish lakes.

The use of artificial substrates that favor pioneer taxa may have led to the unexpectedly low number of Orthocladiinae. Among the entire chironomid community recorded from Lake Abbaye (adults, pupal exuviae and larvae, Verneaux, 1996), this subfamily accounted for nearly 20% of total species richness and 11% of chironomid density. Yet artificial substrates detect orthoclads as only 16% of chironomid taxa and < 2% of density. Although many orthoclads show a preference for hard, mineral or plant substrates (Pinder, 1986), in this study samplers were placed only over soft sediment. Francis & Kane (1995) found that natural substrates influenced not only the composition of benthic communities sampled *in situ*, but also that of the communities collected by artificial substrates.

Discrepancies between ambient communities and those collected by artificial substrates have been reported and have called into question the credibility of the method (Rosenberg & Resh, 1982). Artificial substrates produced higher densities and greater species richness than did sediment samples (Verneaux, 1996). On the other hand, the use of artificial substrates allowed us to eliminate the bathymetric effect on the nature of the substrate and assess the impact of the seasonal variations in environmental conditions on the chironomid communities.

Spatial and temporal pattern of Chironomids

The bathymetric distribution of Chironomids in Lake Abbaye showed several trends.

The seasonal variability of artificial substrate communities clearly increased with increasing depths, suggesting that temporal effects on chironomid community composition were more pronounced in the deeper zones than at shallower depths. Heinis et al. (1989) explained the bathymetric migration of *Tanytarsus* sp. by a decrease in oxygen concentration; *Sergentia coracina* (K.) larvae have been shown to migrate in relation to larval development (Wülker, 1961). No larval migrations were observed in Lake Abbaye, due either to poor environmental conditions or to larval development (Verneaux, 1996). In this lake, summer oxygen depletion may have increased the mortality of some taxa. Although transformation of data excluded the effects of density variation, it is likely that emergence losses from preferred depths have influenced our interpretations of bathymetric affinities. On the other hand, the decrease in oxygen content at the sediment–water interface, which is considered a major factor in chironomid distribution (Brundin, 1951; Prat, 1978; Verneaux et al., 1995), induced the disappearance of many taxa in July at − 20 m and in September at − 15 m. These losses of taxa might induce higher variability in the profundal samples than in the shallower ones because the profundal communities have lower species-richness. However, this higher variability in profundal samples also resulted from greater seasonal variations of the affinities of taxa at these depths. In the deepest zone (− 20 m), three months of severe hypoxic conditions (August, September, October) led to the survival of only *Chironomus anthracinus*. This species copes well with oxygen depletion as it contains haemoglobin, has an anaerobic metabolism, and can enter dormancy (Hamburger et al., 1994; 1995).

We have demonstrated that the depth effect strongly dominates seasonal variation in structuring the larval chironomid community. This result is interesting for the use of chironomid bathymetric distribution in characterizing lakes because it emphasizes the capacity of chironomid larvae to integrate seasonal environmental modifications. Moreover, this predominant effect of depth implies that the depth of emerging species is the same as that of larvae. This conditioned the obtention of the bathymetric distribution of chironomid species with a multi-seasonal sampling.

Although in Lake Abbaye in particular, chironomid communities reveal a great dominance of *C. anthracinus* from the depth − 10 m, avoiding a progressive distribution of the sampled depths comprised between − 5 m and − 20 m, the succession of species (positioned at the mean of depths and relative each to the others) suggests the existence of a bathymetric *continuum*. By analogy with the river typology proposed by Verneaux (1973), based on a species *continuum* of macrobenthos, it is likely that the bathymetric pattern of chironomid species in Lake Abbaye is part of a more generalized *continuum* that might be established through enhancing the number of bathymetric samples collected from a great number of lakes.

Application of Brundin's (1949) and Saether's (1975; 1979) typologies, based on analysis of only profundal chironomid communities, did not give a satisfying differentiation of nine lakes in the Jura (Verneaux, 1996). Moreover the trophic status of these lakes does not seem to be related to the profundal Chironomids collected. In our study, conducted only in Lake Abbaye, the mean bathymetric pattern of chironomid species seemed to be related to the mean annual environmental conditions at the sediment-water interface. This result suggests that comparison of the bathymetric pattern of chironomid species between lakes could give a scheme for lake classification regardless of their trophic status. We recently demonstrated (Verneaux & Aleya, 1997, in press), from ten French lakes, that these bathymetric patterns are related with lake ability to assimilate organic matter. Further studies are needed to understand the behaviour of the allogenic and/or autogenic organic matter that arrives to the sediments at different depths. This must be done simultaneously with an identification of the environmental conditions that contribute to an optimized assimilation of organic matter.

Acknowledgements

We thank Prof. J. Verneaux for coordinating the multidisciplinary research team and Dr P.H. Langton for his help in the determination of some chironomid pupal exuviae. We express our thanks to the anonymous reviewers who provided helpful comments on the manuscript and to Mesdames Gill and Rose for the revision of the English text of the paper.

References

Aagaard, K., 1986. The chironomid fauna of North Norwegian lakes, with a discussion on methods of community classification. Holarct. Ecol. 9: 1–12.

Brinkhurst, O. R., 1974. The benthos of lakes. Mc Millan Press, London and Basingstoke, 190 pp.

Brundin, L., 1949. Chironomiden und andere Bodentiere der Südschwedischen Urgebirgsseen. Rep. Inst. Freshwat. Res. Drottningholm 30: 1–914.

Brundin, L., 1951. The relation of O2-microstratification at the mud surface to the ecology of the profundal bottom fauna. Rep. Inst. Freshwat. Res. Drottningholm 32: 32–42.

Chessel, D., J. Thioulouse & S. Doledec, 1995. ADE-4 program library, multivariate and graphical display of environmental data, 6 vol., Univ. Lyon, 520 pp.

Cranston, P. S., 1995. Introduction to the Chironomidae. In P. Armitage, P.S. Cranston & L.C.V. Pinder (eds), The Chironomidae. The Biology and Ecology of Non-Biting Midges. Chapman & Hall, London, Weinheim, New York, Tokyo, Melbourne, Madras: 1–5.

Dolédec, S. & D. Chessel, 1989. Rythmes saisonniers et composantes stationnelles en milieu aquatique. II – Prise en compte et élimination d'effets dans un tableau faunistique. Acta Oecologica Oecol. gener. 10: 207–232.

Evrard, M., 1995. Quelques aspects méconnus des Chironomidae (Diptera) de Belgique. Nat. belg. 76: 33–50.

Forsyth, D. J., 1978. Benthic macroinvertebrates in seven New Zealand lakes. New Zealand J. of marine and Freshwat. Res. 12: 41–49.

Francis, D. R. & T. C. Kane, 1995. Effects of substrate on colonization of experimental ponds by Chironomidae (Diptera). J. of Freshwat. Ecol. 10: 57–63.

Hamburger, K., P. C. Dall & C. Lindegaard, 1994. Energy metabolism of *Chironomus anthracinus* (Diptera: Chironomidae) from the profundal zone of Lake Esrom, Denmark, as a function of body size, temperature and oxygen concentration. Hydrobiologia 294: 43–50.

Hamburger, K., P. C. Dall & C. Lindegaard, 1995. Effects of oxygen deficiency on survival and glycogen content of *Chironomus anthracinus* (Diptera: Chironomidae) under laboratory conditions. Hydrobiologia 297: 187–200.

Heinis, F., W. J. Van De Bund & C. Davids, 1989. Avoidance of low oxygen and food concentrations by the larvae of *Tanytarsus* sp. Acta Biol. Debr. Oecol. Hung. 3: 141–145.

Laville, H., 1971. Recherches sur les Chironomides (Diptera) lacustres du massif de Néouvielle (Hautes-Pyrénées). Première partie: Systématique, écologie, phénologie. Ann. Limnol. 7: 173–332.

Lindegaard, C., 1992. Zoobenthos ecology of Thingvallavatn: vertical distribution, abundance, population dynamics and production. Oïkos 64: 257–304.

Lindegaard, C., 1995. Classification of water-bodies and pollution. In Armitage P., P. S. Cranston & L. C. V. Pinder (eds), The Chironomidae. The Biology and Ecology of Non-Biting Midges. Chapman & Hall, London, Weinheim, New York, Tokyo, Melbourne, Madras: 385–404.

Pinder, L. C. V., 1986. Biology of freshwater Chironomidae. Ann. Rev. Entomol. 31: 1–23.

Prat, N., 1978. Benthos typology of spanish reservoirs. Verh. Int. Ver. Limnol. 20: 1647–1651.

Real, M. & N. Prat, 1992. Factors influencing the distribution of Chironomids and Oligochaetes in profundal areas of spanish reservoirs. Neth. J. aquat. Ecol. 26: 405–410.

Rosenberg, D. M. & V. H. Resh, 1982. The use of artificial substrates in the study of freshwater benthic macroinvertebrates. In Cairns, J. (ed.), Artificial Substrates. Ann. Arbor. Sci. 175–235.

Saether, O. A., 1975. Nearctic chironomids as indicators of lake typology. Verh. int. Ver. Limnol. 19: 3127–3133.

Saether, O. A., 1979. Chironomid communities as water quality indicators. Holarct. Ecol. 2: 65–74.

Serra-Tosio, B.& C. Gay, 1978. Les Diptères Chironomidés et Chaoboridés du lac de Petichet (Isère). Trav. Lab. Hydrobiol. 69–70: 97–105.

Serra-Tosio, B. & H. Laville, 1991. Liste annotée des Diptères Chironomidés de France continentale et de Corse. Ann. Limnol. 27: 37–74.

Stahl, J. B., 1986. A six-year study of abundance and voltinism of Chironomidae (Diptera) in an Illinois cooling reservoir. Hydrobiologia 134: 67–79.

Thienemann, A., 1913. Der Zusammenhang zwischen dem Sauerstoffgehalt des Tiefenwassers und der Zusammensetzung der Tiefenfauna unserer Seen. int. Rev. ges. Hydrobiol. Hydrogr. 6: 243–249.

Timms, B. V., 1978. The benthos of seven lakes in Tasmania. Arch. Hydrobiol. 81: 422–444.

Verneaux, J., 1973. Cours d'eau de Franche-Comté (massif du Jura). Recherches écologiques sur le réseau hydrographique du Doubs. Essai de biotypologie. Mem. thèse Etat, 257 pp.

Verneaux, J. & J. P. Vergon, 1974. Faune dulçaquicole de Franche-Comté. Sixième partie: les diptères chironomides. Ann. Sci. Univ. Besançon. 3: 179–198.

Verneaux, J., A. Schmitt & V. Verneaux, 1995. Classification biologique des lacs jurassiens à l'aide d'une nouvelle méthode d'analyse des peuplements benthiques. III. Relation entre données biologiques et variables du milieu. Ann. Limnol. 31: 277–286.

Verneaux, J., V. Verneaux & A. Guyard, 1993. Classification biologique des lacs jurassiens à l'aide d'une nouvelle méthode d'analyse des peuplements benthiques. I. Variété et densité de la faune. Ann. Limnol. 29: 59–77.

Verneaux, J., A. Vidonne, F. Remy & A. Guyard, 1991. Particules organiques et rapport C/N des sédiments des lacs du Jura. Ann. Limnol. 27: 175–190.

Verneaux, V., 1996. Structure, dynamique spatiale et temporelle du peuplement chironomidien du lac de l'Abbaye (massif du Jura) et approche typologique. Mem. thèse Univ. Franche-Comté, 187 pp.

Verneaux, V. & L. Aleya, 1996. Diptères chironomidés et caractérisation des lacs. Ann. biol. 35: 220–235.

Verneaux, V., J. Verneaux & A. Guyard, 1993. Classification biologique des lacs jurassiens à l'aide d'une nouvelle méthode d'analyse des peuplements benthiques. II. Nature de la faune. Ann. Limnol. 29: 383–393.

Wiederholm, T., 1980. Chironomids as indicators of water quality in Swedish lakes. Acta Univ. Carolinae Biol. 1978: 275–283.

Wülker, W., 1961. Studien zur Morphologie, Biologie und Verbreitung der Gattung *Sergentia* Kieff. (Dipt., Chironomidae). Arch. Hydrobiol. 25: 307–331.

Hydrobiologia **373/374:** 181–191, 1998.
J.-C. Amiard, B. Le Rouzic, B. Berthet & G. Bertru (eds),
Oceans, Rivers and Lakes: Energy and Substance Transfers at Interfaces.

A wetland managed for agriculture as an interface between the Rhône river and the Vaccarès lagoon (Camargue, France): transfers of water and nutrients

P. Chauvelon
Station Biologique de la Tour du Valat, Le Sambuc, F-13200 Arles, France

Key words: Camargue, Rhône river, Vaccarès lagoon, rice cultivation, nutrients, wetland

Abstract

With the development of rice cultivation in the 50's, a great part of the Isle of Camargue, the central area of the Rhône river delta, was polderized. The Vaccarès lagoon system (110 km^2, 65 km^2 for the Vaccarès lagoon itself) represents the main part of the 'Réserve Nationale de Camargue'. Without human intervention, it is hydraulically isolated from the Rhône river and the Mediterranean sea. It receives drainage water from precipitation and flooded ricefields from its relict catchment.

Our hydrological study concerned the Fumemorte catchment (65 km^2, with 20 km^2 of ricefields). Irrigation for rice, in 1993 and 1994, was estimated at 56.2 and 52.5 Mm^3 respectively. There is great variability in water consumption from one sector to another: 1 400 mm to 5 500 mm. The drainage volume was 45 Mm^3 in 1993 and 53 Mm^3 in 1994, a wetter year. Annual fluxes of 400 tons of nitrogen (N) and 150 tons of phosphorus (P) entered the catchment. With a scenario of maximum transfer, only 5.5% of entering N and 4% of P were exported from the Fumemorte catchment to the Vaccarès lagoon. Organic matter exported by the drainage network represents 5 to 6 times the quantity introduced with Rhône water. We attempted to quantify the processes of consumption, storage and loss which occur in the catchment. From preliminary results on the production of a seagrass bed in the Vaccarès, we found that, at an annual scale, nutrient fluxes from the catchment seem minor compared to other sources (lagoon sediments, atmosphere).

Introduction

Rice crop production increased in the Rhône river delta, or Camargue, after World War II. From 11 km^2 in 1946, the area of rice cultivation reached 160 km^2 in 1960. Intensive rice cultivation radically changed the hydrological regime of the Great Camargue, the central part of the delta, between the two branches of the Rhône. Large quantities of irrigation water from the Rhône river are pumped in the ricefields from May to September. As a consequence, on the base of existing drainage basins and networks, the major part of the agricultural catchment was equipped in the mid 50's with pumping stations for drainage. Since then, 55% of the potential agricultural catchment area of the lagoons has been polderized (see Figure 1). The Fumemorte catchment (65 km^2) represents 55% of the area actually drained to the lagoons, although some hydraulic transfers may occur from the polderized basins to the Vaccarès during the crop season (Heurteaux, 1994; Chauvelon, 1996). It was attempted to estimate the hydrological balance of the system (Heurteaux, 1992, 1994), with relatively sparse data concerning hydraulic management of the system. Research was conducted at the field scale in the Camargue, to quantify processes affecting the nitrogen cycle in ricefields (Bonetto et al., 1988; Minzoni et al., 1988; Golterman et al., 1988). Other work tackled the quality of irrigation and drainage waters (Godin, 1990; De Groot, 1992). Nevertheless, due to discrete sampling and the absence of drainage discharge data, fluxes to the Vaccarès lagoon system could not be estimated accurately.

The objectives of the work presented here were to estimate the major nutrient inputs and outputs for the catchment, and not to analyse the biogeochemical processes affecting the transfer of nutrients and sediments. Thus, the bases for an estimation of fluxes entering the lagoons and their effects on primary production and eutrophication could be determined. In a recent paper (Chauvelon et al., 1996), some results about the impact of hydro-agricultural management on dissolved and solid fluxes and fish population dynamics in the Camargue were presented. In this issue, with the presentation of results on fluxes, we attempted to quantify some processes affecting nitrogen and phosphorus on the catchment: biomass production, sinks and losses. Moreover, the estimates of N and P fluxes entering the Vaccarès were compared (Chauvelon, 1996) to quantities assumed to be necessary for biomass production in the lagoon, using data from literature and initial results on macrophyte biomass for the Vaccarès (Auby, unpubl. data).

Description of sites studied

The Rhône River delta or Camargue, is located on the French mediterranean coast, 80 km West of Marseille. It is characterized by a mediterranean climate, globally warm and dry. Winds are frequent all year long, with a net predominancy of N–NW winds (called 'mistral'). Potential evaporation and evapotranspiration are high (1 300 mm yr^{-1}) due to high summer temperature, solar radiation and windy conditions. Precipitations are not abundant: 590 mm yr^{-1} on average of a 50-year period of observations (Chauvelon, 1996).

In the Camargue, fields managed for rice cropping are relatively small (1.10^4 m^2 to 3.10^4 m^2), and they are carefully leveled and delimited by bunds. The fields are flooded from mid April or May to September and drained by ditches. Cut-offs are made in the bunds, in order to manage some rough weirs used to control the water levels according to different crop stages and to allow the dewatering of the ricefields. In 1994, ricefields covered 145 km^2 in the Great Camargue, from this total, 40 km^2 were on the actual drainage basin of the lagoons, with half being on the Fumemorte catchment.

Since 1869, the Isle of Camargue has been completely diked (see Figure 1). So, it is hydraulically isolated from the Rhône river and the sea without human intervention. At the S-E of the Great Camargue, salines were also isolated from the lagoon system, in order to control salinity. The central lagoons of the Camargue (Vaccarès and 'Etangs inférieurs', or coastal lagoons) extend over an area of 110 km^2 but are shallow (1 m deep on average). The Vaccarès is a permanently flooded lagoon of 1.5 m mean depth and 64 km^2 area. Intermittent connections between the lagoon and the sea are controlled by means of sluices through the seadyke. The Fumemorte drainage canal, which gave its name to the basin, is collecting drainage waters from a complex network 400 km in length.

Due to the fluctuations of rice cultivated area and the irregularity of annual precipitation, salinity of the Vaccarès lagoon varied between 5 g l^{-1} and 38 g l^{-1} during the last 50 years (Heurteaux, 1992). Mean salinity of the Rhône river is less than 1 g l^{-1}, as is generally the case for the Fumemorte canal at the outlet (Chauvelon, 1996). If the raisable barrier at the outlet is opened, wind induced transient salt intrusion from the Vaccarès may occur in the drainage canal (salinity reaching 5 g l^{-1} near the outlet).

Materials and methods

In order to develop an organized data base, a geo-referenced approach was used. It was implemented using digitized maps made from aerial photographs, and land use was updated using satellite remote sensing (Chauvelon, 1996; Sandoz, 1996). All parcels were gathered in larger agricultural units attributed to their own irrigation station (Chauvelon, op. cit.).

Concentrations of total suspended material in the Rhône river were obtained on a daily basis for 1993 and 1994 (Pont, unpubl. data). Concentrations of nitrogen (N), phophorus (P) and organic carbon were calculated using the data obtained by El Habr & Golterman (1987), concerning concentrations and discharge of the Rhône (Chauvelon, 1996). The volume introduced for the irrigation of ricefields was obtained using the distributed approach described by Chauvelon (op. cit.). Each pumping station was monitored (pumping duration and gauging for different stages of functionment) in order to calculate the irrigation volume. The geo-referenced data base allowed us to make a repartition of the volume on the actual irrigated surface corresponding to each pumping station.

A sampling concerning 86% and 70% of the areas devoted to rice and wheat cropping, respectively, was made. From this enquiry, the nature and quan-

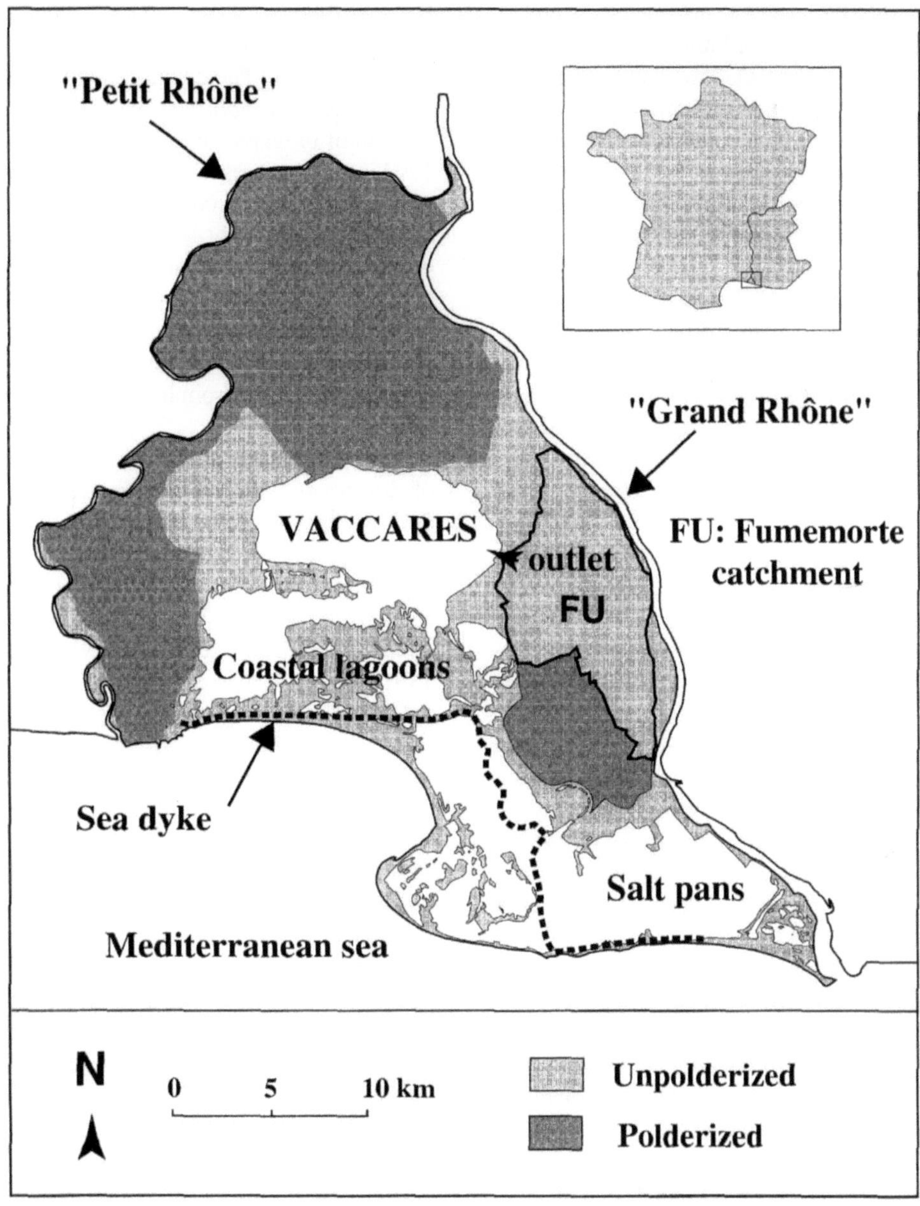

Figure 1. The drainage basins and the lagoons in the 'Grande Camargue', France.

tity of fertilizers used for rice and wheat crops on the Fumemorte basin was obtained. Using the geo-referenced data base it was possible to calculate fertilizer input on a discrete manner, taking into account the spatial variability of fertilization practices. For the farms not sampled, fertilization applications recommended by the French Rice Center were used (Chauvelon, op. cit.): 15 g N m^{-2}. The used form of nitrogen is ammonium sulphate or prilled urea. Phosphorus (under the form of P_2O_5) is applied with the first N application (6 to 8 g P m^{-2}). For the fertilization of wheat, the values of 12 g N m^{-2} and 8 g P m^{-2} were retained (PO_4 is the form used by farmers), (Chauvelon, op. cit.).

In order to estimate direct atmospheric inputs of N and P with rain, data from the literature were used with our precipitation measurements (Chauvelon, op. cit.). For N, we retained the value of 0.5 mg l^{-1}, measured in the west mediterranean coast (Loye-Pilot et al., 1990). For the concentration of phosphorus we used the value of 0.04 mg P l^{-1} recorded by Auby et al. (1994) in the South West of France.

A ultrasonic open channel flow gauge ('Utraflux 2100 CO') was used to obtain the discharge at the outlet of the catchment and an automatic sampler ('Epic 1011') was installed nearby (Figure 1). In order to stop the biological activity, before their collection and refrigeration, 10 ml of a solution (20 g l^{-1}) of mercuric chloride were introduced before the sampling cycles in each flask. Moreover, before each sampling, the suction pipe was automatically flushed out by air forcing. The chemical analysis were carried out according to the French norms for water analysis (AFNOR): ortho-phosphate (and total phosphorus after acid hydrolosis) using molybdate-antimony, ammonia using indophenol, total nitrogen using Kjeldhal method. Water samples were filtered (GF/C filters). The filters were dried at 105 °C to obtain total suspended material (TSM), than passed to an oven at 500 °C to derive the organic suspended material (OSM).

To obtain a rough estimate of vegetation production and its consumption of N and P, we used data from literature. Stutterheim (1994) indicated an apparent nitrogen recovery (ANR = N applied with fertilizer versus N recovered in the crop) of 20% for rice in the Camargue, with a production of 1 500 g DM m^{-2} yr^{-1} (DM: Dry Matter). For wheat we retained a maximum nitrogen recovery of 60%, with a production of 1 000 g DM m^{-2} yr^{-1}. As far as P is concerned, we used a percentage of 0.25% DM for both rice and wheat.

For reedbeds and grass, which colonize edges of agricultural ditches and are burnt each year, we used data from literature (Table 1). An attempt was also made to quantify the algal production in the ricefields. In May and June, in the ricefields of the Camargue, Vaquer (1984a, 1984b) estimated the production of *Chara sp* and phytoplankton (Table 1), assuming that C/N/P = 100/15/1. From our observations (Chauvelon, 1996), the production of benthic cyanophycea in the ricefields of Camargue during the crop season can be estimated to 40 g DM m^{-2} yr^{-1}. An indicator of eutrophication in the drainage network is the presence of submerged macrophytes (essentially *Myriophyllum spicatum* in the Camargue). To quantify their impact on the consumption of N and P, we retained, from literature (Table 1) a biomass of 400 g DM m^{-2}, with 3% DM (N) and 0.25% DM (P).

To compare the annual fluxes of N and P from the catchment to the Vaccarès with the primary production of the lagoon, we attempted to quantify the production of its seagrass bed, using data from literature (Table 2). Because there were not enough data concerning phytoplankton, we only took into account the biomass of macrophytes. The annual production of leaves for seagrass *Zostera noltii* was calculated from values of maximum biomass, measured at the end of summer 1994 in the Vaccarès. The biomass was estimated from three sampling sites on a bathymetric gradient where seagrass developped (Auby, unpubl. data). The value of maximum biomass for leaves was multiplied by a turn over rate (Table 2) in order to estimate the annual leaves production of the seagrass bed. To estimate the annual production of rhizomes, we used another turn over rate (Table 2), from literature concerning *Zostera marina*.

The percentages of P and N in leaves and rhizomes of *Zostera sp* vary during the year, with a maximum during winter and a minimum in July or August (Auby, pers. comm.). We used the average of minimum and maximum values obtained by several authors (Table 2) for *Zostera marina* (Van Lent & Verschuure, 1994) and *Zostera noltii* (Auby, 1991; Perez-Llorens et al., 1991; Perez-Llorens & Niell, 1993). The following values were retained for N and P, in percentage of dry matter (DM): 3% DM (N) and 0.46% DM (P) for the leaves; 1.36% DM (N) and 0.27% DM (P) for rhizomes and roots.

Using maps of the seagrass bed established in 1984 (Vaquer & Heurteaux, 1989) and in 1994 (Réserve Nationale de Camargue, pers. com.), we quantified the evolution of the covered area in the Vaccarès lagoon. For 1994, we extrapolated the biomass of the sampling sites to the total area of the seagrass bed, in order to calculate biomass production and associated consumptions or fixation of nitrogen and phosphorus.

The relative position of the different functional compartments (crops, associated vegetation of the agricultural area, vegetation from the drainage network) within the hydrosystem, and the dominant processes relatively to nutrients dynamics are summarized in Figure 2.

Results

Nutrient and sediment inputs – consumptions for biomass production on the catchment

In 1993 and 1994 the total volume imported from the Rhône to the Fumemorte catchment reached 56.2 10^6 m^3 and 52.5 10^6 m^3, respectively. The irrigation inflow calculations showed that there was important

Table 1. Biomass production (g DM m^2 yr^{-1}), N and P content (% of DM), for vegetation in the agricultural area and drainage network

Type of vegetation	Biomass	N	P	Site	Source
Reedbeds	1 500	2.6	0.14	Camargue	Duncan (1992), Mesléard (1995)
Grass	600	2.5	0.25	Camargue	Mesléard (1995)
Chara sp in ricefields	12 g C m^{-2}	–	–	Camargue	Vaquer (1984a, 1984b)
Phytoplancton in ricefields	26 g C m^{-2}	–	–	Camargue	Vaquer (1984a, 1984b)
Cyanophycea in ricefields	40	–	–	Camargue	Chauvelon (1996)
Myriophyllum sp	400	3	0.25	Camargue, UK, USA	Best & Mantai (1978), Hill (1979), Grillas (1990), Hearne et al. (1994) Van Wijck et al. (1994)

Table 2. Biomass production (g DM m^2 yr^{-1}), turnover rates, N and P content (% of DM), for the seagrass bed

Type of vegetation	Biomass	Turnover	N	P	Site	Source
Zostera noltii leaves	84.9	–	–	–	Camargue	Auby (unpubl. data)
Zostera noltii rhizomes	76.9	–	–	–	Camargue	Auby (unpubl. data)
Zostera marina leaves	–	2–3	–	–	Netherlands	Vermaat et al. (1987)
Zostera marina rhizomes	–	1.8	–	–	Britanny	Jacobs (1979)
Zostera marina leaves	–	–	1.2–4.7	0.2–0.6	Netherlands	Van Lent & Verschuure (1994)
Zostera marina rhizomes	–	–	0.6–2.1	0.1–0.3	Netherlands	Van Lent & Verschuure (1994)
Zostera noltii leaves	–	–	1.8–2.6	0.25–0.4	Arcachon, France	Auby (1991)
	–	–	3–4	0.3–0.5	Netherlands	Perez-Llorens et al. (1991)
	–	–	3–4	0.7–0.9	Spain	Perez-Llorens & Niell (1993)
Zostera noltii rhizomes	–	–	1.25	0.25	Arcachon, France	Auby (1991)
	–	–	1.5	0.25–1.25	Netherlands	Perez-Llorens et al. (1991)
	–	–	1.2–3	0.4–0.8	Spain	Perez-Llorens & Niell (1993)

variability between irrigation basins, as far as water delivery was concerned. Actually, these consumptions varied from 1 400 mm to 5 500 mm during the irrigation period (Chauvelon, 1996). These differences were due to different irrigation practices, physical properties of the areas (soils permeability, drainage density) and hydraulic management constraints.

The associated importations of suspended particles (TSM) were estimated to be more than 2 000 tons per year (Table 3). These particles were mainly mineral, as the inputs of OSM were less than a hundred tons per year.

The calculated inputs of nitrogen and phosphorus with irrigation water were of the same order for the two years (as shown in Table 3). It is worth mentionning that relationships between concentrations and

Table 3. Importation of irrigation water (10^6 m^3 yr^{-1}), sediments, (Total Suspended Materials (TSM) and Organic Suspended Material (OSM)), nitrogen (N) and phosphorus (P) (10^3 kg yr^{-1}) from the Rhône to the Fumemorte catchment

Year	IRRIG.	TSM	OSM	N	P
1993	56.2	2 300	69	69	9.5
1994	52.5	2 200	80	65	9

discharge of the Rhône, which were used for the calculations of fluxes, should be updated, following recent results obtained by Moutin et al. (1996). As could be expected, fertilizers for rice and wheat represented the most important inputs of N and P on the catchment,

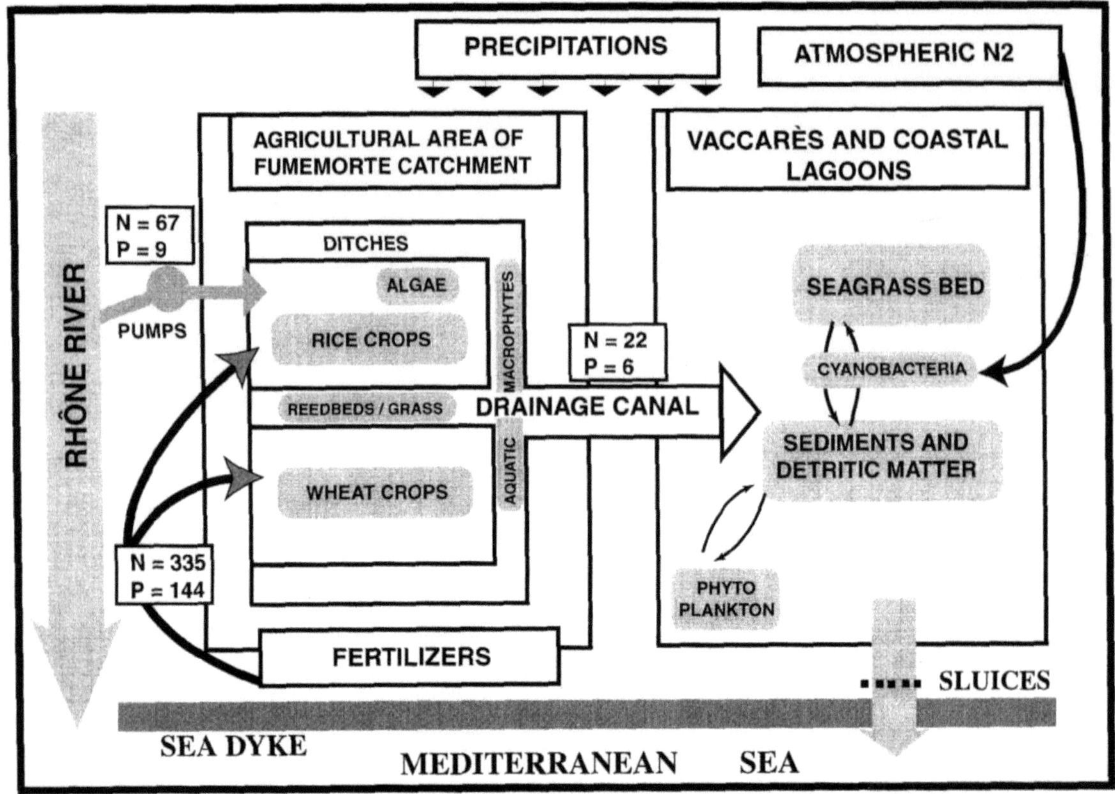

Figure 2. Schematic map of the studied hydrosystem (in light grey, the vegetation compartments consuming or fixing N and P during transfers). Some of the major transfer number for nitrogen (N) and phosphorus (P), are indicated (tons per year).

Table 4. Importation of N and P with fertilizers (10^3 kg yr^{-1}) on the Fumemorte catchment for rice and wheat crops

Year	N rice	P rice	N wheat	P wheat
1993	302	122	40	26
1994	295	117	33	22

with more than 3 10^5 kg of nitrogen and 1 10^5 kg of phosphorus each year (as shown in Table 4). The average quantities of fertilizers applied for rice were: 14.8 g N m^{-2} (range: 12–17 g N m^{-2}) and 5.8 g P m^{-2} (range: 0–9 g P m^{-2}). Seventy five percent of N applied was in the form of ammonia, and 25% was urea.

The calculated atmospheric inputs associated with rain were not negligible as far as nitrogen is concerned: 16 10^3 kg N yr^{-1} and 12 10^3 kg N yr^{-1} for 1993 and 1994 respectively. Using data from literature, atmospheric inputs of phosphorus could reach 1 10^3 kg P yr^{-1}. Combining all possible sources, on average for the two years, we suggest that annual loads of 402 tons of nitrogen and 153 tons of phosphorus were introduced on the Fumemorte catchment.

To compare these quantities of nutrients with macrophyte biomass from the Vaccarès lagoon, we focused on 1994. From the 2.95 10^5 kg of nitrogen introduced for rice fertilization, assuming an applied nitrogen recovery of 20% (Stutterheim, 1994), we suggest that only 6 10^4 kg were effectively assimilated for rice biomass production. Including the production of rice and wheat, we can say that from the agricultural point of view, only 7.6 10^4 kg yr^{-1} of the N fertilizer input were consumed. In the case of phosphorus, total agricultural consumption could reach reach 8.3 10^4 kg yr^{-1}.

For the 0.5 km^2 of channel reedbeds and 0.2 km^2 of grass from ricefields bunds, which are renewed each year, we estimated a production of 8.3 10^5 kg DM, corresponding to consumptions of 2.1 10^4 kg N and 1.3 10^3 kg P. The production of macro-algae, phytoplankton and blue-green algae could reach 4.8 10^5 kg

Table 5. Annual drainage discharge (10^6 m^3 yr^{-1}) and sediments (Total Suspended Materials (TSM) and Organic Suspended Material (OSM)), nitrogen (N) and phosphorus (P) fluxes (10^3 kg yr^{-1}), from the Fumemorte catchment

Year	DRAIN.	TSM	OSM	N	P
1993	45.4	1 500	390	21	6
1994	53.4	1 940	536	23	6

of carbon, thus corresponding to consumptions of 7.1 10^4 kg N and 5 10^3 kg P.

The maximum cumulated biomass of submerged macrophytes in the drainage network of the Fumemorte catchment was estimated to 2.7 10^4 kg DM (Chauvelon, 1996). Despite the spectacular visual effect of this vegetation development in some sectors of the drainage network, these plants only consume 800 kg N and 70 kg P.

Outputs from the catchment – consumptions for biomass production in the lagoon

Description and modeling of hydrological processes are presented elsewhere (Chauvelon, 1996). As could be expected for a catchment with a low hydraulic gradient, there was no relationship between concentrations and discharge at the outlet (Chauvelon, op. cit.), except for TSM during the autumn and winter when discharge was greater than 1 m^3 s^{-1}. For the two years, range and mean values (mg l^{-1}) for TSM and OSM at the outlet were (9–70, 33) and (4–23, 10), respectively. When no precipitations occurred, the drainage of the 20 km^2 of irrigated ricefields corresponded to a discharge of 3 m^3 s^{-1} at the outlet. The total drainage volume (and the associated fluxes) was greater in 1994 than in 1993, (see Table 5) due to a particularly rainy autumn. Finally, only 5.5% of total nitrogen input and 4% of total phosphorus input reached the outlet of the catchment in 1994.

To extrapolate the values of fluxes to the whole catchment of the Vaccarès lagoon, we must take into account a supplement of 61.2 km^2 of actually drained area. On this area, land use was characterized in 1994, by 19 km^2 of ricefields and 1.1 km^2 of fields cultivated for wheat (Chauvelon, 1996). Using the average values obtained on the Fumemorte catchment, we calculated N and P inputs (fertilizer and irrigation water) in the cultivated area, and atmospheric input to the whole catchment. From this, we obtained the following total inputs: 3.70 10^5 kg N, 1.29 10^5 kg P. Considering that the same transfer rates as those of the Fumemorte catchment could be applied (5.5% and 4% for N and P, respectively), the additional nutrient input to the lagoon were therefore 2.04 10^4 kg N and 5.1 10^3 kg P. Using the possibilities of interconnections between the two drainage network (polderized and unpolderized) some water transfers should take place (Heurteaux, 1994). In 1994, during the crop season, a transfer of drainage water (estimated to 13 10^6 m^3) from the normally polderized basin to the Vaccarès occured (Chauvelon, 1996). If we make the assumption that drainage water quality was identical to that of the Fumemorte, this implies additional inputs of 5.8 10^3 kg N and 1.6 10^3 kg P for the Vaccarès. To sum up, this means that up to 4.92 10^4 kg N and 1.31 10^4 kg P entered the Vaccarès lagoon from its drainage basin in 1994. From literature data, the direct inputs with rain water to the lagoon may be estimated as 2.4 10^4 kg N and 2 10^3 kg P.

The area covered by *Zostera noltii* in the Vaccarès increased from 30 km^2 in 1984, to 43 km^2 in 1994 (Chauvelon, 1996). For 1994, the biomass of leaves and rhizomes of *Zostera noltii* were estimated respectively to 84.9 g DM m^{-2} and 76 g DM m^{-2} (Auby, unpubl. data). Using turn over rates of 2 (for leaves) and 1.8 (for rhizomes), the average annual production of the seagrass bed can be estimated: 310 g DM m^{-2} yr^{-1}. Using these preliminary results, we estimated the production of the seagrass bed for 1994 around 1.36 10^7 kg DM yr^{-1}. This means that consumptions of 3 10^5 kg N yr^{-1} and 5.6 10^4 kg P yr^{-1} were necessary for this biomass production.

Discussion

The average irrigation depth used for rice cultivation in the Camargue was estimated from 2 400 mm yr^{-1} (private sector of irrigation) to 3 300 mm yr^{-1} (collective irrigation) (Heurteaux, 1994). Our results are in agreement with these values, but it is worth mentioning that the range is 1 400–4 000 mm yr^{-1} (exceptionnally 5 500 mm yr^{-1}), because this obviously had different environmental impacts downstream the ricefields. In the Ebro delta (NW Spain), the irrigation depth used for rice irrigation (flooding as in the Camargue) is estimated to 2 700 mm from April to September, but there are also 1 300 mm consumed from October to December (Loaso, unpubl. data) in order to avoid salinization of soils.

As irrigation increases, runoff, solutes leaching through soils and bunds, organic matter exportation increase too. Ricefields, although acting as filters for suspended materials (Forès, 1989), are also exporting organic matter in the aquatic environment, as shown by the ratio (1/6) between irrigation input and organic matter output from the Fumemorte catchment. In the Ebro delta, (Forès & Comín, 1992) chemical fertilizers are applied at lower rates than in the Camargue: 11 g N m^{-2} and 20 g P m^{-2}. We quantified some biomass production for vegetation in the catchment, in order to calculate the associated nutrient consumptions. It is worth mentioning that this biomass, although possibly exported from the system (crops, burned vegetation) was stocked, as was the case for algal biomass in the ricefields. As concerned estimates of the N and P ouputs at the outlet of the catchment, it is difficult to give a magnitude of error, because of this, these values are to be considered as maxima. A number of water analysis results were below the lowest detectable values with the methods used by the laboratory. In these cases, we retained the limit values for the calculation of fluxes.

Our purpose was not to review or discuss N and P biogeochemical processes in aquatic environment, which has been done accurately by other authors (Golterman, 1984; Reddy et al., 1989; Picot et al., 1990; De Groot, 1990; De Groot, 1992; Rader & Richardson, 1992; Craft et al., 1995; Madden & Kemp, 1996). In the particular case of N in flooded ricefields, the intensity of nitrification–denitrication processes was already demonstrated in experimental plots in the Camargue (Golterman et al., 1988; Minzoni et al., 1988; El Habr & Golterman, 1990), or other places (Reddy & Patrick, 1986; Forès & Comín, 1987; Frenay et al., 1990). Boisserie (1990) reported that a quantity up to 10% of applied N could be lost by volatilization in Camargue ricefields. Some authors emphasized the fact that precipitation under the form of calcium phosphate ($CaPO_4$) was the major process affecting phosphate in the irrigation and drainage waters in the Camargue (Godin, 1990; De Groot, 1992). This means that the major part of phosphorus, unconsumed by vegetation, is stocked in sediments, in the catchment. Due to low hydraulic gradient, solid transport, and therefore nutrient exportations are not important, even in case of floods, the water velocity in the main drainage canal remaining below 0.4 m s^{-1}. Taking into account nutrient intputs from the whole catchment, and comparing to other Mediterranean catchment-wetland systems (Comín et al., 1990; Lieutaud et al., 1992; Moutin et al., 1993; Forès et al., 1994; Cornwell et al., 1996) it seems that the Vaccarès system could be classified as poorly fed by its catchment, as far as N and P are concerned, at an annual scale.

Concerning macrophyte biomass production in the Vaccarès lagoon, experimental studies are ongoing, including relationships between biomass production with salinity, suspended material dynamics and light attenuation.

Comparing the seagrass bed needs for N and P to the inputs from the catchment, it appears that internal sources and processes in the lagoon were predominant. This has been already suggested in other mediterranean lagoon systems (Comín et al., 1990; Lieutaud et al., 1992; Pichot et al., 1994; Forès et al., 1994).

The role of sediment in the phosphorus and nitrogen cycles in lakes and lagoons has been discussed by several authors (Golterman, 1984; Nowicki & Nixon, 1985; Boström et al., 1988; De Groot, 1990; De Groot & Golterman, 1993; Moutin et al., 1993; Gomez, 1996). The sediments and detritic matter of the Vaccarès lagoon probably constitute the major stock and source of P for primary production (De Groot, 1992). A possible source of N for the aquatic macrophytes is the indirect fixation of molecular nitrogen N_2 (Auby et al., 1994). This is due to bacterial populations, whose presence is facilitated by detritic matter from macrophytes, capable of fixation and transformation in ammonia, thus available for aquatic macrophytes. Capone (1982) estimated this fixation for *Zostera marina* beds between 5 and 8 10^{-3} g N m^{-2} d^{-1}.

In the Bassin d'Arcachon, south-west France, N_2 fixation in *Zostera noltii* beds was estimated (Welsh et al., 1996): minimum values in winter 1–2 10^{-4} g N m^{-2} d^{-1}, maximum values in summer 2–7 10^{-3} g N m^{-2} d^{-1}. These values are consistent with initial experimental results obtained for *Zostera noltii* in the Vaccarès in July 1996 (Auby, unpubl. data): 3 10^{-5} g N m^{-2} h^{-1} (dark), 1.6 10^{-4} g N m^{-2} h^{-1} (light).

Accepting the value of 8 10^{-3} g N m^{-2} d^{-1}, the fixation of nitrogen could be more than 10^5 kg N yr^{-1} for the seagrass bed of the Vaccarès lagoon. In fact, it seems that every input of organic matter in the aquatic environment leads to an increase of atmospheric nitrogen fixation, as demonstrated by Patriquin & Knowles (1972) for the marine environment, and by Seitzinger (1988) in lakes. So, the inputs of organic matter to the Vaccarès, which can be estimated, (on the base of Fumemorte catchment results) to 10^6 kg DM yr^{-1} should favour this N_2 fixation.

Conclusion

Great quantities of fertilizers are used for rice cultivation in the Camargue. Moreover, the used irrigation volume is contributing to nutrient enrichment of the agricultural catchment. Importation of sediments with irrigation water globally compensates for the erosion of the Fumemorte catchment. Annual fluxes at the outlet of the Fumemorte catchment represent only 4 to 6% of nitrogen and phosphorus inputs. This is probably mainly due to denitrification and volatilization of nitrogen, and phosphate precipitation in the catchment. Extrapolating to the whole catchment of the lagoons, and comparing to biomass production of the seagrass bed in the Vaccarès, it is shown that a great part of nitrogen and phosphorus used by the seagrass bed comes from the lagoon sedimentary and detritic stocks, and from N_2 fixation.

As a matter of fact, loads entering the Vaccarès lagoon seemed to be 'acceptable'. But in the long term, there is risk of eutrophication. It would also be necessary to investigate what could be the exchanges of nutrients between the sea and the lagoons, and to estimate the rates of nitrification/denitrification in the lagoons themselves. There is also a lack of data about the biomass dynamic of phytoplankton within the freshwater and brackish hydrosystem in the Camargue.

We hope that this work contributes to the knowledge of agricultural landscape nutrient budget and impact, like other recent publications (Giraud, 1992; Heathwaite & Johnes, 1996; Johnes, 1996). It also highlights the subject of the complex relationships between lowland rice cultivation, hydrological processes and water resources management. Studies were conducted at the field scale (Wopereis et al., 1994; ten Berge et al., 1995), but they were rarely attempted at the catchment scale (Heurteaux, 1994; Wilber et al., 1996).

This work focused on water, nutrient and sediments fluxes from the catchment to the lagoons of the 'Réserve Nationale de Camargue'. A potentially important aspect of water quality in the Camargue concerns the fluxes of pesticides, heavy metals and radio-nucleides coming from the Rhône or introduced to the agricultural systems.

Further studies related to these fluxes need to be done, in order to establish an ecological diagnosis of this protected area under anthropic influence. Certainly, best agricultural and hydraulic management practices should be applied in the Camargue, as was suggested in other important protected wetlands downstream of intensive farming areas, such as the Everglades of South Florida (Snyder & Davidson, 1994; Izuno & Capone, 1995; Harvell et al., 1996).

References

Auby, I., 1991. Contribution à l'étude des herbiers de *Zostera noltii* dans le bassin d'Arcachon: Dynamique, production et dégradation, macrofaune associée. Thèse Doct., Univ. Bordeaux I, 234 pp.

Auby, I., F. Manaud, D. Maurer & G. Trut, 1994. Etude de la prolifération des algues vertes dans le bassin d'Arcachon. Rapport Ifremer-Cemagref-Ssa-Sabarc, 163 pp.

Best, M. D. & K. E. Mantai, 1978. Growth of *Myriophyllum*: sediment or lake water as the source of nitrogen and phosphorus. Ecology 59: 1075–1080.

Boisserie, H., 1990. Etude des pertes d'azote par volatilisation dans les rizières de Camargue. Thèse Doct Univ. Sci. Tech. Languedoc, Montpellier, 221 pp.

Bonetto, C., F. Minzoni & H. L. Golterman, 1988. The nitrogen cycle in shallow water sediment system of ricefields. 2. Fractionation and bioavailaibility of organic nitrogen compounds. Hydrobiologia 159: 203–210.

Boström, N., J. M. Andersen, S. Fleisher & M. Jansson, 1988. Exchange of phosphorus accross the sediment–water interface. In G. Persson & M. Jansson (eds), Phophorus in Freshwater Ecosystems. Developments in Hydrobiology 48. Kluwer Academic Publishers, Dordrecht: 229–244. Reprinted from Hydrobiologia 170.

Capone, D. G., 1982. Nitrogen fixation (acetylene reduction) by the rhizosphere sediments of the eelgrass *Zostera marina*. Mar. Ecol. Prog. Ser. 10: 67–75.

Chauvelon, P., 1996. Hydrologie quantitative d'une zone humide méditerranéenne aménagée: le bassin de Fumemorte en Grande Camargue, delta du Rhône. Thèse Doct. Univ. Sci. Tech. Languedoc, Montpellier, 254 pp.

Chauvelon, P., A. Sandoz, G. Poizat & A. J. Crivelli, 1996. Hydroagricultural management in the Rhone river delta, France: consequences on dissolved and solid fluxes, potential impact on fish populations. Proceedings of the 2nd International Symposium on Habitat Hydraulics, Québec, June 1996, Vol. A: 167–176.

Comín, F. A., M. M. Menéndez & J. R. Lucena, 1990. Proposals for macrophyte restoration in eutrophic coastal lagoons. Hydrobiologia 200/201: 427–436.

Cornwell, J. C., D. J. Conley, M. Owens & J. C. Stevenson, 1996. A sediment chronology of the eutrophication of Chesapeake Bay. Estuaries 19: 488–499.

Craft, C. B., J. Vyzamal & J. Richardson, 1995. Response of Everglades plant communities to nitrogen and phosphorus addition. Wetlands 15: 258–271.

De Groot, C. J., 1990. Some remarks on the presence of organic phosphate in the sediment. In D. J. Bonin & H. L. Golterman (eds), Fluxes Between Trophic Levels and Through the Water–Sediment Interface. Developments in Hydrobiology 62. Kluwer Academic Publishers, Dordrecht: 303–311. Reprinted from Hydrobiologia 207.

De Groot, C. J., 1992. Flux et transformations du phosphate et de l'azote dans les zones humides: La Camargue (delta du Rhône). Thèse Univ. C. Bernard Lyon I. 68 pp.

De Groot, C. J. & H. L. Golterman, 1993. On the presence of organic phosphate in some Camargue sediments: evidence for the importance of phytate. In H. L. 'Golterman (ed.), Sediment–Water Interaction 2. Hydrobiologia 252: 117–126.

Duncan, P., 1992. Horses and grasses. The nutritional ecology of equids and their impact on the Camargue. Ecological Studies 87. Springer-Verlag Ed. 287 pp.

El Habr, H. & H. L. Golterman, 1987. Input of nutrients and suspended matter into the Golfe du Lion and the Camargue by the River Rhône. Rev. Sci. Eau 6: 393–402.

El Habr, H. & H. L. Golterman, 1990. *In vitro* and *in situ* studies on nitrate disapearance in water-sediment systems of the Camargue (Southern France). Hydrobiologia 192: 223–232.

Forès, E. & F. A. Comín, 1987. Chemical characteristics of the water in the ricefields of the Ebro delta (N.E. Spain). Arch. Hydrobiol. 111: 15–24.

Forès, E., 1989. Ricefields as filters. Arch. Hydrobiol. 116: 517–527.

Forès, E. & F. A. Comín, 1992. Ricefields, a limnological perspective. Limnetica 8: 101–109.

Forès, E., R. R. Christian, F. A. Comín & M. Menendez, 1994. Network analysis on nitrogen cycling in a coastal lagoon. Mar. Ecol. Prog. Ser. 106: 283–290.

Frenay, J. R., A. C. F. Trevitt, S. K. De Datta, N. Obcema & J. G. Real, 1990. The interdependence of ammonia volatilization and denitrification as nitrogen loss processes in flooded rice fields in the Philippines. Biol. Fertil. Soils 9: 31–36.

Giraud, F., 1992. Modélisation hydrologique d'une zone humide agricole: perspectives pour l'étude du transport d'azote et de phosphore. Cas du marais de Moèze, Charente-Maritime, France. Thèse Doct., Université de Rennes, 225 pp.

Godin, L., 1990. Impact de l'irrigation pour la riziculture sur l'hydrologie et la chimie des eaux en Camargue. Approche à partir du bassin d'irrigation de l'Aube de Bouic. Thèse Doct. Univ. Franche Comté, 172 pp.

Golterman, H. L., 1984. Sediments, modifying and equilibrating factors in the chemistry of freshwater. Verh. int. ver. Limnol. 22: 23–59.

Golterman, H. L., C. Bonetto & F. Minzoni, 1988. The nitrogen cycle in shallow water sediments systems of ricefields Part III: The influence of N-application on the yield of rice. Hydrobiologia 159: 211–217.

Gomez, E., 1996. Role des sédiments dans l'eutrophisation d'uune lagune littorale méditerranéenne (Etang du Méjean – Hérault): Echanges du phosphate entre le sédiment et l'eau en fonction des conditions environnementales. Thèse Doct., Sci. Enviro. et Santé Publique, Univ. Montpellier 1, Faculté de Pharmacie, 200 pp.

Grillas P., 1990. Distribution of submerged macrophytes in the Camargue in relation to environmental factors. J. Vegetat. Sci. 1: 393–402.

Harwell, M. A., J. F. Long, A. M. Bartuska, J. H. Gentile, C. C. Harwell, V. Myers & J. C. Ogden, 1996. Ecosystem management to achieve ecological sustainability: the case of South Florida. Envir. Mgmt. 20: 497–521.

Hearne J., I. Johnson & P. Armitage, 1994. Determination of ecologically acceptable flows in rivers with seasonnal changes in the density of macrophyte. Regul. Riv. 9: 177–184.

Heathwaite, A. L. & P. J. Johnes, 1996. Contribution of nitrogen species and phosphorus fractions to stream water quality in agricultural catchments. Hydrol. Process. 10: 971–983.

Heurteaux, P., 1992. Modifications du régime hydrique et salin des étangs du système Vaccarès (Camargue, France) liées aux perturbations anthropiques des cinquantes dernières années. Annls Limnol. 28: 157–174.

Heurteaux, P., 1994. Essai de quantification des termes du bilan hydrique des étangs du système Vaccarès (Camargue, France). Annls Limnol. 30: 131–144.

Hill, B. H., 1979. Uptake and release of nutrients by aquatic macrophytes. Aquat. Bot. 7: 87–93. Izuno, F. T. & L. T. Capone, 1995. Strategies for protecting Florida's Everglades: The best management practice approach. Wat. Sci. Technol. 31: 123–131.

Jacobs, P. P. W. M., 1979. Distribution and aspects of the production of biomass of eelgrass, Zostera marina L., at Roscoo, France. Aquat. Bot. 7: 151–172.

Johnes, P. J., 1996. Evaluation and management of the impact of land use change on the nitrogen and phosphorus load delivered to surface waters: the export coefficient modelling approach. J. Hydrol. 183: 323–349.

Lieutaud, A., M. C. Ximenes & T. Moutin, 1992. Lagoon eutrophication assessment for rehabilitation purposes: nitrogen and phosphorus loadings in different compartments. In C. M. Finlayson, G. E. Hollis & T. J. Davis (eds), Managing Mediterranean Wetlands and their Birds. Proc. Symp., Grado, Italy, 1991. IWRB Spec. Publ. n° 20, Slimbridge, UK, 285 pp.

Loye-Pilot, M. D., J. M. Martin, J. Morelli, 1990. Atmospheric input of organic nitrogen to the western mediterranean. Biochemistry 9: 117–134.

Madden, C. J. & W. M. Kemp, 1996. Ecosystem model of an estuarine submersed plant community: calibration and simulation of eutrophication responses. Estuaries 19: 457–474.

Mesléard, F., 1995. Restoration of seasonnally-flooded marshes in abandonned ricefields: the value for grazing and wildlife. Proceedings of the Vth International Rangeland Congress. Salt Lake City, July 1995, Vol. I, 651 pp.

Minzoni, F. C., C. Bonetto & H. L. Golterman, 1988. The nitrogen cycle in shallow water sediments systems of ricefields Part I: The denitrification process. Hydrobiologia 159: 189–202.

Moutin, T., B. Picot, M. C. Ximenes & J. Bontoux, 1993. Seasonal variations of P compounds and their concentrations in coastal lagoons (Herault, France). Hydrobiologia 252: 45–59.

Nowicki, B. L. & S. W. Nixon, 1985. Benthic nutrient remineralization in a coastal lagoon ecosystem. Estuaries. Vol. 8, n° 2B: 182–190.

Moutin, T., P. Raimbault & B. Coste, 1996. Rhône river nutrients inputs into the mediterranean sea: annual observations and comparison with earlier data. Paper presented at the Third International Congress Limnology-Oceanography; 'Interfaces – Oceans rivers and lakes: energy and substance transfers at interfaces'; Nantes, France, 7–9 Oct. 1996.

Patriquin D. & R. Knowles, 1972. Nitrogen fixation in the rhizosphere of marine angiosperms. Mar. Biol. 16: 49–58.

Perez-Llorens J. L., M. Muchtar, P. H. Nienhuis & F. X. Niell, 1991. Particulate organic carbon, nitrogen and phosphorus in roots, rhizomes and differently aged leaves of *Zostera noltii* Hornem. in Oosterschelde (SW Netherlands). Bot. Mar. 34: 319–322.

Perez-Llorens, J. L. & F. X. Niell, 1993. Seasonnal dynamics of biomass and nutrient content in the intertidal seagrass *Zostera noltii* Hornem. from Palmones River estuary, Spain. Aquat. Bot. 46: 49–66.

Pichot, P., M. C. Ximenes, J. M. Deslous-Paoli & C. Juge, 1994. Bilan de l'azote et du phosphore dans le système bassin versant – lagune de Thau. IFREMER, R. int. Del/94.11/Sète, France, 84 pp.

Picot, B., G. Pena, C. Casellas, P. Mancini, D. Bondon & J. Bontoux, 1990. Interpretation of the seasonal variations of nutrients in a mediterranean lagoon: étang de Thau. Hydrobiologia 207: 105–114.

Rader, R. B. & C. J. Richardson. 1992. The effects of nutrient enrichment on algae and macroinvertebrates in the Everglades: a review. Wetlands 12: 121–135.

Reddy, K. R. & W. H. Jr Patrick, 1986. Denitrification losses in flooded ricefields. In S. De Datta & W. H. Jr Patrick (eds), Nitrogen Economy of Flooded Rice Soils. 99–116, Dordrecht, 186 pp.

Reddy, K. R., W. H. Patrick Jr & C. W. Lindau, 1989. Nitrification-denitrification at the plant root–sediment interface in wetlands. Limnol. Oceanogr. 34: 1004–1013.

Sandoz, A., 1996. Proposition d'une méthodologie adaptée au suivi de l'occupation du sol d'une zone humide aménagée: application au bassin du Fumemorte (Grande Camargue, France). Thèse Doct. Univ. Aix-Marseille I, 154 pp.

Seitzinger, S. P., 1988. Denitrification in freshwater and coastal marine ecosystems: ecological and geochemical significance. Limnol. Oceanogr. 33: 702–724.

Snyder, G. H. & J. M. Davidson, 1994. Everglades agriculture: past, present, and future. In S. M. Davis & J. C. Ogden (eds), Everglades, the Ecosystem and its Restoration, St Lucie Press, Boca Raton, FL (USA): 85–115.

Stutterheim, N. C., 1994. Towards higher nitrogen efficiency in European rice cultivation; a case study for the Camargue, South of France. Thesis Wageningen. ISBN 90-9007963-7, 93 pp.

ten Berge, H. F. M., K. Metselaar, M. J. W. Jansen, E. M. de San Augustin & T. Woodhead, 1995. The SAWAH riceland hydrology model. Water Resour. Res. 31: 2721–2732.

Tuong, T. P., M. C. S. Wopereis, J. A. Marquez, & M. J. Kropff, 1994. Mechanisms and control of percolation losses in irrigated puddled rice fields. Soil Sci. Soc. Am. J. 58: 1794–1803.

Van Lent, F. & J. M. Verschurre, 1994. Intraspecific variability of *Zostera marina* L. (eelgrass) in the estuaries and lagoons of the southwestern Netherlands. II. Relation with environmental factors. Aquat. Bot. 48: 59–75.

Van Wijck, C., P. Grillas, C. J. De Groot & L. Tan Ham, 1994. A comparison between the biomass production of *Potamogeton pectinatus* L. and *Myriophyllum spicatum* L. in the Camargue (southern France) in relation to salinity and sediment characteristics. Vegetatio 113: 171–180.

Vaquer, A., 1984a. La production algale dans les rizières de Camargue pendant la période de submersion. Verh. int. ver. Limnol. 22: 1615–1654.

Vaquer, A., 1984b. Biomasse et production des characées dans les rizières de Camargue (France) et leur importance écologique. Oecol. Plant. 5: 299–313.

Vaquer, A. & P. Heurteaux, 1989. Modifications récentes de la végétation aquatique de l'étang du Vaccarès (Camargue, France) liées aux perturbations anthropiques. Annls Limol. 25: 25–38.

Vermaat, J. E., M. J. M. Hootsmans & P. H. Nienhuis, 1987. Seasonal dynamics and leaf growth of *Zostera noltii* Hornem., a perennial intertidal seagrass. Aquat. Bot. 28: 287–299.

Welsh, D. T., S. Bourghès, R. de Wit & R. A. Herbert, 1996. Seasonal variations in nitrogen-fixation (acetylene reduction) and sulplhate-reduction rates in the rhizosphere of *Zostera noltii*: nitrogen fixation by sulphate-reducing bacteria. Mar. Biol. 125: 619–628.

Wilber, D. H., R. E. Tighe, & L. J. O'Neil, 1996. Associations between changes in Agriculture and hydrology in the Cache river basin, Arkansas, USA. Wetlands 16: 366–378.

Wopereis, M. C. S., B. A. M. Bouman, M. J. Kropff, H. F. M. ten Berge & A. R. Maligaya, 1994. Water use efficiency of flooded rice fields. I. Validation of the soil–water balance model SAWAH. Agric. Water Mgmt 26: 277–289.

Hydrobiologia **373/374:** 193–201, 1998.
J. C. Amiard, B. Le Rouzic, B. Berthet & G. Bertru (eds),
Oceans, Rivers and Lakes: Energy and Substance Transfers at Interfaces.

Sediment–water exchanges of ammonium and phosphate in intertidal and subtidal areas of a mesotidal coastal lagoon (Ria Formosa)

Manuela Falcão & Carlos Vale
Instituto de Investigação das Pescas e do Mar. Av. Brasília 1400 Lisboa, Portugal

Key words: ammonium, phosphate, sediment-water interactions, tidal flushing, molecular diffusion

Abstract

We measured the ammonium and phosphate concentrations in near-bottom and pore waters collected across one year, at several stations of the subtidal and intertidal areas of Ria Formosa. In addition, the concentrations were determined in flood water collected during the first 20 min of inundation of the extensive intertidal areas. Ammonium and phosphates were higher in the pore waters than in near-bottom waters, and the predicted flux of NH_4^+ and PO_4^{3-} by molecular diffusion varied seasonally, and was higher in intertidal than in subtidal sediments.

Transport induced by the short-term mixing of intertidal sediments were comparable to the predicted diffusive transports between low tides. The NH_4^+ flux was production-rate controlled, while PO_4^{3-} flux was slowed down by chemical interactions with solids. NH_4^+ was released preferentially to PO_4^{3-}; but the stoichiometry of decomposition/transport varied seasonally. Removal of NH_4^+ from the intertidal sediments was independent of season. As a yearly average, the amounts of NH_4^+ and PO_4^{3-} flushed out of the intertidal area of Ria Formosa were one-two orders of magnitude higher than the quantities released from the subtidal area.

Introduction

There is a basic difference between subtidal and intertidal coastal sediments. Subtidal sediments are always submerged and mass exchanges with the overlying water are ruled by molecular diffusion (Berner, 1980), macrofauna (Aller, 1980), pressure variation on a permeable bottom generated by gravity waves and bottom currents (Shum & Sundby, 1996), and episodic wind-induced resuspension of sediments (Fanning et al., 1982; De Jonge & Van Beusekom, 1995). Intertidal sediments are alternatively exposed inundated. The flood water arriving with the tide forces a flow through the upper sediments, which may cause a renewal of pore water (Kerner & Wallmann, 1992), brings fresh supplies of dissolved oxygen, removes the products of community metabolism (Steele, 1970; Falcão & Vale, 1995), and changes the concentrations of highly reactive species on short-time scales (Caetano, 1977). These exchanges are more important in tide dominated environments.

Since benthic fluxes are influenced by temperature and other induced changes (Aller, 1980), and vary with sediment grain size (Nowicki & Nixon, 1985; Van Raaphorst et al., 1992; Kristensen, 1993), there is a spatial and temporal variability in the sediment-water exchange in most coastal environments (Forja et al., 1994), and dissolved constituents in pore water often differ markedly from steady-state conditions (Lerat et al., 1990; Vidal & Morguí, 1995). Differences of ammonium in sediment should reflect the balance between ammonium production as organic matter decomposes (Nowichi, 1994), nitrification (Lohse et al., 1993), excretion by benthic organisms (Lomstein et al., 1989), and consumption by primary producers living near the sediment–water interface that take advantage of this periodic exposure to light (Nian-zhi & Rong, 1994). Reversible ion-exchange on sediment particles is usually the dominant reaction of this minor cation in marine sediments (Mackin & Aller, 1984) and, thus, the NH_4^+ sorption equilibrium may be modified in minutes by the physical alterations associated with tidal flushing (Falcão & Vale, 1995). Phosphorus reacts with a wide variety of surfaces, being taken up and released from biogenic and abiogenic particles (Van Raaphorst et al., 1994; Prastka, 1995). Of particular importance in marine sediments is the sorption of dissolved phosphorus to iron oxides (Fox, 1993; Sundby et al., 1992; De Jonge et al., 1993) that may be released during anoxia (Krom & Berner, 1981). These sorption reactions slow down the flux of phosphate from the sediment (Aller, 1980) and may present a seasonal variation (Forja et al., 1994).

This paper reports on concentrations of ammonium and phosphate in near-bottom and pore waters of subtidal and intertidal sediments of Ria Formosa and compares their transports to the overlying water, estimated by molecular diffusion and measured directly when intertidal sediments are flooded by the tide.

Study area

Ria Formosa is a shallow coastal lagoon on the south coast of Portugal, covering a wet area of about 105 km^2 (Figure 1). The lagoon has several channels and an extensive intertidal area, around 50% of which is constituted by sand, muddy sand-flats and salt marshes (Bettencourt, 1994). Fresh water input is negligible and salinity remains close to 36‰, except during sporadic periods of winter run-off (Falcão et al, 1992). The tidal elevation ranges from 1 to 3.5 m and, at spring tide around 70% of the lagoon water is renewed.

Sampling and methods

Near-bottom water and surface sediments were collected in March, May, July, September and December of 1993, at four cross sections of the lagoon: A, B, C and D (Figure 1). At each cross section, sediment cores were taken from the subtidal sandy area (stations As, Bs, Cs and Ds), and from the intertidal muddy area (stations Am, Bm, Cm and Dm). At each station, near-bottom water was collected at low tide, directly with a syringe, and frozen in polyethylene vials. Sandy sediments were collected by hand with plastic tubes of 2 cm length and 4 cm diameter. Each sampling comprised 10 short cores from an area of 1 m^2. Box-cores ($15 \times 15\ cm^2$) were collected in the muddy area and the first 2 cm were sliced in 0.5-cm layers. The slicing of the cores was done in loco within a few minutes. Polyethylene vials were filled up with samples from each sediment layer, avoiding air inside the vials. The sediment was centrifuged in polyethylene vials (3500 ppm, 20 min) and the pore water separated for determination of ammonium and phosphate. Temperature of the sediment surface was measured *in situ*. Organic matter (loss on ignition, LOI) was evaluated from weight loss after ignition at 450 °C for 3 h. Porosity was calculated from weight loss at 105 °C.

To evaluate the importance of physical perturbation on the intertidal sediments as water spreads over the tidal flats, independent surveys were performed, in February and July (1993). Flood water was sampled 2, 10, 15 and 20 min after the inundation had started at stations E, F, G, H and I (Figure 1). Flood waters were collected 1 cm above the sediment surface in precleaned syringes, filtered through 0.45 m Nuclepore membranes, and dissolved ammonium and phosphate concentrations measured. Water temperature was measured *in situ*.

Ammonium and phosphate were analysed using a Chemlab autoanalyser according to Technicon Industrial Systems (Grasshoff, 1976). A non-parametric Wilcoxon-Mann-Whitney test was applied to determine the significance of differences between samples (Scherrer, 1984).

Results

Ammonium and phosphate in pore waters and near-bottom waters of subtidal sands

Porosity of sandy sediments from stations As, Bs, Ds and Es varied within a narrow interval (0.2–0.3), and organic matter content was low, between 0.3 a 0.8%. Temperature of the sediment varied from 10 °C in December to 26 °C in July (Figure 2). Concentrations of NH_4^+ and PO_4^{3-} are presented in Table 1. Ammonium concentrations in near-bottom waters were lower than 3.9 μM, while in pore waters levels they were significantly higher ($p \leq 0.01$), ranging from 5.8 to 79 μM; differences between stations, either in water column and pore water changed seasonally, near-bottom water values being lower in May ($p \leq 0.05$) and pore water levels higher in July ($p \leq 0.05$). Phosphate concentrations in near-bottom waters changed from 0.4 to 2.3 μM, while in pore waters were significantly higher ($p \leq 0.01$), between 1.1 and 51 μM; as for ammonium, no statistical differences were recorded among stations. Phosphate levels in near-bottom waters were not significantly different during the sampling period; phosphate in pore waters was higher in July ($p \leq 0.01$).

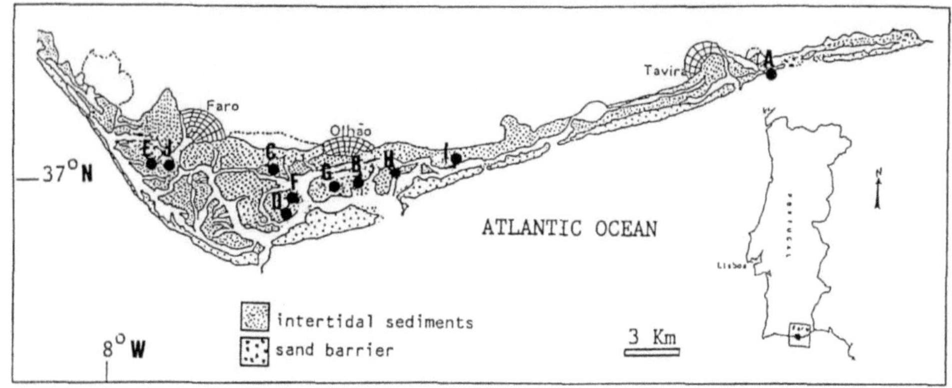

Figure 1. Ria Formosa and the location of the sampling sites A, B, C, D, E, F, G, H, I and J.

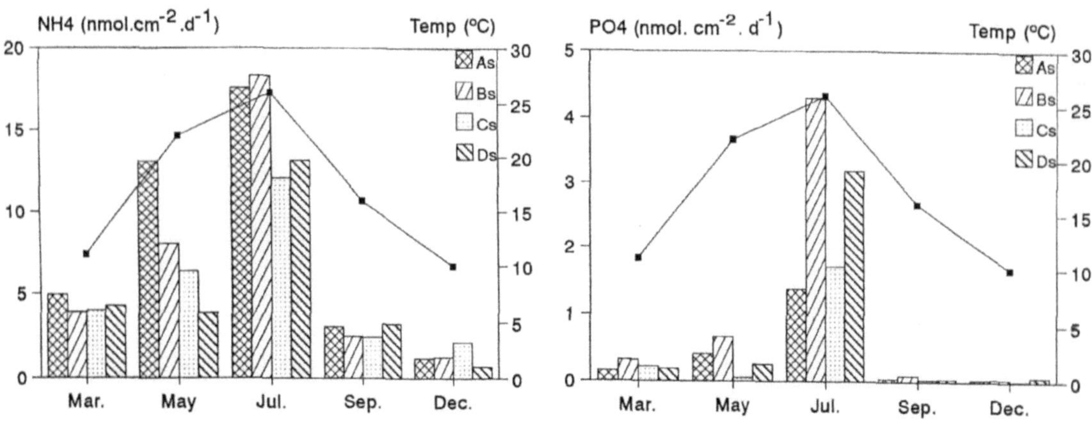

Figure 2. Transport of ammonium and phosphate estimated by the Fick's law of diffusion (nmol cm^{-2} d^{-1}), at stations As, Bs, Cs and Ds of subtidal sandy areas; March, May, July, September and December (1993); mean daily temperature of surface sediments.

Flux of NH_4^+ and PO_4^{3-} between subtidal sediment and overlying water

On the basis on the values given above, the ammonium and phosphate fluxes (*F*), across the sediment–water interface, can be predicted by Fick's first law of diffusion (Berner, 1980):

$$F = -\phi D_s(C_o - C_p) / (\Delta x), \quad (1)$$

where ϕ = porosity (0.2–0.3), $D_s = 14 \times 10^{-6}$ to 22×10^{-6} cm^2. s^{-1} for NH_4^+ and 5×10^{-6} to 8×10^{-6} cm^2. s^{-1} for PO_4^{3-}, considering a temperature range of 10–26 °C (Aller, 1980); $C_o = [NH_4^+]$ or $[PO_4^{3-}]$ in near-bottom water; $C_p = [NH_4^+]$ or $[PO_4^{3-}]$ in pore water; Δx = distance across the interface where concentrations were measured (2 cm).

The calculated fluxes of ammonium and phosphate for each subtidal stations along the annual period of study are presented in Figure 2. In general, lower ammonium fluxes were estimated in September and December (< 3.3 nmol cm^{-2} d^{-1}) and higher fluxes for July (> 12 nmol cm^{-2} d^{-1}). The estimated fluxes of phosphate presented a broader variation: in December (10 °C), they were less than 0.33 nmol cm^{-2} d^{-1}, while in July (26 °C) they were one order of magnitude higher.

Ammonium and phosphate in pore waters and near-bottom waters at submerged intertidal sediments

Porosity of intertidal sandy and muddy sediments from stations Am, Bm, Dm and Em varied from 0.3 to 0.5, and organic matter content ranged between 4% and 6%. Mean temperature of the upper sediment layer changed from 13 °C in winter to 28 °C in July (Figure 3). Around low tide, concentrations of NH_4^+ and PO_4^{3-}, in near-bottom waters and in pore waters of 0.5-

Table 1. Concentrations of ammonium and phosphate (M) in near-bottom waters and pore waters, at stations As, Bs, Cs and Ds of subtidal sandy areas; March, May, July, September and December (1993).

	NH_4^+ Near-bottom water (μM)				NH_4^+ Pore + water (0–2 cm) (μM)			
	As	Bs	Cs	Ds	As	Bs	Cs	Ds
March	2.7	1.5	1.7	2.0	35.3	27.2	28.3	30.3
May	0.5	0.5	0.3	0.8	55.5	34.2	27.0	17.1
July	1.5	2.2	1.1	2.3	75.6	79.4	51.8	57.5
September	3.9	1.4	1.4	2.1	16.9	12	11.9	15.8
December	5	2.3	0.8	1.1	13	10.7	15.1	5.8
March	1.4	1.3	2.1	1.8	4.4	7.3	6.3	5.4
May	2.1	1.0	0.4	1.1	6.9	8.8	1.1	4.1
July	0.7	1	1.1	0.7	16.5	50.8	20.9	37.5
September	2.3	1.3	1.2	1.3	7.9	3.2	4.9	6
December	1.1	1.5	1.3	0.5	1.7	2.3	1.5	1.8

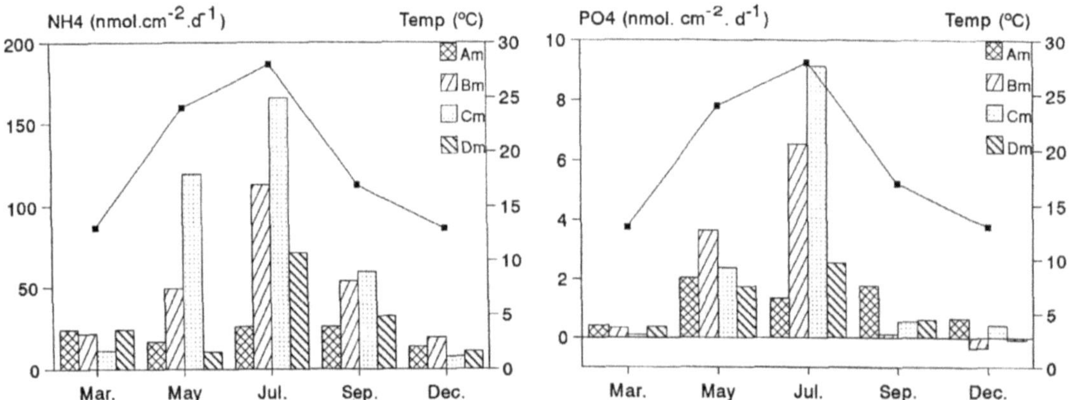

Figure 3. Transport of ammonium and phosphate estimated by the Fick's law of diffusion (nmol cm^{-2} d^{-1}), at stations Am, Bm, Cm and Dm of intertidal muddy areas; March, May, July, September and December (1993); mean daily temperature of surface sediments.

cm sediment layers, are shown in Table 2. Ammonium in near-bottom waters varied from 0.8 to 34.5 μM, while in pore waters they were one order of magnitude higher, between 36 and 335 μM; in near-bottom water and pore water, differences among stations were not statistically significant; in most situations, ammonium concentration increased with depth in the 2-cm upper sediment layer; as observed in the subtidal sandy sediments, ammonium concentrations in pore waters showed an increase in July ($p \leq 0.05$), which was not recorded in the near-bottom waters. Phosphate in near-bottom waters changed from 0.8 to 10.2 μM, while in pore waters they were also one order of magnitude higher, reaching 54.2 μM in the topmost layer and occasionally more in deeper sediments; in the upper 2-cm layer levels they did not show a clear trend with depth; again, no statistical differences were recorded among stations, in water column or pore water; phosphate concentrations in the near-bottom waters were not significantly different during the sampling period, however, values in pore water in May and July (temperature between 24 °C and 28 °C) were significantly higher ($p \leq 0.05$) than in periods of lower temperature (13 °C to 17 °C).

Flux of NH_4^+ and PO_4^{3-} between intertidal sediment and overlying water

On the basis of the values of Table 2, ammonium and phosphate fluxes across the sediment-water interface, were estimated by Fick's first law of diffusion. In equation (1), we assumed for the term Δx the value of 1.3 cm, since the spatial resolution in the intertidal sediment was smaller (0.5 cm). Because the intertidal sediment remained exposed to the atmosphere for a few hours around low tide in each semi-diurnal cycle, and this period changes with tidal amplitude, the fluxes

Table 2. Concentrations of ammonium and phosphate (M) in near-bottom waters and pore waters, at stations Am, Bm, Cm and Dm of intertida muddy areas; March, May, July, September and December (1993).

NH_4^+	March				May				July				September				December			
(μM)	Am	Bm	Cm	Dm	Am	Bm	Cm	Dm	Am	Bm	Cm	Dm	Am	Bm	Cm	Dm	Am	Bm	Cm	Dm
Near-bott w.	18.8	0.8	7.8	2.5	3.0	4.8	10.8	34.5	11.2	15.1	12.2	4.2	16.3	7.4	3.9	8.2	6.2	22.2	4.6	11.4
(0–0.5 cm)	92.7	66.7	42.7	76.0	36.0	100.8	243.0	55.0	61.4	235.2	334.9	142.8	66.2	112.0	120.0	70.8	48.1	81.9	27.7	44.6
(0.5–1 cm)	149.1	78.7	87.3	97.9	66.9	171.5	114.0	73.3	51.7	151.2	249.9	310.2	129.7	121.0	110.0	77.8	123.5	137.2	61.0	54.0
(1–1.5 cm)	104.6	135.0	80.1	105.9	237.7	151.6	87.9	84.0	67.6	147.4	155.6	216.3	116.6	92.0	85.2	87.2	66.9	134.6	42.4	66.6
(1.5–2 cm)	58.7	78.7	66.9	113.4	–	–	–	–	129.2	81.8	14.0	69.4	66.6	84.8	78.8	100.8	100.2	72.4	60.4	72.0
PO_4^{3-}	March				May				July				September				December			
(μM)	Am	Bm	Cm	Dm	Am	Bm	Cm	Dm	Am	Bm	Cm	Dm	Am	Bm	Cm	Dm	Am	Bm	Cm	Dm
Near-bott w.	4.5	2.1	0.8	2.8	3.2	5.8	10.2	5.4	4.3	8.3	5.3	1.8	1.9	2.6	2.2	1.6	2.5	3.3	0.9	1.4
(0–0.5 cm)	8.1	5.0	1.7	6.0	14.1	25.3	23.0	14.7	11.5	43.2	54.2	15.4	1.3	3.3	5.1	4.9	8.1	0.4	4.5	0.7
(0.5–1 cm)	29.1	1.0	10.8	6.4	21.0	32.0	18.0	25.0	11.1	28.1	28.6	50.9	38.8	0.2	20.3	4.2	16.5	11.4	4.0	4.6
(1–1.5 cm)	2.6	5.9	3.0	3.3	18.4	30.2	28.4	19.7	29.2	44.6	17.1	118.6	2.8	9.0	9.3	4.6	20.6	11.9	2.9	1.8
(1.5–2 cm)	9.2	3.9	4.8	8.0	–	–	–	–	45.2	12.9	13.9	168.4	14.3	12.5	10.1	17.5	20.5	8.6	3.0	2.1

should be calculated for the number of hours that sediment is submerged. Assuming that, on average neap-spring tidal cycle, intertidal stations are exposed for three hours per day (average over the neap-spring tidal cycle), the daily fluxes should be estimated for 21 h. The fluxes in intertidal stations predicted by equation (1) are presented in Figure 3. As found at subtidal sandy areas, lower fluxes of ammonium were estimated in March and December (< 24 nmol cm^{-2} d^{-1}), and higher quantities in July (≈ 100 nmol cm^{-2} d^{-1}). Again, the seasonal variation of phosphate fluxes was better defined; lower than 0.66 nmol cm^{-2} d^{-1} in March and December, while in July it reached 9 nmol cm^{-2} d^{-1}. However, the range of the estimated fluxes in the intertidal stations was not so broad as in the subtidal area.

Tidal flushing of ammonium and phosphate from intertidal areas

The concentrations of ammonium and phosphate in flood waters at sites F, G, H, I and J, at different inundation times, in two periods of February and June, are represented in Figure 4. Two minutes after inundation, the ammonium concentrations at the stations varied from 0.8 to 6.8 μM in February, and between 2.2 and 10.3 μM in June; the differences among stations in the following minutes became smaller, and levels between the two periods were not statistically different. However, at each station the concentrations decreased rapidly with time ($p \leq 0.05$), and twenty minutes after inundation, corresponding to approximately 25 cm of water column, levels became similar at all stations (0.4–1.4 μM) and comparable to overlying water. The phosphate pattern was different: concentrations in June changed between 0.7 and 2.3 μM, being statistically higher ($p \leq 0.05$) than those obtained in February (0.2–1.6 μM); the decrease along the 20-min inundation period was not so clearly defined as for ammonium; between stations, phosphate concentrations in June varied in a broader interval.

Associated with tidal inundation of intertidal areas, there is transport of ammonium and phosphate to the water column (T), that may be calculated by:

$$T[\text{nmolcm}^{-2}] = \Sigma(C_i - C_{i-1}).(h_i + h_{i-1})/2, \quad (2)$$

where $i = 2, 10, 15, 20$; $C_i = [NH_4^+]$ and $[PO_4^{3-}]$ in overlying water (μM) at minutes i = 2, 10, 15, 20; h_i = water depth (cm).

On the basis of the direct measurements of concentrations in flood waters at different times of inundation (C_i) one may calculate the 20-min transport at stations F, G, H I and J. Intertidal stations were inundated twice a day, and the daily removal of ammonium and phosphate was estimated multiplying the quantity T by two (Table 3): 30 to 142 nmol cm^{-2} d^{-1} for ammonium and 0.2 to 31 nmol cm^{-2} d^{-1} for phosphate.

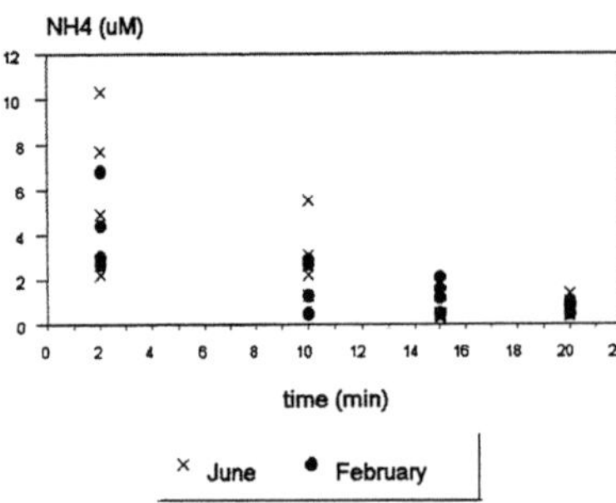

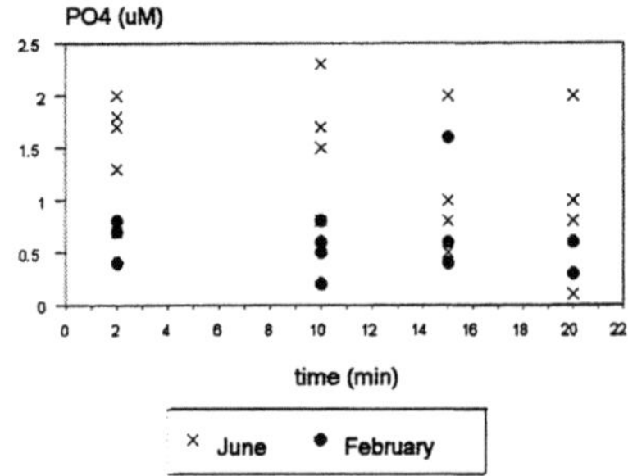

Figure 4. Concentration of ammonium and phosphate (M) in flooding waters during the first 20 min of tidal inundation at stations F, G, H, I and J; February and June surveys (1993).

Table 3. Transport of dissolved ammonium and phosphates (nmol cm^{-2} d^{-1}) by short-term mixing, at stations F, G, H, I and J; June and February surveys (1993).

Stations	NH_4^+ nmol cm^{-2} d^{-1})		PO_4^{3-} nmol cm^{-2} d^{-1})	
	June	February	June	February
E	6.70	81.0	10.0	4.5
F	77.0	78.0	1.5	0.2
G	32.0	–	23.0	–
H	142.0	52.0	31.0	1.2
I	102.0	30.0	20.0	22.0

Discussion

This study reports contrasting amounts of ammonium and phosphate regenerated in subtidal and intertidal sediments of Ria Formosa, their transport to the overlying water as predicted by Fick's first law of diffusion and as measured directly when the tide floods the intertidal zone, and the seasonal variation of NH_4^+/PO_4^{3-} in solute transport.

Higher concentrations of ammonium and phosphate in pore waters of subtidal and intertidal sediments than in near-bottom waters indicate regeneration in the upper sediment layers. The different concentrations across the sediment-water interface imply a flux by molecular diffusion from the sediment to the overlying water column. This has been classically observed in coastal environments (Aller, 1980; Nowicki & Nixon, 1985; Lerat et al., 1990), and reinforces the importance of sediments on the biogeochemical cycling of nutrients in Ria Formosa (Falcão & Vale, 1990). Levels were higher in pore waters of muddy sediments than of sand, which means that the predicted exchanges by molecular diffusion in subtidal areas, whose bottom is essentially constituted by sand, are less important than at intertidal areas. The lower concentrations in sands may be attributed to the lower quantity of organic matter accumulated in these areas and, possibly, to a more efficient renewal of pore water containing the products of organic matter decomposition (Shum & Sundby, 1996). Exposure of intertidal mud of Ria Formosa to the atmosphere results in a rise of temperature and the direct light effect may also contributes to accelerate the mineralization processes of the organic material settled in these areas (Vale et al., 1992). As found in other environments (Nowicki & Nixon, 1985; Forja et al., 1994), the calculated transport of ammonium by molecular diffusion was one or two orders of magnitude higher than that of phosphate, which means that ammonium is more available to the phytoplankton than phosphate.

The flux of NH_4^+ and PO_4^{3-} out of the bottom, including subtidal and intertidal areas, varied seasonally and was highest in summer, lowest in winter. Based on near-bottom and pore water concentrations, together with Fick's first law, the predicted transports increased sharply in July and presumably remained high during the warmer period. Higher benthic fluxes during summer were also recorded in other environments and were interpreted as an intensification in the anaerobic decomposition pathways of organic matter (Forja et al., 1994). A similar explanation could apply to the results obtained in Ria Formosa. Because most decomposition takes place near the sediment–water interface, the fluxes of these constituents are not supposed to be greatly influenced by burrow construction and can be adequately predicted by molecular diffusion (Aller, 1980). However, some intertidal areas of Ria Formosa are used as clam-growth banks, and here we should expect an increment due to excretion and bioturbation (Yamada & Kayama, 1987). For further informa-

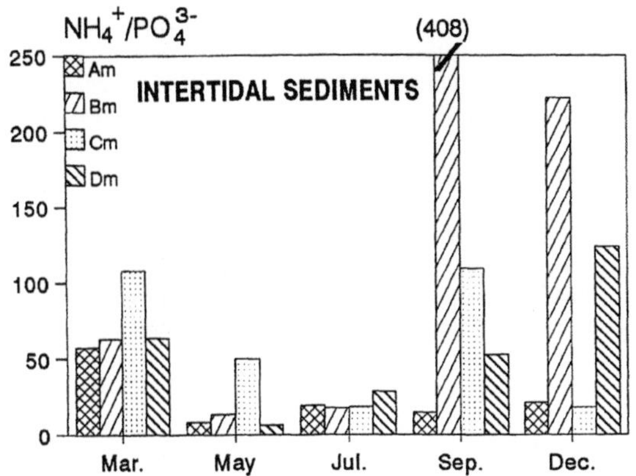

Figure 5. Seasonal variation of the stoichiometry of ammonium and phosphate fluxes (NH_4^+/PO_4^{3-}); March, May, July, September and December surveys (1993).

tion we refer to a previous publication (Falcão & Vale, 1990a).

Moreover, Forja et al. (1994) found that seasonal variations of nutrient fluxes involved alterations in N/P stoichiometry. On the basis of predicted flux of ammonium and phosphates in subtidal and intertidal areas of Ria Formosa, the N/P ratio was calculated (Figure 5). Lower values were obtained in May and July, comparable to the N/P Redfield ratios (16:1), while ratios for September and December fluxes were in the order of hundreds. Consequently, the N/P presented a seasonal variation, and NH_4^+ was released preferentially to PO_4^{3-} during early diagenesis. These results confirm previous conclusions obtained from water column data that suggested phosphorus limitation in Ria Formosa (Falcão & Vale, 1990). They are in agreement with Nowicki & Nixon (1985), who showed a lesser phosphorus availability relative to inorganic nitrogen in a coastal lagoon. The high N/P ratios of nutrient fluxes during autumn/winter in Ria Formosa is mainly due to weaker transport of phosphorus by molecular diffusion across the sediment–water interface. During fall and winter, PO_4^{3-} fluxes may be lowered due to adsorption on Fe-oxides at the sediment–water interface (Aller, 1980). Phosphorus dynamics in sediments are strongly influenced by the cycles of iron and sulphur (Sundby et al., 1992) and these elements are very reactive when flooding water spreads over the sediment of the lagoon (Caetano et al., 1977). Possibly, phosphorus accumulates in solid phase during winter, to be released to pore water and transferred to the overlying water of the lagoon in periods of higher temperature (Ohtake et al., 1984; Kennedy et al., 1984). The well-defined seasonal variation of temperature occurring in Ria Formosa favours contrasting phosphorus exchanges.

The intertidal areas of these shallow meso-tidal environments are characterised by a semi-diurnal pulse mechanism associated with the arrival of the flood water (Falcão & Vale, 1995; Rocha, in press; Caetano et al., 1977). In Ria Formosa, during 20 min of inundation, the quantities of ammonium and phosphate transported to the overlying water were comparable to the predicted diffusive fluxes during the time that sediment remained submerged between low tides. The toplayer sediments of intertidal zones are physically perturbed by the tidal inundiation, which may lead to a renewal of pore water by percolation (Kerner & Wallmann, 1992), and the physical and chemical tide-driven modifications in the oxic and sub-oxic layers may take place in the first minutes of flooding (Caetano et al., in press). This mechanism induces a duplication of the export of ammonium and phosphate from the intertidal area, and reinforces its importance in the cycling of nutrients in the lagoon. However, two situations were observed. In spite of the temperature effect on ammonium production in sediments (Vale et al., 1992), the flushing of ammonium from the intertidal sediments was similar in February and June. Experimental studies demonstrated that ammonium is easily removed from the solid phase by physical processes (Falcão & Vale, 1995), which are not dependent on temperature. Mackin & Aller (1984) demonstrated that dissolved NH_4^+ distribution in marine sediments is substantially affected by reversible sorption, and presumably the physical perturbations caused by flooding water are sufficient to affect this process. In the case of PO_4^{3-}, the transport was again smaller in February, which agrees with the seasonal variation of phosphate association with the particles, that the physical perturbation of the sediment did not modify.

In summary, this study showed that: the NH_4^+ flux is production-rate controlled, while PO_4^{3-} flux is slowed down by chemical interactions with solids; NH_4^+ is released preferentially to PO_4^{3-}; but the stoichiometry of decomposition/transport varies seasonally; flooding of intertidal sediments removes large quantities of NH_4^+ and PO_4^{3-}, but only NH_4^+ release is independent of season; on a yearly basis, the amounts of NH_4^+ and PO_4^{3-} released from the intertidal area of Ria Formosa are two orders of magnitude higher than quantities released from subtidal areas. They are, apparently, capable of supplying most of the daily N and P requirements of phytoplankton in the overlying water.

References

Aller, R. C. & Y. Yingst, 1985. Effect of the marine deposit feeders *Heteromastus filiformis* (Polychaeta), *Macoma baltica* (Bivalvia) and *Tellina texana* (Bivalvia) on averaged sedimentary solute transport, reaction rates and microbial distributions. J. mar. Res. 43: 615–645.

Aller, R. C., 1980. Diagenetic processes near the sediment- water interface of Long Island Sound. Decomposition and nutrient element geochemistry (S, N, P). Adv. Geophys. 22: 237–350.

Berner, R. A., 1980. Early diagenesis – A Theorical Approach. University press, Princeton Series in Geochemistry: 241 pp.

Bettencourt, P., 1994. Les environnements sedimentaires de la côte sotavento (Algarve, sud Portugal) et leur évolution holocéne et actuelle.Thèse en Geologie Marine. Université de Bordeaux I: 494 pp.

Caetano, M., M. Falcão, C. Vale & M. J. Bebianno, 1997. Tidal flushing of ammonium, iron and manganese from inter-tidal sediment pore waters. Mar. Chem. 1475 c, 9 pp.

Caetano M., M. Falcão, C. Vale, & M. J. Bebianno, (1996). Tidal flushing of ammonium, iron and manganese from inter-tidal sediment pore waters. Marine Chemistry (in press).

De Jonge, V. N., M. M. Engelkes & J. F. Bakker, 1993. Bioavailability of phosphorus in sediments of the western Dutch Wadden Sea. Hydrobiologia 253: 151–163.

De Jonge, V. N. & J. E. E. Van Beusekom, 1995. Wind and tide induced resuspension of sediment and microphybenthos from tidal flats in the Ems estuary. Limnol. Oceanogr. 40: 766–778.

Falcão, M. & C. Vale, 1990. Study of the Ria Formosa ecosystem: benthic nutrient remineralization and tidal variability of nutrients in the water. Hydrobiologia 207: 137–146.

Falcão, M. & C. Vale, 1990a. Fluxos de nutrientes em viveiros de ameijoa da Ria Formosa. VI Simpósio Ibérico de Estudo do Bentos Mar. Bent. 6: 145–152.

Falcão, M. & C. Vale, 1995. Tidal flushing of ammonium from intertidal sediments of Ria Formosa, Portugal. Neth. J. aquat. Ecol. 29: 239–244.

Falcão, M., J. Pissarra & H. Cavaco, 1992. Características quimico-biológicas da Ria Formosa: Análise de um ciclo anual. Bol. INIP n° 16: 5–21.

Fanning K. A., Carder K. L & P. R. Betzer, 1982. Sediment resuspension by coastal waters: a potential mechanism for nutrient recycling in the ocean's margins. Deep-Sea Res. 29 (8A): 953–965.

Forja, J. M., J. Blasco & Gómez-Parra, 1994. Spatial and seasonal variation of *in situ* benthic fluxes in the Bay of Cadiz. (South-west Spain). Est. Coast. Shelf Sci. 39: 127–141.

Fox, L. E., 1993. The chemistry of aquatic phosphate: inorganic processes in rivers. Hydrobiologia 253: 1–16.

Gardner, W. S., S. P. Seitzinger & J. M. Malczyk, 1991. The effects of sea salts on the forms of nitrogen released from estuarine and freshwater sediments: does ion pairing affect ammonium flux? Estuaries 14: 157–166.

Grasshoff, K., 1976. Methods of Seawater Analysis. Verlag Chemie-New York, 317 pp.

Kennedy, R., H., W. F. James, R. H. Montgomery & J. Nix, 1984. The influence of sediments on the nutrient status of DeGray Lake, Arkansas. Sediment and water interactions. Peter G. Sly Springer-Verlag. Proceedings of the third international symposium on interaction between sediments and water, 517 pp.

Kerner, M. & K. Wallmann, 1992. Remobilization events involving Cd and Zn from intertdal flats sediments in the Elbe Estuary during the tidal cycle. Est. Coast. Shelf Sci. 35: 371–393.

Kerner, M. & K. Wallmann, 1992. Remobilization events involving Cd and Zn from intertdal flats sediments in the Elbe Estuary during the tidal cycle. Est. Coast. Shelf Sci. 35: 371–393.

Klump, J., V. & C. S. Martens, 1981. Biogeochemical cycling in an organic-rich coastal marine basin. II. Nutrient sediment-water exchange processes. Geochim. Cosmochim. Acta 45: 101–121.

Kristensen, E., 1993. Seasonal variations in benthic community metabolism and nitrogen dynamics in a shallow, organic-poor Danish lagoon. Est. Coast. Shelf Sci. 36: 565–586.

Krom, M. D. & R. A. Berner., 1981. The diagenesis of phosphorus in a nearshore marine sediment. Geochim. Cosmochim. Acta 45: 207–216.

Lerat, Y., P. Lasserre & P. Corre, 1990. Seasonal changes in pore water concentrations of nutrients and their diffusive fluxes at the sediment-water interface. J. exp. mar. Biol. Ecol. 135: 135–160.

Lohse, L., J. F. P. Malschaert, C. P. Slomp, W. Helder & W. Van Raaphorst, 1993. Nitrogen cycling in North Sea sediments: interaction of denitrification and nitrification in offshore and coastal areas. Mar. Ecol. Prog. Ser. 101: 283–296.

Lomstein, B. A., T. H. Blackburn & K. Henriksen, 1989. Aspects of nitrogen and carbon cycling in the northen Bering Shelf sediments. I. The significance of urea turnover in the mineralization of NH_4^+. Mar. Ecol. Prog. Ser. 57: 237–247.

Mackin, J. E. & R. C. Aller, 1984. Ammonium adsorption in marine sediments. Limnol. Oceanogr. 29: 250–257.

Nian-Zhi, J. & R. Wang, 1994. Ammonium uptake and regeneration fluxes of the microplankton communities in Jiaozhou Bay. Chin. J. Oceanol. Limnol. 12: 163–174.

Nixon, S. W., 1981. Remineralization and nutrient cycling in coastal marine ecosystems. Estuaries and Nutrients. Humana Press, Clifton, New Jersey: 111–138.

Nowicki, B. L. & S. W. Nixon, 1985. Benthic community metabolism in a coastal lagoon ecosystem. Mar. Ecol. Prog. Ser. 22: 21–30.

Nowicki, B. L., 1994. The effect of temperature, oxygen, salinity and nutrient enrichment on estuarine denitrification rates measured with a modified nitrogen gas flux technique. Est. Coast. Shelf Sci. 38: 137–156.

Ohtake, H., S. Kato & Y. Date, 1984. The role of tides in removing phosphorus from Lake Nakanoumi, a coastal lagoon in Japan. Arch. Hydrobiol. 99: 443–462.

Prastka, K. e. & T. D. Jickells, 1995. Sediment geochemistry of phosphorus at two intertidal sites on the great ouse estuary, S. E. England. Neth. J. aquat. Ecol. 29: 245–255.

Pregnall, A. M. & S. L. Miller, 1988. Flux of ammonium from surf-zone and nearshore sediments in Nahant Bay, Massachusetts, USA in relation to free-living Pilayella littoralis. Mar. Ecol. Prog. Ser. 50: 161–167.

Rocha, C. Rhythmic ammonium regeneration and flushing in intertidal sediments of warm climates. Limnology and Oceanography (in press).

Rocha, C., G. Cabeçadas & M. J. Broguera, 1995. On the importance of sediment-water exchange processes of ammonia to primary production in shallow areas of the Sado estuary (Portugal). Neth. J. aquat. Ecol. 29: 265–273.

Scherrer, B., 1984. Biostatistique. Gaëtan Morin. Canada-Québec, 850 pp.

Shum K. T. & B. Sundby, 1996. organic matter processing in continental shelf sediments – the subtidal pump revisited. Mar. Chem. 53: 81–87.

Steele, J. H., A. L. S. Munro & G. S. Giese, 1970. Environmental factors controlling the epipsammic flora on beach and sublitoral sands. J. Mar. Biol. Assoc.

Stumm, W. & J. D. Leckie, 1970. Phosphate exchange with sediments: Its role in the productivity of surface waters. Adv. Wat. Poll. Res. 2: 1–26.

Sundby, B., C. Gobeil, N. Siverberg & A. Mucci, 1992. The phosphorus cycle in coastal marine sediments. Linmol. Oceanogr. 37: 1129–1145.

Vale, C., J. Hussenot, M. Falcão & J-L. Martin, 1992. Dawn-dusk chemical variations near the sediment-water interface of a clam-growth bank of Ria Formosa. Bol. INIP, n° 19: 71–79.

Van Noort, 1992. Nitrogen cycling in two types of sediments of the southern field data and mesocosm results. Neth. J. Sea Res. 28: 293–316.

Van Raaphorst, W. & H. T. Kloosterhuis, 1994. Phosphate sorption in superficial intertidal sediments. Mar. Chem. 48: 1–16.

Van Raaphorst, W., H. T. Kloosterhuis, E. M. Berghuis, A. J. M. Gieles, J. F. P. Malschaert & G. J. Van Raaphorst, W., P. Ruardij & A. G. Brinkman, 1988. The assessment of benthic phosphorus regeneration in a marine ecosystem model. Neth. J. Sea Res. 22: 23–36.

Vidal, M. & J. A. Morguí, 1995. Short-term pore water ammonium variability coupled to benthic boundary layer dynamics in Alfacs bay, Spain (Ebro Delta, NW Mediterranean). Mar. Ecol. Prog. Ser. 118: 229–236.

Yamada, H. & M. Kayama, 1987. Liberation of nitrogenous componds from bottom sediments and effect of bioturbation by small bivalve Theora lata (Hinds). Est. Coast. Shelf. Sci. 24: 539–555.

Yentsch, C. S. & D. W. Menzel, 1963. A method for determination of phytoplankton chlorophyll and phaeophytin by fluorescence. Deep-Sea Res. 10: 221–251.

Yurkovskis, A., F. Wulff, L. Rahm, A. Andruzaitis & M. Rodriguez-Medina, 1993. A nutrient budget of Gulf of Riga; Baltic Sea. Est. Coast. Shelf Sci. 37: 113–127.

Zwolsman, J. G., 1994. Seasonal variability and biogeochemistry of phosphorus in the Scheldt Estuary, South-west Netherlands. Est. Coast. Shelf Sci. 39: 227–248.

Hydrobiologia **373/374:** 203–216, 1998.
J.-C. Amiard, B. Le Rouzic, B. Berthet & G. Bertru (eds),
Oceans, Rivers and Lakes: Energy and Substance Transfers at Interfaces.

Phosphate mobility at the sediment–water interface of a Mediterranean lagoon (etang du Méjean), seasonal phosphate variation

E. Gomez[1], M. Fillit[2], M. C. Ximenes[2] & B. Picot[1]
[1] *Département Sciences de l'Environnement et Santé Publique, UMR CNRS 5556, Ecosystèmes lagunaires, Faculté de Pharmacie. 15 Av. C. Flahault, 34060 Montpellier, France*
[2] *CEMAGREF, BP 5095, Montpellier, France*

Key words: Phosphate, sediments, coastal lagoon, seasonal variations, F fractions

Abstract

The Méjean Lagoon (France) is a confined, shallow system, 0.7 m in depth, with a surface area of 747 ha. Its sediments have a fine granulometry and are evenly distributed. The bathymetry and hydrodynamic behaviour of the basin create two zones with restricted exchanges between them.

The western part (60% of the total surface area) is rich in dissolved phosphate (> 1 mg l^{-1} in summer) and receives the majority of the phosphate (P) inputs from the watershed. The heavy macroalgal population consists of *Gracilaria* throughout the year and *Ulva* in summer and autumn. Phytoplankton levels sometimes reach 100 μg l^{-1} of chlorophyll *a*.

The sediment-water phosphate exchange varies with environmental conditions and macroalgal growth. P mobility was studied on the basis of these characteristics.

Algal biomass and water characteristics were measured at 15-day intervals throughout 1993. A seasonal study was conducted in 1994 to investigate which P fractions in the sediment were involved in the exchanges. The spring-time rise in temperature reactivated decomposition of the macroalgal biomass that had accumulated in autumn. The redox potential (E_H) fell as a result of this biological activity, leading to a decrease in the inorganic P fractions.

This P release accounts for the higher P water concentrations observed in the lagoon in summer.

The mobility of P fractions shows that the P stored in the sediments plays an active role in the dynamics of the overlying water. Seasonal variations in these fractions explain the patterns of P storage and mobilization.

Introduction

As human activity in the watersheds increases, an increasing amount of waste from the continental land mass enters the sea. Coastal lagoons act as an intermediate recipient, protecting the marine environment. They act as a sink for the pollutants. When large amounts of nutrients (N and P) enter these stretches of water, repeated algal blooms are observed and the lagoons become eutrophic.

Coastal lagoons occupy a significant part (13%) of the world's shorelines (Nixon, 1982). These shallow ecosystems are usually very productive and play an important role in the economic development of the coastal zones. The decomposition of the large quantities of algae can produce anoxic conditions in the water column. The oxygen depletion results in widespread mortality in the fish and farmed shellfish populations. Massive algal growth has been observed in various different eutrophic bays and lagoons: Saint Brieuc Bay, the Po River Delta, the Venice Lagoon, the Salses-Leucate Lagoon, the Prévost Lagoon, etc. (Lieutaud et al., 1992; Pirou & Menesguen, 1992; Viaroli et al, 1991; Sfriso et al., 1987; Boutière et al., 1982).

To understand the internal functioning of these ecosystems, two sites where nutrients are preserved can be taken into consideration: the sediment and the algae (macrophytes and phytoplankton).

Table 1. General characteristics of the environment

Location	French Mediterranean coast (Languedoc Roussillon)
Watershed name	Lez-Mosson
Watershed area (km^2)	574
Inhabitant/km^2 (watershed)	588
Total area (ha)	747
Average depth (m)	0.7
Water salinity (‰)	11–36
P input (t/year)	12

Algal growth varies during the year. Seasonal changes in species composition, biomass and the production of macroalgal communities in coastal waters depend on the interaction of physical, chemical and biological factors (Thom & Albright, 1990). There is a close link between the growth and decomposition of *Ulva* in particular and phosphate and nitrogen processes (Viaroli et al., 1991). Phytoplankton appears in short successive blooms when input nutrients are high and there is no competition from macrophytes (Vaulot & Frisoni, 1986).

Sediment plays a major role in the coastal ecosystems because of the lagoons' shallowness, which means that it mixes readily with the overlying water. It can act either as a sink or as a source of phosphate. Sedimentary phosphate dynamics involve a large number of mechanisms. Different approaches have been adopted to study phosphate dynamics in sediment (López Laseras, 1991): the relationships between the different descriptors and the established vertical and horizontal variations, phosphate release rates and the factors controlling them, and studies of the chemical fractions of phosphate in sediments.

This study was carried out for the Programme National d'Océanographie Cotière (PNOC) with the aim of understanding the nature of the eutrophication problem before restorative action is taken. The Méjean Lagoon (Figure 1), one of the shallow coastal lagoons bordering the Mediterranean, was chosen.

The study attempts to identify mechanisms explaining sedimentary phosphate dynamics in terms of seasonal variations in algal growth. This necessitated simultaneous monitoring of the sedimentary phosphate fractions and the water conditions.

Materials and methods

Site

The study was carried out using the sediment of the Méjean Lagoon, a Mediterranean coastal lagoon in the South of France which receives pollution mainly from urban waste water. The lagoon has a total surface area of 747 ha and a mean depth of < 1 m. Variable inputs of fresh water (with rainfall) and sea water cause the water's salinity to range from 9 to 31‰ in the course of a year. The rate of water renewal is very low in dry periods (0.01% per day of the lagoon's volume) but can reach significantly higher values in periods of rainfall. The general characteristics of the lagoon are summarized in Table 1.

Two different sampling points, point 1 in the eastern (Pérols) part and point 2 in the western (Méjean) part, were selected for the removal of sediment samples. The treated waste water from the Montpellier region (300 000 habitants) is disposed of in the Méjean area. The Pérols area does not receive any waste directly from the treatment plant. The hydrodynamics of the lagoon create little mixing of the water masses from the Méjean area and the Pérols area. The choice is the result of the spatial variation study of Gomez (1996) which showed the existence of two different parts in this lagoon.

The periods chosen for sediment sampling reflect the different phases of algal growth.

Sampling techniques

Sediments and pore water were sampled at the same time as the flux measurements, at both points in April, June, August and November 1994. The chemical extractions of phosphate, as well as determination of pH and E_H profiles were carried out in February 1994, in order to have a winter reference value for the sediment.

Samples from undisturbed sediments were taken with a manual corer (Rofés & Savary, 1981). The cores were cut into 5 cm slices (0–5 cm and 5–10 cm), stored in a plastic flask, kept at 4 °C and treated the same day. 5 cm slices were chosen due to the high proportion of shells in the sediments which means thinner slices would be difficult to cut and rather selective. However, the measurement of pH and E_H can be carried out *in situ* at a higher resolution by using microelectrodes placed directly in the sediments. Sediment characteristics were determined after sieving with a 2 mm sieve.

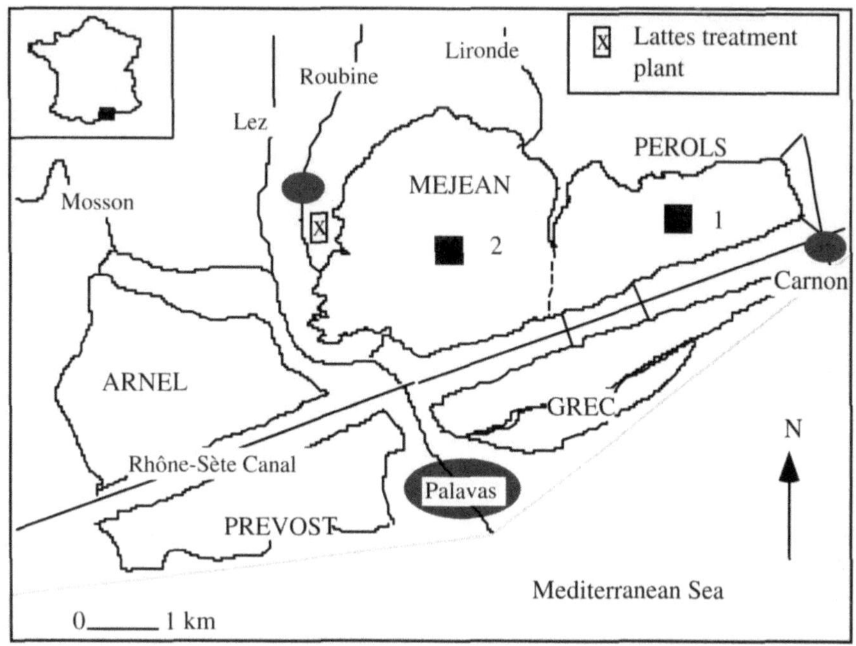

Figure 1. Location of the sampling points in the Méjean Lagoon (France).

The removal of the sediments for extraction of pore waters was carried out by coring in the absence of air, in order to avoid any contact with oxygen.These sediments were then placed in a glove box containing nitrogen. The porewaters were sampled by the press method (Reeburgh, 1967), using the pressure of nitrogen (Mesnage et al., 1993).

Water characteristics

pH, temperature and O_2 were measured in the field directly after sampling. pH was measured with a WTW pH meter and redox potential with an Ingold Pt 4800-M5 electrode. Results were corrected for the standard-H_2 electrode (Bühler & Galster, 1980). O_2 was measured with a combined oxygen-temperature WTW92 meter. Soluble reactive phosphate (SRP) was measured using the standard method (Standard Methods, 1992). Each macroalgal sample was collected over an area of 0.25 m^2. The samples were sieved, sorted into species, and the dry weight of each species was measured after drying at 110 °C for 24 h. Biomass was measured in grams dry weight per square metre (g dw m^{-2}). Phytoplankton was measured as $\mu g\ l^{-1}$ of chlorophyll *a* (acetone extraction and spectrofluorimetry).

Sediment characteristics

One part of the sediment was freeze-dried to determine total organic carbon, total nitrogen, and $CaCO_3$ content. Another part was used to determine porosity ($Ø = Vol_{water}/Vol_{total}$) and water content ($H$ = water weight/total weight of wet sediment).

Granulometric distribution in the sediment samples was determined by laser granulometry according to Duchaufour's pedologic nomenclature (1988).

Total phosphate was analyzed, after persulphate mineralization, as o-P and SRP, according to the Standard Methods (1992): ammonium molybdate and potassium antimonyl tartrate react in acid medium with orthophosphate to form a heteropolyacid-phosphomolybdic acid, that is reduced to intensely coloured molybdenum blue by ascorbic acid. Total iron was determined after Golterman et al. (1978). Phosphate fractionation was carried out using the Golterman & Booman (1988)/De Groot & Golterman (1990) method, which was appropriate for our sediments (Mesnage & Picot, 1993). After a water extraction of weakly adsorbed phosphate (o-P), chelating agents such as NTA or EDTA were used to extract iron-bound and calcium-bound phosphate (Fe(OOH)-P and $CaCO_3$-P). H_2SO_4 0.5M was used to extract acid-soluble organic phosphate (ASOP) and persulphate digestion to extract residual organic phosphate (ROP).

pH profiles and redox potential were measured *in situ*, with the electrode introduced into the sediment every 2 cm in depth.

Statistical analysis

Statistical analysis (Anova, Statview 4) was carried out to evaluate spatial and temporal variations in phosphate concentration in sediments, with three replicates per measurement.

Abbreviations:

SRP: Soluble Reactive Phosphate in water
o-P: Sediment adsorbed phosphate
Fe(OOH)-P: Ferric hydroxide-bound phosphate
$CaCO_3$-P: Calcium carbonate-bound phosphate
ASOP: Acid-soluble organic phosphate
ROP: Residual organic phosphate
Tot-P: Total P in sediment
Tot-N: Kjeldhahl N (including NH3)
Tot-C: Total carbon;
Corg: Organic carbon;
Cmineral; mineral carbon.

Results

Water column

Water column characteristics are shown in Table 2. The temperatures recorded varied between 7.7 °C in February and 28.2 °C in June. The overlying water was always sursaturated in dissolved oxygen (measured near the surface) in the Méjean part and quite high (at least 77%) in the Pérols part. The highest values were recorded in August. The redox potential values were always over 400 mV, in accordance with the continual presence of oxygen in the water at the time of sampling. These results are a indication of continual photosynthesis activity. More frequent monitoring in 1993 showed brief periods of anoxia in early April, late June and mid-August, always in the Méjean part (Fillit, 1995).

The mean pH of the water was 8.5 ± 0.3 in the Pérols part and 8.6 ± 0.4 in the Méjean part. Salinity varied between 4.2‰ (Pérols) in November and 27‰ (Pérols) in June. The lowest values coincided with periods of heavy rainfall (February, October and November), while high salinity coincided with periods of high evaporation (June and August).

The graph of SRP over an annual cycle (Figure 2) shows higher concentrations over the summer period (June to October). The lagoon appears to be divided into two parts, with higher concentrations in the Méjean than in the Pérols part.

The SRP for the seasonal study varied from close to zero to 5.16 μM (or ≈ 0 to 160 $\mu g\ l^{-1}$) in Pérols and to 29.68 μM (or ≈ 0 to 920 $\mu g\ l^{-1}$) in Méjean. As in the preceding year, the values were highest in summer.

The changes in macroalgal and phytoplankton biomass over an annual cycle (Fillit, 1995) show differences in growth between the two sectors of the lagoon. The P stored in phytoplankton was calculated using measured chlorophyl *a* values, to which the following conversions were applied: carbon/chlorophyl *a* = 60 (Strickland, 1960), carbon/nitrogen = 5.5 (Goldman et Mann, 1980), nitrogen/phosphorus = 16.7 (Harris et Riley, 1956). The P stored in macroalgae was calculated using the biomass and phosphate content measured at each sampling (Fillit, 1995).

– The eastern sector (or Pérols part) is characterized by strong phytoplankton growth. The P stored in the phytoplankton thus represents about 0.5 tonnes of P throughout the year, except for the June to September period when it is very slight. The macroalgae in this sector develop from July onwards and their biomass represents, at the maximum, 1.4 tonnes of P in October. This level then declines.

– The western sector (or Méjean part) is the area where most macroalgal growth occurred. Only extremely small amounts of phytoplankton are present while the macroalgae develop to a massive extent between April and October. Their biomass reaches a maximum value of 4.5 tonnes of P.

The western sector can consequently be regarded as an area of macroalgal growth. The eastern sector, on the other hand, can be regarded as an area of phytoplankton growth.

The seasonal survey of algal growth in 1994 (Table 2) showed the continual presence of phytoplankton for the Pérols part (34.8 to 144.7 $\mu g\ l^{-1}$) and later macroalgal growth. The latter were present in minimal quantities except in November (5.3 g m^{-2}). Macroalgal growth also occurred later than in the previous year in the Méjean part, from June to November (between 12.4 and 302.4 g $m^{-2)}$. Phytoplankton was present in lower concentrations than in the Pérols part in the four samples taken at this point (2.5 to 106.5 $\mu g\ l^{-1}$). The

Table 2. Water characteristics at time of sampling during 1994 seasonal survey: (×) below the detection limit, (–) not determined.

Méjean part					
Month	February	April	June	August	November
Temperature (°C)	8.5	11.2	27.7	24.6	12.5
O_2 (% saturation)	104	95	93	126	77
pH	8.6	–	8.9	8.0	8.9
E_H (mV)			418	421	454
Salinity (g l^{-1})	12.5	17.5	23.0	26.0	4.9
Conduct. (mS cm^{-1})	–	–	36.6	47.5	70.2
o-P (μM)	0.32	3.23	20.00	29.68	×
Chl *a* (μg l^{-1})	–	6.1	2.5	24.6	107
Algae (g m^{-2})	–	×	12	302	147
Perols part					
Month	February	April	June	August	November
Temperature (°C)	7.7	11.0	28.2	25.5	14.3
O_2 (% saturation)	134	111	140	142	123
pH	8.1	8.2	8.7	8.5	8.8
E_H (mV)	667		415	440	463
Salinity (g l^{-1})	8.9	18	27	26.5	4.2
Conduct. (mS cm^{-1})	–	–	39.7	48.5	58.5
o-P (μM)	0.65	1.61	5.16	4.84	×
Chl *a* (μg l^{-1})	–	98	35	145	60
Algae (g m^{-2})	–	×	×	×	5.3

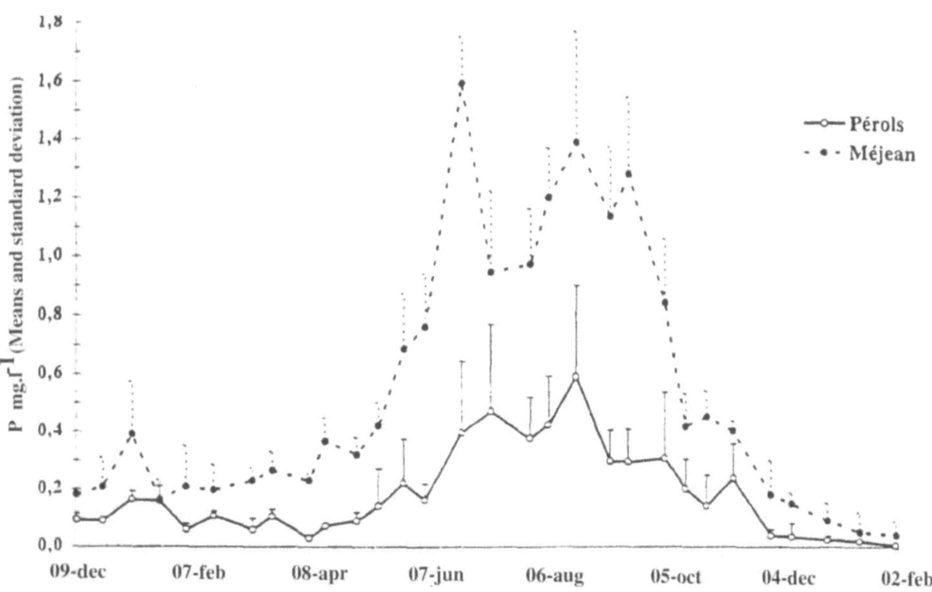

Figure 2. Changes in phosphate concentrations (SRP) in the Méjean Lagoon in 1993 (Fillit, 1995). Means and standard deviations in Méjean and Pérols parths from December to February.

Table 3. Sediment characteristics

0–5 cm	PEROLS part					MEJEAN part				
month	April	June	August	November	mean	April	June	August	November	mean
Silt < 50 μm	–	68	67	68	68 ± 0.6	–	75	77	82	78 ± 4
Fine sand 50–200 μm	–	9	9	8	9 ± 0.6	–	7	7	7	7
Coarse sand 200 μm–2 mm	–	23	24	24	24 ± 0.6	–	18	16	11	15 ± 4
Water content (%)	64	62	62	61	62 ± 1.3	65	57	62	72	64 ± 6
Porosity (%)	84	78	79	81	80 ± 2.6	81	71	80	87	80 ± 7
$CaCO_3$ (%)	–	57	56	56	56 ± 0.6	–	58	53	48	53 ± 5
Tot-P $\mu g\ g^{-1}$	511 ± 25	610 ± 22	580 ± 10	512 ± 44	553 ± 50	701 ± 45	603 ± 24	714 ± 35	901 ± 7	741 ± 110
Norg mg g^{-1}	4.2	4.2	4.2	4.3	4.2 ± 0.05	5.1	5.6	4.8	5.9	5.3 ± 0.5
Tot-C mg g^{-1}	92.3	95.5	94.0	99.6	95 ± 3	93.0	92.5	94.0	98.0	94 ± 2
Cmineral mg g^{-1}	31.3	38.8	45.2	57.5	43 ± 11	27.4	32.2	39.0	45.8	36 ± 8
Corg mg g^{-1}	61.0	56.7	48.8	42.1	52 ± 8	65.6	60.7	55.0	52.2	58 ± 6
Corg/Norg	14.5	13.5	11.6	9.8	12 ± 2	12.9	10.8	11.5	8.9	11 ± 2
5–10 cm	**PEROLS part**					**MEJEAN part**				
month	April	June	August	November	mean	April	June	August	November	mean
Silt < 50 μm	–	72	68	83	74 ± 8	–	73	73	72	73 ± 0.6
Fine sand 50-200 μm	–	7	12	3	7 ± 4	–	7	5	6	6 ± 1
Coarse sand 200 μm–2 mm	–	21	20	14	18 ± 4	–	20	22	22	21 ± 1
Water content (%)	62	62	60	61	61 ± 1	62	61	58	60	60 ± 2
Porosity (%)	82	79	81	79	80 ± 10	83	77	77	75	78 ± 3
$CaCO_3$ %	–	54	55	48	52 ± 4	–	62	59	58	60 ± 2
Tot-P $\mu g\ g^{-1}$	382 ± 4	388 ± 3	350 ± 31	319 ± 24	360 ± 32	543 ± 15	478 ± 31	471 ± 10	494 ± 12	496 ± 32
Norg mg g^{-1}	3.9	3.7	3.9	4.2	3.9 ± 0.2	3.6	4.4	3.5	3.8	3.8 ± 0.4
Tot-C mg g^{-1}	82.0	91.8	92.2	89.4	88 ± 5	92.0	94.0	94.7	98.0	95 ± 2
Cmineral mg g^{-1}	16.6	35.0	39.9	43.2	34 ± 11	36.9	34.6	49.1	55.1	44 ± 10
Corg mg g^{-1}	69.4	56.8	52.3	46.2	56 ± 9	55.1	59.4	38.3	42.9	49 ± 10
Corg/Norg	17.8	15.4	13.4	11.0	14 ± 3	15.3	13.5	11.0	11.3	13 ± 2

values obtained for these four periods showed a trend similar to that observed the previous year.

Sediment characteristics

The main sediment characteristics are shown in Table 3.

The sediment was very fine, with over two thirds in the < 63 μm fraction (clay and silt). The < 63 μm fraction represented 79% at Méjean and 68% at Pérols for the 0–5 cm layer; 74% of the sediment in the 5–10 cm layer consisted of this fraction at both sampling points. The water content (62% of wet weight for point 1 and 64% for point 2) and the porosity (80%) of the sediment in the 0–5 cm layer were fairly constant throughout the year. Values were of the same order for the lower layer (5–10 cm). The sediments were calcareous (carbonate calcium accounts for 55% of dry weight on average).

Most of the sediment's total carbon was organic. There were higher levels of organic carbon in the Méjean part (65.6 to 52.2 mg g^{-1}) than in the Pérols part (61.0 to 42.1 mg g^{-1}). This was due to the anthropic input (quite considerable in this part of the lagoon) and the macroalgal biomass that developed at this sampling point. The decrease observed in organic carbon levels from April to November in both sectors of the lagoon indicates that a mineralization of organic matter was enhanced by an increase in temperature. There was little variation in the levels of organic nitrogen.

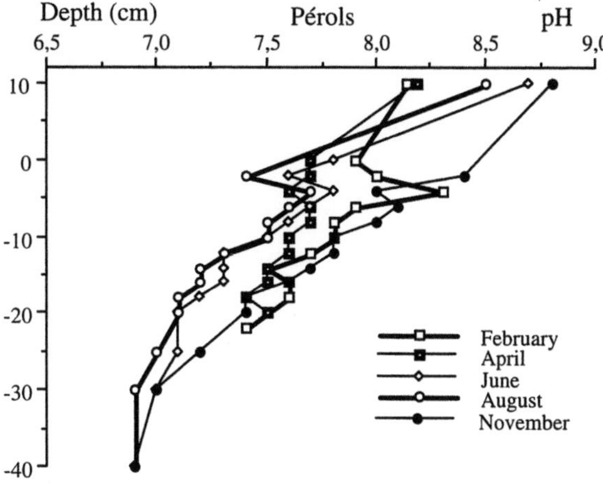

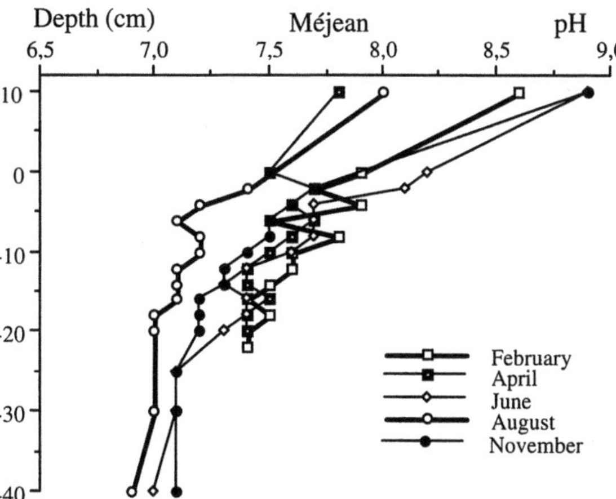

Figure 3. Seasonal variation in the pH of the sediment and the overlying water for the two sampling points studied.

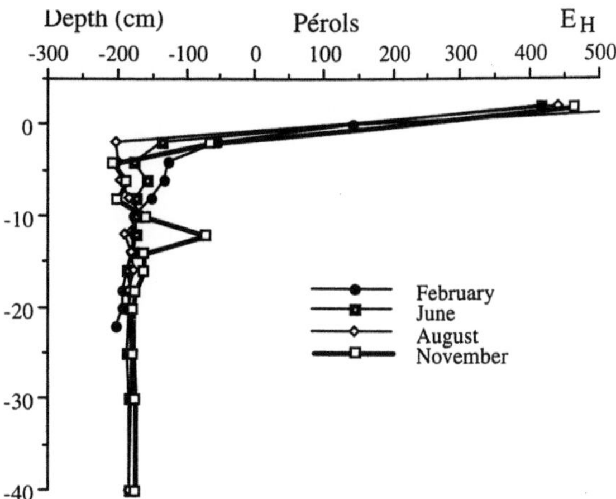

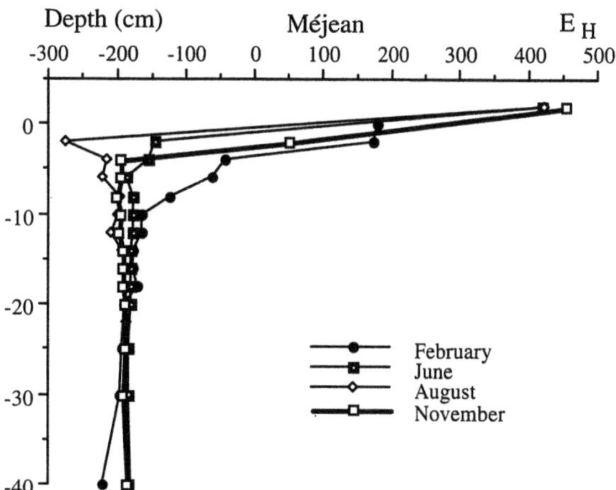

Figure 4. Seasonal variation in the E_H of the sediment and the overlying water for the two sampling points studied.

which decreased with depth: 4.2 to 5.9 mg g^{-1} at the surface and 3.5 to 4.4 mg g^{-1} at depth. As nitrogen levels were fairly constant, the Corg/Norg ratio (between 11 and 17) decreased from April to November as a result of the reduction in organic carbon.

The pH values (Figure 3) of the sediment were lower than those of the water in all seasons. Sediment pH decreased with depth, ranging from 7.7 ± 0.2 (Pérols) and 7.8 ± 0.4 (Méjean) at –2 cm to 7.3 ± 0.2 for both sampling points at –20 cm. The pH profile for August showed the lowest pH values at both sampling points.

The redox potential (Figure 4) in the Méjean Lagoon only varied over the first few centimeters of sediment. After the first two centimeters, it was between –150 and –200 mV, irrespective of season. The E_H was positive at the surface in February, April and November, indicating oxic conditions at the interface. This was not the case in June and August, when the sediment's anoxic state made it difficult to measure the E_H at the interface. The lowest values for the year at –2 cm (in August: –203 mV for the Pérols part and –276 mV for the Méjean part) were recorded during the hottest months and provide information about the state of the interface.

Decomposition of algae led to a major deposition of organic matter on the sediment, and the carbon content was always over 40 mg g^{-1} in the first 10 centimeters. This major deposit of organic matter mineralizing in the sediment led to a reduction in pH with depth and continuing very low redox potential values.

Total phosphate and P fractions

The sediments had a higher P load (Table 3) in the Méjean (603 ± 24 to 901 ± 7 $\mu g\ g^{-1}$) than in the Pérols part (511 ± 25 to 610 ± 22 $\mu g\ g^{-1}$). The values for the 5–10 cm layer were 30 to 40% lower than those at the surface (471 ± 10 to 543 ± 15 $\mu g\ g^{-1}$ for the Méjean part and 319 ± 24 to 388 ± 3 $\mu g\ g^{-1}$ for the Pérols part).

Seasonal changes in total P were quite marked at the surface (0–5 cm) and different for the two sectors studied: the maximum values occurred in the Méjean part in spring and in the Pérols part in winter. The variations in the 5–10 cm layer are not significant.

Sedimentary phosphate fraction data is given in Tables 4 (content) and 5 (percentages of total P).

The o-P fraction represented at most 2.7% of total P. Maximum values were observed in November.

The ferric hydroxide-bound phosphate fraction (Fe(OOH)-P), which represented between 7 and 31% of total P, behaved differently in each sector of the lagoon. The Fe(OOH)-P value of the sediment from the eastern (Pérols) part of the lagoon was highest in August (103 ± 2) and lowest in November (67 ± 6) and February (47 ± 2). In the Méjean part, it was highest in April (216 ± 14) and lowest during the hot period: June (112 ± 20) and August (92 ± 2).

The calcium-bound phosphate fraction ($CaCO_3$-P) was the largest inorganic fraction. The highest values were observed in the same season in both parts of the lagoon, in February for the surface sediments (260 $\mu g\ g^{-1}$ Pérols, 418 $\mu g\ g^{-1}$ Méjean) and in April for the sediments of the 5–10 cm layer (188 $\mu g\ g^{-1}$ Pérols, 275 $\mu g\ g^{-1}$ Méjean). These values were not measured in February. The lowest values appeared at different seasons in the two sectors of the lagoon. They were measured in November for the Pérols part (169 $\mu g\ g^{-1}$ at the surface and 106 $\mu g\ g^{-1}$ at 5–10 cm) and in June for the Méjean part (205 $\mu g\ g^{-1}$ at the surface and 141 $\mu g\ g^{-1}$ at 5–10 cm).

The acid-soluble organic P fraction (ASOP) represented between 7 and 18% of total sedimentary P. It was the only fraction whose behaviour was independent of the sector concerned: lowest in spring (April) and highest in summer (June–August) regardless of the depth.

The measured residual organic phosphate fraction (ROP) represented between 28 and 56% of total sedimentary P. Like the calcium-bound fraction, it was a major phosphate fraction in the sediment. The highest levels for the Pérols part were observed in June (257 $\mu g\ g^{-1}$ at the surface, 161 $\mu g\ g^{-1}$ at 5–10 cm). The highest levels for the Méjean part were observed in November at the surface(393 $\mu g\ g^{-1}$) and in June at 5–10 cm (197 $\mu g\ g^{-1}$).

Discussion

Analysis of the exchange situation during the main algal growth phases

The overall results indicate an influence of the inputs from the watershed, which were partly responsible for the high dissolved P values in the Méjean part. The mean net input (inflow–outflow) was 870 ± 160 kg of P per month (Burtin 1994 in Fillit 1995). Virtually all of it came from the tailings of the neighbouring treatment plants (Lattes – population 6 500 and Montpellier – population 330 000). Under 10% of inputs are of more distant origin (upstream Lez, Mosson).

Four phosphate exchange situations were observed in the west (Méjean part) of the lagoon, where macroalgal growth is dominant.

In April, when there was an absence of phytoplankton and macroalgae in the water column (Table 2), conditions were favourable to storage: continuous inputs from the watershed, no P requirement by the water column, an oxic layer on the surface of the sediment and an oxygenated environment. Storage behaviour was in agreement with the experiments conducted in the laboratory in which all the conditions mentioned were present. The same sediment, when in suspension (aproximately 5 g l^{-1} d.w.) in well-oxygenated water without algae, had adsorbed 300 μg of phosphate addition of 350 $\mu g\ g^{-1}$ (Gomez et al., 1998). Others sediments show the same behaviour (Furumai & Ohgaki, 1988; Istanovics, 1994). On the other hand, total P values (Table 3) show a major reduction in the sedimentary load: from 871 ± 20 $\mu g\ g^{-1}$ in February to 701 ± 45 $\mu g\ g^{-1}$ in April. This reduction should have led to an increase in the concentration of dissolved P in the water column. The concentration was, however, only 3.23 μm. There are only two possible explanations for the loss of sedimentary P: either re-suspension of the sediment (and of the P it contained) or its dilution. Re-suspension can occur because of wind. In 1993 this phenomenon led to a noticeable increase in suspended matter, which exceeded 0.46 g l^{-1} (Fillit, 1995). Dilution can be due to high rainfall, which is capable of doubling the initial depth of the water column. There was a rainy

Table 4. Means of phosphate fractions measured during seasonal survey (phosphate in $\mu g\ g^{-1}$) and standard deviations

$\mu g\ g^{-1}$	Month	o-P	Fe(OOH)-P	Fe(OOH)	CaCO3-P	ASOP	ROP	Fe/P
	February	4 ± 0.5	47 ± 2	891 ± 68	260 ± 15	33 ± 2	151 ± 3	19 ± 0.8
Pérols	April	2 ± 0.1	75 ± 10	2054 ± 68	247 ± 17	38 ± 7	148 ± 17	27 ± 2.8
	June	2 ± 0.1	85 ± 5	1193 ± 53	175 ± 3	91 ± 1	257 ± 21	14 ± 0.1
0–5 cm	August	4 ± 0.3	103 ± 2	1377 ± 54	192 ± 5	74 ± 1	208 ± 18	13 ± 0.2
	November	5 ± 0.3	67 ± 6	1153 ± 42	169 ± 8	70 ± 10	201 ± 24	17 ± 0.9
	April	2 ± 0.1	59 ± 4	1135 ± 26	188 ± 7	12 ± 3	121 ± 5	19 ± 0.9
Pérols	June	1 ± 0.7	44 ± 2	1284 ± 69	111 ± 11	70 ± 2	161 ± 13	30 ± 2.8
	August	2 ± 0.1	53 ± 4	677 ± 45	130 ± 5	54 ± 1	111 ± 32	13 ± 0.1
5–10 cm	November	7 ± 0.3	41 ± 1	477 ± 12	106 ± 8	49 ± 2	116 ± 16	12 ± 0.3
	February	8 ± 0.9	156 ± 10	1721 ± 56	418 ± 3	75 ± 15	214 ± 9	11 ± 0.8
Méjean	April	2 ± 0.6	216 ± 14	4058 ± 176	219 ± 24	61 ± 10	203 ± 38	19 ± 0.4
	June	1 ± 0.1	112 ± 20	2310 ± 3	205 ± 1	140 ± 22	144 ± 19	21 ± 3.7
0–5 cm	August	8 ± 0.2	92 ± 2	1693 ± 186	216 ± 2	121 ± 19	277 ± 13	18 ± 2.3
	November	35 ± 8	116 ± 5	1761 ± 131	256 ± 5	101 ± 15	393 ± 2	15 ± 0.5
	April	1 ± 0.3	101 ± 1	3151 ± 1	275 ± 213	20 ± 14	146 ± 4	31 ± 27
Méjean	June	3 ± 0.2	45 ± 1	516 ± 18	141 ± 24	92 ± 33	197 ± 21	11 ± 0.1
	August	3 ± 0.3	78 ± 1	1094 ± 29	165 ± 12	96 ± 3	128 ± 20	14 ± 0.5
5–10 cm	November	8 ± 0.1	74 ± 1	1103 ± 28	195 ± 9	62 ± 9	155 ± 17	15 ± 0.3

Table 5. Percentages of phosphate fractions measured during seasonal survey

%	Month	o-P	Fe(OOH)	CaCO3-P	ASOP	ROP
	February	0.8	10	52	7	31
Pérols	April	0.3	15	48	7	29
	June	0.2	10	20	10	42
0–5 cm	August	0.5	13	24	9	36
	November	0.7	9	24	10	56
	April	0.4	16	49	3	32
Pérols	June	0.2	8	20	13	42
	August	0.4	11	29	12	31
5–10 cm	November	1.5	10	24	11	36
	February	1.0	18	48	9	25
Méjean	April	0.2	31	31	9	29
	June	0.2	19	34	23	24
0–5 cm	August	1.1	13	30	17	39
	November	4.0	13	28	11	44
	April	0.2	19	51	4	27
Méjean	June	0.4	7	21	13	41
	August	0.5	13	28	16	27
5–10 cm	November	1.2	11	30	10	31

period in February 1994 which might have been responsible for the loss of considerable quantities of P from the lagoon. During a similar rainfall, 33% of the total water volume of the lagoon was renewed in less than six days. This would explain why concentration increased by only 2.90 μm during the period under consideration whereas it should have been higher in view of the inputs (from the watershed and release by the sediment).

In June, the algal biomass in the water was low (2.5 $\mu g\ l^{-1}$ of Chla and 12.4 g m^{-2} of macroalgal biomass) and the concentration of P in the water very high (20 μm). The dissolved P came from the sediment, and the load diminished between April and June from 701 $\pm$ 45 to 603 $\pm$ 24 $\mu g\ g^{-1}$, which was statistically significant. The increase in temperature compared to that of April was very considerable (from 11.2 to 27.7 °C). Biological activity should therefore have been stimulated both in the water, which had a pH of 8.9, and in the sediment. Phosphate release can be caused by mineralization of the organic matter by bacterial activity, as Hasson (1989) observed under anaerobic conditions. This release was increased by the establishment of reduced conditions at the water–sediment interface.

In August 1994, as in the previous year, algal growth was very heavy (302 g m^{-2} of macroalgal biomass) with much sunshine and high temperatures. Despite the increased sedimentary load, P release appears to have continued after June, as the concentration of dissolved P in the water (29.68 μm) increased. There was a simultaneous growth of macroalgae and degradation of part of the biomass during this period of algal growth. The observed increase in the sedimentary load was caused by the accumulation of dead biomass in the sediment.

In November, there was a reduction in macroalgal biomass compared to the August level, but it was still considerable (147 g m^{-2} of macroalgal biomass). Phytoplankton bloom was observed during this sampling period (106 $\mu g\ l^{-1}$ of Chl*a*). The heavy algal presence was responsible for the disappearance of P in the water (value below the detection limit). The fall in temperature from 24.6 to 12.5 °C slowed down biological activity in the sediment and enabled the sediment surface to return to oxic conditions, favouring storage. The load of sedimentary P increased again between August and November, due to the deposition of decomposing macroalgae.

The annual pattern of changes was completely different in the Pérols part.

The algal population consisted solely of phytoplankton growth (Table 2), with macroalgae present only when the November samples were taken and even then in small quantities (5.3 g m^{-2}). The rate of phytoplankton renewal was much more rapid that for macroalgae, which explains its appearance in the form of brief, successive blooms. Fillit (1995) had shown a continuous existence of phytoplankton in this part of the lagoon in 1993, with high chlorophyll *a* values in winter and low ones in the summer (September). In 1994 phytoplankton growth occupied the stretch of water, with a maximum level in August, in the absence of competition from macroalgae.

Seasonal changes in the load of sedimentary P were smaller and the reverse of those observed in the Méjean part. Only the increase observed between April and June and the reduction between August and November are significant. The P accumulating in the sediment between April and June could not have come from the macroalgae, which were absent in this sector of the lagoon, or from the inputs from the watershed, which remained fairly constant over time. The temperature increase led to a change to reduced conditions in the sediment from July onwards. The resulting release is indicated by the reduction in the sedimentary load between August and November.

Sediment

The high primary production of the Méjean Lagoon led to major inputs of organic matter in the sediment, which were reflected in low redox values, in the order of –150 mV (Figure 4). The results obtained for the E_H profiles in this eutrophic environment are similar to those found in summer in the oyster-farming area of the Thau Lagoon (Mesnage et al., 1994; Mesnage & Picot, 1995). Several authors place the iron oxidoreduction, controlling the availability of some of the phosphate trapped by the sediment within this range of observed redox values (Whitfield, 1969; Bagander & Niemisto, 1978).

The P fractions in the Méjean Lagoon showed a mobility capable of creating fluxes at the water–sediment interface as a function of the environmental variates.

Macroalgal growth and mortality were also involved in the phosphate exchanges in the Méjean part.

The presence of an oxic layer on the sediment surface favoured adsorption of the P by the ferric hydroxide-bound fraction. Adsorption was confirmed by the increase in Fe(OOH)-P between February and April. The increase in this fraction under quite high redox conditions agrees with the observations of numerous authors, both in the field (Detenbeck & Brezonik, 1991; Nair & Balchand, 1993), in laboratory experiments (Furumai & Ohgaki, 1989) and in our own experiments (Gomez et al., 1998). In June the temperature increase activated bacterial degradation of organic matter (Organic carbon decrease, Table 3). This phenomenon caused a fall in E_H (–150 mV) over the first five centimeters of sediment and provoked the release of phosphate and iron (1.7 mg g^{-1} of Fe) from the Fe(OOH)-P fraction. At the same time, the pH on the surface sediment increased (0.8 units). Numerous authors, including Boström et al. (1988) and Seitzinger (1991), have shown that an increase in pH also reduces the adsorption capacity of ferric hydroxides.

The calcium-bound phosphate fraction is sensitive to pH variations in the surface layer: it declined between February and April because of a 1-unit fall in pH. Its mobilization in the water explains the reduction in the load of sedimentary P observed between February and April. $CaCO_3$-P increased again between August and November at the same time as pH. Variations in this fraction are significant only in the two situations mentioned. Variations in pH do not produce the same results systematically as they are combined with variations in other variates (Jensen & Andersen, 1992). In addition, laboratory experiments with calcium salts confirm that the latter play an active part in P retention when there is a high pH. Stumm & Morgan (1981) show the formation of hydroxyapatite at high pH and other authors suggest increased precipitation of P and calcite when the pH rises (Ishikawa & Ishikuni, 1981; Lopez & Morgui, 1992). Precipitation of P with calcite occurs very rapidly (10 to 100 minutes), and is followed by a slow phase (Otsuki & Wetzel, 1972; House & Donalson, 1986; House et al., 1986; Suzuki et al., 1986). Work by Martens (1970) and Guldbransen et al. (1984) has shown, however, that precipitation with calcium is slower, in the presence of magnesium, in marine environments.

The organic phosphate (ROP) in the surface sediment increased after degradation of the lower layers of macroalgae, which were deposited on the sediment from August onwards. The organic carbon measured during this period decreased, despite the deposition of macroalgal debris. This indicates a more rapid decomposition of carbon compounds than of phosphate compounds.

The difference in concentration observed in November between the pore water (Table 4) and the overlying water (values below the detection limit) may be responsible for a diffusive P flux at the interface (Enell & Löfgren, 1988). The flux will tend to equilibrate the two concentrations at an equilibrium value such as has been measured by McCallister & Logan (1978), Froelich (1988) and Oliya & Reddy (1993). The consumption of P by the phytoplankton accounts for the continuing gradient at the interface.

The load of total P increased in June in the Pérols part (Table 3) because of the organic fractions. The origin of the organic P cannot be explained on the basis of the data collected for this sector (absence of macroalgal bloom). A transfer of dissolved P from the Méjean sector is conceivable, given the considerable difference in concentration (20 μM in Méjean, 5.16 μM in Pérols).

The organic fractions declined between June and August because of bacterial degradation, which increased with temperature (Table 2). In aerobic environments bacterial activity is responsible for an increased rate of P release (Sinke & Cappenberg, 1988). Release may occur by oxidation of the organic forms of P or by dissolution of the $CaCO_3$-P following a drop in pH. In anaerobic environments, bacteria are regarded as catalysts for iron hydroxide reduction which results in the release of iron-bound P, in other words as an indirect source of P through the use and release of internal P accumulated in aerobic conditions (DeMontigny et al., 1993). In the present case, ferric hydroxide reduction cannot be considered as the redox potential remained unchanged (~ -180 mV). In addition, the Fe(OOH)-P fraction increased, even if only slightly. It is therefore likely that mineralization of the organic fractions is a relevant factor in explaining the release. The ASOP fraction is considered to be an available P fraction (Golterman & Booman, 1988); it is therefore logical that this fraction should be mobilized. The same is also true, however, for one part of the ROP fraction, whereas it had initially been regarded as non-degradable. Research by Mesnage (1994) shows, however, that this fraction can be involved in phosphate exchanges. De Groot & Golterman (1993) show the presence of phytate in this fraction on paddy field sediments.

The inorganic fraction was involved when the degradation of organic matter creates strong reducing conditions in the surface sediments during August (Figure 4). Between the measurement in August, when the sediment did not have an oxic layer at the surface, and November, the load of sedimentary P declined because of the Fe(OOH)-P and $CaCO_3$-P fractions. The iron-bound fraction was mobilized at the same time as the ferric hydroxides in reducing conditions, which corresponds to the traditional theory of P release by iron reduction (Mortimer, 1941, 1942; Patrick & Khalid, 1974; Sundby et al., 1992).

The strong phytoplankton presence maintains high dissolved oxygen concentrations and a high pH in the water in winter. Because of these two conditions, both phosphate adsorption by iron and phosphate precipitation with calcium are possible. The sediment load is replenished by sedimentation of inorganic P.

Conclusion

The Méjean Lagoon is a coastal lagoon subject to severe eutrophication, reflected in a proliferation of macroalgae and phytoplankton over the course of the year. The seasonal cycle of the dissolved P in this shallow lagoon is determined not by inputs from the watershed, which are relatively regular, but by factors internal to the lagoon itself: the biological cycle of the macroalgae, and release from the sediment.

A climatic factor, temperature, regulates the behaviour of the sedimentary phosphate in the Méjean Lagoon. The high temperatures of the hottest months in the region trigger the release flux following degradation of the sediment's organic matter. The November fall in temperature slows down biological activity in the sediment and, by the same token, phosphate release.

Replenishment of the sediment's phosphate stock starts when macroalgal mortality reaches major proportions and continues throughout winter. Macroalgal mortality also has the effect of prolonging the phenomenon of sedimentary phosphate release by decomposition of the deposited algal biomass.

In the absence of macroalgae in the sector characterized by phytoplankton growth, there is no lengthening of the autumn release and the sedimentary load of organic phosphate can hardly be replenished in autumn. The phosphate load is replenished in winter by an increase in the inorganic fractions caused by the oxic conditions prevailing in the environment and by a high level of phytoplankton activity in this sector.

Macroalgal growth appears to be a phenomenon unconnected with phosphate release from the sediment. The inputs from the watershed in fact give rise to sufficiently high concentrations of dissolved P for algal growth. Fillit (1995) has shown that in the Méjean Lagoon this element is never limiting macroalgal growth and concludes that there must be limitation by light and temperature.

The phosphate released by the sediment is not essential to phytoplankton growth, as the highest chlorophyll *a* values are recorded in winter, when there is no release. The phytoplankton develops on the strength of the inputs from the watershed and takes advantage of the phosphate supplied by the sediment in periods of release.

Phosphate inputs arriving in the Méjean Lagoon come mainly from the tailings of the treatment plants. The first step towards the rehabilitation of the lagoon is for inputs from the watershed to be reduced or even stopped.

If inputs stopped completely, the supply of phosphate for the water column and hence for macroalgal growth would depend for a long time on the release of P by the sediment. As the amount of phosphate released is greater than the amount necessary for macroalgal growth, phosphate would not be a growth-restricting factor.

It is likely that the same amount of biomass will be produced as at present in the area in which macroalgae are the dominant feature. The deposition of this biomass on the sediment would replenish the sedimentary load of P. The system would lose phosphate only through the process of water renewal, which allows dissolved P to depart.

A relatively long period thus appears necessary before the phosphate starts to restrict algal growth and, by the same token, before any hypothetical rehabilitation of the lagoon can be effective.

Acknowledgements

This research was carried out with the financial support of the Programme National d'Océanographie Côtière.

References

Bagander, L. E. & L. Niemisto, 1978. An evaluation of the use of redox measurements for characterizing recent sediments. Estuar coast. Shelf Sci. 6: 127–134.

Boström, B., J. M. Anderson, S. Fleischer & M. Jansson, 1988. Exchange of phosphorus across the sediment–water interface. Hydrobiologia 170: 229–244.

Boutière H., F. deBovée, D. Delille, M. Fiala, C. Gros, G. Jacques, M. Knoepffler, J. P. Labat, M. Panouse & J. Soyer, 1982. Effet d'une crise dystrophique dans l'étang de Salses-Leucate. Oceanol. Acta, SP: 231–142.

Bühler, H. & H. Galster, 1980. Mesures redox, bases et problèmes. Ingold Suisse (Ed. et Publ.) – Zürich. 23 pp.

Burtin, C., 1994. Apports en azote et phosphore d'un bassin versant méditerranéen – Impact sur le complexe lagunaire récepteur. Mémoire de fin d'étude d'ingéniorat, Ecole National du Génie de l'Eau et de l'Environnement de Strasbourg/Cemagref, groupement de Montpellier, division Aquaculture et Pêche. Mémoire (144p) et Annexes (188 pp).

De Groot, C. J. & H. L. Golterman, 1990. Sequential fractionation of sediment phosphate. Hydrobiologia 192: 143–148.

De Groot, C. J. & H. L. Golterman, 1993. On the presence of organic phosphates in some Camargue sediments: evidence for the importance of the phytate. Hydrobiologia 252: 117–126.

De Montigny, C., Y. T. Prairie, P. C. M. Boers, T. E. Cappenberg & W. Van Raaphorst, 1993. The relative importance of biological and chemical processes in the release of phosphorus from a highly organic sediments. Hydrobiologia 253: 141–150.

Detenbeck, N. E. & P. L. Brezonik, 1991. Phosphorus sorption by sediments from a soft-water seepage lake1. An evaluation of kinetic and equilibrium models. Envir. Sci. Technol. 25: 395–403.

Duchaufour, P., 1988. Pédologie. 2nd Abrégés Masson, Paris: 224 pp.

Enell, M. & S. Löfgren, 1988. Phosphorus in interstitial water: methods and dynamics. Hydrobiologia 170: 103–132.

Fillit, M., 1995 Eutrophisation et macroalgues dans une lagune littorale méditerranéenne (Etang de Méjean-Pérols): L'étang change, les ulves prospèrent. Ph.D., Montpellier II University: 199 pp.

Froelich, P. N., 1988. Kinetic control of dissolved phosphate in natural rivers and estuaries: A primer on the phosphate buffer mechanism. Limnol. Oceanogr. 33: 649–667.

Furumai, H. & S. Ohgaki, 1988. Radiochemical analysis of phosphorus exchange kinetics between sediments and water under aerobic conditions. J. Envir. Qual. 17: 205–212.

Furumai, H. & S. Ohgaki, 1989. Adsorption-desorption of phosphorus by lake sediments under anaerobic conditions. Wat. Res. 23: 677–683.

Goldman, J. C. & R. Mann, 1980. Temperature-influenced variations in speciation and chemical composition of marine phytoplancton in outdor mass cultures. J. exp. mar. Biol. Ecol. 46: 29–39.

Golterman, H. L. & A. Booman, 1988. Sequential extraction of iron-phosphate and calcium-phosphate from sediments by chelating agents. Verh. int. Ver. Limnol. 23: 904–909.

Golterman, H. L., R. Clymo & M. A. M. Ohnstad, 1978. Methods for physical and chemical analysis of fresh waters, IBP Manual N°8. Blackwell Scientific Publishers, Oxford, 213 pp.

Gomez, E., 1996. Rôle des sédiments dans l'eutrophisation d'une lagune littorale (Etang du Méjean, Hérault): échanges du phosphate entre le sédiment et l'eau en fonction des conditions environnementales. Ph. D., Montpellier I University, 198 pp.

Gomez, E., D. Durillon, G. Rofés & B. Picot, 1998. Phosphate adsorption and release from sediment of brackish lagoon: pH, O2 and loading influence. Wat. Res., submitted.

Guldbrandsen, R. A., C. E. Roberson & E. T. Neil, 1984. Time and the crystallization of apatite in seawater. Geoch. Cosmoch. Acta 48: 213–218.

Hansson, L. A., 1989. The influence of a periphytic biolayer on phosphorus exchange between substrate and water. Arch. Hydrobiol. 115: 21–26.

Harris, E. & G. A. Riley, 1956. Oceanography of Long Island Sound, 1952–1954. VIII: Chemical composition of the plankton. Bull. Bingham Oceanog. Coll. 15: 315–323.

House, W. A. & L. Donaldson, 1986. Adsorption and coprecipitation of phosphate on calcite. J. Colloid. Interf. Sci. 112: 309–323.

House, W. A., H. Casey, L. Donaldson & S. Smith, 1986. Factors affecting the coprecipitation of inorganic phosphate with calcite. Wat. Res. 20: 917–922.

Ishikawa, M. & M. Ishikuni, 1981. Coprecipitation of phsophate with calcite. Geochem. J. 28: 283–288.

Istanovics, V., 1994. Fractional composition, adsorption and release of sediment phosphorus in the Kis-Balaton reservoir. Wat. Res. 28: 717–726.

Jensen, H. S. & F. O. Andersen, 1992. Importance of temperature , nitrate and pH for phosphate release from aerobic sediments of four shallow eutrophic lakes. Limnol. Oceanogr. 37: 577–589.

Lieutaud, A., M. C. Ximenes & T. Moutin, 1992. Lagoon eutrophication assessment for rehabilitation purposes: nitrogen and phosphorus loadings in different compartments. In Managing mediterranean wetlands and their birds. I.W.R.B. Spec. Publ. 20, Slimbridge, U.K. 285: 147–153.

Lopez, P. & J. A. Morgui, 1992. Phosphate and calcium carbonate saturation in a stratified coastal lagoon. Hydrobiologia 228: 55–63.

Lopez Laseras, P., 1991. Sedimentary phosphorus dynamics in epicontinental systems: a literature review. Oceanol. Aquat. 10: 113–125.

Martens, C. S. & R. C. Harris, 1970. Inhibition of apatite precipitation in the marine environment by magnesium ions. Geoch. Cosmoch. Acta 34: 621–625.

McCallister, D. L. & T. J. Logan, 1978. Phosphate adsorption-desorption characteristics of soils and bottom sediments in the Maumee River Basin of Ohio. J. Envir. Qual. 7: 87–92.

Mesnage, V., 1994. L'étude de la mobilité des formes de phosphate à l'interface eau-sédiment dans les écosystèmes lagunaires. Ph.D., Montpellier II University, 253 pp.

Mesnage, V., J. M. Deslous-Paoli, B. Picot, S. Landrein & C. Jege, 1993. L'eau interstitielle du sédiment d'une lagune méditerranéenne: comparaison de méthodes de prélèvement. Seminaire européen sur les technologies marines et environnement en Méditerranée. Santa Margherita (Italie): 167–171.

Mesnage, V. & B. Picot, 1995. The distribution of phosphate in sediments and its relation with eutrophication of a Mediterranean coastal lagoon. Hydrobiologia 297: 29–41.

Mesnage, V. & B. Picot, 1993. Comparaison de deux méthodes d'extraction sélectives pour évaluer la mobilité des phosphates dans des sédiments de lagunes littorales. Journal Français d'Hydrologie 24: 55–68.

Mesnage, V., B. Picot, E. Gomez, T. Moutin & J. Bontoux, 1994. Study of coastal mediterranean lagoons: phosphate speciation mobility in sediments. 1994. Proc. 4th Intern. Conf. on Wetland Systems for Water Pollution Control, organised by IAWQ, 6–10 november Guanghou (China): 503–512.

Mortimer, C. H., 1941. The exchange of dissolved substances between mud and water in lakes (I). J. Ecol. 29: 280–329.

Mortimer, C. H., 1942. The exchange of dissolved substances between mud and water in lakes (II). J. Ecol. 30: 280–201.

Nair, S. M. & A. N. Balchand, 1993. Phosphate–phosphorus adsorption characteristics of sediments from a very high productive coastal zone. Toxic. Envir. Chem. 39: 81–95.

Nixon, S. W., 1982. Nutrient dynamics, primary production and fisheries yields of lagoons. Oceanol. Acta, Proceedings International Symposium on Coaltal Lagoons. SCOR/IABO/UNESCO, Bordeaux, 8–14 septembre 1981: 357–371.

Oliya, O. G. & K. R. Reddy, 1993. Phosphorus sorption characteristics of sediments in shallow eutrophic lakes of Florida. Arch. Hydrobiol. 129: 45–65.

Otsuki, A. & R. G. Wetzel, 1972. Coprecipitation of phosphate with carbonates in a marl lake. Limnol. Oceanogr. 17: 763–767.

Patrick, J. & R. A. Khalid, 1974. Phosphate release and sorption by soils and sediments: effect of aerobic and anaerobic conditions. Science 186: 53–55.

Pirou, J. Y. & A. Ménesguen, 1992. Environmental factors controlling the Ulva sp. blooms in Brittany (France). In G. Colombo, I. Ferrara, V. U. Cecherelli & R. Rossi (eds), Marine Eutrophication and Polution Dynamics. 25th European Marine Biology Symposium. Olsen & Olsen, Fredensborg.

Reeburg, W. X., 1967. An improved interstitial water sampler. Limnol. Oceanogr. 12: 163–165.

Rofes, G. & M. Savary, 1981. Description d'un nouveau modèle de carottier pour sédiments fins. Bull. Franç. Piscicult. 283: 102–113.

Seitzinger, S. P., 1991. The effect of pH on the release of phosphorus from Potomac estuary. Estuar. coast. Shelf. Sci. 33: 409–418.

Sinke, A. J. C. & T. E. Cappenberg, 1988. Influence of bacterial processes on the phosphorus release from sediments in the eutrophic Loosdrecht Lakes, the Netherlands. Arch. Hydrobiol. Beih. Ergebn. Limnol. 30: 5–13.

Sfriso, A., A. Marcomini & B. Pavoni, 1987. Relationship between macroalgal biomass and nutrient concentrations in a hypertrophic area of the Venice lagoon. Mar. Envir. Res. 22: 297–312.

Standard Methods for the examination of water and watewater, 1992. 18th edition APHA, AWWA, WEF. Edited by A. E. Greenberg, L. S. Clesceri & A. D. Eaton.

Strickland, J. D. H., 1960. Measuring the production of marine phytoplancton. Bull. Fish. Res. Bd Can. 122: 1–172.

Stumm, W. & J. J. Morgan, 1981. Aquatic chemistry, an introduction emphasizing chemical equilibria in natural waters. 2nd edn. Wiley-Interscience Publication, New York, 780 pp.

Sundby, B., C. Gobeil, N. Silverberg & A. Mucci, 1992. The phosphorus cycle in coastal marine sediments. Limnol. Oceanogr. 37: 1733–1743.

Suzuki, T., S. Inomata & K. Sawada, 1986. Adsorption of phosphate on calcite. J. Chem. Soc. Faraday Trans. 82: 1733–1743.

Thom, R. M. & R. G. Albright, 1990. Dynamics of benthic vegetation standing-stock, irradiance and water properties in central Puget Sound. Mar. Biol. 104: 129–141.

Vaulot, D. & G. F. Frisoni, 1986. Phytoplanctonic productivity and nutrients in five Mediterranean lagoons. Oceanol. Acta 9: 57–63.

Viaroli, P., A. Pugnetti & I. Ferrari, 1991. Ulva rigida growth and decomposition processes and related effects on nitrogen and phosphorus cycles in a coastal lagoon (Sacca di Goro, Po river delta). In G. Colombo, I. Ferrara, V. U. Cecherelli & R. Rossi (eds), Marine Eutrophication and Polution dynamics. 25th European Marine Biology Symposium. Olsen & Olsen, Fredensborg.

Whitfiel, M., 1969. E_H as an operational parameter in estuarine studies. Limnol. Oceanogr. 14: 547–558.

Hydrobiologia **373/374:** 217–235, 1998.
J.-C. Amiard, B. Le Rouzic, B. Berthet & G. Bertru (eds),
Oceans, Rivers and Lakes: Energy and Substance Transfers at Interfaces.

Wetland effects on water quality: input-output studies of suspended particulate matter, nitrogen (N) and phosphorus (P) in Grand-Lieu, a natural plain lake

Loïc Marion & Luc Brient
Laboratoire d'Evolution des Systèmes Naturels et Modifiés, UMR CNRS Ecobio 6553, Museum National d'Histoire Naturelle & Université de Rennes, Campus Beaulieu, 35042 RENNES cedex, France

Key words: wetlands, lake, water quality, sediment, nitrogen, phosphorus, macrophytes, nutrient retention

Abstract

The role of wetland on water quality of rivers was estimated in a natural lake, Grand-Lieu (5600 ha), discharging to the Loire estuary. Inputs of its two tributaries, budgets and retention within the lake for particulate matter (PM), nitrogen (N) and phosphorus (P) were compared during the inflow period (Oct–May), by an input-output study, in two hydrologically contrasted years, 1993–94 with high inflow (292 10^6 m^3), and 1995–96 with low inflow (76 10^6 m^3). Globally the loads per ha were similar for the two tributaries for the same year, with higher values at the beginning of the flows, and total inputs markedly higher in 1993–94. During this year, average loads for the main tributary were 154 kg ha^{-1} PM, 40 kg ha^{-1} total N and 1.35 kg ha^{-1} total P. In the two tributaries, NO_3 represented 80% of total N for the two years, and PO_4 65% and 44% of total P. Total inputs, total outputs and storages are highly related to annual inflow, with large differences between elements. The highest change of inputs occur for NO_2, PO_4 and PM, and the lowest for NH_4. Storage and outputs of the lake were also much higher in 1993–94, the most important annual differences concerning P storage, and outputs of PM and all forms of nitrogen except NH_4. However, only the retention rate of total P doubled with the high outflow of 1993–94 (40% against 18%), while those of PO_4 and NH_4 were equivalent (79–72% and 72–66% respectively). In contrast the retention rate of all the other elements was lower with the large flow of 1993–94: 61 against 86% for NO_3, 85–90% for NO_2, 32–60% for total N, and 14–20% for PM. Globally, this wetland received important discharge of inorganic nitrogen from its catchment area, trapping or converting most of it (62–85%) into organic matter, while it exported a large amount of dissolved and particulate organic nitrogen (16 g m^{-2} y^{-1} in 1993–94 and 4.5 in 1995–96), 2.6 and 1.9 times more than it received. The exportation of organic N per m^2 and retention of total N (9.9–14.5 g m^{-2} y^{-1}) represent record values. This seems mainly due to the importance of water flow, juxtaposition of habitats with different degree of closure, plant biomass and resuspended endogenous, organic sediments in exportation. The lake is not able to counterbalance the dramatic increase of agricultural and sewage inputs, that induce its eutrophication and silting up.

Introduction

Wetlands, including shallow lakes, play an important role in water quality of rivers by storing sediments and nutrients, even if this action is often controversial (Lee et al., 1975; Jones et al., 1976; Gersberg et al., 1986; Johnston et al., 1990; Johnston, 1991). Numerous studies on wetlands give concentrations of nutrients in water or standing stocks, but according to Johnston (1991) and Hopkinson (1992), studies on fluxes in wetlands are less common, in comparison with rivers (Dolan et al., 1981; Meybeck, 1982, 1993; Golterman et al., 1983; Probst, 1985; Seux et al., 1985; Walling, 1988; Dorioz et al., 1989; Kronvang, 1992). We measured input and output fluxes of particulate matter, nitrogen and phosphorus on the major

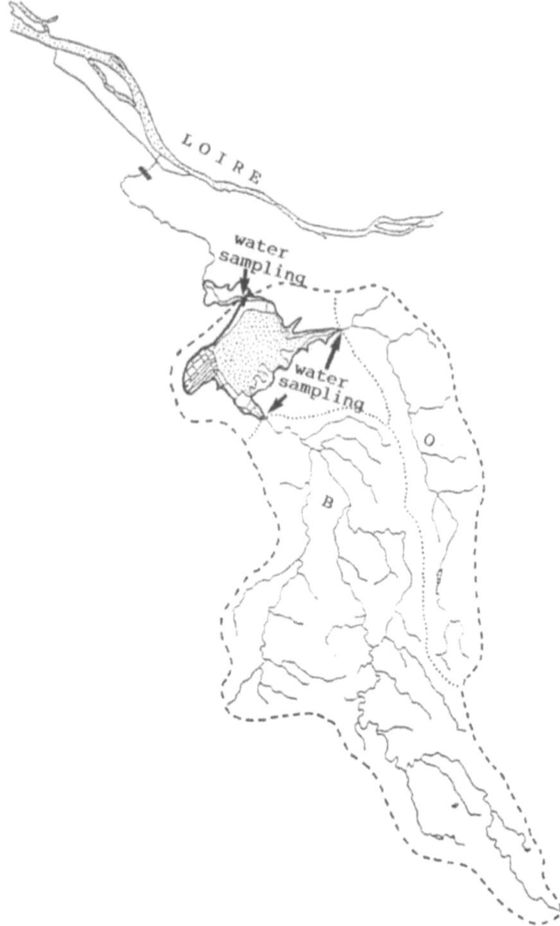

Figure 1. Map of the watershed of the lake (B = Boulogne, O = Ognon), with water samples points.

and oldest natural French floodplain lake, Grand-Lieu (5600 ha), that flows into the Loire estuary. Although most of the discharge in and out of the lake occurs in winter, its small catchment area (670 km^2) and its important biomass of algae, macrophytes and trees suggested that it could play an important role in water quality, in agreement with Odum's theory (1969) on higher retention in old succession stages' systems. We tested this role by a triple approach: comparisons between the two tributaries, between inputs and outputs of the lake, and between two hydrologically opposite winters, 1993–94 with one of the highest annual inflows since 1946 and 1995–96 with one of the lowest annual inflows. This paper focuses on the retention in the lake rather than on the watershed fluxes.

Description of site studied

The lake of Grand-Lieu is situated 23 km from the Atlantic Ocean, 25 km upstream of the Loire estuary (Figure 1). It is a natural lake formed by subsidence between the Tertiary and Pliocene periods. Several sea transgressions and regressions occurred until the Flandrian, the sea and an Ypresian river (50 million years) depositing sand in parts of the lake and of the catchment area. The lake covers about 4000 ha during summer, and 6300 ha during winter, by flooding adjacent peaty marsh grasslands used as pastures (Figure 2). Within the summer wet area, about 2000 ha of peat bog, that has progressively appeared in the last 7000 y, border a 20 km long shore, with *Phragmites*, *Salix* and *Alnus*. Parts of this peat bog are floating, while others are flooded from October to July. Most of the permanently flooded area of the lake is covered from April to October or November by about 1400 ha of floating macrophytes (mainly *Nymphea alba*, *Nuphar lutea*, *Trapa natans*, *Nymphoides peltata*), and more scarcely *Scirpus lacustris*. This macrophyte area has a muddy bottom, and its water depth is about 0.7 m in July. Other 600 ha, with rocky, sandy and/or clay bottom, has no floating macrophytes and extends to the eastern shore, bordered by a narrow vegetation belt of emerging macrophytes (*Scirpus lacustris*, *Typha angustifolia*, *Phragmites communis*). In July, depth varies from 1.2–1.7 m in the central part to 0.1 m on the edge. The rocky and sandy bottom subsides toward the West, to reach a pit, originally 15–20 m deep and now entirely full of muddy sediments. Organic sediment and peat (without marsh grass lands) represent about 160 10^6 m^3. For more details see Marion & Marion (1975).

Two rivers flow into the lake: the Boulogne (88 km long, average slope 0.01%) and the Ognon (45 km long, average slope 0.1%), with catchment areas of 485 and 185 km^2 respectively (Figure 1). These catchments are mainly under cultivation (cereals, maize, vine, grasslands with hedgerows, small woods). Soils are mainly sandy and clayey. The outlet, the Acheneau (25 km long), flows to the Loire estuary. The hydroperiod of the wetland with contrasted high and low water levels, is regulated by a sluicegate. This closes the outlet canal from the end of the flooding of the adjacent pastures (April) until the following winter, when a new flood must be evacuated (generally from December onwards). This sluicegate is situated on a ditch created in the 19th century, that separates the lake in two parts (5600 ha above, including 4000 ha of

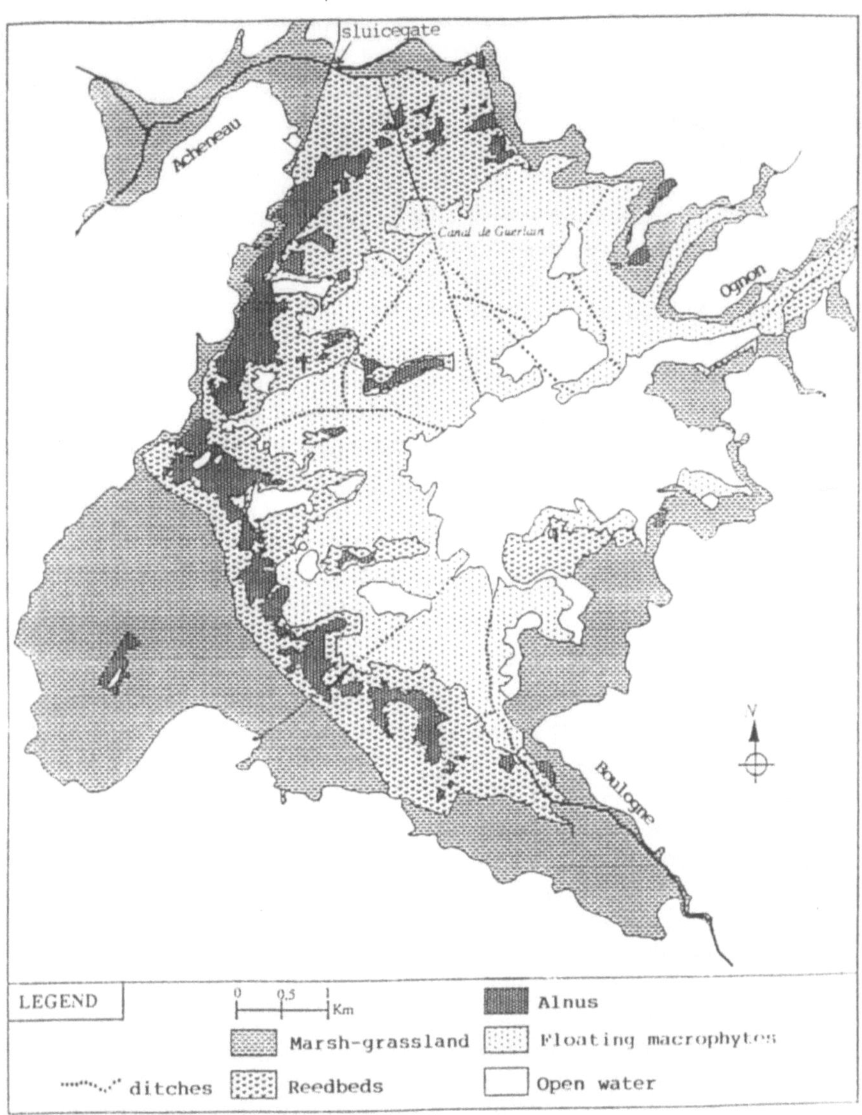

Figure 2. Map of the lake of Grand-Lieu with the different vegetation areas. The summer limit of water excludes the marsh grass land, that is seasonally flooded (Nov./Dec.–April/May). The reedbeds with Alnus are intermittently exposed (August–Sept.), and macrophytes with open water (central area) are permanently flooded.

permanent lake area with the peat bog and 1600 ha of adjacent marsh flooding grasslands, and 700 ha below, with only marsh flooding grasslands, see Figure 2). During the closing of the sluicegate, inflow from the catchment is generally nil, and rainfall on the lake (5600 ha = 16 10^6 m^3) does not counterbalance evapotranspiration (Penman method: 29 10^6 m^3) and pumping (3 10^6 m^3). So a 0.4 m decrease of the lake level is noted from May to September. Accordingly to these two hydraulic periods, the volume of water in the lake (before the sluicegate) varies during the year from 25 to 120 10^6 m^3. Most of the annual mean inflow of 168 10^6 m^3 (N = 45 y: 1946–90) occurs from December or January to March, with few peaks floods. During winter (October to April), mean direct input by rain represents only 24 10^6 m^3 and direct loss by evapotranspiration 12 10^6 m^3. Average net input from the atmosphere represents only 7% of total flow entering the lake during this period. There is no exchange with groundwater. Nutrient inputs from rivers are mainly due to agriculture, secondary to sewage. The lake was increasingly eutrophied since the 1970's (Marion et al., 1994). Water is saturated with oxygen during winter and oversaturared in the central area during the day in summer, while O_2 concentration in the macrophytes is about 4–5 mg l^{-1} at the surface of water and about 1 mg l^{-1} at – 50 cm (Marion et al., 1986). The redox potential is high in water (rH = about 32 MV) and from 12.5 to 17 MV in sediments in September (Rofès, 1992). pH is about 6 in the peat bog, about 7 in the macrophyte area and in the central area in winter, but can increase to 11 in summer (Marion et al., 1992). The Alkalinity (about 0.8 meq l^{-1}) and carbonates (< 1%) are low, and the conductivity increased from about 250 μS cm^{-1} in 1982 (Marion et al., 1986) to about 500 μS cm^{-1} in 1991 (Rofès, 1992).

Materials and methods

Flows and input – output concentrations of PM, N and P measurement

For simplicity, we name the 1993–94 and 1995–96 winters 1994 and 1996, respectively. During the inflow periods, water samples were taken once a week at the junction of the two tributaries with the lake, manually from early January to the end of May 1994, and by using automatic refrigerated sampling pumps every two hours (samples cumulated per day) in 1996 from December 24 to May 31. This last method was also used in the outlet draining channel during the two whole outflow periods (from December to May). Daily inflows were automatically measured every day in the two tributaries during the two years. Daily outflows of the draining channel were estimated for 1994 from the inputs (including net rain input) and daily variations of water level in the lake, and automatically measured in 1996. The water samples were analyzed in the laboratory for a range of variables including: particulate matter (PM), NO_2-N, NO_3-N, NH_4-N and PO_4-P after filtration on GF/C, and total N and total P on crude water. Data for ouputs give concentrations of water samples taken just before the sluicegate, even when the latter was closed (in this case output is nil although positive concentrations). PM was measured by filtering 250-750 ml water through precombusted and preweighted 0.7 m glass-fibre filters. Filters were dried for at least 1 h at 103 to 105°C, cooled in a dessicator and weighed. At the exit of the lake a turbidimeter (Ponselle Mesure) collected data every 15 mn when water output occurred. Concentrations of orthophosphate P were determined using the colorimetric technique with ascorbic acid reduction according to Murphy & Riley (1962). According to Golterman et al. (1978), total P and total N were determined using persulphate as oxidizing agent at 120°C during 1/2 h, NO_3 by the reduction cadmium method, NH_4 by the phenate method and NO_2 by diazoting with sulfamide and coupling with N-(1 BNnaphtyl)-ethylediamine to form a colored azodye.

Modelling of PM, N and P budgets and retention in the lake

Fluxes from January to May for each element were calculated by multiplying concentrations by water flow. To approach annual fluxes, we simulated fluxes from October to December, using daily inflows effectively measured in each tributary during each year, and average concentrations obtained during the January–May periods, increased by 7% for N and 14% for P and PM. These corrective factors are based on data obtained on a yearly basis in 1981–82 and 1990–91 (Marion et al., 1994). This estimate for October–December represents 34% of the global water inflow of the 1994 winter (October-May), and only 11% for 1996. Input from June to September was ignored, because it represented only 0.02% of the annual input of nitrogen and 0.1% of phosphorus in the 1990-91 study (Marion et al., 1994). Also, precipitation during the winter period was ignored, because it only contained 0.1 mg l^{-1} of NO_3-N, 0.04 mg l^{-1} of NH_4-N and no P (1996 data), due to the proximity of the sea from which most of wind and clouds come. For the whole winter, precipitation only countains about 2.4 tons of NO_3-N and 0.96 ton of NH_4-N. We also do not take into account inputs by birds, which represented 0.4% of total N and 6.6% of total P in the 1990–91 study (Marion et al., 1994).

Losses of PM and nutrients from the catchment area of the two rivers were obtained by dividing inputs arriving to the lake by the surface of each catchment. Real losses per ha are higher because only part reaches the lake. Retention rates for PM and nutrients within the lake were obtained by the difference of the input and the output, without supposing that output concerns the same particles. Also, retention per m^2 was obtained by dividing the total retention by 5 600 ha (= the area situated before the sluicegate), without supposing that real accumulation is regular in this area (see Discussion). We did not take into account dredging at the exit of the lake during the 1994 and 1996 winters (Marion, 1997).

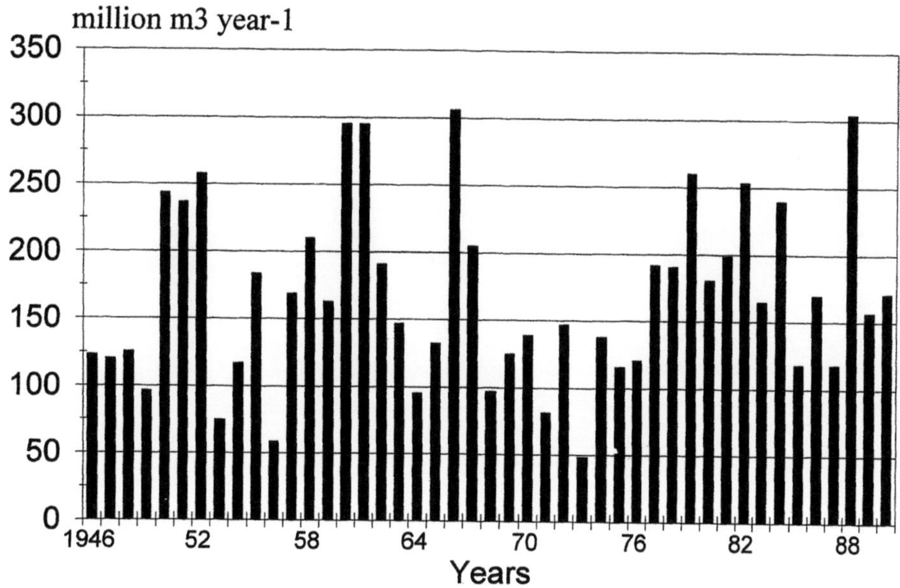

Figure 3. Annual inflow from adjacent streams entering the lake of Grand-Lieu from 1946 to 1990.

We shall take into condideration the summer period in the discussion about retention because each new winter input is first diluted in the volume of the lake, whose quality is influenced by processes occurring within the lake during its closing stage (spring-autumn).

Results

Inflow and outflow of water

1994 winter

During winter 1994, rivers discharge was the third highest since 1946: 292 10^6 m^3 against 306 in 1966 and 304 in 1988 (Figure 3). After a limited flow in October, the inflow really began on December 5 and practically stopped mid April. Most of the flow occurred between mid December and mid February. The Boulogne contributed 74% of the inflow, with 9 flood peaks (Figure 4A). The most important flows occurred during the first week of January (up to 8 10^6 m^3 during 24 h) and the first week of February (up to 10 10^6 m^3). Run-off varied during the winter from 0.90 to 119 m^3 s^{-1} ($\bar{x}=10.26$ m^3 s^{-1}). The Ognon displayed simultaneous peak flows but at a lower level (< 3 10^6 m^3 per day). During the winter, its run-off varied from 0.25 to 36 m^3 s^{-1} ($\bar{x}=3.66$ m^3 s^{-1}).

The outflow of the lake represented 288 10^6 m^3. It started on December 13 but was important only at the end of December, then decreased regularly until the beginning of May (Figure 4B). The outflow varied from 1.24 to 59.78 m^3 s^{-1} ($\bar{x}=19.05$ m^3 s^{-1}). By reference to the minimum volume of the lake (in September), the residence time of water was 0.09 y.

1996 winter

During winter 1996 the discharge was only 75.5 10^6 m^3. Only two years have had a lower discharge since 1946 (48 10^6 of m^3 in 1973, 58 in 1956). The discharge began at the end of December and practically stopped at the end of March, most of it occurring before the end of February, with seven small flow peaks (all with about 1 10^6 m^3 for the Boulogne except the last one with 2 10^6 m^3, Figure 5A). This river contributed 68% of the winter inflow with a run-off varying from 0.10 to 29 m^3 s^{-1} ($\bar{x}=2.79$ m^3 s^{-1}). Run-off of the Ognon varied from 0.10 to 16.60 m^3 s^{-1} ($\bar{x}=1.31$ m^3 s^{-1}).

The outflow of the lake was lower than the inflow by only 58 10^6 m^3 (Figure 5B). It started on January 17, stopped temporarily the three last days of February, from March 30 to April 27, and from April 28 to May 3, and stopped definitely on May 30. When the sluicegate was open, the outflow varied from 0.32 to 12.44 m^3 s^{-1} ($\bar{x}=4.88$ m^3 s^{-1}). The residence time of water was 0.33 y.

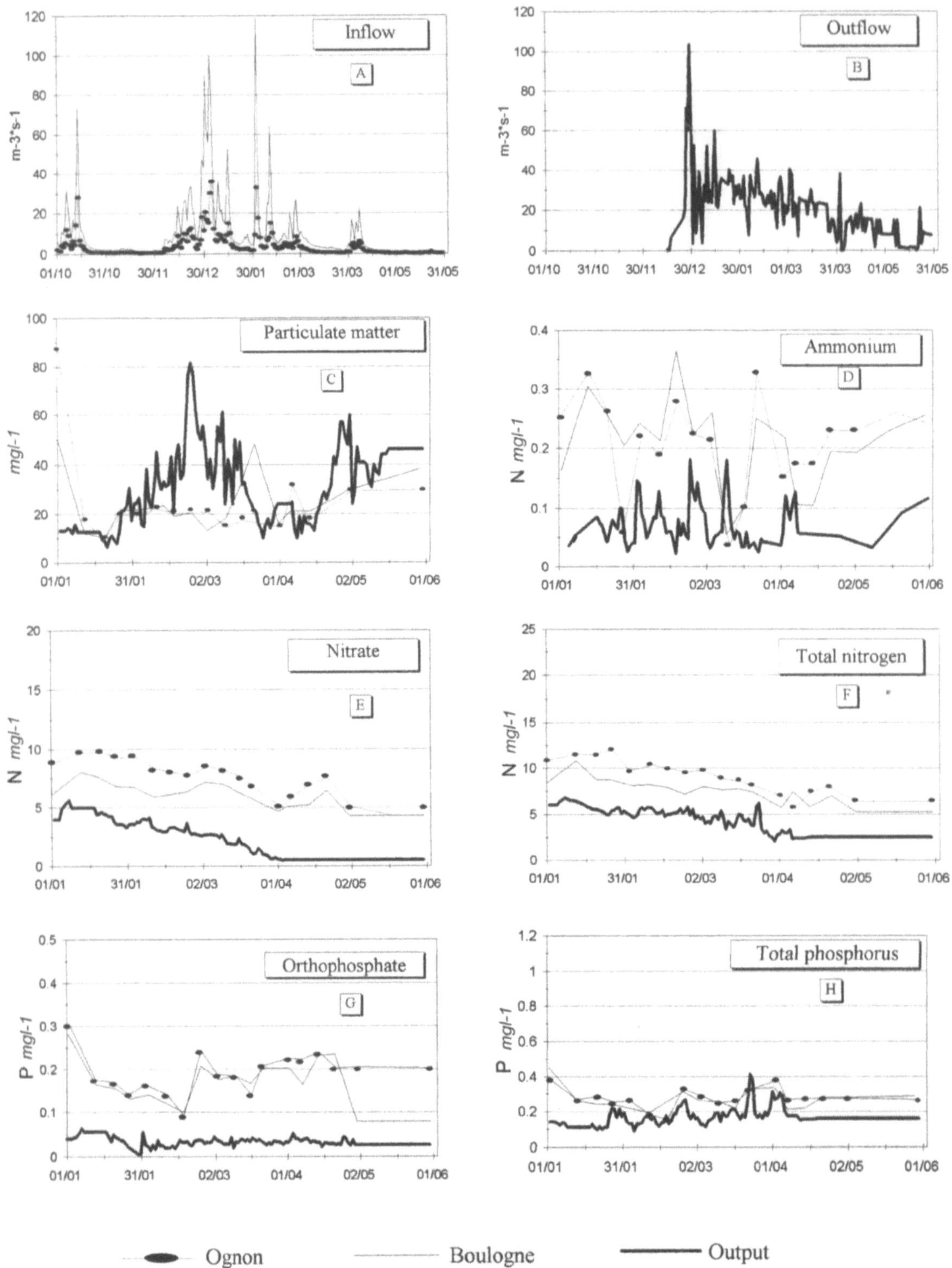

Figure 4. Comparison between inflow from the two tributaries (Boulogne and Ognon), outflow and concentrations of PM and nutrients in 1994.

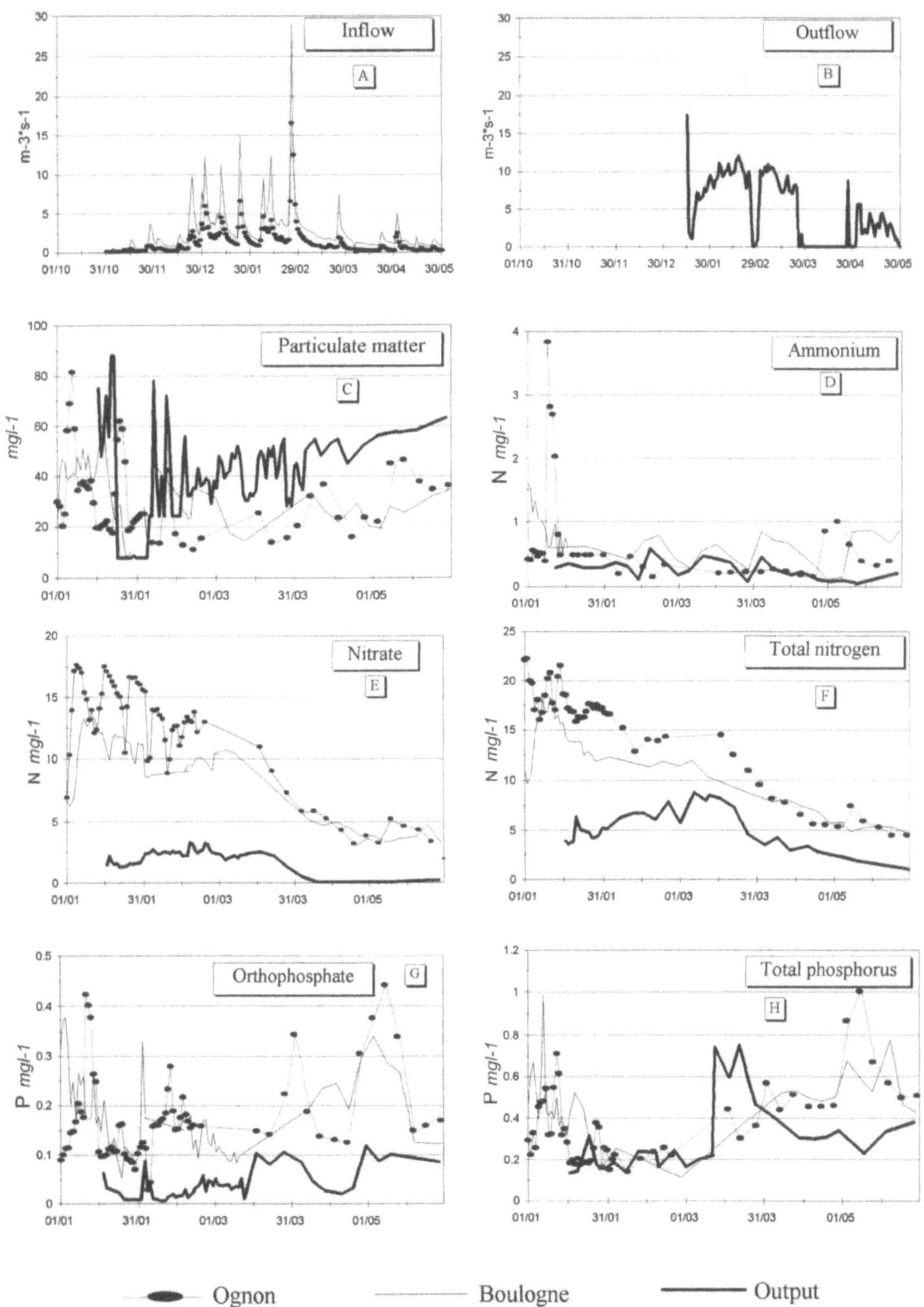

Figure 5. Comparison between inflow from the two tributaries (Boulogne and Ognon), outflow and concentrations of PM and nutrients in 1996.

Particulate matter (PM)

PM is largely represented by sediments during the winter inflow and outflow, whereas algae become more important in April and May (Marion et al., 1992).

1994 winter

PM concentration varied from 5 to 87 mg l^{-1} ($\bar{x}$=24.44 ± SD 5.08 mg l^{-1}) in the Boulogne and from 5 to 49 mg l^{-1} ($\bar{x}$=25.22 ± 7.81 mg l^{-1}) in the Ognon (Figure 4C). The highest values for both were observed during the first flow of January, whereas the main flow peak (February 3) did not cause a peak of PM (21 mg l^{-1} in the Boulogne and 12 mg l^{-1} in the Ognon). Estimation of the PM input to the lake for the whole period October–May was about 9153 tons (69% from the Boulogne), and represented losses of 154 and 130 kg ha^{-1} for the catchment of the Boulogne and Ognon.

The PM output of the lake varied from 6 to 82 mg l^{-1} ($\bar{x}$=30.95 ± 3.23 mg l^{-1}, Figure 4C). During the whole winter, it totaled 7 913 tons, representing a retention rate of 14% within the lake (22 g m^{-2} y^{-1}).

1996 winter

PM concentrations varied from 9 to 62 mg l^{-1} ($\bar{x}$=32.58 ± 3.10 mg l^{-1}) in the Boulogne and from 11 to 82 mg l^{-1} ($\bar{x}$=31.03 ± 4.02 mg l^{-1}) in the Ognon (Figure 5C). As in 1994, the highest values occurred during the first flow of the rivers, for example 45 mg l^{-1} on January 2 in the Boulogne when entered 1 10^6 m^3 of water, against only 15 mg l^{-1} with 1.4 10^6 m^3 on January 24 or 34 mg l^{-1} during the major flow peak of 2.5 10^6 m^3 on February 26. The Ognon showed 82 mg l^{-1} on January 5 and only 37 mg l^{-1} on February 26. The estimate for the whole winter was 2526 tons (71% from the Boulogne), and represented only 40 and 37 kg ha^{-1} for each catchment.

The output of the lake varied from 8 to 88 mg l^{-1} ($\bar{x}$=36.51 ± 3.83 mg l^{-1}). During winter, it totaled 2019 tons, representing a retention rate of 20% (9 g m^{-2} y^{-1}).

Comparison between the two winters

In 1994, the PM input of the lake was 3.6 times and the output 3.9 times higher than in 1996, mainly due to the difference of inflow (3.9 times) and outflow (5 times), and consequently to the difference of speed of current in the rivers. However, the difference in total retention in the lake is less important (2.5 times), and the retention rate was higher in 1996, in relation with the longer highest residence time of water.

Nitrogen

1994 winter

Total N concentrations varied from 5.50 to 10.82 mg l^{-1} ($\bar{x}$=7.61 ± 0.64 mg l^{-1}) in the Boulogne and from 5.80 to 12.10 mg l^{-1} ($\bar{x}$=9.06 ± 0.90 mg l^{-1}) in the Ognon. In the two rivers concentrations decreased by about 50% from January to April, with comparatively higher values in the Ognon most of the time (Figure 4F). The estimate for the whole winter was about 2543 tons (70% from the Boulogne), representing losses of 41 and 37 kg ha^{-1} for the Boulogne and Ognon.

The output from the lake varied from 2.1 to 6.8 mg l^{-1} ($\bar{x}$=4.61 ± 0.26 mg l^{-1}) and decreased during winter by about 65% (Figure 4F). It totaled 1731 tons for the winter, resulting in a total N retention in the lake of 32% of the input (14.50 g m^{-2} y^{-1}).

Most nitrogen fluxed concerned nitrate, at concentrations from 4.50 to 7.76 mg l^{-1} of NO_3-N ($\bar{x}$=6.18 ± 0.49 mg l^{-1}) in the Boulogne and from 5.00 to 9.47 mg l^{-1} ($\bar{x}$=7.67 ± 0.75 mg l^{-1}) in the Ognon (Figure 4E). During winter, input to the lake was estimated to about 2040 tons (68% from the Boulogne), representing losses of 29 and 35 kg ha^{-1} from the river catchments.

The output from the lake varied from 0.53 to 5.63 mg l^{-1} ($\bar{x}$=2.71 ± 0.33 mg l^{-1}), with a steeper decrease during winter than for total N. During winter, 804 tons of nitrate were exported, or a retention rate of 61% (22.08 g m^{-2} y^{-1}).

Nitrite concentrations were not important and varied from 0.02 to 0.80 mg l^{-1} of NO_2-N ($\bar{x}$=0.33 ± 0.13 mg l^{-1}) in the Boulogne, and from 0.03 to 0.67 mg l^{-1} ($\bar{x}$=0.32 ± 0.12 mg l^{-1}) in the Ognon, with similar evolution in both rivers: increase in January, relatively low values until the major peak in March, and decrease thereafter. Input to the lake for the whole winter was about 94 tons (74% from the Boulogne). Output from the lake was lower and varied from 0.009 to 0.37 mg l^{-1} ($\bar{x}$=0.048 ± 0.009 mg l^{-1}). It totaled 14 tons, representing a retention rate of 85% (1.44 g m^{-2} y^{-1}).

Ammonium concentrations were generally similar in both rivers and fluctuated from 0.06 to 0.37 mg l^{-1} of NH_4-N in the Boulogne ($\bar{x}$=0.20 ± 0.04 mg l^{-1}) and 0.04 to 0.33 mg l^{-1} ($\bar{x}$=0.20 ± 0.04 mg l^{-1}) in the Ognon (Figure 4D). They were lower in the outputs from the lake ($\bar{x}$ = 0.07 mg l^{-1} ± 0.02) and largely fluctuating. The total input to the lake was only 63 tons (72% from the Boulogne), with an output of 18 tons (retention in the lake: 72% = 0.80 g m^{-2} y^{-1}).

1996 winter

During winter 1996, total N concentrations varied from 5.7 to 18.3 mg l^{-1} ($\bar{x}$ = 11.81 ± 1.00 mg l^{-1}) in the Boulogne and from 4.5 to 22.8 mg l^{-1} ($\bar{x}$ = 14.74 ± 1.40 mg l^{-1}) in the Ognon. They increased strongly with flow at the beginning of January in the Boulogne (from 9.9 mg l^{-1} on December 24 to 18.34 on January 9), and more quickly and strongly in the Ognon (Figure 5F). They decreased by about 75% from mid January to April, when concentrations began to be similar in both rivers. During the whole winter, the estimated input to the lake only reached about 930 tons (61% from the Boulogne), representing losses of 12 and 20 kg ha^{-1} from the catchment areas.

The output from the lake varied from 2.00 to 8.82 mg l^{-1} ($\bar{x}$ = 5.12 ± 0.56 mg l^{-1}), with increasing values until mid March and decreasing values after (Figure 5F). During winter, output totaled 376 tons (retention: 60% = 9.88 g m^{-2} y^{-1}).

Nitrate concentrations varied from 2.75 to 13.27 mg l^{-1} NO_3-N ($\bar{x}$ = 9.22 ± 0.62 mg l^{-1}) in the Boulogne and from 2.92 to 17.65 mg l^{-1} ($\bar{x}$ = 11.79 ± 1.03 mg l^{-1}) in the Ognon. Concentrations also largely increased from only 6.32 mg l^{-1} in the Boulogne and 6.43 in the Ognon on December 29 to reach the highest values in the flow of January (Figure 5E). They fluctuated until mid February in relation with flow, with highest values in the Ognon. Concentrations began to be similar in both rivers during the second period of the decrease. During winter, total input was estimated to 742 tons (60% from the Boulogne), representing losses of 9 and 16 kg ha^{-1} respectively for the two catchment areas.

The output from the lake varied from 0.06 to 3.29 mg l^{-1} ($\bar{x}$ = 1.98 ± 0.19 mg l^{-1}). In contrast to 1994, it increased during the first half period before decreasing to reach virtually nil in April (Figure 5E). For the whole winter, the output was only 105 tons (retention: 86% = 11.39 g m^{-2} y^{-1}).

Nitrites were negligible in 1996, varying from 0.01 to 0.11 mg l^{-1} NO_2-N ($\bar{x}$ = 0.06 ± 0.01 mg l^{-1}) in the Boulogne, and from 0.003 to 0.18 mg l^{-1} ($\bar{x}$ = 0.09 + 0.01 mg l^{-1}) in the Ognon. The input of the lake for the winter is estimated to only 5 tons (59% from the Boulogne). The output from the lake was virtually nil and varied from 0.005 to 0.06 mg l^{-1} ($\bar{x}$ = 0.009 ± 0.004 mg l^{-1}). It totaled 0.52 tons, representing a retention rate of 90% (0.08 g m^{-2} y^{-1}).

Ammonium concentrations were much higher than in 1994 and fluctuated from 0.12 to 1.76 mg l^{-1} in the Boulogne ($\bar{x}$ = 0.87 ± 0.14 mg l^{-1}) and 0.15 to 3.84 mg l^{-1} ($\bar{x}$ = 0.62 ± 0.18 mg l^{-1}) in the Ognon (Figure 4D). Peaks of concentrations occurred in the Boulogne during the first flow on December 24 (out of Figure), and on January 9 in the Ognon, also with the first storm run-off, while following concentrations were always relatively low even during storms. Total input was estimated at 51 tons (76% from the Boulogne). Concentrations were lower in the outputs from the lake but higher than in 1996 ($\bar{x}$ = 0.28 ± 0.04), involving a similar total output of 17 tons (retention: 66% = 0.60 g m^{-2} y^{-1}).

Comparison between the two winters

On the one hand, the total N and nitrate inputs appeared relatively similar during both years. However, the decrease in 1996 was half as important as in 1994. On the other hand, the outputs differed largely, with a regular decrease in 1994 against an increase followed by a decrease in 1996.

Differences in fluxes between years (Figure 6) were less important for total N input (× 2.8 times in 1994) than for inflow of water (× 3.9), but the output basically fitted the outflow (× 4.6 and 5). However, the difference of global retention in the lake (Figure 7) was not very important (× 1.5), and the retention rate was almost twice higher in 1996 (Figure 8). Nitrates displayed a similar trend with a difference of 2.8 times for inputs and only 1.9 for retention, although the difference between outputs was important (× 7.7). So, the retention rate was also more important in 1996 (86% against 61%). Effectively, while nitrates represented the same proportion of total N in inputs for the two years (80%), they represented for outputs 46% of total N in 1994 and only 28% in 1996. Like for PM, the long residence time of water in 1996 probably favoured loss of nitrates. The annual difference of nitrites was the highest of all the studied elements: the input of 1994 was 18 times more important than in 1996, the output 27 times and the global retention 17 times. Thus the retention rate was also more important in 1996 (Figure 8). Inversely, ammonium was the least sensitive to run-off (Figure 6), the difference between years being only 1.2 times for inputs, while outputs and retention rates were similar, in spite of the large difference of flows.

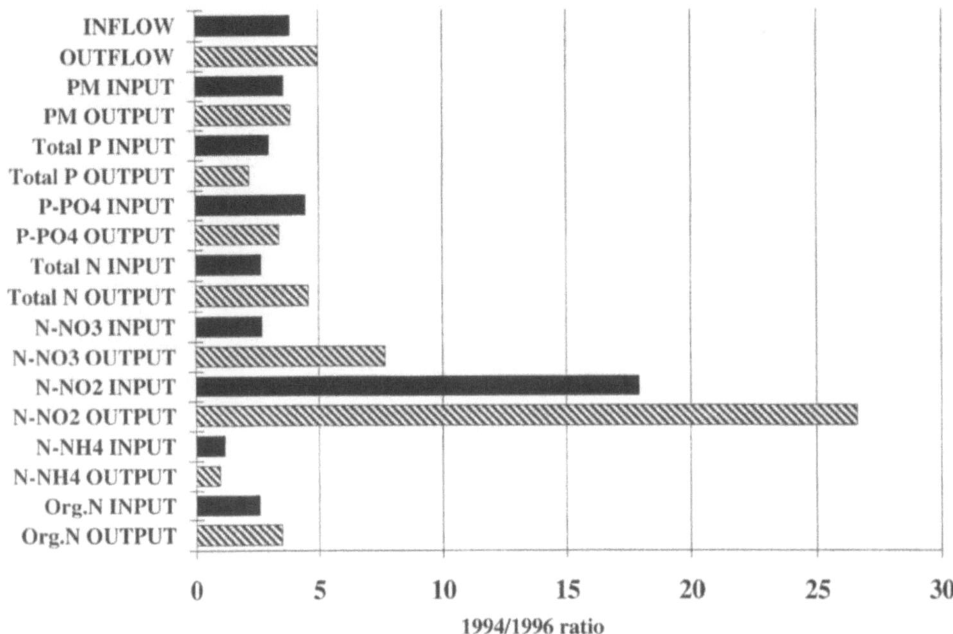

Figure 6. Annual differences of input and output of flow, PM, phosphorus and nitrogen (ratio of fluxes 1994/1996).

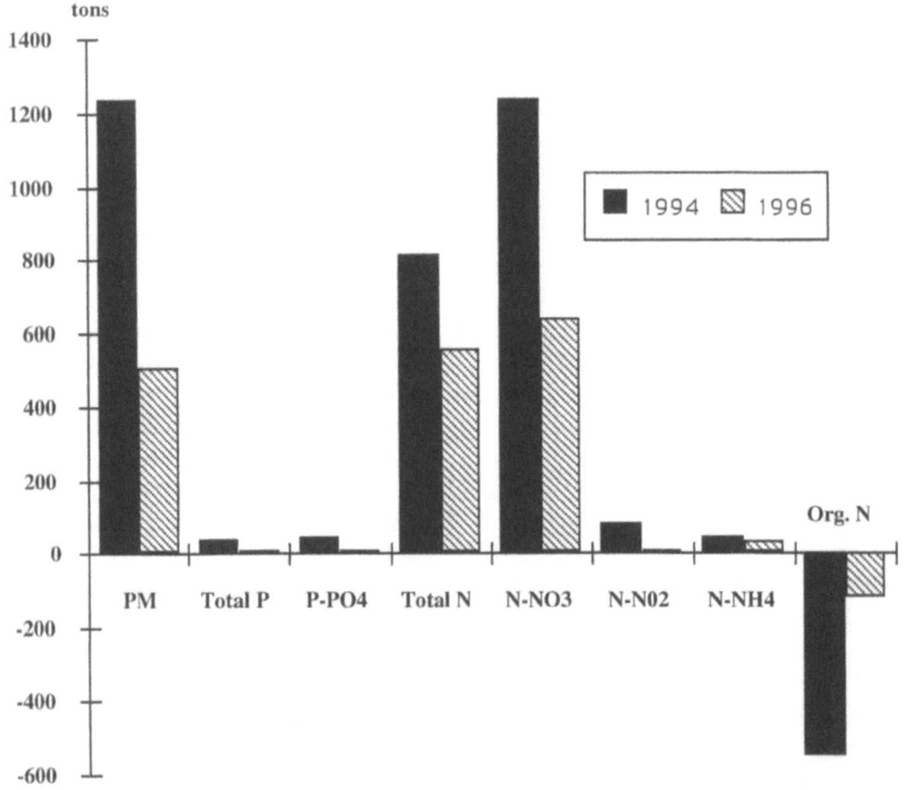

Figure 7. Annual differences of total retention (input–output in tons) between 1994 and 1996 for PM, P and N.

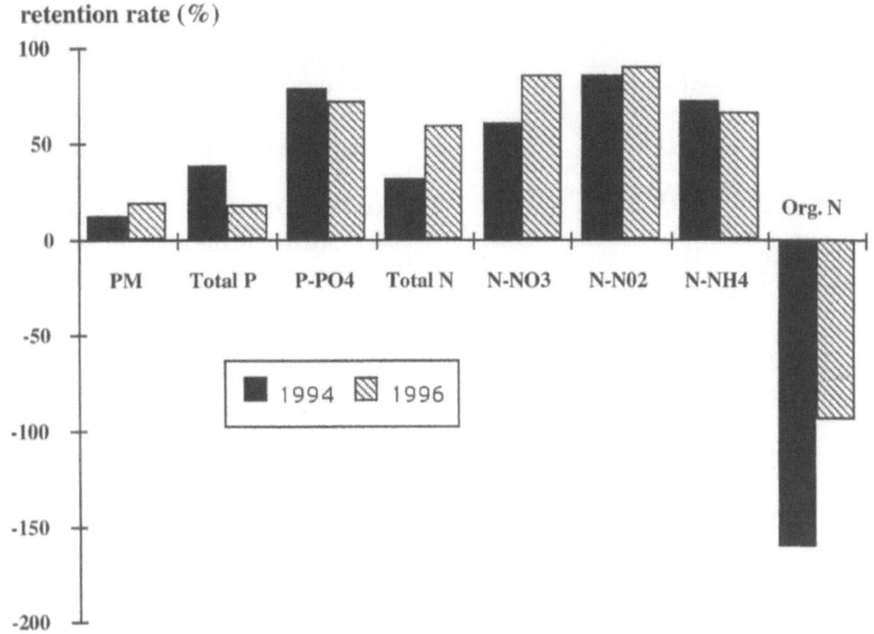

Figure 8. Annual differences of net retention rates (input–output/input) between 1994 and 1996 for PM, P and N.

Phosphorus

1994 winter

Total P concentrations varied from 143 to 450 mg m^{-3} ($\bar{x} = 260 \pm 33$ mg m^{-3}) in the Boulogne and from 153 to 380 mg m^{-3} ($\bar{x} = 275 \pm 28$ mg m^{-3}) in the Ognon. They largely fluctuated during the study, similarly for both rivers, with highest values at the beginning of January and April (Figure 4H). During the whole period, the input of the lake was estimated to about 88 tons (74% from the Boulogne), representing losses of 1.35 and 1.24 kg ha^{-1} for the catchment areas of each river.

The output from the lake varied from 89 to 415 mg m^{-3} ($\bar{x} = 170 \pm 12$ mg m^{-3}). During the whole period, it represented 53 tons, resulting in total P retention in the lake of 40% (0.63 g m^{-2} y^{-1}).

Concentrations of orthophosphate varied from 75 to 278 mg m^{-3} PO_4-P ($\bar{x} = 160 \pm 23$) in the Boulogne and from 88 to 300 mg m^{-3} ($\bar{x} = 190+ 23$) in the Ognon, with similar evolution during the studied period for both rivers, with a decrease until mid February and higher concentrations after (Figure 4G). The estimated input of the lake reached about 57 tons (72% from the Boulogne), corresponding to losses of 0.84 and 0.88 g m^{-2} y^{-1} for the two catchments.

The output from the lake was low and varied from 5 to 63 mg m^{-3} ($\bar{x} = 30 \pm 2$ mg m^{-3}), with unimportant changes during the season. During the winter, it totaled only 12 tons (retention in the lake: 79% = 0.80 g m^{-2} y^{-1}).

1996 winter

Total P concentrations varied from 109 to 776 mg m^{-3} ($\bar{x} = 460 \pm 52$ mg m^{-3}) in the Boulogne and from 143 to 1010 mg m^{-3} ($\bar{x} = 370 \pm 50$ mg m^{-3}) in the Ognon. They showed several peaks corresponding to the storm run-off of the beginning of January, but less important than for nitrogen (Figure 5H). They then decreased to reach the lowest values of the studied period (about 200 mg m^{-3}) from end of January to mid March, in spite of a major flow peak of February 26 (2.5 10^6 m^3 and only 170 mg m^{-3} P in the Boulogne). Then concentrations increased again to reach the highest values ever for the Ognon (up to 1010 mg m^{-3} on May 8), but the flow was at this time low. Globally the curve of total P was inverse to that of nitrogen, except during the first storm flows. For the whole winter, the estimated input reached 29 tons (67% from the Boulogne), representing 0.40 and 0.51 kg ha^{-1} for the catchment of each river.

The output varied from 140 to 741 mg m^{-3} ($\bar{x} = 300 \pm 63$ mg m^{-3}). Concentrations were relatively low (about 200 mg m^{-3}) until mid March, similar to the inputs, then increased during one weak and regularly decreased after (Figure 5H). For the

Table 1. Comparison of N and P inputs in the Lake of Grand-Lieu for 4 y (concentrations averaged on fluxes).

	1980–81	1990–91	1993–94	1995–96
Inflow million m^3	270	135	292	76
N input in tons	780	2184	2543	930
Average concentration mg l^{-1}	2.89	16.18	8.69	12.24
P input in tons	82	36	88	29
Average concentration mg l^{-1}	0.30	0.27	0.30	0.38

whole winter, the output reached 24 tons, resulting in total P retention of 18% (0.09 g m^{-2} y^{-1}).

Orthophosphate was relatively similar in both rivers, with concentrations varying from 52 mg to 390 mg m^{-3} in the Boulogne ($\bar{x} = 180 \pm 20$) and from 83 to 440 mg m^{-3} ($\bar{x} = 170 \pm 20$) in the Ognon. Trend was globally similar to that for total P (Figure 5G). Estimate of the input to the lake was about 13 tons (66% from the Boulogne), and represented losses of 0.17 and 0.24 kg ha^{-1} for the two catchments.

The output varied irregularly during the period from 5 to 122 mg m^{-3} ($\bar{x} = 40 \pm 7$ mg m^{-3}). For the whole winter, it was only 3.6 tons (retention in the lake: 72% = 0.16 g m^{-2} y^{-1}).

Comparison between winters

Like for nitrogen, differences in fluxes (Figure 6) between years were less important for total P inputs (3 times) and output (× 2.3) than for inflow and outflow of water (× 3.9 and 5) or PM (× 3.6 and 3.9), but the annual differences in retention (× 6.6) and in retention rate (× 2.2) were the highest of all elements studied, except nitrites (Figure 7 and 8). The input of orthophosphate was 4.5 times more important in 1994, representing 65% of total P against 44% in 1996. Even if its proportion in total P outputs was also higher in 1994 (23% against 15%), the difference was lower and global retention was 4.9 times more important in 1994, despite a relatively similar retention rate (79% against 72%).

Discussion

Even though the flows of the two years (292 and 76 10^6 m^3) were at the extremes side of the known range of this lake, the nutrient inputs observed in the present study were largely higher than normal, due to the dramatic increase in nutrients inputs from sewage and agricultural run-off since the 1960's (Marion et al., 1994). In 1980–81 total P concentrations were already as high as in the 1990's, but the situation for N has degraded since this period (Table 1). Annual total N losses from the watershed (14 kg ha^{-1} for 1996, 39 for 1994) are within the range of values given in literature for agricultural watersheds. The same is true for total P loss for 1996 (0.43 kg ha^{-1}), but the 1994 P value was higher (1.32 kg ha^{-1} instead of a range of 0.18–1.0 reported by Jones et al., 1976).

The four annual budgets (Table 1) like in other lakes (Meybeck, 1995), show that total inputs are highly related to inflow, but considerable differences occur between elements. A comparison between 1994 and 1996 (Figure 6) shows that the largest annual differences were in nitrites, orthophosphate and PM, and the lowest in ammonium. Also, large inflow involved a dilution of most elements (by about 30–70% for PM, total N, NO_3 and total P, and 3–4 times for NH_4), except for orthophosphate (same average concentrations), and above all nitrites, average concentrations of which were 4.3 times lower in 1996. Such higher concentration with flow may be due to overflowing sewage storage tanks (of towns and cattle farms). Respective fluxes from both tributaries were also highly related to annual flow, but less in 1996. In 1994, the part of the Boulogne flux in total input to the lake was relatively similar, for all elements (68 to 74%), to its inflow (74%) and to the surface of its catchment (72%), the lowest proportions concerning nitrate and PM. In 1996, values were more heterogeneous, part of total N, nitrate and nitrites fluxes provided by the Boulogne being lower (61 to 59%) than its inflow (68%), those of ammonium and PM being higher to 1994, and only total P and PO_4 being similar to the contribution of its discharge flow.

As for inputs, the outputs from the lake of elements were higher during the larger flow of 1994, confirming Mitsch & Gosselink (1986) and Hopkinson (1992), who found that increased water inputs generally reflect increased nutrients loading. For Johnston (1991), fluxes are more involved than storage compartments in the control of the net effect of a wetland on water quality. Here again there are important differences between elements (Figure 6), the higher annual differences between outputs concerning PM and all forms of nitrogen except ammonium. Global retention in the lake also reflects the importance of inflow for all elements, but mainly for nitrites, total

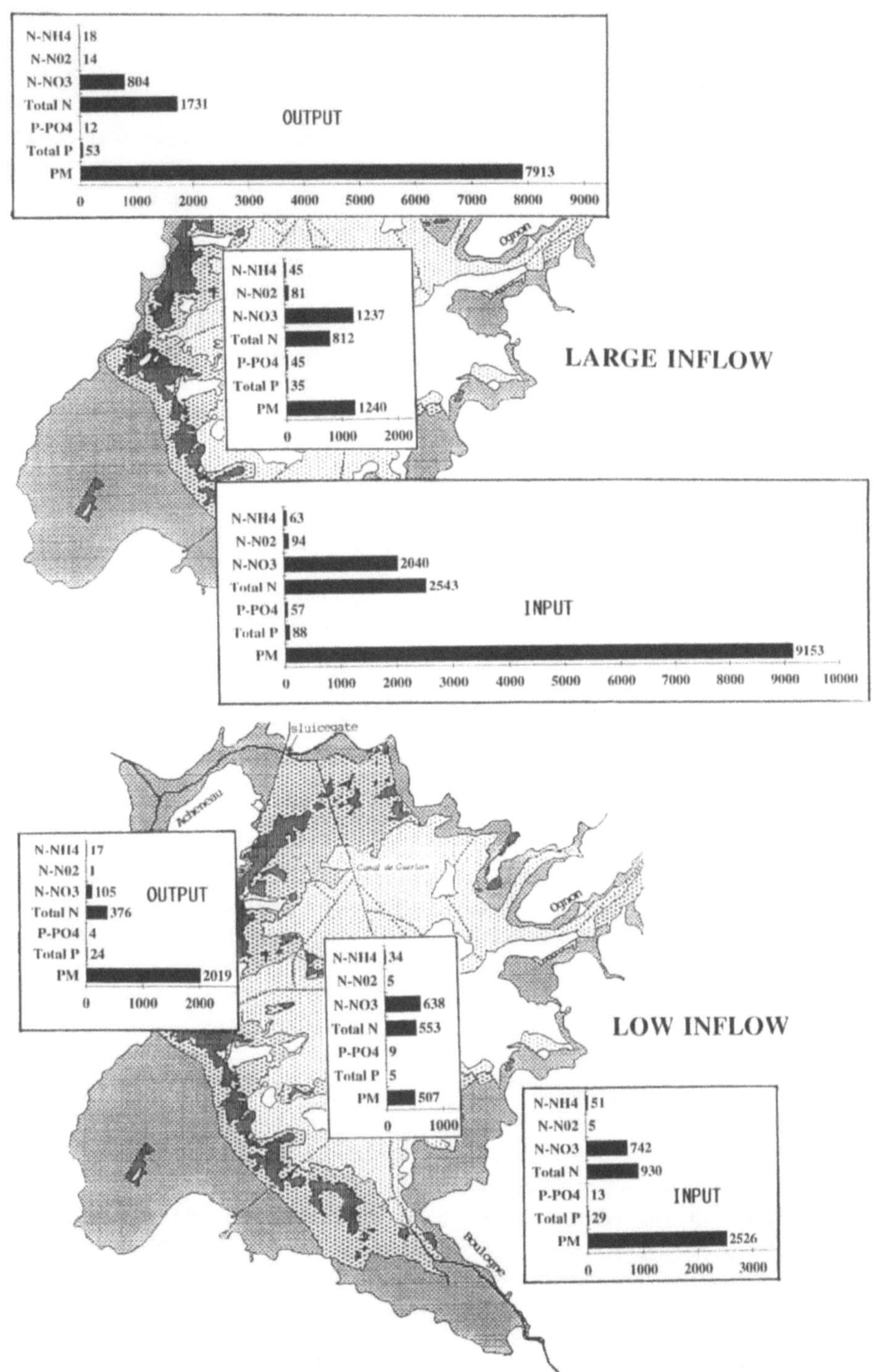

Figure 9. Overview of global retention (in tons) of PM and nutrients within the lake (central box), compared to inputs and outputs during the two opposite inflows in 1994 (large) and 1996 (low).

P, and orthophosphate (Figure 7). Retention rates do not respect this rule (except for total P), 1996 showing higher values for PM and all forms of nitrogen except ammonium (Figure 8). Globally, retention in the lake was less influenced by flow than input and output (Figure 9).

Particulate Matter

The small difference between retention rates of PM during the two years (14-20%), in spite of the different flows and amounts of sediments transported and deposited (from 2.5 to 5 times more important in 1994), could suggest that this retention rate is usual for this lake. It is lower than reported in literature from input-output studies, varying from 23 to 93% for 18 wetlands, with a mean of 44% (Johnston, 1991, after Phillips, 1989). The accumulation rate of allochthonous PM in Grand-Lieu (9–22 g m^{-2} y^{-1}) is naturally much lower than that of Alpine lakes (650–16000 g m^{-2} y^{-1}, Campy & Meybeck, 1995), and Grand-Lieu should be compared to shallow lakes with a low residence time, that limits sedimentation (Meybeck, 1995). Other factors explain this relatively low retention. Although recent agricultural practices increase erosion (maize crop and vineyards without protecting vegetation during winter, drainage, destruction of hedgerows), such erosion is not important because the watershed is situated in an old, eroded basin (Tertiary), with a smooth relief. Only the PM yields of the two rivers observed during the major flow of 1994 (130 and 154 kg ha^{-1} y^{-1}) are of the same magnitude as shown by Kronwang (1992) for two Danish rivers (143 and 284 kg ha^{-1}), by Ongley (1973) for Canadian tributaries to Lake Ontario, or by Walling (1978) and Foster et al. (1985) for two British rivers. The PM yields that we observed in 1996 were much lower (37 and 40 kg ha^{-1}); Webb & Walling (1984) reported similar values during a nine-year study of a British river. Jigorel (1992) showed that erosion input into the Lake of Grand-Lieu was mainly clay and limon particles (average particle size: 3.6 μ for the Boulogne, 8.8 μ for the Ognon), with relatively little particulate organic matter (17% for the Boulogne, 23% for the Ognon), and many mineral (quartz) and Diatoms skeletal. This input would represent about 32% in 1994 and only 9% in 1996 of the estimated annual production of sediments within the lake, obtained by a model and sedimentation traps in 1990–91 (28800 tons, Marion et al., 1992). Net retention (input minus output) would only represent 4.3% and 1.8% of this internal production, respectively.

The dynamics of PM inputs conforms to those of other lakes, particularly shallow lakes such as Balaton where wind plays a major role (Sebestyen, 1950; Campy & Meybeck, 1995). Sediments rapidly deposit in the proximate eastern parts of the lake, and small particles are partly resuspended by wind, transported to the center, and finally the western part of the lake (Jigorel, 1992). During winter, when the sluicegate is opened, outputs of sediments mainly concern autochthonous production of the lake, with a mean size of 2.0 μ and much of particulate organic matter (26-39%). Most of this output is sediment resuspended by wind in the western part of the lake, i.e. the macrophyte area, where organic matter represents from 23 to 40% of D.W. (Rofès, 1992). Thus, even if the global retention of sediments in the lake is small, the real sedimentation of PM input is probably higher that it appears here, partly due to its higher density (168 g l^{-1}) than autochthonous sediments (110 g l^{-1} at the surface of sediment), and because concentrations in inflows are higher at the beginning of inflow, when the exit is still closed.

Preliminary results from a recent study based on ^{210}Pb profiles indicate an annual accumulation of about 0.4 cm y^{-1} in the macrophyte area (Creach & Marion, in prep.). A first model concluded that the accretion rate was 1.4 cm y^{-1}, assuming that accumulation occurred on about 1100 ha (Marion et al., 1992), a probable underestimate. Both values confirm that net retention in the lake from inputs from the watershed (input minus output) is considerably lower than internal production. Total accumulation obtained with ^{210}Pb method is similar to that of Lac des Allemands in Louisiana (0.44-0.81 cm y^{-1}, Stow et al., 1985), but lower than that in other American lakes: 1.7 cm y^{-1} in Big Lake (Eckblad et al., 1977), 2.00 in lake Ellyn (Striegl, 1987) and even 2.6 for Capitol Lake (DeLaune et al., 1989). These are the highest values reported for wetlands, after Johnston (1991) who cited a mean of 0.69 cm y^{-1} (N = 29 wetlands with mineral sediments). Sedimentation can be even higher in ditches in freshwater marshes in Western France (2-6 cm y^{-1}, Giraud, 1992; Feunteun, 1992). Grand-Lieu represents a record of sedimentation if we only consider wetlands with organic soils (Johnston, 1991), with a mean of 0.12 cm y^{-1} (N = 13).

It is important to consider here, for PM and for nutrient retention (see below), the degree of closure of systems (related to the velocity of flow through wetlands), that increases along the four succession stages of the lake. The retentions mentioned in the present study are assumed to be an average at the scale of the lake, since effective retentions are certainly different in each area. The central area, directly connected to tributaries and sensitive to wind, is the most open and the most permanent in aspect throughout year. Its water quality follows that of tributaries (Marion et al., 1986). The floating macrophyte area has the same characteristics as the latter in winter, but becomes increasingly closed during spring. Due to decreasing water level, the peat bog is completely closed from the end of spring to autumn, being partially opened with winter flooding (Marion et al., 1994). The fourth area, the marsh grassland, is a man-created area from the peat bog by cutting all the trees. It is entirely disconnected from the lake when water level is low from spring to autumn, and even during floodings its water is filtered by the peat bog. There is no wind-resuspension in this area. Accumulation of sediment is important in the macrophyte area and in the adjacent peat bog, but nil in the centre (despite high turbidity) and in the grassland areas (no turbidity). Parts of peat bog that are close to the macrophytes probably receive most sediment, because reedbeds and trees suppress waves and currents (Prentki et al., 1979). This sedimentation decreases from the central area to the grassland area.

The water level was lower in 1996, decreasing both level and time of flooding and thus linkage of central water and the peat bog. These conditions may explain part of the differences in global retention between years, and the differences in nutrient retention.

Nitrogen

There are few studies about input-output of nitrogen in large wetlands crossed by streams, for which total N retention is comparable to Grand-Lieu: for example 39% and 50% in two swamps, and about 50% in a bog in the USA, although with only 0.68, 1.33 and 0.66 g m^{-2} y^{-1}, respectively (Johnston, 1991), much lower than in Grand-Lieu (about 10–15 g). This result is surprising because swamps and bogs are more closed than Grand-Lieu, and the percentage of nutrient inputs retained by a system is strongly related to its degree of closure (Hopkinson, 1992).

In contrast to the retention rate of sediments, the retention rate of total N in 1996 was twice more important than in 1994 (60–32%), related to the large difference in inputs, while loss was less variable. We consider than Grand-Lieu was a more closed system in 1996 than in 1994, due to a longer water residence time. Even though flow occurs in this lake during winter, retention of nitrates is important (61% in 1994, 86% in 1996), although retention per m^2 was higher in 1994 (22 g against 11). In 1996, higher retention rate gave lower concentrations in outputs in spite of higher concentrations in inputs.

Apart from the role of the flow and of residence time of water on nutrient retention, it is difficult to interpret all these 'black-box' input-output differences with internal lake processes because of the complexity of the N cycle in wetlands, for which Johnston (1991, adapted from Nixon & Lee, 1986) give seven kinds of standing stocks and 28 ways of fluxes. However, nitrogen fluxes seem essentially concerned at lake Grand-Lieu by inputs and outputs by rivers, consumption by vegetation, storage in sediments and denitrification, because of absence of exchanges with groundwater, negligible flux with precipitation (0.14 mg l^{-1}), and probably low fixation. Even though asymbiotic fixation, mainly from Cyanobacteria, can be important in eutrophic lakes (Labroue et al., 1995), it could partly be inhibited in the center of this lake by several factors (high levels of inorganic N, O^2, pH and turbididy). Symbiotic fixation by the very localized *Alnus* in the peatbog does not occur in winter and is moreover incorporated in wood and only released with decay. Before direct accumulation in sediments, the main retention factor during winter seems consumption by algae and denitrification, that could explain the strong decrease of nitrates from January to the end of March (Figure 4E and 5E), even if algae density is much lower (about 36 mg l^{-1} chlorophyll a in March) than during spring and summer (about 380 mg l^{-1} in June). From April onwards, concentration of nitrate in the water of the lake is very low, partly due to adding consumption by growing macrophytes and their periphyton. Between April and August, the *Nymphea alba-Nuphar lutea* area is the most depleted in total N (47 mg l^{-1}), followed by the less nitrophilous *Trapa natans* (110 mg l^{-1}), while proximate areas without macrophytes have the highest concentrations (137 mg l^{-1}, Marion et al., 1986). By adding *Limnanthemum peltata*, the standing stock of these floating macrophytes reached 105 tons of N (12 g m^{-2}) in summer 1996, representing about 19% of the net input of the lake during the preceding winter, a high level if we consider that these plants only represent about 25% of the total macrophytes biomass in this lake (Marion et al., 1989). Nevertheless, if we refer to literature, important part of this stock is directly taken in sediments and, except for annual *Trapa*, by retranslocation from roots, before partly returning in them by translocation before decay (Wetzel, 1990; Pieczynska, 1993). The total N stock in the first 10 cm of sediments represents in September at Grand-Lieu about 155 g m^{-2} in the macrophytes area and about 227 g m^{-2} in the sandy open water without macrophytes (after Rofès, 1992), totaling about 3 500 tons for the lake (without the peatbog area). While important, this stock only represents 4 to 6 times the annual net input of the lake, and the lower level in the macrophytes area could illustrate plant consumption, because these organic soils generally average twice as much N as mineral soil (Johnston, 1991). Macrophytes are supposed to pump nutrients from sediment, increasing nutrients concentration in water (Wetzel & Manny, 1972; Testard, 1995), but this opinion is controversial (Labroue et al., 1995) and seems to apply to submerged macrophytes. We consider here that the part of nutrients directly taken from the water by floating macrophytes and associated periphyton is probably not negligible and may contribute to decrease the nitrogen stock in water, even after decay of leaves. Jigorel (1992) showed that accumulation of total N in the deposited sediment in this lake was higher under the macrophytes during all the spring and summer, and reached 1.2% of total deposition. Yet, part of the latter is exported during winter by wind resuspension.

Denitrification was not measured and could occur mainly in the macrophyte area, at the water-sediment interface, and within the peatbog, where alternating aerobic and anaerobic conditions exist in space and time, a favourable situation for denitrification (Smith & Patrick, 1983). However, all muddy sediments under macrophytes are anoxic even at the surface, while the water column is generally aerobic even in summer (except in hot shallow areas, Marion, 1995), and always in winter (Marion, 1992), conditions that can limit denitrification (Reddy et al., 1980). Also, nitrates required for denitrification are provided by rivers only during winter, and their strong decrease before May could limit denitrification in such a lake. Labroue et al. (1995) reported a similar case of an eutrophic gravel pit lake (St-Caprais) where a strong decrease of nitrate occurs from March to June, when water has about 50 mg l^{-1} chlorophyll a before increasing up to 150 mg l^{-1} in summer. In this lake denitrification only occurs from December to May. At Grand-Lieu, the part of nitrates due to nitrification of ammonium from the inputs is probably relatively low during winter, because of its low level (ammonium only represented 2.5% and 5.5% of total N input in 1994 and 1996), and its low retention by the lake (16-18 times less than nitrates in flux). Nitrification is probably mainly supported by direct upward diffusion of NH_4 from sediments.

Finally, all these processes of N cycling during the closing watergate period involve low concentrations of total N in water of the lake (0.95 mg l^{-1} from April to September, Marion et al., 1994), before the beginning of new input from catchment area. This depletion in concentrations corresponds to a higher decrease in N stock of the water column because the simultaneous decrease in water level by 40 cm (about 33% of the volume of April), due to evapotranspiration, concentrates nutrients. Globally, this wetland receives an important discharge of inorganic nitrogen from its catchment area, and keeps most of it (62% in 1994 and 85% in 1996), but exports a large amount of dissolved and particulate organic nitrogen (254 tons = 4.54 g m^{-2} y^{-1} in 1996, 896 tons = 16 g m^{-2} y^{-1} in 1994), 1.9 to 2.6 times more than it receives (Figure 7 and 8). These characteristics are mainly related to the importance of vegetation in this hypereutrophic lake, and of wind resuspended organic sediments in outputs. In his review of input-output studies, Johnston (1991) only reported one wetland (Okefenokee Swamp, Blood, 1981) with a negative retention of organic N, moreover with considerable lower value (only -9%) than for Grand-Lieu. More recently Hopkinson (1992) mentioned net exportation of organic N for the same wetland, with higher rates (-22%). Grand Lieu is both an opened and a closed system, favourizing retention of nutrients. Thus, its peat bog is relatively similar to swamps, with all the characteristics of high closure of old succession stages: high vegetation density that increases physical deposition of sediments during flooding, high biomass and storage capacities for N (long term or definitely stocked in wood and peat). However because of its proximity of the central area of the lake and of periodic important floods, this area is also largely more opened than swamps which receives most of their water budget from rain (77%, Hopkinson, 1992). The importance of vegetation on nutrient budgets of Lake Grand-Lieu is confirmed by organic C concentrations of outputs, higher than inputs (about 25 mg against 10), as for other bog lakes for which dissolved organic C can reach 40 mg l^{-1} in flooded areas (Gorham et al., 1985). In the sediments within the lake, all C is organic (Rofès, 1992). Grand-Lieu is mainly an autotrophic system, converting most of inorganic nutrients inputs to organic matter, which is only partly exported.

Phosphorus

Most considerations on nitrogen also apply to phosphorus. The review by Johnston (1991) about retention in wetlands mentioned similar retention rates for N and P in three studies in which both nutrients were studied, and a P retention varying from 4% to 80% (average 45%) for 11 wetlands where input-output studies were done. Only one lake was concerned, for which retention was only 10%, however with similar retention per m^2 than at Grand-Lieu in 1994 (0.56). The retention per m^2 at Grand-Lieu in 1996 (0.09 g) was the minimum cited by Johnston (1991), but it was higher than those observed in 7 major lakes on Iowa streams (0.03 g after Jones et al., 1976). Only one study (Kuenzler et al., 1980) compared two years with different flows, with a retention rate twice as high during large flow (57%) as during low flow (30%), due to increased sedimentation and uptake by algae. It is the same magnitude than that we observed at Grand-Lieu (40–18%), with also large difference in PM input, sedimentation (22 g m^{-2} against 9) and algae production (Le Rouzic et al., in prep.). This is not surprising, because this relationship of phosphorus with PM is known (Walling & Kane, 1982; Probst, 1985). However, Kronwang (1992) showed that there was no difference in transport of total P between two Danish rivers although a strong difference (twice more) in PM transport. In our study, even inputs of orthophosphate, although known to be related to storm run-off, had a lower difference between years than nitrites and total N. As for these elements, difference must be done between global relation concerning water flow and fluxes of nutrients, related to increase of loading from land, and on the other hand the punctual increase of P with first flow in rivers, partly due to storage of sewage P during low flow, that can reach 70% according to Dorioz et al. (1989).

In the present paper, we do not take into account net inputs by bird droppings, retention rates being only calculated from inputs-outputs of rivers. In fact, real total P retention rates of this lake would increase by about 1.7% in 1994 and 6.5% (total retention rate = 25% instead of 18%) in 1996, according to the study made on this site in 1990–91 (about 2.5 tons P, Marion et al., 1994). Birds do not really change the N input.

The lake's retention of P is obviously more important for orthophosphate (79–72%) than for total P. However, contrary to N, total P concentrations do not really decrease in spring and summer in the lake (see Marion et al., 1994). These similar concentrations correspond in fact to a decrease of the P stock because of the simultaneously decrease of water volume (by 33%, see above for N). This limited loss compared to N may be partly due to release of phosphorus by sediments under anaerobic conditions on the water-sediment interface (Boström et al., 1982, Wetzel, 1983). The total P stock in the first 10 cm represents in September at Grand-Lieu about 12 g m^{-2} in the macrophyte area and about 20 g m^{-2} in the sandy open water (mainly non apatitic inorganic phorphorus, the more easily used by plants), that represents without the peatbog area about 284 tons, 8 (1994) to 53 (1996) times the annual net input into the lake. Such a concentration in sediment exceeds the range cited by Labroue et al. (1995) for lakes (2–10 g m^{-2}). Like for nitrogen, the lower sediment concentration in the macrophytes area could illustrate plant uptake, because P concentrations are generally compararable for both organic and mineral soils (Johnston, 1991). The standing stock of the floating macrophytes mentioned above for N reached 13.5 tons of P (1.6 g m^{-2}) in summer 1996, representing about 2.5 times the net input of the lake that year but 39% in 1994. Probably algae play a major role in P retention from the water, rather than macrophytes that take most of their P in sediment. In contrast to N, there is effectively no difference from April to August in P water concentrations between *Trapa* area, *Nymphea-Nuphar* area and open water without microphytes. This can also be due to release from sediments and from regular decay of leaves, like in other lakes (Wetzel, 1990; Pieczinska, 1993; Labroue et al., 1995). The final and important decay of whole macrophytes in autumn is not noticeable in water P compared to autumnal inputs from the watershed.

Contrary to N, there is no loss of P by atmospheric exchange, so its stock increases continually in the system due to increasingly net input. In the 1960's, before the tremendous increase of sewage and agriculture run-off, inputs of P was very low and the lake was oligotrophic, probably receiving up to 95% of its P input from birds (Marion et al., 1994). Retention is probably lower now, because the actual input exceeds its retention capacities and makes the lake endangered. Such a decrease of retention rate with eutrophication has been demonstrated for lac Léman between the 1960's and the 1980's (Jacquet, 1985). The very low nutrients input occurring in 1996 involved a low turbidity in spring and summer due to low density of algae, and a return of submerged macrophytes, that disappeared after 1984. Water was also better oxygenated, due to higher water level because botulism also disappeared this year, after a dramatic outbreak in 1995 that killed 30 000 to 50 000 birds (Marion, 1995). This spectacular change illustrates the sensibility of this system to winter nutrient input and water level, and the multiplying effect of additional nutrients on the eutrophication processes when oxygen concentration drops in the water and favours mobilization of nutrients stocked in the sediment.

Acknowledgements

This study was supported by grants from the Life Grand-Lieu Program (European Community, Ministère de l'Environnement, Agence de l'Eau Loire-Bretagne, Région Pays de la Loire, Département de Loire Atlantique), directed by L. Marion for the Société Nationale de Protection de la Nature and the Centre National de la Recherche Scientifique. We thank the SEMA Pays de Loire for measures of runoff of tributaries, P. Boret for its help on the field, V. Bouchard, J. Cl. Lefeuvre, E. Feunteun and B. Le Rouzic for improvement of the manuscript.

References

Blood, E. R., 1981. Environmental impact of upland streams on the Okefenokee Swamp. University of Georgia Institute of Ecology Rep. N^o ERC 06-81, Athens, GA.

Boström, B., M. Jansson & C. Forsberg, 1982. Phosphorus release from lake sediments. Arch. Hydrobiol. Ergebn. Limnol. 18: 5–59.

Campy, M. & M. Meybeck, 1995. Les sédiments lacustres. In Pourriot, R. & M. Meybeck (eds), Limnologie générale. Masson, Paris: 185–226.

DeLaune, R. D., J. H. Whitcomb, W. H. Patrick, J. H. Pardue & S. R. Pezeshki, 1989. Accretion and canal impacts in a rapidly subsiding wetland. I. ^{137}Cs and ^{210}Pb techniques. Estuaries 12: 247.

Dolan, D. M., A. K. Yui & R. D. Geist, 1981. Evaluation of river load estimation methods for total phosphorus. J. Grt Lakes Res. 7: 207–214.

Dorioz, J. M., E. Pilleboue & A. Ferhi, 1989. Phosphorus dynamics in watersheds: role of trapping processes in sediments. Wat. Res. 23: 147–158.

Eckblad, J. W., L. L. Peterson, K. Ostlie & A. Temte, 1977. The morphometry, benthos and sedimentation rates of a floodplain lake in Pool 9 of the upper Mississippi River. Am. Midl. Nat. 97: 433–443.

Feunteun, E., 1994. Le peuplement piscicole du marais littoral endigué de Bougneuf-Machecoul (France, Loire Atlantique). Thèse Univ. Rennes, 256 pp.

Foster, I. D. L., J. A. Dearing, A. Simpson, A. D. Carter & P.G. Appleby, 1985. Lake catchment based studies of erosion and denudation in the Merevale catchment, Warwickshire, UK. Eth Surf. Proc. Land Forms 10: 45–68.

Gersberg, R. M., B. V. Elkins, S. R. Lyon & C. R. Goldman, 1986. Role of aquatic plants in wastewater treatment by artificial wetlands. Water Res. 20: 363–388.

Giraud, F., 1992. Modélisation hydrologique d'une zone humide agricole: perspective pour l'étude du transport d'azote et de phosphore dans le réseau hydraulique. Cas du marais de Moëze. Thèse Univ. Rennes, 224 pp.

Golterman, H. L., R. S. Clymo & M. A. N. Ohnstad, 1978. Methods for physical and chemical analysis of fresh waters. I.B.P. Handbook n° 8, Blackwell Scientific Publications.

Golterman, H. L., P. G. Sly & R. L. Thomas, 1983. Study of the relationship between water quality and sediment transport. A guide for the collection and interpretation of sediment quality data. UNESCO, Paris, 231 pp.

Gorham, E., S. J. Eisenreich, J. Ford & M. V. Santelman, 1985. The chemistry of bog waters. In W. Stumm (ed.), Chemical Processes in Lakes: 339–363.

Hopkinson, C. S., 1992. A Comparison of Ecosystem Dynamics in Freshwater Wetlands. Estuaries 15: 549–562.

Jacquet, J. M., 1985. Bilan et modèle d'entrées-sorties du phosphore total dans le Léman. Annls Limnol. 21: 177–189.

Jigorel, A., 1992. Etude sédimentologique du lac de Grand-Lieu. Rapport de l'Institut National des Sciences Appliquées de Rennes, 83 pp.

Johnston, C. A., 1991. Sediment and nutrient retention by freshwater wetlands: effects on surface water quality. Crit. Rev. Envir. Contr. 21: 491–565.

Johnston, C. A., N. E. Detenbeck & G. J. Niemi, 1990. The cumulative effect of wetlands on stream water quality and quantity. A landscape approach. Biogeochemistry 10: 105–141.

Jones, J. R., B. P. Borofka & R. W. Bachmann, 1976. Factors affecting nutrient loads in some Iowa streams. Wat. Res. 10: 117–122.

Kronwang, B., 1992. The export of particulate matter, particulate phosphorus and dissolved phosphorus from two agricultural river basins: implications on estimating the non-point phosphorus load. Wat. Res. 26: 1347–1358.

Kuenzler, E. J., P. J. Mulholland, L. A. Yarbro & L. A. Smock, 1980. Distributions and budgets of carbon, phosphorus, iron and manganese in a floodplain swamp ecosystem. Water Resources Research Institute. University of North Carolina, Raleigh, Report n° 157.

Labroue, L., J. Capblancq & A. Dauta, 1995. Cycle des nutriments: l'azote et le phosphore. In Pourriot, R. & M. Meybeck (eds), Limnologie générale. Masson, Paris: 727–764.

Lee, G. F., E. Bentley & R. Amundson, 1975. Effects of marshes on water quality. In Halser, A. D. (ed.), Coupling of Land and Water Systems. Ecol. Stud. 10, Springer-Verlag, Berlin: 105-127.

Marion, L. 1992. Productivité des macrophytes et du phytoplancton du lac de Grand-Lieu. SESLG-Univ. Rennes, 57 pp.

Marion, L., 1995. La catastrophe écologique du lac de Grand-Lieu: des dizaines de milliers d'oiseaux meurent de botulisme. La Recherche 283: 28–29.

Marion, L., 1997. Désenvasement: une expérience de génie écologique à grande échelle sur le lac de Grand-Lieu. In Forum des Gestionnaires 'la gestion de l'eau pour les zones humides', Ministère de l'Environnement, RNF, ENF, Paris (in press).

Marion, L., P. Marion, A. Bocquenet, L. Brient, 1986. L'eutrophisation des eaux du lac de Grand-Lieu. Ministère de l'Environnement-SESLG-LESNM-Univ. Rennes, 109 pp.

Marion, L., G. Bertru, L. Brient, M. Danais, A. Desgrée, A. Durbec, A. Dutartre, G. Galea, A. Jigorel, G. Rofès & P. Marion, 1992. Causes de l'envasement du lac de Grand-Lieu et mesures de sauvetage. Rapport de synthèse. Ministère Environnement, 72 pp.

Marion, L. , P. Clergeau, L. Brient & G. Bertru, 1994. The importance of avian-contributed nitrogen (N) and phosphorus (P) to Lake Grand-Lieu, France. Hydrobiologia 279/280: 133–147.

Marion, L. & P. Marion, 1975. Contribution à l'étude écologique du lac de Grand-Lieu. Bull. H.S. Soc. Sc. Nat. Ouest France, 611 pp.

Marion, L., P. Marion & Ph. Clergeau, 1989. Causes and consequences of accelerated eutrophic evolution of the lake of Grand-Lieu, France. Proc. Third Intern. Conf. Wetland, Rennes September 1988: 219–220.

Meybeck, M., 1982. Carbon, nitrogen and phosphorus transport by world rivers. Am. J. Sci. 282: 401–450.

Meybeck, M., 1993. C, N, P and S in rivers: from sources to global inputs. In: Wollast, R., F. T. Machenzie & L. Chou (eds), Interaction of C, N, P and S Biogeochemical Cycles and Global Change. Springer Verlag, Berlin: 163–193.

Meybeck, M., 1995. Les lacs et leurs bassins. In R.Pourriot & M. Meybeck (eds), Limnologie générale. Masson, Paris: 6–59.

Mitsch, W. J. & J. G. Gosselink, 1986. Wetlands. Van Nostrand Reinhold, New York, 534 pp.

Murphy, J. & J. P. Riley, 1962. A modified single-solution method for the determination of phosphate in natural waters. Analyt. Chim. Acta 27: 31–36.

Nixon, S. W. & V. Lee, 1986. Wetlands and water quality, Wetlands Research program. Tech. Rep. Y-86-2, U.S. Army Engineer Waterways Experiment Station, Vicksburg, MS.

Odum, E. P., 1969. The strategy of ecosystem development. Science 164: 262–270.

Ongley, E. D., 1973. Sediment discharge from Canadian basins into lake Ontario. Can. J. Eth. Sci. 10: 146–156.

Phillips, J. D., 1989. Fluvial sediment storage in wetlands. Water Resour. Bull. 25: 867.

Pieczynska, E., 1993. Detritus and nutrient dynamics in the shore zone of lakes: a review. Hydrobiologia 251: 49–58.

Prentki, R. T., M. S. Adams, S. R. Carpenter, A. Gasith, C. S. Smith & P. R. Weiler, 1979. The role of submersed weedbeds in internal loading and interception of allochthonous materials in Lake Wingra, Wisconsin, USA. Arch. Hydrobiol. Suppl. 57: 221–250.

Probst, J. L., 1985. Nitrogen and phosphorus exportation in the Garonne Basin (France). J. Hydrol. 76: 281–305.

Reddy, K. R., P. D. Sacco & D. A. Graetz, 1980. Nitrate reduction in an organic soil-water system. J. Envir. Qual. 9: 283–288.

Rofès, G., 1992. Lac de Grand-Lieu: étude sédimentologique. Rapport du CEMAGREF de Lyon, 64 pp.

Seux, R., B. Sorelard, P. Boutin & J. P. Bechac, 1985. Azote et phosphore dans un cours d'eau du nord de la Bretagne: origine des apports, estimation des flux. Trib. Cebedeau 38: 11–24.

Smith, C. J. & W. H. Patrick, 1983. Nitrous oxide emission as affected by alternate anaerobic and aerobic conditions from soil suspensions enriched with ammonium sulfate. Soil Biol. Biochem. 15: 683.

Stow, C. A., R. D. DeLaune & W. H. Patrick, 1985. Nutrient fluxes in a eutrophic coastal Louisiana freshwater lake. Envir. Mgmt 9: 243.

Striegl, R. G., 1987. Suspended sediment and metals removal from urban runoff by a small lake. Wat. Res. Bull. 23: 985.

Testard, P. 1995. Rôle des macrophytes littoraux dans le fonctionnement des écosystèmes lacustres. In Pourriot, R. & M. Meybeck (eds), Limnologie générale. Masson, Paris: 296–326.

Walling, D. E., 1978. Suspended sediment and solute response characteristics of the the River Exe, Devon, England. In Research in Fluvial Geomorphology, Geo-Abstracts, Norwich: 169–197.

Walling, D. E., 1988. Erosion and sediment yield research – some recent perspectives. J. Hydrol. 100: 113–141.

Walling, D. E. & P. Kane, 1982. Temperal variation of suspended sediment properties. In Recent Developments in the Explanation and Prediction of Erosion and Sediment Yields, Proceedings of the Exeter Symposium. IAHS, Publ. n° 137: 409–419.

Webb, B. W. & D. E. Walling, 1984. Magnitude and frequency characteristics of suspended sediment transport in Devon Rivers. In Catchement Experiments in fluvial Geomorphology, Geobookds, Norwich: 399–415.

Wetzel, R. G. 1983. Limnology. Saunders College Publ., Philadelphia, 760 pp.

Wetzel, R. G. 1990. Land-water interfaces: Metabolic and limnological regulators. Verh. int. Ver. Limnol. 24: 6–24.

Wetzel, R. G. & B. A. Manny, 1972. Secretion of dissolved organic carbon and nitrogen by aquatic macrophytes. Verh. int. Ver. Limnol. 18: 162–170.

Hydrobiologia **373/374:** 237–246, 1998.
J. C. Amiard, B. Le Rouzic, B. Berthet & G. Bertru (eds),
Oceans, Rivers and Lakes: Energy and Substance Transfers at Interfaces.

The input of nutrients by the Rhône river into the Mediterranean Sea: recent observations and comparison with earlier data

T. Moutin, P. Raimbault, H. L. Golterman* & B. Coste
Laboratoire d'Océanographie et de Biogéochimie, Centre d'Océanologie de Marseille, Campus de Luminy, case 901, F-13 288 Marseille cédex 09, France
* *Mas dou Bòchi, Chemin de Tintarlot, F-13 200 Arles, France*

This paper is dedicated to the memory of Kees De Groot

Key words: nutrients, Rhône river, Gulf of Lions (Golfe du Lion), Mediterranean Sea

Abstract

The nutrient input by the Rhône river into the Mediterranean Sea was measured from June 1994 to May 1995. The various fractions of N and P (dissolved and particulate, organic and inorganic) as well as chlorophyll and dissolved silicate concentrations were measured twice a month at Arles, 40 km upstream of the river mouth, in this period. In addition, some samples were taken when the river was in high flood.

The study shows that nitrogen is mainly present as nitrate (76%). Dissolved and particulate organic nitrogen represent 8 and 9%, ammonium 5% and nitrite 1% of total nitrogen respectively. Almost half of the phosphate is particulate phosphate, the largest part of which is calcium-bound phosphate. Orthophosphate represents 31% of total phosphate.

We estimated the total input of nitrogen, phosphate and silicate as 115–127 kt y^{-1} (N), 6.5–12.2 kt y^{-1} (P) and 135–139 kt y^{-1} (Si) by different methods. For nitrogen an annual input of 92.3 to 96.1 kt of nitrate is found, 1.3 to 1.5 kt of nitrite, 6.3 to 6.7 kt of ammonium, 9.7 to 9.8 kt of dissolved organic nitrogen and 5.3 to 12.7 kt of particulate nitrogen; for phosphate the annual imput was 2.7 to 3.0 kt of orthophosphate, 1.3 to 1.7 kt of dissolved organic phosphate & polyphosphate and 2.5 to 7.5 kt of particulate phosphate. While the N-input was mainly in the form of nitrate, a large part of the phosphate input was particulate-P. Comparisons to previous estimations show that the mean annual nitrate concentration in the Rhône has increased by about 50% during the last two decades.

Introduction

Nutrient input by rivers contributes considerably to the nutrient balance of the Mediterranean Sea. The Rhône is the main river flowing into the western basin of the Mediterranean Sea. Its mean flow rate of 1690 $m^3 s^{-1}$, estimated over a 30 y period, is three times higher than the sum of that of the Ebro (200 $m^3 s^{-1}$), the Arno (103 $m^3 s^{-1}$) and the Tiber (234 $m^3 s^{-1}$) (Martin and Saliot, 1992).

The input of nutrients has a direct influence on primary production in the Golfe du Lion. On an annual basis, around 50% of the primary production in the Gulf can be attributed to terrigenous input (Coste, 1974, Morel et al., 1990). The nitrogen or phosphate concentration is generally considered to be a limiting factor for algal production in the Mediterranean (Berland et al., 1980). An increase in the input may cause a corresponding increase in primary production leading to changes in the functioning of the Mediterranean ecosystem. Therefore it is necessary to know the long-term variability of the input.

The aim of this work is to specify the concentrations of the various fractions (dissolved or particulate, organic or inorganic) of nitrogen and phosphate in the Rhône near the estuary, to estimate their input into

Figure 1. Map of the study area. A = Arles.

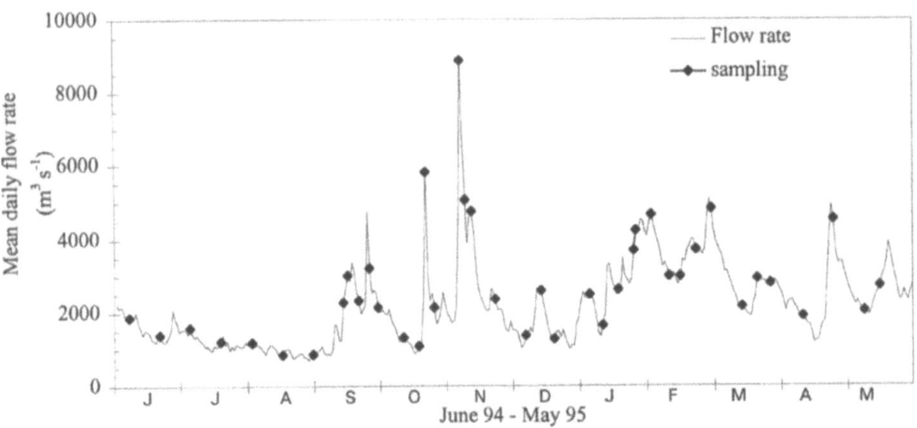

Figure 2. Mean daily flow rate and sampling date.

the Mediterranean Sea and to discuss their long term changes.

Material and methods

The study of the nutrient input by the Rhône into the Mediterranean Sea was carried out from 01/06/1994 to 31/05/1995. The various fractions of nitrogen and phosphate as well as dissolved silicate were measured twice a month in Arles (Figure 1). Additional results obtained from discrete sampling during flood are indicated in Figure 2. Chlorophyll *a* and phaeopigment concentrations were also measured. Samples were kept cooled and, if necessary, filtered over GF/F membranes and/or frozen at – 20 °C before analysis in the laboratory.

Chemical analyses

The methods used to determine nutrient (nitrate, nitrite, ammonium, orthophosphate and dissolved silicate) concentrations are described in Tréguer & Le Corre (1975). Particulate and dissolved organic nitrogen and phosphate are measured according to Pujo-Pay and Raimbault (1994). Tot-N and Tot-P correspond to the sum of these various fractions. Chlorophyll and phaeopigment concentrations are determined according to the fluorometric method described by Herbland et al. (1984).

An analysis of the various fractions of particulate phosphate has also been carried out on the suspended matter collected during four three-day sampling periods (14, 15 and 16 Oct. 92, March, May and June 1993). The protocol used is that of Golterman & Booman (1988) and of De Groot & Golterman (1990). Concentrations are expressed as mg l^{-1} (N, Si) or μg l^{-1} (P). N and P concentrations in particulate matter are expressed as mg g^{-1}. Loadings are expressed as t y^{-1} (N, P or Si).

Input calculation

The input into the Mediterranean Sea corresponds to the mass fluxes entering the estuary area (in tons per year). They have been estimated by the two most classical (Walling & Webb, 1985; Thomas & Meybeck, 1992) and two other methods ($*$) which only differ from the precedent in the estimation of the particulate fluxes.

Method 1: based on the calculation of weighted average

The fluxes are calculated from the concentrations C_i (mg f^{-1}) and c_i (mg g^{-1}) of the different variables listed above, the concomitant flow rates Q_i and suspended matter concentrations S_i, and the total water and solid flow rates over the year studied (respectively $Q°$ et $S°$). The latter were estimated, from records of mean daily water and solid flow rates, at 70 10^9 m^3 and 14 10^6 t (Pont & Bombled, 1996).

The method amounts to the estimation of the yearly dissolved flux F_d from a water discharge weighted average concentration CE_d, and of the yearly particulate flux F_p from a sediment discharge weighted average concentration CE_p.

For dissolved compounds:

$$CE_d = \frac{\sum_{i=1}^{n} Q_i \cdot C_i}{\sum_{i=1}^{n} Q_i}$$

For particulate compounds:

$$CE_p = \frac{\sum_{i=1}^{n} S_i \cdot c_i}{\sum_{i=1}^{n} S_i}$$

with n: number of samples and $F_d = Q^o \cdot CE_d$; $F_p = S^o \cdot CE_p$.

*Method 1**: As method 1 but considering particulate fluxes as dissolved fluxes.

Method 2: based on a linear interpolation

The annual flux is the sum, over the year, of the product for each period j between two samples, of the mean value for two successively measured concentrations $\overline{C}_{dj}$ (or $\overline{c}_{pj}$) by the mean value of the mean daily flow rate $\overline{Q}_i$ (or solid flow rate $\overline{S}_j$) measured during this time interval:($k_j - 1$) days (i.e. extrapolation of concentration).

For dissolved compounds:

$$F_d = \sum_{j=1}^{n-1} \overline{C_{d}j}, \overline{Q_j}(k_j - 1)$$

For particulate compounds:

$$F_p = \sum_{j=1}^{n-1} \overline{c_{p}j}, \overline{S_j}(k_j - 1)$$

Method 2:* As method 2 but considering particulate fluxes as dissolved fluxes.

For methods 1 or 2, the total input of a compound is equal to the sum of its dissolved and particulate fluxes: $F = \Sigma F_d + \Sigma F_p$

Methods 1* & 2* do not take into account the state (dissolved or particulate) of the compounds studied. The fluxes are calculated by adding up the fluxes of the various compounds all calculated as if they were dissolved coumpounds.

Results and discussion

Hydrology

The mean flow rate of the Rhône river calculated for the period 1961–1994 is 1690 $m^3 s^{-1}$ (s.d. = 320, n = 34). the minimum and maximum annual discharge measured being 1064 and 2388 $m^3 s^{-1}$ respectively.

The annual cycle studied is characterized by a mean annual flow rate of 2220 $m^3 s^{-1}$ which corresponds to a high flow rate, by a minimum flow rate of about 1000 $m^3 s^{-1}$ and by numerous floods, one of which was exceptional on the 6th of November 1994: the probability of this event is once in 30 yr. Figure 2 gives daily flow rate and times of sampling.

The yearly inputs are calculated from concentrations and flow rates. Concentrations are measured at Arles and flows at Beaucaire, the distance between the two places is about 20 km (Figure 1). The branching off of the Rhône river near the mouth is considered to have no influence on the input, as the water flow is divided but the concentrations are similar in the 'Petit Rhône' and in the 'Grand Rhône'.

Several reservoirs are located even in the lower part of the Rhône basin, which influence the relation between flow rate and suspended matter. The lake of Geneva which intercepts about 7000 km^2 of the drainage area retains the major part of the suspended matter of glacier origin.

Otherwise, inter-annual variations of the flow rate are significant: the mean annual flow rate can double as the values of the minimum and maximum show during the period 1961–1994. This means that extrapolation to other years is only allowed if the relation between concentration and flow rate is known.

Concentration analysis:

The results obtained for the different variables measured between June 1994 and May 1995 are given in Table 1. They concern arithmetic means as well as standard deviations, minimum and maximum values, weighted average concentrations, medians and quantiles, and total number of analyses. The discharge-weighted averages of Tot-N, Tot-P and dissolved silicate are respectively 1.73 mg l^{-1} (N), 0.12 mg l^{-1}(P) and 1.99 mg l^{-1}(Si). Nitrogen is mainly present as nitrate (76%). Org-N_{diss} and N_{part} account respectively for 8% and 9% of Tot-N, ammonium for 5% and nitrite for 1%. The major part of the phosphate is P_{part}. o-P represents only 31% of Tot-P.

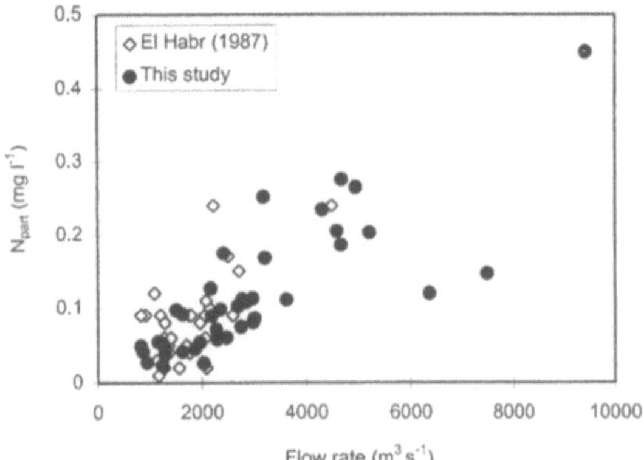

Figure 3. N_{part} concentration versus Q. Data from El Habr (1987) and from this study.

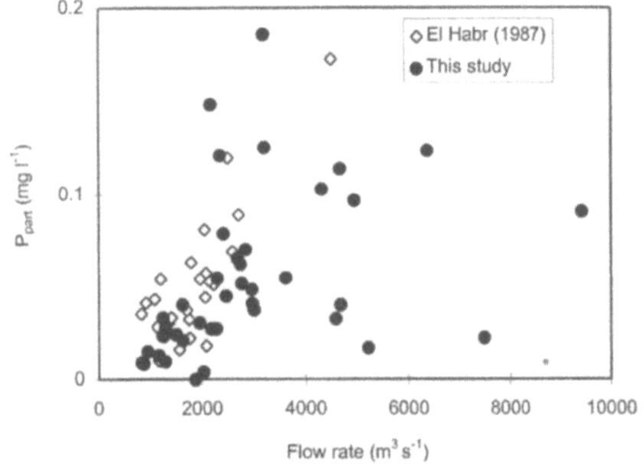

Figure 4. N_{part} concentration versus Q. Data from El Habr (1987) and from this study.

Considering the importance of P_{part}, it seems essential to specify the various forms present, in order to estimate their fate in the Golfe du Lion. Table 2 gives the results of four analyses of the P-fractions carried out on the suspended matter sampled between Oct. 92 and June 1993. All flow rates were somewhat below the average, but not exceptionally. Iron-bound P ranged between 5–34%, Ca-bound P between 43–56%, Acid Soluble Organic P between 6–28% and NaOH Soluble Organic P between 11–13%. Ca-bound P is therefore the largest fraction, which is due to $CaCO_3$ being formed in the Rhône. The ratio between Iron-bound P and Iron extracted can not be explained: two values fall in the usual range (1:10–1:20), but especially the very high value in June can not be explained. It seems likely that the system is not yet in chemical equilibrium. Many more data are needed in order to predict the quantity of bio-available P transported by the Rhône. In

Table 1. Characteristics of variable measured in the Rhone river during the period June 1994 – May 1995.

Variables	Units	Arithmetic mean	Standard deviation	Minimum	Maximum	Water-discharge weighted average concentration	Sediment-discharge weighted average concentration	Median and quantiles					n
								Q_{10}	Q_{25}	Q_{50}	Q_{75}	Q_{90}	
Instantaneous fow rate	$m^3\ S^{-1}$	2847	1702	836	9431			1238	1635	2448	3420	4707	40
Suspended matter	$mg\ l^{-1}$	311	848	8.5	5200	683		14	31	70	285	694	39
Nitrate	$mg\ l^{-1}$ (N)	1.34	0.43	0.82	2.26	1.32		0.90	0.99	1.15	1.72	1.96	40
Nitrite	$mg\ l^{-1}$ (N)	0.020	0.005	0.010	0.034	0.019		0.015	0.016	0.020	0.023	0.026	39
Ammonium	$mg\ l^{-1}$ (N)	0.09	0.05	0.004	0.18	0.091		0.020	0.061	0.097	0.134	0.148	39
org-N_{diss}	$mg\ l^{-1}$ (N)	0.14	0.08	0.00	0.51	0.14		0.07	0.09	0.13	0.17	0.21	38
N_{part}	$mg\ l^{-1}$ (N)	0.12	0.09	0.02	0.45	0.16		0.04	0.05	0.09	0.16	0.24	40
N_{part}	$mg\ g^{-1}$ (N)	1.69	1.51	0.09	7.47		0.38	0.25	0.76	1.27	2.08	3.28	39
Tot-N (Sum)	$mg\ l^{-1}$ (N)	1.70	0.49	1.04	3.01	1.73		1.16	1.32	1.53	2.10	2.40	37
o-P	$\mu g\ l^{-1}$ (P)	44	29	9	195	38		19	29	41	53	61	40
org-p_{diss} & poly P	$\mu g\ l^{-1}$ (P)	23	12	2	55	19		9	13	22	31	36	36
P_{part}	$\mu g\ l^{-1}$ (P)	55	43	4	186	67		12	24	40	76	121	40
P_{part}	$mg\ g^{-1}$ (P)	0.76	0.51	0.02	2.17		0.18	0.07	0.39	0.72	0.97	1.34	39
Tot-P (Sum)	$\mu g\ l^{-1}$ (P)	117	50	43	231	124		63	80	110	143	195	34
Silicate	$mg\ l^{-1}$ (Si)	1.89	0.50	0.84	2.74	1.99		1.20	1.59	2.06	2.25	2.46	39
Chlorophyll *a*	$\mu g\ l^{-1}$	2.2	2.1	0.4	13.3			0.7	1.2	1.8	2.7	3.6	40
Phaeopigments	$\mu g\ l^{-1}$	1.5	1.4	0.1	8.4			0.3	0.6	1.3	1.8	2.6	40

Table 2. Characteristics of 4 samples of suspended matter collected in 1992/1993, with special attention to fractionation of P_{part}.

Variables	Units	Oct 92	March 93	May 93	June 93
Flow rate	$m^3\ s^{-1}$	1100	780	1350	1590
Suspended matter	$mg\ l^{-1}$	400	230	182	179
org-C	$mg\ g^{-1}$	4.2	8.2	29.0	17.0
org-N	$mg\ g^{-1}$	0.3	0.9	2.8	2.1
Tot-Fe	$mg\ g^{-1}$	4.6	7.4	15.0	23.0
Tot-P	$\mu g\ g^{-1}$	200	490	454	1200
$Fe(OOH)_{extr}$	$\mu g\ g^{-1}$	400	770	40	510
$Fe(OOH) \approx P$	$\mu g\ g^{-1}$	47	30	20	405
$CaCo_3 \approx P$	$\mu g\ g^{-1}$	90	210	250	570
Acid Soluble Organic P (1)	$\mu g\ g^{-1}$			123	73
Residual Organic P (2)	$\mu g\ g^{-1}$			50	150
(1) + (2)	$\mu g\ g^{-1}$	70	150	173	223
Sum of fractions	$\mu g\ g^{-1}$	207	390	443	1198

the Golfe du Lion, a shift may occur from iron-bound P to calcium-bound P due to the greater salinity and high calcium concentration and higher pH of the sea water, while most of the incoming o-P may precipitate as both inorganic forms.

The nutrient concentrations varied during the period studied. The maximum factor of variation is 3 for Tot-N, 6 for Tot-P and 3 for dissolved silicate. These fluctuations are much greater if one considers the various fractions of nitrogen and phosphate, the maximal fluctuations concerning the dissolved and particulate organic fractions. The concentrations of nitrate show a maximum in winter and a minimum in summer whereas there is no marked seasonal variation in the other nitrogen compounds as was already noticed by Coste & Raimbault (1993). The other variables measured during the period of study do not show significant seasonal variations either. Furthermore, there is no linear relationship between the mean daily flow rate and any of the variables studied, except for N_{part} which showed a significant relationship. Figures 3 and 4 show the concentrations of N_{part} and P_{part} as functions of the flow rate (data of El Habr (1987) and from this study). With a linear regression of the type $y = ax$, a slope of 3.7 to 4.5 10^{-5} is obtained for N_{part} and of 1.5 to 2.1 10^{-5} for P_{part} at the 95% confidence threshold ($R^2 = 0.60$ for nitrogen ($n = 69$) and 0.06 for phosphate ($n = 66$)). The relation obtained for P_{part} is not significant, due to the fact that P_{part} comprises several organic and inorganic compounds, while N_{part} is mainly organic.

Comparison with other rivers

Table 3 indicates the median value of the frequency distributions of concentrations found in other rivers. It shows that the concentrations of dissolved nutrients in the Rhône are clearly higher than those measured in 60 non-polluted major rivers (Meybeck, 1995), particularly as far as nitrogen is concerned (12, 20 and 7 times higher for nitrate, nitrite and ammonium, respectively). The median value is 4 times higher for o-P. On the contrary, for particulate fractions, nitrogen and phosphate are about 2 times less concentrated in the turbid Rhône river. This is partly related to the fact that N_{part} and P_{part} median values are actually much different from the sediment discharge averages because of increasing sampling during flood period.

Input in the Mediterranean Sea

Table 4 presents the mass fluxes estimated according to the four methods proposed. Considering the maximum and minimum values obtained by these methods of estimation for each variable, an annual N-input of 92.3 to 96.1 kt as nitrate is found, 1.3 to 1.5 kt as nitrite, 6.3 to 6.7 kt as ammonium, 9.7 to 9.8 kt as org-N_{diss} and 5.3 to 12.7 kt as N_{part}; an annual P-input of 2.7 to 3.0 kt as o-P, 1.3 to 1.7 kt as org-P_{diss}/poly-P and 2.5 to 7.5 kt as P_{part}.

The input of Tot-N is 115-127 kt and the input of Tot-N is 6.5-12.2 kt for the year studied. The nitrate input represents about 80% of the Tot-N input while that of o-P is only about 25-45% of the Tot-P. There-

Table 3. Medians of frequency distributions of nutrient concentrations or contents for: (1) 60 non-polluted rivers (Watershed > 100 000 km^2) (Maybeck & Helmer 1989 in Meybeck, 1995). (2) References in Meybeck 1993; (3) This study.

	Nitrate mg l^{-1}	Nitrite mg l^{-1}	Ammonia mg l^{-1}	org-N_{diss} mg l^{-1}	N_{part} mg l^{-1}	N_{part} mg g^{-1}	o-P mg l^{-1}	org-P_{diss} & poly P mg l^{-1}	P_{part} mg l^{-1}	P_{part} mg g^{-1}	Silicate* mg l^{-1}
Non-polutted rivers (a)	0.1	[0.001]	0.015	0.26	0.35	2.3	0.01	[0.015]	0.17	1.1	5.0
Mackenzie (2)	0.1			0.1	0.2						1.4
Amazone (2)	0.17		<0.02	0.2	0.32		0.024	[0.015]	[0.16]		3.2
Seine (2)	4.3		1	0.5	0.3		0.40				
Rhône (3)	1.15	0.02	0.10	0.13	0.09	1.3	0.041	0.022	0.040	0.6	2.06

* Transformed in mg l^{-1} of Si in this table.

Table 4. Estimation of nutrient input (kt of N, P or Si per year) by the Rhône river into the Mediterranean Sea using different calculation methods: methods 1* & 2* as method 1 & 2 but assuming the same transport model for particulate and dissolved compounds (see method section).

	Nitrate	Nitrite	Ammonia	org-N_{diss}	N_{part}	Tot-N	o-P	org-P_{diss} & poly P	P_{part}	Tot-P	Silicate
Method 1	92.3	1.3	6.3	9.7	5.3	115	2.7	1.3	2.5	6.5	139
Method 1*					11.2	121			4.7	8.7	
Method 2	96.1	1.5	6.7	9.8	12.7	127	3.0	1.7	7.5	12.2	135
Method 2*					8.7	123			3.8	8.5	
Minimum	92.3	1.3	6.3	9.7	5.3	115	2.7	1.3	2.5	6.5	135
Maximum	96.1	1.5	6.7	9.8	12.7	127	3.0	1.7	7.5	12.2	139

fore the o-P flux represents less than half of the Tot-P input into the Mediterranean Sea. Given the rapidity of the exchanges between the particulate and the dissolved fractions of phosphate (De Groot, 1992; Moutin, 1992), it is necessary to calculate the Tot-P input as the sum of all fractions.

The two methods used to estimate the dissolved compounds fluxes give essentially equivalent results (Table 4). This is mainly related to the relatively dense sampling which makes the choice of the method to estimate the fluxes less crucial. On the contrary, for the particulate compounds, there is a difference of about a factor 3 between the maximum and minimum values obtained by the four methods of estimation. This is awkward in the case of phosphate for which the particulate fraction is far from negligible. The transport is different for dissolved and particulate compounds which led us to use two different calculations to evaluate their respective fluxes: one based on the evaluation of the total water flow rate over the year and the other based on the evaluation of the total solid flow rate (method 1). In the second case, it implies that there is a constant relationship between the variables studied and the concentration of suspended matter. That is true when the concentration of suspended matter does not exceeds 150 mg l^{-1}. In this case, N_{part} concentration in suspended matter is equal to 2.28 mg g^{-1} with a confidence interval of 0.34 and the P_{part} concentration in suspended matter is equal to 1.16 mg g^{-1} with a confidence interval of 0.11 ($\alpha = 0.05$). On the contrary, when the concentration of suspended matter exceed 150 mg l^{-1}, it seems that it is poor in nitrogen (Figure 5) and the relationship with P_{part} is not linear (Figure 6): method 1 probably underestimates N_{part} and P_{part} inputs. The fact that P_{part} & N_{part} concentrations (mg.l^{-1}) increase more or less linearly with flow (Figures 3 and 4) lead us to prefer methods 1* or 2* to method 1 to estimate particulate nutrient fluxes. Method 2, which permits to take into account the variability of the fluxes of suspended matter, seems to be the most appropriate for the quantification of particulate fluxes (N_{part} & P_{part}) and total fluxes (Tot-N & Tot-P). For dissolved compounds, good results are obtained with method 1 which is of easy use.

Long-term pattern of nitrate input

Figure 7 represents the nitrate input by the Rhône river into the Mediterranean Sea:

(1) calculated by Coste & Raimbault (1993) at four periods (1968, 1984, 1989, 1990),

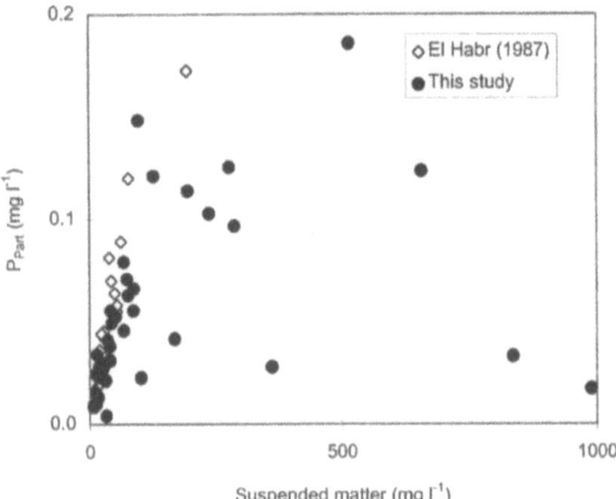

Figure 5. N_{part} concentration versus suspended matter concentration. Data from El Habr (1987) and from this study, one sample omitted ($x=5200$; $y=0.45$).

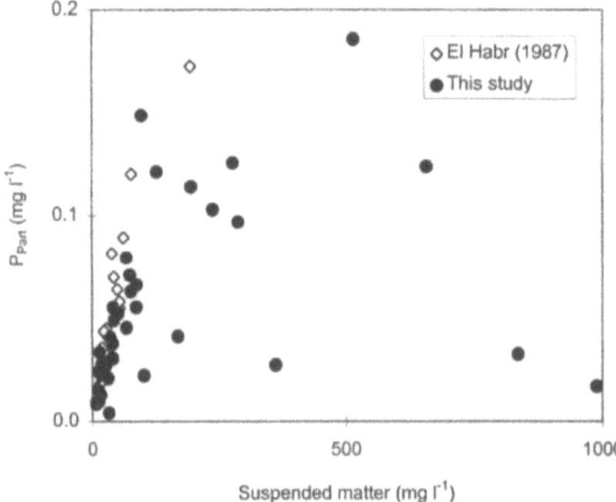

Figure 6. N_{part} concentration versus suspended matter concentration. Data from El Habr (1987) and from this study, one sample omitted ($x=5200$; $y=0.09$).

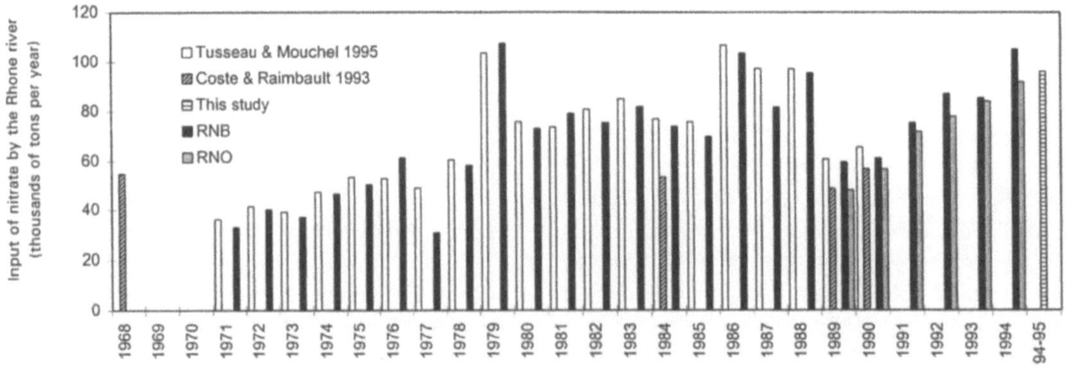

Figure 7. Input of nitrate by the Rhône river into the Mediterranean Sea since 1968.

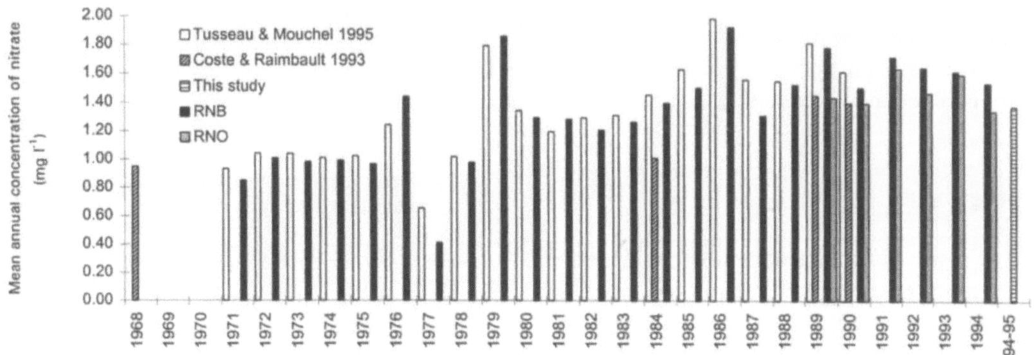

Figure 8. Mean annual concentration of nitrate in the Rhône river near the estuary since 1968.

(2) estimated by Tusseau and Mouchel (1995) from a statistical method and using the data provided by the Agence de l'Eau Rhône-Méditerranée-Corse for the period 1971–1990 (Réseau National de Bassin (RNB), a French river water survey),

(3) calculated with method 1 from the data of this study,

(4) calculated with method 1 from the data provided by the Agence de l'Eau Rhône-Méditerranée-Corse for the period 1971–1990 (RNB),

(5) calculated with method 1 from data obtained from sampling near the estuary for the period 1989–1994 (Réseau National d'Observation (RNO), a French coastal water survey).

Coste & Raimbault (1993) mentioned that there is no significant increase in the nitrate flux during their study period (1968–1990). Tousseau & Mouchel concluded, from the analysis of their data, that the input increased. These contradictory conclusions are based on different approaches and data sets, more precisely on a different number of observations due to a different sampling frequency.

To evaluate the possible increase of the nitrate input, we calculated the mean annual concentration of nitrate in the Rhône as the mass flux calculated by method 1 divided by the mean annual flow rate. This in order to get rid of the inter-annual variability of the flow rates. In all cases, it seems that the mean annual concentration of nitrate in the Rhône has increased during the two last decades (Figure 8). The presence of a trend was tested with a non parametric permutation test using a Wald and Volfowitz's statistic (Good, 1994): the hypothesis of an absence of trend has to be rejected ($\alpha = 0.0022$). The least square estimate of the increase rate of N is 0.029 mg l^{-1} $y^{-1}$with a standard deviation of 0.006. The steady state noted by Coste & Raimbault (1993) must be related to the particular hydrological features of the years included in their study.

Comparison to the total inputs in the Mediterranean Sea

According to Helmer (1977), the annual load carried by rivers into the Mediterranean Sea in the early 1970's was 600–1000 kt of N and 200–400 kt of P. The input by the Rhône river may represent about 10–20% of the Tot-N and 2–6% of the Tot-P. These fluxes are considerable with regard to the global nutrient balance and primary production in the Mediterranean Sea, more especially as Helmer's estimation is based on data from a UNEP report (1977) which seem to be overestimated (Coste et al., 1988).

Conclusion

Input of Tot-N, Tot-P and dissolved silicate by the Rhône river into the Mediterranean Sea was evaluated during a one-year period to 115–127 kt of N, 6.5–12.2 kt of P and 135–139 kt of dissolved silicate (Si). The N-input is mainly as nitrate (92.3–96.1 kt y^{-1}) while that of P is for an important part in particulate form. These quantities can play an essential role in the nutrient balances and the primary production in the Golfe du Lion and a non-negligible role at the scale of the Mediterranean Sea.

The analysis of our data and of the previous estimations leads us to conclude that the mean concentrations of nitrate in the Rhône increased during the two last decades by about 50%. It is necessary to prolong these studies and to establish the relation between the nutrient concentration and flow rate more precisely, and to study the influence of this increase in nitrate input on the primary production of the Golfe du Lion.

Acknowledgements

We want to thank B. Bombled, N. Garcia, D. Pont coordinator of the program, M.C. Bonin and Mrs J. C. Golterman-Hardenberg for help in preparation of the manuscript and M. Meybeck and J.P. Durbec for helpful comments. Financial support was provided by the Agence de l'Eau Rhône-Méditerranée-Corse and the Agence Régionale pour l'Environnement Provence-Alpes-Côte d'Azur.

References

Berland, B., D. J. Bonin & S. Y. Maestrini, 1980. Azote ou phosphore? Considérations sur le paradoxe nutritionnel de la Méditerranée. Oceanologica Acta 3: 135–142.

Coste, B., 1974. Rôle des apports nutritifs minéraux rhodaniens sur la production organique des eaux du Golfe du Lion, Téthys 6: 727–740.

Coste, B., P. Le Corre & H. J. Minas, 1988. Re-evaluation of the nutrient exchanges in the strait of Gibraltar. Deep-Sea Res. 35: 767–775.

Coste, B. & P. Raimbault, 1993. Recent data on the nutrient input into the Mediterranean Sea by the Rhône river. In Martin J.-M. & H. Barth (eds), EROS 2000. NERC, Plymouth, U. K., Wat. Pollut. Res. Rep. 30: 47–50.

De Groot, C. J., 1992. Flux et transformations du phosphate et de l'azote dans les zones humides: la Camargue (Delta du Rhône). Thèse de l'Université Claude Bernard (Lyon I). Thèse et articles de la thèse.

De Groot, C. J. & H. L. Golterman, 1990. Sequential extraction of sediment phosphate. Hydrobiologia 192: 143–148.

El Habr, H., 1987. Les éléments nutritifs du Rhône; leur devenir dans les canaux d'irrigation et les marais en Camargue. Thèse de l'Université Claude Bernard (Lyon I), 150 pp.

Golterman, H. L. & A. Booman, 1988. Sequential extraction of iron-phosphate and calcium-phosphate from sediments by chelating agents. Verh. int. Ver. Limnol. 23: 904–909.

Good, P., 1994. Permutation tests. A practical guide to resampling methods for testing hypotheses. Springer series in statistics. Springler-Verlag New York, 228 pp.

Helmer, R., 1977. Pollutants from land-based sources in the Mediterranean. Ambio 6: 312–316.

Herbland, A., A. LeBouteiller & P. Raimbault, 1984. Size structure of phytoplankton biomass in the Equatorial Atlantic Ocean. Deep-Sea Res. 32: 819–836.

Martin, J. M. & A. Saliot, 1992. Bilan des apports fluviatiles et atmosphériques d'éléments et composés chimiques en Méditerranée Occidentale. 3ème rencontre de l'Agence Régionale pour l'Environnement Provence-Alpes-Côte d'Azur, 24–27 septembre 1991: 61–66.

Meybeck, M., 1995. Les lacs et leur bassin. In Pourriot, R. & M. Meybeck (eds), Limnologie générale. Masson, Paris: 6–59.

Meybeck, M., 1993. C, N, P and S in rivers: from sources to global inputs. In Wollast R., F. T. Mackenzie & L. Chou (eds), Interaction of C, N, P and S Biogeochemical Cycles and Global Change. Springer Verlag: 163–193.

Meybeck, M. & R. Helmer, 1989. The quality of rivers: from pristine stage to global pollution. Global Planet. Change 75: 283–309.

Morel, A., A. Bricaud, J. M. André & J. Pelaez-Hudlet, 1990. Spatial-temporal evolution of the Rhône river plume as seen by CZCS imagery: Consequences upon primary productions in the Gulf of Lions. In Martin J. M. & H. Barth (eds), EROS 2000. NERC, Plymouth, U. K., Wat. Pollut. Res. Rep. 20: 45–62.

Moutin, T., 1992. Contribution à l'étude du cycle du phosphate dans les écosystèmes lagunaires. Thèse de l'Université de Montpellier II, 251 pp.

Pont, D. & B. Bombled, 1996. Estimation du débit solide. In Evaluation des charges polluantes du Rhône à la Méditerranée. Rapport Agence de l'Eau Rhône-Méditerranée-Corse, 15 pp.

Pujo-Pay, M. & P. Raimbault, 1994. Improvement of the wet-oxidation procedure for simultaneous determination of particulate organic nitrogen and phosphorus collected on filters. Mar. Ecol. Prog. Ser. 105: 203–207.

Thomas, R. L., & M. Meybeck, 1992. The use of particulate material in Water Quality and Assessments, D. Chapman (ed.), Chapman and Hall, London: 121–170.

Tusseau, M. H. & J. M. Mouchel, 1995. Nitrogen inputs to the gulf of Lions via the Rhône river. In Martin J.-M. & H. Barth (eds), EROS 2000. NERC, Plymouth, U.K., Water Pollution Research Report 32: 49–60.

Tréguer, P. & P. LeCorre, 1975. Manuel d'analyse des sels nutritifs dans l'eau de mer (Utilisation de l'autoAnalyseur II). 2ème ed., Laboratoire d'Océanographie chimique, Université de Bretagne Occidentale, Brest, 110 pp.

Walling, D. E. & B. W. Webb, 1985. Estimating the discharge of contaminants to coastal waters by rivers: some cautionary comments. Marine Poll. Bul. 16: 488–492.

UNEP (United Nations Environnement Programme), 1977. Preliminary report on the state of pollution of the Mediterranean Sea. UNEP Preliminary Report IG.11/INF. 4.

Hydrobiologia **373/374:** 247–258, 1998.
J.-C. Amiard, B. Le Rouzic, B. Berthet & G. Bertru (eds),
Oceans, Rivers and Lakes: Energy and Substance Transfers at Interfaces.

Seasonal and spatial trends of nitrogen and phosphorus loads to the upper catchment of the river Vilaine (Brittany): relationships with land use

Samuel Moreau[1], Georges Bertru[1] & Christian Buson[2]
[1] *Laboratoire d'Evolution des Systèmes Naturels et Modifiés, Université Rennes 1*
[2] *Institut Scientifique et Technique de l'Environnement*

Key words: Nitrogen, phosphorus, catchment, water quality, river, discharges

Abstract

Influences of urban and agricultural activities on river water quality were analysed on the upper catchment of the River Vilaine (902 km^2), France. Agricultural land covered 78% of the study area. Corn, grass and cereals were the main crops. Total Nitrogen (TN) and Total Phosphorus (TP) loads were determined for individual subcatchments (from 10 to 194 km^2) during the hydrological year 1994/95.

Relationships between water quality and catchment management were studied. Non-point source produced 95% of TN and 90% of TP of the whole river basin. Significant variation in TN and TP occurred, in time and space. The greatest part was discharged in January and February. Annual river exportation rates ranged from 20 to 65 kg TN ha^{-1} and from 1.15 to 3.05 kg TP ha^{-1}. Reservoirs will held almost 15% of TN and TP loads. Cyanobacteria blooms occurred in summer and autumn, with significant production of organic matter here, and downstream, in the rivers.

Greatest TP river retention rate were determined downstream a town of 15 000 people, and highest TN loss occurred in the subcatchment with the highest density of cattle and the most important percent of permanent grassland.

Introduction

Freshwater quality surveys have developed during the last 30 years, in parallel with an increase of human activities and the need for new water sources. In western Europe, the trophic level of rivers has risen significantly, and now a major part of them is eutrophic (Meybeck, 1996). Numerous studies of river basins have been carried out at different scales, from few km^2 (Omernik et al., 1981; Burt et al., 1988; Kronvang, 1992) to thousands of km^2 (Edwards et al., 1990; Hunsaker et al., 1995). One of our aims was to establish relationships between catchment characteristics and nutrient concentrations in their streams. The Upper Catchment of the River Vilaine, an agricultural catchment of eastern Brittany, has been under investigation since 1994. The size of this study area (902 km^2) enables the results to be used for catchment management. Indeed, in this agricultural region different subcatchments with various anthropogenic pressures exist, which cause different nutrient losses from the catchment.

In this paper, we discuss the loads of Total Nitrogen (TN) and Total Phosphorus (TP), for the hydrologic year 1994/95, in subcatchments of the River. TN and TP river exportation or loss (in kg ha^{-1}) are compared with their anthropogenic and natural characteristics.

Site description

The River Vilaine, above Rennes, is 90 km long (stream order 5) and drains a catchment area of 902 km^2, with 78% of the land use being agricultural (Figure 1 and Table 1). The density of population is

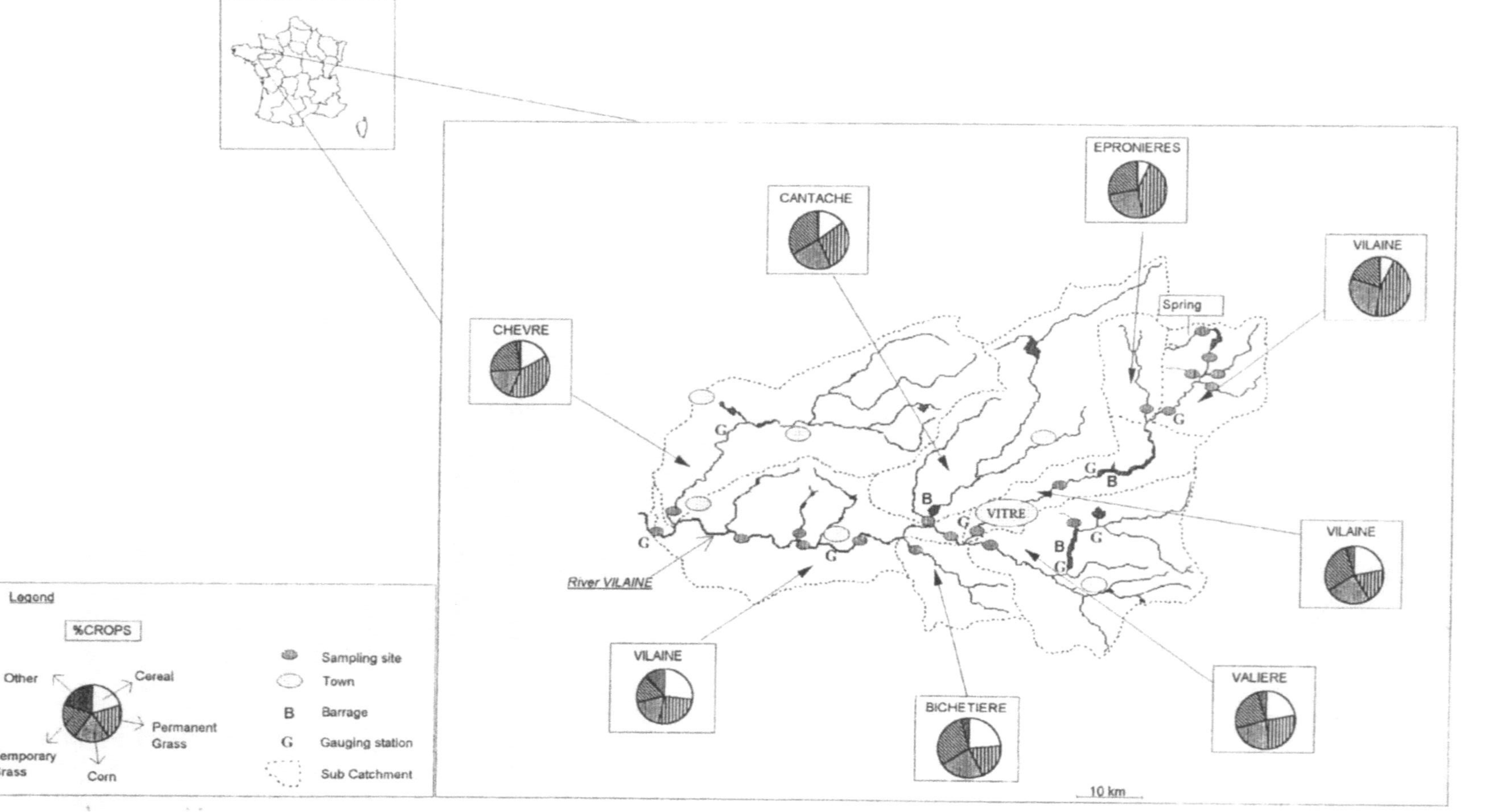

Figure 1. Upper catchment of the River Vilaine.

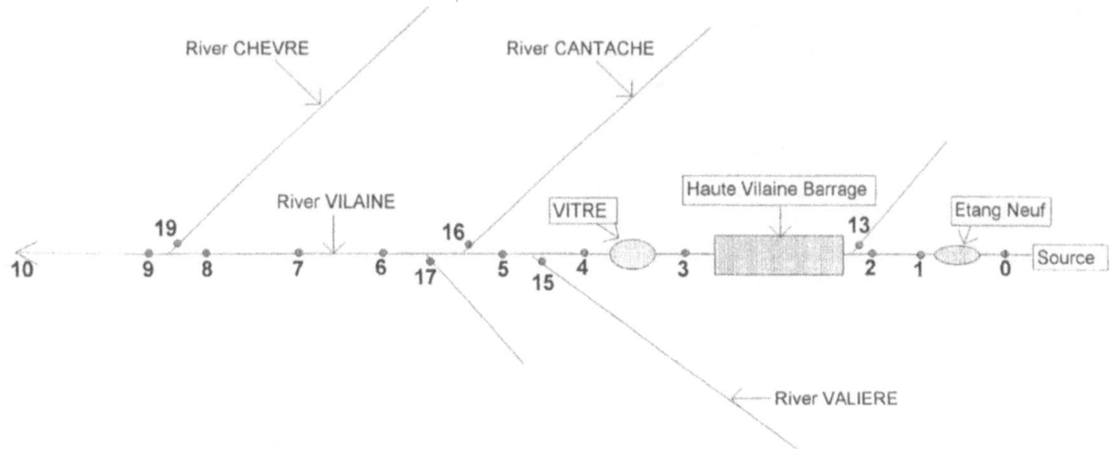

Scheme 1. Main sampling sites on the Upper River Vilaine Catchment.

Table 1. Main characteristics by subcatchment

Site	Stream order	River	Sewage plant municipal/	Industrial	Area km^2	% of agri-cultural land
1	2	Vilaine	1	0	12	85
2	3	Vilaine	0	0	61	87
3	4	Vilaine	0	0	134	90
4	4	Vilaine	1	2	151	89
5	5	Vilaine	1	0	307	84
6	5	Vilaine	0	0	548	84
8	5	Vilaine	3	1	651	83
9	5	Vilaine	3	0	902	78
13	3	Epronières	1	0	35	92
15	4	Valière	5	0	144	78
16	4	Cantache	2	0	176	84
17	3	Bichetière	1	0	39	83
19	4	Chevré	2	1	194	64

about 70 persons km^2 with the majority of the 60 000 habitants living in 40 towns and villages; the biggest town, Vitré, has 15 000 people. Twenty municipal and four industrial sewage treatment plants are the major point sources to the streams. This region is one of the best producers of cattle in France (1,93 cows ha^{-1} in 1988), and consequently cereal grain, corn and grass production predominate.

The study area is typical of eastern Brittany, with low river flows during summer/autumn and a short period of high discharge. Between December and March, 70% of the annual flow occurs, and discharge ranges from 1 to 100 $m^3 s^{-1}$. Three reservoirs were built on the river's course, Valière (5.7 Mm^3) in 1979, Haute Vilaine (7 Mm^3) in 1983 and Villaumur (7 Mm^3) in 1993, to reduce water shortage in summer and to regulate discharge.

Discharge and rainfall were measured at eight different sites by the DDE Flow Survey Network (Figure 1). Mean annual rainfall is 1 000 mm for the headwaters of the basin, reducing to 750 mm in the west. Altitude ranges from 190 metres at the source of the river to 20 metres at the lowest point, in Cesson (site 9).

Materials and methods

Water analysis

Seasonal trends in N and P loads were examined during 1994/1995. Water samples were collected every two weeks when river flow was low, and more frequently during periods of high discharge (32 samples), at 20 different sites (Figure 1). Water was analyzed within four hours of being collected. Detemination of Total Nitrogen and Total Phosphorus was performed on unfiltered samples. Subsamples for dissolved nitrogen and phosphorus forms were filtered through 0.45 μm Millipore filters. All analysis were carried out colorimetrically. For ammonia, Indophenol Blue method was used (Stickland & Parsons, 1968). Nitrate was reduced to NO_2^- (which gives a diazo compound with sulfanilamid by passing through a cadmium column) and analysed using a Technicon auto analyzer

(Barnes & Kolkard, 1951). Total Dissolved Nitrogen (TDN) and Total Nitrogen (TN) were determined as NO_3^- after oxidation with $NaOH/K_2S_2O_8$ (Koroleff, 1972).

Subsamples for Total Dissolved Phosphorus (TDP) and Total Phosphorus (TP) were digested with $H_2SO_4/K_2S_2O_8$, which converted all P forms into Molybdate Reactive Phosphorus (MRP) (Langler & Hendrix, 1982). The PO_4^{3-} (MRP) was analyzed using the molybdenum blue procedure of Murphy & Riley (1962).

Load calculation

Six gauging stations were used to calculate TN and TP loads of the main subcatchments studied (Scheme 1). Flow measurements were available every six minutes.

When there was no gauging station at a sampling site (A), discharge was calculated with the nearest station (B).

Discharge at the site A = Discharge B
∗ area A/catchment area B.

Daily discharges were used to determine N and P loads, and rainfall was considered to contribute nutrient concentrations to discharge day by day. Monthly nitrate loads, from 1990 to 1995 at site seven, were determined in the same way.

N and P input

Atmospheric deposition, agricultural practice, import from livestock, atmospheric N_2 fixation, and sewage are the main sources of N and P influx to a water course. Wet deposition was determined at two gauging stations, near the source (site 2) and at the mouth (site 9) of the catchment. Daily and weekly surveys, which collected discharge data from municipal and industrial wastewater treatment plants, were used to estimate the contribution of the major point sources to TN and TP loads. French Agricultural Statistics (RGA, 1988) collect land use, livestock and crop data for each city. The portion of the area of each council, for each subcatchment was calculated, and used to determine agricultural practice by subcatchment. A comparison with a recent agricultural study (SEEGT, 1994) on the subcatchment of the River Cantache (site 16), conducted in 1994, suggested there has been no significant change in land use since 1988.

Nitrate background

Before discussing the 1994/95 results, we present the evolution of nitrate in this eutrophic river since 1971 (Figure 2), based on three different nitrate survey Networks. The R.N.B (National Basin Network) controls monthly the water quality of the Vilaine, six km downstream (site 10) the lower point of our study area (site 9). Another survey of nitrate has been carried out (DDASS Network) on the River Vilaine at Chateaubourg (Site 7, middle of the catchment), to assess quality for use as drinking water since 1980. When the concentration of nitrate increased, daily measurements were made. During our own study, nitrate concentration was measured since September 1994. The spatial range of nitrate concentration was less than 4% between sites 7 and 10, making it possible to describe nitrate trends with these three different data bases.

The main feature is the sinusoidal pattern of nitrate concentration, with high values in winter and low values in summer/autumn. This is common for rivers under similar climatic conditions (Burt et al., 1988; Roberts et al., 1984). The most influential factor is the variability of annual rainfall, because high nitrate concentrations are associated with leaching from soil during high rainfall winters. A significant increase in mean nitrate concentration, from 1.5 to 8 mg NO_3-N l^{-1} has been observed since 1971.

In 1991, a gauging station was set up near the point of the drinking water plant (site 7), and we used its daily data with the DDASS Nitrate Survey to determine evolution of NO_3^- loads since 1990. Figure 3 presents the monthly NO_3^- loads at Chateaubourg. A relationship with runoff and consequently with winter rainfall (November to March), when soils are saturated with water is clear (Table 2).

The dry winter of 1991/92 produced low loads (141 tonnes NO_3^--N). Wetter winters like 1990/91 and 1992/93, with about 245 mm of rainfall, gave loads of 843 and 1100 tonnes NO_3^--N, respectively. The highest mean and maximum concentrations of NO_3^- occurred during these two hydrologic years. The winter 1993/94 was very wet (320 mm), the concentrations of NO_3^- decreased but the high discharges led to the highest load in five years. The winter of 1994/95 was the wettest, with 475 mm of rainfall. Despite this, the loads were lower than in 1993/94. The concentrations decreased because of the dilution effect of the high discharges, and the excess water which caused flooding decreased the efficiency of nitrate leaching.

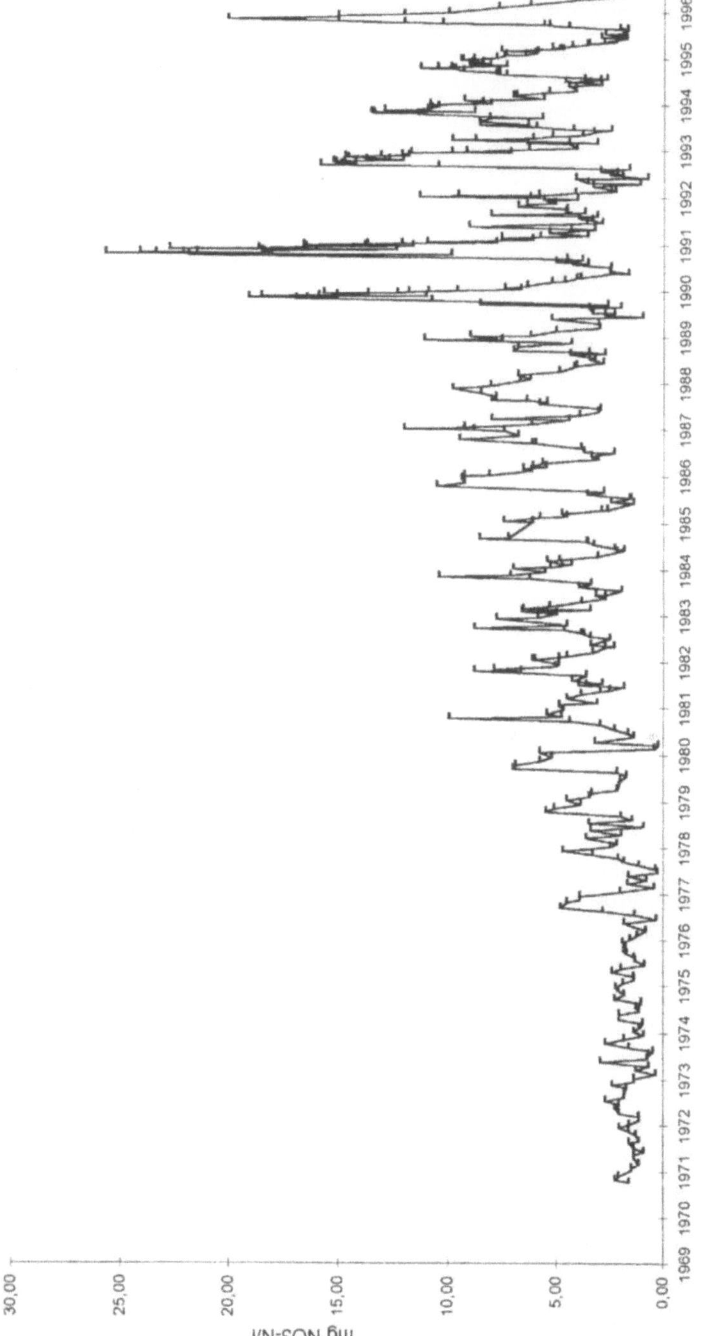

Figure 2. Evolution of nitrate concentration in the lower area of the catchment (1971–1996).

Table 2. Rainfall, effective Runoff, nitrate concentrations and nitrate loads in site 7 (1990–1995)

Hydrologic year	Rainfall (mm) Nov to March	Effective Runoff (10^6 m^3) Nov–March	Annual	Note	Mean and Max of NO_3-N (mg/l)		Annual load in NO_3-N tonnes
1990/1991	245	46	65		12	25	843
1991/1992	180	18	30		7	11	141
1992/1993	245	68	96	Wet April and June	10	18	1108
1993/1994	320	146	214		9	14	2285
1994/1995	475	217	252		6	11	2007

Sixty percent of the annual load occurred during two to four weeks, between December and February. This is the main period to carry out management to reduce nitrate load. But, if rainfall is the first variable to describe annual evolution of nitrate (Correl et al., 1991; William et al., 1994), it is not sufficient to explain the increase in annual mean concentrations since 1971.

Thus, we studied the flows of TN and TP through the hydrological system of the River Vilaine for the year 1994/95. Our aim was to determine the role of subcatchments (10 to 194 km^2 in area) in the transfer of nutrients to the Upper Basin of the River Vilaine.

Results

Agricultural situation

We used the French Agricultural Statistical Data Base (R.G.A., 1988) to characterize agricultural practice in this catchment. This was the most appropriate, due to a lack of recent data and the size of the study area (902 km^2), which made field inquiry impossible (Table 3). The main crops of the catchment are corn, cereals and grass (temporary and permanent), but the proportion varies according to subcatchment. The head of the basin was characterized by a low percentage of cereal (less than 7% of the agricultural land), more than 40% of permanent grass and around 30% of corn. In the Cantache (site 16) and in the Chevré (site 19) catchments, the area of cereals was around 15%. Moving down the catchment, the percentage contribution of cereals increased to an average of 24% in that for maize decreased. Permanent grass land ranged from 14 to 45% of the agricultural area, while temporary grassland ranged from 16 to 32% without spatial pattern.

A spatial trend existed in the three main animals produced in the catchment (cow, pigs and poultry). The east had the highest concentration of cattle and poultry, with more than two cows and more than 10 poultry per ha of agricultural land. The distribution of pigs was more homogeneous, although fewer lived at the head of the basin. Two catchments (Sites 15 and 17) had a cattle density over the mean (three) of the whole basin. Three other subcatchments at the head of the basin had a density of poultry and cow greater than the mean for the entire catchment. Finally, the middle eastern part of the upper basin had the greatest density of all.

Nitrogen

Nitrate accounted for 86% of TN load and for more than 91% of TDN (Table 4). Figure 4a summarizes the spatial trend of TN loads on the main river. Downstream of sites 1 and 3, there was a decrease in load which indicates that N was retained in ponds and reservoirs. One hundred and forty tonnes of TN were retained in the reservoir of the Haute Vilaine (on the River Vilaine). This involved a loss of –25.1 kg TN ha^{-1} in the area drained by this river, between sites 2 and site 13 upstream, and site 3 downstream of the reservoir.

The region which comprises the biggest town, Vitré, showed an increase of 45 tonnes of TN, i.e almost 1.5% of the annual load of the entire catchment. The annual discharge of the three sewage treatment plants of the region gave a result of 16 tonnes of TN.

To compare different areas, additional area (between two sites on the river Vilaine) and subcatchment exportation rates were calculated (Table 3). A wide range of nitrogen loss from 25.1 to 64.9 kg TN ha^{-1} was noticed. The greatest N loss on an area basis (64.9 kg TN ha^{-1}) was to the Vilaine (site 2), at the head of the basin, due to an important supply of N from three agricultural tributaries on the left bank. In contrast, the

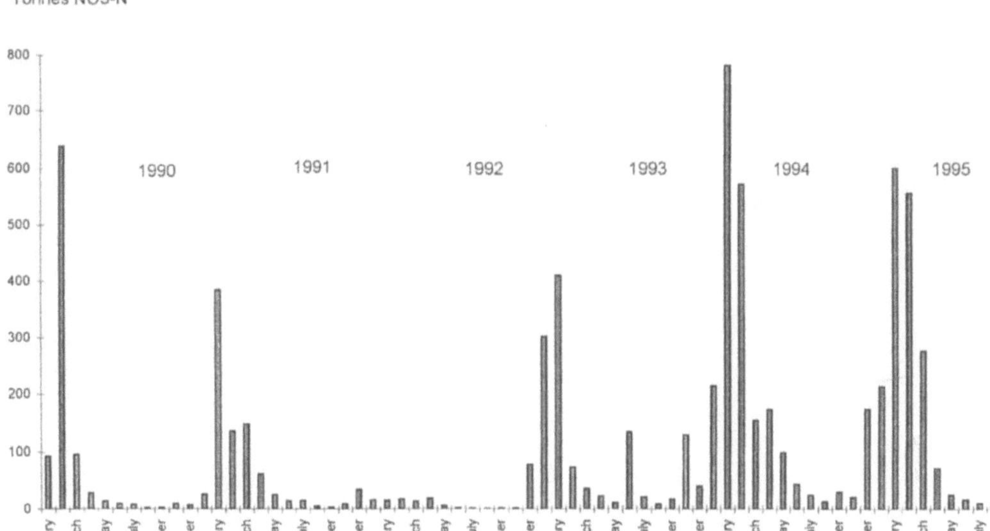

Figure 3. Monthly nitrate loads in site 7 (1990–1995).

Table 3. Land use and cattle density in 1988, and TN and TP losses (1994/95), between two sites on the River Vilaine and for each main tributary basin

	Animal ha^{-1} of agricultural land			Crops as % of agricultural land				Loss (kg ha^{-1})	
Site	COW	PIG	POULTRY	Cereal	Permanent grassland	Corn	Temporary grassland	Nitrogen	Phosphorus
	River Vilaine								
1	2.52	0.61	10.5	6	42	30	21	36.0	1.42
2	2.34	0.71	18.9	7	45	27	20	64.9	1.46
3	1.79	1.91	6.01	14	39	20	25	–25.1	–1.10
4	2.03	2.65	10.05	22	19	25	29	25.9	3.03
5	2.22	3.08	17.52	28	15	24	27	27.1	1.75
6	1.57	2.83	9.09	27	26	18	17	25.0	1.26
8	1.31	1.4	17.55	30	17	18	26	23.3	1.20
9	2.19	3.72	5.82	31	21	20	22	23.3	1.20
	Tributaries								
13	2.13	0.87	10.81	7	39	26	27	20.2	1.50
15	1.95	2.77	18.85	22	26	22	25	43.7	1.15
16	2.24	3.09	7.13	14	28	24	32	42.5	1.23
17	2.14	3.47	16.92	24	19	23	30	45.5	1.51
19	1.85	5.52	4.04	17	39	18	23	34.6	1.21
	Average for the whole Catchment								
	1.93	2.71	11.10	19	29	21	26	32.5	1.14

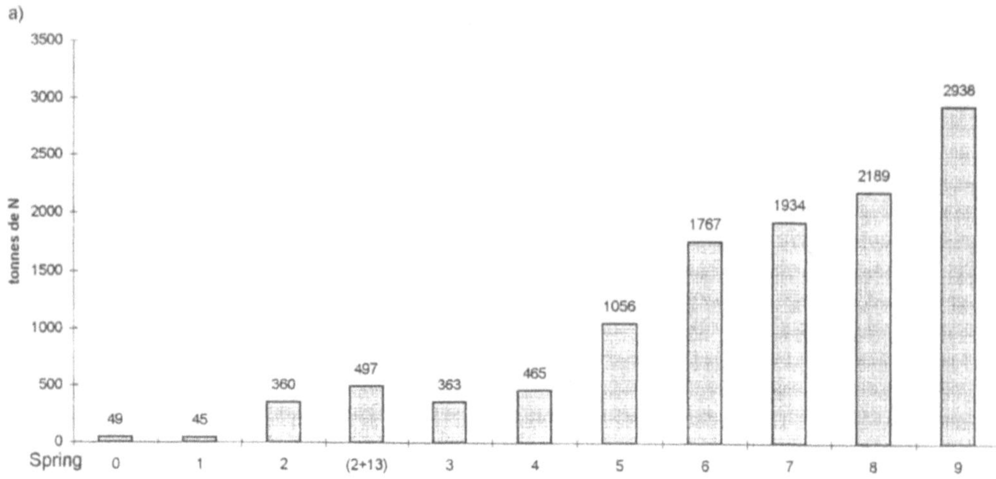

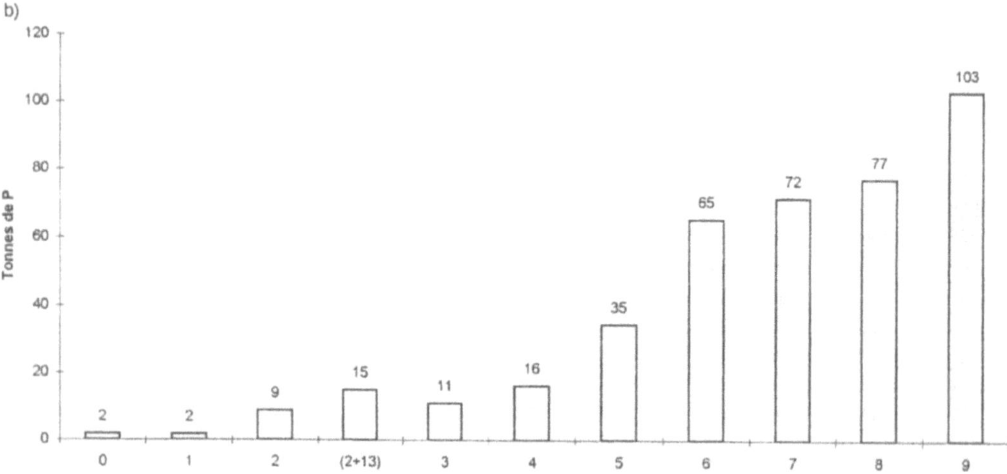

Figure 4. (a) Spatial trend of TN in the River Vilaine; (b) Spatial trend of TP in the River Vilaine.

Epronières stream (Site 13) contributed only one third of this amount (20.1 kg TN ha^{-1}). Its catchment is 35 km^2 in area. The river Bichetière (Site 17), which drains 39 km^2, also discharged significant amounts of TN, including the greatest concentration of nitrate of the region (Max 20.5 mg NO3-N l^{-1} in 1994/95).

The seasonal trend at four sampling sites on the main river is summarized in Figure 5a. As discussed above, almost 80% of the load is associated with the winter period, when the highest annual flows occur.

Loss to the entire River Vilaine Basin (site 9) was 32.5 kg TN ha^{-1} during this hydrologic year, which corresponds to 2938 tonnes of N (of which 2665 was as NO_3^--N).

Phosphorus

The spatial evolution of TP is similar to that of TN on the main stream (figure 4b). One of the main features of P behavior is the buffer power of the Haute Vilaine reservoir, with 28% of TP load retained, or a decrease of TP load from 15 to 11 tonnes. Buffer power was less efficient in Etang Neuf and the load at its mouth was two tonnes of TP. This load increased to 15 tonnes at the Haute Vilaine reservoir. The impact of Vitré on TP loads is greater than with TN. The increase of TP load after this town was about five tonnes. The calculation of TP derived from industrial and urban sewage plants of Vitré gave a value of 1.8 tonne, which was not sufficient to explain this great increase. Other point sources may be involved near this town.

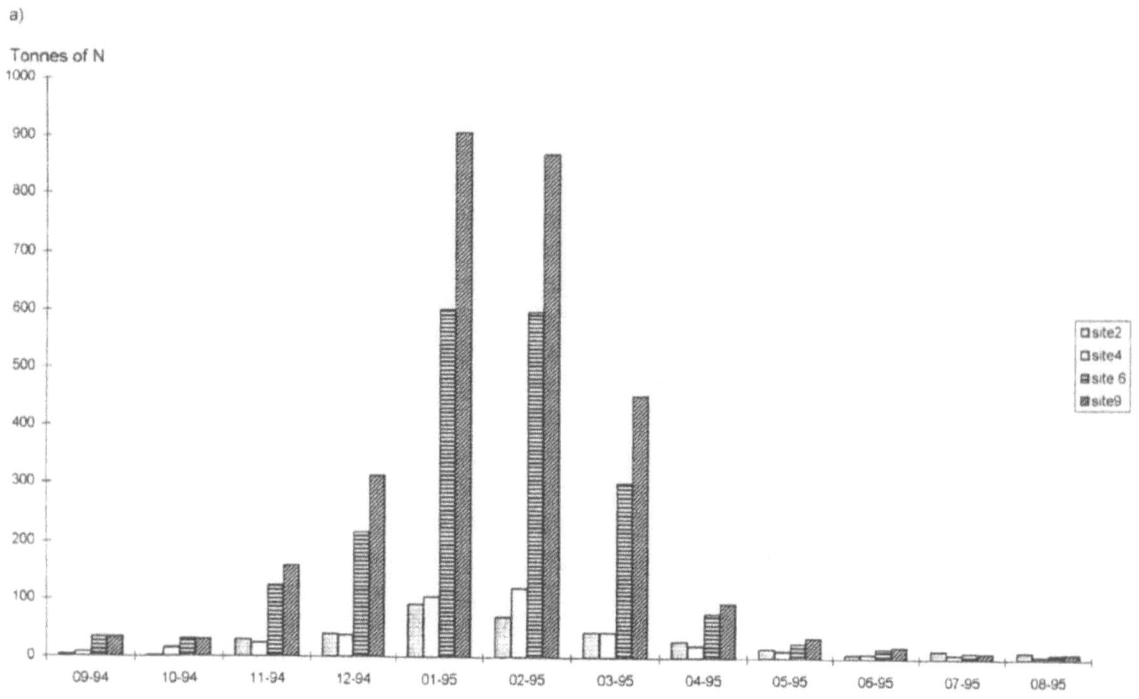

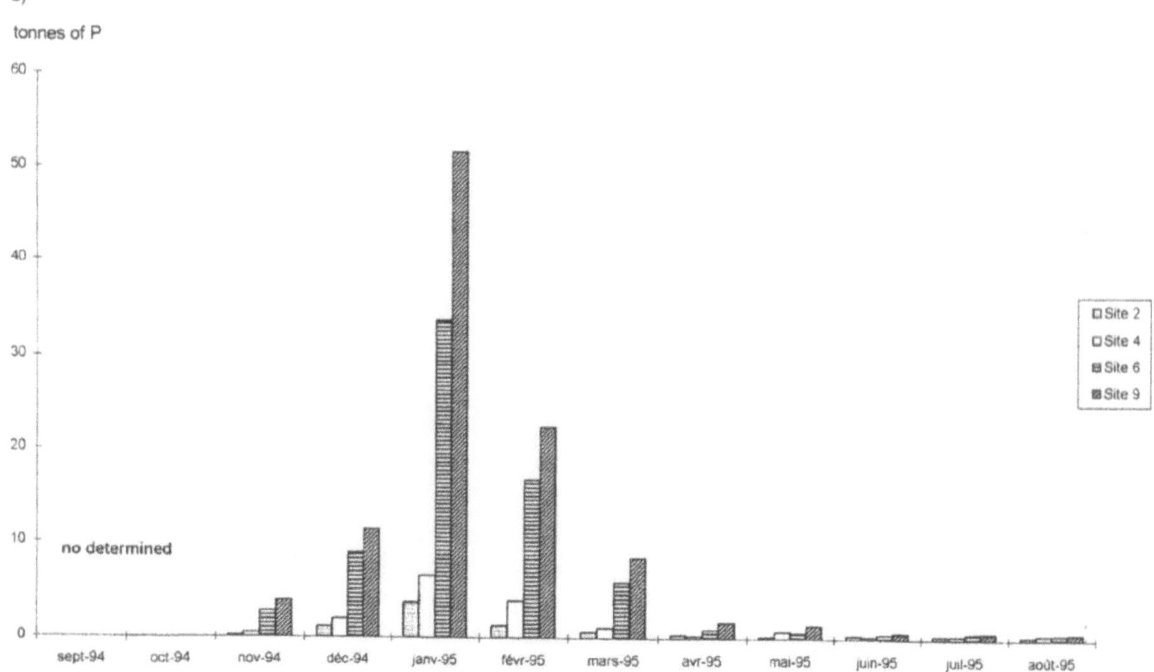

Figure 5. (a) Seasonal trend of TN loads in the river Vilaine; (b) Seasonal trend of TP loads in the river Vilaine.

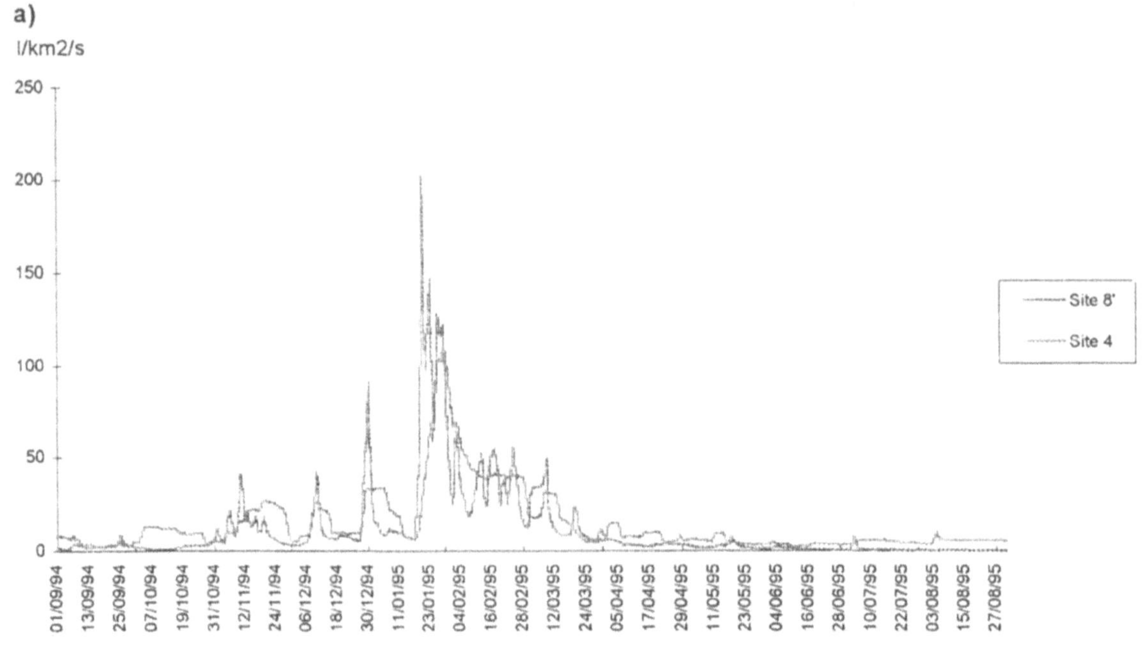

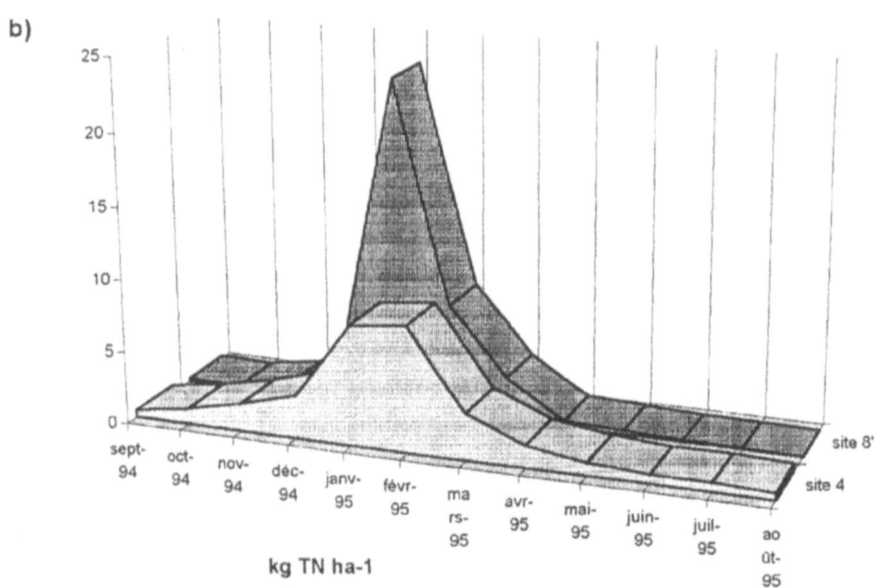

Figure 6. (a) Discharge trends in site 4 and 19 (1994/95); (b) Monthly TN loads in site 4 and 19 (1994/95).

Table 4. Percentage of N forms loading for four sites on the River Vilaine

Sampling Site	2	3	4	9
Nitrate	85.3	86.2	78.5	86.0
Ammonia	2.0	2.1	4.1	2.5
Nitrite	0.2	0.3	0.3	0.3
D.O.N	3.4	4.4	5.3	2.3
Particulate N	9.1	7.0	11.9	9.0

Downstream, P loss increased until site 6 (1.3 kg TP ha^{-1}) and decreased to 1.15 kg TP ha^{-1} at the lowest point of the catchment (site 9). Approximately 100 tons of TP were exported from this catchment. Figure 5b shows the seasonal trend for four sites. Similarly to TN, the main load occurs during high flows.

Discussion

It is known that TN and TP loading to streams is linked with catchment land use, and particularly agricultural practices (Omernik et al., 1981; Nielsen & Jensen, 1990; Jaworski et al., 1992). In this study, we presented annual loads for individual subcatchments which cover a large part of this agricultural basin, and have a high production of pig, poultry, and to a lesser extent, cattle.

Reservoirs and lakes retain large amounts of TN. Our estimation gives an amount of 500 tonnes of TN, held presumably through assimilation and denitrification, in the main water bodies of the upper river Vilaine, equivalent to 15% of the total N load. At the same time, the reservoirs also retain about 15% of the whole amount of TP. Reservoirs are thus important buffer zones for P and N forms. However, as a result algal blooms occurred in these water bodies, with cyanobacteria (*Oscillatoria agardhii*, *Microcystis aeruginosa* and *Gomphosphaeria sp.*) dominating. We also noticed a production of dissolved organic carbon in the reservoir of Haute Vilaine (65 tonnes of dissolved Carbon for the whole reservoir).

Different management measures involved different losses of TN and TP. Agricultural practices, urbanization and stream management vary for each subcatchment and their effect on the trend of TN and TP loads were different.

We found relationships between TN loss and animal density, and with agricultural practices. The highest loss of TN occurred in subcatchments having cattle densities over the mean for the whole catchment (sites 2, 16 and 17). The second agricultural aspect was the increase in cereal area and decrease in grassland from the head of the catchment to the plains in the west. It is interesting to note that the greatest loss of nutrients occurred in subcatchments with the highest percent of permanent grassland. The biggest loss of TN also occurred in a subcatchment without sludge treatment plant and with a low population density (site 2). Non-point sources of TN and TP were dominant (respectively > 95% and 90%) during 1994/95.

The regulation of flow by reservoirs changed the amount of individual stream nutrient loading. Indeed, the River Vilaine and its tributaries were in high flood during January 1995. Discharge was efficiently regulated by the Haute Vilaine reservoir. A comparison between TN loads at site 4 (downstream of this barrage) and at site 19, on the river Chevré (without barrage), illustrates the role of this reservoir in controlling TN loads (Figure 7). At the end, cumulated losses in TN dowstream of Vitré (site 4) were only about 31 kg ha^{-1} against 34.6 kg ha^{-1} at the mouth of the River Chevré, the third main tributary of the Vilaine. A catchment with a relatively low land use density (64% agricultural land) and without reservoir lost similar amounts of TN and TP than a more agricultural basin (80% agricultural area) but with a reservoir. This retention of nutrients by stagnant waterbodies induced an increase in organic matter production and an appearance of algal blooms donwstream.

The spatial range in TP loss is less than for TN, with 1.45 kg TP ha^{-1} at the head of the basin and 1.2 kg TP ha^{-1} for higher order stream. In streams with no sewage treatment works, the loss of TP increased with a decrease of stream order probably reflecting instream sorption processes between sediment and water (Bostrom et al., 1988; Froelich, 1988). When the River Vilaine crossed Vitré, loads rose by about five tonnes and represented the greatest increase in TP into the river. Less than two tonnes came from the three sewage treatment plants, which means other sources of P were involved and/or significant releases of P from the sediment occurred in this area.

Conclusion

This simple holistic study of a catchment and its water quality allowed us to evaluate the role of individual subcatchments with different properties, in the transfer of nutrients, under similar climatic conditions. An initial characterization of agricultural practices was made for each subcatchment using agricultural statistics from 1988. But we have to quantify processes involved with non point sources of nutrients. We need to build a more accurate data base of cattle density to find good relationships between catchment management and water quality.

The second point we have to develop is the transfer of P from sediment to water in the stream. Sorption mechanisms are well known but it is still difficult to quantify the P load released or adsorbed by this process (Pilleboue-Baptendier, 1987). At same time, we have to identify the bioavailable P in this sink of P, in a river system where cyanobacterial bloomd occur every year.

Acknowledgements

This work was supported in part by the Institut Scientifique et Technique de l'Environnement and the Société Vitréenne d'Abattage (S.V.A).

References

Barnes, H. & A. R. Kolkard, 1951. The determination of nitrites. National Council for Exploration of the Sea, Charlottenlund.

Bostrom, B., J. M. Andersen, S. Fleisher & M. Jansson, 1988. Exchange of phosphorus across the sediment-water interface. Hydrobiologia 170: 229–244.

Burt, T. P., S. T. Trudgill & D. E. Walling, 1988. Stream nitrate levels in a small catchment in south west England over a period of 15 years (1970–1985). Hydro. Proc. 2: 267–284.

Correll, D. L., E. J. Thomas & D. E. Weller, 1992. Nutrient flux in a landscape: effects of coastal land use and terrestrial community mosaic on nutrient transport to coastal waters. Estuaries 15: 431–442.

Edwards, A. C., K. Pugh, G. Wright & A. H. Sinclair, 1990. Nitrate status of two major rivers in N.E Scotland with respect to land use and fertiliser additions. Chem. Ecol. 4: 97–107.

Froelich, P. N., 1988. Kinetic control of dissolved phosphate in natural rivers and estuaries: A primer on the phosphate buffer mechanism. Limnol. Oceanogr. 33: 649–668.

Hunsaker, C. T. & D. A. Levine, 1995. Hierarchical approaches to the study of water quality in rivers. Bioscience 45: 193–203.

Jaworski, N. A., P. M. Groffman, A. A. Keller & J. C. Prager, 1992. A watershed nitrogen and phosphorus balance: the upper Potomac river basin. Estuaries 15: 83–85.

Koroleff, F., 1977. Simultaneous persulfate oxidation of phosphorus and nitrogen compounds in water. In report of the Baltic Intercalibration Worshop: 52–53.

Kronvang, B., 1992. The export of particulate matter, particulate Phosphorus and dissolved Phosphorus from two agricultural river basins: Implication on estimating the non-point phosphorus load. Wat. Res.l 26: 1347–1358.

Langler, C. L. & P. F. Hendrix, 1982. Evaluation of a persulfate digestion method for particulate nitrogen and phosphorus. Wat. Res. 16: 1451–1454.

Meybeck, M., 1996. River quality. Global ranges, time and space variabilities, proposal for some redefinitions. Verh. int. Ver. Limnol. 26: 91–96.

Murphy, J. & J. P. Riley, 1962. A modified single-solution method for the determination of phosphate in natural waters. Anal. Chim. Acta. 27: 31–36.

Nielsen, N. E. & H. F. Jensen, 1990. Nitrate leaching from a loam soil as affected by crop rotatio and nitrogen fertilizer application. Fert. Res. 26: 197–207.

Omernik, J. M., A. R. Abernathy & L. M. Male, 1981. Stream nutrient levels and proximity of agricultural and forest land to stream: Some relationships. J. soil & water 36: 227–231.

Pilleboue-Baptendier, E., 1987. Origines, bilans et mécanismes de transfert du phosphore et de l'azote d'un bassin versant vers un lac. Thèse de Doctorat Paris 6.

Roberts, G., J. A. Hudson & J. R. Blackie, 1984. Nutrient input and outputs in a forested and grassland catchment at Plynlimon, mid Wales. Agricult. Water Mgmt 9: 177–191.

R.G.A., 1988. Recensement général Agricole, inventaire communaux. 1988. Ille et Vilaine et Mayenne. Ministère de l'Agriculture et des Forêts.

SEEGT, 1994. Etude de qualité de la Cantache. Département d'Ille et Vilaine

Strickland, J. H. & T. R. Parson, 1968. A Practical Handbook of Seawater Analysis, 2nd edn. Bull. Fish. Res. Bd Can.: 167.

William, T. S., A. Jenkins & A. Eastherall, 1994. A simple model of stream nitrate concentrations in forested and deforested catchments in mid-Wales. J. Hydrol. 158: 61–78.

Wright, G. G., A. C. Edwards, J. G. Morrice & K. Pugh, 1991. North east Scotland river catchment nitrate loading in relation to agricultural intensity. Chem. Ecol. 5: 263–281.

Hydrobiologia **373/374:** 259–279, 1998.
J.-C. Amiard, B. Le Rouzic, B. Berthet & G. Bertru (eds),
Oceans, Rivers and Lakes: Energy and Substance Transfers at Interfaces.

Fate and effects of micropollutants in the Gironde estuary, France: a multidisciplinary approach

C. Amiard-Triquet[1], S. Altmann[2], J. C. Amiard[1], C. Ballan-Dufrançais[1], P. Baumard[1], H. Budzinski[1], C. Crouzet[2], P. Garrigues[1], E. His[3], A. Y. Jeantet[1], R. Menasria[1], P. Mora[1], C. Mouneyrac[1], J. F. Narbonne[1] & J. F. Pavillon[1]
[1] *CNRS, GDR1117 Écotoxicologie et Chimie Marines, Université de Nantes, 1 Rue Gaston Veil, 44035 Nantes Cedex, France*
[2] *BRGM, DR/PCI Orléans, France*
[3] *IFREMER, Arcachon, France*

Key words: sediment-bound pollutants, sediment toxicity tests, bioavailability, bivalves, translocation experiments, biomarkers

Abstract

Important Cd contamination has been observed in the Gironde estuary, France, but other metallic and organic pollutants are also present. Since sediment is well-known as a major compartment for the storage of numerous pollutants in aquatic environments, its contribution as a source of pollutants for the biota has been examined according to different methodologies. Geochemical studies have provided estimates of Cd mobility particularly with decreasing pHs and increasing salinity, a result in agreement with the relative abundance of exchangeable Cd and Cd bound to carbonates shown by sequential extraction. When *in vitro* assays tended to be more realistic with regard to the digestion process in bivalves, Cd extraction at low pH was lowered but was still important compared to Cu or Zn. Moreover, toxicity of Gironde sediments to copepod and sea urchin larval stages (not to oyster larvae) brought indirect evidence of the bioavailability of sediment-bound pollutants. Except the aromatic compound perylene, bioaccumulation in bivalves concerns mainly metals. In oysters they were found preferentially in the soluble phase and, in the cytosol, a strong relationship between cadmium and metallothionein-like proteins has been shown, suggesting a detoxication process in this species. This hypothesis is in agreement with the fact that neither cytopathological effects in gills and digestive glands nor marked changes of condition index were observed. On the other hand, no changes in MTLP levels in mussels were induced by metal accumulation in individuals transplanted from a comparatively unpolluted area (Bay of Bourgneuf). The stability of GST activity may be related to a poor accumulation of aromatic compounds. Changes in MDA concentration, AChE and catalase activities are discussed. Mobility, bioavailability of pollutants, significant responses of biomarkers suggest a potential environmental hazard. However, since interspecific differences occur, is the risk at the level of the whole estuarine ecosystem equilibrium or is it limited to a small number of species?

Introduction

In aquatic environments, a great number of pollutants bind to sedimentary particles whether deposited or in suspension. In the case of benthic organisms, either filter – or deposit-feeders, food ingestion systematically involves sediment uptake. Thus an essential issue raised for aquatic environments is either the sediment is a source or a sink for pollutants. The different modes to assess bioavailability in field or laboratory conditions have been reviewed (Amiard, 1992). It is of interest to find out under which physico-chemical forms metals are involved in the sediment since these forms condition the degree of degradability or mobility of sediment-bound pollutants. Thus several patterns of sequential extraction have been proposed to determine the different geochemical fractions binding metals in suspended or deposited sediments (Kersten & Förstner, 1989). However, extractions cannot simulate the biological systems implicated in the processes of resorption and extraction of the trace elements adsorbed on the particles. A more thorough interpretation of sequential extractions in terms of bioavailability requires a methodology incorporating a knowledge of the physico-chemical conditions undergone by particles during digestion. Digestion in the oyster *Crassostrea gigas* has been precisely studied (Boucaud-Camou et al., 1983; Lebesnerais, 1985). Based upon these biological data, *in vitro* desorption tests have been proposed to mimic the digestive processus in molluscs (Amiard et al., 1995; Ettajani et al., 1996). Even if mobility or potential bioavailability are evidenced through such procedures, accumulation strategies of invertebrates can interfere as shown by these authors for lead and arsenic. The assessment of sediment contamination by laboratory bioassay testing has the advantage of reflecting the active dose of a contaminant. In field sediments contaminated with complex mixtures of chemicals, bioassays are currently the only way to assess potential toxicity (Hill et al., 1993).

Important Cd-contamination has been observed in the Gironde estuary, France, but other metallic (Cu, Zn and to a lesser degree Cr, Hg) and organic pollutants are also present (RNO, 1995).

The contamination of living organisms can occur via the aqueous phase or the particulate phase. In the Gironde estuary, Cd concentrations in water are rather high, rising occasionally above 300 ng l^{-1} (Boutier et al., 1989). Metal concentrations in sediments are moderate and similar from the upstream to the downstream parts of the estuary (Crouzet & Altmann, 1996). However, the gross concentrations of contaminants in any compartment of the environment have only a low ecotoxicological significance.

The sedimentary compartment is subdivided into two sub-compartments: deposited sediment and suspended matter. In turbulent environments such as tidal coastal areas and estuaries, exchanges between both of these sub-compartments occur continuously. This is particularly true in the Gironde estuary. Consequently, in the present study, superficial sediments easy to collect in sufficient quantities were used to assess metal mobility either directly using remobilization measurements or sequential extractions. *In vitro* studies were designed to mimic digestive conditions in bivalves in order to estimate directly metal bioavailability. Indirect evidence of toxicant bioavailability was obtained using toxicity tests with larval stages of invertebrate (copepods, bivalves, sea urchins). In addition to these laboratory experiments, mussels and oysters were also transplanted from a clean site (Bay of Bourgneuf, Fr. to the Gironde estuary. Condition indices, contaminant uptake, biochemical variables (metallothionein and MDA concentrations, AChE, GST and catalase activities), and ultrastructural changes were examined in bivalves maintained at the clean site or transplanted and in oysters of the Gironde estuary.

Materials and methods

Experiments in the field (Figure 1)

Mussels (*Mytilus edulis*) and oysters (*Crassostrea gigas*) from a relatively clean coastal area (Bay of Bourgneuf) were translocated to sites on the left bank of the Gironde estuary, La Fosse and Le Cheyzin from May to August 1994 and September to November 1995 respectively. Additional populations were studied: mussels and oysters living in the Bay of Bourgneuf during the same periods, used as controls and oysters growing naturally in the Gironde estuary which will be referred to in the following text a 'residents'.

Sediment sampling

Superficial sediment was scraped using a plastic blade and placed immediately in a plastic box which wa filled brim-full in order to eliminate air bubbles. Samples were then transported to the laboratory with ice packs in an isothermic container. For ecotoxicity tests sediments (May, June and August 1994) were stored in the dark, at 4 °C and used within a week. Metal remobilization measurements, sequential extraction and *in vitro* simulation of digestion were carried out in sediments stored at –20 °C until use. Concerning cadmium, this part of the study was also carried out with samples of 1994. Additional studies for copper and zinc were carried out with sediments sampled in November 1995 at upstream (Phare de Richard) and downstream (La Chambrette) sites (Figure 1).

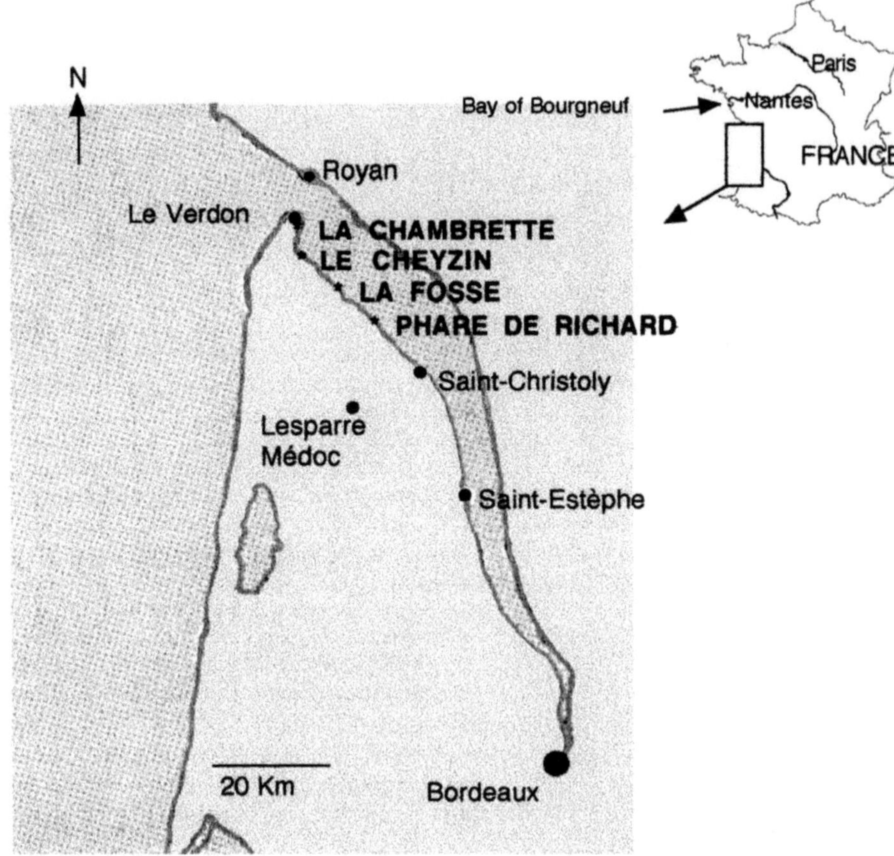

Figure 1. Sampling sites.

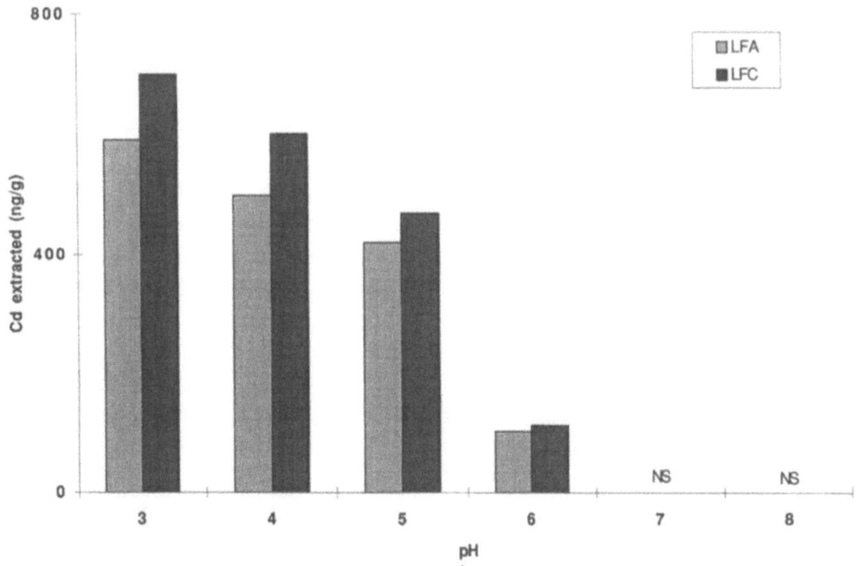

Figure 2. Cd extraction from sediment (La Fosse; A-May 1994; C-August 1994) at different pHs. Duration of assays: 48 h. Natural pH in sediment: 8.

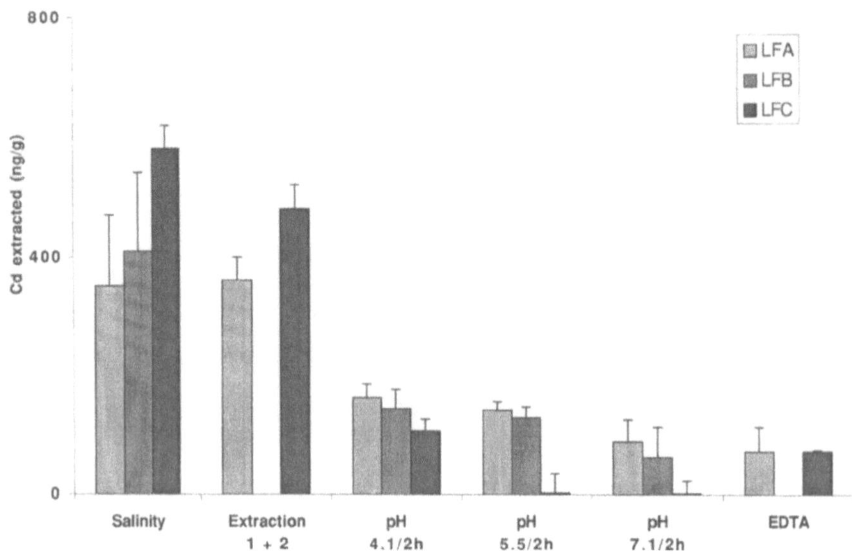

Figure 3. Cd extraction from sediment (La Fosse; A-May, B-June, C-August, 1994). Influence of pH (duration of assays: 2 h), EDTA, salinity. Comparison with Cd recovered in phases 1 and 2 after sequential extractions.

Mobility of sediment-bound Cd

The influence of pH, salinity, EDTA, and Cd partitioning among geochemical fractions, were studied according to the procedures described below:

* Cd release as function of pH

0.8 g of wet sediment was placed in a polypropylene flask with 80 ml of 0.01 M NaCl and the suspension was agitated for 24 hours. Then, the pH was measured and adjusted to the desired pH (from 3 to 8) using suprapure HCl and, for pH 5, a CO_2 flow, several times during 48 hours.

Released Cd concentration in the filtered (0.45 μm) solution was determined by flameless atomic absorption spectrometry (AAS – 300 Zeeman VARIAN).

* Cd release with salinity (33‰)

0.5 g of sediments was agitated during 24 hours in 40 ml of 33‰ saline solution. Cd concentration in the sediments was determined after separation by centrifugation (1000 g for 10 minutes).

* Cd release in the presence of an organic ligand (EDTA)

0.8 g of wet sediments was in contact with 80 ml of 10^{-2}M EDTA (Na_2 EDTA, Merck 8418) in a 100 ml polypropylene tube and agitated for 24 hours. Released Cd was determined in the filtrated solution by flameless AAS.

* Cd partitioning in the sediments

Sequential chemical extractions were used to estimate the distribution of cadmium in various operationally defined metal-solid associations. A combination of chemical extractions were used in order to distinguish:

1. exchangeable metal (Salomons and Forstner, 1980);
2. carbonate associated metal (Tessier et al., 1979);
3. metal associated with Fe or Mn oxyhydroxides (Tessier et al., 1979);
4. metal complexed with organic matter (Tessier et al., 1979);
5. metal in the residual phase (mineral matrix) (Shannon and White, 1991).

Bioavailability of sediment-bound Cd

In vitro desorption tests were designed by Amiard et al. (1995) to mimic the digestive processus in bivalves. In comparison with the work of these authors, the simulation of bivalve digestion was limited to the effect of physiological pHs (4.1, 5.5, 7.1). All desorption tests were carried out in duplicate. Each sediment sample (50 mg) was dispersed into buffer (acetic acid 1%) and adjusted to adequate pHs by adding suprapure ammonia. The digestion time was 2 h at ambient temperature (19–21 °C) since oysters are poikilothermic organisms. Then sediments were recovered by centrifugation and rinsed with the buffer at pHs identical to those of the tests. Trace metal analysis before and after treatments was carried out by atomic absorption spectrophotometry (AAS) after digestion carried ac-

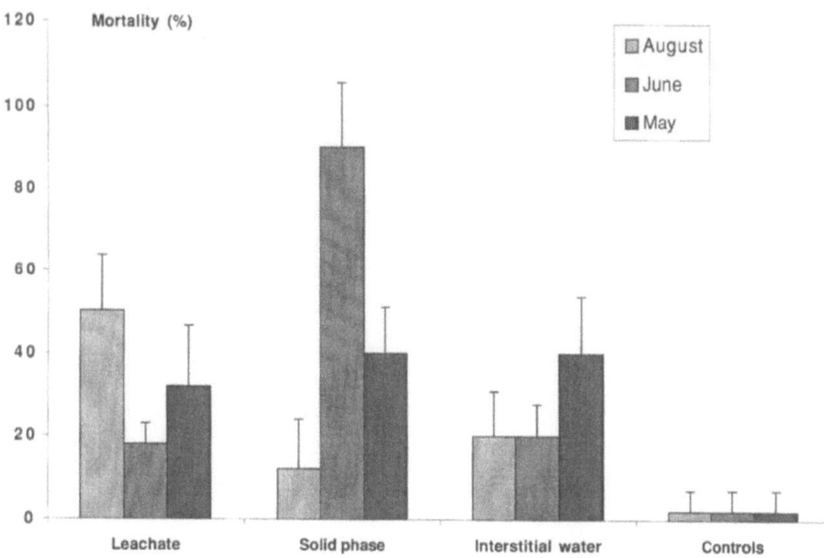

Figure 4. Toxicity of sediments (La Fosse) to the copepod *Tigriopus brevicornis* at different sampling dates and according to the mode of exposure. Mean ($n = 5$) and confidence interval at 95%.

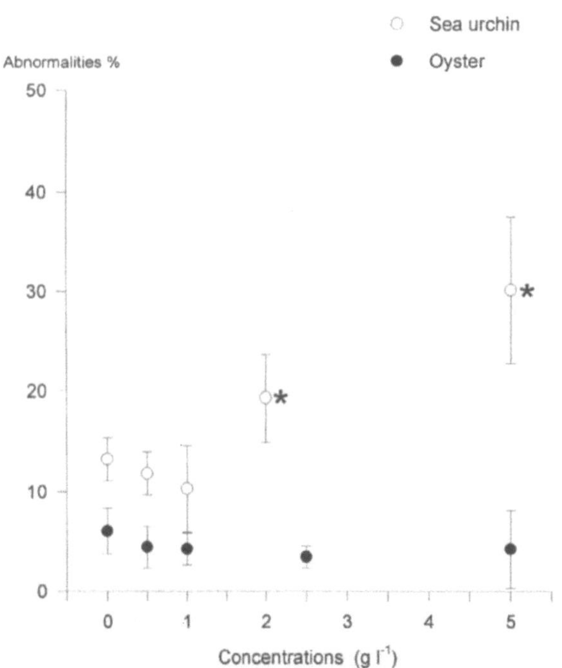

Figure 5. Toxicity of sediments (La Fosse, August 1994) to embryos of oyster (*Crassostrea gigas*) and sea urchin (*Paracentrotus lividus*). X: concentrations of solid sediment in incubation medium. Y: mean ($n = 5$) percentage of larval abnormalities and confidence interval. Asterisks note significant differences with controls.

cording to the recommendations of Charlou & Joanny (1983). The percentage of Cd desorption was calculated from the concentrations of this trace element in initial sediment samples and in sediment samples submitted to the different desorption tests.

Ecotoxicity tests

Test organisms were larval stages of the copepod *Tigriopus brevicornis* , the oyster *Crassostrea gigas* and the sea urchin *Paracentrotus lividus*.

T. brevicornis were collected in a relatively unpolluted area (Le Croisic, South of Brittany) and cultivated in the laboratory. Tests were carried out in 25 ml crystallising dishes, each containing 10 individuals selected according to their stage of development (Copepodite, 1st to 7th day). Copepods were starved all over the exposure period (8 days). The temperature was 20 °C and the photoperiod 12 h/12 h. Five replicates were made for each treatment and cumulated mortalities over 8 d were registered. These treatments were: i. control test with reference seawater collected off Roscoff, France; ii. in leachate test, one volume of reference seawater was agitated with one volume of tested sediment for 20 mn then water was recovered for testing by using centrifugation (750 rpm for 10 mn); iii. in solid-phase test, reference water was poured over the tested sediment, the height of each medium being 1 cm; iv. interstitial water test.

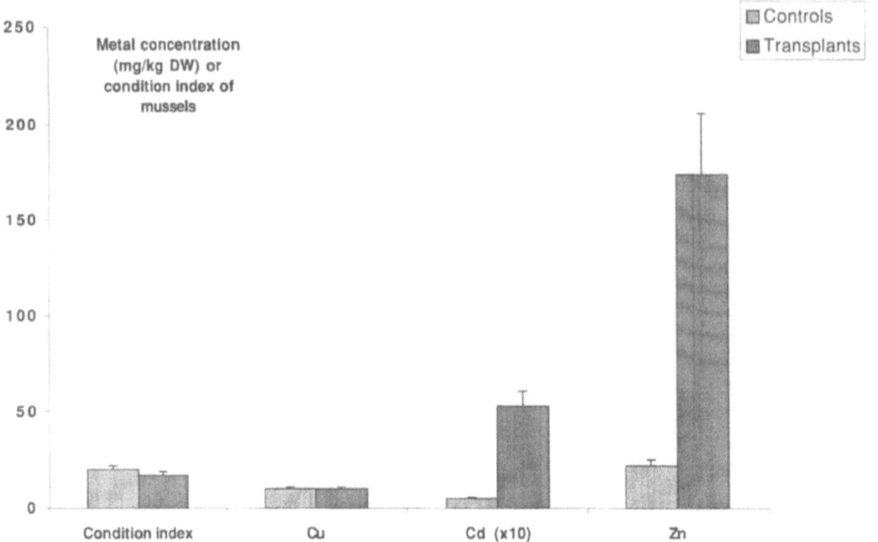

Figure 6. Differences in condition index and metal concentrations in mussels (*Mytilus edulis*) of the Bay of Bourgneuf (controls) and individuals originating of this site and translocated to La Fosse, Gironde estuary, for 3 months. Mean ($n = 20$) and confidence interval at 95%.

Adult *C. gigas* oysters were induced to spawn by thermal stimulation. Freshly collected eggs and sperm were suspended in 0.2 μm filtered seawater (FSW) in sterile glass beakers to achieve fertilization, facilitated by gentle stirring.

Adult *P. lividus* sea urchins were collected from the wild (Arcachon, SW France) and induced to spawn by intracœlomic injection of 0.5 ml KCl at 35 g l^{-1}. Dense sperm from male sea urchins was diluted (1:50) in FSW prior to fertilization.

After 15 min, fertilized eggs from both species were counted under the microscope (taking 4 samples from a measuring cylinder agitated with a plunger) and delivered into 25 ml polyethylene Coulter Counter cuvettes (30 per ml) filled with the different treatments. These treatments were: i. sediment concentrations ranging from 0 to 10 g l^{-1} for the solid phase bioassays, and ii. elutriates of 0 to 200 g l^{-1} obtained following the methods described by Beiras & His (1995) for the suspended phase bioassays. Five replicates per treatment were incubated either at 24 °C for 24 h (oysters) or at 21 °C for 48 h (sea urchins).

After incubation 100 μl of 4% buffered formalin were added to each cuvette and 100 randomly chosen individuals per cuvette were observed under an inverted microscope in order to discriminate the well formed larvae from abnormal larvae or earlier stages of development, according to the criteria of His et al. (1997) for oysters and Klöckner et al. (1985) for sea urchins.

Bioaccumulation of metals and condition index

Gross concentrations were measured in bivalves sampled in May, June and August 1994 at La Fosse. Twenty specimens were examined for each species and each condition (controls, transplants and residents). They were weighed (total weight) then the soft tissues were separated from the shell, mopped up on absorbent paper, weighed (wet weight), dried at 85 °C to constant weight and digested with suprapure nitric acid (Amiard et al., 1987) with a view to metal analysis by AAS.

The condition index (C.I.) was determined according to the recommendation of the French Association for Standardization (AFNOR, NF V45056, Septembre 1985):

$$\text{C.I.} = \text{Drained weight of soft tissues} \; . \; 100 \; . \; \text{Total weight}^{-1}$$

Metal compartmentation was carried out in autochthonous oysters collected in November 1995 at Le Cheyzin. Gills, digestive gland and remainder were examined separately in 16 specimens. Soft tissues were homogenized in 50 mM Tris-NaCl (pH = 8.6) buffer (4:1, v:w) in an ice bath, then centrifuged at 21 250 g for 1 h at 4 °C. Tris contained mercaptoethanol to avoid the formation of disulfide bridges between MT molecules.

A fraction of the supernatant was reserved with a view to metallothionein-like proteins – MTLP – analy

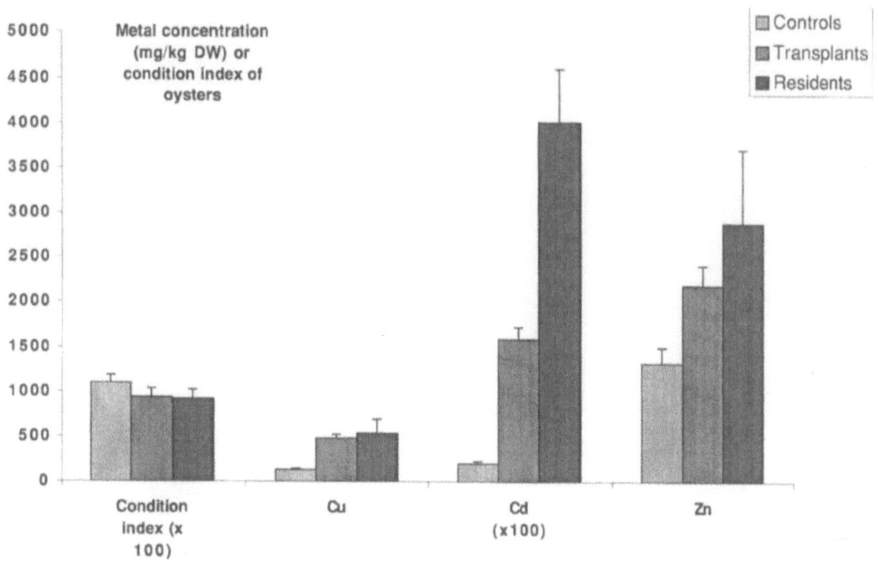

Figure 7. Differences in condition index and metal concentrations in oysters (*Crassostrea gigas*) of the Bay of Bourgneuf (controls), individuals originating of this site and translocated to La Fosse, Gironde estuary, for 3 months and autochthonous specimens of the latter (residents). Mean ($n = 20$) and confidence interval at 95%.

sis (see below). The remainder as well as the pellets were digested with suprapure nitric acid with a view to metal analysis by AAS.

Trace metal analysis

In acid solutions resulting from the digestion of sediments or biological samples, metals were quantified by flame or electrothermal AAS with Zeeman effect for the correction of background (Hitachi Z 8200). The analytical method has been validated through international intercalibration exercises. The results of the most recent one (Coquery & Horvat, 1996) are shown in Table 1.

Study of aromatic compounds: preparation of sediment and mussel tissue extracts

The sediment sampled at La Fosse in August 1994 was freeze-dried and then sieved at 2 mm. It (1 g) was extracted using microwave assisted extraction as described elsewhere (Budzinski et al., 1995; Letellier et al., 1996) using the dichloromethane as solvent of extraction. The organic extract was desulphurized using activated copper. It was then purified on a micro-column containing alumina and fractionated on a micro-column containing silica in order to collect separately the saturated and aromatic compounds according to a procedure adapted from Behar et al. (1989). The aromatic fraction was analyzed by Gas Chromatography / Mass Spectrometry.

Among mussels collected in August 1994, 19 individuals of the Bay of Bourgneuf and 60 individuals of La Fosse were devoted to the determination of aromatic compounds. Soft tissues were separated from the shell, freeze-dried and homogenized. Assay samples (2 g) were extracted according to a procedure described by Baumard et al. (1996). They were saponified for 4 hours with an ethanolic solution of potassium hydroxide (EtOH, KOH, 4N). The digest was liquid-liquid extracted with pentane. The final extract was then purified and fractionated into an aliphatic and an aromatic fraction on alumina and silica micro-columns (Baumard et al., 1996). The aromatic fraction was analyzed by Gas Chromatography/Mass Spectrometry.

Gas Chromatography/Mass Spectrometry (GC/MS) conditions

Aromatic fractions were analysed by GC/MS using a HP 5890 series II gas chromatograph (splitless injection, purge delay: 1 min and purge flow: 60 ml min^{-1}; injector temperature: 270 °C; interface temperature: 290 °C; oven temperature: from 50 °C (2 min) to 290 °C (20 min) at 2 °C min^{-1}) coupled to a HP MSD 5972 mass spectrometer (electron impact: 70 eV; voltage: 2000 V; source temperature: 150 °C) operated under the SIM (selected ion monitoring) mode using

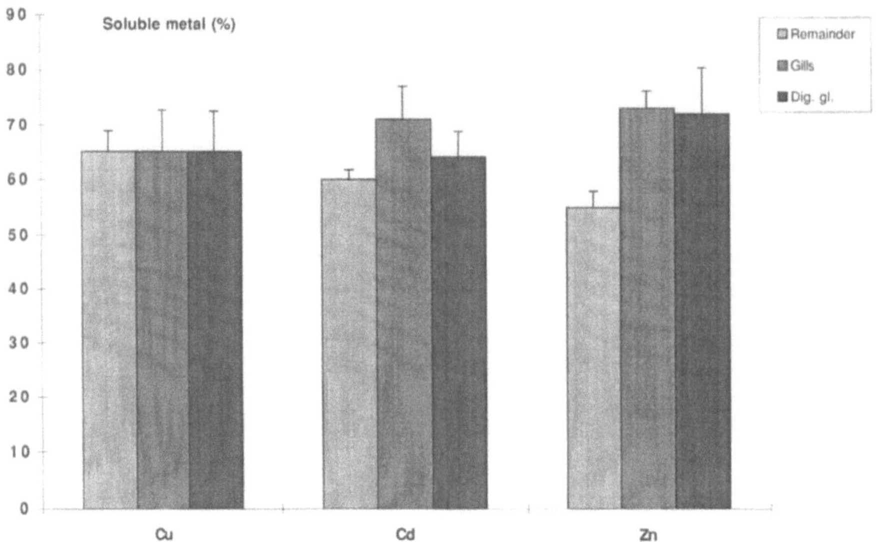

Figure 8. Compartmentation of metals in oysters. Percentages of Cu, Cd and Zn present as soluble forms in different organs or groups of organs. Mean ($n = 16$) and confidence interval at 95%.

Table 1. Intercalibration exercise (Coquery and Horvat, 1996, IAEA – MED-POL): determination of Cd, Cu and Zn in sediment and biota (mean and confidence interval at 95%, mg kg^{-1} dw)

	Cd	Cu	Zn
Marine sediment			
SD-MEDPOL-1/TM			
Certified value	0.59 (0.10)	25.1 (3.8)	191 (17)
Our value	0.70 (0.06)	23.9 (1.4)	189 (2)
Fish homogenate			
MA-MEDPOL-1/TM			
Certified value	0.015 (0.012)	0.62 (0.12)	16.80 (0.48)
Our value	0.0021 (0.0007)	0.66 (0.10)	17.3 (0.6)

the molecular ions of the compounds being studied at 1.2 scan s^{-1}. A 60 m $\times$ 0.25 mm i. d. $\times$ 0.25 μm film thickness capillary column, coated with phenyl (5%)-methyl-polysiloxanne (HP-5, Hewlett Packard, Les Ulis, France), was used with helium as the carrier gas at a constant flow of 1 ml min^{-1}.

Some coelutions have to be noted between structural isomers: (i) chrysene coelutes with triphenylene. Their sum will be noted Chrys$*$; (ii) benzo(b)fluoranthene coelutes with benzo(k)fluoranthene and benzo(j) fluoranthene. Their sum will be noted BF; (iii) dibenz(ah)anthracene coelutes with dibenz(ac)anthracene. Their sum will be noted DaA.

Perdeuterated aromatic compounds were used as internal standards for quantitative analyses in each aromatic series. They were added prior to the extraction. The response factors of the PAHs selected for study relative to the perdeuterated PAHs were regularly determined by the use of the Standard Reference Material, SRM 2260 (NIST, Gaithersburg, MD, USA).

In order to test the accuracy of the quantification method, the SRM 2260 was regularly run on the GC/MS system. Moreover, the entire analytical procedure was applied several times to the certified marine sediment, SRM 1941a (NIST, Gaitherburg, MD, USA). The recoveries, for five replicates on SRM 1941a, were between 95 and 103% and the repro-

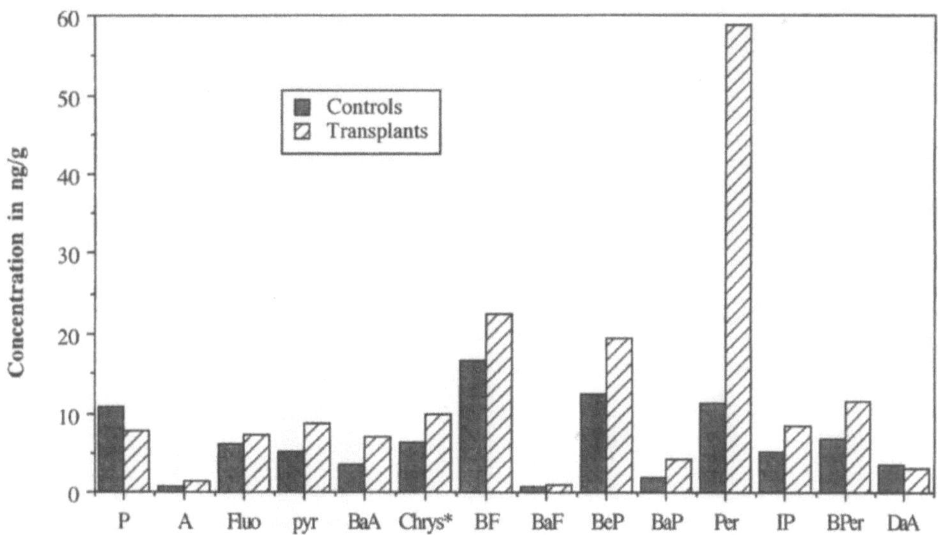

Figure 9. Aromatic compounds in mussels (*Mytilus edulis*) of the Bay of Bourgneuf (controls) and individuals originating from this site and translocated to La Fosse, Gironde estuary, for 3 months. Soft tissues of 19 (Bourgneuf) to 60 (Gironde) specimens were pooled. Phenanthrene (P), anthracene (A), fluoranthene (Fluo), pyrene (Py), benz(a)anthracene (BaA), chrysene (Chrys), triphenylene (Trip), benzo(b)fluoranthene (BbF), benzo(k)fluoranthene (BkF), benzo(j)fluoranthene (BjF), benzo(a)fluoranthene (BaF), benzo(e)pyrene (BeP), benzo(a)pyrene (BaP), perylene (Per), indeno(1,2,3-cd)pyrene (IP), benzo(g,h,i)perylene (BPer), dibenz(a,h)anthracene (DahA), dibenz(a,c)anthracene (DacA).

ducibility of the technique ranged from 5% to 10% depending on the compounds. The same approach was applied to certified mussel tissue sample, SRM 1974a (NIST, Gaithersburg, MD, USA). The recoveries were $90 \pm 16\%$.

Biochemical measurements

Cholinesterase (ChE), glutathione-S-transferase (GST), catalase activities and malonedialdehyde (MDA) concentration were measured in mussels collected in August 1994. For controls and transplants, 12 samples of gills (each combining the tissues of 2 individuals) and 4 or 5 samples of digestive glands (combining the tissues of 5 to 7 individuals) were analysed. Subcellular tissue fractions were prepared as follow.

The digestive glands were weighed and homogenized in two volumes of 100 mM phosphate buffer, pH 7.4, using a potter homogenizer (ELVEHJEM). Homogenates were centrifuged (20 minutes, 9000 g), and the resulting supernatant (post-mitochondrial fraction or S9) was centrifuged for 1 hour at 105000 g yielding the supernatant cytosol and the microsomal (pellet) fractions. Cytosol was stored at –30 °C. Microsomes were resuspended in a 100 mM phosphate buffer containing 20% glycerol and 1 mM EDTA, pH 7.4, to a final concentration of about 30 mg protein ml^{-1} and stored at –80 °C prior to analysis. All experiments were carried out at 4 °C.

The gills were also weighed and homogenized in three volumes of 100 mM phosphate buffer, pH 7.4, for 20 seconds per 750 mg, with the ultra-turrax homogenizer at 13500 t min^{-1}. The homogenates were centrifuged (20 minutes, 9000 g) to obtain the S9.

Enzymatic activities were measured in the post-mitochondrial fraction S9, in gills for AChE and GST, in the digestive gland for catalase by using spectrophotometric techniques. AChE activity was determined by measuring the kinetic of propionylthiocholine hydrolysis (used as substrate) to thiocholine. The reaction of thiocholine with 5-thio-2-nitrobenzoate gives a yellow chromophore determined at 412 nm (Ellman, 1961).

GST activity was determined by measuring the kinetic occurrence of the 1-chloro-2-4-dinitrobenzene-glutathione complex at 340 nm (Habig et al., 1974). Catalase activity was determined by measuring the disappearance of hydrogen peroxide at 240 nm by a method adapted from Claiborne (1985). The concentration of MDA was measured in the cytosol of digestive glands after complexation with thiobarbituric acid producing a pink chromophore determined by colorimetry at 532 nm (Buege & Aust, 1978). All these data were expressed in n mole mg^{-1} proteins, the measurement of which was carried out according to the method of Lowry et al. (1951).

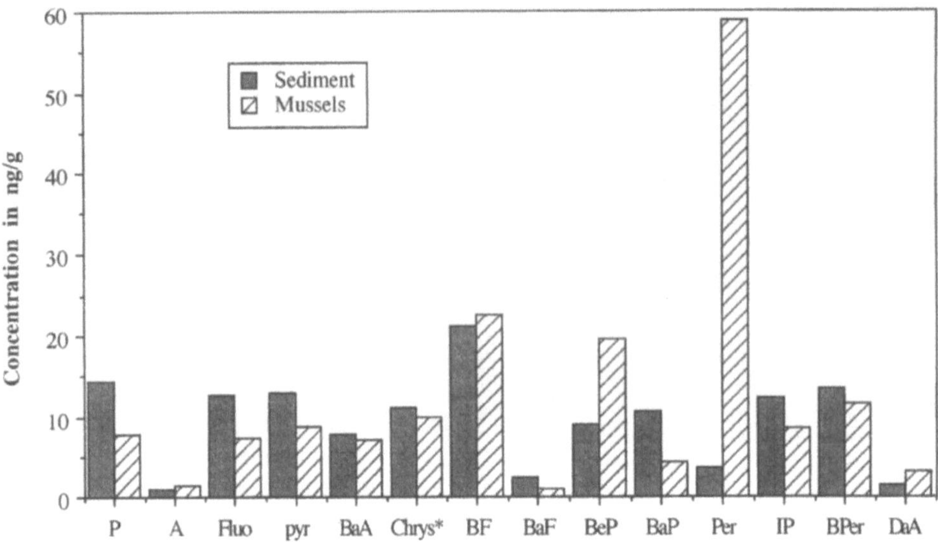

Figure 10. Comparison of aromatic compound levels in mussels translocated for 3 months (pool: $n = 60$) at La Fosse, Gironde estuary and sediment from the same site. For abbreviations see Figure 9.

MTLP were quantified in both species of bivalves sampled in November 1995 (controls originating of the Bay of Bourgneuf, specimens translocated towards the Gironde estuary, residents in the Gironde estuary). The first part of the procedure was common with metal compartmentation (see above). Then a fraction of the supernatant was heat-denaturated (70 °C for 15 min) to precipitate thermolabile compounds which were subsequently separated from heat-stable compounds (including MTLP) by centrifugation (12 000 g; 10 min). Differential pulse polarography was used to determine MTLP in the supernatant according to the method described by Thompson and Cosson (1984). To limit the size of the present paper, MTLP concentrations recalculated for the whole soft tissues are often shown rather than data obtained for separated groups of organs.

Electron microscopy and electron probe microanalysis

A cytological study was carried out in oysters sampled in the Gironde estuary at La Fosse (5 transplants and 5 residents) in June 1994. Gills and digestive glands were selected for this study, the first one by reason of its direct exposure to micropollutants in the medium, the last one for its role in biotransformations. Pieces of these tissues were fixed in 3% glutaraldehyde buffered with cacodylate 0.2 M at pH 7.4. For ultrastructural study, specimens were post fixed in 1% osmium tetroxyde in water. All the tissue samples were embedded into Epon-Araldite. Thin section (50–80 nm) were collected on copper grids and observed on a Phillips 300. The electron probe microanalysis (EPMA) of non osmicated and unstained material, thin sectionned (80–100 nm) and collected on titanum carbon coated grids was performed on a Cameca MBX (Camebax). The probe diameter was 500 nm for a beam intensity of 150 nA at 45 KV. Intensities analysis were 100 sec. counting periods. Significative values were retained at 2σ of confidence interval.

Results

Mobility of metals

The results of remobilization measurements according to pHs are shown in Figure 2. After 48 h contact, extracted Cd increased with increasing acidity of the medium reaching about 70% of total sediment Cd at the lowest pH tested. This is typical of metallic cation mobilization by desorption from acidic functional groups on oxyhydroxydes and organic matter. The same tendency was observed after 2 h contact at physiological pHs (Figure 3). Two pHs almost equal were tested in both experiments: at pH 7 or 7.1, no Cd was extracted whatever the conditions. At pH 4 or 4.1, Cd mobility reached about 60% corresponding to about 0.6 μg Cd extracted per g of sediment after 48 h contact, whereas after 2 h contact, the Cd quan-

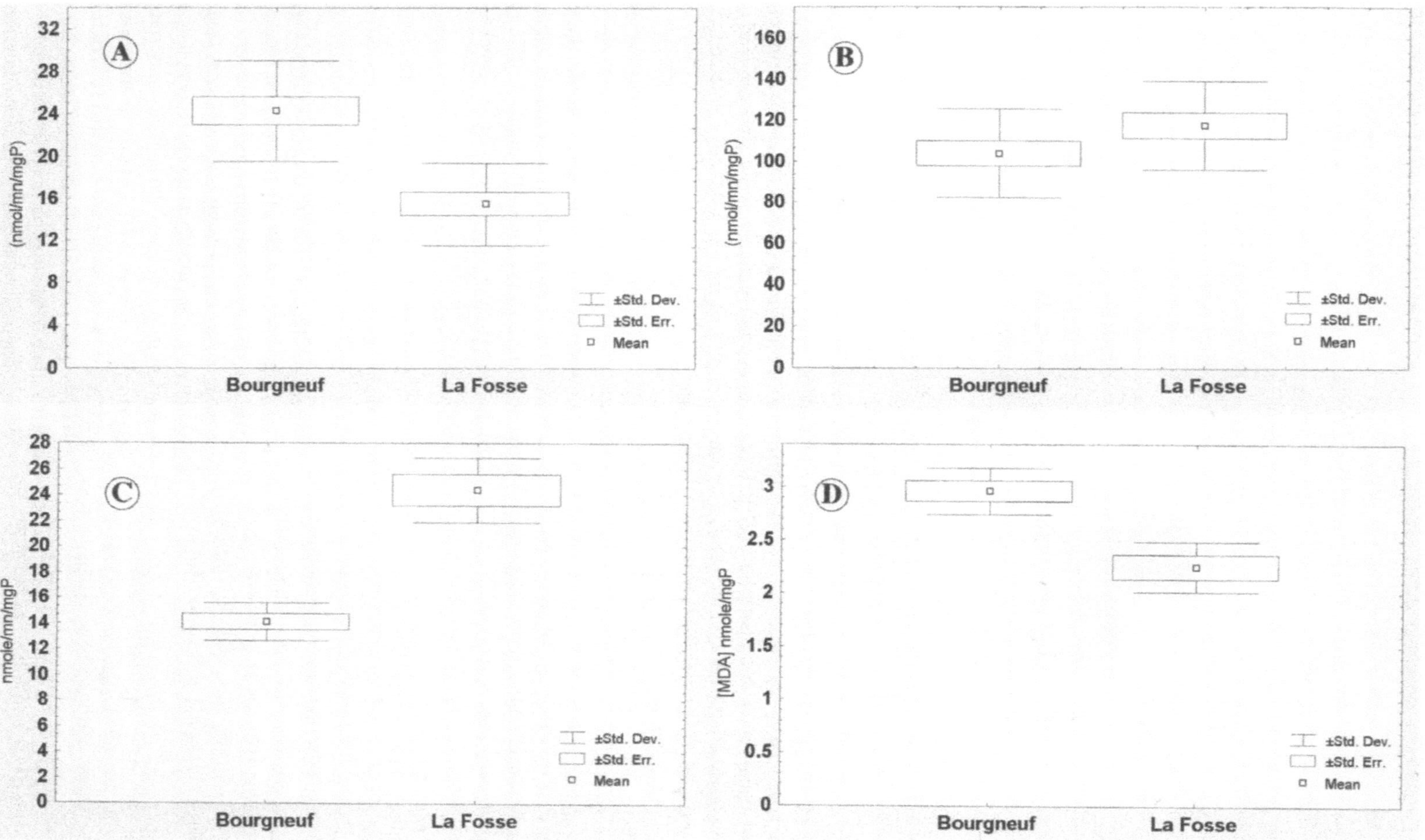

Figure 11. Comparison of AChE (A) and GST (B) activities in gills, catalase activities (C) and MDA concentrations (D) in the digestive glands of mussels (*Mytilus edulis*) from the Bay of Bourgneuf (controls) and individuals originating from this site and translocated to La Fosse, Gironde estuary, for 3 months. Enzymatic activities (A, B, C) in nmol per mn per mg of proteins, concentrations of MDA (D) in nmol per mg of proteins. Gills: $n = 12$, each sample combining tissues of 2 specimens; digestive gland: $n = 4$ or 5, each sample combining the tissues of 5 to 7 specimens.

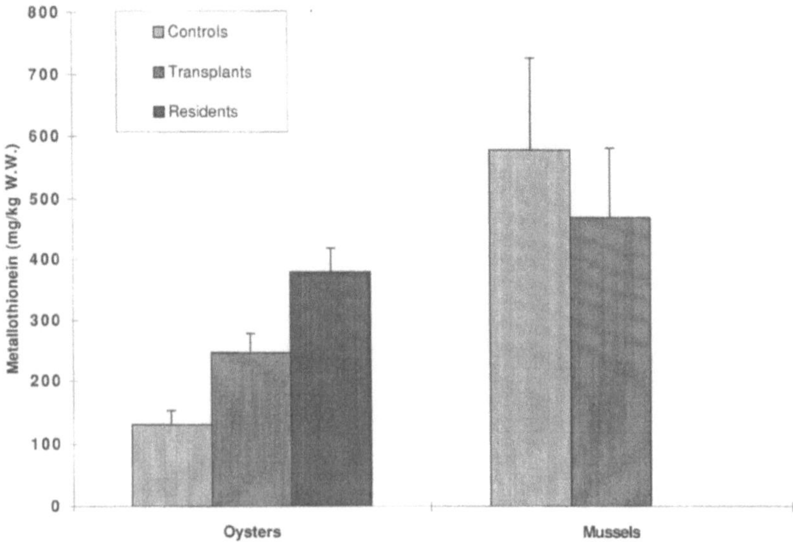

Figure 12. Metallothionein-like protein concentrations in bivalves of the bay of Bourgneuf (controls), bivalves originating of this site and translocated to Cheyzin, Gironde estuary, for 2 months and autochthonous oysters of the latter (residents). Mean (n = 8) and confidence interval at 95%.

tity extracted varied from roughly 0.1 to 0.2 μg g^{-1} depending on the sampling date.

Remobilization of Cu and Zn was not assayed with sediments from La Fosse but due to the homogeneity of sediments in the Gironde, data obtained upstream (Phare de Richard) or downstream (La Chambrette) may be considered. Moreover, Cd desorption was similar to the one determined at La Fosse (Table 2). In the same conditions of pH, desorption of Cu and Zn was not as important as for Cd. At all the physiological pHs tested, there was a significant difference between copper concentration before and after treatment whereas this difference was significant only at the lowest pH tested (4.1) for zinc (Table 2).

During their transport towards the Atlantic Ocean, sedimentary particles follow a gradient of salinity. At salinity nearing that of seawater, Cd bound to sediment from the brackish site of La Fosse was transferred to the soluble phase. Cd extraction was more important in summer than in spring (Figure 3). However, the full chemical analysis did not show striking differences between sediments collected in May and August (Table 3).

Sequential extractions have allowed to determine the distribution of cadmium in five geochemical fractions (Table 4). This distribution was very similar in samples of both months. Exchangeable Cd and Cd bound to carbonates which are considered as the most liable to be bioaccumulated together represented 42 and 47% of total cadmium in May and August respectively. Cadmium extracted in these fractions (1 + 2) is quantitatively equivalent to Cd remobilized at seawater salinity (Figure 3). Cadmium extracted in these fractions and in the fraction 'Fe-Mn oxyhydroxides' is closed to the quantity of Cd remobilized at pH 3 (proton exchanges, carbonate dissolution and Cd desorption).

Sediment Cd is only slightly solubilized by the 10^{-2} m EDTA (7% of total sediment Cd). Significant amounts of iron and calcium (7% and 80–90% of total content, respectively) contained in the sediments, were also extracted by EDTA.

Ecotoxicity tests

The results obtained with the copepod *Tigriopus brevicornis* are shown in Figure 4. Going into the details, differences may be described according to the date of sampling or to the procedure of exposure (leachate, solid phase or interstitial water) but the most striking feature is that in all cases (except the solid phase in August), the exposure of copepods to sediments induced a significant mortality compared to controls. In half of the cases shown in Figure 4, mortality can reach 40% or more of the experimental population.

Percentages of abnormal larvae have been angular transformed (arcsine) prior to calculate the 95% con-

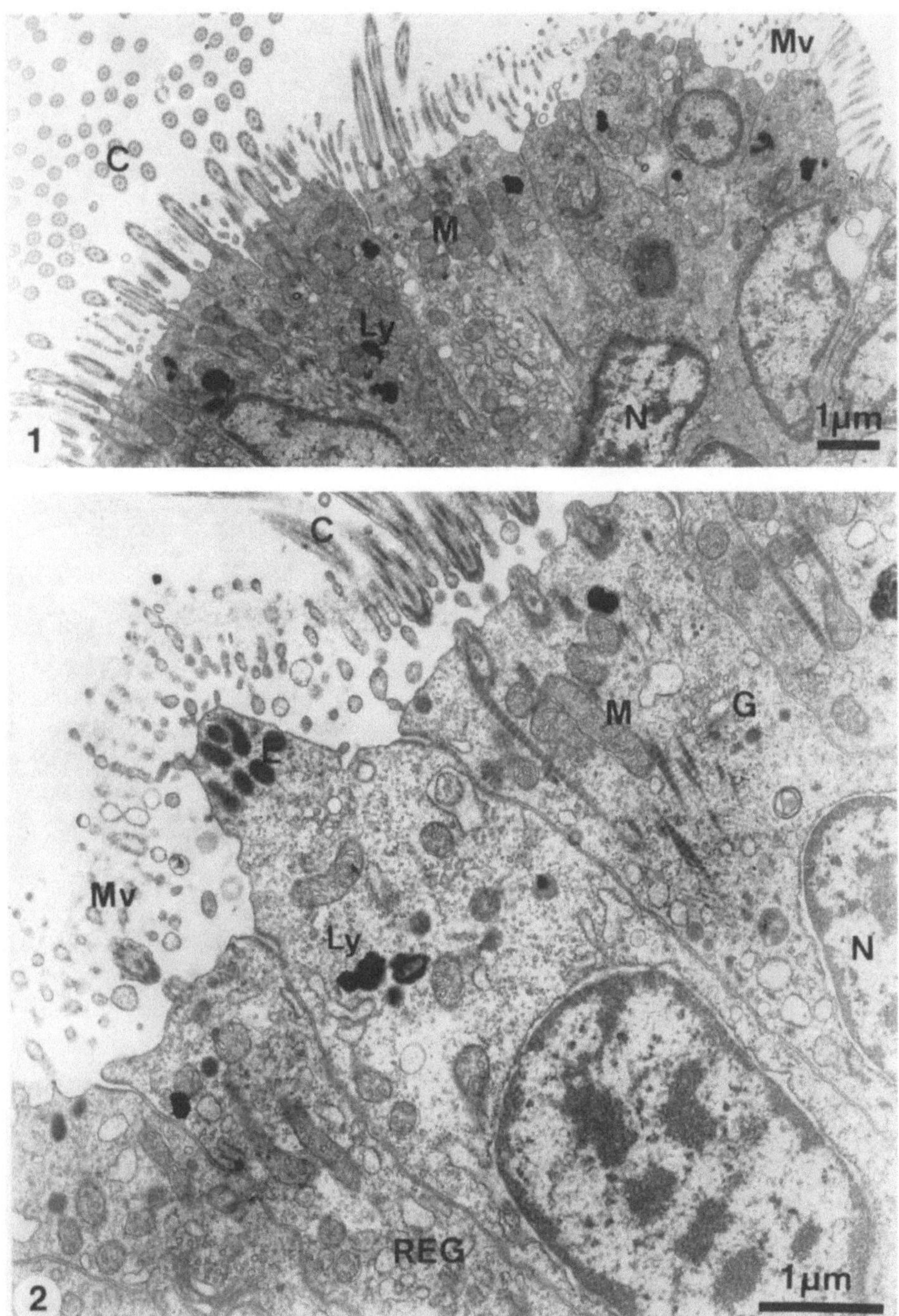

Figure 13. Top: Gill of an oyster originating from La Fosse (Gironde estuary). Cadmium was not detected in the lysosomes of branchial cells. G. × 9000. Bottom: Gill of an oyster originating from Bourgneuf and transplanted to La Fosse. The epithelium did not show any cytopathological aspect and cadmium did not occur anywhere. G × 20 000. C = cilia; E = endosome; G = Golgi apparatus; Ly = lysosome; M = mitochondria; Mv = apical microvilli; N = nucleus; REG = granular endoplasmic reticulum.

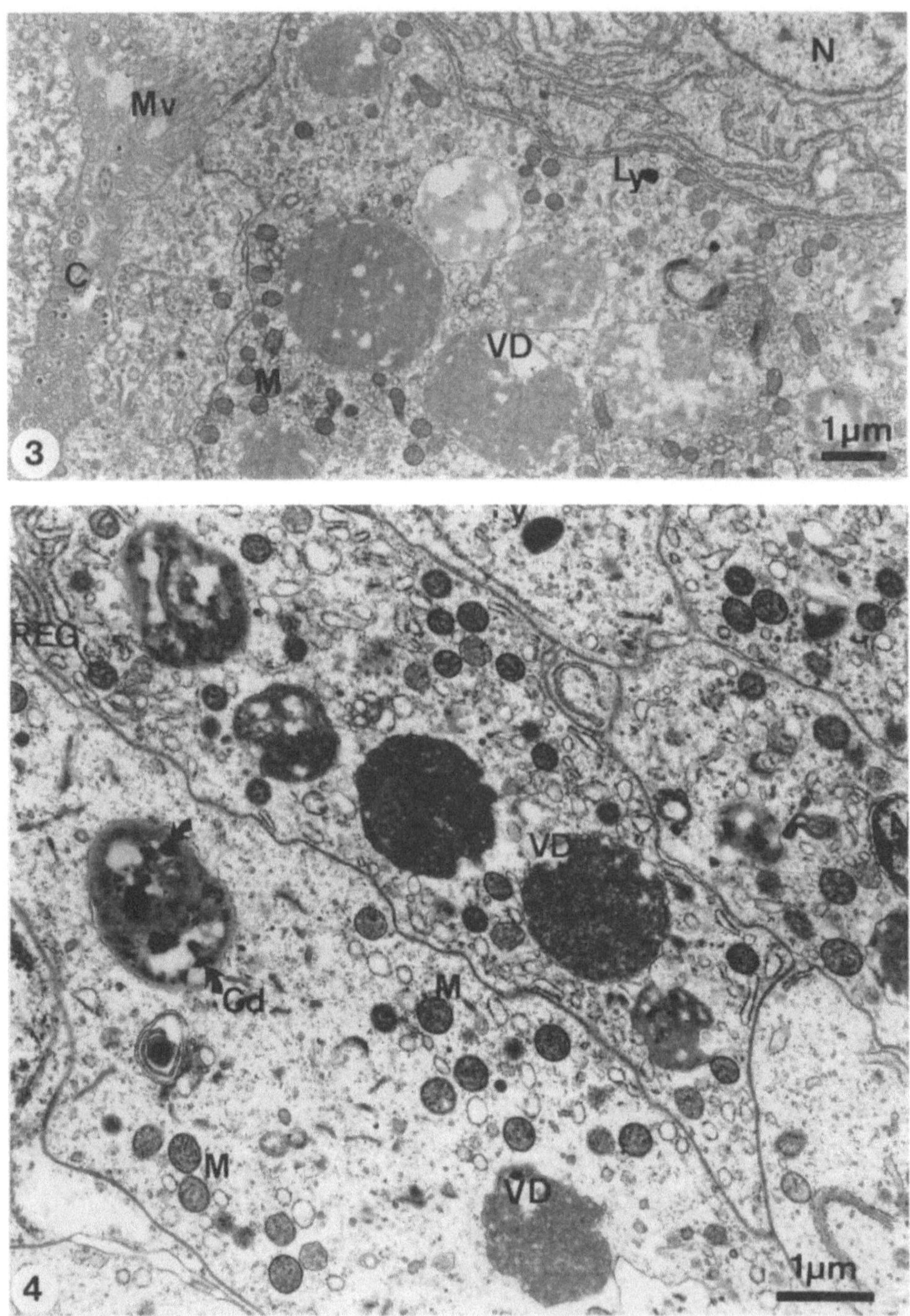

Figure 14. Top: Digestive gland of an oyster transplanted from Bourgneuf to La Fosse. As in branchial epithelium, cells didn't show any cytopathological aspect. Cadmium was not detected anywhere. G. × 9000. Bottom: Digestive gland of an oyster originating from La Fosse. Cadmium was detected in lysosomes occurring in some of the digestive vacuoles (arrow). G × 14 000. C = cilia; Ly = lysosome; M = mitochondria; Mv = microvilli; N = nucleus; VD = digestive vacuole.

Table 2. Metal desorption from sediments sampled in the Gironde estuary at physiological pHs. Upstream site: Phare de Richard (PR); downstream site: La Chambrette (LC)

	Cadmium		Copper		Zinc	
	LC	PR	LC	PR	LC	PR
% extracted at pHs*						
4.1	40 ± 11	44 ± 3	13 ± 3	23 ± 5	10 ± 2	15 ± 2
5.5	28 ± 8	28 ± 5	12 ± 6	14 ± 2	NS	NS
7.1	19 ± 6	NS	11 ± 1	11 ± 2	NS	NS
Concentration in raw sediment (μg g^{-1})	0.443	0.405	14.3	14.3	110	100
Metal extracted at pH 4.1 (μg g^{-1})	0.177	0.178	3.1	5.4	11	54

* Mean ± confidence interval at the level of 95%
NS: no significant difference between metal concentration before and after treatment.

Table 3. Full chemical analysis of sediments collected at La Fosse in May (A) or August (C). Data in percents

Sample	PP 450 °C	Organic C	Mineral C	Total S	Si	Al	Fe
A	3.31	0.45	1.35	0.05	28.8	6.1	2.4
C	4.57	0.55	1.15	0.07	27.0	6.0	2.4
	Ca	Mg	K	Mn	P		
A	3.6	0.81	2.37	0.08	701		
C	3.7	0.90	2.24	0.05	672		

Table 4. Distribution of cadmium among geochemical fractions in sediments collected at La Fosse in May (A) and August (C)

	A		C	
	μg g^{-1}	%	μg g^{-1}	%
1 – Exchangeable	0.13	15	0.11	11
2 – Carbonates	0.23	27	0.37	36
3 – Fe hydroxydes and Mn oxides	0.26	30	0.31	31
4 – Organic matter	0.08	9	0.06	6
5 – Residual	0.16	19	0.16	16

fidence interval of the means (Sokal & Rohlf, 1981), and back-calculated for presentation.

Low percentages of abnormalities were found in all oyster embryo bioassays (ranging from 2.8 to 7.2%) with independence of the sediment concentration and phase assayed, and no significant differences with the controls were obtained in any treatment (Table 5). In contrast, sea urchin embryogenesis success was significantly (95% confidence level) reduced at concentrations of solid sediment $\geq$ 2 g l^{-1} (Table 5, Figure 5).

Bioaccumulation

In mussels and oysters translocated from the Bay of Bourgneuf to the Gironde estuary, Cd and Zn concentrations increased (Figure 6 and 7). However, for oysters even after 3 months of exposure, the level of Cd in transplants had not reached the level in residents whereas the level of Zn is approaching. Translocation did not affect the level of Cu in mussels whereas the level of Cu in oysters was increased but was no longer significantly different between transplants and residents after 3 months (Figure 6 and 7).

In all the organs or groups of organs of oysters, Cd, Cu and Zn were present mainly in the cytosol (Figure 8). Thus MTLP levels were determined in gills

Table 5. Mean percentage abnormalities (95% confidence intervals, $n = 5$) in *Crassostrea gigas* and *Paracentrotus lividus* embryogenesis in the presence of different concentrations of sediments (Solid phase) or elutriates from sediments (Suspended phase) sampled from La Fosse. n.t. = not tested. *significant difference from control

	Solid phase				Suspended phase
	C. gigas		*P. lividus*		*C. gigas*
$g\,l^{-1}$	25 May	24 August	24 August	$g\,l^{-1}$	24 August
0	4.0 (2.8, 5.3)	6.0 (3.3, 8.6)	13.0 (10.8, 15.3)	0	4.2 (2.5, 6.1)
0.5	2.8 (1.8, 3.9)	4.4 (2.3, 6.7)	11.8 (9.4, 14.3)	50	5.4 (2.1, 9.3)
1	4.6 (3.0, 6.3)	4.2 (2.7, 5.9)	10.2 (6.2, 14.6)	100	5.4 (2.4, 8.8)
2	6.8 (4.7, 9.1)	n.t.	19.4* (15.4, 23.5)	200	5.4 (3.3, 7.6)
2.5	n.t.	3.4 (2.3, 4.6)	n.t.		
5	5.6 (4.2, 7.1)	4.2 (2.9, 5.6)	30.0* (22.7, 37.6)		
10	n.t.	7.2 (3.7, 11.1)	n.t.		

(235 ± 31 mg kg^{-1}), digestive gland (749 ± 79 mg kg^{-1}) and remainder (145 ± 23 mg kg^{-1}). In order to see the combined effect of metal concentrations and weight on MTLP concentrations, multiple regression analysis was performed. MTLP concentrations can be expressed as a linear model:

$$[\text{MTLP}] = \alpha_0 + \alpha_1\,[\text{Cd}] + \alpha_2\,[\text{Cu}] + \alpha_3\,[\text{Zn}] + \alpha_4\,\text{weight}$$

with concentrations (MTLP and metals) recalculated for the whole soft tissues and [Cd], [Cu] and [Zn] expressed in μM. The coefficients $\alpha_0, \ldots, \alpha_4$ were calculated from concentrations (MTLP and metals) determined individually in 16 oysters. When we express MTLP (dependent variable) as a function of metal concentrations, and weight (independent variables), we need to standardize the independent variables to avoid the problem of unit. The proposed method is to transform natural variables x_i (Cd, Cu, Zn and weight) to coded variables X_i.

$$\text{MTLP} = \alpha_0 + \Sigma \alpha_i X_i$$

$X_i = +1$ for the maximum value of a variable and $X_i = -1$ for the minimum one. The passage from natural variables to coded variables is given by the following equation:

$$X_i = \frac{x_i - (x_{i\,\max} + x_{i\,\min})/2}{(x_{i\,\max} - x_{i\,\min})/2}$$

where $X_{i\,\min}$ = the lowest value of the natural variable, and $X_{i\,\max}$ = the highest value of the natural variable.

The coefficients ($\alpha_0, \ldots, \alpha_4$) were tested by using Student's t-test.

MTLP = –0.414 + 0.385 [Cu] – 0.448 [Zn] + 0.508 [Cd] – 0.498 weight.

($t = 1.48$) ($t = 1.78$) ($t = 2.10$) ($t = 2.49$)
($p = 0.17$) ($p = 0.10$) ($p = 0.06$) ($p = 0.03$).

This model is highly significant with a high R^2-value (0.872). It seems that MTLP concentration in oysters at this site depends mainly upon weight of and Cd concentration in soft tissues because of the high coefficients obtained and the levels of probabilities.

Concerning the insoluble fraction, the electron probe microanalysis (EPMA) showed Cd only in a few digestive vacuoles of autochthonous oysters. No Cd was detected elsewhere, whereas Al, Cu, Fe, Zn, P and S were present in the lysosomes. The quantity of Cd was at the limit of the sensitivity of the Camebax, making impossible the research of the molecular form in which it occurred.

Bioaccumulation of aromatic compounds

Figure 9 allows a comparison between the concentrations of aromatic compounds in mussels of the Bay of Bourgneuf and those determined in specimens translocated in the Gironde estuary. Differences are moderate except for perylene which was highly accumulated in transplants.

Figure 10 allows a comparison between the concentrations of aromatic compounds in sediments sampled at La Fosse and those in mussels exposed *in situ* at the same site. Concentrations were generally of the same order of magnitude in sediment and biota except for perylene, the level of which is very low in sediment compared to mussels.

Biological effects

In mussels, changes have been observed at the biochemical level (Figure 11). Acetylcholinesterase and GST activities were determined in gills since they exhibited the highest activities among the tissues of these bivalves. A depletion of AChE activity ($p = 0.00006$) was observed in mussels translocated to the Gironde estuary whereas the GST activity was not significantly modified ($p = 0.14$). MDA concentration in the digestive gland of translocated mussels was also depleted ($p = 0.002$ for cytosolic MDA, Figure 11; $p = 0.001$ for MDA measured in microsomes, not shown), whereas catalase activity was enhanced ($p = 0.01$).

In mussels, MTLP levels did not differ significantly in transplants vs controls (Figure 12). On the other hand, in oysters, MTLP levels were significantly higher in transplants than in controls but after two months of exposure in the Gironde estuary they had not reached the concentrations determined in residents (Figure 12).

Cytopathological effects have been investigated in gills and digestive glands of oysters translocated from the Bay of Bourgneuf to the Gironde estuary (La Fosse) as well as in residents. Results obtained in both tissues and both temporary and chronic exposures are shown in Figure 13 and 14. No deleterious effects were observed.

In both species of bivalves, a more general marker of potential effects has been determined: the condition index (Figure 6 and 7). Changes observed at this level were negligible.

Discussion

The geochemical characterization of sediments sampled in the Gironde estuary has shown that relatively important quantities of Cd were present in the exchangeable phase or bound to carbonates. These fractions are considered to be the most available for bioaccumulation. Moreover, Cd, Cu and Zn are observed to desorb from sediments with decreasing pH or increasing salinity; conditions encountered in the field when sedimentary particles are ingested by living organisms or during their transport towards the sea. EDTA was used as a model of biological compounds which may exert a binding ability towards metals during digestion. The fact that only a relatively small amount of Cd was solubilized by EDTA, compared to that extracted in the first 4 extractions (7% versus 80%), is likely to be due to competition by Ca, Fe and other sediment elements. These elements are present at much higher total concentration in the sediments relative to Cd (Table 3). Their ability to compete against Cd will depend both on their relative total concentrations and their EDTA complexation constants. Iron is both present in high concentration (24 mg g^{-1}) and has a high constant ($10^{27.7}$, Morel, 1993). Ca (36 mg g^{-1}) has a constant of $10^{12.4}$ (Morel, 1993). It is interesting to note that EDTA solubilized 70% of the iron mobilized in the first 4 sequential extractions.

According to Campbell et al. (1988) 'the nature of most of the geochemical and biological work that has been done in the past makes it difficult to relate geochemical to biological data, much less estimate the related bioavailability of metals in sediments'.

In freshwater bivalves, bioaccumulation of Cu, Pb and Zn is well correlated with the initial stages of sequential extractions. In marine invertebrates (*Macoma balthica, Scrobicularia plana, Nereis diversicolor*) a good correlation has been shown between the concentrations of different metals in organisms and those extracted with HCl 1M for Ag, Cd, Co and Zn or with HNO_3 (15M) for Cu, Hg and Pb (review by Amiard, 1992). However, the correlation may be indirect since metal bioavailability may be controlled by chelating agents such as iron hydroxide present in the sediment (Allen & Hall, 1980; Langston, 1980; Tessier et al., 1984).

A more thorough interpretation of geochemical data requires an integrated knowledge of the processes undergone by particles during digestion. In the present study, we have shown that a decrease in the length of digestive transit modelled by contact duration (2 h vs 48 h) between particles and acid solutions led to a decrease of Cd – desorption.

Complementary to geochemical studies, toxicity tests with field contamined sediments may provide indirect evidence of bioavailability considering that only bioaccumulated toxicants can exert a biological effect. Sediments collected in the Gironde estuary (or extracts of these) induced noxious effects at early stages of life in the copepod *Tigriopus brevicornis* and the sea-urchin *Paracentrotus lividus* but not in oyster embryos (*Crassostrea gigas*). This resistance of oyster embryos may be associated with the existence of MTs in embryos of *Crassostrea virginica*. In this species, concentrations as high as 20 μg Cd l^{-1} did not perturb embryogenesis (Ringwood et al., 1995).

Sediment bioassays take into account the joint action of sediment-bound materials. Techniques of

Toxicity Identification Evaluation have been proposed to localize toxicity in particular sediment fractions and ultimately for identifying which substances are contributing to the overall toxic effect (Hill et al., 1993). Since the Gironde estuary has given rise to extensive monitoring (RNO, 1995), we have preferred to focus our research on metal and HAP bioaccumulation.

In mussels transplanted from the Bay of Bourgneuf, considered as a clean site, to the Gironde estuary, no important accumulations were observed except for perylene a compound resulting from combustion processes at high temperature (cars, industries, etc. ...). Sediments of the Gironde estuary (La Fosse) sampled concomitantly with mussels did not show high concentrations of this particular compound. However, we have observed high levels of perylene in some sediments at certain periods in the year. Thus, the preferential bioaccumulation of perylene in mussels may be due to their exposure to freshwater fluxes before (March to June) our sampling date (August) whereas in summer, at lowest water level, the marine influence was preponderant at this site (La Fosse) near the mouth of the river. Moreover, the preferential storage of perylene may be re-inforced as a consequence of its hydrophobic nature, limiting its metabolization.

Metal bioaccumulation in mussels and oysters translocated from the Bay of Bourgneuf to the Gironde estuary differed between metals within the same species and differed also between species when a particular metal was considered. Such observations are abundant in the literature and have been attributed to the biological role of different metals (essential such as Cu and Zn versus toxic such as Cd) or to specific strategies of accumulation and storage (Rainbow, 1993). Cd may to be stored in lysosomes of gills, digestive gland and kidney cells of oysters after experimental treatment with Cd Cl_2, but was not detected previously in natural conditions in the Gironde estuary (Jeantet et al., 1985). The storage of Cd, Cu and Zn in intracellular structures have been extensively studied in mussels and various species of oysters, highlighting interspecific differences responsible for the higher levels of bioaccumulation in oysters compared to mussels (George et al., 1978; George et al., 1982; Pirie et al., 1984; Thomson et al., 1985; Jeantet et al., 1985; Jeantet et al., 1997).

Moreover, in the present study, interspecific differences were demonstrated in the case of one particular metal-binding protein, namely metallothionein (or MT-like protein). MTLP levels were higher in mussels than in oysters and translocation from a clean to a metal-contaminated site did not induce identical responses in both species.

Metal compartmentation studies in oysters have shown that metals were found mainly in the soluble phase. In the cytosol, a strong relationship between Cd and MTLP levels has been demonstrated. Given the probable metabolic role of MTs (Cosson et al., 1991; Carpene, 1993; Dallinger, 1993), these observations suggest that a detoxication process may be at work in oysters of the Gironde estuary. This hypothesis is in agreement with the lack of observed physiological effects at the organism level (condition index) as well as cytopathological effects in target organs (gills, digestive gland).

Some authors have assumed that AChE activities are specific indicators of organophosphorus or carbamate-insecticides exposure (Hill & Fleming, 1982; Galgani & Bocquené, 1989; Day & Scott, 1990; Bocquené & Galgani, 1991). In the case of the Gironde estuary, monitoring programs have not revealed such type of pollution (RNO, 1995). On the other hand, both *in vitro* and *in vivo* studies with mammals, fish or invertebrates point to AChE inhibitory effects of metals including cadmium (Galzigna et al., 1969; Heywood et al., 1978; Olson & Christensen, 1980; Suszkiw et al., 1984; Schmidt & Ibrahim, 1994; Labrot et al., 1996). Thus the decrease of AChE activity in the gills of mussels translocated from the Bay of Bourgneuf to the Gironde estuary may be due to the exposure to Cd which has also induced an important incorporation of this toxicant.

No significant change was observed in GST activity in gills of mussels transplanted from Bourgneuf to the Gironde estuary. The GST activity is generally considered as a biomarker of phase II metabolism of lipophylic organic contaminants (Narbonne, 1991). Previous results from the Baltic sea coastal areas have shown a correlation between GST activity in the gills of mussels and the logarithm of PAH concentration in sediment (Narbonne & Hansen, 1996). Compared with Bourgneuf, the site of La Fosse has a slightly higher level of total PAH sediment contamination (logarithm about 10% more), whereas no significant increase of GST activity was observed. However, the mussels *Mytilus edulis* are known to have multiple forms of GST with overlapping specificities (Fitzpatrick & Sheehan, 1993) which are not measured using only CDNB as substrate. Thus, it is not possible to exclude variations in the pattern of isoforms, even if the mean activities with the used substrate are similar (Regoli, 1995).

Catalase is a protective enzyme against oxidative stress (Winston et al., 1990). It catalyses the dismutation of hydrogen peroxide to molecular oxygen and water. In the blue mussel, exposure to paraquat and benzo(a) pyrene have led to increased catalase activity (Garcia-Martinez & Livingstone, 1995). Catalase activity can be a biomarker of the presence of chemicals inducing oxidative stress but it is also sensitive to abiotic stress (Pellerin-Massicotte, 1994). Moreover, in a site devoid of pollution, Pellerin-Massicotte (1997) has shown an increase of catalase activity which may be related to a physiological stress like repeated spawning.

When oxidative stress overwhelms the protective forces such as catalase, a deleterious effect on membranes may be shown through an increase of MDA concentrations since MDA is an end-product of lipid peroxidation. Such increases have been shown in the digestive gland of mussels following exposure to paraquat and heavy metals (Viarengo et al., 1990).

Thus, according to Pellerin-Massicotte (1997) 'catalase may be more sensitive to subtle variations in the environment while MDA could be more representative of a stressful condition not counteracted by anti-oxidant forces'. This is at least partly the case in our study since catalase activity was increased in the digestive gland of mussels transplanted from the Bay of Bourgneuf to the Gironde estuary. Increase of catalase activity without increase of MDA levels could therefore be interprated as a production of free radicals efficiently counteracted, thus without physiological disturbances. In fact MDA levels were not only constant in transplanted mussels vs controls but they even decreased. It must be duly noted that ecological and physiological conditions may interact with pollution factors. A seasonal trend in MDA level was detected in *M. edulis* from the St Lawrence estuary and the Magdalen Islands (North Atlantic Ocean) and variations of MDA levels in *M. galloprovincialis* have been related to their changing metabolic status (Viarengo et al., 1991; Pellerin-Massicotte, 1997).

Conclusion

Sediment bound metal mobility, assessed using chemical extractions (pH variation, organic ligand, sequential extractions), bioavailability of sediment bound pollutants, significant responses of biomarkers examined in bivalves transplanted from a relatively clean area to the Gironde estuary suggest a potential environmental hazard in this area. However, interspecific differences have been demonstrated with a particular tolerance of oysters *Crassostrea gigas* at different stages of their life cycles. These results suggest several questions for the future. The fact that a tolerant species is also as a strong accumulator presents hygiene authorities with a serious problem. Oyster culture has been limited in the Gironde estuary in order to meet the regulations (Journal Officiel de la République Française, 1995) but the standards could be improved by studying the physico-chemical forms of metabolic storage of toxicants, and their potential risk of transfer to the consumer. Interspecific differences raise the question if the risk is limited to a few species, or is it at the level of the equilibrium of the whole estuarine ecosystem ? Thus a major aim for the future of ecotoxicology is to develop firm bonds between chemistry, toxicology and benthic ecology.

References

Allen, H. E., R. H. Hall & T. D. Brisbin, 1980. Metal speciation: effects on aquatic toxicity. Envir. Sci. Technol. 14: 441–447.

Amiard, J. C., 1992. Bioavailability of sediment-bound metals for benthic aquatic organisms. In J. P. Vernet (eds), Impact of Heavy Metals on the Environment, Elsevier, Amsterdam: 183–202.

Amiard, J. C., A. Pineau, H. L. Boiteau, C. Métayer & C. Amiard-Triquet, 1987. Application of atomic absorption spectrophotometry using zeeman effect to the determination of eight trace elements (Ag, Cd, Cr, Cu, Mn, Ni, Pb and Se) in biological materials. Wat. Res. 21: 693–697.

Amiard, J. C., H. Ettajani, A. Y. Jeantet, C. Ballan-Dufrançais & C. Amiard-Triquet, 1995. Bioavailability and toxicity of sediment-bound lead to a filter-feeder bivalve *Crassostrea gigas* (Thunberg). BioMetals 8: 280–289.

Baumard, P., H. Budzinski & P. Garrigues, 1997. Analytical procedure for the analysis of PAHs in biological tissues by gas chromatography coupled to mass spectrometry: application to mussels. Fresenius J. Anal. Chem. (in press).

Behar, F., C. Leblond & C. Saint Paul, 1989. Analyse quantitative des effluents de pyrolyse en milieu ouvert et fermé. Rev. Inst. Français Pét. 44: 387–411.

Beiras, R. & E. His, 1995. Toxicity of freeze-dried hydrocarbon polluted sediments to *Crassostrea gigas* embryos. Mar. Pollut. Bull. 30: 47–49.

Bocquené, G. & F. Galgani, 1991. L'acétylcholinestérase chez les organismes marins, outils de surveillance des effets des pesticides organophosphorés et carbamates. Oceanis 17: 439–448.

Boucaud-Camou, E., C. Lebesnerais, P. Lubet & I. Lihrmann, 1983. Dynamique et enzymologie de la digestion chez l'huître *Crassostrea gigas* (Thunberg). In IFREMER (ed.), Bases biologiques de l'Aquaculture. Montpellier, Actes de Colloques 1: 75–96.

Boutier, B., J. F. Chiffoleau, J. M. Jouanneau, C. Latouche & I. Philipps, 1989. La contamination de la Gironde par le cadmium: origine, extension, importance. IFREMER, Rapp. Sci. Tech. 14, 105 pp.

Budzinski, H., A. Papineau, P. Baumard & P. Garrigues, 1995. Extraction assistée par chauffage micro-ondes focalisées (MOF) à pression ambiante des composés organiques dans les matrices naturelles: application à l'analyse des composés aromatiques. C.R. Acad. Sci. Paris, 321 (Série IIb): 69–76.

Buege, J. A. & S. D. Aust, 1978. Microsomal lipid peroxydation. Meth. Enzymol. 50: 302–310.

Campbell, P. G. C., A. G. Lewis, P. M. Chapman, A. A. Crowder, W. K. Fletcher, B. Imber, S. N. Luoma, P. M. Stokes & M. Winfrey, 1988. Biologically available metals in sediments. Public. NRCC, Ottawa (Canada) Pub. N° NRCC 27694, 298 pp.

Carpene, E., 1993. Metallothionein in marine molluscs. In R. Dallinger & P. S. Rainbow (eds), Ecotoxicology of Metals in Invertebrates. Lewis Publishers, Boca Raton: 55–72.

Charlou, J. L. & M. Joanny, 1983. Dosage du mercure et d'autres métaux (Pb, Zn, Cu, Cd, Co, Ni, Cr, Mn) dans les sédiments marins par absorption atomique. In A. Aminot & M. Chaussepied (eds), Manuel des Analyses Chimiques en Milieu Marin, CNEXO, Brest: 285–295.

Claiborne, A., 1985. Catalase activity. In R. A. Greenwald (ed.), Handbook for Oxygen Radical Research, C.R.C. Press, Boca Raton, Florida: 283–284.

Coquery, M. & M. Horvat, 1996. The analytical performance study for the *MED POL* Area: determination of trace elements in marine sediment *SD-MEDPOL-1/TM* and in fish homogenate *MA-MEDPOL-1/TM*. In: International Atomic Energy Agency Marine Environment Laboratory, Monaco.

Cosson, R. P., C. Amiard-Triquet & J. C. Amiard, 1991. Metallothioneins and detoxification. Is the use of detoxication protein for MTs a language abuse? Wat. Air Soil Pollut. 57–58: 555–567.

Crouzet, C. & S. Altmann, 1996. Comparaison de diverses méthodologies pour estimer le pouvoir toxique de sédiments. Publically available BRGM report – R38894.

Dallinger, R., 1993. Strategies of metal detoxification in terrestrial invertebrates. In R. Dallinger & P. S. Rainbow (eds), Ecotoxicology of Metals in Invertebrates. Lewis Publishers, Boca Raton: 245–289.

Day, K. E. & I. M. Scott, 1990. Use of acetylcholinesterase activity to detect sublethal toxicity in stream invertebrates exposed to low concentrations of organophosphate insecticides. Aquat. Toxicol. 18: 101–114.

Ellman, G. L., K. D. Courtney, V. Andres & R. M. Featherstone, 1961. A new and rapid colorimetric determination of acetylcholinesterase activity. Biochem. Pharmac. 7: 88–95.

Ettajani, H., C. Amiard-Triquet, A. Y. Jeantet & J. C. Amiard, 1996. Fate and effects of soluble or sediment-bound arsenic in oysters (*Crassostrea gigas* Thun.). Arch. envir. Contam. Toxicol. 31: 38–46.

Fitzpatrick, P. J. & D. Sheehan, 1993. Separation of multiple forms of glutathione S-transferases from the blue mussel, *Mytilus edulis*. Xenobiotica 23: 851–861.

Galgani, F. & G. Bocquené, 1989. A method for routine determination of organosphosphates and carbamates in sea water. Envir. Technol. Lett. 10: 311–322.

Galzigna, L., G. S. Corsi, B. Saia & A. A. Rizzoli, 1969. Inhibitory effects of triethyl lead on serum cholinesterase *in vitro*. Clin. Chem. Acta 26: 391–393.

Garcia-Martinez, P., G. W. Winston, C. Metash-Dickey, S. C. M. O'Hara & D. R. Livingstone, 1995. Nitrofurantoin-stimulated reactive oxygen species production and genotoxicity in digestive gland microsomes and cytosol of the common mussel (*Mytilus edulis* L.). Toxicol. Appl. Pharmacol. 131: 332–341.

George, S. G. & J. M. Frazier, 1982. Some aspects of the relationship between tolerance to heavy metal pollution and metabolisms of Cd, Cu and Zn in oysters. Thalassia Jugosl. 18: 203–219.

George, S. G., B. J. S. Pirie, A. R. Cheyne, T. L. Coombs & P. T. Grant, 1978. Detoxication of metals by marine bivalves: an ultrastructural study of the compartmentation of copper and zinc in the oyster *Ostrea edulis*. Mar. Biol. 45: 147–156.

Habig, W. H., M. J. Pabst & W. B. Jakoby, 1974. The first enzymatic step in mercapturic acid formation. J. Biol. Chem. 249: 7130–7139.

Heywood, R., R. W. James, A. E. Street, C. J. Rudd & P. S. I. Barry, 1976. *In vitro* and *in vivo* study of the ability of tetraethyl lead to inhibit cholinesterase activity. Toxicol. Lett. 2: 11–16.

Hill, E. F. & W. J. Fleming, 1982. Anticholinesterase poisoning of birds: field monitoring and diagnosis of acute poisoning. Envir. Toxicol. Chem. 1: 27–38.

Hill, I. R., P. Matthiessen & F. Heimbach (eds), 1993. Guidance document on sediment toxicity tests and bioassays for freshwater and marine environments. Proc. SETAC Europe Workshop, Renesse, Netherlands.

His, E., M. N. L. Seaman & R. Beiras, 1997. A simplification of the bivalve embryogenesis and larval development bioassay method for water quality assessment. Wat. Res. 31: 351–355.

Jeantet, A. Y., C. Ballan-Dufrançais & J. L. Martin, 1985. Recherche des mécanismes de détoxication du cadmium par l'huître *Crassostrea gigas* (Mollusque Bivalve). II. Sites intracellulaires d'accumulation du métal dans les organes absorbants et excréteurs. C. R. Acad. Sci., Paris 301: 177–182.

Jeantet, A. Y., C. Ballan-Dufrançais & A. Anglo, 1997. Atteintes cytologiques chez les bivalves marins. In L. Lagadic, T. Caquet, J. C. Amiard & F. Ramade (eds), Biomarqueurs en Ecotoxicologie – Aspects Fondamentaux. Masson, Paris: 315–354.

Journal Officiel de la République Française, 1995. Arrêté du 25 Juillet 1995 relatif au classement de salubrité et à la surveillance des zones de production et des zones de reparcage des coquillages vivants. Publ. on the 1st of Sept. 1995. République Française, Paris.

Kersten, M. & U. Förstner, 1989. Speciation of trace elements in sediments. In G.E. Batley (ed.), Trace Element Speciation: Analytical Methods and Problems, CRC Press, Inc. Bota Raton, Florida: 245–309.

Klöckner, K., H. Rosental & J. Willführ, 1985. Invertebrate bioassays with North Sea water samples. Structural effects on embryos and larvae of serpulids, oysters and sea urchins. Helgoländer Meeresunters 30: 1–19.

Labrot, F., D. Ribera, M. Saint Denis & J. F. Narbonne, 1996. *In vitro* and *in vivo* studies of potential biomarkers of lead and uranium contamination: lipid peroxidation, acetylcholinesterase, catalase and glutathione peroxidase activities in three non-mammalian species. Biomarkers 1: 21–28.

Langston, W. J., 1980. Arsenic in U.K. estuarine sediment and its availability to benthic organisms. J. mar. biol. Ass. U.K. 60: 869–881.

Lebesnerais, C., 1985. Etude expérimentale de la digestion chez l'huître japonaise *Crassostrea gigas* (Thunberg). Thèse de doctorat, Université de Caen, Fr.

Letellier, M., H. Budzinski, P. Garrigues & S. Wise, 1996/1997. Focused microwave assisted extraction of polycyclic aromatic hydrocarbons in open cell from reference materials (sediment, soil, air particulates). Spectroscopy: An International Journal 13: 71–80.

Lowry, O. H., N. J. Rosenbrough, A. I. Farr & R. J. Randall, 1951. Protein measurement with the folin phenol reagent. J. Biol. Chem. 193: 265–275.

Morel, F. M. M. & J. G. Hering, 1993. Principles and applications of aquatic chemistry. John Wiley and Sons. Inc.

Narbonne, J. F., 1991. Mécanismes de biotransformation des polluants organiques chez les animaux marins. Océanis 17: 449–458.

Narbonne, J. F. & P. D. Hansen, 1996. Recent studies on biomarkers on coastal areas: assessment of ecosystem health – ecotoxicological studies. Proceedings of international conference on animal physiology, Münster/Westphalia, May 31-June 2, 1996.

Olson, D. & G. M. Christensen, 1980. Effects of water pollutants and other chemicals on fish acetylcholinesterase (*in vitro*). Envir. Res. 21: 327–335.

Pellerin-Massicotte, J., 1997. Influence of elevated temperature and air-exposure on MDA levels and catalase activities in digestive glands of the blue mussel (*Mytilus edulis* L.). J. Rech. Océanogr. 22: 91–98.

Pellerin-Massicotte, J., 1994. Oxidative processes as indicators of chemical stress in marine bivalves. J. aquat. Ecos. Health 3: 101–111.

Pirie, B. J. S., S. G. George, D. G. Lytton & J. D. Thomson, 1984. Metal-containing blood cells of oysters: ultrastructure, histochemistry and X-ray microanalysis. J. mar. biol. Ass. U.K. 64: 115–123.

R.N.O., 1995. Surveillance du milieu marin. Ministère de l'Environnement, 32 pp.

Rainbow P. S., 1993. The significance of trace metal concentrations in marine Invertebrates. In R. Dallinger & P. S. Rainbow (eds), Ecotoxicology of Metals in Invertebrates. Lewis Publishers, Boca Raton: 3–23.

Regoli, F. & G. Principato, 1995. Glutathione, glutathione-dependent and antioxidant enzymes in mussel, *Mytilus galloprovincialis*, exposed to metals under field and laboratory conditions: implications for the use of biochemical markers. Aquat. Toxicol. 31: 143–164.

Ringwood, A. H. & M. Brouwer, 1995. Patterns of metallothionein expression in oyster embryos. Mar. Envir. Res. 39: 101–105.

Salomons, W. & U. Forstner, 1980. Trace metals analysis on polluted sediments. Envir. Techn. Lett. 1: 506–517.

Shannon, R. D. & J. R. White, 1991. The selectivity of a sequential extraction procedure for the determination of iron oxyhydroxides and iron sulfides in lake sediments. Biogeochem. 14: 193–208.

Schmidt, G. H. & N. M. M. Ibrahim, 1994. Heavy metal content (Hg^{2+}, Pb^{2+}) in various body parts: its impact on cholinesterase activity and binding glycoproteins in the grasshopper *Aiolopus thalassinus* adults. Ecotox. envir. Saf. 29: 148–164.

Sokal, R. R. & F. J. Rohlf, 1981. Biometry. Freeman, New York.

Suszkiw, J., G. Toth, M. Murawsky & G. P. Cooper, 1984. Effects of Pb^{2+} and Cd^{2+} on acetylcholine release and Ca^{2+} movements in synaptosomes and subcellular fractions from rat brain and *Torpedo* Electric Organ. Brain Res. 323: 31–46.

Tessier, A., P. G. C. Campbell & M. Bisson, 1979. Sequential extraction procedure for the speciation of particulate trace metals. Anal. Chem. 51: 844–851.

Tessier, A., P. G. C. Campbell, J. C. Auclair & M. Bisson, 1984. Relationships between the partitioning of trace metals in sediments and their accumulation in the tissues of the freshwater mollusc *Elliptio complanata* in a mining area. Can. J. Fish. aquat. Sci. 41: 1463–1472.

Thompson, J. A. J. & R. P. Cosson, 1984. An improved electrochemical method for the quantification of metallothioneins in marine organisms. Mar. Envir. Res. 11: 137–152.

Thompson, J. D., B. J. S. Pirie & S. G. George, 1985. Cellular metal distribution in the Pacific Oyster, *Crassostrea gigas* (Thun.) determined by quantitative X-ray microprobe analysis. J. exp. mar. Biol. Ecol. 85: 37–45.

Viarengo, A., L. Canesi, M. Pertica & D. R. Livingstone, 1991. Seasonal variations in the antioxidant defence systems and lipid peroxidation of the digestive gland of mussels. Comp. Biochem. Physiol. 100C: 187–190.

Viarengo, A., L. Canesi, M. Pertica, G. Poli, M. N. Moore & M. Orunesu, 1990. Heavy metal effects on lipids peroxidation in the tissues of *Mytilus galloprovincialis* Lam. Comp. Biochem. Physiol. 97C: 37–42.

Hydrobiologia **373/374:** 281–295, 1998.
J.-C. Amiard, B. Le Rouzic, B. Berthet & G. Bertru (eds),
Oceans, Rivers and Lakes: Energy and Substance Transfers at Interfaces.

Particulate organic matter composition in the Estuarine Turbidity Maxima (ETM) of the Seine River estuary

Laurent Bodineau*, Guy Thoumelin, Vincent Béghin & Michel Wartel
Laboratoire de Chimie Analytique et Marine, ELICO EP1750 Cité scientifique, Bat. C8 2ème étage, 59655 Villeneuve d'Ascq, France (e-mail: bodineau@univ-lille1.fr)
* *corresponding author*

Key words: suspended particulate matter, organic matter, fatty acids, sterols, Estuarine Turbidity Maximum, Seine River estuary

Abstract

Organic matter (OM) was investigated in particles collected in the Estuarine Turbidity Maximum (ETM) zone of the Seine River estuary, for two different hydrodynamic regimes in February and September 1995. During ebb resuspension, occurrence of the ETM was documented two to four hours before ebb. The discharge regime greatly influence the position and importance of the ETM. It occurs more prominently during high flow than during low flow. Suspended particulate matter (SPM) and particulate organic carbon (POC) maxima, as well as surface-bottom concentration gradients, decrease when river flow is enhanced. During flood resuspension a second peak was observed, but only for bottom waters. This indicates decoupling of surface (fresh waters) and bottom layers (brackish/marine waters) at the beginning of the flood. Organic matter content in SPM remained almost unchanged (2 to 3%) and only a slight decrease was noticed during the ETM.

Recent particulate organic matter sources such as terrestrial plant, phytoplankton, bacteria and domestic sewage were differentiated by the use of lipidic biomarkers (fatty acids and sterols). In the ETM, terrestrial plant, sewage and bacterial inputs are at maximum. On the other hand, phytoplankton abundance is minimum in ETM but it may be important during slack periods. This is the case in September, when evidence is given for a diatom bloom during high tide slack.

Introduction

Estuaries are complex systems including land and adjacent coastal ocean, where fresh and saline waters mix in various proportions depending on river discharge and tidal energy. In macrotidal estuaries, transport of water and suspended particles from rivers through estuaries to the sea is complicated by the broad zone of abnormally high particle concentrations known as Estuarine Turbidity Maxima (ETM). The ETM is formed by tidal dynamics, residual circulation and resuspension-deposition cycles of the fine sediment (Festa & Hansen, 1978; Nichols, 1986; Jay & Musiak, 1994). Usually, the ETM acts as a particle trap and consequently particles travel slowly between different parts of the estuary. Much of the estuarine particulate matter comes from the continental inputs, but marine particles can be also concentrated in the ETM as shown, for example, in the Seine River estuary by Dupont et al. (1994).

q Estuarine organic matter originates from the complex mixing of abundant and diverse sources, including inputs of allochthonous and autochthonous origin: higher plants derived from the surrounding watershed and delivered by river inflow, phytoplankton, benthic algae and higher plants. To a lesser extent, heterotrophic organisms and sewage wastes, also significantly influence overall composition of organic matter. In macrotidal estuaries, generally, this organic fraction represents less than 15% of total suspended matter (Laane, 1982; Meybeck et al., 1988; Chiffoleau et al., 1994).

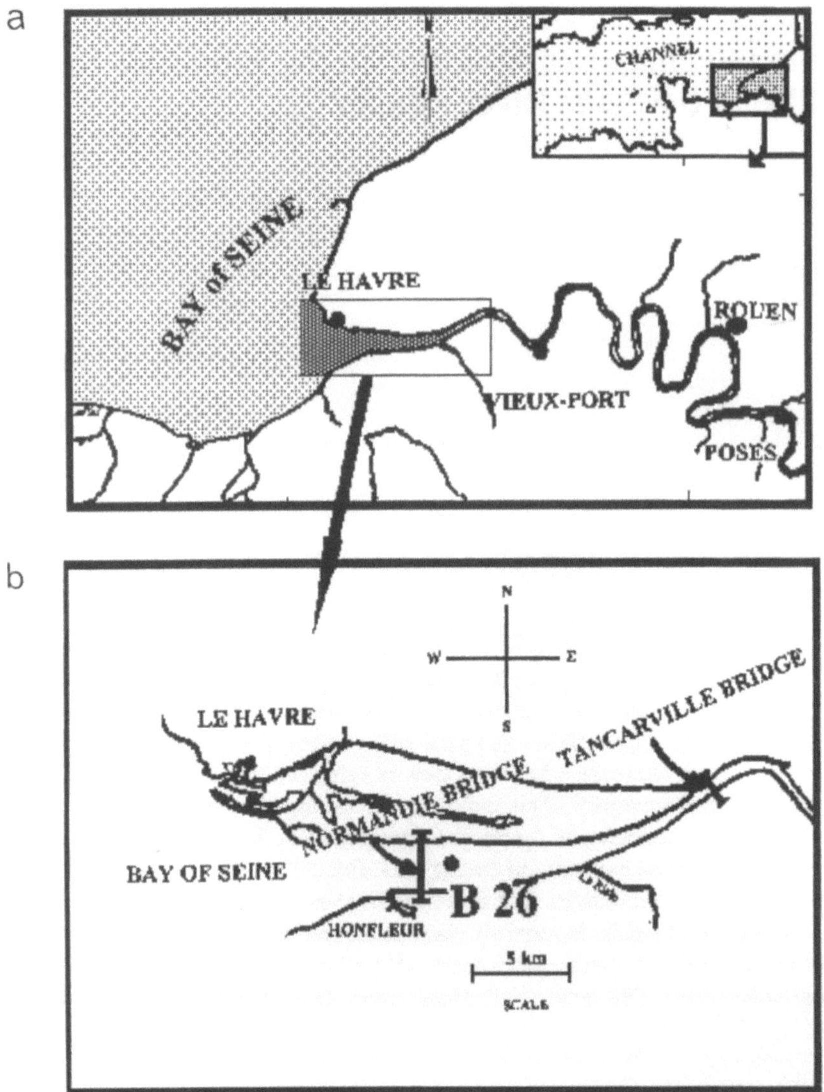

Figure 1. Map of the Bay of Seine in the northeastern part of France (a) with location of the fixed buoy 26 (b) situated near the Normandie bridge in the inner Seine estuary.

In order to understand the biogeochemical roles of organic matter, it is important to characterize it and to identify its major sources. In some macrotidal estuarine studies, the lipid biomarker approach has been widely used, together with gross organic content determination (Saliot et al., 1984, 1988; Tronczynski et al., 1986; Quéméneur & Marty, 1992; Gearing et al., 1994 and Yunker et al., 1995), which is usually examined through its distribution along estuarine longitudinal gradient. Meybeck et al. (1988), for example, have postulated that vertical gradients are negligible in comparison with longitudinal variations. However, temporal changes are present in short time scales as minutes to hours, e.g. as a response to the tidal resuspension or advection (Canuel et al., 1995). As far as we know, only few papers (Sicre et al., 1993; Simenstad et al., 1994; Prahl & Coble, 1994; Reed & Donovan, 1994) have focused on fluctuations of bulk organic matter on the tidal-time scale.

The goal of our study was to discrimate short-term shifts in organic matter and to estimate the importance of its different sources. We examined the intensity and the position of the ETM in the Seine River estuary under opposite hydrodynamic regimes: high and low

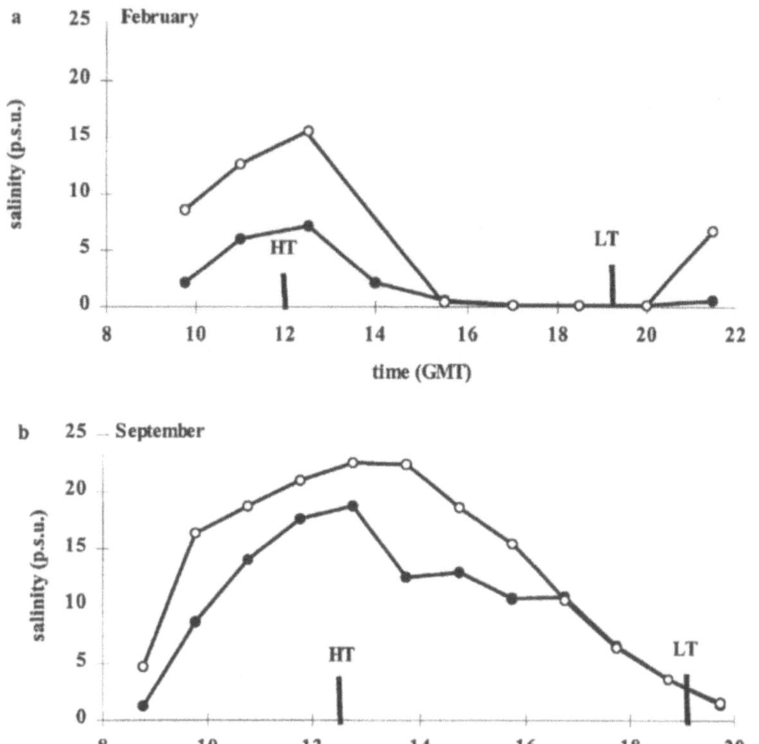

Figure 2. Salinity (p.s.u.) of surface (—●—) and bottom (—○—) waters under high (2074 m^3 s^{-1}) and low (298 m^3 s^{-1}) discharges, respectively in February (a) and September (b) 1995 at the B26 site in the Seine River estuary. (HT: high tide; LT: low tide)

river flow. By the use of pigments (chlorophyll *a* and phaeopigments) and biomarkers (fatty acids (FA) and sterols (ST)), we quantified changes in the particulate organic matter composition during complete tidal cycles.

Study locations descriptions

The Seine River's catchment area drains about 75 000 km^2, encompassing 20% of the cultivated areas, 30% of the population and 40% of the economic activities of France (Cabioch, 1986). The estuary is the main outlet of the Seine Bay to the English Channel. Its configuration is completely artificial; submerged embankments channelize the main part of the river flow (Figure 1a).

The average freshwater runoff is about 450 m^3 s^{-1}, with considerable variation between low (100–200 m^3 s^{-1}) and high discharge (1500–2000 m^3 s^{-1}). During 1995, the average flow at the Poses Dam was about 745 m^3 s^{-1}, with lower and upper limits from 139 and 2153 m^3 s^{-1}, respectively. Tidal range can reach 8 m in the spring tide and this macrotidal estuary is well-mixed (Le Floch, 1961).

The Seine River estuary is about 160 km long and can be divided into three zones: the fluvio-estuarine part from Poses to Vieux-Port with only dynamic tide; the inner estuary between Vieux-Port and Le Havre, characterized by the mixing of fresh and marine waters and the excursion of the ETM; and the outer estuary off Le Havre.

Since the 1980's, several studies have been conducted in the Seine estuarine system to improve our knowledge about the human activities impact, including sediment transport (Avoine, 1981; Dupont et al., 1994) and chemical contamination (Chiffoleau et al., 1994). However, few studies have addressed organic matter (Texier et al., 1993).

The study location (Figure 1b) was the fixed Buoy 26 (B26) located between the Normandie Bridge and the Tancarville Bridge, within the inner estuary. Two cruises were conducted in February (SEAV B) and September (SEAV D) 1995, within spring tide cycles

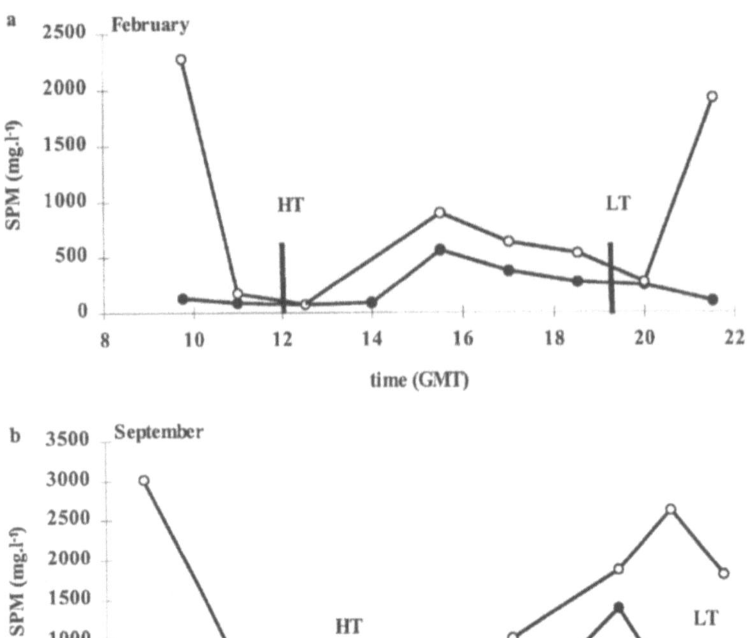

Figure 3. Suspended particulate matter (SPM, mg l^{-1}) of surface (—●—) and bottom (—○—) waters under high (2074 m^3 s^{-1}) and low (298 m^3 s^{-1}) discharges, respectively in February (a) and September (b) 1995 at the B26 site in the Seine River estuary.

but under different hydrodynamic regimes (Table 1). During SEAV B, the river flow at Poses Dam was 2074 m^3 s^{-1} and corresponds to a high discharge peak. SEAV D occurred after a drought of three months when flow Poses Dam only 298 m^3 s^{-1}. These two opposite situations in terms of river flow correspond to different residence times from the Poses Dam to the Normandie Bridge, estimated to be between 2 and 3 days, 10 and 15 days, respectively (A. Ficht, pers. comm.).

Table 1. Synopsis of mean characteristics during the two surve (SEAV B and SEAV D) conducted under different river flows with me: spring-tide

Cruise name	SEAV B	SEAV D
Date	February 1st 1995	September 12th 1995
River flow (m^3 s^{-1})	2074	298
Tidal coefficient:		
AM	104	94
PM	104	89

Materials and methods

Sampling

Sampling was performed from the *R/V Côte d'Aquitaine* or *Côte de Normandie* anchored at site B26, using Niskin bottles, Downing optical backscatter sensor and salinity (conductivity-temperature-depth) probe (Seabird CTD). During each cruise, both surface (one metre below the surface) and bottom (one metre above the bottom) waters were collected, approximatively every hour, during a complete tidal cycle.

Samples of between 0.1 to 1 liter were filtered immediately through polycarbonate filters (0.45 μm size) for suspended particulate matter (SPM), GF/C Whatman glass fiber filters (1.2 μm) for photosynthetic pigments and pre-ignited (450 °C, overnight) GF/F Whatman glass fiber filters (0.7 μm) for lipids, particulate organic carbon and nitrogen (POC and PON).

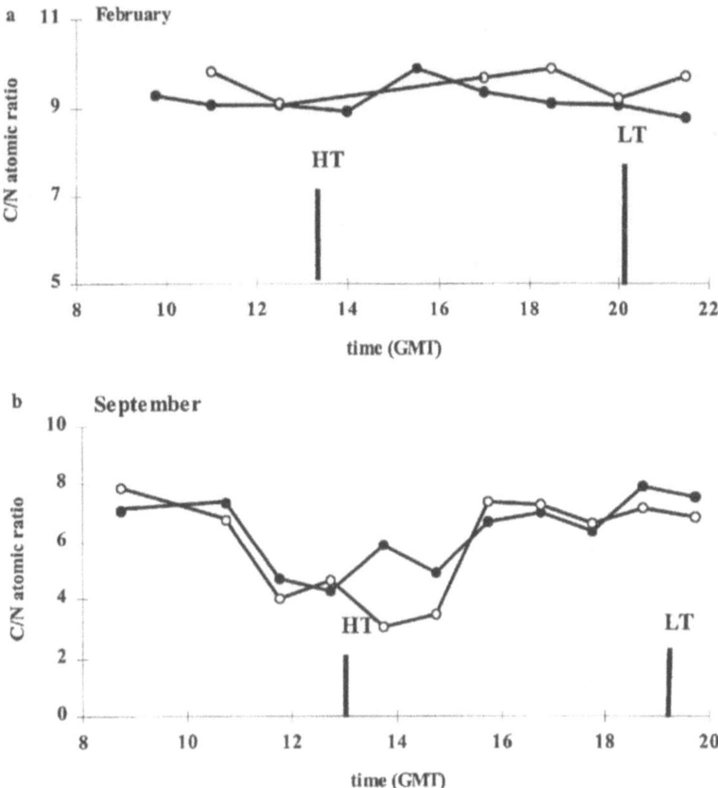

Figure 4. C/N atomic ratio of surface (—●—) and bottom (—○—) waters under high (2074 m^3 s^{-1}) and low (298 m^3 s^{-1}) discharges, respectively in February (a) and September (b) 1995 at the B26 site in the Seine River estuary.

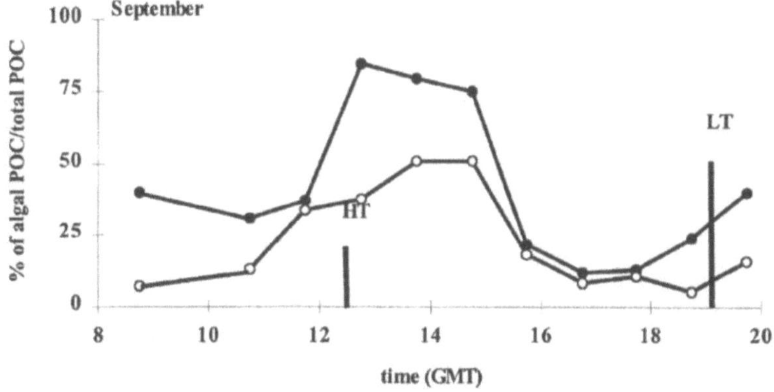

Figure 5. Relative abundances of algal (alive and detritic) organic carbon (in % of POC) variations observed in surface (—●—) and bottom (—○—) under low (298 m^3 s^{-1}) discharge (September 1995) at the B26 site in the Seine River estuary.

Unfortunatly, only during the SEAV D low flow survey, four samples of estuarine material were collected every three hours using a 60 μm mesh net, from which phytoplankton was isolated by prefiltration through a 200 μm sieve and stored at –20 °C.

POC and PON filters were stored at –20 °C until analysis. After addition of magnesium hydrogenocarbonate in the end of the filtration, pigment filters were kept in dark and stored at –20 °C. Lipid sample filters were stored in methanol and in darkness at –20 °C.

Analysis

SPM was determined gravimetrically. POC and PON were measured by means of elemental CHNS analyzer

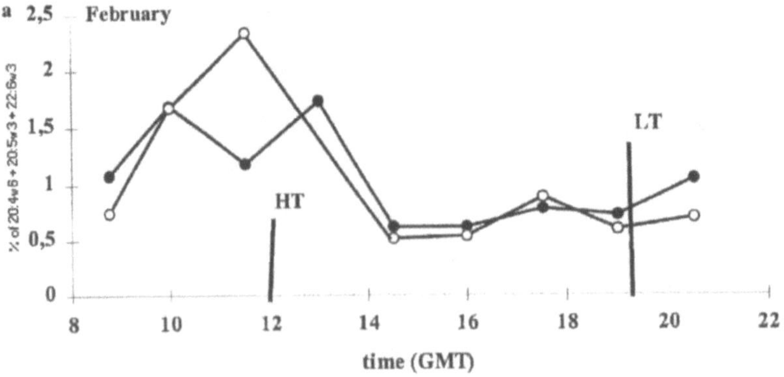

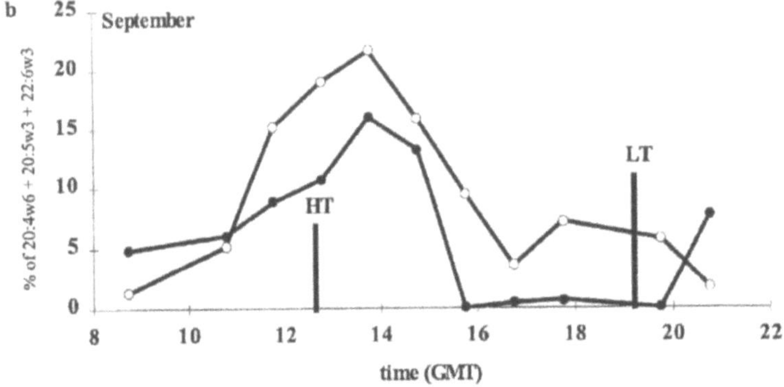

Figure 6. Relative abundances of unsaturated fatty acids 20:4ω6, 20:5ω3 and 22:6ω3 (expressed as% of total FA) variations in surface (—●—) and bottom (—○—) waters under high (2074 $m^3 s^{-1}$) and low (298 $m^3 s^{-1}$) discharges, respectively in February (a) and September (b) 1995 at the B26 site in the Seine River estuary.

(Leco 932) after inorganic carbon removal by HCl fumes.

Chlorophyll *a* and phaeopigments were extracted overnight at 4 °C in 90% acetone. Spectrophotometric measurements were done before and after acidification, following Lorenzen (1967).

Lipids were extracted following Bligh & Dyer (1959) using one-phase mixture of $CHCl_3$/MeOH/H_2O (1:2:1, v/v) by means of ultrasonication and formation of two-layers mixture. Extraction was done twice and rotary-evaporated lipid extracts were combined. After drying the lipid fraction under gentle nitrogen stream, transmethylation was performed by the solution of 14% BF_3 in methanol under reflux for 10 min, following the procedure of Metcalfe & Schmitz (1961). Fatty acid methyl esters (FA) and sterols (ST) were separated by high performance liquid chromatography (Thoumelin, 1988) and then analysed by gas chromatography (GC).

FA and ST trimethylsilyl derivatives (TMS-ST) were analysed using a Hewlett Packard 5890 Series II GC equipped with a flame ionization detector (FID) and an on-column injector. Hydrogen was used as carrier gas. FA and ST were respectively separated in a capillary column (25 m × 0.32 mm i.d) coated with a polar Acid Modified Polyethylene Glycol phase (HP-FFAP) and an apolar cross-linked methyl/phenyl (95:5) silicone (HP-5) fused silica capillary column (30 m × 0.25 mm i.d). Samples were injected at 60 °C. The oven was temperature programmed for the range of 60–150 °C at a rate of 30 °C min^{-1}, then at 3 °C min^{-1} to final temperatures of 230 for FA and 290 °C for ST. Identification was made by comparison of the retention times with those of known standards. Cholestane and tricosanoïc acid methyl ester (23:0) were used as internal standards for ST and FA quantifications. Additionally, GC/MS (Hewlett Packard 5890 Series II linked with a 5972 Mass Selective Detector) was used to confirm the structure of both, ST and FA compounds using an apolar HP5 ultra 2 column (0.2 mm i.d., 30 m, 0.1 μm film phase) with helium as carrier gas. The standard deviation for triplicate sam-

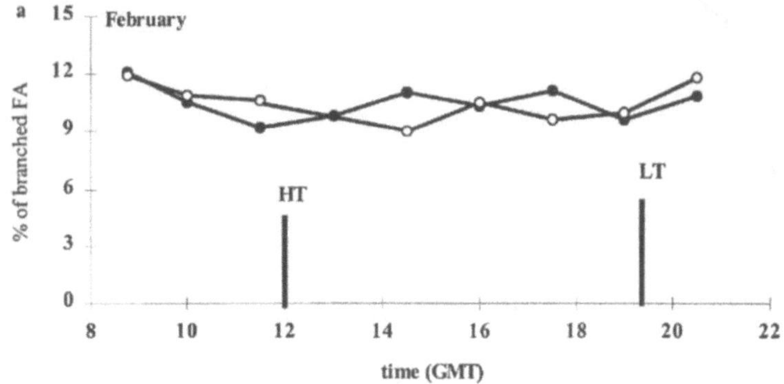

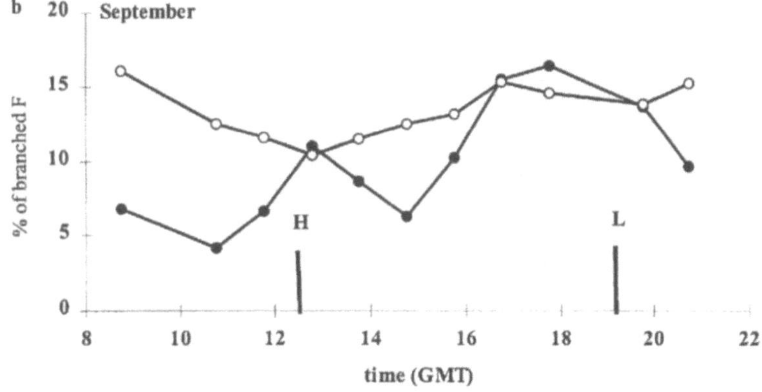

Figure 7. Relative abundances of branched fatty acids (as iso and anteiso 15:0 and 17:0, expressed as% of total FA) variations in surface (—●—) and bottom (—○—) waters under high (2074 m^3 s^{-1}) and low (298 m^3 s^{-1}) discharges, respectively in February (a) and September (b) 1995 at the B26 site in the Seine River estuary.

ples was < 5% for total amounts of major components of each class of lipids.

FA are described by a n:pωx formula, where n is number of carbon atoms in the aliphatic chain, p the number of double bonds and the first double bond is located between carbons x and $x + 1$ numbered from the terminal methyl group. Other possible double bonds are methylene-interrupted. Iso and anteiso prefixes indicate the position of a methyl group at the penultimate and antepenultimate carbon position, respectively. ST are designated by their systematic name. For ST with 28 and 29 carbon atoms, two epimers at C_{24} may be present in samples, but they were not separated by the analytical procedure used in our work.

Results

Salinity

Patterns in temporal variation of salinity at the two sampling depths during the cruises in February and September, give a rough insight into the fresh and marine waters mixing under high and low river inflows (Figures 2a and b). Vertical stratification was present at all periods around the flood-slack, as shown with the significant enhancement of salinity through the water column. These gradients are higher during February under high river discharge than during low flow in September. However, surface salinities are maximum for the latter period and exceed 15 p.s.u. High bottom water salinities indicate salt intrusions that last longer in September. Flood-slack periods increase between February and September as a result of river flow decrease. During ebb periods and at low tide, stratification is completely destroyed and estuarine waters become well-mixed, as it is indicated by disappearance of any salinity gradient.

ETM identification

Suspended particulate matter (SPM) concentrations vary from 30 to 3000 mg l^{-1} for all surface and bottom samples (Tables 2 and 3), which is in good agreement

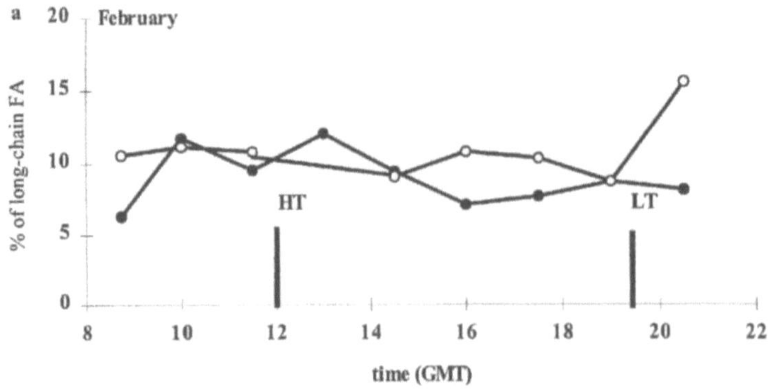

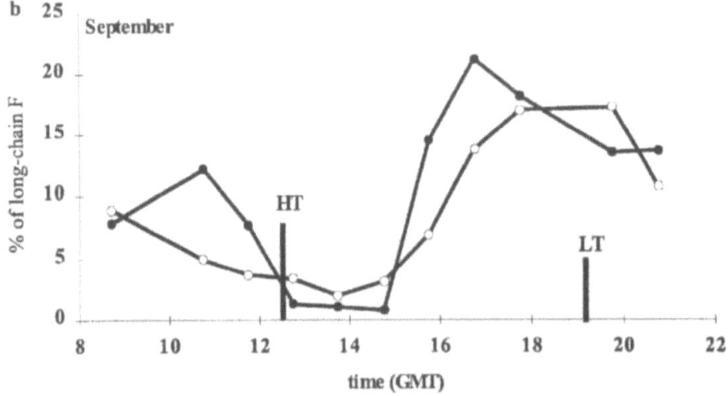

Figure 8. Relative abundances of even long-chain fatty acids (from 24:0 to 32:0, expressed as % of total FA) variations in surface (—●—) and bottom (—○—) waters under high (2074 $m^3 s^{-1}$) and low (298 $m^3 s^{-1}$) discharges, respectively in February (a) and September (b) 1995 at the B26 site in the Seine River estuary.

with previously reported results (Dupont et al., 1994; Thoumelin et al., 1997). During ebb, maximum concentrations were observed in both surface and bottom waters (when collections were possible) between one and four hours before low tides (Figures 3a and b). This indicates the ETM presence caused by the resuspended material under ebb flows. In February when river flow is higher, maximum concentrations in ETM of 500 mg l^{-1} and of 1000 mg l^{-1} were observed in surface and in bottom waters (Figure 3a). Higher maximum concentrations were observed in September (Figure 3b), especially in bottom waters (> 2.5 g l^{-1}). Thus, in drought conditions, a much more developed ETM was observed, suggesting that the rate of SPM export towards the outer estuary is probably small. In contrast, under high discharge (February), a much more important part of ETM is flushed out to the external estuary (Avoine, 1981).

We also detected an ETM pulse during the flood. However, it was observed only in bottom waters and, therefore, it induces strong SPM gradient, decreasing from bottom towards surface waters. This indicates that decoupling between bottom and surface waters takes place particularly during flood ETM. Finally, lowest concentrations were observed during the slack after flood.

Particulate organic carbon

POC concentrations varied significantly between 1 and 73 mg l^{-1}, and covaried with SPM, during each cruise (Tables 2 and 3). Similar variations were observed for POC and SPM. Linear correlations between POC versus SPM were significant ($\rho < 0.001$) indicating that 2–3% of SPM was organic carbon.

C/N atomic ratio

In February, the C/N ratio remained relatively constant between 9 and 11 through the entire tidal cycle (Figure 4a and Table 2). These high values are typical of winter conditions, when continental organic inputs predominate over primary production. Ebb and flood

Table 2. Tidal-time changes observed for SPM, POC, C/N atomic ratio, particulate total fatty acids and sterols in February 1995. Summary of some source-specific lipid constituents (individual fatty acid and sterol expressed as % of total fatty acids and sterols)

	Time (GMT)		09:45 Flood ETM	11:00	12:30	14:00	15:30 Ebb ETM	17:00	18:30	20:00 ...ES...	21:30 Flood ETM
				........FS........							
	SPM (mg l^{-1})										
		s	131	86	76	91	561	380	279	249	100
		b	2265	167	73		896	640	540	279	1918
	POC (mg l^{-1})										
		s	3.7	2.9	3.0	2.7	13.5	9.8	7.2	7.1	3.1
		b		5.4	2.6			12.4	11.4	8.6	35.4
	C/N atomic ratio										
		s	9.3	9.1	9.1	8.9	9.9	9.4	9.1	9.1	8.8
		b		9.8	9.1			9.7	9.9	9.2	9.7
	Total FA (μg l^{-1})										
		s	34.6	20.2	18.6	21.3	124.5	83.5	83.0	74.0	32.6
		b	423.8	38.7	22.3		135.7	126.9	91.1	100.9	290.2
	Total ST (μg l^{-1})										
		s	11.6	6.4	4.4	6.6	20.6	24.7	24.9	18.1	16.6
		b	174.2	18.8	8.0		36.7	43.2	30.5	42.6	77.3
Phytoplanktonic	16:2, 16:3										
		s	1.8	1.9	1.9	1.6	1.1	1.0	1.1	1.2	1.2
		b	0.0	0.3	0.3		0.0	0.1	0.2	0.2	0.0
	18:2w6, 18:3w3, 18:3w6										
		s	4.2	1.3	1.2	1.5	1.0	1.3	1.5	1.5	1.5
		b	0.3	1.8	5.1		0.3	0.5	1.3	1.3	0.3
	20:4w6, 20:5w3, 22:6w3										
		s	1.1	1.7	1.2	1.7	0.6	0.6	0.8	0.7	1.0
		b	0.7	1.7	2.3		0.5	0.5	0.9	0.6	0.7
Bacterial	iso + antéiso 15:0 + 17:0										
		s	12,1	10.5	9.2	9.8	11.0	10.3	11.1	9.6	10.8
		b	11.9	10.8	10.6		9.1	10.5	9.6	10.0	11.8
Higher plants	Sum of even FA (24:0–32:0)										
		s	6.3	11.7	9.5	12.0	9.4	7.1	7.6	8.7	8.1
		b	10.5	11.1	10.7		9.0	10.7	10.2	8.7	15.7
Domestic sewage	Sum of 5β-stanols										
		s	13.3	20.4	13.6	12.4	15.0	13.8	26.6	21.6	6.4
		b	14.9	7.1	7.0		6.0	9.2	9.4	10.6	7.8

(s: surface; b: bottom)
FS: slack after flood-tide
ES: slack after ebb-tide

Table 3. Tidal-time changes observed for SPM, POC, C/N atomic ratio, particulate total fatty acids and sterols in September 1995. Summary of some source-specific lipid constituents (individual fatty acid and sterol expressed as % of total fatty acids and sterols)

	Time (GMT)		08:45	10:45	11:45	12:45	13:45	14:45	15:45	16:45	17:45	18:45	19:45
			Flood ETM		........FS........						Ebb ETM		...ES...
	SPM (mg l^{-1})												
		s	128	105	113	69	86	41	88		1370	380	229
		b	3017	447	273	178	198	155	1016		1864	2612	1800
	POC (mg l^{-1})												
		s	3.8	3.5	2.9	2.0	2.0	1.6	3.5	28.2	29.0	8.4	5.2
		b	63.1	18.8	6.5	4.5	2.6	5.9	22.8	40.3	38.6	72.3	23.3
	C/N atomic ratio												
		s	7.0	7.3	4.7	4.3	5.9	4.9	6.7	7.0	6.4	7.9	7.5
		b	7.8	6.7	4.0	4.7	3.0	3.5	7.4	7.3	6.6	7.2	6.9
	Total FA ($\mu g\ l^{-1}$)												
		s	49.7	33.6	35.9	68.9	75.6	73.7	14.4	167.8	185.3	153.0	59.7
		b	355.5	195.1	109.5	85.5	49.8	85.4	190.1	396.3	501.6	692.6	588.6
	Total ST ($\mu g\ l^{-1}$)												
		s	4.4	3.4	1.5	2.0	2.5	2.4	3.2	23.5	24.3	16.4	5.8
		b	25.3	22.2	11.8	9.8	3.7	6.6	29.9	55.9	54.8	113.6	50.8
	Chlorophyll *a* ($\mu g\ l^{-1}$)												
		s	5.7	2.7	3.1	37.8	26.1	15.2	0.9	42.6	43.1	10.7	31.5
		b	61.4	13.6	5.9	23.5	7.6	22.7	21.7	5.3	4.1	14.4	30.8
	Phaeopigment ($\mu g\ l^{-1}$)												
		s	53.9	39.2	39.3	30.6	37.3	31.9	29.1	85.6	101.1	67.8	50.5
		b	117.3	79.3	80.4	44.2	44.0	96.3	137.9	123.3	153.1	135.5	111.0
Phytoplanktonic	16:2, 16:3												
		s	5.0	0.0	0.0	17.3	14.0	11.0	0.0	0.4	0.5	0.0	3.0
		b	3.9	7.2	9.5	8.2	8.5	9.5	6.7	4.4	6.0	6.0	5.0
	18:2w6, 18:3w3, 18:3w6												
		s	2.0	3.0	4.1	3.6	4.9	4.1	0.0	0.4	0.6	0.6	1.6
		b	2.9	4.0	4.4	4.1	4.9	5.4	3.6	3.3	3.6	4.1	3.4
	20:4w6, 20:5w3, 22:6w3												
		s	4.8	6.0	8.8	10.7	16.0	13.2	0.0	0.4	0.7	0.0	7.7
		b	1.3	5.1	15.2	19.0	21.6	15.8	9.4	3.5	7.1	5.7	1.8
	Dinosterol												
		s	0.0	0.0	7.2	7.7	7.2	7.0	4.0	5.9	3.8	0.0	0.0
		b	0.0	0.0	3.7	3.2	7.4	8.4	3.2	4.8	4.5	4.2	5.7
Bacterial	iso + antéiso 15:0 + 17:0												
		s	5.7	3.6	5.7	7.5	5.3	2.6	9.1	11.9	12.9	12.2	8.6
		b	13.3	9.5	7.2	6.4	6.9	7.6	9.5	12.3	10.4	9.8	12.5
Higher plants	Sum of even FA (24:0–32:0)												
		s	7.9	12.1	7.8	1.3	1.1	0.8	14.5	21.2	18.1	13.4	13.6
		b	8.9	5.0	3.7	3.4	2.0	3.2	7.0	13.7	17.0	17.2	10.7
Domestic sewage	Sum of 5β-stanols												
		s	10.7	11.1	11.0	5.0	5.4	2.9	15.7	22.3	22.2	15.1	15.2
		b	13.1	9.4	11.2	6.5	6.1	0.0	7.7	12.6	13.1	10.2	17.0

(s: surface; b: bottom)
FS: slack after flood-tide
ES: slack after ebb-tide

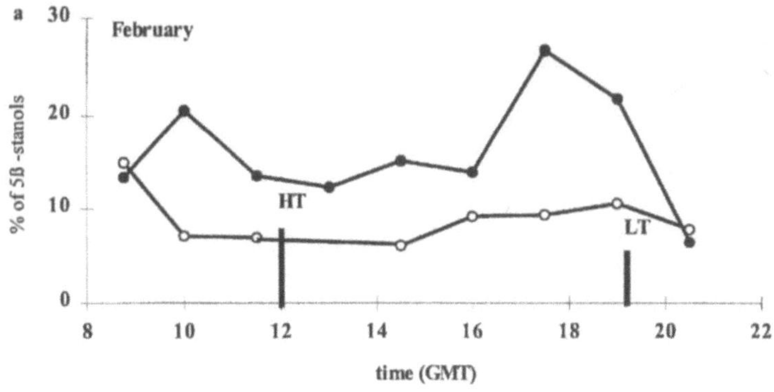

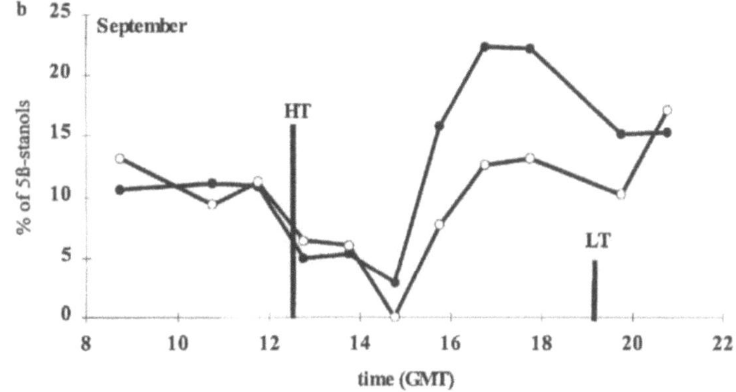

Figure 9. Relative abundances of 5β-stanols (as the sum of 5β(H)-cholestan-3β-ol, 24-methyl-5β(H)-cholestan-3β-ol, 24-ethyl-5β(H)-cholestan-3β-ol, expressed as% of total sterol) variations in surface (—●—) and bottom (—○—) waters under high (2074 m^3 s^{-1}) and low (298 m^3 s^{-1}) discharges, respectively in February (a) and September (b) 1995 at the B26 site in the Seine River estuary.

ETM events induce no particular changes. However, in September, lowest values of C/N ratio were observed (3 to 8) and decreased during the slack after flood tide (Figure 4b). The low values which are then observed are typically indicative of planktonic organisms.

Pigments

A conservative estimate of total POC as phytoplankton component was calculated by using the average POC/(chlorophyll *a* and phaeopigment) ratio of 25.3 as determined from discrete plankton samples. This value is similar to that found in other rivers (Relexans et al., 1988). In this way, the contributions of algae to total POC, corresponding to photosynthetic and degraded plankton cells, for both surface and bottom waters were estimated for September (Figure 5). Greatest algal POC fractions were observed in surface waters after the flood-tide slack, whereas lowest values were connected with both ebb and flood ETM events. Around high tide, maxima contribution of algal POC to total POC of 80–90% and 40–50% occurred in surface and bottom waters, respectively.

Particulate biomarkers variations

In order to assess the organic matter composition during the two different river flow regimes (February and September), we used specific lipid biomarkers (Table 4) (Mayzaud et al., 1989; Scribe et al., 1991). For each organic source, we evaluated fluctuations in relative abundance of appropriate indicators (Figures 6 to 9).

Among polyunsaturated fatty acids, which are usually used as plankton indicators (Chuecas & Riley 1969), 22:6ω3, 20:5ω3 and 20:4ω6 predominated (Figure 6 and Tables 2 and 3) especially in the September survey. Percentages of these plankton indicators fluctuated between 0 to 22% of total FA with higher maximum values in September (Figure 6b) than in February (Figure 6a). This was particularly the case at the flood-tide slack in September. During this pe-

Table 4. Typical source assignments

Source	Individual fatty acid and sterol components
Phytoplankton	– *diatom:* 16:2ω7, 16:3ω4, 20:5ω3, 20:4ω6, 22:6ω3 – *chlorophyceae:* 18:3ω3, 16:2ω6, 16:3ω3, – *dinoflagellate:* dinosterol (4α-23,24-trimethyl-5α-cholest-22-ène-3β-ol)
Higher plants	Even-numbered saturated fatty acids with chain lenghts > *n*–24:0
Bacteria	iso- and anteiso-15:0 and 17:0 saturated fatty acids
Domestic sewage	– coprostanol (5β(H)-cholestan-3β-ol) – 24-methylcoprostanol, (24-methyl-5β(H)-cholestan-3β-ol) – 24-ethylcoprostanol (24-ethyl-5β(H)-cholestan-3β-ol)

riod, maximum percentages of C_{18}, and especially C_{16} polyunsaturated FA, were also observed (Table 3). For C_{16} FA, higher values were observed in surface waters than in bottom waters while the contrary was evident for 20:4ω6 + 22:5ω3 + 22:6ω3. On another, highest percentages of dinosterol, a compound typically found in dinoflagellates (Joseph, 1993), were also observed during the slack after high tide. Low or nil proportions of polyunsaturated FA are detected during the ebb ETM, especially in surface waters. This is not the case for dinosterol except in surface waters during low tide. This compound was also not detected during the flood ETM.

In February, lower proportions of phytoplanktonic polyunsaturated FA were observed (Table 2 and Figure 6a). For example, percentage of 20:4ω6 + 22:5ω3 + 22:6ω3 varied between 0.7 and 2.3% and dinosterol was not observed. The variations of the proportion of 20:4ω6 + 22:5ω3 + 22:6ω3 are roughly similar to those observed during September (Figure 6b), with maximum values around high tide and minimum values around low tide. However, more homogeneity was observed in the water column, especially during ebb.

Bacterial contribution to POC in estuarine particles can be detected with the use of iso and anteiso 15:0 and 17:0 (hereafter called branched FA, Figure 7), to which a bacterial origin is generally attributed (Perry et al., 1979). In February, rather constant proportions are observed for surface and bottom waters (10–12% of total FA, Table 2 and Figure 7a). Greater variation occurred in September (Figure 7b): the highest values were detected for surface and bottom waters during the ebb ETM and also in bottom waters during the flood ETM. Minimum values were observed at high tide in bottom waters at the same time a secondary maximum was observed in surface waters.

Continental inputs derived from higher plants (Kolattukudy, 1976) can be recognized by the appearance of even long-chain saturated FA (from 24:0 to 32:0). Rather constant relative abundances of these long-chain FA were observed in February uniformly in both surface and bottom waters (Table 2 and Figure 8a). In September, much more variation was evident and the highest values were associated with the ebb ETM and minimal ones around high tide (Table 3 and Figure 8b).

The urban sewage input may be depicted by three sterols (5β(H)-cholestan-3β-ol, 24-methyl-5β(H)-cholestan-3β-ol and 24-ethyl-5β(H) -cholestan-3β-ol, hereafter called 5β-stanols, Figure 9), which are widely used as successful and sensitive indicators of faecal pollution (Grimalt et al., 1990). In February, the highest proportions of the 5β-stanols were not associated with the ebb ETM. Maximum values were observed in surface waters before low tide (Figure 9a). Except at the beginning and end of the sampling series, proportions in surface waters were higher than those in bottom waters. In September, we observed similar patterns to variation continental inputs, with maxima and minima obtained during ebb and around the slack after flood-tide, respectively (Table 3 and Figure 9b).

Discussion

The influence of river flow was clearly shown by the salinity variations (Figure 2). The marine water input was more prominent in September when the river flow was low. Ebb ETM events observed during the spring tide, either under high or low flow conditions, were associated with the stratification decay. SPM concentrations were the highest at the beginning of the flood, but only in bottom waters. These SPM maxima are due to strong resuspension of sediment material close to the bed. This resuspension is probably due to the strong pulse of marine water intruding along the bottom at the beginning of the flood.

We observed insignificant seasonal shifts in POC, shown as percentage of the SPM in surface and bottom water samples. This relative uniformity in the organic carbon content of estuarine particles collected in the ETM zone is not related to river flow or seasonal changes. Low absolute POC concentrations obtained in February were probably governed by export of suspended material in the outer estuary at each tidal cycle and by a low biological activity. In September, however, low flow conditions probably induced prominent development of the ETM that reduces the light penetration and, therefore, had a negative impact on photosynthetic production of POC. Generally, POC percentages remained unchanged during the tidal cycle, except during ETM events which induced slight decreases in the SPM organic carbon content. Higher percentage of organic matter in resuspended particles during the flood ETM (9.7%) than during the ebb (4.7%), was found by Reed & Donovan (1994) in the Columbia River estuary. This was not the case in the Seine River estuary, where particles were relatively poor in organic carbon in comparison with other estuarine studies (Laane, 1982; Saliot et al., 1988). In February, conjunction of high river flow and low biologic activity in the estuary probably contributed to a global homogeneization in organic matter content., as shown by low C/N variations. On the contrary, in September low river flow as well as higher biological activity induced variations in organic matter content.

Particulate organic matter changes during tidal cycle

In September, we found evidence of a phytoplankton bloom in the estuary, with high values of chl*a*, especially in surface waters, high calculated algal POC/total POC ratios and low C/N values (Table 3) around high-tide slack. This was confirmed by maximum proportions of polyunsaturated planktonic FA at the same time. Trapping of algal organic matter and/or resuspension of material previously trapped in the ebb ETM were shown by rather high percentages of polyunsaturated FA in bottom waters. This algal organic matter comes from the marine part of the estuary and possibly from freshwaters. Among polyunsaturated FA high percentages of C_{16}, C_{20} and C_{22} polyunsaturated FA, especially 16:2, 16:3 and 20:5ω3, indicate a diatom predominance. This has also been shown by microscopic observations (Dupont, pers. comm.). However, chlorophyceae and dinoflagellate species are also present as it is suggested by C_{18} polyunsaturated FA and dinosterol respectively. During February, in contrast, low sterols and fatty acid contributions indicated a lower plankton input due to high river runoff, less salinity intrusion and low plankton production. Dinoflagellates were absent because dinosterol is not found.

In September, branched fatty acids varied between surface and bottom waters (Figure 7b). The bacterial fingerprint was not closely associated with phytoplankton, even if a secondary maximum of branched fatty acids was observed at high tide in the surface waters (Figure 7b). The maximum proportions of these compounds were observed during the ebb ETM. This suggests that heterotrophic activity is enhanced in the presence of organic matter trapped in the ETM. But in February, branched fatty acids contributions were relatively constant, and similar in surface and bottom waters throughout the tidal cycle. This is probably due to the fact that the high winter river flows produced strong mixing of particulate organic matter, the greater part of it being directly exported out of the estuary instead of being trapped in the ETM.

As with bacterial markers, terrestrial and faecal organic markers varied more in September than in February (Figures 8 and 9). In September, the highest percentages of the terrestrial and faecal biomarkers were observed in the ebb ETM, and their proportions were higher at low tide than at high tide (Figures 8 and 9). Ebb ETM events appeared to accumulate material of continental origin which was also predominantly brought by freshwaters. Terrestrial-derived material is closely associated with rather large particles which are trapped in the ETM and submitted to sedimentation/resuspension cycles. In contrast, during flood, preferential sedimentation of continental and detrital particles occurred as shown by the lowest abundances of long-chain fatty acids. By comparing flood and ebb variations of proportions and absolute concentrations, we also suggest a seaward transport of these fractions. Particle sizes were probably different for these two types of organic matter. Relatively to terrestrial-derived organic matter, faecal organic matter was probably associated with more easily resuspended particles, as indicated by the highest percentages in surface waters during ebb compared to bottom ones during the flood ETM. In February, as already been seen for bacterial markers, the river flow has induced an homogeneization in faecal and terrestrial organic matter proportions throughout the tidal cycle. However, as in September, faecal organic matter was found in higher proportions in surface than in bottom waters but at that time during the complete tidal cycle.

Consistent proportions of faecal organic matter in the estuarine waters suggest constant inputs from the urban effluent which supplements the introduction of labile organic material to the estuary and, therefore, contributes to oxygen consumption by heterotroph organisms. Nevertheless, as suggested by Relexans et al. (1988) in Loire biogeochemical studies, the domestic input decrease due to the better plant treatment, would generally have a diminishing effect on the POC content in the ebb ETM and in the estuary.

River flow has an important effect on ETM events. During high discharge conditions, suspended material is exported out of the estuary and this process reduces ETM development. Consequently, particulate organic matter composition appears comparative in nature in both time and vertical structure. The recent identified sources were terrestrial plants, bacteria and sewage effluents. Moreover, due to the seasonal depression in phytoplankton, contributions in February are minimal. In September, favourable summer conditions enhanced phytoplankton development in the inner estuary. Due to low river flow, the ETM was well-developed in the inner estuary. Evidence is then given for accumulation of particulate organic matter. Consequently phytoplankton organic matter is trapped in ETM. However, even if it is accumulated in ETM, a net seaward transport of organic matter from terrigenous and anthropogenic is also observed under low flow. Finally, to increase our knowledge of such complex ecological system as ETM, future studies will be needed with special attention to tidal (daily, monthly,. . .) cycles in order to document more accurately the particle trapping mechanism.

Acknowledgements

We are grateful to Jean-Paul Dupont and Robert Lafite from the Geological Department of the Rouen University for the collection of SPM and the excellent cruise monitoring. We are also indebted to the crews of the *R/V Côte d'Aquitaine* and *R/V Côte de Normandie* for their cooperation in field sampling. The CNRS and Région Nord-Pas de Calais are gratefully acknowledged for their grant according to one of us (LB). This work has been undertaken within the SEINE-AVAL program financially supported by the Régions basse et haute Normandie, IFREMER and Agence de l'eau du bassin Seine-Normandie. Finally, we sincerely thanks the two anonymous reviewers for helpful comments.

References

Avoine, J., 1981. L'estuaire de la Seine: sédiments et dynamique sédimentaire. PhD, University of Caen, France, 236 pp.

Bligh, E. G. & W. J. Dyer, 1959. A rapid method of total lipid extraction and purification. Can. J. Biochem. Physiol. 37: 911–917.

Cabioch, L., 1986. Introduction. In La Baie de Seine, GRECO MANCHE. Actes de Colloques 4, Caen 24–26 avril 1985, Caen: 4–8.

Canuel, E. A., J. E. Cloern, D. B. Ringelberg, J. B. Guckert & G. H. Rau, 1995. Molecular and isotopic tracers used to examine sources of organic matter and its incorporation into the food webs of san Francisco Bay. Limnol. Oceanogr. 40: 67–81.

Chiffoleau, J. F., D. Cossa, D. Auger & I. Truquet, 1994. Trace metal distribution, partition and fluxes in the Seine estuary (France) in low discharge regime. Mar. Chem. 47: 145–158.

Chuecas, L. & J. P. Riley, 1969. Components of fatty acids of the total lipids of some marine phytoplankton. J. mar. Biol. Ass. U.K. 49: 97–116.

Dupont, J. P., R. Lafite, M. F. Huault, P. Hommeril & R. Meyer, 1994. Continental/marine ratio changes in suspended and settled matter across a macrotidal estuary (the Seine estuary, northwestern France). Mar. Geol. 120: 27–40.

Festa, J. F. & D. V. Hansen, 1978. Turbidity maxima in partially mixed estuaries: a two-dimensional numerical model. Estuar. coast. Mar. Sci. 7: 347–359.

Gearing, J. N., J. Tronczynski & S. A. Macko. 1994. Particulate organic matter in the St. Lawrence Estuary: anthropogenic and natural sources. In K. R. Dyer & R. J. Orth (eds), Changes in Fluxes in Estuaries: Implications from Sciences to Management. Olsen & Olsen, Fredensberg: 125–130.

Grimalt, J. O., P. Fernandez, J. P. Bayona & J. Albaiges, 1990. Assessment of fecal sterols and ketones as indicators of urban sewage inputs to coastal waters. Envir. Sci. Tech. 24: 357–363.

Jay, D. A. & J. D. Musiak, 1994. Particle trapping in estuarine tidal flows. J. Geophys. Res. 99: 445–461.

Joseph, J. D., 1993. Octadecapentaenoic acid: a biomarker for toxic marine phytoplankton ? International News Fats, Oils Related Matter. 4: 534.

Kollatukudy, P. E., 1976. Chemistry and Biochemistry of Natural Waxes. New-York: Elsevier.

Laane, R. W. P. M., 1982. Chemical characteristics of the organic matter in the waterphase of the Ems-Dollart estuary. Thèse Doct., University of Groningen, No. 6.

Le Floch, J. F., 1961. Propagation de la marée dynamique dans l'estuaire de la Seine Maritime. PhD, Paris, 509 pp.

Lorenzen, C. J., 1967. Determination of chlorophyll and pheopigments: spectrophotometric equations. Limnol. Oceanogr. 12: 343–346.

Mayzaud, P., J. P. Chanut & R. G. Ackman, 1989. Seasonal changes of the biochemical composition of marine particulate matter with special reference to fatty acids and sterols. Mar. Ecol. Prog. Ser. 56: 189–204.

Metcalfe, L. D. & A. A. Schmitz, 1961. The rapid preparation of fatty acid esters for gas chromatographic analysis. Anal. Chem. 33: 363–364.

Meybeck, M., G. Cauwet, S. Dessery, M. Somville, D. Gouleau & G. Billen, 1988. Nutrients (Organic C, P, N, Si) in the eutrophic river Loire, (France) and its estuary. Estuar. coast. Shelf Sci. 27: 595–624.

Nichols, P. D., 1986. Formation of the turbidity maximum: a review. EBSA Symposium on the dynamics of turbid coastal environments. Plymouth, UK.

Perry, G. J., J. F. Volkman, R. B. Johns & Jr. H. J. Bavor, 1979. Fatty acids of bacterial origin in contemporary marine sediments. Geochim. Cosmochim. Acta 43: 1715–1725.

Prahl, F. G. & P. G. Coble, 1994. Input and behavior of dissolved organic carbon in the Columbia River Estuar. In K. R. Dyer & R. J. Orth (eds), Changes in Fluxes in Estuaries: Implications from Sciences to Management. Olsen & Olsen, Fredensberg: 451–458.

Quéméneur, M. & Y. Marty, 1992. Sewage influence in a macrotidal estuary: fatty acid and sterol distributions. Estuar. coast. Shelf Sci. 34: 347–363.

Reed, D. J. & J. Donovan, 1994. The character and composition of the Columbia River estuarine turbidity maximum. In K. R. Dyer & R. J. Orth (eds), Changes in Fluxes in Estuaries: Implications from Sciences to Management. Olsen & Olsen, Fredensberg 445–450.

Relexans, J. C., M. Meybeck, G. Billen, M. Brugeaille, H. Etcheber & M. Somville. 1988. Algal and microbial processes involved in particulate organic natter dynamics in the Loire estuary. Estuar. coast. Shelf Sci. 27: 625–644.

Saliot, A., A. Lorre, J. C. Marty, P. Scribe, J. Tronczynsky, M. Meybeck, S. Dessery, M. Marchand, J. C. Caprais, G. Cauwet, H. Etcheber, J. C. Relexans, M. Ewald, P. Berger, C. Belin, D. Gouleau, G. Billen & M. Somville-Groupe De Géochimie Organique Du GRECO I.C.O., 1984. Biogéochimie de la matière organique en milieu estuarien: stratégies d'échantillonnage et de recherche élaborées en Loire, France. Oceanol. Acta. 7: 191–207.

Saliot, A., J. Tronczynski, P. Scribe & R. Letolle, 1988. The application of isotopic and biogeochemical markers to the study of the biochemistry of organic matter in a macrotidal estuary, The Loire, France. Estuar. coast. Shelf Sci. 27: 645–669.

Scribe, P., J. Fillaux, J. Laureillard, V. Denant & A. Saliot, 1991. Fatty acids as biomarkers of plankton inputs in the stratified estuary of the Krka river, Adriatic Sea: relationship with pigments. Mar. Chem. 32: 299–312.

Sicre, M-A., I. Broyelle, A. Lorre & A. Saliot, 1993. Sources and transport of particulate hydrocarbons in the meso-tidal Changjiang estuary. Estuar. coast. Shelf Sci. 37: 557–573.

Simenstad, C. A., D. J. Reed, D. A. Jay, J. A. Baross, F. G. Prahl & L. F. Small, 1994. Land-Margin Ecosystem Research in the Columbia River Estuary: an interdisciplinary approach to investigating couplings between hydrological, geochemical and ecological processes within estuarine turbidity maxima. In K. R. Dyer & R. J. Orth (eds), Changes in Fluxes in Estuaries: Implications from Sciences to Management. Olsen & Olsen, Fredensberg: 437–444.

Texier, H., R. Lafite, J. P. Dupont, V. Firmin, F. Gadel, L. Serve & B. Charrière, 1993. Organic matter transport across a macrotidal estuary gradient: the Seine estuary, France. Neth. J. Aquat. Ecol. 27: 405–413.

Thoumelin, G., 1988. Décomposition de la matière organique dans les eaux côtières : Etude en laboratoire de la biodégradation d'effluents urbains. Thèse de doctorat, Université de Bretagne Occidentale: pp. 181.

Thoumelin, G., L. Bodineau & M. Wartel, 1997. Origin and transport of organic matter across the macrotidal Seine estuary: fatty acids and sterols variations, Mar. Chem. 58: 59–71.

ronczynski, J., J. C. Marty, P. Scribe & A. Saliot, 1986. Dissolved and particulate hydrocarbons in the Loire estuary, from the riverine zone to the external estuary: budget at different seasons. International Journal of Env. Analysis Chemistry, 23: 169–187.

Yunker, M. B., R. W. Macdonald, D. J. Veltkamp & W. J. Cretney, 1995. Terrestrial and marine biomarkers in a seasonally ice-covered Arctic estuary-integration of multivariate and biomarker approaches. Mar. Chem. 49: 1–50.

Hydrobiologia **373/374:** 297–310, 1998.
J.-C. Amiard, B. Le Rouzic, B. Berthet & G. Bertru (eds),
Oceans, Rivers and Lakes: Energy and Substance Transfers at Interfaces.

Influence of the level of oxygenation in sediment and water on copper bioavailability to marine bivalves: laboratory experiments and translocation experiments in the field

Herman Hummel[1], Paolo Magni[1,*], Claude Amiard-Triquet[2], Florence Rainglet[2], Roel Modderman[1], Yvette van Duijn[3], Marcel Herssevoort[3], Johan de Jong[3], Linda Snitsevorg[4], Maaike Ytsma[4], RoelofBogaards[1] & Lein de Wolf[1]

[1] *Netherlands Institute of Ecology, Centre for Estuarine and Coastal Ecology, Vierstraat 28, 4401 EA Yerseke, The Netherlands*

[2] *Université de Nantes, CNRS, Faculté de Pharmacie, Laboratoire d' Ecotoxicologie, 1, Rue Gaston-Veil, 44035 Nantes Cedex, France*

[3] *Polytechnical Faculty, Hogeschool Rotterdam & Omstreken, Kluyverweg 4, 2629 HT Delft, The Netherlands*

[4] *Van Hall Institute, Hereweg 99, 9721 AA Groningen, The Netherlands*

* *Present address: Department of Bioresource Science, Kagawa University, 761-07 Miki, Kagawa, Japan*

Key words: bivalves, oxygen, condition, copper, bioavailability, translocation, sediment, silt fraction

Abstract

The effects of differences in the level of oxygenation of sediment or water on the condition and copper content of two bivalves, the Baltic clam *Macoma balthica* and the cockle *Cerastoderma edule*, were assessed. Specimens from four intertidal flats in the Netherlands and France were compared, translocated and exposed to different levels of oxygen in the laboratory. Cockles showed no significant differences in condition and copper content between animals from light (= more oxygenated) and dark (= less oxygenated) sediments. Baltic clams also showed no differences in condition, but the clams had a higher copper content (concentration as well as body burden) in dark than in light sediments. During the translocation experiments no significant changes occurred. In the laboratory experiments the level of oxygen had no effect on the condition or copper content of the Baltic clam. The only factor affecting the copper content of Baltic clams was the addition of copper to the water or sediment. The copper, organic carbon and silt fraction ($< 16\ \mu m$) was higher in dark sediments than in light sediments. The copper content in the sediment was positively related to the silt and organic carbon content. We argue that the relation between coloration (= degree of oxygenation) of sediments and the copper content of Baltic clams could be indirect: due to a higher silt fraction and/or organic content at some places on a tidal flat, these places are more hypoxic and therefore darker, whereas simultaneously these places have a higher copper concentration because of more copper-complexing sites (and surface), whereby the higher copper concentration in the sediment relates to a higher copper concentration in the clams.

Introduction

Variations of metal content in bivalves can be related to several environmental and internal ecophysiological factors, such as temperature, salinity, sediment grain size or tissue weight (Bryan, 1980, 1984; Luoma, 1983, 1989; McLusky et al., 1986; Forstner, 1987, 1990; Millward, 1995). The influence of the oxygenation level in sediment and water is mostly referred to with much reservation, because of the complex interrelation of oxygen-levels with e.g. acidity and sulphide levels (Theede, 1985; Forstner, 1990; Di Toro et al., 1992; Ankley et al., 1994). At lower oxygen levels in water the carbon-dioxide levels will increase, resulting in a change of the bicarbonate system, giving a shift to a more acidic pH. This may lead to more uncomplexed, more bioavailable, forms of some heavy metals, such as copper (Neuhoff, 1983; Theede, 1985). In anoxic sediments with metal-sulphides, increased fluxes of oxygen or successive reduction-oxidation events near organisms (because of their food intake and inhalation processes) stimulate the mobilization of some metals, among others copper, which may become potentially available to organisms (Westerlund et al., 1986; Forstner, 1990; Gerringa, 1990, 1991). Although in anaerobic sediments the acid volatile sulfide (AVS) fraction may reduce the metal activity (i.e. bioavailability) (Di Toro et al., 1992; Ankley et al., 1994), in summary, the bio-availability of metals was mostly found to be increased in water and sediments with decreasing oxygen tensions. Likewise, it was suggested that anoxic conditions enhanced the bioavailability of copper to the clams *Macoma balthica* and *Scrobicularia plana* (Luoma & Bryan, 1982; Luoma et al., 1985). In anoxic mudflats, and during periods of anoxia, higher concentrations of copper were observed in these clams. In both studies, the occurrence of higher copper concentrations in animals living under anoxic conditions were incidental (episodic) observations, noted because they were deviating from the commonly observed relations with e.g. grain size or body weight.

In this study we present field-data on copper levels in two bivalve species, the Baltic clam *M. balthica* and the cockle *Cerastoderma edule*, from locations specifically chosen for their difference in the degree of oxygenation of the sediment. Moreover, experiments were carried out, i.e. translocations between the selected field locations and laboratory experiments with exposure to different degrees of oxygenation.

Copper is studied because, although it is an essential element, it can be (sub)lethal at low concentrations (Flemming & Trevors, 1989; Bryan & Langston, 1992). Therefore, a higher bioavailability of copper in more hypoxic, e.g. eutrophicated, sediments might be of crucial importance to biota.

The two bivalve species were chosen because they are important, often dominant, members of the benthos, and because of their differences in life-strategies. *M. balthica* is a facultative deposit-feeder and lives 3 to 8 cm deep in the sediment, whereas *C. edule* is an obligate suspension-feeder living near the surface (0–2 cm deep). Comparison of these two species may yield additional information on the impact of the level of oxygenation in sediments: the impact of differences in the level of oxygenation of the sediment should be most noticeable in the deeper burrowing Baltic clam.

Figure 1. Location of the sampling stations (Dort = Dortsman, Oosterschelde sea-arm; Paul = Paulina, Westerschelde estuary; Somm = Somme estuary; Stro = Stroodorpe, Oosterschelde sea-arm).

Material and methods

Field sampling

Specimens of *M. balthica* and *C. edule* were sampled from dark and light sediments at a distance of less than 200 m from each other at the tidal flats of Dortsman and Stroodorpe (Oosterschelde sea-arm), Paulina (Westerschelde estuary) and Somme (Figure 1). The sampling sites were between mean tidal level (MTL) and low water level (LWL) and at a salinity of 25 to 30 ppt.

The difference between the two types of sediment was based on the depth of the light-brownish coloured (oxygenized) upper layer. In all sediment cores a clear distinction could be made between the upper light-brownish sediment-layer, and the lower dark-greyish

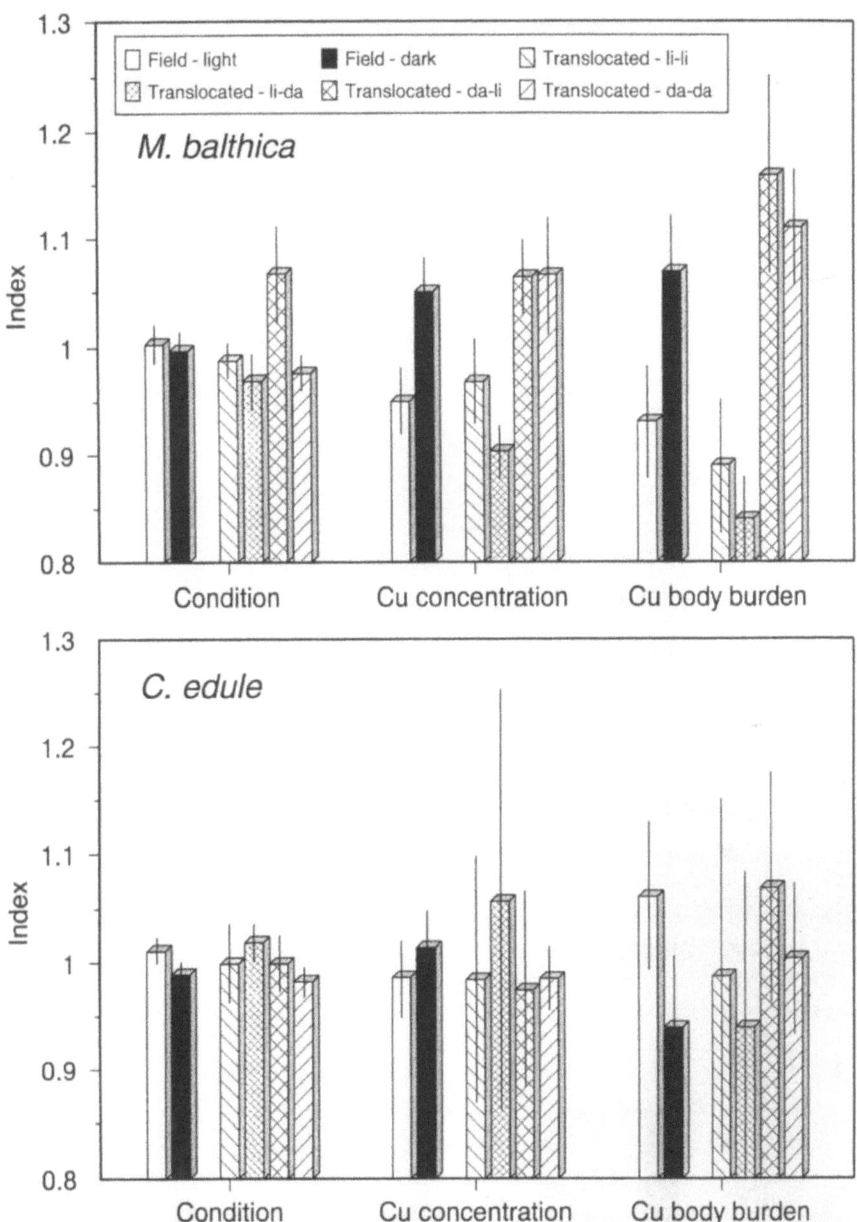

Figure 2. Average indices (and standard error) for the condition, copper concentration and copper body burden in the bivalves *M. balthica* and *C. edule* as found for specimens from light (li) and dark (da) sediments in the field and after translocation. The indices were calculated as follows. Firstly, for each station an average was calculated of each group of observations obtained during 1 day, being during field surveys 2 data, i.e. li and da, or after a translocation experiment 4 data, i.e. those from light sediment brought to light (li-li) or dark (li-da) sediment and those from dark brought to light (da-li) or dark (da-da) sediment. Secondly, the individual data of a day at a station were then divided by the average.

Table 1. Geochemical data at the sampling stations (Stro = Stroodorpe, Dort = Dortsman, Paul = Paulina, Somm = Somme; li = light sediment-type, da = dark sediment-type; L/D depth = depth of border between light-brownish sediment-layers and dark (black-grey) sediment-layers; Copper = copper concentration in ppm on dry weight sediment; TOC = % total organic carbon on dry weight sediment; sampling months in 1993: 1 = January, 11 = November)

Station	Sediment type	Sampling depth	L/D depth (cm)	Copper (ppm)	TOC (%)	Silt < 16 μm (%)		Median grain size (μm)	
(sampling months)			1/11	1	11	1	11	1	11
Paul	li	0–2 mm		4.9	4.3	0.63	12.7	89.0	85.1
Paul	li	0–10 cm	5		12.0	0.69	21.5	106.9	71.5
Paul	da	0–2 mm		7.2	4.5	0.87	13.4	138.5	78.6
Paul	da	0–10 cm	2		5.2	0.94	14.0	99.4	68.0
Dort	li	0–2 mm		1.4	0.4	0.15	1.4	135.5	145.9
Dort	li	0–10 cm	6	1.2	0.7	0.14	1.1	131.2	146.6
Dort	da	0–2 mm		1.4	0.4	0.19	2.0	134.9	145.6
Dort	da	0–10 cm	2.5	3.1	0.8	0.21	1.6	130.0	146.6
Stro	li	0–2 mm		1.5		0.30		120.5	
Stro	li	0–10 cm	3	2.2		0.56		128.1	
Stro	da	0–2 mm		1.1		0.40		110.5	
Stro	da	0–10 cm	1	2.4		0.72		129.2	
Somm	li	0–2 mm			0.7		2.2		235.6
Somm	li	0–10 cm	8		1.4		3.7		198.4
Somm	da	0–2 mm			2.1		5.4		173.0
Somm	da	0–10 cm	2		2.0		5.9		156.9

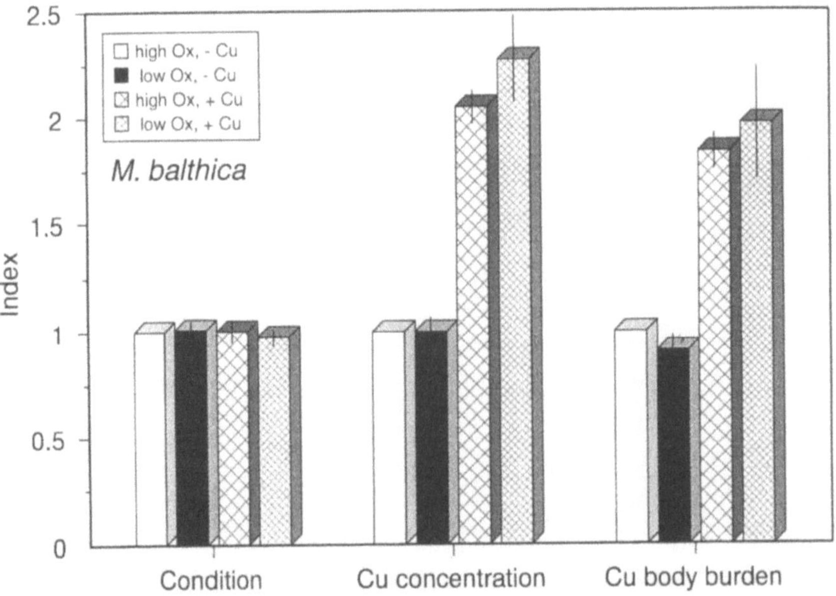

Figure 3. Average indices (and standard error) for the condition, copper concentration and copper body burden in the bivalve *M. balthica* as found in the laboratory experiments. The indices were calculated as follows. The original data of the treatment with 100% oxygen and no copper additions were given index values 1. The data of other treatments of the same experiment within the same week were divided by the original data with index value 1. Then, for each treatment, all indices were averaged. The basic data are given in Table 5.

Table 2. Experimental conditions

	Experiment 1				Experiment 2			
Animals from	Paulina				Paulina			
Period (month/year)	8/92				10/92			
Temperature (°C)	16				16			
Salinity (ppt)	30				30			
Sediment	sand				sand			
Treatment*	a	b	c	d	a	b	c	d
Oxygen level in water (%)	100	40	100	40	100	25	100	45
Cu in water (ppb)	3	5	30	37	4	2	36	31
Cu in sediment (ppm)	1.6	1.3	1.4	1.9	1.1	1.0	2.2	1.6

	Experiment 3											
Animals from	Dortsman											
Period (month/year)	2/94											
Temperature (°C)	5											
Salinity (ppt)	30											
Sediment	sand				mix				mud			
Silt fraction (% < 16 μm)	0.2				6.1				34.5			
Treatment*	a	b	c	d	a	b	c	d	a	b	c	d
Oxygen level in water (%)	100	45	100	45	100	45	100	45	100	45	100	45
Cu in sediment (ppm)**	1	1	154	154	2	2	123	123	13	13	276	276

* Types of treatments: a = high oxygen level, no copper, b = low oxygen level, no copper, c = high oxygen level, copper added, d = low oxygen level, copper added

** Average of measures carried out during experiments (copper additions for experiment 1 and 2 in water at a nominal level of 50 ppb, for experiment 3 in the sediment at a nominal level of 200 ppm DW).

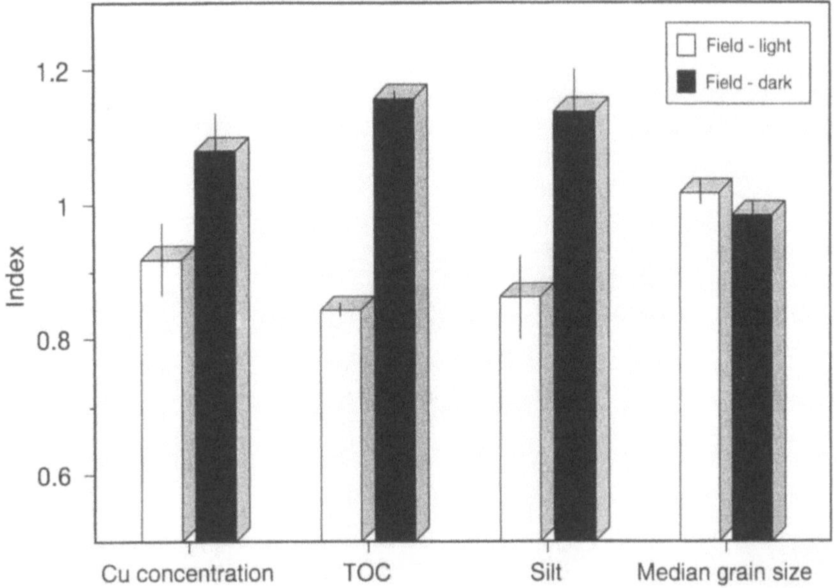

Figure 4. Indices for the median grain size and the contents of copper, total organic (TOC) and silt (fraction < 16 μm) in light and dark sediments (for calculation of the indices see Figure 2; basic data in Table 1).

Table 3. Basic data on the samples from the field and from the translocation experiments to assess the differential effect of dark versus light sediments on the condition (Condit), copper concentration (Cu-conc in ppm DW) and copper body burden (Cu-bb) of the bivalves *M. balthica* and *C. edule* (Stro = Stroodorpe, Dort = Dortsman, Paul = Paulina, Somm = Somme; Start = situation as found in the field (for those used in the translocation experiment it is the situation before translocation), Transloc = situation during the translocation experiments (except: f = field sample, not used for translocation experiments, s = field sample, used as start of translocation experiments): li = animals from light sediment, da = animals from dark sediment)

Species	Year	Month	Station	Sediment Start	Transloc	Number	Condit. mg/cm^3	Cu-conc ppm	Cu-bb μg/ind
M. balthica	92	Dec.	Stro	li	f	20	10.31	52.20	1.80
M. balthica	92	Dec.	Stro	da	f	20	11.31	52.60	1.95
M. balthica	92	Dec.	Dort	li	f	20	9.46	22.50	0.59
M. balthica	92	Dec.	Dort	da	f	20	9.32	24.70	0.68
M. balthica	92	Dec.	Paul	li	f	20	8.05	20.15	0.50
M. balthica	92	Dec.	Paul	da	f	19	8.59	33.15	1.13
C. edule	92	Dec.	Stro	li	f	10	10.10	5.67	0.66
C. edule	92	Dec.	Stro	da	f	20	9.58	4.44	0.52
C. edule	92	Dec.	Dort	li	f	19		3.93	0.38
C. edule	92	Dec.	Dort	da	f	10	12.27	4.23	0.39
M. balthica	93	Jan.	Paul	li	s	40	8.17	19.73	0.56
M. balthica	93	Jan.	Paul	da	s	40	7.01	19.58	0.52
M. balthica	93	Jan.	Paul	li	li	40	7.73	21.63	0.44
M. balthica	93	Jan.	Paul	li	da	40	6.60	21.45	0.45
M. balthica	93	Jan.	Paul	da	li	40	7.72	23.75	0.36
M. balthica	93	Jan.	Paul	da	da	40	6.89	21.93	0.47
M. balthica	93	Jan.	Dort	li	s	40	8.14	15.88	0.30
M. balthica	93	Jan.	Dort	da	s	40	7.62	26.18	0.62
M. balthica	93	Jan.	Dort	li	li	40	7.69	21.13	0.28
M. balthica	93	Jan.	Dort	li	da	40	8.26	17.08	0.29
M. balthica	93	Jan.	Dort	da	li	40	8.12	18.50	0.33
M. balthica	93	Jan.	Dort	da	da	40	8.06	26.15	0.42
M. balthica	93	Jan.	Stro	li	f	20	10.57	53.45	1.82
M. balthica	93	Jan.	Stro	da	f	20	9.64	52.75	1.36
M. balthica	93	Jan.	Dort	li	f	20	8.88	23.00	0.56
M. balthica	93	Jan.	Dort	da	f	20	9.21	21.70	0.57
M. balthica	93	Jan.	Paul	li	f	19	9.49	20.30	0.46
M. balthica	93	Jan.	Paul	da	f	20	7.31	24.55	0.61
C. edule	93	Jan.	Stro	li	f	10	10.45	8.84	0.73
C. edule	93	Jan.	Stro	da	f	10	9.19	13.23	0.85
C. edule	93	Jan.	Dort	li	f	10	12.28	4.50	0.43
C. edule	93	Jan.	Dort	da	f	10	11.92	5.43	0.68
M. balthica	93	Oct.	Dort	li	s	40	13.64	25.40	0.79
M. balthica	93	Oct.	Dort	da	s	40	14.59	33.50	1.29
M. balthica	93	Oct.	Dort	li	li	30	12.73	16.73	0.48
M. balthica	93	Oct.	Dort	da	li	30	13.79	16.80	0.60
M. balthica	93	Oct.	Dort	da	da	30	12.99	14.10	0.42
M. balthica	93	Oct.	Dort	li	da	40	13.80	13.03	0.37
C. edule	93	Oct.	Dort	li	s	40	13.79	4.87	0.61
C. edule	93	Oct.	Dort	da	s	40	15.24	4.95	0.51
C. edule	93	Oct.	Dort	li	li	16	13.44	3.27	0.41
C. edule	93	Oct.	Dort	da	li	20	11.30	3.90	0.45
C. edule	93	Oct.	Dort	da	da	40	11.98	3.66	0.44
C. edule	93	Oct.	Dort	li	da	30	11.92	3.49	0.36

Table 3. (continued)

Species	Year	Month	Station	Sediment Start	Sediment Transloc	Number	Condit. mg/cm^3	Cu-conc ppm	Cu-bb μg/ind
M. balthica	93	Oct.	Paul	li	s	40	13.82	18.35	0.51
M. balthica	93	Oct.	Paul	da	s	40	13.28	19.25	0.80
M. balthica	93	Oct.	Paul	li	li	40	12.86	14.73	0.37
M. balthica	93	Oct.	Paul	da	li	40	12.85	17.65	0.77
M. balthica	93	Oct.	Paul	da	da	30	12.90	16.27	0.67
M. balthica	93	Oct.	Paul	li	da	20	12.33	15.55	0.41
C. edule	93	Oct.	Paul	li	s	40	17.17	5.61	1.41
C. edule	93	Oct.	Paul	da	s	40	17.44	5.49	1.93
C. edule	93	Oct.	Paul	li	li	40	15.25	3.89	0.92
C. edule	93	Oct.	Paul	da	li	40	15.28	4.59	1.66
C. edule	93	Oct.	Paul	da	da	30	15.52	3.98	1.44
C. edule	93	Oct.	Paul	li	da	12	15.65	3.70	0.86
M. balthica	93	Nov.	Somm	li	f	20	16.34	12.16	1.31
M. balthica	93	Nov.	Somm	da	f	20	16.81	8.39	0.77
C. edule	93	Nov.	Somm	li	f	20	15.76	2.96	1.67
C. edule	93	Nov.	Somm	da	f	20	16.08	2.76	1.09
M. balthica	94	March	Dort	li	s	40	13.51	10.10	0.28
M. balthica	94	March	Dort	da	s	40	12.64	9.78	0.22
M. balthica	94	March	Dort	li	li	24	18.36	11.80	0.68
M. balthica	94	March	Dort	da	li	24	18.79	13.02	0.74
M. balthica	94	March	Dort	da	da	7	18.92	12.05	0.70
M. balthica	94	March	Dort	li	da	10	18.85	11.06	0.60
C. edule	94	March	Dort	li	s	40	11.37	5.10	0.58
C. edule	94	March	Dort	da	s	40	10.59	4.78	0.32
C. edule	94	March	Dort	li	li	16	14.56	3.54	0.63
C. edule	94	March	Dort	da	li	1	15.45	3.38	0.81
C. edule	94	March	Dort	da	da	7	14.92	4.16	0.74
C. edule	94	March	Dort	li	da	23	15.28	7.53	1.13
M. balthica	94	March	Paul	li	s	40	12.10	19.05	0.63
M. balthica	94	March	Paul	da	s	40	16.79	23.05	0.63
M. balthica	94	March	Paul	li	li	36	11.09	14.08	0.72
M. balthica	94	March	Paul	da	li	36	14.85	19.64	1.26
M. balthica	94	March	Paul	da	da	16	10.51	20.88	1.08
M. balthica	94	March	Paul	li	da	8	10.01	15.26	0.65
C. edule	94	March	Paul	li	s	40	12.07	4.78	1.63
C. edule	94	March	Paul	da	s	40	11.78	4.88	1.00
C. edule	94	March	Paul	li	li	30	13.29	6.50	2.40
C. edule	94	March	Paul	da	li	20	14.92	4.72	1.41
C. edule	94	March	Paul	da	da	4	13.46	5.19	1.46
C. edule	94	March	Paul	li	da	12	15.15	3.60	1.35

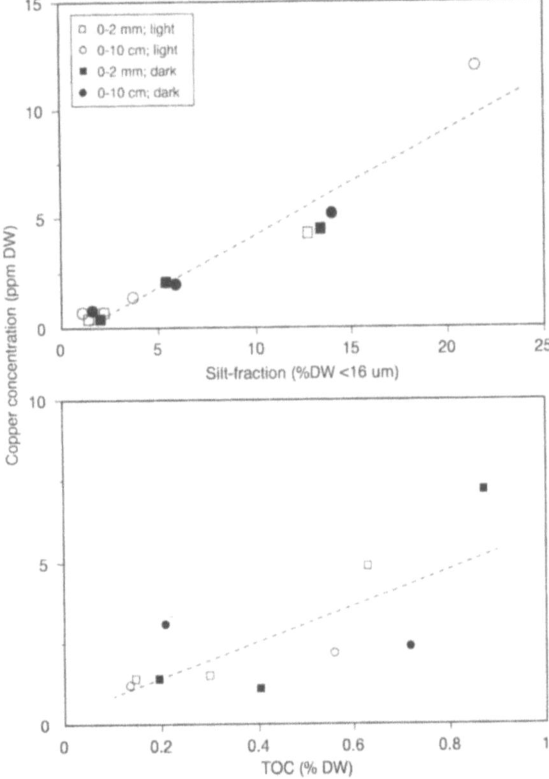

Figure 5. Relation between the silt-fraction (fraction < 16 μm as % of the dry weight) or total organic carbon (% of the dry weight) and the copper concentration in light and dark sediments, sampled to a depth of 2 mm or 10 cm (lines are best fitting linear regressions; with silt: $r^2 = 0.91$, $n = 12$, $p < 0.001$; with TOC: $r^2 = 0.56$, $n = 10$, $p = 0.04$).

or black (anoxic) layers. Light sediments (indicated by li) were those with light sediment-layers deeper than 3 cm (Table 1). The dark sediments (da) were those with light sediment-layers less than 3 cm deep (Table 1). Of each species, 20 to 40 animals were analysed immediately after sampling and depuration during 6 h in filtered sea water. In case of translocation experiments, part of the sampled animals were labeled and either put back to the sediment-type from which they originated (control: li-li, or da-da), or translocated to the differently coloured sediment-type on the same tidal flat (li-da = translocated from light to dark, or da-li = from dark to light). After 6 weeks, 10 to 40 of the translocated animals were recovered, and brought to the laboratory for further analyses.

Laboratory experiments

The experiments lasted 4 to 5 weeks each and were conducted in aquaria of 20 l. In the aquaria, a central box (14 cm ϕ and 10 cm high) was placed with a sediment-layer of approximately 8 cm deep. The water was recirculated at a rate of 4.5 liters per hour by a peristaltic pump. Water was changed two times per week with 0.2 μm filtered Oosterschelde sea-water (salinity 30 ppt). Three types of sediment were used: sand (from dunes), mud (from a tidal flat in the Oosterschelde) and a mixture of sand and mud (Table 2). Before use, the sediment was dried at 220 °C for 24 hours, and heated at 500 °C for 2 hours to remove organic matter. The organic matter was removed in order to have during the experiment no disturbance by additonal oxygen demand from the break-down of organic matter.

Adult *Macoma balthica* of 3 to 7 years old (shell-length 12.5 to 17.5 mm) were collected at Paulina (Figure 1), acclimatized and depurated in filtered Oosterschelde water at ambient temperatures during 1 day before starting the experiment. Forty specimens were introduced in each aquarium, and no food was given to avoid copper uptake and adsorption by algae (Absil et al., 1996).

In all experiments a combination was made between different oxygen concentrations and copper additions. For oxygen, a high (100% saturation) or low level (25 to 45%) was maintained. For copper, no additions were made, or copper was added to a nominal concentration of 50 ppb in water (experiment 1 and 2) or to a nominal concentration of 200 ppm dry weight in sediment (experiment 3). These sediments were loaded as follows: they were mixed during 12 hours with a copper-chloride solution, then water was taken away and sediments once rinsed with normal Oosterschelde seawater to eliminate unbounded copper. Measured oxygen and copper concentrations (averages) in experimental conditions are given in Table 2).

The required oxygen concentrations were adjusted – through air bubbles supply (air saturated aquaria) or by an appropriate air and nitrogen mixture (aquaria with low oxygen level). The oxygen percentage was continuously measured by means of oxygen probes (YSI 5700).

Condition index

Animals were depurated during 4 to 6 hours in filtered sea water. The shell length of the animals was determined. The soft tissues were separated from the shell, freeze-dried and weighed (dry weight). The condition index is defined as the ratio of the soft tissue dry weight and the shell volume. The shell volume was estimated on basis of shell length, by (shell length)^3 (Beukema & de Bruin, 1977).

Determination of copper concentrations and silt fraction

The soft tissues of each individual specimen were digested with HNO_3 suprapur, and copper concentrations were determined with flame atomic absorption spectrometry (AAS) with Zeeman background correction according to Amiard et al. (1987), Nieuwenhuize et al. (1988) and Nieuwenhuize & Poley-Vos (1989). The analytical technique was validated through international intercalibration (Coquery & Horvat, 1996). In standard materials MA-MEDPOL-1/TM (fish homogenate) and SD-MEDPOL-1/TM (marine sediment) our values (reference of the lab: 12) were 0.66 (±0.10) and 23.9 (±1.4) mg kg^{-1} DW, respectively, compared to certified values: 0.62 (±0.12) and 25.1 (±3.8) mg kg^{-1} DW. All Cu concentrations were based on lyophilized dry weight.

Water was sampled weekly in pre-cleaned 250 ml poly-ethylene bottles just before and 6 hours after changing the water. Immediately after sampling 200 μl HCl 30% suprapur (at least pH 2) was added. After the organic matter was destroyed during 4 hours by UV-irradiation, the copper concentration was determined by Potentiometric Stripping Analysis (PSA) according to Kramer (1986).

Sediments in the field were sampled in two ways: by scraping the top 2 mm (from which the clams feed) at 10 randomly chosen spots around the sampling station or by taking sediment cores of 10 cm depth (till which depths clams live) with a plexy-glass cylinder, 2 cm diameter, from 4 randomly chosen spots. In the experiments 2 cores of 8 cm were taken. The samples were freeze-dried. One part was used for particle size analysis with a Malvern particle sizer, the other part ground in an agate grinder, digested with aqua regia (12 ml of 37% HCl and 4 ml of 65% HNO_3 suprapur), and copper concentrations determined with atomic absorption spectrometry (AAS) with Zeeman background correction according to Nieuwenhuize et al. (1991).

Indices and statistical analysis

Copper content and condition may vary considerably between seasons and stations. In order to minimize this variation due to season or station, and to accentuate and assess only the differences related to dark and light sediments, some indices were calculated. To this end, for each station an average was calculated of each group of observations obtained during 1 day, being during field surveys 2 data, i.e. li and da, or after a translocation experiment 4 data, i.e. those from light sediment brought to light (li-li) or dark (li-da) sediment and those from dark brought to light (da-li) or dark (da-da) sediment. The individual data of a day at a station were then divided by the average. In this way, for the field surveys and translocation experiments, average indices (and standard errors) were calculated for the condition, copper concentration and copper body burden in the bivalves *M. balthica* and *C. edule* as found for specimens from light (li) and dark (da) sediments.

In the laboratory experiments the indices were calculated as follows. The original values found for the treatments with 100% oxygen and no copper additions were given index value 1. The values of other treatments within the same experiment and of the same week were divided by the original values belonging to index 1 (i.e. 100% Ox, no Cu). Then, for each treatment, all indices were averaged.

To analyse the significance of the influence of sediment-type, oxygen level or copper additions, ANOVA (Sokal & Rohlf, 1995) was performed with the SYSTAT statistical program.

Results

Field data

The values measured for condition, copper concentration or copper body burden vary strongly depending on species, season, and sampling station (Table 3). To assess, beside the variation caused by season or station, the differences related to dark and light sediments, indices were calculated (see Material and Methods). For both bivalve species no significant differences were found in the condition of animals from light and dark sediments (Figure 2, Table 4A). For the copper content, cockles showed no differences between light and dark sediments, whereas Baltic clams did show significant differences (Figure 2, Table 4A). In dark sediments Baltic clams had a higher copper concentration than in light sediments.

Table 4. Significance of the effect related to light and dark sediments, or high and low oxygen concentrations, on the condition (Condit), copper concentration (Cu-conc) and copper content (body burden; Cu-bb) of bivalves as found in the field, after translocation in the field and after laboratory experiments (ANOVA, *: $p < 0.05$, **: $p < 0.01$, –: n.s.)

A) Field-situation: comparison of animals from light and dark sediments (all data from Table 3 marked with an f or s in the column under Sediment-Transloc, thus excluding translocated samples)

Species	Effect on	Effect related to Light-Dark difference in the field
M. balthica	Condit.	–
	Cu-conc.	*
	Cu-b.b.	–
C. edule	Condit.	–
	Cu-conc.	–
	Cu-b.b.	–

B) Experimental translocation in the field: testing if the type of sediment (light or dark) before and/or during the translocation had any effect (data from Table 3, excluding those marked with an f or s in the column Sediment-Transloc) (Start = situation before the start of the experiment, Transloc = situation during the experiment, Start*Transloc = interaction of effects before and during the experiment)

Species	Effect on	Effect by Light-Dark difference		
		at Start	Transloc	Start*Transloc
M. balthica	Condit.	–	–	–
	Cu-conc.	**	–	–
	Cu-b.b.	**	–	–
C. edule	Condit.	–	–	–
	Cu-conc.	–	–	–
	Cu-b.b.	–	–	–

C) Laboratory experiments: testing if the degree of oxygenation had influence (data from Table 5, effect of sediment-type only tested for experiment 3).

Species	Effect on	Effect by		
		Oxygen	Added Cu	Sediment-type
M. balthica	Condit.	–	–	–
	Cu-conc.	–	**	–
	Cu-b.b.	–	**	–

Table 5. Basic data on the samples from the laboratory experiments to indicate the effect of the degree of oxygenation of the water and the addition of copper on the condition (Condit), copper concentration (Cu-conc) and copper body burden (Cu-bb) of the bivalve *M. balthica*. (Exp. = experiment 1, 2 or 3; Expos. period = exposure period in wk; Sed. = type of sediment used; Ox = oxygen concentration used, Cu = copper added (+) or not (−); background data on Exp., Sed., Ox and Cu in Table 2)

Exp.	Expos. period (wk)	Sed.	Ox	Cu	Number	Condit mg/cm^3	Cu-conc ppm	Cu-bb μg/ind
M. balthica								
1	0		start	–	20	13.5	16.1	0.62
1	2	sand	high	–	10	11.7	17.9	0.87
1	4	sand	high	–	10	11.4	22.4	0.80
1	2	sand	low	–	10	12.7	19.5	0.69
1	4	sand	low	–	10	13.7	18.8	0.69
1	2	sand	high	+	10	14.1	33.9	1.31
1	4	sand	high	+	10	11.8	53.2	1.67
1	2	sand	low	+	10	11.6	37.4	1.16
2	0		start	–	20	12.4	24.9	0.84
2	2	sand	high	–	10	10.1	21.5	0.65
2	3	sand	high	–	10	12.1	27.8	1.00
2	4	sand	high	–	10	11.4	24.2	0.71
2	2	sand	low	–	10	9.9	24.8	0.72
2	3	sand	low	–	10	10.5	26.0	0.73
2	2	sand	high	+	10	11.0	43.3	1.26
2	3	sand	high	+	10	9.7	56.6	1.63
2	4	sand	high	+	10	10.9	49.2	1.52
2	2	sand	low	+	10	11.1	56.9	1.59
2	3	sand	low	+	10	10.6	64.7	1.89
3	0		start		20	10.4	33.4	0.81
3	5	mud	high	–	10	10.9	26.8	0.65
3	5	sand	high	–	10	10.0	21.2	0.49
3	5	mix	high	–	10	10.6	26.6	0.62
3	5	mud	low	–	10	10.2	21.4	0.53
3	5	sand	low	–	10	9.9	27.1	0.52
3	5	mix	low	–	10	10.9	31.6	0.71
3	5	mud	high	+	10	10.5	57.3	1.30
3	5	mix	high	+	10	10.4	49.2	1.11
3	5	mud	low	+	10	10.6	42.1	1.00
3	5	mix	low	+	10	10.0	72.5	1.65

During the translocation experiments no significant changes occurred (Table 4B). Significant differences in the translocated Baltic clams are only connected to the differences already present at the start of the experiment, i.e. animals originating from dark sediments have a higher copper content (concentration as well as body burden) (Figure 2, Table 4B).

Laboratory experiments

In all three experiments, the increase of the copper content in copper exposed Baltic clams was equivalent (Table 5), reaching a concentration of 50 to 70 ppm Cu within 3 to 5 weeks, irrespective if the copper was added to the water (exp. 1 and 2) or in the sediment (exp. 3). The sediment-type, i.e. sand or mud, or the oxygen level had no significant effect on the condition or copper content of the Baltic clam (Figure 3, Table 4C, 5). The only factor affecting significantly the copper content of the Baltic clams was the addition of copper to the water or sediment.

Geochemistry of sediment

The two types of sediment, light and dark, showed significant differences (t-test, Sokal & Rohlf, 1995): the copper ($p = 0.05$), organic carbon ($p < 0.01$) and silt fraction ($< 16\ \mu$m; $p = 0.01$) was higher in dark sediments than in light sediments (Figure 4). Irrespective the level of oxygenation in the sediment, the silt fraction and copper content were strongly related to each other, as were also, to a less extent, the organic carbon and copper content (Figure 5).

Discussion

The strong relation between copper concentration and the silt-fraction, and to a lesser extent between copper and the organic content, of the sediment, corroborates the results of many studies (see reviews of Luoma, 1989; Forstner, 1987, 1990; Flemming & Trevors, 1989). A higher fraction of silt and organic carbon coincides, because of its relative higher surface area and higher number of complexing sites, with a higher concentration of copper. A higher fraction of silt and higher organic content may however also result in a lower permeation of oxygen and higher use of oxygen in the sediment, thus resulting in a more hypoxic situation. A higher copper concentration and a lower oxygen content in the darker sediments could thus be indirectly interrelated through the higher silt and/or organics content.

As generally found for deposit-feeding clams (Bryan & Langston, 1992), for the Baltic clam a positive relation is found between the copper concentration in the sediment and that in the clams (Cain & Luoma, 1990; Absil, 1993; Hummel et al., 1997). Therefore, it is not surprising that also in this study, the higher copper concentration of the dark sediments (on average 17.7% higher when compared to the lighter sediments) went together with a corresponding higher copper concentration in Baltic clams (on average 17.3% higher than those in lighter sediments). Thus, also the relation between the oxygenation of sediments and the copper content of clams could be indirect, as is illustrated in the tentative scheme of Table 6. Due to a higher silt fraction and/or organic content some places at a tidal flat are more hypoxic, and by that darker coloured, than at other places. At these darker places, also because of the higher silt and/or organic content, higher copper contents can be found in the sediment and consequently also in the endofaunal, deposit-feeding Baltic clams.

A higher bioavailability of copper under more hypoxic, reduced, sediment conditions, thus could in this study not be proven for Baltic clams. The previous indications for an effect of the level of oxygenation on the copper content of clams (Luoma & Bryan, 1982; Luoma et al., 1985) were based on incidental observations that were strongly deviating from normal values (expected on basis of standard relations with weight or grain size or from average seasonal fluctuations). Anoxic conditions in the environment were then *post hoc* suggested as a possible cause. Other indications that a low oxygen content would lead to a relatively higher uptake of copper were derived from exposure of clams (without sediments) to low copper concentrations in combination with low oxygen concentrations (Neuhoff, 1983; Theede, 1985). However, at closer observation of the data of Neuhoff and Theede, it is clear that they exposed their clams, which were kept under low and normal oxygen concentrations, to very different copper concentrations (8 or 15 ppb and 37 or 86 ppb Cu, respectively). Moreover, they found the higher copper concentrations in clams kept under the normoxic conditions (and thus highest copper exposures). The physiological characters they measured in clams, such as oxygen consumption, glycogen content or Adenylate Energy Charge, showed indeed a worse performance of the animals kept under low oxygen condition, but such can not be related to a

Table 6. Tentative scheme explaining the indirect relations between coloration, oxygen levels and copper content in sediment and Baltic clams

higher fraction of silt (32%) and organic carbon (37%) in sediment	
↓↓↓	↓↓↓
relative higher surface area	lower permeation of oxygen
↓↓↓	+
higher number of complexing sites	higher oxygen consumption
↓↓↓	↓↓↓
higher concentration of copper in sediment (17.7%)	stronger hypoxic situation
↓↓↓	↓↓↓
higher concentration of copper in clams (17.3%)	darker colour

higher copper concentration in the environment or in clams. Therefore summarizing, significant higher copper concentrations in clams because of low levels of oxygenation were in fact not shown in the studies of Neuhoff (1983) and Theede (1985).

Cockles did not show the (indirect) relationships as found for Baltic clams, probably because of their life strategy, i.e. living near the sediment surface, and for their food intake fully dependent on the water column. Moreover, cockles may have a physiological regulatory mechanism to control their copper concentration, whereas Baltic clams probably lack such a regulation (Bryan & Hummerstone, 1977; Langston & Zhou, 1987; Hummel et al., 1997).

Then, taking all results together, we infer that the level of oxygenation, as indicated in this study by the coloration of sediments, was only indirectly related to, and not directly influencing, the copper content of bivalves (Table 6).

Acknowledgements

The research of P.M. was supported by an EC-COMETT II post-graduate fellowship no. 8\91. The sampling in France was supported by a fund from the committee for French-Dutch cooperation via the Netherlands Foundation of Sea Research (SOZ; presently Geoscience Foundation GOA). We thank Jaap van Liere, Joop Nieuwenhuize, Jan Sinke, Julian van Oostrum for help in sampling and analyses. Publication N100 2412, Centre for Estuarine and Coastal Ecology.

References

Absil, M. C. P., 1993. Bioavailability of heavy metals for the deposit feeder *Macoma balthica* with special emphasis on copper. Ph.D. thesis, Agricultural University Wageningen, 171 pp.

Absil, M. C. P., M. Berntssen & L. J. A. Gerringa, 1996. The influence of sediment, food and organic ligands on the uptake of copper by sediment-dwelling bivalves. Aquat. Toxicol. 34: 13–29.

Amiard, J. C., C. Amiard-Triquet, B. Berthet & C. Metayer, 1987. Comparative study of the patterns of bioaccumulation of essential (Cu, Zn) and non-essential (Cd, Pb) trace metals in various estuarine and coastal organisms. J. exp. mar. Biol. Ecol. 106: 73–89.

Ankley, G. T., N. A. Thomas, D. M. Di Toro, D. J. Hansen, J. D. Mahony, W. J. Berry, R. C. Swartz, R. A. Hoke, A. W. Garrison, H. E. Allen & C. S. Zarba, 1994. Assessing potential bioavailability of metals in sediments: A proposed approach. Envir. Mgmt 18: 331–337.

Beukema, J. J. & W. de Bruin, 1977. Seasonal changes in dry weight and chemical composition of the soft parts of the tellinid bivalve *Macoma balthica* in the Dutch Wadden Sea. Neth. J. Sea Res. 11: 42–55.

Bryan, G. W., 1980. Recent trends in research on heavy-metal contamination in the sea. Helgol. Meeresunters. 33: 6–25.

Bryan, G. W., 1984. Pollution due to heavy metals and their compounds. In O. Kinne (ed.), Marine Ecology, Vol. 5, Part 3. John Wiley & Sons Ltd, Chichester: 1289–1431.

Bryan, G. W. & L. G. Hummerstone, 1977. Indicators of heavy-metal contamination in the Looe estuary (Cornwall) with particular regard to silver and lead. J. mar. biol. Ass. U.K. 57: 75–92.

Bryan, G. W. & W. J. Langston, 1992. Bioavailability, accumulation and effects of heavy metals in sediments with special reference to United Kingdom estuaries: a review. Envir. Poll. 76: 89–131.

Cain, D. J. & S. N. Luoma, 1990. Influence of seasonal growth, age and environmental exposure on Cu and Ag in a bivalve indicator, *Macoma balthica*, in San Francisco Bay. Mar. Ecol. Prog. Ser. 60: 45–55.

Coquery, M. & M. Horvat, 1996. The analytical performance study for the MED POL area: determination of trace elements in marine sediment SD-MED POL-1/TM and in fish homogenate MA-MED POL-1/TM. Intern. Atom. Energy Agency, 85 pp.

Di Toro, D. M., J. D. Mahony, D. J. Hansen, K. J. Scott, A. R. Carlson & G. T. Ankley, 1992. Acid volatile sulfide predicts the acute toxicity of cadmium and nickel in sediments. Envir. Sci. Technol. 26: 96–101.

Flemming, C. A. & J. T. Trevors, 1989. Copper toxicity and chemistry in the environment: a review. Wat. Air Soil Pollut. 44: 143–158.

Förstner, U., 1987. Sediment-associated contaminants – an overview of scientific bases for developing remedial options. Hydrobiologia 149: 221–246.

Förstner, U., 1990. Inorganic sediment chemistry and elemental speciation. In R. Baudo, J. P. Giesy & H. Muntau (eds), Sediments: Chemistry and Toxicity of in-place Pollutants. Lewis Publishers Inc., Ann Arbor: 61–105.

Gerringa, L. J. A., 1990. Aerobic degradation of organic matter and the mobility of Cu, Cd, Ni, Pb, Zn, Fe and Mn in marine sediment slurries. Mar. Chem. 29: 355–374.

Gerringa, L. J. A., 1991. Mobility of Cu, Cd, Ni, Pb, Zn, Fe and Mn in marine sediment slurries under anaerobic conditions and at 20% air saturation. Neth. J. Sea Res. 27: 145–156.

Hummel, H., R. Modderman, C. Amiard-Triquet, F. Rainglet, Y. van Duijn, M. Herssevoort, J. de Jong, R. Bogaards, G. Bachelet, M. Desprez, J. Marchand, B. Sylvand, J.-C. Amiard, H. Rybarczyk & L. de Wolf, 1997. A comparative study on the relation between copper and condition in marine bivalves and the relation with copper in the sediment. Aquat. Toxicol. 38: 165–181.

Kramer, C. J. M., 1986. Apparent copper complexation capacity and conditional stability constants in North Atlantic waters. Mar. Chem. 18: 335–349.

Langston, W. J. & M. Zhou, 1987. Cadmium accumulation, distribution and elimination in the bivalve *Macoma balthica*: Neither metallothionein nor metallothionein-like proteins are involved. Mar. Envir. Res. 21: 225–237.

Luoma, S. N., 1983. Bioavailability of trace metals to aquatic organisms – a review. Sc. Total Envir. 28: 1–22.

Luoma, S. N., 1989. Can we determine the biological availability of sediment-bound trace elements? Hydrobiologia 176: 379–398.

Luoma, S. N. & G. W. Bryan, 1982. A statistical study of environmental factors controlling concentrations of heavy metals in the burrowing bivalve *Scrobicularia plana* and the polychaete *Nereis diversicolor*. Estuar. coast. Shelf Sci. 15: 95–108.

Luoma, S. N., D. Cain & C. Johansson, 1985. Temporal fluctuations of silver, copper and zinc in the bivalve *Macoma balthica* at five stations in South San Francisco Bay. Hydrobiologia 129: 109–120.

McLusky, D. S., V. Bryant & R. Campbell, 1986. The effects of temperature and salinity on the toxicity of heavy metals to marine and estuarine invertebrates. Oceanogr. Mar. Biol. Ann. Rev. 24: 481–520.

Millward, G. E., 1995. Processes affecting trace element speciation in estuaries. A review. Analyst 120: 609–614.

Neuhoff, H. G., 1983. Synergistic physiological effects of low copper and various oxygen concentrations on *Macoma balthica*. Mar. Biol. 77: 39–48.

Nieuwenhuize, J., C. H. Poley-Vos, A. Goud & M. A. Hemminga, 1988. Determination of Cd, Cu, Pb and Zn in small marine insect larvae with a microdigestion method. Atomic Spectroscopy 9: 204–206.

Nieuwenhuize, J. & C. H. Poley-Vos, 1989. A rapid microwave dissolution method for the determination of trace and minor elements in lyophilized plant material. Atomic Spectroscopy 10: 148–153.

Nieuwenhuize, J., C. H. Poley-Vos, A. H. van den Akker & W. van Delft, 1991. Comparison of microwave and conventional extraction techniques for the determination of metals in soil, sediment and sludge samples by atomic spectrometry. Analyst 116: 347–351.

Sokal, R. R. & F. J. Rohlf, 1995. Biometry. Freeman & Company, New York, 887 pp.

Theede, H., 1985. Some problems of pollution with heavy metals under the environmental conditions of the Baltic Sea. Symp. Biol. Hung. 29: 223–238.

Westerlund, S. F. G., L. G. Anderson, P. O. J. Hall, A. Iverfeldt, M. M. Rutgers van der Loeff & B. Sundby, 1986. Benthic fluxes of cadmium, copper, nickel, zinc and lead in the coastal environment. Geochim. Cosmochim. Acta 50: 1289–1296.

Hydrobiologia **373/374:** 311–316, 1998.
J.-C. Amiard, B. Le Rouzic, B. Berthet & G. Bertru (eds),
Oceans, Rivers and Lakes: Energy and Substance Transfers at Interfaces.

Characterisation of marine zooplankton alkaline phosphatase activity in relation to water quality

Dominique Jamet & Gérard Boge
Université de Toulon et du Var, Département de biologie appliquée, Laboratoire d'Hydrobiologie, L.E.P.I., B.P.132. F-83957 La Garde cedex, France

Key words: alkaline phosphatase activity, zooplankton Ecology, marine pollution

Abstract

In this study, we measured the non-specific zooplanktonic phosphate activity inside and outside Toulon bay. These activities were compared in relation to substrate and phosphate concentrations, pH and sea water composition. Great differences were observed between the two sites. Phosphatase activities were higher inside Toulon bay, where high zooplankton density and high protein concentrations were observed. Concerning enzyme activities, the intersite differences increased at the lowest substrate concentrations. Moreover, zooplankton alkaline phosphatase activity showed a good adaptation to the natural medium conditions (pHs and ionic strength) and is likely directly liberated to the water, where it contributes to an enrichment in mineral phosphorus from organic material.

Introduction

Since the 1960s, intensifying industrial and domestic activities in the Mediterranean basin have produced a regular increase of sea water nutrient levels, disturbing the natural biological equilibrium (Bethoux & Copin-Montegut, 1988). Enclosed or semi enclosed areas such as estuaries, bays, sea-ports, lagoons and coastal pools, are affected by this phenomenon, which is characterised by the development of planktonic organisms associated with toxic pollution. The Toulon-Bay is one such sensitive zone (Belin et al., 1995).

In order to gain deeper insight into the planktonic evolution in this area, we have undertaken a study of the zooplanktonic community during an annual cycle. Preliminary results have shown an intense development of these organisms together with elevated chlorophyll concentrations (Jamet et al., 1996). We also investigated the role of zooplankton in production of inorganic phosphate from organic phosphoric esters. For this purpose, we measured phosphatase activities which are likely to hydrolyse organic compounds originating from biological material or from anthropogenic inputs (Kobori & Taga, 1979). Activities in areas inside and outside Toulon Bay were compared, in relation to protein, substrate and phosphate concentrations, pH and sea water composition.

Material and methods

Study sites

Two sampling stations located at the French Mediterranean coast were chosen:

Station 1: Toulon bay (43°5′ N, 6°0′ E). Zooplankton were collected at the 'petite rade' of Toulon between 8 and 20 m depth. This site is characterised by a heavy maritime traffic due to military and commercial activities. The REPHY data (French Phytoplankton Monitoring Network) indicated high levels of PAH, PCB, and heavy metals (mercury, lead) (Cossa 1994, IFREMER 1993). The presence of artificial breakwaters reduces the influence of the dominant West and East winds and the surface and under-currents (Paillard et al., 1993).

Station 2: Le Niel bay. This station is located in Giens, 20 km from Toulon, in front of Porquerolles Island. The quality of the water is improved by an abundance of *Posidonia oceanica*. This station is protected from West winds and is influenced by North-East under-currents (Paillard et al., 1993). Zooplankton were collected between 4 and 14 m depth.

Field sampling

Zooplankton were collected in May-June using a net of 88 μm mesh. The volume of filtered water was measured by a flow meter (model General Oceanic 2030 R). Size class filtrations (1180, 950, 710, 600, 500, 400, 300, 250, 180, 132 and 90 nm) were performed and the organisms collected in clean sea water taken from the less polluted site (Le Niel Bay).

Zooplankton analysis

Countings were achieved on plankton filtered onto 1180–90 nm filters and stored in buffered (CaCO3) 5% sea water formalin.

Phosphatase activity

Enzyme activity was measured spectrophotometrically with paranitrophenyl phosphate (PNPP) as substrate (Reichardt et al., 1967; Jamet et al., 1995). Plankton was centrifuged (10 minutes at 6000 g) and the supernatant used as enzyme source. Substrate and enzyme blanks were carried out using prefiltrated sea water. The incubation medium consisted of sea water from the non polluted site (Le Niel Bay). When specific media were used, their composition are presented in the text.

Protein determinations

Particulate protein concentrations were estimated on plankton filtered onto 90 nm filters according to Lowry et al. (1951), with an albumin (BSA) standard.

Results

Intersite comparison

Zooplankton biomass: The quantities of proteins per m3 of sea water were much higher in Toulon Bay than in the Niel Bay (Table 1).

Phosphatase activities: For substrate concentration of 15 mM, phosphatase activity related to the filtered sea water volume or to the protein amount was higher in Toulon bay (Table 1). This activity showed a linear evolution with time during five minutes indicating that no limitation factor existed in the incubation medium over this period.

Table 1. Protein levels and phosphatase activity in Toulon Bay and Le Niel Bay

Site	Toulon Bay	Le Niel Bay
Protein concentrations (mg m^{-3})	15.1	0.50
Phosphatase activity:		
Nanomoles PNP min^{-1} m^{-3} seawater	113	1.45
Nanomoles PNP min^{-1} mg^{-1} prot	7.46	2.9

At these two sites, the maximal activity corresponded to 120–250 μm size class organisms. Microscopic observations and countings confirmed that this enzyme activity was essentially supported by zooplankton since the density of these organisms was much higher at these sizes (Figure 1).

Substrate concentration dependence

The role of PNPP concentration on planktonic phosphatase activity has been studied. The results have been plotted according to Eadie Hofstee (Activity vs Activity/Substrate concentration) (Figure 2).

Theoretically, a straight line with decreasing slope is characteristic of pure enzyme activity. With zooplankton, more complex curves were obtained, especially in Toulon Bay, which suggests a possible involvement of several enzymes with different kinetic constants. Their K_m can be estimated from the tangents drawn at the extremities of the curves. For the high affinity system, they were almost the same inside and outside Toulon Bay. For the lowest affinity system, the slope was too low and the values could not be determinated with enough accuracy.

These results suggested that the intersite differences in plankton activities could be essentially quantitative. However, the ratio of inside to outside activity revealed that these differences increased as substrate concentration decreased (Figure 3). This is an indication that qualitative differences may also exist.

pH dependence

This study was undertaken to detect if several enzyme activities with distinct pH requirements could be involved in PNPP hydrolysis. pH levels were prepared

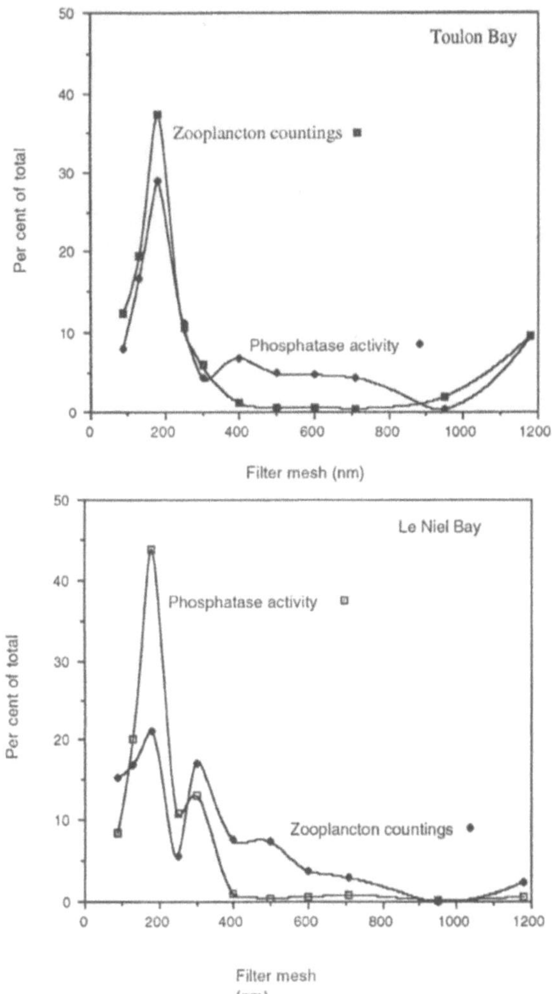

Figure 1. Phosphatase activity related to different size classes of zooplankton in Toulon Bay and in Le Niel Bay.

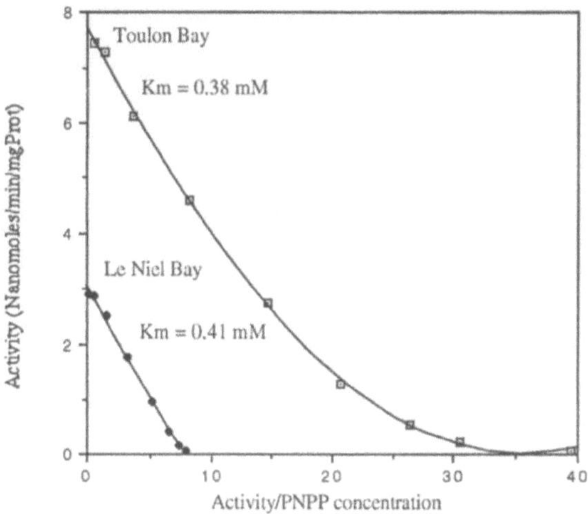

Figure 2. Eadie-Hoffstee plots of the phosphatase activity of Toulon and Niel Bay zooplankton (Per cent of number).

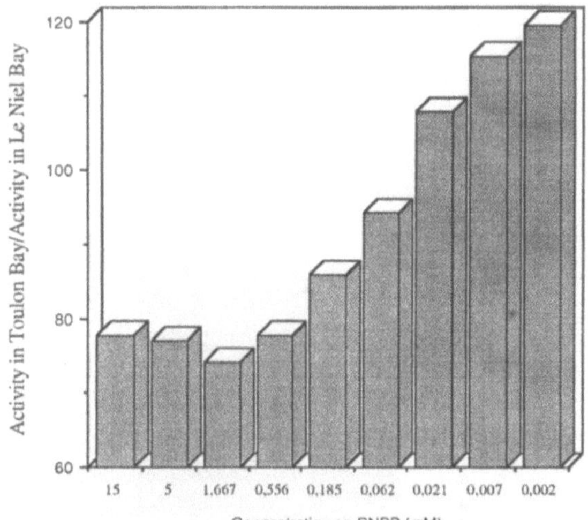

Figure 3. Ratio of the phosphatase activities in zooplankton collected in Toulon Bay and Le Niel bays.

with acetate buffer (up to 7) and glycine buffer (from 7 to 10) and two PNPP concentrations were used (1 and 10 mM). In parallel, bovine pure alkaline phosphatase was checked for comparison (Figure 4).

For the two substrate concentrations, only one optimum pH has been observed with zooplankton samples, which suggests that one enzyme activity could be responsible for the hydrolysis of PNPP. No intersite differences were found in this pH dependence, whereas the pH optimum of bovine alkaline phosphatase was much higher.

Role of sea water

In order to detect a possible influence of sea water components on planktonic enzyme activity, the extracts were prepared in distilled water adjusted to pH 8 with glycine buffer. They were incubated in 10 mm PNPP, either in distilled water or in sea water. The same experiment was also carried out using pure bovine alkaline phosphatase. Taking the values obtained in distilled water as basic activity, the data demonstrated that sea water stimulated plankton enzyme activity more intensively than bovine activity (Figure 5).

Role of phosphate concentrations

Alkaline phosphatase activity is typically repressed by phosphate ions (Jansson et al., 1988).The influence of

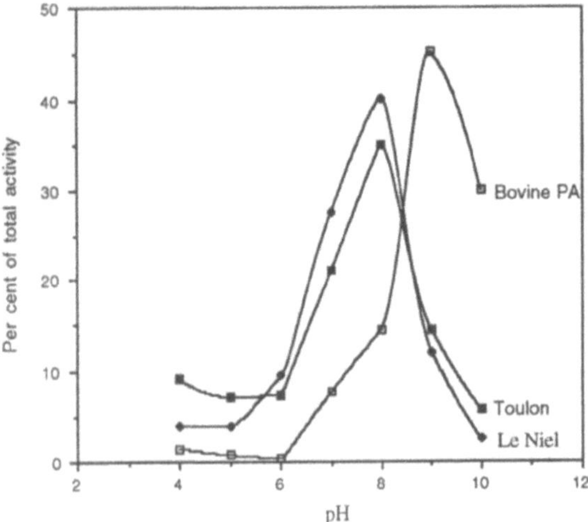

Figure 4. pH curves of the phosphatase activities in zooplankton collected in Le Niel Bay and Toulon Bay. Comparison with bovine alkaline phosphatase activity.

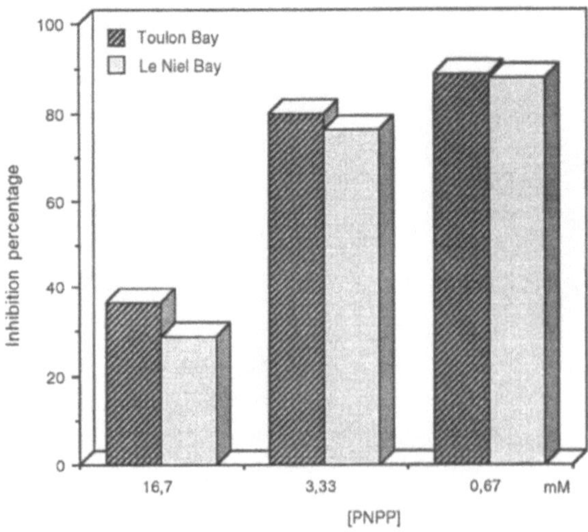

Figure 6. The effect of inorganic phosphate ions on the phosphatase activity of zooplankton in both Toulon and Le Niel Bays.

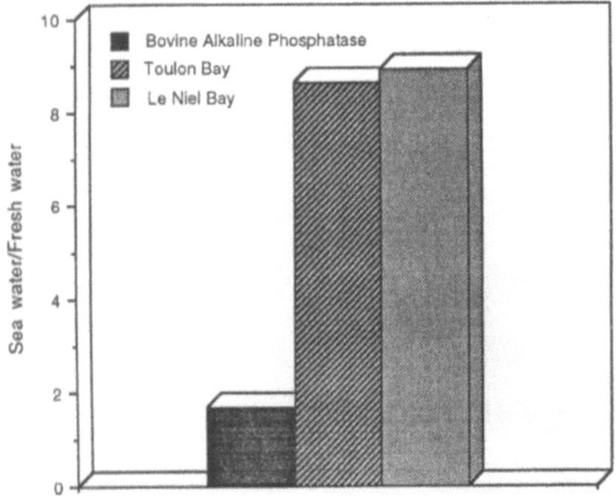

Figure 5. The effect of salt water on the phosphatase activity of zooplankton in both Toulon and Le Niel Bays.

a high concentration of orthophosphate (10 mM) on the hydrolysis of two PNPP concentrations (0.67 mM and 3.33 mM) by zooplankton was studied. Na_2HPO_4 was used, but owing to its low solubility in sea water, distilled water was chosen as incubation medium instead of sea water. pH was adjusted to 8 (50 mM glycine buffer).

A notable inhibition of the plankton enzyme activity from Toulon Bay or Le Niel Bay was observed, especially for 0.67 mm PNPP (Figure 6). These results support the hypothesis that a specific alkaline phosphatase activity was responsible of the hydrolysis of this substrate.

Discussion

Major differences have been observed between the inside and outside of Toulon Bay concerning zooplankton biomass and the activity of non-specific phosphatases. The particulate protein concentrations were about 30 times higher inside Toulon Bay. This observation can be dependent to the density of zooplanktonic organisms and more especially to copepods (Jamet et al., 1996). As to phosphatase activities related to water units or protein concentrations, they were also higher in the Toulon Bay.

Two different enzymes with high and low affinites probably contributed to total activity, since non linear Eadie Hofstee curves were observed (Figure 2). Electrophoretic characterisation of this activity will need to be carried out to know if different enzymes were really involved in the total activity. Such enzyme activities may originate from different zooplankton species. Countings carried out in parallel to these enzyme determinations showed that at least 16 distinct species of copepods were present in zooplankton collected in Toulon bay, with these species abundant (*Oithona nana, Euterpina acutifrons, Acartia sp*) (Jamet et al., 1996). Phytoplankton or bacteria could also have a role in this total enzyme activity, but except for fixed and ingested organisms, they are not retained by the 90 nm mesh filters. Concerning the possible activity dissolved in sea water, preliminary determinations have shown that no such measurable activity existed following short term incubations.

The correlation between phosphatase activity and the abundance of zooplankton ($p = 0.05$) offers a possibility of using this enzyme to assess zooplankton biomass or its activity, as has been proposed by several authors (Bonin & Travers, 1992). More accurate estimations of biomass would seem possible at high substrate concentrations (at least 10 mM PNPP), since the influence of medium components would be lowered, as was shown with phosphate (Figure 6). Contrasting results exist relative to this sensitivity to phosphate ions. Jansson (1976) concluded that zooplankton enzymes were insensitive, whereas Boavida (1984) showed that enzymes of cultured *Daphnia* were sensitive. Our results demonstrate that this sensitivity is highly dependent upon substrate concentration (Figure 6), and suggest a possible competitive-like effect with substrate on the active site (see Petterson 1979, 1980).

This sensitivity to phosphorus ions and the increased activity at alkaline pH indicates that alkaline phosphatase could be responsible of the hydrolysis of PNPP and other phosphoric esters by zooplankton. This activity showed a good adaptation to the natural medium since its optimum pH and its ionic requirements were well covered in sea water (Figures 4 and 5). In contrast, the dependence of vertebrate enzyme toward extra physiological alkaline pH has often been questioned (Bogé et al., 1992).

Alkaline phosphatase, essentially bacterial enzyme, is adaptive. Its activity is controled by dissolved phosphate, and is higher in phosphorus-depleted organisms (Petterson, 1979; Siuda, 1984; Jansson et al., 1988). In Toulon Bay, higher differences in activity than Le Niel Bay were found for the lowest substrate concentrations (Figure 3). This indicates that qualitative differences also exist. They could be ascribed to differences in the specific richness of the zooplankton. But they could also be due to the fact that zooplankton organisms were more depleted in phosphorus and had optimised their phosphatase activity. Preliminary measurements of the phosphate concentrations showed that they are lower than 10 μgPO4 l^{-1} in Toulon Bay and slightly higher than outside (personal observations).

The present data suggest that alkaline phosphatase could be released directly to the natural medium. Indeed, this activity was always assayed in the supernatant following the centrifuging of a concentrated suspension of zooplanktonic organisms (see materials and methods). Moreover a 0.45 μm filtration of the supernatant had no effect on total activity (data not shown). In aquatic organisms, alkaline phosphatase can also be bound to cellular membranes, especially in bacteria and in phytoplankton. In fresh water copepods, phosphatase is liberated directly in water and contributes to the enrichment in mineral phosphorus from organic material (Boavida & Heath, 1984; Jansson, 1976). The present results suggest that this could be also the case with marine copepods. This release would make possible its direct determination in sea water. In fresh water this has been carried out using PNPP as substrate (Reichardt et al., 1967). However, in such conditions, zooplankton activity could not be easily distinguished from bacterial or phytolankton activities. Moreover, owing to the lower activities in sea water, long time incubations would be necessary (probably several days), or the utilisation of fluorimetric methods, which are more sensitive (Perry, 1972; Hoppe, 1983).

The present results raise the question of the contribution of the zooplankton to the total activity in sea medium. In fresh water, authors give it the same importance as bacteria or phytoplankton (Jansson, 1976). Assays are in progress to distinguish the specific contribution of each community in Toulon Bay sea water.

References

Belin, C., B. Beliaff, B. Raffin, M. Rabia & F. Ibanez, 1995. Phytoplanktonic time-series data of the french phyto-plankton monitoring network: toxic and dominant species. In P. Lassus, G. Arzul, E. Arard, P. Gentien & C. Marcaillou (eds), Harmful Marine Algal Blooms, 771–776. Techn. & Doc., Lavoisier, Intercept Ltd.

Bethoux, J. P & G. Copin-Montegut, 1988. Phosphorus and nitrogen in the Mediterranean Sea: specificities and forecasting. In H. J. Minas & P. Nival (eds), Oceanologica Acta nESP, océanographie pélagique méditerranéenne: 75–78.

Boavida, M. J & R. T. Heath, 1984. Are the phosphatases released by Daphnia magna components of its food? Limnol. Oceanogr. 29: 641–645.

Bogé, G., M. Leydet & D. Houvet, 1992. The effects of hexavalent chromium on the activity of alkaline phosphatase in the intestine of rainbow trout (*Oncorhynchus mykiss*). Aquat. Toxicol. 23: 247–260.

Bonin, D. J. & M. Travers, 1992. Examen critique des méthodes d'estimation de la biomasse et de l'activité des microorganismes dans les systèmes aquatiques. Mar. life 2: 1–29.

Cossa, D., 1994. Le mercure en milieu marin: le cas du littoral français dans le contexte d'une contamination à l'échelle planétaire Equinoxe: 47, 48–52.

IFREMER, 1993. Qualité du milieu marin littoral. IFREMER et Ministère de l'enseignement supérieur et de la recherche, Ministère de l'agriculture et de la pèche, Ministère de l'équipement, des transports et du tourisme, Ministère de l'environnement, 241 pp.

Hope, H. G., 1985. Significance of exoenzymatic activities in the ecology of brackish water: measurements by means of methylumbelliferyl substrates. Mar. Ecol. Prog. Ser. 11: 299–308.

Jamet, D., L. Aleya & J. Dévaux, 1995. Diel changes in the alkaline phosphatase activity of bacteria and phytoplancton in the hypereutrophic Villerest reservoir (Roanne, France). Hydrobiologia 300/301: 49–56.

Jamet, J. L., A. S. Ferec-corbel, S. Richard, C. Geneys & G. Bogé, 1996. The zooplanktonic diversity of a perturbed littoral marine ecosystem: the Toulon Bay (Var, France), in press.

Jansson, M., 1976. Phosphatases in lake water: characterization of enzymes from phytoplancton and zooplancton by gel filtration Science.

Jansson, M., H. Olsson & K. Petterson, 1988. Phosphatases: characteristics and function in lakes. Hydrobiologia 170: 157–175.

Kobori, H. & N. Taga, 1979. Phosphatase activity and its role in the mineralization of organic phosphorus in coastal sea water. J. Explor. mar. Biol. Ecol. 36: 23–39.

Lacaze, J. C., 1993. La dégradation de l'environnement côtier: Conséquences écologiques Masson eds.

Lowry, O. H., N. J. Rosebrough, A. L. Farr & R. J. Randall, 1951. Protein measurement with the Folin phenol reagent. J. Biol. Chem. 193: 265–275.

Paillard, M., V. Gravez, P. Clabaut, P. Walker, J. J. Blanc, C. F. Boudouresque, T. Belsher, F. Urscheler, F. Poydenot, J. M. Sinnassamy, C. Augris, J. P. Peyronnet, M. Kessler, J. M. Augustin, E. Le Drezen, C. Prudhomme, J. M. Raillard, G. Pergent, A. Hoareau & E. Charbonnel, 1993. Cartographie de l'herbier de Posidonie et des fonds marins environnants de Toulon à Hyères (Var – France). Reconnaissance par sonar latéral et photographie aérienne. Notice de présentation. IFREMER & GIS POSIDONIE Publ., Fr.: 1–36 + 3 cartes annexes.

Olson, H., 1983. Origin and production of phosphatases in the acid Lake Gädsjön. Hydrobiologia 101: 49–58.

Petterson, K., 1980. Alkaline phosphatase activity and algal surplus phosphorus as phosphorus-deficiency indicators in Lake Erken. Hydrobiologia 89: 54–87.

Petterson K. & M. Jansson, 1978. Determination of phosphatase avctivity in lake water-astudy of methods Verh. int. Ver. Limnol. 20: 1226–1230.

Reichardt, W., J. Overbeck & L. Steubing, 1967. Free Dissolved Enzymes in Lake Waters. Nature 216: 1345–1347.

Siuda, W., 1984. Phosphatases and their role in organic phosphorus transformation in natural waters. A review. Pol. Arch. Hydrobiol. 31: 207–233.

Hydrobiologia **373/374:** 317–332, 1998.
J. C. Amiard, B. Le Rouzic, B. Berthet & G. Bertru (eds),
Oceans, Rivers and Lakes: Energy and Substance Transfers at Interfaces.

The lead content of atmospheric aerosols above the eastern channel: seasonal variability and solubility in a coastal seawater

Marie-Laure Bertho, Karine Deboudt*, Pascal Flament & Emile Puskaric
Université du Littoral – Côte d'Opale, Laboratoire de Chimie Analytique Appliquée à l'Environnement Naturel (EP CNRS 1750) Station Marine, BP 80, F-62930 Wimereux, France
*(*author for correspondence)*

Key words: atmospheric aerosols, lead, time serie, solubility, seawater, eastern channel

Abstract

In order to describe the seasonal variability of the atmospheric lead concentrations above the Eastern Channel, aerosols samples have been collected by bulk filtration at the Cape Gris-Nez (Dover Strait) during one year. A statistical analysis of the obtained time-series from lead concentrations and rain amounts illustrates the geochemical implications of a scavenging process: when precipitations become lower than the average rain events, the lead concentrations become higher than the mean values and reciprocally. On a global scale, the lead concentrations over the Eastern Channel and the Southern Bight of the North Sea have decreased by about one order of magnitude since fifteen years. Lead concentrations have been compared to the origin of the collected air masses. The use of the back-trajectories permits us to classify the origins of the air masses among five geographical sectors relevant to the position of our sampling site: Atlantic Ocean and English Channel; North Sea; British Islands; Southern and Western Europe; Scandinavia and Central Europe. Continental sectors show greater lead concentrations than marine and semi-continental sectors, by a factor of about 3 to 7. These results are in a good agreement with the assumption according to which the atmospheric lead is mainly associated with emissions from heavy industrialised and urbanised areas, especially from automotive origin. Under our experimental conditions, the solubility in a coastal seawater for 'continental' lead is different from the one of 'maritime' lead. These dissolution experiments are developed in order to ultimately answer to the questions 'how does the dry fall-out scatter into the upper layer of the seawater mass, regarding as a chemical system?' and more particularly 'what is the part of the heavy metals transported via the atmosphere which is rapidly transferred into the dissolved phase, i.e. under a labile form potentially available for the interfacial organisms?'

Introduction

Despite uncertainties on the measurement of dry and wet deposition fluxes, it is now well recognized that the atmosphere is an important pathway for pollution transfer to the sea (Duce et al., 1991). The extent to which the atmospheric input influences both biological processes and chemical cycling is more particularly related to our knowledge about the processes controlling the removal of atmospheric particulate matter and gases specially, under high latitudes, by the wet deposition mode (Bergametti et al., 1989; Lim et al., 1994). About that, current uncertainties in the evaluation of dry and wet deposition fluxes, over the North Sea and adjacent seas, represent a good illustration for the need in a monitoring approach, on a global scale, including emission data, accurate determination of the deposition rate of airborne particulate matter and studies on the physical mechanisms affecting the deposition onto the sea surface (Injuk & Van Grieken, 1995). But in the case, for example, of heavy-metals such as lead, a global measurement strategy for the evaluation of the real impact of atmospheric particulate matter on marine biogeochemical cycles implies an assessment of both kinetic and thermodynamic effects during the transfer of the particle across the air-sea interface.

Only few studies approach the description of dissolution, precipitation and/or adsorption processes affecting the behaviour of trace metals at the air–sea interface. Kersten et al. (1991) studied the solubility of atmospherically derived trace metals into seawater originating from Helgoland Bay and underlined the opposition between the solubility of cadmium (95% of metal released in seawater) and lead (16%) in presence of artificial suspended matter. Our work focused on this last element essentially for two reasons: first, the transport of lead to the sea, via the atmosphere, represents the major pathway for this metal to the ocean reservoir (Chester, 1990), specially for the North Sea (North Sea Task Force, 1993). On the other hand, lead can be considered as a tracer for the presence of pollution aerosols, regarding to its specific anthropogenic origin (Buat-Menard, 1993). For Giusti et al. (1993), release kinetics in seawater show a bimodal pattern with a fraction of lead (and other metals) solubilized in a few minutes and another fraction released over a longer leaching period. For Hodge et al. (1978), lead collected in an urban area exhibits a higher solubility (39%) than lead from rural samples (13%), in a coastal zone. On a general manner, the solubility of lead associated to urban-rich aerosols is higher than the one for lead related to crust-rich aerosols (Chester et al., 1993).

So as to establish a possible relation between the origin of lead associated with atmospheric aerosols and its solubility in a coastal seawater, atmospheric particulate matter has been collected during one year on the coastline of the Dover Strait (Eastern Channel). This work is a part of a study performed, on one hand in order to determine the elemental concentrations of heavy metals in aerosols by the means of a continuous sampling program based upon daily samples collection, on the other hand in order to tackle the behaviour of these dry fall-out in surface waters. Solubility measurements have been carried out by Differential Pulse Anodic Stripping Voltammetry (DPASV); the origin of atmospheric lead is estimated from air-masses back trajectories.

Materials and methods

Sampling method

Atmospheric aerosols have been sampled at Cape Gris-Nez (Eastern Channel: 50°52′ N; 1°36′ E; see Figure 1) from August '95 to August '96. Atmospheric particulate matter has been collected by air filtration onto cellulosic filters (Whatman® , grade 41; 7 cm diameter). Continuous sampling was performed by the means of an automatic filter-turn off, once a day.

The cellulosic filter, in a PVC filter-holder, is run at the top of a 12 meter-high mast, completed by a PVC head, in order to avoid any metallic contamination from the mast. More details on the sampling procedure are given in Flament et al. (1996). The mean filtered volume is 108 ± 24 m^3 for 24 h (172 ± 29 m^3 during the weekend, for a 36 h collection period). Two systems are run in parallel in order to achieve an analytical blank in the same conditions of sampling.

Determination of lead and aluminium in aerosols

In order to calculate both the abundance of lead in aerosols and the enrichment factors EF_{crust} (Rahn, 1976) derived from the composition of the terrestrial crust (Taylor, 1964), Pb and Al are measured by GFAAS. Mineralisation of atmospheric particulate matter is performed by acid digestion ($HNO_3 + HClO_4 + HF$, Ultrapure Grade Prolabo®) at 120 °C, for 2 h. After evaporation, nitric acid is added (50 μl) and the residue is solubilized in ultra-pure water ($\rho = 18$ MΩ cm) and then transferred into a 25 ml polyethylene flask. More details are also given in Flament et al. (1996). We used a Perkin – Elmer Model 1100 Spectrophotometer equipped with an HGA 500 graphite furnace monitor and an A.S. 40 auto-sampler. Results are expressed in nanograms per cubic meter air.

The accuracy of this operating procedure is evaluated by the acid digestion and the analysis of a standard reference material from the National Institute of Standards and Technology (Urban Particulate Matter ref. SRM 1648). The obtained recovery is respectively 95% for lead and 104% for aluminium.

dvips hydrja11

Dissolution experiments in seawater

We used for this study a coastal seawater provided by the National Research Council of Canada (CASS-3) with a certified value for the Pb concentration

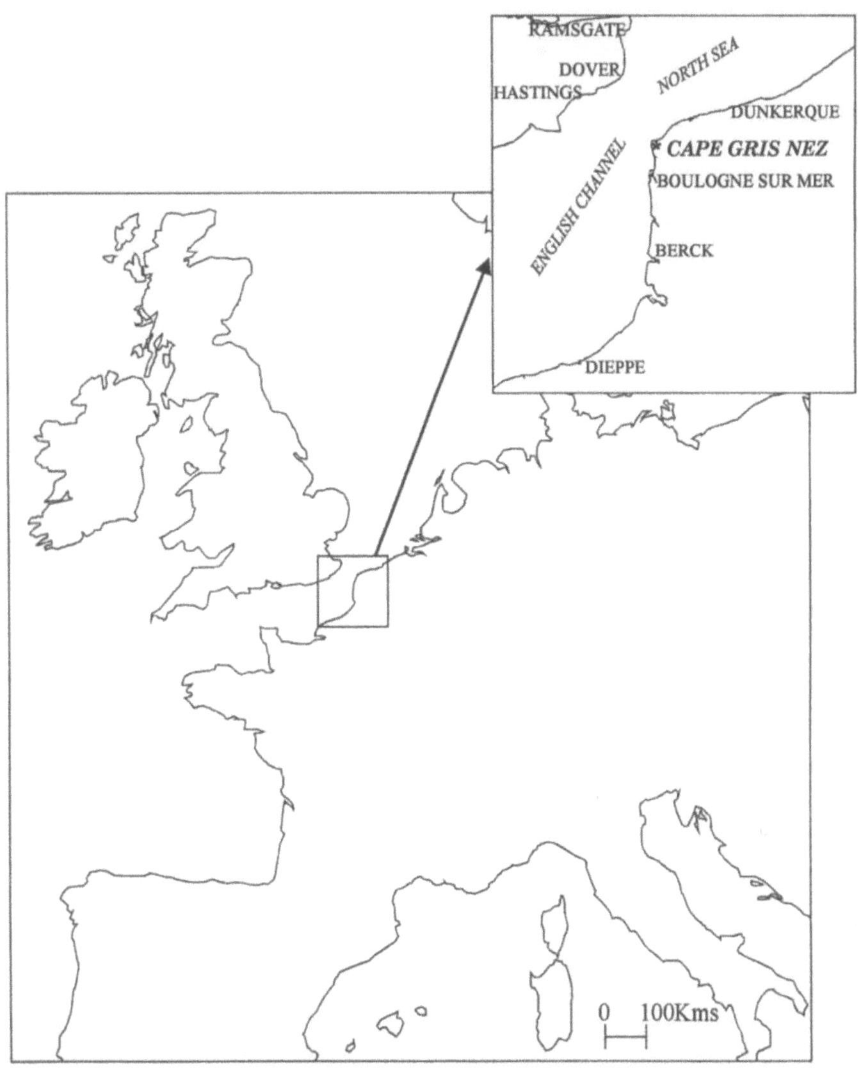

Figure 1. Cape Gris-Nez sampling site.

(12 ± 4 ng l^{-1}), very lower than the concentrations measured during dissolution experiments (10^3 and 5.10^4 ng l^{-1}) between. The pH of this seawater, initially at 1.6, was adjusted at 8 with ammonia solutions (Suprapur Grade, Merck®). Lead measurements were performed by DPASV, into a 20 ml Teflon (PTFE) double compartment cell maintained, between experiments, into an aqueous solution (1 M) of nitric acid (Suprapur Grade, Merck®). All the manipulations were performed under a laminar-flow hood (Class 100, Federal Standard 209a) using polyethylene gloves. We utilised a POL 150 Voltammeter equipped with a voltammetric stand MDE 150, a rotating glassy carbon electrode EDI 101T (Radiometer Analytical®) and an Ag/AgCl, KCl (3M) reference electrode to which all reported potentials are referred. Operating conditions are optimised for lead measurements. Before each analysis, solutions were deoxygenated for 5 min by nitrogen bubbling and the electrode was rotated (3000 rpm) and maintained at − 150 mV for 30 s. Metal deposition was performed by electrolysis (at − 900 mV for 5 min) of seawater, in which $Hg(NO_3)_2$ was added (10^{-5} M) while the electrode was rotated at 3000 rpm. After 30 s at rest, the measurement was started by using a differential pulsed voltammetric scan carried out in the positive direction (scan time 0.2 s, scan amplitude

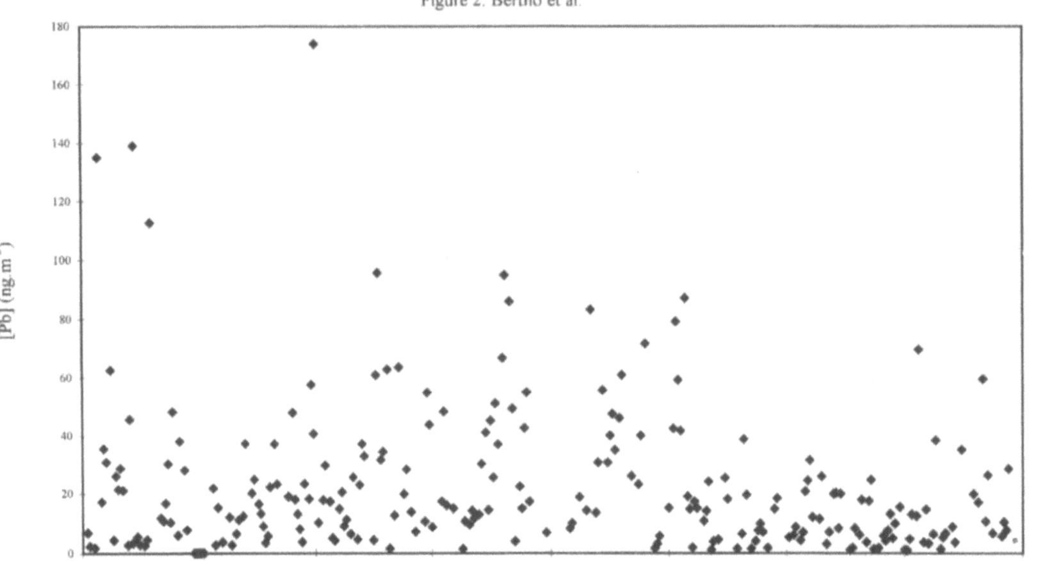

Figure 2. Concentrations of atmospheric lead ([Pb]) during the one year survey, from August '95 to August '96, at Cape Gris-Nez (Sampling dates in Julian days).

2 mV, pulse time 40 ms, pulse amplitude 50 mV). The final potential was set at – 150 mV. Lead concentrations were determined from standardisation straight line which was built at pH $\approx$ 2 with CASS-3 seawater. At this pH value, we can consider that the totality of soluble lead is electroactive (Capodaglio et al., 1995), since this seawater has a low organic matter content (the total dissolved organic carbon concentration is about 0.5 mg l^{-1}).

Electroactive lead concentrations were thus followed up from the contact ($t=0$) of the filter which support aerosols with seawater to 48 h, the cell being maintained under N_2 atmosphere during this survey.

Low modifications of the ASV signal intensity were observed after polishing the electrode surface. So, it was regularly polished at a low rotation rate (500 rpm) and the error on the lead concentration was estimated at ± 10%.

Results

Lead concentrations and Enrichment Factors

Figure 2 presents the concentrations of particulate lead (in nanograms per m^3 air) obtained in the lower atmosphere at Cape Gris-Nez. Due to the meteorological conditions (rain events), 257 samples out of 339 have been analysed. The extent to which the earth crust and anthropogenic emissions contribute to the input of particulate trace metals to the atmosphere can be assessed using the enrichment factor EF_{crust} (Rahn, 1976) calculated as follow:

$$EF_{crust} = (Pb/Al)_{air}/(Pb/Al)_{crust}$$

in which $(Pb/Al)_{air}$ is the lead concentration over the aluminium concentration in the aerosol and $(Pb/Al)_{crust}$ is the same ratio but in the average crustal rocks (Taylor, 1964). Obtained EF_{crust} values are listed in the Figure 3.

For the lead concentrations and the enrichment factors, the geometric means (respectively 10.9 ng m^{-3} and 1401) are lower than the arithmetic means (22.1 ng m^{-3} and 4046) but are close to the medians (14.4 ng m^{-3} and 1367); the distributions may approximate a log-normal one. For the lead concentrations ([Pb]; standard deviation = 25.0 ng m^{-3}) a Kolmogorov-Smirnov-Lilliefors test, carried out in order to determine the log-normality of the distribution (Lilliefors, 1967), shows that this one is not log-normal. Nevertheless, the geometric mean is privileged with respect to the arithmetic mean, because it gives less importance to episodically high values, which are not representative (Flament et al., 1996). The geometric mean of the lead concentrations, used by many

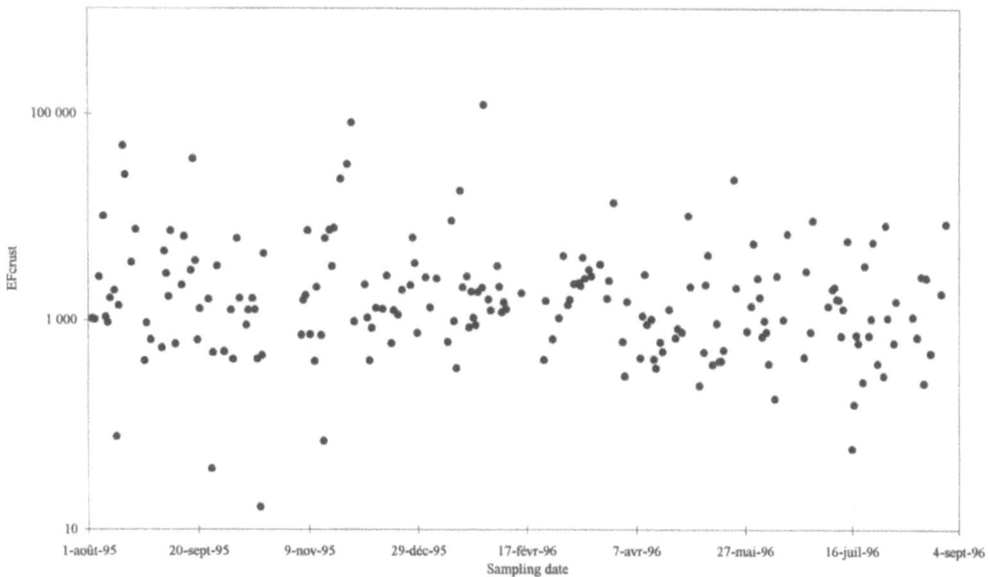

Figure 3. Enrichment factors EF_{crust} during the one year survey, from August '95 to August '96, at Cape Gris-Nez (Sampling dates in Julian days).

Table 1. Concentrations of atmospheric lead ([Pb]) above the English Channel and the North Sea from 1980 to 1996 (this work). See text for the definition of the local wind sectors. (*) In brackets are listed, for each local wind sectors (our results), the geometric mean of the daily rainfall (in mm).

Authors	Site	Local wind sectors (see text)	Sampling period	[Pb] (ng m^{-3})
Baeyens and Dedeurwaerder (1991)	Southern bight of the North Sea	1	1980/85	12.6
		2	ibid	152.9
		3	ibid	30.9
		4	ibid	99.5
Flament (1985)	Eastern Channel	1	1982/84	42
		2	ibid	179
		3	ibid	41
		4	ibid	113
Otten et al. (1989)	Southern North Sea	All sectors	1984/88	62
Chester and Bradshaw (1991)	Southern bight of the North Sea	All sectors	1988/89	20
Van Jaarsveld (1993)	Southern bight of the North Sea	All sectors	1989	22
Dannecker et al. (1994)	North Sea	All sectors	1991	4.5
Flament et al. (1996)	Wimereux	All sectors	1994	8
This work (*)	Cape Gris Nez	1 (0.2)	1995/96	6.4
		2 (0.2)	ibid	15.0
		3 (0.2)	ibid	7.7
		3 (0.2)	ibid	6.8

authors, is compared, in Table 1, to those reported in other works around the Channel and the North Sea.

To investigate whether a possible relationship between the element concentrations in aerosols and the wind directions exists, each sample has been classified

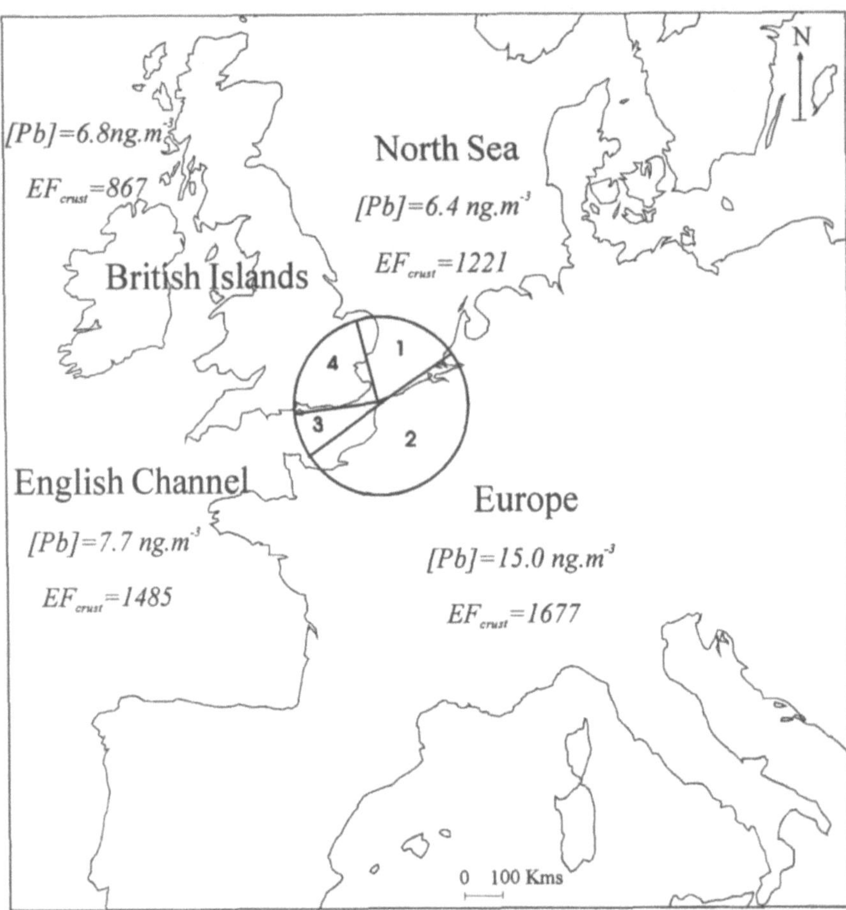

Figure 4. Definition of the local wind sectors, average lead concentrations and enrichment factors in each sector (geometric means; this work).

according to the local wind sectors, determined geographically as follow (see Figure 4): Sector 1 (345°–55°); Sector 2 (55°–235°); Sector 3 (235°–260°) and Sector 4 (260°–345°).

The geometric mean for [Pb] and EF_{Crust}, corresponding to 'continental' (Europe; Sector 2), 'semi-continental' (British Islands: Sector 4) and 'marine' (North Sea and Eastern Channel; Sectors 1 and 3) samples were calculated (see Figure 4). The obtained values for [Pb] are compared in Table 1 to the results from other workers and more particularly from authors who have chosen this classification according to the local winds.

In order to better understand the abundance of lead in aerosols at Cape Gris Nez, a classification of the collected air-masses, with respect to the emission areas, has been performed by using the air-masses history. This is the back trajectories during the previous 48 h before sampling, calculated by the Peridot model (Meteo-France). By this way, we assumed that the aerosols displacements are associated to those of the air-masses, on a large scale. No indication is available on the rain events during the displacements of the air-mass. But, additional 3D trajectories might allow us to obtain more indications on the origin of the collected airborne particulate material. The back trajectories furnished us extra informations; indeed, two samples collected under similar local meteorological conditions (same wind direction) may have different air-masses histories. Two calculated back trajectories are ended at the site each day (0h00 and 12h00 TU). Consequently, each 24 hours-sample was associated with 3 retrospective air-masses tracings (4 during the week end). Five sectors have been distinguished: 'Atlantic-English Channel', ' North Sea', 'British Islands', 'South-western Europe' and 'Central Europe and Scandinavia'

Table 2. Concentrations of atmospheric lead ([Pb]) and Enrichment Factors EF_{Crust} versus the origin of the air-masses. Listed values (our results) for [Pb], EF_{Crust} and *h* are the geometric means.

This work Sampling period: 1995–96 Sector	Number of sampling	EF_{Crust}	[Pb] ($ng\ m^{-3}$)	h ($mm\ day^{-1}$)	(a): Chester and Bradshaw (1991) (b): Ottley and Harrison (1993) Sampling period: 1988–89 Sector	[Pb] ($ng\ m^{-3}$)
Atlantic–English Channel	24	1631	7.1	0.4		
North Sea	17	700	3.7	0.2	North Sea	(a): 3.6–6.8 (b): 3.5
British Islands	45	1036	6.7	0.2	British Islands	(a): 22–52 (b): 32.5
Southwestern Europe	16	1782	22.7	0.1	Europe	(a): 38–52–189 (b): 63.0
Central Europe and Sandinavia	9	1661	25.0	0.1		

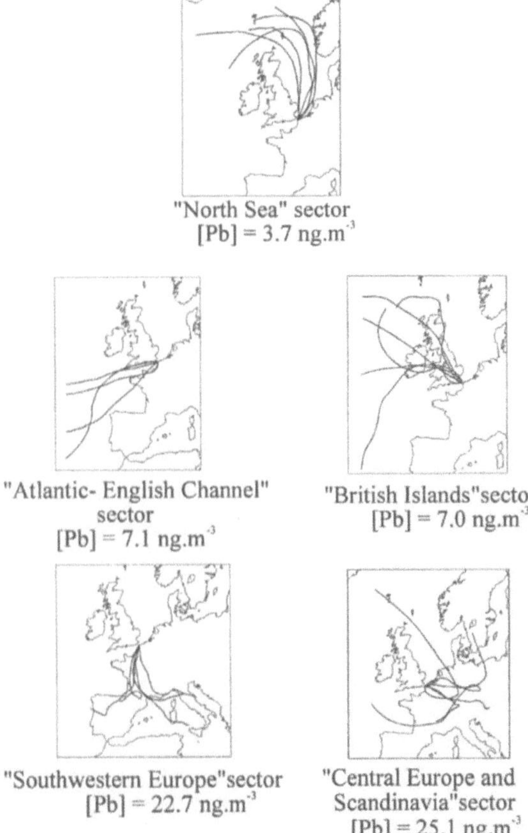

Figure 5. Definition of the geographical sectors based upon the air-masses back trajectories, with some examples of calculations; concentrations of atmospheric lead [Pb] in each sector (geometric mean).

(see Figure 5, including some examples of calculated back-trajectories). Among the 257 analysed samples, 126 have been separated from this classification, because the origin of the air mass cannot be univocally associated with any sector. By comparison with the classification according to the local winds, for the 131 remaining samples, only 58% have been classified in the same sector, 33% have been classified in an adjoining sector and 9% in a distant sector. That shows the insufficiency of the previous classification. The geometric means of the lead concentrations and the enrichment factors have been calculated in each sector. These results are listed in Table 2.

Dissolution of atmospheric aerosols in seawater

The kinetic of the dissolution of atmospheric lead in seawater has been investigated by the study of five filters on which were collected aerosols from different origins. Characteristics of these samples are presented in Table 3.

A maximal concentration of dissolved lead in seawater $[Pb]_{MD}$ is deduced from the total mass collected on the filter. The percentage of dissolved lead from aerosols can be calculated as the concentration of electroactive lead in the experimental cell (blank not subtracted) over $[Pb]_{MD}$ (see Figure 6 and Figure 7). These results have to be compared with the concentration of dissolved lead arising from each 'blank' filter (contamination lead); the average mass of Pb on a blank filters is 335 ± 160 ng and 35.5% of this contamination lead is rapidly dissolved in seawater at the natural pH. Thus, the lead concentrations obtained from the dissolution of atmospheric aerosols originated from the North Sea and the Atlantic–English Channel sectors are in the same order of magnitude that their 'blank' concentra-

Table 3. Concentrations of atmospheric lead [Pb] and maximal concentration of dissolved lead in seawater $[Pb]_{MD}$ for the five samples processed in the dissolution experiments.

Sampling date	Origin of the air mass	EF $_{crust}$	[Pb] in ng m^{-3} air	$[Pb]_{MD}$ in $\mu g\ l^{-1}$
4–6/08/95	North Sea	1020	22	13.51
13–15/10/95	South-Western Europe	1250	24.8	47.54
12–13/03/96	Central Europe	2290	30.7	38.05
2–3/06/96	Atlantic–English Channel	2550	4.3	5.1
9–10/07/96	British Islands	1616	7.7	18.9

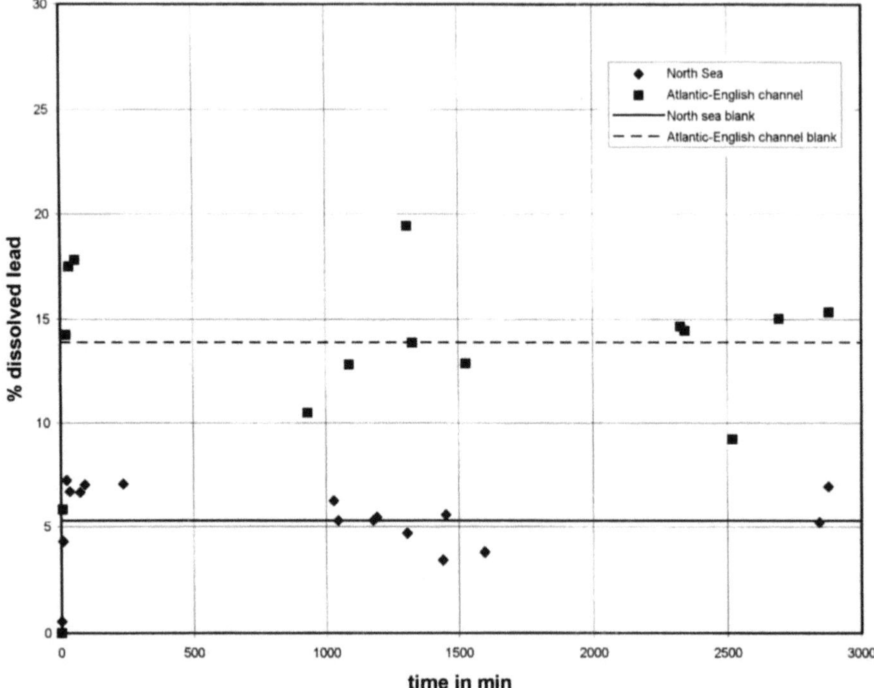

Figure 6. Evolution of the percentage of dissolved lead with time during the dissolution experiments: ◆ aerosols from the North Sea sector, — North Sea blank; ■ aerosols from the Atlantic/English Channel sector, - - - Atlantic/English Channel blank.

tions (see Figure 6). On the other hand, during the dissolution experiment for aerosols from continental and semi-continental sectors, contamination lead can be neglected (see Figure 7).

Davison et al. (1987) have shown that the ASV peak current is depressed by increasing pH conditions due to sorptive losses. Sorption of heavy metals onto the cell-walls and onto the filter can significantly affect measurements in solution and has to be taken into account. About that, the evolution of electroactive lead in a spiked seawater at 5 $\mu g\ l^{-1}$ was observed for 5 h with and without the presence of a virgin filter (see Figure 8). We can note that after 5 h, at natural pH, about 50% of the initially dissolved lead are lost by sorption onto the walls of the experimental cell and more than 60% are lost by sorption onto the walls and the filter.

Discussion

General trend

Table 1 shows the evolution of the atmospheric lead concentrations since 1980 above the Eastern Channel and the Southern Bight of the North Sea. Concentra-

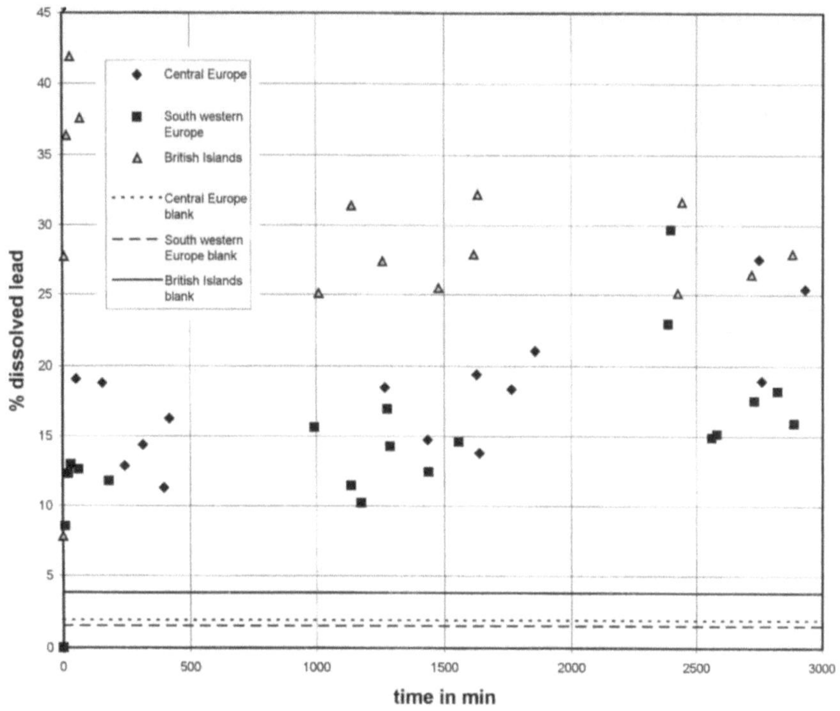

Figure 7. Evolution of the percentage of dissolved lead with time during the dissolution experiments: ■ aerosols from the South-western Europe sector, - - - South-western Europe blank; ♦ aerosols from the Central Europe sector, ... Central Europe blank; ▲ aerosols from the British Islands sector, — British Islands blank.

tions have decreased from about 100 ng m^{-3} in 1980 to about 10 ng m^{-3} in 1995.

These values must be considered as an order of magnitude, because sampling location and duration are different according to the authors. Furthermore, on the 82 not analysed filters, 61% have been ruined because of leaching by rainwater and 39% destroyed during the sampling operations. This leads to an underestimation of the scavenging effects. Major changes in the long term evolution of the rainfalls on the Eastern Channel and the Southern Bight of the North Sea could as well affect the expected trend in the airborne concentrations of lead and must be considered. About that, data from the French meteorological authority (Meteo France) over the past fifteen years show that the rainfalls at Dunkerque and Boulogne sur Mer (Figure 1) are higher in 1982–84 (about 690 mm yr^{-1}) than in 1995–96 (540 mm yr^{-1}). The conjunction of the two phenomena (loss of many rainy-day samples in 1995–96 and decrease of the precipitations between 1982–84 and today) should lead, with the assumption that the scavenging ratios (Church et al., 1992) remain unchanging, to an underestimation of the real evolution of the extent to which the regional transient sources contribute to the airborne lead accumulation in the troposphere, between the mid-eighties and today. Concerning more particularly the evaluation of major modifications in the lead content of aerosols over the Eastern Channel by integration (see Table 1) of the data from Flament (1985), meaningful changes in the local wind speeds and frequencies could affect the comparison. Nevertheless, comparing a 1995–96 windrose (Figure 9a) and a ten years (1986–95) windrose (Figure 9b) at Boulogne sur Mer (15 km from our sampling site in the south – Figure 1), it is possible to affirm that our meteorological conditions are representative of the mainly observed ones from the past ten years. Despite the lack of a long duration study which could bring us an integrated view of the problem, the diminution of the atmospheric lead concentrations over the last decade, already shown by other workers in some continental (Bouchereau, 1992) and coastal (Grimaldi et al., 1993; Migon et al., 1994) areas, seems confirmed. According to the previous authors and an earlier work on the Eastern Channel (Flament et al., 1996), this evolution is explained by the reduction of the lead

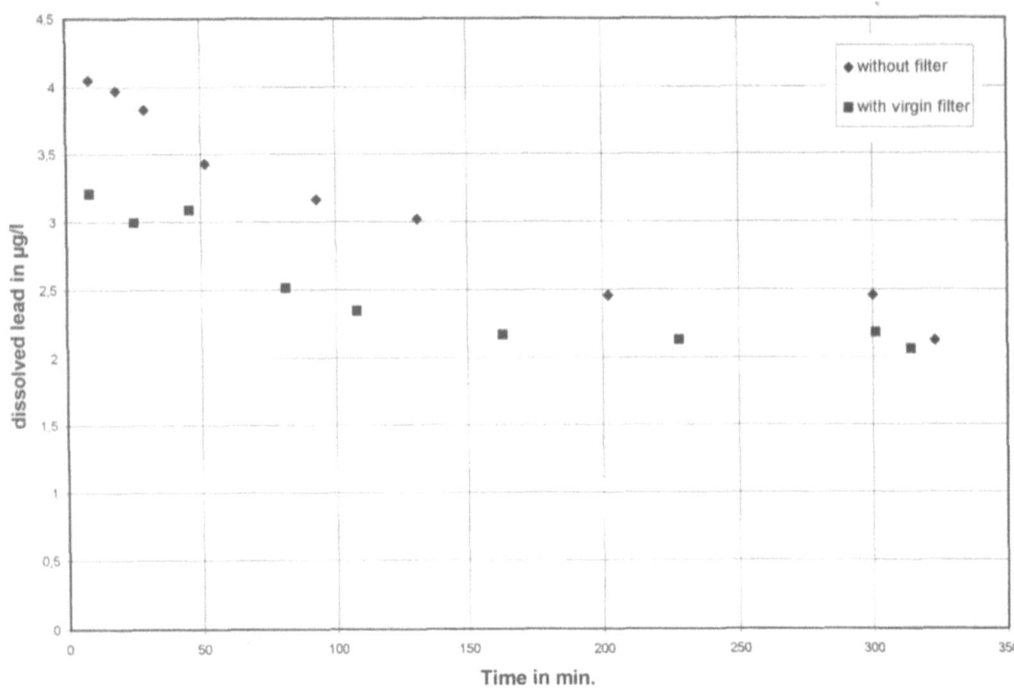

Figure 8. Sorption of lead from a spiked seawater by the walls of the electrochemical cell (◆) and evolution of the lead concentrations in the presence of a virgin filter (■) during the dissolution experiments.

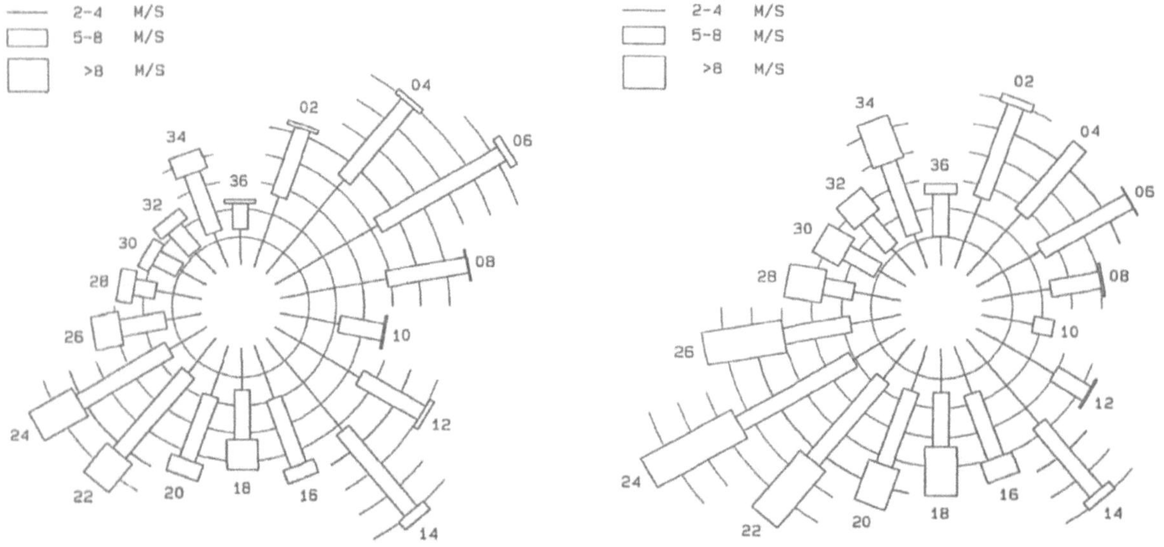

Figure 9. Annual and ten years windroses at Boulogne sur Mer. The distance between two consecutive circles represents 1% of the observed cases (eight records a day) from August '95 to July '96 (on the left) and from January '86 to December '95 (on the right). Width of the segments are related to the wind – speed (see figure).

automotive emissions, due to the new policies (1987 in France) which promote the use of unleaded gasoline. With regard to the enrichment factors EF_{Crust}, the evolution is not so evident. Even if the lead concentrations have decreased, EF_{Crust} values remain important (close by 1400). The origin of Pb in aerosols appears

Table 4. Correlation coefficients (Spearman's coefficients) calculated from the relationship degree existing between, on one hand the rank of one experimental value into the ordinal distribution of all the experimental values of the same form (i.e. [Pb], EF_{crust} or H) and, on the other hand, the sampling date. If the coefficient is meaningful with respect to a critical value which depends of the size of the data base (here approximately – 0.100), the probability (at least 99% in our statistical test) of an existent trend is accepted; if not, this probability is rejected. Negative values for the Spearman's coefficients show here a negative trend.

Variable	Spearman's coefficient	Probability of an existen trend > 99%
[Pb] (ng m^{-3}	- 0,191	Accepted
EF_{crust}	- 0.084	Rejected
H (mm)	- 0,398	Accepted

to be primarily anthropogenic and atmospheric lead can be always considered as an abnormally enriched element, in comparison with the main natural sources of atmospheric particulate matter. Our work is now focused (Veron et al., in press) on the use of the stable lead isotopes in order to trace the anthropogenic emission sources of this element in northwestern France.

Temporal variability of the lead concentrations and enrichment factors

Except the origin of the sampled air masses, rain events are among the environmental factors which strongly influence the trace-elements concentrations in the atmosphere (Migon, 1988; Jaffrezo & Colin, 1988; Bergametti et al., 1989). By nature, rain events present a seasonal variability. The rain event frequencies and their magnitude during the transport of the sampled air masses are not indicated by the Peridot model used here. Nevertheless, the daily rain amounts H (in mm) at our sampling site are known. It is then possible, at least, to undertake the study of the influence of the local scavenging on the lead concentrations measured at Cape Gris-Nez. In order to account for the effect of this factor on our results, the temporal variability of the lead concentrations and enrichment factors have been analysed by considering our data as a time-series (see Figures 2 and 3). The general trend of the series has been assessed by making a calculation of linear regression on the rank of each value in the corresponding population ([Pb], EF_{crust} and rain amount H) versus time (Spearman's correlation; Siegel & Castellan, 1988). The obtained coefficient is not significant for EF_{crust} and indicative of a slight diminution of [Pb] and rain amount (H) versus time (see Table 4). Our survey consists to a continuous sampling based upon daily collections. By this way, the 'day n' filter may collect a part of a 'batch' of lead associated to an air mass, immediately before the turn-off, while the 'day $n+1$' filter will collect the end of the batch. By the way of a 'memory' effect, an autocorrelation of the series may appear and edge off the temporal analysis of the obtained values. This autocorrelation is assessed for the lead concentrations [Pb] after extraction of the general trend, performed by the variate-difference method (Kendall, 1976). The application of the Spearman's correlation test to the new calculated series (noted as Δ[Pb]) confirms the absence of trend after extraction (Spearman's coefficient = – 0.084; probability of an existent trend > 99% rejected). The results of the calculation of the autocorrelation $r_{yy}(k)$ (Legendre & Legendre, 1984) for Δ[Pb] shows no autocorrelation ($r_{yy}(k) = -0.031$ after two days). For the variable $EF_{crust}(r_{yy}(k) = 0.008$ after two days) no memory effect is observable too. The first step of the time-series analysis on the variable [Pb] permits us to extract the trend of the series in order to show the absence of autocorrelation on this variable. We are, so, authorised to examine the influence of the rain events on the temporal variability of the lead concentrations. That is processed by the means of the cumulative-sums method (Ibanez et al., 1993). The obtained series by this method are represented on Figures 10 and 11. The definition of six time-periods (noted as 1 to 6 on the two graphs) in which, for example, the rain amount is lower than the average amount (negative local trend) while the concentration of lead becomes higher than the average one (positive local trend) and reciprocally, is possible. Such evolution represents a good illustration of the elimination of particulate matter from the atmosphere by the rain, called as the 'wash-out' process (Jaffrezo & Colin, 1988) and of the accumulation of particles in the air during the dry periods.

Local winds influence

The geometric means for the lead concentrations in each sector defining from the local wind direction are shown in Figure 4. Continental winds (sector 2) are associated with higher concentrations than marine (sectors 1 and 3) or semi-continental winds (sector 4). The geometric means of the precipitations being identical for each sector (Table 1), this abundance distribution can't be explained by some differences in the

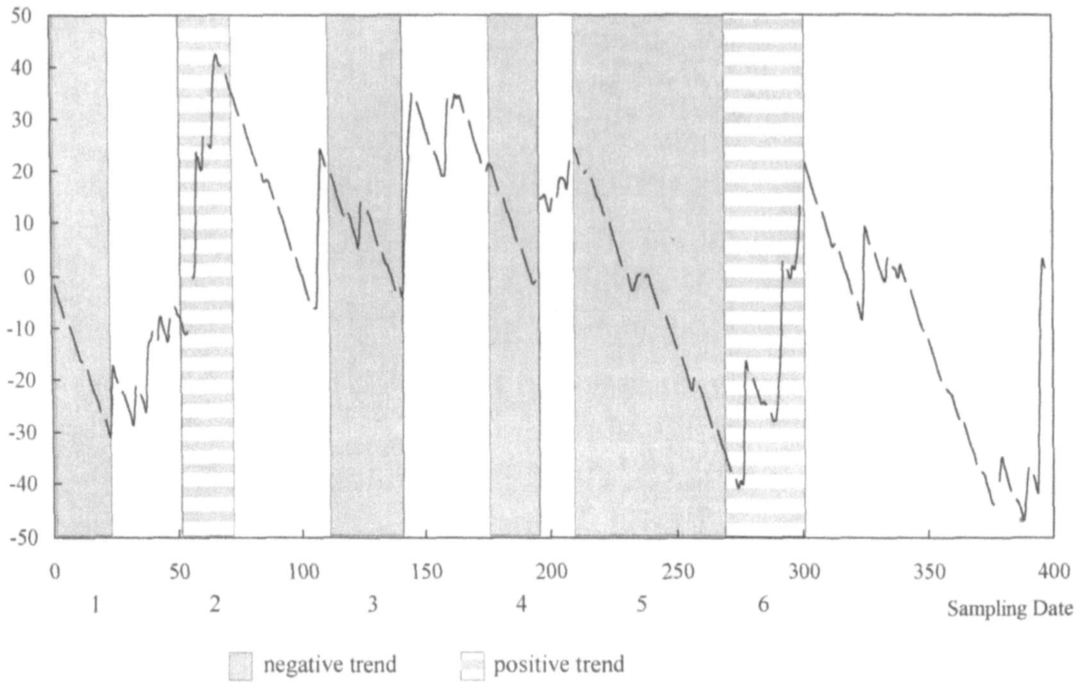

Figure 10. Cumulative – sums diagram for the daily rain amount H (mm).

Table 5. Matrix of the probabilities of heterogeneity ('*U*' Mann-Withney test; local wind sectors classification): if the probability is less than 0.010 (0.009 for example, in the comparison between the sectors 2 and 3), the averages [Pb] for the two sectors are statistically different. Sectors numbering as noted in the text.

Sector	1	2	3	4
1	1.000			
2	0.001	1.000		
3	0.760	0.009	1.000	
4	0.971	0.000	0.726	1.000

magnitude of the local wash-out related to the previous sectors. The abundance of lead in aerosols seems to be related to the emissions of the urbanised and industrialised areas from northwestern Europe. This observation is consequent with the anthropogenic character of lead. However, the classification according to the local wind mean direction is not sufficient in order to affirm the existence of a gradient between the distribution of lead in continental and marine air masses. Indeed, the mean concentration ([Pb] = 6.8 ng m^{-3}) for the sector 'British Islands' (semi-continental area), is not intermediate between the English Channel's (marine sector: [Pb] = 7.7 ng m^{-3}) and continental Europe's ([Pb] = 15.0 ng m^{-3}) values. Then, the crustal enrichment factors for the marine sectors (Figure 4) are stronger than the ones for the sector 'British Islands'. In order to establish, nevertheless, that local winds directions allow, at least, the discrimination of the large source-areas surrounding our sampling station, statistical tests of heterogeneity were applied on each sectorial values sub-population. The Kruskal-Wallis test (Scherrer, 1984), applied to our data with a 5% significance limit, shows that the four sub-populations form a whole heterogeneous group. From this result, the sectors have been compared two by two with the '*U*' Mann-Whitney test (Mann & Whitney, 1947), in order to extract the sectors which are responsible of the heterogeneity (see Table 5). In this matrix, two populations are considered to be different when the calculated probability is less than 1%. Finally, the sectors corresponding to the North Sea, the Atlantic Ocean-Channel

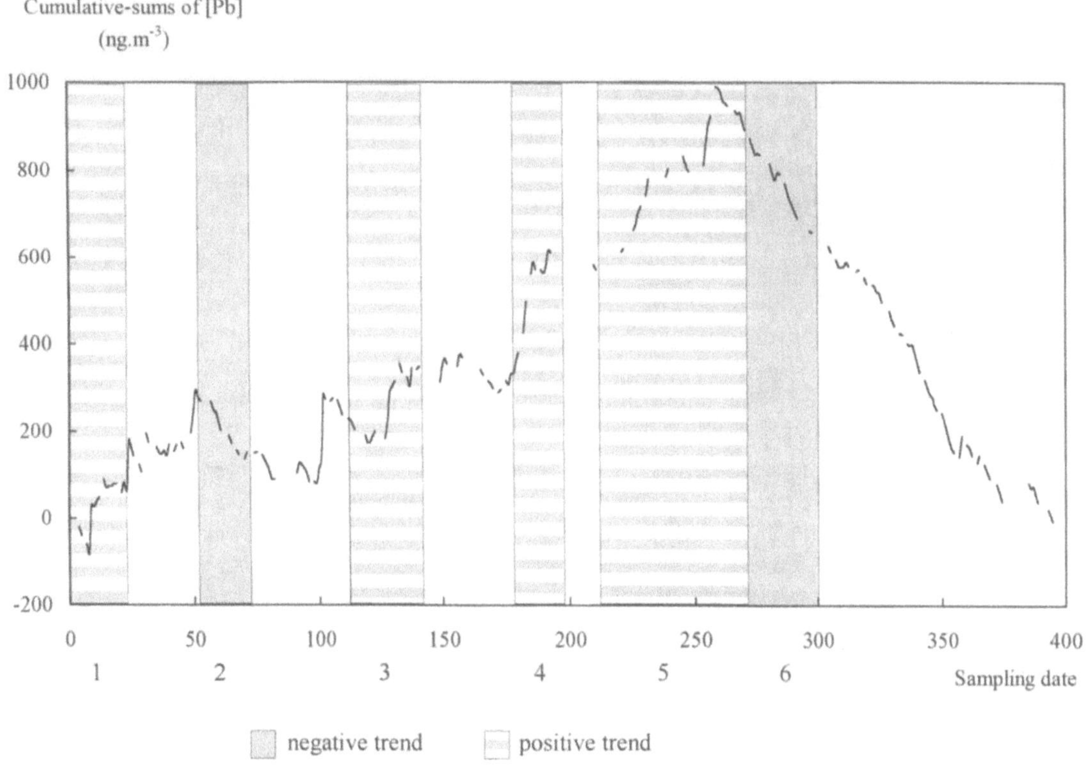

Figure 11. Cumulative – sums diagram for the concentration of lead [Pb] (ng m^{-3}).

zone and the British Islands are statistically different from the European sector.

As shown in Table 1, Flament (1985) and Baeyens & Dedeurwaerder (1991) have used the same classification according to the wind directions. Their sampling locations were respectively Wimereux (Eastern Channel) and the West Hinder lightship (Southern Bight of the North Sea), close to our sampling site. In that way, the different sectors can be directly compared and represented on the same location on Figure 4, according to the used scale. First, these results are consequent with the assumed anthropogenic origin of lead since sectors 2, which correspond to the localisation of the greatest part of urban and industrial areas, have the most important average concentrations. Secondly, considering that the precipitations on the French coast of the Strait of Dover are superior in 1982–84 than in 1995–96 (see section 4.1), the diminution of the atmospheric lead concentrations is confirmed since the mid-eighties in all sectors. Specially, for the European sector, the average lead concentration decreases from about 180 ng m^{-3} in 1982 to 15 ng m^{-3} in 1995; for the British sector, it decreases from 113 ng m^{-3} to 6.8 ng m^{-3} in 1995. As established elsewhere in some coastal areas, this diminution is probably linked to the use of unleaded gasoline (Migon et al., 1993; Flament et al., 1996). For the North Sea sector (sector 1) and the English Channel one (sector 3), concentrations decrease from about 40 ng m^{-3} in 1982 to respectively 6.4 ng m^{-3} and 7.7 ng m^{-3} in 1995. These less important evolutions could be explained by the fact that these sectors fit with remote continental sources.

Lead concentrations versus air-masses back trajectories

The geometric means for the lead concentrations obtained according to the classification with back trajectories (see Figure 5 and Table 2) confirm that the atmospheric concentrations depend on the origin of the air-mass and are higher when this one is a continental one. It is the case for the comparison between the mean values from the South-western and Central Europe zones (respectively 22.7 and 25.0 ng m^{-3}, with

Table 6. Matrix of the probabilities of heterogeneity ('*U*' Mann-Withney test; air-masses back trajectories classification): if the probability is less than 0.003, the averages [Pb] for the two compared sectors are statistically different.

Sector	A	B	C	D	E
Atlantic–English Channel (A)	1.000				
North Sea (B)	0.131	1.000			
British Islands (C)	0.628	0.140	1.000		
South-Western Europe (D)	0.000	0.002	0.000	1.000	
Central Europe and Scandinavia (E)	0.000	0.001	0.000	0.129	1.000

high crustal enrichment factors) and those originating from the North Sea (3.7 ng m^{-3}, with lower crustal enrichment factors). The mean of lead concentrations and crustal enrichment factors obtained for the British Islands sector are relatively low, compared to the continental sectors. The insular nature of this zone could be responsible of this fact, the transit time of the air-mass over the continent being short and then, the accumulation of anthropogenic aerosols should be limited. These results could also be explained by an important scavenging of the air-mass during its transport over this sector, that could be checked by the analysis of the precipitation events along the back-trajectories. Concerning the 'Atlantic–English Channel' sector one, the relatively high crustal enrichment factors and lead concentrations can be explained by the epicontinental character of this area. Indeed, considering the narrowness of this sector, the limited precision of the calculation of the back-trajectories and the dispersion of the aerosols in the atmosphere, the atmospheric particular matter can be contaminated by the aerosols from the bordering European sources and this sector can be assimilated as a semi-continental emission area.

The statistical tests used in the study on the discrimination of [Pb] according to the local wind directions have been achieved on this classification. The Kruskal-Wallis test, applied to these data with a 5% significance limit, shows an heterogeneous group for the five sub-populations. The '*U*' Mann Whitney test (see Table 6) converges to the same conclusions as in the previous section, but with a more exacting significance limit (0.3% against 1%); the sectors corresponding to the North Sea, the Atlantic & English Channel zone and to the British Islands are statistically different from the two European sectors. In conclusion, the air-masses back trajectories allow us to obtain a more discriminating classification than the precedent one (based upon wind directions).

This classification has already been used by other workers (Chester & Bradshaw, 1991; Ottley & Harrison, 1993; see Table 2). In the North Sea sector, the mean concentration (3.7 ng m^{-3}) is, in 1995, close by those obtained in the previous works (respectively from 3.6 to 6.8 and 3.5 ng m^{-3}) in 1988. Elsewhere, semi-continental (British Islands) and continental (Europe) concentrations are lower in 1995 than in 1988, with respectively a diminution from about 35 to 6.7 ng m^{-3} and from about 85 to 25 ng m^{-3}. This decrease, particularly significant regarding to the continental sources, has been previously explained by the diminution of the automotive component of atmospheric lead.

Dissolution of atmospheric aerosols in seawater

Whatever the origin of the atmospheric particulate matter collected onto the filter is, the maximal concentration of electroactive lead, so the maximal percentage of dissolved lead, is obtained in the first minutes after the contact of the aerosols with the coastal seawater (See Figure 6 and 7). After this laps of time, a low decrease of the lead concentrations is generally observed and then any sharp tendency can be distinguished. Since any variation of the seawater pH during the dissolution experiments was discerned, this evolution could be explained by sorption phenomena onto the walls of the electrochemical cell and onto the filter, which were already observed elsewhere (Giusti et al., 1994; Diaz-Cruz et al., 1992). On the other hand, this low decrease of the lead concentrations during the dissolution experiments can also be explained by sorption onto insoluble atmospheric particulate matter or by complexation with soluble organic matter issued from the aerosols. In spite of these possible artefacts, we can note, as a first result, that the atmospheric lead originated from the North Sea and the Atlantic–English Channel sectors is weakly or not soluble in seawater. At the opposite, atmospheric lead originated from continental or semi-continental sectors seems more soluble, since the maximal percentage of dissolved lead obtained in the first minutes is about 40% for the British Island sector, 20% for the Central Europe sector and 15% for South-western Europe sector.

Conclusion

A one-year survey has been processed in order to describe the global evolution of the atmospheric lead concentrations above the Eastern Channel. This work must be also considered as a first part of a global study on the behaviour into the surface waters of the pollutant metals associated to the atmospheric particulate matter.

From a general point of view, lead concentrations [Pb] have decreased by about one order of magnitude over the last decade. This time evolution is explained by the diminution of the automotive and industrial components of atmospheric lead. Nevertheless, this element appears always abnormally enriched in comparison with the crustal sources of aerosols. The occurrence of local rain events generates a seasonal variability of [Pb] by the elimination of particulate matter from the atmosphere, probably by a wash-out process.

An other source of variability is due to the wind influence. Local wind directions can be used as a factor in order to explain the variations of [Pb], specially in terms of an opposition between continental and purely marine winds influences. But a more discriminating classification, according to the origin of the collected aerosols, is obtained by the analysis of the back trajectories of the sampled air-masses.

Whatever its origin, a fast dissolution episode occurs during the first minutes after introduction of the atmospheric lead in a coastal seawater. After reaching a (quasi)equilibrium, it appears that lead with a continental origin is more soluble than lead originated from a marine area.

Acknowledgements

The authors thank the two reviewers for their helpful comments. We are grateful to the technical team from the 'Station Marine de Wimereux' and expressly to Dominique Menu, Daniel Hilde and Michel Priem, for their assistance during this work. Particular thanks to Lucie Courcot-Deram and Erik Loiseau for their help during the preparation of the manuscript. This work was partly supported by the National Centre for Scientific Research (CNRS) and the Nord - Pas de Calais County (program DYSCOP II). MLB thanks the Pas de Calais District Council for its financial support.

References

Baeyens, W. & H. Dedeurwaerder, 1991. Particulate trace metals above the southern bight of the North Sea. II- Origin and behaviour of the trace metals. Atmos. Envir. 25A: 1077–1092.

Bergametti, G., A. L. Dutot, P. Buat-Ménard, R. Losno & E. Remoudaki, 1989. Seasonal variability of the elemental composition of atmospheric aerosol particles over the northwestern Mediterranean. Tellus, 41B: 353–361.

Bouchereau, J. M., 1992. Estimation des émissions atmosphériques de métaux lourds en France pour le Cr, Cu, Ni, Pb et Zn. C.I.T.E.P.A., Paris: 82 pp.

Buat-Menard, P., 1993. Global Changes in Atmospheric Metal Cycles. In Hewitt, C. N. & W. T. Sturges (eds), Global Atmospheric Chemical Change. Chapmann & Hall, London: 271–312.

Capodaglio, G., G. Scarponi, G. Toscano, C. Barbante & P. Cescon, 1995. Speciation of trace metals in seawater by anodic stripping voltammetry: Critical analytical steps. Fresenius J. Anal. Chem. 351: 386–392.

Chester, R., 1990. Marine Geochemistry, Chapman & Hall, London, 680 pp.

Chester, R. & G. F. Bradshaw, 1991. Source control on the distribution of particulate trace metals in the North Sea atmosphere. Mar. Poll. Bull. 22: 30–36.

Chester, R., K. T. J. Murphy, J. F. Lin, A. S. Berry, G. A. Bradshaw & P. Corcoran, 1993. Factors controlling the solubilities of trace metals from non-remote aerosols deposited to the sea surface by the 'dry' deposition mode, Mar. Chem. 42: 107–126.

Church, T., A. Veron, C. Patterson & D. Settle, 1992. Trace Metal Scavenging from the North Atlantic Troposphere. In Schwartz, S. E. & W. G. N. Slinn (eds), Proceedings of the 5th Int. Conf. on Precipitation, Scavenging and Atmosphere-Surface Exchange, Vol. 1, Hemisphere Publish. Corp., Washington: 433–446.

Dannecker, W., H. Hinzpeter, H.J. Kirzel, H. Luthardt, M. Kriews, K. Naumann, M. Schulz, M. Schwikowski, M. Steiger & U. Terzenbach, 1994. Atmospheric transport of contaminants, their ambient concentration and input into the North Sea. In Sündermann, J. (ed), Circulation and contaminant fluxes in the North Sea, Springer-Verlag, Berlin: 138–189.

Davison, W., S. J. De Mora, R. M. Harrison & S. Wilson, 1987. pH and ionic strength dependence of the ASV response of cadmium, lead and zinc in solutions with simulate natural waters. Sci. Total Envir. 60: 35–44.

Diaz-Cruz, J. M., M. Esteban, M. A. Van den Hoop & H. P. Van Leeuwen, 1992. Stripping voltammetry of metal complexes: interferences from adsorption onto cell components. Anal. Chem. 64: 1769–1776.

Duce R. A., P. S. Liss, J. T. Merill, E. L. Atlas, P. Buat Ménard, B. B. Hicks, J. M. Miller, J. M. Prospero, R. Arimoto, T. M. Church, W. Ellis, J. N. Galloway, L. Hansen, T. D. Jickells, A. H. Knapp, K. H. Reinhardt, B. Schneider, A. Soudine, J. J. Tokos, S. Tsunogai, R. Wollast & M. Zhou, 1991. The atmospheric input of trace species to the world ocean. Global Biogeochem. Cycles 5: 193–259.

Flament, P., 1985. Les métaux traces associés aux aérosols atmosphériques: apport au milieu marin du littoral Nord-Pas de Calais. Thèse de 3ème cycle, Lille, 175 pp.

Flament, P., M.-L. Bertho, K. Deboudt & E. Puskaric, 1996. Changes in the lead content of atmospheric aerosols above the Eastern Channel between 1982–83 and 1994. Sci. Total Envir. 192: 193–206.

Giusti, L., Y.-L. Yang, C. N. Hewitt, J. Hamilton Taylor & W. Davison, 1993. The solubility and partitioning of atmospherically derived trace metals in artificial and natural waters: a review. Atmos. Envir. 27A: 1567–1578.

Giusti, L., J. Hamilton Taylor, W. Davison & C. N. Hewitt, 1994. Artefacts in sorption experiments with trace metals. Sci. Total Envir. 152: 227–238.

Grimaldi F., H. Bascou, A. Viala, E. Muls, N. Esberard & S. Casabianca, 1993. Action du comité Marseille Provence de l'APPA en matière d'étude des polluants métalliques de l'atmosphère. Pollut. Atmos. 139: 98–103.

Hodge, V., S. R. Johnson & E. D. Goldberg, 1978. Influence of atmospherically transported aerosols on surface ocean water composition. Geochem. J. 12: 7–20.

Ibanez, F., J. M. Fromentin & J. Castel, 1993. Application de la méthode des sommes cumulées à l'analyse des séries chronologiques océanographiques. C.R. Acad. Sci. Paris, Sciences de la Vie, 316: 745–748.

Injuk, J. & R. Van Grieken, 1995. Atmospheric concentrations and deposition of heavy metals over the North Sea: a litterature review. J. Atmos. Chem. 20: 1567–1578.

Jaffrezo, J. L. & J. L. Colin, 1988. Rain aerosol coupling in urban area: scavenging ratio measurement and identification of some transfer processes. Atmos. Envir. 22: 929–935.

Kendall, M., 1976. Time series, Charles Griffin & Co Ltd, London, 197 pp.

Kersten, M., M. Kriews & U. Förstner, 1991. Partitioning of trace metals released from polluted marine aerosols in coastal seawater. Mar. Chem. 36: 165–182.

Legendre, L. & P. Legendre, 1984. Ecologie numérique, 2, Masson, Presses de l'Université du Québec, Paris, Québec, 335 pp.

Lilliefors, H. W., 1967. The Kolmogorov-Smirnov test for normality with mean and variance unknown. J. Am. Stat. Assoc. 62: 399–402.

Lim B., T. D. Jickells, J. L. Colin & R. Losno, 1994. Solubilities of Al, Pb, Cu and Zn in rain sampled in the marine environment over the North Atlantic Ocean and Mediterranean Sea. Global Biogeochem. Cycles 8: 349–362.

Mann, H. R. & D. R. Whitney, 1947. On a test whether one of two random variables is stochastically larger than the other. Ann. Math. Stat. 18: 52–54.

Migon, C., 1988. Etude de l'apport atmosphérique en métaux traces et sels nutritifs au milieu côtier méditerranéen: implications biogéochimiques. Thèse de Doctorat, Nice, 217 pp.

Migon, C., L. Alleman, N. Leblond & E. Nicolas, 1993. Evolution of atmospheric lead over the northwestern Mediterranean Sea between 1986 and 1992. Atmos. Envir. 27A: 2161–2167.

Migon, C., E. Jourdan, E. Nicolas & B. Gentili, 1994. Effects of reduced leaded fuel consumption on atmospheric lead behaviour. Chemosphere 28: 139–144.

North Sea Task Force, 1993. North Sea Quality Status Report. Oslo and Paris Commissions, London. Olsen & Olsen, Fredensborg, Denmark, 132 pp.

Otten, P., H. Storms, C. Xhoffer & R. Van Grieken, 1989. Chemical composition, source identification and quantification of the atmospheric input into the North Sea. In Pichot, G. (ed), Progress in Belgian Oceanographic Research. Prime Minister's Services of Science Policy Office & Ministry of Public Health and Environment, Brussels: 413–422.

Ottley, C. J. & R. M. Harrison, 1993. Atmospheric dry deposition flux of metallic species to the North Sea. Atmos. Envir. 27A: 685–695.

Rahn, K. A., 1976. The chemical composition of the atmosphere aerosol. Technical Report. University of Rhode Island, 265 pp.

Scherrer, B., 1984. Biostatistique, Gaëtan Morin Editeur, Québec, 850 pp.

Siegel, S. & N. J. Castellan, 1988. Nonparametric statistics, Mac Graw-Hill, New-York, 399 pp.

Taylor, S. R., 1964. Abundance of chemical elements in the continental crust: a new table. Geochim. Cosmochim. Acta 28: 1273–1285.

Van Jaarsveld, J. A., 1993. Atmospheric inputs to the North Sea and coastal waters. J. IWEM, 7: 24–31.

Veron, A., P. Flament, M.L. Bertho, L. Alleman, R. Flegal & B. Hamelin. Lead isotopic characterisation of transient anthropogenic emissions to the European Troposphere. Atmos. Envir. (in press).

Hydrobiologia **373/374:** 333–347, 1998.
J.-C. Amiard, B. Le Rouzic, B. Berthet & G. Bertru (eds),
Oceans, Rivers and Lakes: Energy and Substance Transfers at Interfaces.

Biological components from the Seine estuary: first results

Pascal Mouny[1], Jean Claude Dauvin[4], Christophe Bessineton[2], Bernard Elkaim[3] & Serge Simon[2]

[1] *Muséum National d'Histoire Naturelle, Laboratoire de Biologie des Invertébrés Marins et Malacologie, URA CNRS 699, 57 rue Cuvier, F-75231 Paris Cedex 05*

[2] *Cellule de suivi du littoral Haut-Normand, Quartier des Affaires Maritimes, 4 rue du Colonel Fabien, F-76083 Le Havre Cedex, France*

[3] *Université P. & M. Curie, Laboratoire d'Hydrobiologie, 12 rue Cuvier, F-75005 Paris Cedex 05, France*

[4] *Station Marine de Wimereux, Université de Lille 1, CNRS UPRES A ELICO, BP80, 28 Avenue Foch, F-62930 Wimereux, France*

Key words: Seine estuary, macrobenthos, mesozooplankton, suprabenthos, ichtyofauna, trophic chains

Abstract

The Seine estuary plays an important role in the dynamics of the eastern English Channel ecosystem. Nevertheless, its biological compartment is poorly known. This constitutes an important handicap to establish the precise state of the health of this major European estuary. The objectives of this study were to identify the life resources of the estuary: macrobenthos, mesozooplankton, suprabenthos, and fish populations and to define the main trophic links in two parts of the estuary (i.e. the polyhaline and oligohaline zones). There is an impoverishment of the biological diversity from the polyhaline zone to the oligohaline zone. The benthic and pelagic fauna of the Seine estuary is similar to other North-Eastern Atlantic estuaries. But the pelagic fauna, especially the copepod *Eurytemora affinis* and the shrimp *Palaemon longirostris* seemed to be more abundant in the Seine estuary than in other estuaries. Two macrobenthic communities occurred in the estuary: a diversified and abundant *Abra alba-Pectinaria koreni* community in the outer part of the estuary and a *Macoma balthica* community in the inner part. This latter was especially poor in specific richness, density and biomass, in all areas, except on tidal mud flats. Two trophic chains were identified. In the oligohaline zone corresponding to the maximum turbidity zone (high turbidity, low concentration of oxygen), the trophic chain was exclusively planktonic due to the dredging of the estuary which prevented permanent benthic fauna formation. In this zone, the number of fishes was relatively low in spite of high biomass of mesozooplankton and suprabenthos. In the outer part of the estuary, low turbidity and high concentration of oxygen are more favourable to fish populations which feed especially on benthic fauna.

Introduction

The Seine estuary is the largest macrotidal estuary in the English Channel, with an area of about 150 km^2 at high tide. Freshwater input into the estuary is mainly from the Seine river with a drainage area of approximately 78 650 km^2. The discharge varies seasonally from a maximum of 2 000 m^3 s^{-1}in winter to a minimum of 100–200 m^3 s^{-1}in summer. Mean annual discharge ranges from 240 to 625 m^3 s^{-1} with an average of 380 m^3 s^{-1} (Avoine, 1986). It is an important area for the French economy, because 1/4 of the French population and 1/3 of the industry and agriculture are concentred along the Seine river.

Several multidisciplinary research programs have occurred in the Bay of Seine (especially during the beginning of the 1980s; Cabioch, 1986). The distribution and dynamics of marine benthic communities in the eastern part of the Bay of Seine were studied (e.g. Gentil et al., 1986; Dauvin & Gillet, 1991; Thiébaut et al., 1997). Data on mesozooplankton and suprabenthos were given later by Wang & Dauvin (1994) and

Wang et al. (1994, 1995) at the mouth of the estuary. The distribution of the ctenophore *Pleurobrachia pileus* was also studied in spring with several sets of spring plankton and suprabenthos samples (Wang et al., 1995).

Nevertheless, knowledge on the composition and distribution of the fauna in the Seine estuary remained very limited, and concerned only the outer part of the estuary during the 'Schéma d'Aptitude à l'Utilisation de la Mer' (at the end of the 1970s, SAUM, 1980). In 1978–1979, quantitative measurement on the distribution of subtidal macrobenthic communities was conducted from the mouth of the estuary to the upstream limit of the salinity intrusion (Proniewski & Elkaim, 1980). The distribution of intertidal macrobenthic communities was also studied by Desprez (1981). In the inner part of the estuary, the distribution of the Maximum Turbidity Zone and environmental parameters, e.g. concentrations of pollutants and dissolved oxygen are well described (Avoine, 1981).

Moreover, the Seine estuary was the main nursery ground of the English Channel (Duval, 1982). Conversely, the data on the fauna from other main estuaries of French coasts and North-Eastern Atlantic was well documented: e.g. Gironde (Castel, 1981; Sorbe, 1981), Loire (Robineau & Marchand, 1984; Marchand, 1981), Schelde (Mees, 1994; Mees et al., 1995; Soetaert & van Rijswijk, 1993), and Ems (Baretta & Malschaert, 1988).

The present study, as a part of the French Seine-Aval program, was initiated in 1995 to examine the life resources and to define the main trophic links within each part of the Seine estuary in order to: (i) to describe the gross taxonomic composition of the main biological compartments, i.e. macrobenthos, mesozooplankon, suprabenthos and ichtyofauna mainly in two seasons spring and autumn, (ii) to describe the main components of the trophic chains in the inner and outer parts of the estuary.

Study site

To assess the distribution of live resources, samples were collected from the mouth of the estuary (East of 0 of longitude) to the upstream limit of the salinity intrusion upstream Pont de Tancarville (see Figure 1).

Sampling strategy and treatment of the samples

Macrobenthos

Three replicates were collected at each station using a Smith McIntyre grab (0.10 m^2), 14 were taken on November 1993 to estimate the specific composition, density and biomass of the macrofauna from the subtidal zone of the mouth of the estuary (Fosse Nord: five stations, and Fosse Sud: five stations, three from the outer part and two from the inner part), and in the navigation channel along a transect of four stations from the mouth of the Seine estuary to the inner part of the estuary near the Pont de Tancarville. Three transects (Pont de Normandie, Marais du Hode and Pont de Tancarville) with four to six stations were sampled in the inner part of the estuary in May and October 1995: 14 stations in May (3 intertidal and 11 subtidal), and 12 subtidal stations in October. The samples were taken with a benthic corer with a surface 0.0625 m^2 (four replicates at each station) at the intertidal stations. A 'Rallier du Baty' dredge was used in the subtidal stations, so that about 30 to 40 l of sediment were preserved for each dredge sample.

After, the sediment was sieved (1 mm mesh), the retained material was sorted twice. The samples were fixed with 10% neutral formalin and identified to specific level. The dry weight (DW) of each species was determined and results were expressed as number of animals or biomass per m^2.

Mesozooplankton

Mesozooplankton were obtained using a WP2 (200 μm mesh size) plankton net with a TSK flowmeter in the mouth part to measure the volume of filtered water. The plankton net was hauled oblique from bottom to the surface against the tide. To assess the spatial distribution of the mesozooplankton in the estuary, a set of samples was taken during a daytime flood tide along a transect of 5 to 6 stations with a salinity gradient from 25.6 to 0.4 P.S.U. (SAPO32/Pk371 to SAPO37/Pk336 upstream Pont de Tancarville) in May 1994 and 12.1 to 0.4 P.S.U. in September 1995 (SAPO96/Pk365 to SAPO100/Pk336) (see Figure 1, Table 1). Materials collected were fixed with 5% neutral formalin for approximately one week then washed and preserved in 70% alcohol. Planktonic organisms were identified and counted under a dissecting microscope according to Frontier's subsampling method (Frontier, 1972). The ash-fresh dry weight (AFDW) of each species was determined and results were expressed as number of organisms or AFDW biomass m^{-3}.

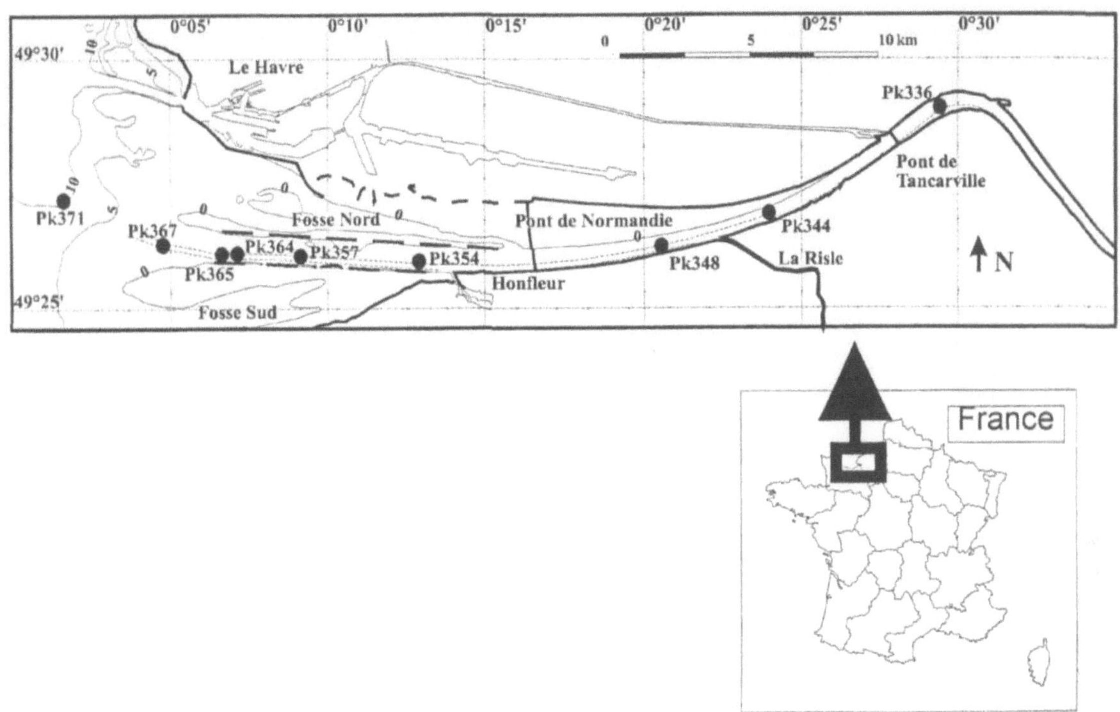

Figure 1. Map of the Seine estuary showing the different sampling stations.

Table 1. Sampling stations for the different areas of the estuary with indication of their environmental characteristic. Nm: number of mesozooplankton sample, Ns: number of suprabenthic sample, Pk: kilometric point (Pk = 0 at Paris), Hr: hour of sampling (HT: High tide at the Havre; May 1994: HT = 15:55, September 1995: HT = 15:55).

	Date	Type	Sampling station				Environmental characteristics	
			Nm	Ns	Pk	Hr (UT+2)	Surface salinity	Temperature (C)
Suprabenthos and Mesozooplankton	May 1994	Transect	SAPO32	SAT11	371	11:00	25.5	13.3
			SAPO33	SAT12	364	12:00	20.3	14.2
			SAPO34	SAT13	357	14:00	12.9	15.5
			SAPO35	SAT14	348	15:00	4.3	16.6
			SAPO36	SAT15	344	16:15	0.7	17.6
			SAPO37	SAT16	336	17:15	0.3	17.8
	September 1995	Transect	SAPO96	SAT44	365	10:30	12.1	17.8
			SAPO97	SAT45	362	11:30	7.7	18.0
			SAPO98	SAT46	354	13:00	4.4	18.2
			SAPO99	SAT47	348	14:00	1.8	18.5
			SAPO100	SAT48	336	15:00	0.4	18.6
Ichtyofauna	May 1995	2 areas: Honfleur	–	–	–	HT	14.0	17.2
		Tancarville	–	–	–	HT	0.5	16.4
	September 1995	3 areas: Bouée Amfard	–	–	367	HT	32.5	17.9
		Honfleur	–	–	–	HT	24.0	16.0
		Tancarville	–	–	–	HT	6.0	16.5

Table 2. Densities of benthic species collected in eight areas from the seine estuary.

Zone	Fosse Sud Outer part	Fosse Sud Inner part	Fosse Nord	Chenal dowstream Honfleur	Chenal upstream Honfleur	Pont de Normandie			Marais du Hode			Pont de Tancarville	
Date	November 93		November 93	November 93	November 93	May 95		October 95	May 95		October 95	May 95	October 95
Number stations	3, subtidal	2, subtidal	5, subtidal	2, subtidal	2, subtidal	1, intertidal	3, subtidal	6, subtidal	2, intertidal	3, subtidal	3, subtidal	6, subtidal	3, subtidal
Species richness	25	13	12	8	4	7	11	14	6	7	9	5	6
N. endobenthic species	23	12	9	4	–	6	5	7	3	2	3	1	2
N. vagile species	2	1	3	4	4	1	6	7	3	5	6	4	4
Endobenthic species													
Abra alba	566	–	–	–	–	–	–	–	–	–	–	–	–
Cerastoderma edule	–	213	13	8	–	–	–	4	–	–	–	–	–
Eteone longa	11	4	4	5	–	10	10	2	–	–	10	–	3
Hediste diversicolor	–	–	4	–	–	615	9	6	42	2	6	16	12
Macoma balthica	3	5	21	9	–	58	18	7	–	–	–	–	–
Nephtys hombergii	66	17	32	1	–	40	11	13	5	2	2	–	–
Owenia fusiformis	345	3	–	–	–	–	–	–	–	–	–	–	–
Pectinaria koreni	344	8	–	–	–	–	–	–	–	–	–	–	–
Other polychaeta	27	8	8	–	–	–	–	8	–	–	–	–	–
Divers	26	7	1	–	–	37	–	–	13	–	–	–	–
Vagile species													
Gastrosaccus spinifer	3	–	–	–	–	–	3	–	–	–	–	–	–
Carcinus maenas	2	–	1	4	–	–	–	1	2	1	–	2	–
Crangon crangon	–	5	2	22	5	–	28	21	–	9	6	–	29
Mesopodopsis slabberi	–	–	4	20	7	–	22	24	–	7	6	–	–
Neomysis integer	–	–	–	–	–	–	8	3	–	33	4	28	25
Palaemon longirostris	–	–	–	–	–	–	17	–	–	22	16	40	3
Amphipoda													
Corophium volutator	–	–	–	–	–	1022	–	–	–	–	–	–	–
Other amphipoda	–	–	–	–	5	–	1	2	3	–	2	2	1
Isopoda	–	–	–	–	4	–	–	–	–	–	–	–	–
Divers	–	–	–	2	–	–	11	12	3	–	–	–	–
Total density m^{-2}	1393	270	90	82	21	2252	138	100	68	76	52	88	73
Total Biomass mg m^{-2}	44961	2366	655	3912	441	6120	1661	1928	221	2878	1206	4545	1928
Endobenthic Biomass mg m^{-2}	44807	2148	389	211	0	5060	193	43	40	40	46	53	43
Vagile Biomasse mg m^{-2}	154	218	266	3701	441	1060	1468	1885	181	2838	1160	4492	1885

Table 3. Densities of mesozooplanktonic species collected along the salinity gradient (oblic samples) in the Seine estuary in May 1994 and September 1994.

Date	May 1994						September 1995				
Samples	SAPO32	SAPO33	SAPO34	SAPO35	SAPO36	SAPO37	SAPO96	SAPO97	SAPO98	SAPO99	SAPO100
Kilometric Point	371	364	357	348	344	336	365	362	354	348	336
Sub-surface Salinity (P.S.U.)	27.6	23.0	17.8	4.2	1.3	0.4	12.1	7.7	4.4	1.8	0.4
Cladocera											
Bosmina spp.	–	–	1	257	218	907	88	–	8	7	2000
Daphnia spp.	–	–	–	16	29	88	–	–	–	–	105
Evadne nordmanni	19	132	174	3	–	–	–	–	–	–	–
Other cladocera	14	3	1	1	21	7	–	7	–	–	158
Copepoda											
Acartia spp.	283	282	6	–	–	–	62941	6897	190	–	–
Eurytemora affinis	–	1	945	47145	157890	1033	853	10414	5337	464	13684
Euterpina acutifrons	5	1	6	–	–	–	853	7	–	–	–
Temora longicornis	56	185	3	–	–	–	–	–	–	–	–
Other copepoda	19	69	3	–	–	–	412	14	24	–	579
Copepodid											
Acartia spp.	274	692	–	–	–	–	1235	62	–	–	–
Eurytemora affinis	–	127	13826	47000	35266	45	–	9172	5159	879	4105
Temora longicornis	27	1104	–	–	–	–	–	–	–	–	–
Other copepodid	7	162	–	–	–	–	–	–	–	–	–
Cirriped larvae	30	812	260	–	10	48	–	–	–	–	–
Mysidacea											
Mesopodopsis slabberi	–	4	437	2	–	–	81	–	–	–	–
Neomysis integer	–	–	–	11	–	–	657	757	104	8	21
Other mysidacea	–	–	–	–	–	–	–	–	–	–	–
Chaetognatha											
Sagitta elegans	–	1	–	–	–	–	35	–	–	–	–
Appendicular											
Oikopleura dioica	65	212	30	–	–	–	–	–	–	–	–
Fish larvae and fishes											
Gobiidae	1	1	3	–	–	–	19	4	8	5	17
Other species	3	1	7	–	–	–	–	–	–	–	–
Decapoda larvae	6	11	7	1	–	–	3	1	1	1	–
Ostracoda	29	113	122	–	–	1	–	7	–	–	–
Amphipoda	–	–	–	–	–	–	–	–	–	–	–
Polychaeta larvae	2	15	3	–	–	–	–	–	–	–	–
Ctenophore						–					
Pleurobrachia pileus	8	10	3	–	–	–	–	–	–	–	–
Other taxa	6	19	27	–	–	–	–	–	–	–	–
Species richness	27	32	25	14	8	14	14	12	9	7	11
Total density (ind m^{-3})	854	3957	15864	94436	193434	2129	67177	27342	10831	1364	20669
Total biomass (mg m^{-3})	5.26	14.97	218.81	536.00	1513.10	11.62	1558.85	1582.24	257.02	21.05	181.52

Suprabenthos

A suprabenthic sledge with a principle similar to the GIROQ version of the MACER sledge was used (Dauvin & Lorgeré, 1989; Dauvin et al., 1995). Four WP2 zooplankton nets (0.5 mm mesh size) were superposed on 0.10–0.40 (net 1), 0.45–0.75 (net 2), 0.80–1.10 (net 3) and 1.15–1.45 m (net 4) above the sea bed. Each net holded a TSK flowmeter to measure the volume of water filtered. Mean trawling speed during the 5-min-long hauls was 1.5 knots against the tide. As for the mesozooplanktonic samples, a set of samples was taken along a transect of six stations (SAT11/Pk371 to SAT16/Pk336) in May 1994, and a transect of five stations in September 1995 (SAT44/pk365 to SAT48/pk336) at the same stations for mesozooplanktonic samples (see Figure 1 and Table 3).

Materials collected were fixed with 10% neutral formalin for approximately one week then the materials were washed and preserved in 70% alcohol. Suprabenthic fauna was examined by a dissecting microscope. The ash fresh dry weight (AFDW) of each species was estimated and the mean of total number of individuals of the four nets or biomass in a haul was standardized to 100 m^3.

Ichthyofauna

Trawling with a prawn trawl (22 mm mesh size) was carried out in May 1995 and in September-October 1995 at stations 'Bouée Amfard/Pk367', Honfleur and Pont de Tancarville (see Figure 1). Fresh fishes were identified, counted and weighted. The weight of fishes sampled were standardized to 1000 m^2.

Trophic chains

To determine the trophic chains, stomach contents of four main fishes collected in the estuary (i.e. the flounder *Platichthys flesus*: 187 stomachs, the sole *Solea vulgaris*: 56 stomachs, the sea perch *Dicentrarchus labrax*: 34 stomachs, the goby *Pomatoschistus microps*: 200 stomachs) and the shrimp *Palaemon longirostris* (420 stomachs) were examined to identify the main prey of these species. Sampling sites were in the meso-oligohaline (downstream Pont de Tancarville) and meso-euhaline (downstream Pont de Normandie) zones. For the goby *P. microps* and the shrimp *P. longirostris*, the number of prey (Np) and the ratio in percentage of number of prey i on the total number of preys (Cn) were calculated. For the demersal fishes the occurrence of prey was only recorded.

Physical environment

The main characteristics of the environment parameters in May and September–October were given in the Table 1. In May the turbidity, near the sea-bed, increased regularly from the outer part of the estuary to the inner part which varied from < 50 mg l^{-1} at Pk371 to 375 mg l^{-1} at Pk344 and then decreased upstream from less than 50 mg l^{-1}. In September there was the same gradient but the Suspended Particule Matter (SPM) concentration exceeded 1000 mg l^{-1} near the Pont de Tancarville (lowest water level of the Seine River).

Results

Macrobenthos

During the autumn (November 1993 and October 1995), 25 species were sampled in the subtidal outer part of the estuary in the 'Fosse Sud' (Table 2). The species richness was lower in Honfleur and 'Marais du Hode' (i.e. 14 and 4 respectively). Total density and biomass showed also a similar pattern which decreased from the outer part of the estuary in the 'Fosse Sud' to the inner part of the estuary in Honfleur. The density was particularly low (< 21 ind m^{-2}) in the channel station near Honfleur; in the other stations the density was near 100 ind m^{-2}. The biomass showed a very high value in the outer part of the 'Fosse Sud' (DW = 45 g m^{-2}) and was lower than 4 g m^{-2} in the other part of the estuary. It was important to notice the contribution of the vagile species (suprabenthic species) at the stations from the navigable channel and the stations upstream Honfleur. In these stations, the biomass of the endobenthic macrofauna was very low (Table 2).

The specific richness, density and biomass at stations of Pont de Normandie, Marais du Hode and Pont de Tancarville showed no seasonal changes.

For the intertidal zone, the 'Pont de Normandie' station showed a low species richness but a high density which was due to the amphipod *Corophium volutator* and the polychaete *Hediste diversicolor* (Table 2). In the 'Marais du Hode' intertidal station, although the density was of the same order of magnitude as the subtidal station, its biomass was very low.

In summary, the benthic macrofauna showed a drastic reduction of species richness, density and biomass from the outer part of the estuary in the 'Fosse

Table 4. Densities of suprabenthic species collected along the salinity gradient in the Seine estuary in May 1994 and September 1995.

Date	May 1994						September 1995				
Samples	SAT11	SAT12	SAT13	SAT14	SAT15	SAT16	SAT44	SAT45	SAT46	SAT47	SAT48
Kilometric Point	371	364	357	348	344	336	365	362	354	348	336
Sub-surface Salinity (P.S.U.)	27.6	23.0	17.8	4.2	1.3	0.4	12.1	7.7	4.4	1.8	0.4
Amphipoda	6	2	1	–	–	–	1300	569	57	12	–
Mysidacea											
Gastrosaccus spinifer	19	1	8	1	–	–	–	–	–	–	–
Mesopodopsis slabberi	1	1	47	522	22	–	34329	12	–	–	–
Neomysis integer	–	–	–	20443	332	4	261	77056	56425	2153	1137
Other Mysids	21	40	11	–	–	–	8	–	–	–	–
Decapoda											
Crangon crangon	81	37	104	1055	43	–	2	555	446	41	–
Palaemon longirostris	–	–	1	157	4	–	–	–	175	1769	776
Other decapoda	1	3	–	–	–	–	116	2	2	–	–
Decapoda larvae	137	34	109	–	–	–	19	69	3	–	–
Cumacea	14	–	–	–	–	–	55	–	–	–	–
Chaetognatha											
Sagitta elegans	1	–	–	–	–	–	71	–	6	–	–
Fishes											
Pomatoschistus microps	649	630	862	1588	26	–	61	59	209	251	992
Pomatoschistus minutus	–	–	–	45	–	–	141	126	435	14	2
Other fishes	–	–	–	–	–	1	2	2	7	1	1
Clupeidae larvae	240	283	58	10	–	–	3	–	–	–	–
Gadidae larvae	36	7	3	1	–	–	–	–	–	–	–
Gobiidae larvae	–	–	–	–	–	–	614	81	24	1	65
Pleuronectidae larvae	5	1	8	5	–	–	–		–	–	–
Soleidae larvae	28	9	1	1	–	–	–		–	–	–
Other larvae	5	1	6	3	–	–	–		–	–	–
Total richness	26	19	18	13	5	2	20	11	15	9	6
Total density (ind 100 m^{-3})	1244	1049	1219	23831	427	5	36982	78531	57789	4242	2973
Total biomass (mg 100 m^{-3})	75749	72892	93922	224631	4139	72	57341	100634	142037	216774	188864

Sud' to the channel especially in the part between Honfleur and Pont de Tancarville. In this zone, the contribution of the vagile fauna was important. No seasonal changes was observed in the channel stations.

Mesozooplankton

For both seasons, spring and autumn, the mesozooplankton community showed the same changes along the salinity gradient (Table 3). The species richness decreased from the outer part to the inner part of the estuary; the maximum value in May (i.e. 32 species) was recorded for a salinity of 23.0 P.S.U. and the minimum value in September (i.e. 14 species) was in a salinity zone of 12.1 P.S.U. The lowest species richness was recorded in May for a salinity of 1.3 (i.e. 8 species) and in September for a salinity of 1.8 P.S.U. (i.e. 7 species). Upstream, the species richness increased with the increasing of freshwater species (cladocerans).

The mesozooplankton community showed three distinct patterns (Table 3). In the polyhaline zone (18–30 P.S.U., Venice system of classification of saline waters; Anonymous, 1959) the community was characterized by marine copepods (*Pseudocalanus elongatus, Paracalanus parvus* and *Temora longicornis*), ostracods and the ctenophore *Pleurobrachia pileus*. In the mesohaline zone (5–18 P.S.U.) the community was characterized by the copepod *Acartia* (64,176 ind m^{-3} in autumn) and the mysid *Neomysis integer*. In

Table 5. Biomass (g fresh weight/1000 m^{-2}) of fishes collected by trawls in three areas of the Seine estuary.

Species	Pont de Tancarville Pk 340 May	Pont de Tancarville Pk 340 October	Honfleur Pk 356 May	Honfleur Pk 356 October	Bouée Amfard Pk 367 September
Anguilla anguilla (L., 1758)	–	–	–	–	206
Agonus cataphractus (L., 1758)	–	–	–	–	13
Buglossidium luteum (Risso, 1810)	–	–	–	–	434
Callionymus lyra L., 1758	–	–	–	–	561
Ciliata mustela (L., 1758)	–	–	–	45	–
Clupea harengus L., 1758	–	2	130	955	710
Dicentrarchus labrax (L., 1758)	99	5	–	105	34
Gobiidae	–	–	–	–	1
Limanda limanda (L., 1758)	–	–	–	–	392
Merlangius merlangus (L., 1758)	–	–	–	843	21
Mugilidae	–	8	–	–	–
Platichthys flesus (L., 1758)	131	228	860	473	239
Pleuronectes platessa L., 1758	–	–	–	–	156
Pomatoschistus microps (Kröyer, 1838)	–	1	35	39	57
Solea vulgaris Quensel, 1806	–	–	–	–	40
Sprattus sprattus (L., 1758)	–	1	206	–	138
Trachinus draco L., 1758	–	–	–	1	13
Trachurus trachurus (L., 1758)	–	–	–	24	307
Trigla lucerna L., 1758	–	–	–	44	1
Trisopterus luscus L., 1758	–	–	639	360	238
Species richness	2	6	5	10	18
Biomass (g fresh weight/1000 m^2)	230	245	1870	2889	3561

the oligohaline zone (0.5–5 P.S.U.), the community was characterized by the dominance of the copepod *Eurytemora affinis* (≅ 190,000 ind m^{-3} in May) and the presence of freshwater cladocerans (*Bosmina* spp.). In the oligohaline zone, the density and biomass reached their maximum in May with the values of 193,434 ind m^{-3} and 1,513 mg m^{-3} respectively, while in the mesohaline zone in September they were 67,177 ind m^{-3} and 1,582 mg m^{-3} respectively. During spring, the biomass in the polyhaline zone was lower than those found in the meso- and oligohaline zones (Table 3).

Suprabenthos

At both seasons, the specific richness decreased as the salinity decreased. The maximum species richness was 26 species in May and 20 in September for salinities of 27.6 P.S.U and 12.1 P.S.U. respectively, while few species were reported in the low salinity zone (i.e. only two species in May and six species in September for a salinity of 0.4 P.S.U. (Table 4).

The suprabenthic community was characterized by six dominant species: the mysids *Mesopodopsis slabberi* and *Neomysis integer*, the decapods *Crangon crangon* and *Palaemon longirostris*, the gobies *Pomatoschistus minutus* and *P. microps*. The quantitative composition of this compartment varied between the mesohaline zone which was dominated by *M. slabberi* (34,330 ind 100 m^{-3} in September), *C. crangon* (104 ind 100 m^{-3} in May) and the young gobies (860 ind 100 m^{-3}) in spring and the oligohaline zone which was characterized by the dominance of *N. integer*, *P. longirostris* and *P. microps*, with maximal densities of 77 000, 1 769 and 991 ind 100 m^{-3} respectively in September (Table 4).

The total density of suprabenthic community was always higher in the oligohaline zone than in the mesohaline zone (Table 4). The biomass was higher in the summer than in the spring. At both seasons the

Table 6. Diet of three size classes of the gobiidae *Pomatoschistus microps* in the oligohaline zone of the Seine estuary in May 1994 and September 1995. Np: Number of prey; CN: Ratio in percentage of number of prey i on the number of total prey; L1: 10–29 mm; L2: 29–39 mm; L3: 39–50 mm.

	May 1994					
	L1		L2		L3	
Number of stomach	34		34		33	
Stomachs with prey	33		34		33	
Total number of prey	308		2029		2766	
Mean number of prey per stomach	9.1	59.7	83.8			
	Np	Cn	Np	Cn	Np	Cn
Copepoda						
Eurytemora affinis	308	100	2023	99.6	2741	99.1
Other copepoda	–	–	–	–	9	0.3
Cladocera						
Bosmina spp.	–	–	4	0.30	3	0.1
Mysidacea						
Neomysis integer	–	–	–	–	6	0.2
Mesopodopsis slabberi	–	–	–	–	3	0.1
Decapoda						
Crangon crangon	–	–	1	0.05	–	–
Other decapoda	–	–	–	–	1	0.04
Decapoda larvae	–	–	–	–	1	0.04
Ostracoda	–	–	1	0.05	–	–
Pleuronectidae larvae	–	–	–	–	1	0.04
Chironomidae larvae	–	–	–	–	1	0.04
	September 1995					
	L1		L2		L3	
Number of stomach	53		44		3	
Stomachs with prey	38		35		3	
Total number of prey	207		192		21	
Mean number of prey per stomach	5.4		5.5		7.0	
	Np	Cn	Np	Cn	Np	Cn
Copepoda						
Eurytemora affinis	186	89.9	162	84.4	9	42.9
Mysidacea						
Neomysis integer	21	10.1	29	15.1	12	57.1
Mesopodopsis slabberi	–	–	1	0.5	–	–

biomass was maximum in the salinity zone around 4 P.S.U. (Table 4).

Ichthyofauna

A total of 20 fish species was identified from all trawls. The major ichtyofauna components of the Seine estuary were the flounder *Platichthys flesus*, the sole *Solea vulgaris*, the sea perch *Dicentrarchus labrax*, the herring *Clupea harengus* and the sprat *Sprattus sprattus* (Table 5).

The species richness in the outer part of the estuary was 5 in May and 10 in October at downstream Honfleur, and 18 in September at Bouée Amfard. By contrast, low species richness as 2 in May and 6 in October at Pont de Tancarville was observed in the inner part of the estuary (Table 5). The biomass showed the same pattern as 3561 and only 245 g of fresh weight.

Table 7. Diet of two size classes of the decapoda *Palaemon longirostris* in the oligohaline zone of the Seine estuary. Np: Number of prey; Cn: ratio in percentage of number of prey i on the number of total prey. L1: 20–29 mm; L2: > 29 mm.

	May 1994			
	L1		L2	
Number of stomach	159		77	
Stomachs with prey	29		16	
Total number of prey	26		17	
Mean number of prey per stomach	0.9		1.1	
	Np	Cn	Np	Cn
Copepoda				
Eurytemora affinis	8	30.8	9	52.9
Other copepoda	4	15.4	–	–
Mysidacea				
Neomysis integer	7	26.7	4	23.5
Other mysid	6	23.1	2	11.8
Decapoda				
Crangon crangon	–	–	1	5.9
Other decapoda	1	0.04	1	5.9
Number of stomach with Detritus	29	–	9	–
	September 1995			
	L1		L2	
Number of stomach	121		63	
Stomachs with prey	42		27	
Total number of prey	145		62	
Mean number of prey per stomach	3.5		2.3	
	Np	Cn	Np	Cn
Cladocera				
Bosmina spp.	3	2.1	–	–
Other cladocera	1	0.7	–	–
Copepoda				
Eurytemora affinis	134	92	54	87.1
Other copepoda	5	3	4	6.5
Mysidacea				
Neomysis integer	–	–	1	1.6
Other mysid	2	1.4	3	5
Number of stomach with detritus	42	–	27	–

1000 m^{-2} for respective by the outer and the inner parts of the estuary in October. Comparatively speaking, species richness and total biomass were higher in autumn than in spring. For example at Honfleur, 10 species were present in autumn for a total biomass of 2889 g of fresh weight 1000 m^{-2} while in spring only 5 species were sampled for a total biomass of 1870g of fresh weight 1000 m^{-2}.

Inner part trophic chain

The stomach contents analysis showed that the copepod *Eurytemora affinis* was the main prey of the goby *Pomatoschistus microps* (Table 6). The diet of the young goby (Length < 30 mm) was based exclusively on this copepod in spring, while some *Neomysis integer* were present in the diet in autumn. The diet of the large *P. microps* (length > 30 mm) was not so specific, including the decapod *Crangon crangon* and cladocerans (Cn: 0.1 to 0.3). In spring, the mean number of preys per stomach in large goby was higher than the small ones (L2: 59.7 and L3: 83.8). However, it was similar for the three size classes in autumn (5.4 to 7.0). In general, the copepod *E. affinis* was a dominant diet component at both seasons, although mysids were also found in autumn (Table 6).

In the decapod *Palaemon longirostris* (Table 7) as in the *P. microps*, the copepod *E. affinis* was also the major food component of the diet (Cn 30.8 to 92.0%). The mysid *N. integer* composed the secondary diet (Cn: 1.6 to 26.7%). Detritus was always present in the stomach of *P. longirostris*. No seasonal change of the diet was observed, but the mean number of prey per stomach was higher in autumn than in spring.

Outer part trophic chain

Table 8 showed the analysis of stomach contents of three main fishes sampled in the lower estuary. An important number of stomachs with prey was recorded: 100% for the three size class of *D. labrax*, 93.3 to 100% for *P. flesus* and 6.3 to 92.3% for *S. vulgaris*. The flatfishes *P. flesus* and *S. vulgaris* showed more benthic diet than *D. labrax*. The suprabenthic Decapoda Caridea were the major diet component of the three species. The diet of *D. labrax* and *P. flesus* were more diverse (respectively 5 and 6 preys) than *S. vulgaris* (3 preys). There was also an intraspecific variability of the diet: the juveniles of *P. flesus* and *D. labrax* presented a more diverse diet than adults which preferred planktonic copepods and mysids. The diet of young *D. labrax* was characterized by mysids (Cn:

68%) while copepods represented 31% of the diet of the young *P. flesus*. Generally, the benthic preys in the diet increased with the size of the fish. For example, benthos formed 33% of the juvenile and 92% of the adult *S. vulgaris* diets.

Discussion

This present study furnished the first quantitative data on the mesozooplankton, suprabenthos and the first quantitative data for the subtidal macrobenthos in the mesohaline and oligohaline areas of the Seine estuary at only two seasons spring and autumn. It gave also supplementary data on fish composition and analysis on 5 dominant suprabenthic species with accomplishment of the food chains in two parts of the estuary. Among the abundant literature on estuaries, data were mainly compared with these available on four other north-eastern Atlantic estuaries, the Ems and the Schelde (The Netherlands), and the Loire and the Gironde (France) (see Heip & Herman, 1995).

Faunistic composition and spatio-temporal distribution

Macrobenthos. The fauna belongs to two distinct benthic communities. (i) In the outer part of the Fosse Sud, it was an *Abra alba-Pectinaria koreni* community which occurred in the eastern part of the Bay of Seine on muddy fine sand (Gentil et al., 1986; Thiébaut et al., 1997). It was dominated by the bivalve *Abra alba*, and the polychaetes *Pectinaria koreni, Owenia fusiformis*, and *Nephtys hombergii* which exhibited high density and biomass (Table 2). (ii) In the other parts of the estuary, it was a *Macoma balthica* community which commonly occurs in the north-eastern Atlantic estuaries (Bachelet, 1979; Robineau & Marchand, 1984; Meire et al., 1991). In the Seine estuary, the *Macoma balthica* community showed a very low number of species in relation to the salinity gradient. Very low density and biomass were also found, except on intertidal stations near the Pont de Normandie where *Hediste diversicolor* and *Corophium volutator* were important species in density. Nevertheless, as a 1 mm mesh size sieve was used, the densities of small species like the Oligochaeta which may constitute very high densities in the less haline zone of the estuaries was not estimated. In the Seine estuary on 1 mm mesh size sieve, the density of the *Macoma balthica* was nearly 100 ind m^{-2} and the biomass ranged from 0.2 to 6 g m^{-2}, while the maximum density and biomass exceeded 10 000 ind m^{-2} and 10 g m^{-2} in the other estuaries (Bachelet, 1979; Robineau & Marchand, 1984; Meire et al., 1991). Our results agreed with the previous study from Proniewski & Elkaim (1980) on the subtidal macrofauna of the Seine estuary, who described both communities and underlined the absence of typical oligohaline fauna in this estuary. The commonest species was the polychaete *Hediste diversicolor* which occurred along the salinity gradient. In the Fosse Nord and Fosse Sud (Figure 1), the community was dominated by the bivalves *Cerastoderma edule* and *Macoma balthica*. In contrast, in the inner part of the estuary, the mysid and decapod suprabenthic species were dominant. This distribution patterns with two distinct zones was due to a strong abiotic limitation of maximum SPM concentration and oxygen deficiency in the inner estuary compared to the outer part with low turbidity and high concentration of oxygen, and to a regular dredging of the channel which prevented the settlement of species in the channel. In the outer part of the Fosse Sud (Figure 1), the *Abra alba-Pectinaria koreni* exhibited a high density and biomass which indicated an eutrophic condition without anoxia (Gentil et al., 1986).

Mesozooplankton. In the Seine estuary, the gross composition of the mesozooplankton was similar to those found in the other north-European estuaries (Castel, 1981; Bakker, 1994; Sautour & Castel, 1995) which included 3 different zones. (i) A polyhaline zone (18 to 30 P.S.U.) where neritic copepods *Temora longicornis*, *Euterpina acutifrons*, *Pseudocalanus elongatus* and *Paracalanus parvus*, and the cladocera *Evadne nordmanni* were common. In spring, copepod abundance was in the same order of magnitude as in other estuaries (maximum values: 2 600 ind m^{-3} in the Seine, 3 800 ind m^{-3} in the Ems, 2 300 ind m^{-3} in the Gironde and 5 100 ind m^{-3} in the Schelde, Sautour & Castel, 1995). (ii) A mesohaline zone, with the salinity between 5 and 18 P.S.U., was dominated by the copepods *Acartia* spp. especially in autumn, but *Eurytemora affinis* could also develop a dense population at both seasons. In spring, copepod abundance reached to 15 000 ind m^{-3} while in autumn it exceeded 62 000 ind m^{-3}. The density of *Acartia* spp. was very low in winter and autumn ($\approx$ 300 ind m^{-3}) (Mouny et al., 1996). In spring, the maximum density of *Acartia* spp. in the Seine estuary (1 000 ind m^{-3}) was in the same order of magnitude as in the Gironde (1 100 ind m^{-3}) but lower than found in the West-

Table 8. Occurrence of prey items in the stomach of 3 size classes of *Dicentrarchus labrax* (D.l.), *Platichthys flesus* (P.l.) and *Solea vulgaris* (S.v.) from May and October hauls in the downstream Seine estuary (Pk 340, Pk 356 and Pk 367) (L1 = 9–15 cm; L2 = 16–21 cm; L3 > 21 cm).

	D.l.			P.f.			S.v.		
	L1	L2	L3	L1	L2	L3	L1	L2	L3
Number of stomachs	15	16	3	137	30	20	16	27	13
% of Stomachs with prey	100	100	100	95.6	93.3	100	6.3	77.7	92.3
Copepod	8	–	–	31	7	–	–	–	–
Mysid	68	9	–	1	4	–	–	–	–
Decapod Caridea	12	57	66	4	40	40	60	26	8
Benthic preys	4	6	–	7	19	20	33	74	92
Fishes	–	8	–	2	4	–	–	–	–
Other preys	12	20	33	22	22	40	7	–	–
Detritus	–	–	–	33	4	–	–	–	–

Table 9. Comparative densities of mesozooplankton and suprabenthos from four North-European Atlantic estuaries (Gironde, Schelde and Ems: Mesozooplankton data: Sautour & Castel, 1995; Suprabenthic data: Mess et al., 1995; Seine: this study).

	Gironde	Seine	Schelde	Ems
Mesozooplankton				
Type of samples	Surface/Bottom	oblique	oblique	oblique
Gyre	WP2 net (200 μm)	WP2 net (200 μm)	WP2 net (200 μm)	WP2 net (200 μm)
Date of sampling	March to June	May	March, April and June	March and May
Maximum spring density (ind m^{-3}) of				
Acartia spp.	1100	1000	2800	18200
Eurytemora affinis	14500	193160	11500	16800
Suprabenthos				
Gyre	Sledge (mesh size: 2–1 mm) (Hamerlynck & Mees, 1989)	Sledge (mesh size: 500 μm) (Dauvin & Lorgeré, 1991)	Sledge (mesh size: 2–1 mm) (Hamerlynck & Mees, 1991)	Sledge (mesh size: 2–1 mm) (Hamerlynck & Mees, 1991)
1991)				
Date of sampling	August	September	August	August
Maximum summer density (ind m^{-2})				
Neomysis integer	30	240	200	20
Palaemon longirostris	224	6000	0	1
Pomatoschistus spp.	6300	4000	5	435

erschelde (2 800 ind m^{-3}) and the Ems (18 200 ind m^{-3}) (Table 9). (iii) An oligohaline zone (0.5 to 5 P.S.U.) which was dominated by the copepod *Eurytemora affinis.* The spring peak of abundance of *E. affinis* in the Seine estuary was more than ten time higher than those found in Gironde, Schelde and Ems estuaries (Table 9). The density of *Eurytemora affinis* remained relatively high during the autumn and the winter (16 000 to 27 000 ind m^{-3}) (Mouny et al., 1996).

Suprabenthos. The suprabenthic fauna was composed of 39 species in May and 34 species in September. There was a decrease in the number of species along the salinity gradient from the polyhaline zone (> 20 species), the mesohaline zone (11–20 species), to the oligohaline zone (< 15 species). In the eu-

Table 10. Comparative fish species richness in four north-european estuaries (fishes sampled with a suprabenthic sledge). Gironde, Schelde & Ems: Mees et al., 1995; Seine: this study.

Material	Gironde Sledge (mesh size: 2–1 mm) (Hamerlynck & Mees, 1991)	Seine Sledge (mesh size: 500 μm) (Dauvin & Lorgeré, 1989)	Schelde Sledge (mesh size: 2–1 mm) (Hamerlynck & Mees, 1991)	Ems Sledge (mesh size: 2–1 mm) (Hamerlynck & Mees, 1991)
Species				
Anguilla anguilla	*	*		
Callionymus lyra		*		
Callionymus reticulatus		*		
Ciliata mustela		*		
Clupea harengus		*	*	*
Dicentrarchus labrax		*		
Gasterosteus aculeatus	*	*	*	
Gobius niger		*		
Gymnocephalus cernuus				*
Lampetra fluviatilis		*		
Limanda limanda				*
Liparis liparis				*
Merlangius merlangus		*		
Osmerus eperlanus	*			*
Platichthys flesus		*		
Pleuronectes platessa				*
Pomatoschistus lozanoi	*		*	*
Pomatoschistus microps	*	*	*	*
Pomatoschistus minutus	*	*	*	*
Solea spp.	*	*		*
Sprattus sprattus	*	*	*	*
Syngnathus spp.	*	*	*	*
Trisopterus luscus				*
Zoarces viviparus				*
Total number of fish species	9	16	7	14

haline zone of the Seine estuary (Wang & Dauvin, 1994), the suprabenthos was more diverse than in the estuarine zones, with 46 species of Decapoda and Peracarida sampled, especially amphipods, mysids, and cumaceans, and 19 other macrozooplankton taxa (Wang et al., 1994). Spatial changes of the suprabenthos fauna were less pronounced than the mesozooplankton because some species, e.g. the decapod caridea *Crangon crangon* and the mysid *Mesopodopsis slabberi* showed seasonal migrations between the bay and the estuary. However, dominant species presented their maximal abundance in the mesohaline zone: the mysids *Mesopodopsis slabberi* and *Neomyis integer*, or in the oligohaline zone: the decapods *Palaemon longirotris* and *Crangon crangon*, and the goby *Pomatoschistus microps*. The maximum density in spring increased from the polyhaline zone to the oligohaline zone (Table 4), but the density of the fauna decreased dramatically if the salinity below 1.0 P.S.U. In the euhaline zone, the suprabenthic fauna was dominated by the ctenophore *Pleurobrachia pileus* and fishes larvae in spring (Wang et al., 1994, 1995). In summer, the abundance of *Palaemon longirostris* in the Seine estuary was higher than found in the Gironde, Schelde and Ems estuaries (Table 9). The abundance of *Neomysis integer* in the Seine was similar to the Schelde estuaries. *Pomatoschistus* spp. was in the same order of magnitude in the Gironde and the Seine estuaries (Table 9).

Ichtyofauna. The fish composition of the Seine estuary was relatively similar to those found in the Loire (Lardeux, 1986), the Forth and the Tyne estuaries (Pomfret et al., 1991), and the Gironde, Schelde and Ems estuaries (Mees et al, 1995). Comparative sampling of fishes by suprabenthic sledge was available for four estuaries (Table 10), the number of species recorded was in the same order of magnitude in the Seine and Ems estuaries, and higher than Schelde and Gironde estuaries.

Trophic chains

In the Seine estuary, two trophic chains were identified.

Upstream of Pont de Normandie, the shrimp *Palaemon longirostris* and the goby *Pomatoschistus microps* preyed on the pelagic compartment exclusively (copepods and mysids) (Tables 6 & 7). In this zone, the dominant copepod *Eurytemora affinis* was the main prey for both species. But, *Pomatoschistus* spp. (in the Ythan estuary, North Sea) and *P. longirostris* (in the Loire estuary) had a diet mainly on benthic preys (Healey, 1972; Marchand, 1981). However, the diet of the shrimp *Palaemon* in the Loire estuary included also pelagic prey (*N. integer*); the role of the detritus which should be an important source of organic materials found in the stomach of the shrimp in the Seine estuary and the Loire estuary should be elucidated. In the Ythan estuary, the diet of the common goby showed seasonal changes with a more diverse diet in spring than in autumn, coinciding with the availability of preys (Healey, 1972). Conversely, no seasonal changes of the diet of the Palaemonidae was observed in the Loire estuary. The difference between the Seine estuary and the other estuaries might be due to the dredging of the navigation channel of the Seine estuary which prevented the formation of a permanent benthic fauna.

The diet of fishes downstream of Honfleur showed a classical change of diet with age: juveniles eating mainly on planktonic prey (copepods, mysids and decapods) and adults eating benthic prey (polychaetes and bivalves). The benthic prey were especially important for *Solea vulgaris*. This result agreed with those of studies that the diet of the demersal fishes especially the flounder *Platichthys flesus* (Arntz, 1977) was mainly based on the benthic fauna. In summary, in the oligohaline zone (high turbidity, low concentration of oxygen) the diversity and density of fishes are relatively low in spite of high biomass of mesozooplankton and suprabenthos. In the outer part of the estuary, low turbidity and high concentration of oxygen are favourable of older fish populations which especially eat benthic fauna. In this area, a high macrozoobenthic biomass and a weak planktonic biomass are found.

Acknowledgements

This study was undertaken within the framework of the Seine-Aval program co-ordinated by the 'Région Haute Normandie'. The authors are grateful to the crew of 'N.O. Pluteus II', and 'N.O. Côte d'Aquitaine', Mrs C. Conq, J.C. Lorgeré and F. Proniewski for their valuable assistance in field sampling and, Mr. E. Thiébaut, and Miss Li-Lian Liu for comments on the manuscript, and two anonymous referees for their very useful remarks on the first version of the manuscript.

References

Anonymous, 1959. Symposium on the classification of brackish waters. Venice 8–14th April 1958. Archivio di oceanografia Limnologia Volume 11. Supplemento (Simposio sulla Classificazione della Acque Salmastre. Venizia 8–14 Aprile, 1958).

Arntz, W. E., 1978. Predation on benthos flounders, *Platichthys flesus* (L.) in the deeper parts of Kiel bay. Ber. dt. wiss. Kommn. Meeresforsch. 26: 70–78.

Avoine, J., 1981. L'estuaire de la Seine: sédiments et dynamique sédimentaire. Thèse 3ème cycle, Université de Caen, 236 pp.

Avoine, J., 1986. Evaluation des apports fluviatiles dans l'estuaire de la Seine. Actes Colloq. IFREMER 4: 117–124.

Bachelet, G., 1979. Les peuplements macrobenthiques de l'estuaire de la Gironde: évolution annuelle des paramètres démographiques. J. Rech. Océanogr. 4: 3–13.

Bakker, C., 1994. Zooplankton species composition in the Oosterschelde (SW Netherlands) before, during and after construction of a storm-surge barrier. Hydrobiologia 282/283: 117–126.

Baretta, J. W. & J. F. P. Malschaert, 1988. Distribution and abundance of zooplankton of the Ems estuary (North sea). Neth. J. Sea. Res. 22: 69–81.

Cabioch, L. (ed), 1986. La baie de Seine. Actes Colloq. IFREMER 4: 1–531.

Castel, J., 1981. Aspects de l'étude écologique du plancton de l'estuaire de la Gironde. Océanis 6: 535–577.

Dauvin, J. C. & P. Gillet, 1991. Spatio-temporal variability of demographic structure of *Owenia fusiformis* Delle Chiaje (Annelida-Polychaeta) from the Bay of Seine (eastern English Channel). J. exp. mar. Biol. Ecol. 152: 105–122.

Dauvin, J. C. & J. C. Lorgeré, 1989. Modifications du traîneau Macer-Giroq pour l'amélioration de l'échantillonnage quantitatif étagé de la faune suprabenthique. J. Rech. Océanogr. 14: 65–67.

Dauvin, J. C., J. C. Sorbe & J. C. Lorgeré, 1995. Benthic boundary layer macrofauna from the upper continental slope and the Cap Ferret Canyon. Oceanol. Acta 18: 113–122.

Desprez, M., 1981. Etude du macrozoobenthos intertidal de l'estuaire de la Seine. Thèse 3ème cycle, Université de Rouen, 186 pp.

Duval, P., 1982. Etude des nurseries littorales de la baie de Seine. Rapport GRPCM/ISTPM, 25 pp.

Frontier, S., 1972. Calcul de l'erreur sur un comptage de zooplancton. J. exp. mar. Biol. Ecol. 10: 121–132.

Gentil, F., J. P. Irlinger, B. Elkaïm & F. Proniewski, 1986. Premières données sur la dynamique du peuplement macrobenthique des sables fins envasés à *Abra alba* de la baie de Seine orientale. Actes Colloq. IFREMER 4: 409–420.

Healey, M. C., 1972. On the population ecology of the common goby in the Ythan estuary. J. Nat. Hist. 6: 133–145.

Heip, C. & P. Herman, 1995 (eds). Major Biological Processes in European Tidal Estuaries. Hydrobiologia 311: 1–7.

Lardeux, F., 1986. Biologie, écologie et dynamique de population de l'éperlan (*Osmerus eperlanus* L.) (Poisson, Clupéiforme, Osméridé) dans l'estuaire de la Loire. Thèse de Doctorat, Université de Nantes, 587 pp.

Marchand, J., 1981. Observations sur l'écologie de *Crangon crangon* (Linné) et *Palaemon longirostris* H. Milne Edwards (Crustacea, Decapoda, Natantia) – Estuaire interne de la Loire (France). Vie Milieu 31: 83–92.

Mees, J., 1994. The hyperbenthos of shallow coastal waters and estuaries: community structure and biology of the dominant species. PhD Thesis, Univ. of Ghent, Belgium, 212 pp.

Mees, J., N. Fockedey & O. Hamerlynck, 1995. Comparative study of the hyperbenthos of three European estuaries. Hydrobiologia 311: 153–174.

Meire, P. M., J. J. Seys, T. J. Ysebaert & J. Coosen, 1991. A comparison of the macrobenthic distribution and community structure between two estuaries in SW Netherlands. In M. Elliott & J. P. Ducrotoy (eds), Estuaries and Coasts: Spatial and Temporal Intercomparisons. Olsen & Olsen, International Symposium Series (Denmark) 221–230.

Mouny, P., Z. Wang & J. C. Dauvin, 1996. Premières données sur la structure spatio-temporelle du mésozooplancton et du suprabenthos de l'estuaire de la Seine. J. Rech. Océanogr. 21: 100–108.

Pomfret J. R., M. Elliott, M. G. O'Reilly & S. Phillips, 1991. Spatial and temporal patterns in the fish communities in two UK North Sea estuaries. In M. Elliott & J.P. Ducrotoy (eds), Estuaries and coasts: spatial and temporal intercomparisons. Olsen & Olsen, International Symposium Series (Denmark) 277–284.

Proniewski, F. & B. Elkaïm, 1980. Les peuplements benthiques subtidaux de l'estuaire de la Seine. Résultats préliminaires. C. R. heb. Acad. Sci., Paris 219, D: 545–547.

Robineau, B. & J. Marchand, 1984. Distribution des communautés macrozoobenthiques des substrats meubles de l'estuaire interne de la Loire (France). J. Rech. Océanogr. 9: 146–149.

S. A. U. M., 1980. Dossier n 4: a – eau et sédiments – M.E.B.S. – M.A.B.N., rapport, 98 pp.

Sautour, B. & J. Castel, 1995. Comparative spring distribution of zooplankton in three macrotidal European estuaries. Hydrobiologia 311: 139–151.

Soetaert, K. & P. Rijswijk van, 1993. Spatial and temporal patterns of the zooplankton in the Westerschelde estuary. Mar. Ecol. Prog. Ser. 97: 47–59.

Sorbe, J. C., 1981. La macrofaune vagile de l'estuaire de la Gironde. Distribution et migration des espèces. Modes de reproduction. Régimes alimentaires. Océanis 6: 579–592.

Thiébaut, E., L. Cabioch, J. C. Dauvin, C. Retière & F. Gentil, 1997. Spatio-temporal stability of the *Abra alba-Pectinaria koreni* muddy-fine sand community of the eastern Bay of Seine (English Channel). J. mar. biol. Ass. U.K. 77: 1165–1185.

Wang, X. & J. C. Dauvin, 1994. The suprabenthic crustacean fauna of the infralittoral fine sand community from the Bay of Seine (eastern English Channel): composition, swimming activity and diurnal variation. Cah. Biol. Mar. 35: 135–155.

Wang, Z., J. C. Dauvin & E. Thiébaut, 1994. Preliminary data on the near-bottom meso- and macrozooplanktonic fauna from the eastern Bay of Seine: faunistic composition, vertical distribution and density variation. Cah. Biol. Mar. 35: 157–176.

Wang, Z., E. Thiébaut & J. C. Dauvin, 1995. Spring abundance and distribution of the ctenophore *Pleurobrachia pileus* in the Seine estuary: advective transport and diel vertical migration. Mar. Biol. 124: 313–324.

Hydrobiologia **373/3374:** 349–359, 1998.
J.-C. Amiard, B. Le Rouzic, B. Berthet & G. Bertru (eds),
Oceans, Rivers and Lakes: Energy and Substance Transfers at Interfaces.

Sediment phosphorus and internal phosphate flux in the hydroelectric reservoir of Bort-les-Orgues, France

V. Ruban & D. Demare
Laboratoire Central des Ponts et Chaussées, B.P. 19, 44340 Bouguenais, France

Key words: reservoir, sediment, phosphate, flux, sequential extraction

Abstract

The hydroelectric reservoir of Bort-les-Orgues (France) plays an important role in regional tourism. However this function is under threat because the lake is likely to become eutrophic. External phosphate load is large (about 250 t y^{-1} of P are discharged to the lake) but internal flux also has to be taken into account. The potential of P release at the sediment-water interface was studied. Not all forms of phosphate are likely to be released, to know which part of the pool can be mobilized we used sequential extraction. In Bort-les-Orgues about 80% of P, mostly iron-bound, was mobile. The percentage of calcium-bound P did not exceed 20%. Phosphate and iron fluxes were measured in a laboratory experiment. As long as oxygen concentration remained above 0.5 mg l^{-1}, no release was observed. Phosphate and iron were released together when oxygen dropped to 0.5 mg l^{-1} and redox potential reached − 350 mV at the sediment-water interface. Such conditions were common during summer, and the corresponding P flux was 18 mg m^{-2} d^{-1} PO_4-P. This internal phosphate source should be taken into account in any project of lake restoration. Sequential extraction of phosphate and flux measurement are thus necessary tools for a better management of eutrophic water bodies.

Introduction

Eutrophication (i.e., the proliferation of algae due to an excess of nutrients) is a major water pollution problem all over the world, with economic consequences (eg., the preparation of drinking water becomes more difficult because of filter clogging by algae). This is recognized by the European Urban Wastewater Directive of 21 May 1991 (EEC 1991) in which eutrophic waters are identified as 'sensitive areas' with stricter standards for phosphorus and nitrogen discharge into receiving waters. Phosphorus (P) is a key factor in freshwater eutrophication (Vollenweider, 1968; OECD, 1982; Golterman & De Oude, 1991; Barroin, 1991; Boers & Van der Mollen, 1993). Phosphorus has two origins, i.e., external and internal. External P originates from point (industrial and domestic effluents) or diffuse sources (weathering, agricultural runoff), but it is also released by sediments, which act as an internal source (Holdren & Armstrong 1980, Logan 1982, Boström & Pettersson 1982, Premazzi & Provini 1985, Boers 1991). Sediment can contribute phosphate to the overlying waters at levels comparable to the external source (Rossi & Premazzi 1991).

The hydroelectric reservoir of Bort-les-Orgues, in the French Massif Central, which has been identified as a 'sensitive area' plays an important role in tourism. However, aquatic activities are under threat because the lake is likely to become eutrophic. The external phosphate load is large (about 250 t of P are discharged into the lake on an annual basis, Ruban & Fournier, unpublished data) but the internal flux also has to be accounted for.

Not all forms of phosphate are likely to be released, and it is important to know which part of the stock can be mobilized (in aquatic environment, the biogeochemical cycle of phosphorus is identical with that of phosphate, (Jahnke, 1988), therefore, 'phosphate' can be used instead of 'phosphorus'). Different forms of phosphate in sediments can be determined by sequen-

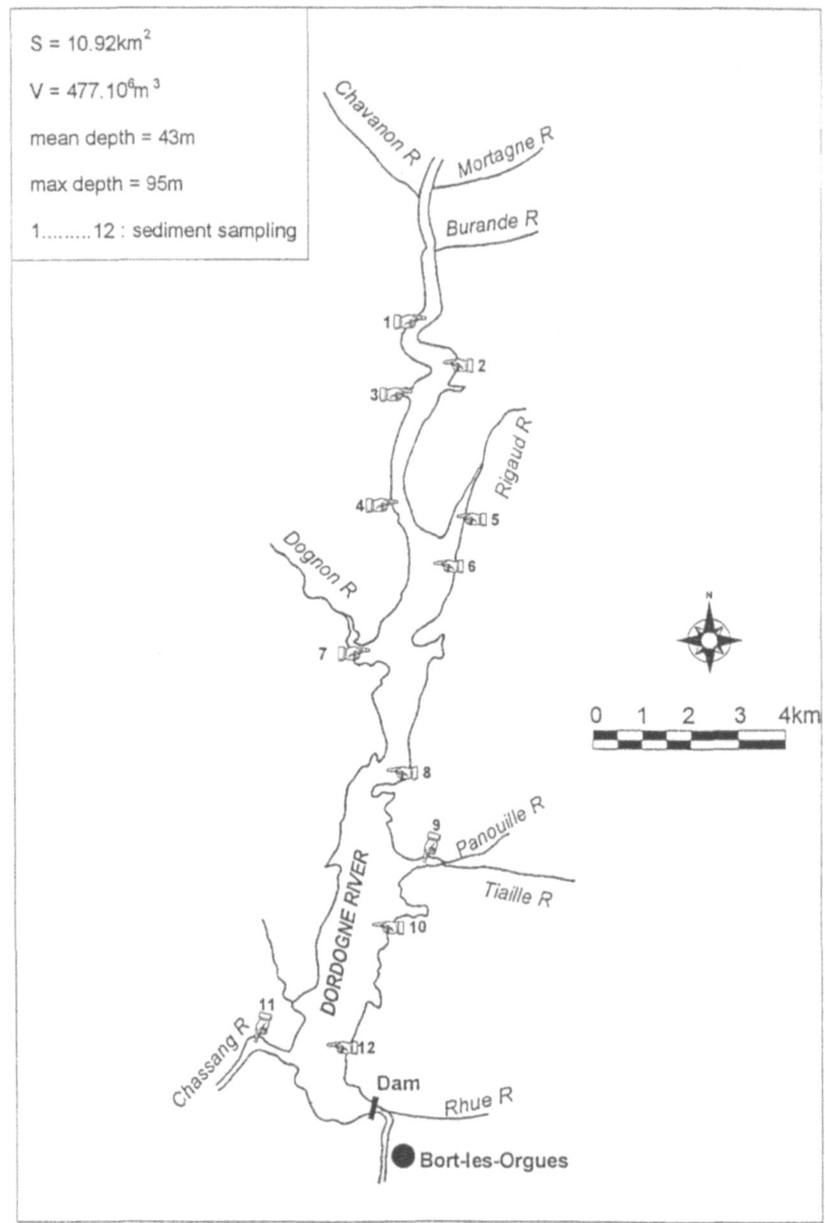

Figure 1. The hydroelectric reservoir of Bort-les-Orgues.

tial extraction. This technique, first developed for soils and then extended to sediments, has been widely used over the last two decades. Most of the schemes are defined by the reagents used to isolate or identify the different forms, i.e. they allow the quantification of phosphorus in a specific phase of a sediment. There are many available schemes, here, the Williams method was chosen (Williams et al., 1971, 1976, 1980; Manning et al., 1982, 1983; Burrus, 1984; Armengol et al., 1986; Armengol & Vidal, 1988; Thomas et al. 1991) because it is relatively simple. It utilizes NaOH to solubilize Fe and HCl to dissolve Ca; this method is not perfect though, and leads to resorption by carbonates in calcareous sediments (The Bort-les-Orgues sediments are not calcareous).

The aim of the present study was to evaluate the potential phosphate release and to quantify the flux at the sediment–water interface in the reservoir o

Bort-les-Orgues, and the measurement of P flux in a laboratory experiment.

Description of the study site

Location and geology

The hydroelectric reservoir of Bort-les-Orgues is located in the upper part of the Dordogne River, not far from its source (Figure 1); it is the largest reservoir of the French Massif Central. The main tributaries are the Chavanon, the Lys and the Dognon on the right bank and the Burande, the Mortagne, the Rhue and the Tialle on the left bank. The watershed area is 1745 km^2 (1010 km^2 for the Dordogne River and 735 km^2 for the Rhue River). The basin consists of volcanic rocks; there are no limestones. The mean slope of the watershed is steep (3%).

Hydrology and climate

The reservoir was commissioned in 1952, its main characteristics are as follows:
- length: 21 km
- max depth: 95 m
- mean depth: 43.7 m
- area: 10.92 km^2
- volume: 477.10^6 m^3
- retention time: 2.5 y

Annual means based on 6 measurements are:
- transparency (Secchi disk): 2.5 m (ranging from 1.1 m in July to 3.3 m in winter)
- mean Total Phosphorus: 60 mg m^{-3} (upstream), 78 mg m^{-3} (downstream) with seasonal variations ranging from 10 mg m^{-3} in winter to 110 mg m^{-3} from July to October.
- mean chlorophyll a: 11.5 mg m^{-3} (upstream), 18 mg m^{-3} (downstream), with a minimum of 3 mg m^{-3} and a maximum of 25 mg m^{-3}.

The mean annual discharge into the Lake is 63 m^3 s^{-1}, with the Rhue River contributing 50%, and the Dordogne 20%.

The climate is temperate, with an average annual rainfall of 900 mm.

Land use and population

Land use in the basin is dominated by agriculture (55% of the area), cattle breeding for milk and meat, and forests, which comprise 28% of the basin. Industry is mainly represented by food processing (dairy products), whose effluents are often untreated. Population is concentrated in four major centres: La Bourboule, Le Mont Dore, Murat-le-Quaire and Riom-es-Montagne, with dephosphatation of the effluents in a sewage treatment plant. Though there is no intensive use of fertilizers, the nutrient load from agriculture is important and represents the major source of nutrient pollution in the Bort-les-Orgues reservoir.

Material and method

Sampling and preservation

Twelve sediment samples were collected all over the reservoir by means of a grab sampler (Figure 1), only the upper 5 cm were sampled. The samples were put into polyethylene bags and stored at 4 °C prior to treatment. The sediment was passed through a 2 mm sieve in order to eliminate coarse plant fragments and then freeze-dried. Though not perfect, this technique induces little change in the sediment and allows reproducible analyses even after several months.

Phosphate flux at the sediment-water interface was measured through a laboratory experiment in which a plexiglass chamber was used. The sediment (approx. 60 kg of the upper 5 cm) was collected by means of a grab sampler in the lower part of the reservoir where P concentrations were the highest, and immediately transferred to the laboratory. Water (approx. 100 l) was sampled at the same location. The sediment was put into a tank and water was poured in gently to avoid resuspension. The system was allowed to settle for 16 h before the plexiglass chamber was put in place. The chamber was 34.6 cm high, with a base of 26.9×26.9 cm. Water samples were collected by means of a peristaltic pump, and the chamber was tight. Dissolved oxygen inside the chamber was controlled (Figure 2). Water conditions i.e., temperature, pH, dissolved oxygen, and Eh were registered by probes inside the plexiglass chamber:
- temperature (Ponselle HP);
- oxygen (YSI);
- pH (Ponselle HP);
- Eh (Tacussel PHN81, Ag/AgCl electrode).

Water samples were filtered (Millipore 0.45 μm pore filter) under nitrogen and acidified with HNO_3 (iron determination) or HCl (phosphate and sulphate determinations) to pH < 2 and stored in polyethylene bottles previously soaked in hydrochloric acid and rinsed with deionized water. The bottles were stored

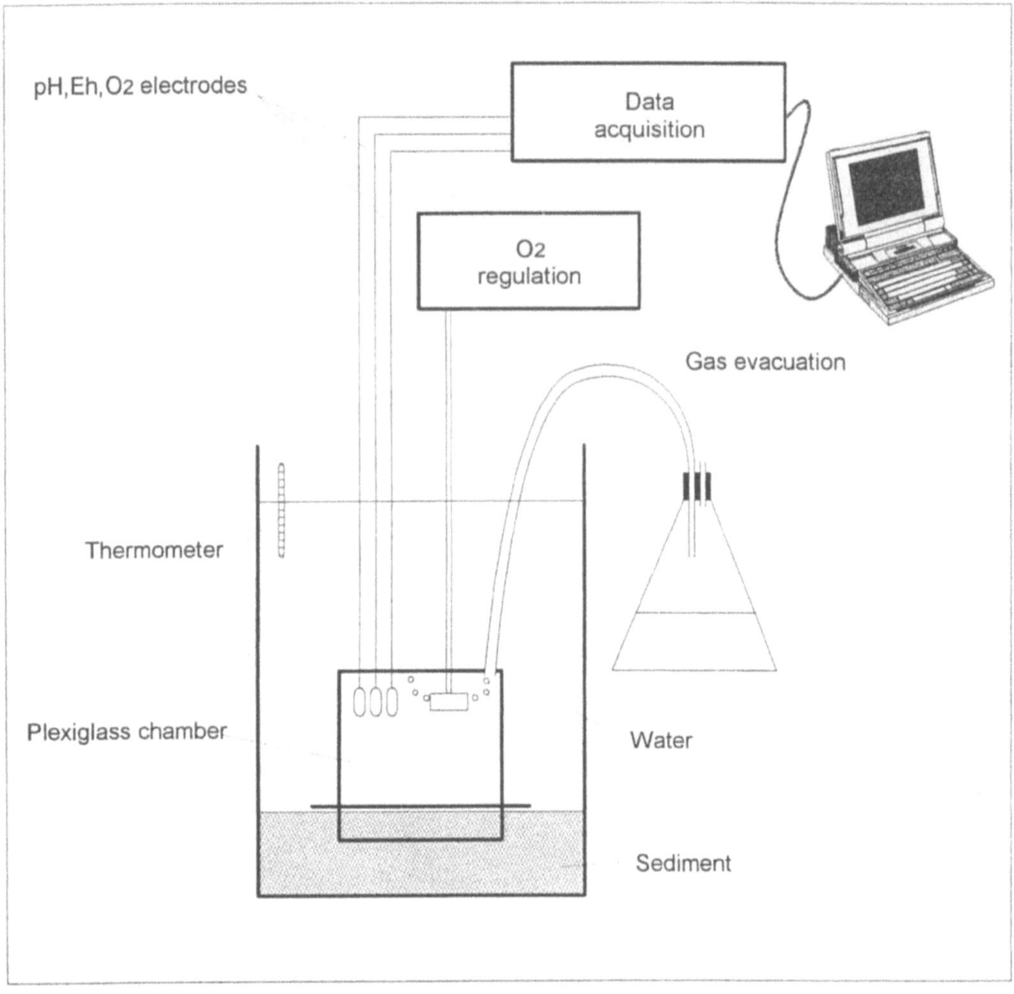

Figure 2. The plexiglass chamber used for the laboratory experiment.

at 4 °C prior to analyses which were performed the same day.

Sample analysis

The following determinations were carried out on the fraction of the sediment of particle size less than 2 mm:

- loss on ignition (calcination at 550 °C for 2 h);
- particle size distribution (laser diffraction, Malvern Mastersizer MS 1005);
- total phosphorus (TP) (persulphate digestion);
- sequential extraction of phosphate according to (Williams et al., 1971, 1976, 1980); the initial scheme included a CDB extraction which was later omitted (Burrus, 1984). This operationally defined method is composed of three steps i.e. 1 M NaOH (step 1), 3.5 M HCl (step 2), and 1 M HCl (step 3) and allows the determination of the following forms:
  - Apatite Phosphorus (AP). This form represents orthophosphate ions involved in the
  - structure of apatite, it is a Ca-bound form;
  - Organic Phosphorus (OP) represents P linked to organic matter either autochtonous (plankton) or allochtonous (wastewaters, plant fragments);
  - Non Apatite Inorganic Phosphorus (NAIP). This form is mostly bound toiron/aluminium oxides or hydroxides, it is of anthropogenic origin (wastewaters, agriculture).

All forms extracted were measured as orthophosphate according to the Murphy & Riley spectrophotometric method (Murphy & Riley 1962), and so were phosphates in water samples;

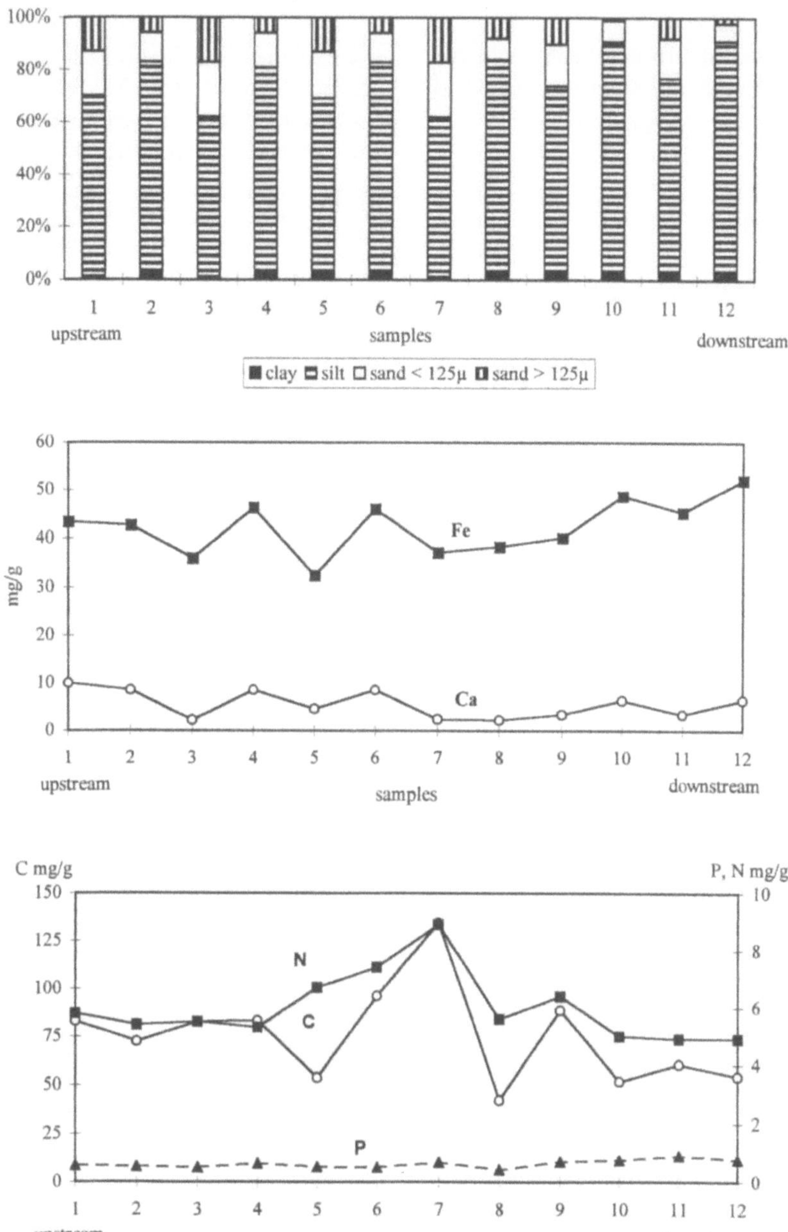

Figure 3. Bort-les-Orgues – Characteristics of the sediment.

– Sulphate. Ion chromatography (Dionex DX 100) was used for sulphate determination.
– Iron. Flame atomic absorption spectrometry (Varian Spectra A 300) was used both for sediment and water analysis of Fe and Ca.
– organic carbon was measured according to the French Standard NF X31 109 (AFNOR 1996).
– Kjeldhal nitrogen (TKN) was measured according to the NF ISO 11 261 standard (AFNOR 1996). TKN was considered as equivalent to organic N since NH_4^+ was negligible in all samples.

All chemicals used were ultra pure. Note that Ca, Al, NO_3, NH_4^+ and Kjeldhal N were also measured in water samples but the results are not reported here. The detection limits were 10 $\mu g\,l^{-1}$ for Fe, PO_4^{3-}, SO_4^{2-} and 0.01 mg g^{-1} for AP, OP, NAIP and TP. The average value of two replicates was taken for each determination. Quantification of phosphate, sulphate and iron was based upon calibration curves of standard solutions.

Results and discussion

Characteristics and origin of the sediment

The main characteristics of the sediment are reported in Figure 3. The sediment of the reservoir was mainly silt (approx. 75%, whereas the percentage of sand and clay were 20–25% and less than 2%, respectively). Loss on ignition was relatively high (17%). The sediment tended to be finer downstream but this tendency, related to the dynamics and morphology of the reservoir, was not as strong is in other reservoirs in the Massif Central, such as St Etienne Cantales (Gandais-Ruban et al., 1993). Note that pollutants (heavy metals, phosphate) preferentially sorb onto these fine sediments.

Calcium concentration was low (around 5 mg g^{-1}), whereas iron concentration was high (45 mg g^{-1}) reflecting the nature of the watershed (Figure 3b). The large carbon content (75 m g^{-1} on average) was mainly due to organic carbon as carbonate concentration was low. There was an increase in carbon from sites 5 to 7 (Figure 3c), which could be explained by input from the small Rigaud stream. The C/N ratio has been utilized by a number of authors to evaluate the source of organic carbon (Kemp et al., 1977; Damiani & Thomas, 1974; Thomas et al., 1991). Sediments of terrestrial origin have a C/N ratio (by weight) around 14–20, while phytoplankton, the major source of autochtonous organic matter in lakes produces ratios from 7 to 9. The C/N ratio averages 14.8 in the sediments of Bort-les-Orgues, attesting to a predominance of terrestrial inputs.

Total P concentration in the sediment was high (1.84 mg g^{-1} DW on average). This phosphate had two main origins:

- domestic and industrial effluents discharged into the lake by the tributaries; these are easily identifiable point sources;
- agricultural runoff, difficult to quantify because it is a diffuse source; effluents from agriculture represented about 2/3 of total P input to the reservoir (Ruban & Fournier, unpublished data). Atmospheric deposition and erosion accounted for less than 1%.

TP increased from upstream (1.1 mg g^{-1} DW) to downstream (2.56 mg g^{-1} DW); this tendency could be explained, as already mentioned, because of finer particles likely to sorb P downstream (the correlation between silt and TP was good $r^2 = 0.92$, $n = 12$) and also by inputs from the Rhue River which contributed about 50% of the TP load reaching the reservoir (128 t y^{-1}, the annual load being 250 t y^{-1}, Ruban & Fournier, unpublished data). It should be noted that TP concentrations in the water column also were higher downstream. The link between P concentration in a sediment and its concentration in water has already been highlighted (Nurnberg 1984, 1987). The higher the P content of the sediment, the greater the risk of eutrophication.

Sequential extraction of phosphate

The determination of TP is not sufficient for a good knowledge of P dynamics in a reservoir. Not all forms play the same role in the development of eutrophication. Figures 4a and 4b show the concentration of several P forms in the sediment. Non-Apatite Inorganic Phosphorus (NAIP) and Organic Phosphorus (OP), which are the forms likely to be released, represent about 80% of total sediment P (Figure 4a). Mean NAIP and OP concentrations were 0.92 and 0.63 mg g^{-1}, respectively (Table 1 and Figure 4b). De Groot (1990) mentioned that the use of aggressive methods (using HCl-NaOH such as in Williams' method) to extract inorganic phosphate from sediment tends to underestimate the organic phosphate pool. This might be the case here but would not alter our results, i.e. the sediments of Bort-les-Orgues have a large potential for P release.

Apatite phosphorus (AP) which has a detrital origin and is not available to algae accounted for 20% of TP (mean concentration 0.29 mg g^{-1}DW). The validity of the extraction sequence and the accuracy with which it has been carried out may be tested by comparing the sum of the single fractions with the total measured separately (Barbanti & Sighinolfi, 1988). Such a comparison (Table 1) shows that $TP - \Sigma(AP + NAIP + OP) = 0.01$, the mean difference is thus negligible.

TP was strongly correlated with Fe ($r^2 = 0.90$, $n = 12$), in agreement with classical theory (Mortimer, 1941; Ostrofsky, 1987; Jensen et al., 1992); the correlation was even better between NAIP (iron-bound P) and Fe ($r^2 = 0.93$, $n = 12$). The importance of iron with regard to P has been highlighted by Sondergaard et al. (1996) who observed that in shallow Danish lakes, TP

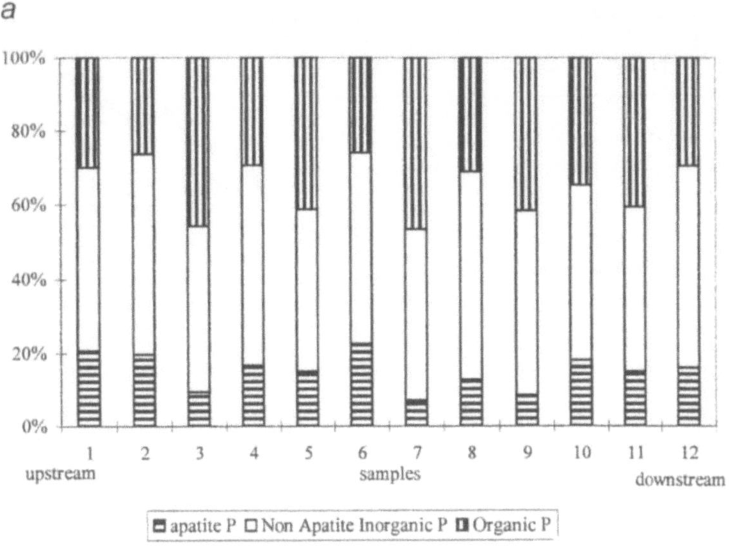

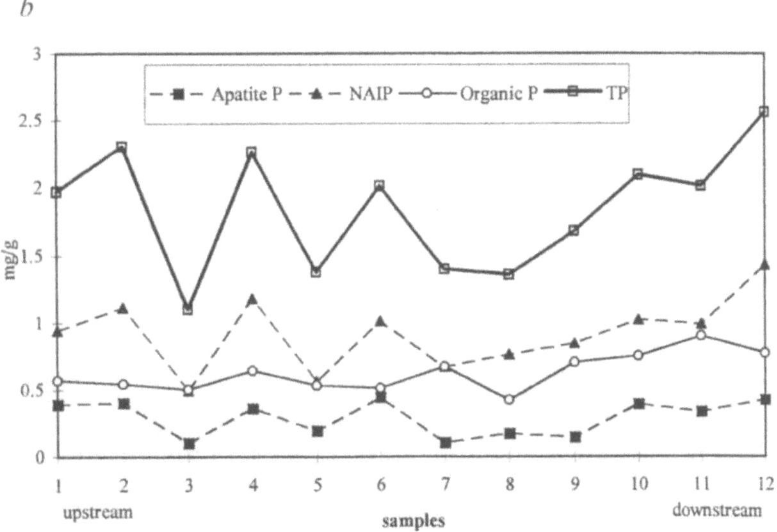

Figure 4. Bort-les-Orgues – Sequential extraction of phosphate in the sediment according to the Williams' method (modified). (a) relative distribution of the P forms. (b) concentration of the different P forms.

Table 1. Forms of phosphate (sequential extraction according to Williams'method modified) in the hydroelectric reservoir of Bort-Les-Orgues.

	NAIP mg g^{-1} (d.w.)	OP mg g^{-1} (d.w.)	AP mg g^{-1} (d.w.)	TP measured mg g^{-1} (d.w.)	Σ AP + NAIP + OP mg g^{-1} (d.w.)	Fe mg g^{-1} (d.w.)
Min	0.49	0.42	0.10	1.10	1.09	32.4
Max	1.43	0.90	0.44	2.56	2.62	52.1
Mean	0.92	0.63	0.29	1.83	1.84	47.8
Std. dev.	0.26	0.13	0.13	0.43	0.44	6.15

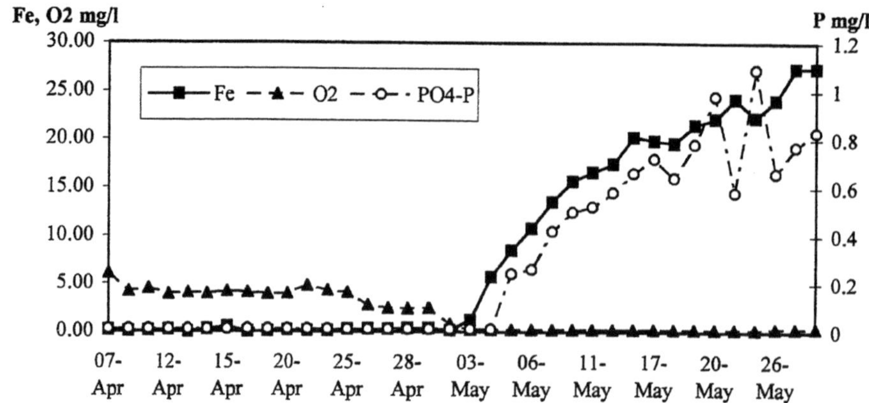

Figure 5. Changes in phosphate, iron and oxygen concentrations in the plexiglass chamber during the experiment.

concentrations were high (> 3 mg P g^{-1} DW) when Fe exceeded 20–30 mg g^{-1}.

Flux at the sediment-water interface

Phosphate, sulphate and iron fluxes were measured in a plexiglass chamber between 7 April and 30 May. Dissolved oxygen, as well as pH and Eh were also measured; the water temperature during the experiment was 20 °C. The change in concentration of orthophosphate, iron and oxygen is shown in Figure 5.

Fluxes were calculated taking into account: the difference of concentration (ΔC) during the period considered (Δt), the volume of the plexiglass chamber (V), and the sediment surface (S). Hence the flux (in mg m^{-2} d^{-1} DW):

$$\Phi = \frac{\Delta C \times V}{\Delta t \times S}$$

Note that though fluxes were not measured *in situ*, other experiments have shown similarity between laboratory and *in situ* fluxes (Cavalcante, 1995; Ruban & Fournier, unpublished data). This allows us to consider the flux measured in the plexiglass chamber as representative of *in situ* conditions in the zones where P concentration in the sediment is close to that used for the experiment (about one half of the reservoir).

As long as oxygen concentration above the sediment remained above 0.5 mg l^{-1} (which was the case during the first part of the experiment when oxic conditions were maintained), no P flux was observed. The release started when oxygen dropped below 0.5 mg l^{-1} and redox potential was low; such conditions occur in the reservoir in the summer months. Note that Fe release started a day earlier than P release but the change in both concentrations in the plexiglass chamber was similar until 20 May (Figure 5). Thereafter, iron concentration continued to increase regularly, whereas P fluctuated before increasing again. Low redox potential and oxygen depletion favoured the reduction of Fe^{3+} to Fe^{2+}; and iron hydroxides in the oxidized surface layer were reduced to soluble Fe^{2+}. Iron diffused to the water column, hence the increase measured in the plexiglass chamber. Phosphates, sorbed onto iron hydroxides, were thus also released. Most of the P released was PO_4^{3-}, with little organic P in the water.

Sulphates had the opposite behaviour (Figure 6): they increased from 3.65 to 5.72 mg l^{-1} until oxygen dropped below 0.5 mg l^{-1} and then decreased to 0.19 mg l^{-1}. This could be due to a precipitation of sulphur during the reduction of sulphates, with Fe^{2+} forming pyrite typical of reduced environments. However, it can be seen from Figure 7 that pH increased from 6.4 in oxic conditions to 6.7 in anoxia, while redox potential varied from + 300 mV to - 300 mV in anoxy. The Eh-pH diagram for the Fe-S-H_2O system (Lowson, 1982) shows that pyrite (FeS_2) is unstable in such conditions, with formation of H_2S. pH should be much lower for pyrite to be stable, while FeS exists at pH 6–7 and Eh < - 500 mV; therefore, the observed decrease in S concentration is most probably due to H_2S formation.

The phosphate flux was 18 mg m^{-2} d^{-1} (PO_4-P), that of iron 337 mg m^{-2} d^{-1}. During the same time, a negative sulphate flux (-68 mg m^{-2} d^{-1}) was observed (Figure 6). The sediment surface Fe:P ratio has been shown to be a useful tool to predict release from aerobic sediments in shallow lakes (Jensen et al., 1992). These authors considered that P release is limited as long as Fe:P is above 15 (by weight). This, however,

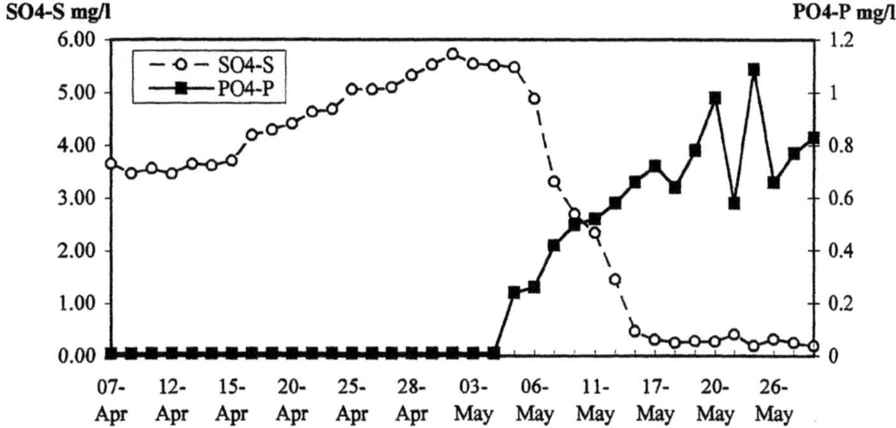

Figure 6. Changes in phosphate and sulphate concentrations in the plexiglass chamber during the experiment.

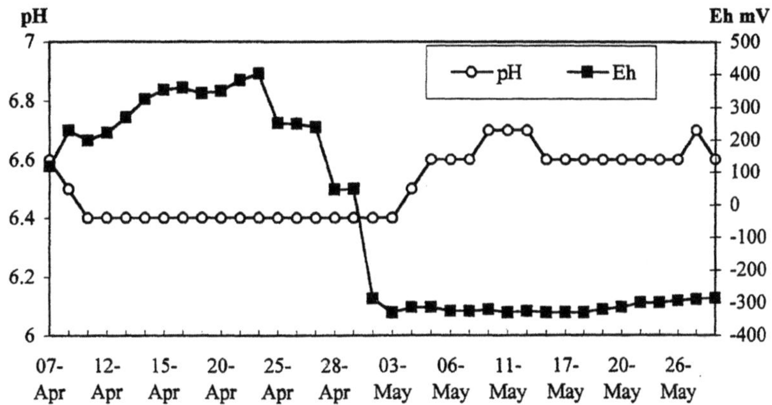

Figure 7. Variations of pH and Eh in the water of the plexiglass chamber during the laboratory experiment.

Table 2. Bort-Les-Orgues, laboratory experiment – P fractions in the sediment at the beginning (Ti) and at the end (Tf) of the experiment.

	Organic P	Apatite P	Non Apatite Inorganic P
Ti	0.64	0.36	1.18
Tf	0.64	0.36	1.16

does not seem to be the case in an anoxic environment since our results showed that P was released at a mean Fe:P ratio of 26. One could suggest that rather the Fe:Fe-P ratio should be used as this would give more information on potential phosphate release because iron-bound P provides the best indication of the mobile P pool. This ratio averages 52 (by weight) in the sediment of Bort-les-Orgues. These sediments are rich in iron, and hence have a strong P-sorption capacity. Phosphate release appeared to be mostly controlled by physical-chemical phenomena (Eh, pH, O2); this is confirmed by the comparison between P fractions at the beginning (T_i) and at the end (T_f) of the experiment (Table 2). The sequential extraction shows no difference between T_i and T_f for OP and AP, whereas NAIP concentrations varied from 1.18 to 1.16 mg g^{-1}. A significative contribution of bacteria would probably have affected OP concentration, which was not the case; however, the influence of bacteria has not been investigated. Also, bioturbation seemed of minor importance (no bioturbation was observed).

An attempt was made to estimate the annual P release in the reservoir. As mentioned, the flux measured in the laboratory was supposed to be a good estimator of the *in situ* flux (Cavalcante, 1995) and it was assumed, on the basis of the topography of the reservoir and the characteristics of the sediment, that release occurred in about half of the lake during 3–4 months (summer, during which oxygen concentration drops below 0.5-1 mg m^{-3} and redox potential is low). The annual P release is then 9–12 t, which needs to be compared to P mass in the water. This mass averages 33 t (mean lake volume 477.10^6 m^3, and mean P concentration 70 mg m^{-3}). Therefore, the quantity of P released at the sediment-water interface is far from negligible and must be taken into account in restoration projects.

Acknowledgements

We thank the Adour-Garonne Water Agency and the County Councils of Cantal, Corrèze and Puy-de-Dôme for their financial support. We are also grateful to EPI-DOR (Etablissement Public Interdépartemental de la Dordogne).

References

AFNOR, 1996.Recueil de normes françaises. qualité des sols.

Armengol, J., M. Crespo, J. A. Morgui & A. Vidal, 1986. Phosphorus budget and forms of phosphorus in the Sau reservoir sediments: an interpretation of the limnological record. Hydrobiologia 143: 331–336.

Armengol, J. & A. Vidal, 1988. The use of different phosphorus fractions for the estimation of the trophic evolution of the Sau reservoir. Arch. Hydrobiol. Beih. Ergebn. Limnol. 30: 61–70.

Barbanti, A. & G. P. Sighinolfi, 1988. Sequential extraction of phosphorus and heavy metals from sediments: methodological considerations. Envir. Technol. Lett. 9: 117–134.

Barroin G., 1991. La réhabilation des plans d'eau. La Recherche 238: 1412–1422.

Boers P., 1991. The release of dissolved phosphorus from lake sediments. Thesis Univ. Wageningen, 131P.

Boers, P. C. M. & D. T. Van der Mollen, 1993. Control of eutrophication in lakes: the state of the art in Europe. Eur. Wat. Poll. Control.: 19–25.

Boström, B. & K. Pettersson, 1982. Different patterns of phosphorus release from lake sediments in laboratory experiments. Hydrobiologia 170: 229–244.

Burrus, D., 1984. Contribution à l'étude du transfert du phosphore dans le Rhône alpin. Thesis Univ. Genève n° 2135, 100 pp.

De Groot, C. J., 1990. Some remarks on the presence of organic phosphate in the sediment. Hydrobiologia 207: 303–309.

Cavalcante, P., 1995. Etude sur la mobilisation du phosphore, des formes azotées et de quelques métaux associés dans les vases eutrophes: expérimentation sur les vases de l'Erdre *in situ* et en laboratoire. Thesis Univ. Nantes, 187 pp.

Damiani, V. & R. L. Thomas, 1974. Mercury in the sediments of the Pallanza Basin. Nature 251: 696–697.

EEC Directive, 1991. Urban Wastewater Directive n° 91/271, 21 May 1991.

Gandais-Ruban, V., D. Demare & A. Fournier, 1993. Etude du phosphore dans une retenue hydroélectrique eutrophe. Cas de St Etienne Cantalès, France. Bull. liaison Labo. P. et Ch. 186: 79–84.

Golterman, H. L. & N. T. De Oude, 1991. Eutrophication of lakes, rivers and coastal seas. The Handbook of Environmental Chemistry, O. Hutzinger (ed.) 5A: 79–124.

Holdren, G. G. & D. E. Armstrong, 1980. Factors affecting phosphorus release from intact sediment cores. Envir. Sci. Technol. 1: 79–87.

Janhke, R. A., 1988. The phosphorus cycle. Chap. 14. In Butchers S. S. (eds), Global Biogeochemical Cycles. Academic Press, Harcourt Brace Jovanovich Pub., London.

Jensen, H. S., P. Kristensen, E. Jeppesen & A. Skytthe, 1992. Iron: phosphorus ratio in surface sediments in shallow lakes. Hydrobiologia 235/236: 731–743.

Kemp, A. L. W., R. L. Thomas, H. K. T. Wong & L. Johnston, 1977. Nitrogen and C/N ratio in the sediments of Lakes Superior, Huron, St Clair and Ontario. Can. J. Earth Sci. 14: 2402–2413.

Logan, T. J., 1982. Mechanisms for release of sediment-bound phosphate to water and the effects of agricultural land management on fluvial transport of particulate and dissolved phosphate. Hydrobiologia 92: 519–530.

Lowson, R.T., 1982. Aqueous oxidation of pyrite by molecular oxygen. Chem. Reviews 5: 461–497.

Manning, P. G. & W. Jones, 1982. The binding capacity of ferric hydroxides for non apatite inorganic phosphorus in sediments of the depositional basins of Lakes Erie and Ontario. Can. Miner. 20: 169–176.

Manning, P. G. & K. R. Lum, 1983. Forms of iron phosphorus and trace metal ions in a layered sediment core from Lake Ontario. Can. Miner. 21: 121–128.

Mayer, T. & E. De Losreyes, 1996. Phosphorus and metal contaminant transport in Two Southern Ontario rivers: the Grand River and its tributary, the Nith River. Wat. Qual. Res. J. Can. 1: 119–151.

Mortimer, C. H., 1941. The exchange of dissolved substances between mud and water in lakes. J. Ecol. 29: 280–339.

Murphy, J.& J.P. Riley, 1962. A modified single solution method for the determination of phosphate in natural waters. Anal. Chim. Acta 27: 31–36.

Nurnberg, G., 1984. The prediction of internal phosphorus load in lakes with anoxic hypolimnia. Limnol. Oceanogr. 1: 111–124.

Nurnberg, G., 1987. Prediction of phosphorus release rates from total and reductant soluble phosphorus in anoxic lake sediments. Can. J. Fish. aquat. Sci. 45: 453–462.

OCDE, 1982. Eutrophisation des eaux. Méthodes de surveillance, d'évaluation et de lutte. Organisation de Coopération et de Développement Economiques. Paris: 164 pp.

Ostrofsky, M. L., 1987. Phosphorus species in the surficial sediments of lakes of Eastern North America. Can. J. Fish aquat. Sci. 44: 960–966.

Premazzi, G. & A. Provini, 1985. Internal P loading in lakes: a different approach to its evaluation. Hydrobologia 120: 23–33.

Rossi, G. & G. Premazzi, 1991. Delay in lake recovery caused by internal loading. Wat. Res. 5: 567–575.

Sondergaard, M., J. Windolf & E. Jeppesen, 1996. Phosphorus fractions and profiles in the sediment of shallow Danish lakes as related to phosphorus load, sediment composition and lake chemistry. Wat. Res. 4: 992–1002.

Thomas, R. L., S. Santiago, V. Gandais, L. Zhang & J.-P. Vernet, 1991. Forms of particulate phosphorus and the carbon/nitrogen ratio as indicators of phosphorus origins in aquatic systems. Wat. Poll. Res. J. Can. 4: 433–451.

Vollenweider, R. A., 1968. Scientific fundamentals of the eutrophication of lakes and flowing waters, with particular reference to nitrogen and phosphorus as factors in eutrophication. Techn. Rep. DAS/CSI, 68:27, OECD, Paris: 192 pp.

Williams, J. D. H., J. K. Syers, S. S. Shukla, R. F. Harris & D. E. Armstrong, 1971. Levels of inorganic and total phosphorus in lake sediments as related to other sediment parameters. Envir. Sci. Technol. 11: 1113–1120.

Williams, J. D. H., J.-M. Jaquet & R.L. Thomas, 1976. Forms of phosphorus in the surficial sediments of lake Erie. J. Fish. Res. Bd Can. 33: 413–429.

Williams, J. D. H., T. Mayer & J. O. Nriagu, 1980. Extractibility of phosphorus from phosphate minerals common in soil and sediments. Soil Sci. Soc. Am. J. 44: 462–465.

Hydrobiologia **373/374:** 361–377, 1998.
J.-C. Amiard, B. Le Rouzic, B. Berthet & G. Bertru (eds),
Oceans, Rivers and Lakes: Energy and Substance Transfers at Interfaces.

Impact of a flood event on the biogeochemical behaviour of a mesotrophic alpine lake: Lake Bourget (Savoy)

B. Vinçon-Leite[1], P.-E. Bournet[1], X. Gayte[2], D. Fontvieille[2] & B. Tassin[1]
[1] *CERGRENE – Ecole Nationale des Ponts et Chausshes, Ecole Nationale du Génie Rural des Eaux et des Forêts, F-77455 Marne la Vallée, Cedex 2, France*
[2] *GRETI (and ESA CNRS n° 5023), Université de Savoie, 73 376 Le Bourget du Lac, France*

Key words: lake, flood, interface, nutrients, bacteria, particles

Abstract

A flood event which occurred during the onset of thermal stratification and of algal growth (March 1996) was studied in mesotrophic Lake Bourget (France). Both physical and biogeochemical processes occurring during this episode were assessed. The dominant effect was a decrease of nutrient concentrations along the river–lake ecotone. This phenomenon seemed mainly linked to biological factors: stimulation of the spring algal and bacterial growth. After the flood, the algae aggregated with allochthonous particles brought by the river and were rapidly eliminated by sedimentation. The effect of this early spring flood, seemed spatially restricted and rather neutral as regards lake water quality. In the lake region crossed by the river plume, the overall effect after the flood was a decrease of P available for the phytoplankton. In the open water, the P concentration was probably not modified by the load supplied by the flood.

Introduction

As pointed out by many authors, lakes exhibit their strongest gradients along the vertical. The order of magnitude of horizontal mixing time scale is the week while the order of magnitude of vertical mixing time scale is the year (Imboden, 1990). Thus, horizontal heterogeneity has been largely neglected and major efforts in limnology are aimed at improving the overall knowledge of lake behaviour by studying the pelagial. The inputs from the watershed are taken into account on large space and time scales but small space and time scale phenomena, including the exchanges of energy, dissolved and particulate species, biotic and abiotic particles in the lake-river interface, are rarely assessed. Nevertheless, during flood events, the interactions between the incoming rivers and the lake may become significant. Field data concerning the physical and chemical processes occurring in the interface region during such periods, and the ensuing biological response, are scarce, the highly random nature of floods preventing the planning of surveys.

Table 1. Mooring trap periods during spring 1996

Dates	Duration	Period n°
March 1st–14th	13 days	1
March 14th–29th	15 days	2
March 29th–April 9th	11 days	3
April 9th–29th	20 days	4

Physical impact of a flood is caused by the density current created by the incoming river. Depending on the density difference between lake and river water, a surface overflow, an intermediate depth interflow or a near-bottom underflow is created (Carmack et al., 1979). The intrusion of the river brings about a modification of the lake water circulation. The turbulence associated with this intrusion originates the entrainment of surrounding lake water. Moreover, a flood occurring in spring, during the set-up of the thermal stratification, can affect the depth of the thermocline (Bournet, 1996). Frequently, in small watersheds,

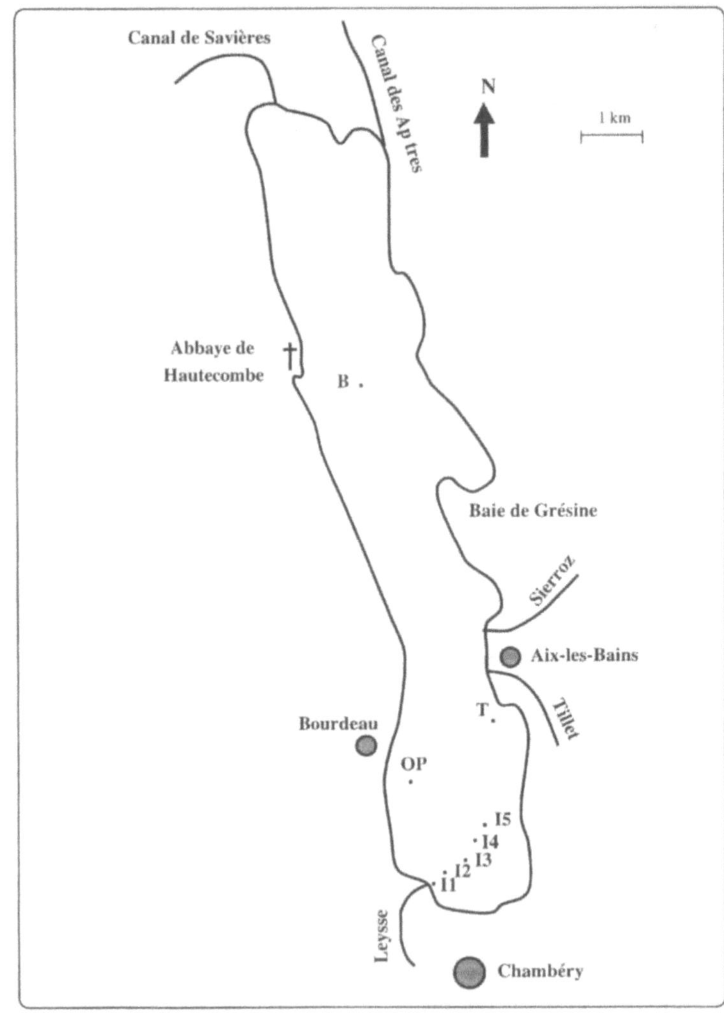

Figure 1. Sampling points during the flood.

flood episodes are associated with meteorological perturbations, especially high wind speeds and cooling of air temperature. Therefore, interactions between the mixing effects driven by meteorological conditions and those caused by the flooding river itself are closely overlapping.

From a biogeochemical point of view, a flood may cause conflicting effects. The suspended solids transported by the turbid inflow into the lake decrease light penetration in the water column and therefore limit algal growth (Tilzer, 1988; Reynolds, 1984). The particulate matter derived from watershed erosion (e.g. small clay particles, metallic oxides) can have a high binding capacity for toxic substances and nutrients (Stabel & Geiger, 1985; Sigg et al., 1987; Froelich, 1988) and decrease the bioavailability of these species. This is particularly important for phosphate, which is the limiting nutrient of primary production in many temperate lakes. Soil leaching also increases dissolved organic matter in river water (Boissier & Fontvieille, 1993). The natural dissolved organic matter (DOM) can present a high complexing capacity and can enhance the adsorption of toxics and nutrients (Zumstein & Buffle, 1989; Buffle & Leppard, 1995).

But a flooding river can also introduce to a lake a large amount of nutrients which increases the nutrient stock in the water column (CEMAGREF, 1985; Balland, 1980; Dorioz et al., 1989). Therefore, flood events are suspected to be responsible of a significant amount of annual nutrient loading. In summer, they may contribute to epilimnion nutrient replenishment. Microbial biota is one of the biological component

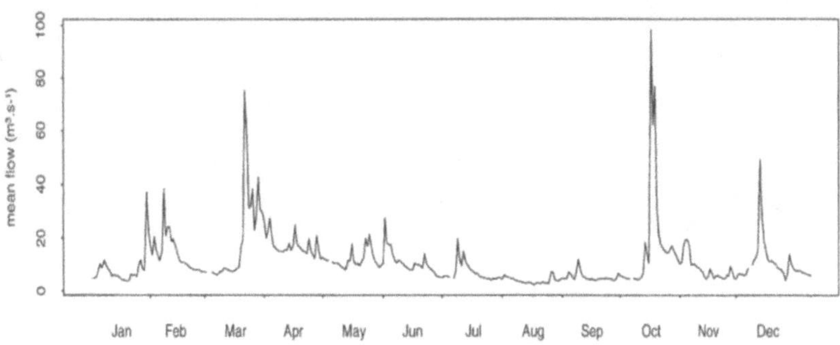

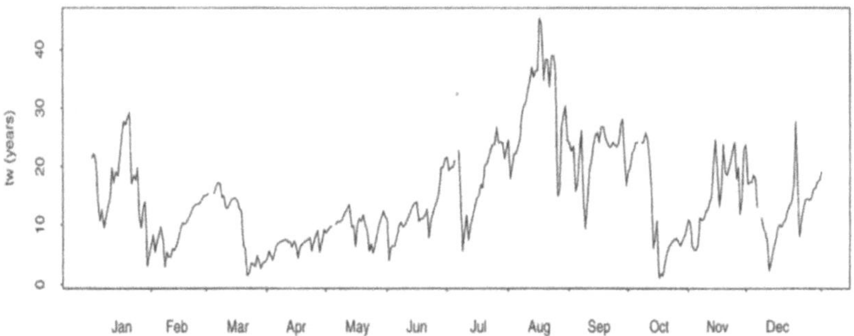

Figure 2. 1988 river Leysse mean flows and instantaneous water residence time.

Table 2. Pre-flood characteristics of river and lake water

	River Leysse (9th February 1996)	Lake Bourget (5th March 1996)
Total P ($\mu g\ l^{-1}$)	48	37
Particulate P ($\mu g\ l^{-1}$)	17	4
Total soluble P ($\mu g\ l^{-1}$)	31	33
Soluble reactive P ($\mu g\ l^{-1}$)	26	26
TSM (mg l^{-1})	5.5	0.8
BDOC (mg l^{-1})	0.54	0.28
RDOC (mg l^{-1})	2.05	2.17
Total number bacteria (cells ml^{-1})	5.88 10^{+5}	9.17.10^{+5}
Bacterial production ($\mu gC\ l^{-1}\ h^{-1}$)	0.71	0.72

highly sensitive to such disturbances (Brock, 1966; Fletcher et al., 1987). If little is known about the influence of flood events on phytoplankton behaviour (Elber & Schanz, 1990), even less is known about the microbial component (Gayte & Fontvieille, submitted).

Finally, the overall impact of a flood event on a lake is difficult to assess as it depends on four main factors: (1) season, (2) the magnitude of the discharge, (3) the characteristics of the lake and the river water and (4) the suspended matter load. Literature conclusions are often conflicting (Burkholder, 1992).

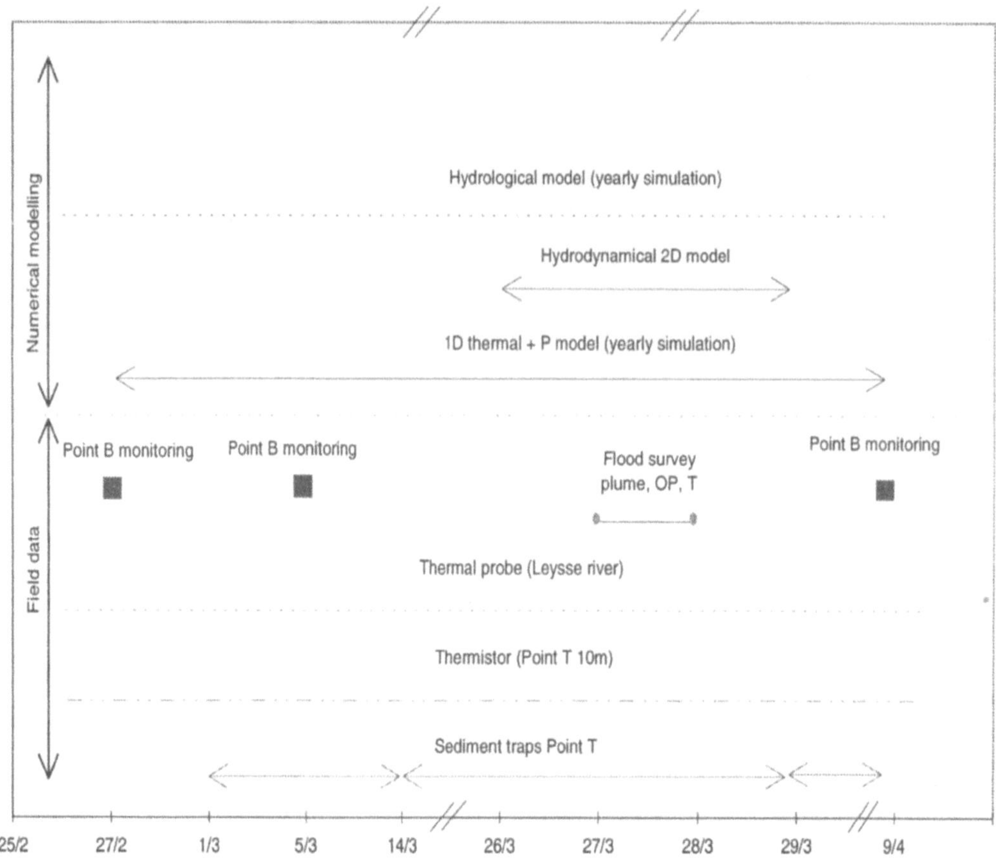

Figure 3. General organization of the flood survey: field measurements and numerical modelling.

Table 3. Na^+, Cl^-, SO_4^{-2} concentrations in river Leysse and lake Bourget surface water

	Na^+ (mg l^{-1})	Cl^- (mg l^{-1})	SO_4^{-2} (mg l^{-1})
Leysse – mean 1996	4.7	6.5	15.3
Leysse (this flood)	2.8	4.15	8.5
Lake (surface) – mean 1996	4.7	6.9	14.9
Lake (this flood)	4.3	6.2	13.2

A significance of floods in Lake Bourget, mainly on the Phosphorus cycle (P-cycle), has been postulated (Groleau, 1994; Vinçon-Leite et al., 1995). A field survey aimed at monitoring the lake-river interface during a flood was conducted in March 1996. The objective of this paper is to highlight the impact of this flood event in Lake Bourget on both physical and biogeochemical aspects. The physical impact was assessed through river and lake water temperature time series analysis and modelling. The biogeochemical effects were estimated through the characterization of the particles: size distribution, morphological observations with scanning electronic microscope (SEM) and semi-quantitative analysis of their composition (EDX). Bacterial biomass and production were also measured, to better understand DOC and bacterial dynamics in the river plume.

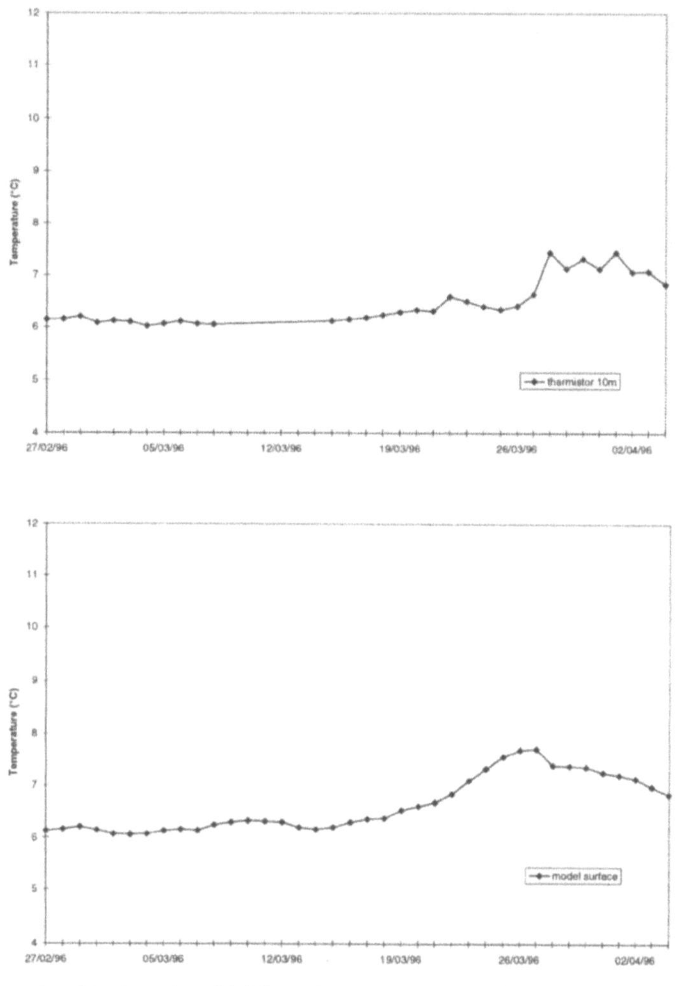

Figure 4. Results of the thermal model (lake surface water) and 10 m depth thermistor temperature data.

Lake Bourget

Lake Bourget is a medium sized lake (surface area: 42 km^2, maximal depth: 145 m, residence time $\approx$ 7 years) located in the French Alps (Figure 1).

The lake watershed area is about 560 km^2. The main tributary, river Leysse, has a 320 km^2 watershed area and its spring is located at 1200 m a.s.l.: 75% of the total water input to Lake Bourget comes from the Leysse. Approximately 30% of the watershed area has an altitude higher than 1000 m and is snow covered from December to the end of February. The Leysse discharges upstream of its entrance in Lake Bourget were gauged until 1988. Afterwards the gauging station was disactivated. Since February 1996, a new station, located at the same point, has been installed. Water levels are measured again but the new gauging curves are not yet available, especially for high flows. Therefore, the general hydrological characteristics were established according to the 1980–88 data.

River Leysse mean discharge rate is 8.5 m^3 s^{-1}, and maximum flow measured on the 1980–88 period is c. 100 m^3 s^{-1}. Floods frequently happen in early spring, during snow melting, and in autumn after storm events. They last generally less than a week. Figure 2 shows the daily mean flow of river Leysse in 1988 as an example highlighting the wide range of the instantaneous residence time of water in the lake caused by the variability of daily mean flow. The threshold considered to define a flood is a discharge rate greater than 20 m^3 s^{-1} (CEMAGREF, 1985).

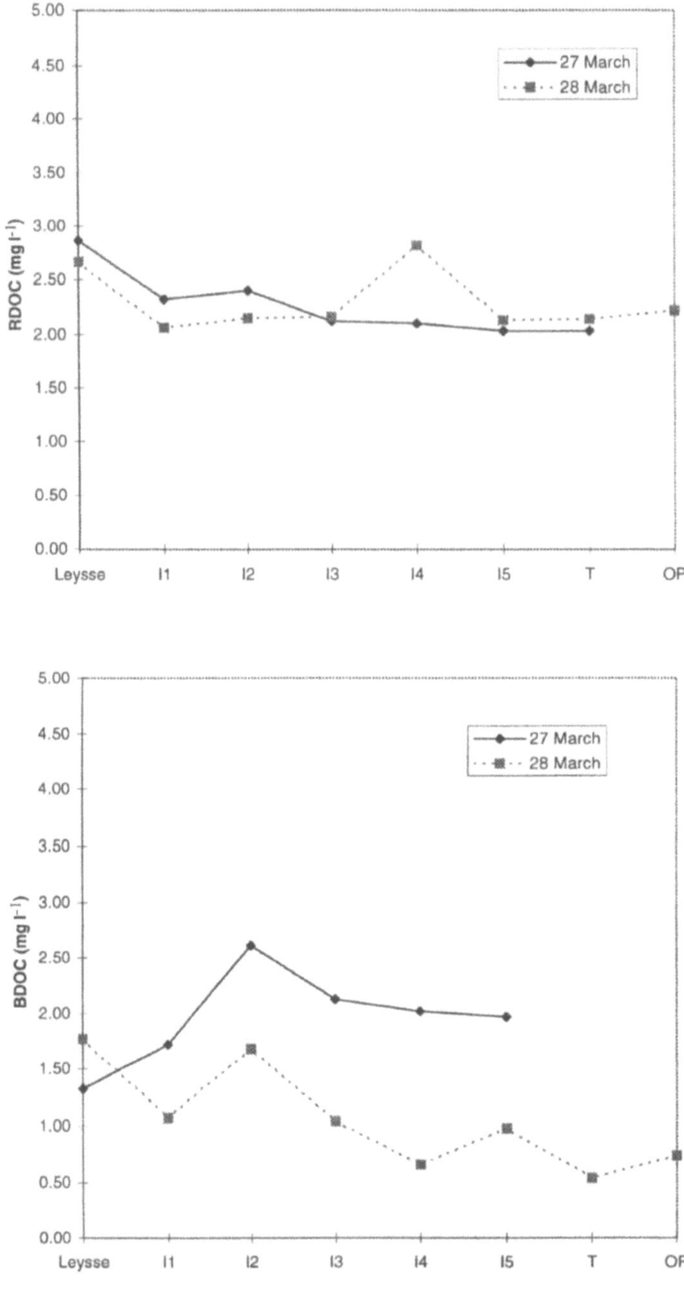

Figure 5. BDOC and RDOC evolution in Lake Bourget on March 27th-28th 1996.

Material and methods

Methodology

Following the analysis of data from different points of the lake gathered between 1969 and 1978 (INRA, 1972; CEMAGREF, 1978), it is assumed that Lake Bourget, on a time scale of days, is horizontally homogeneous and a single point, point B, located at the lake maximum depth (Figure 1) is monthly monitored.

As a flood cannot be forecasted, we can only assess the lake pre-flood conditions through the monthly monitoring. A major difficulty in studying flood events comes from the difference of field sampling scales

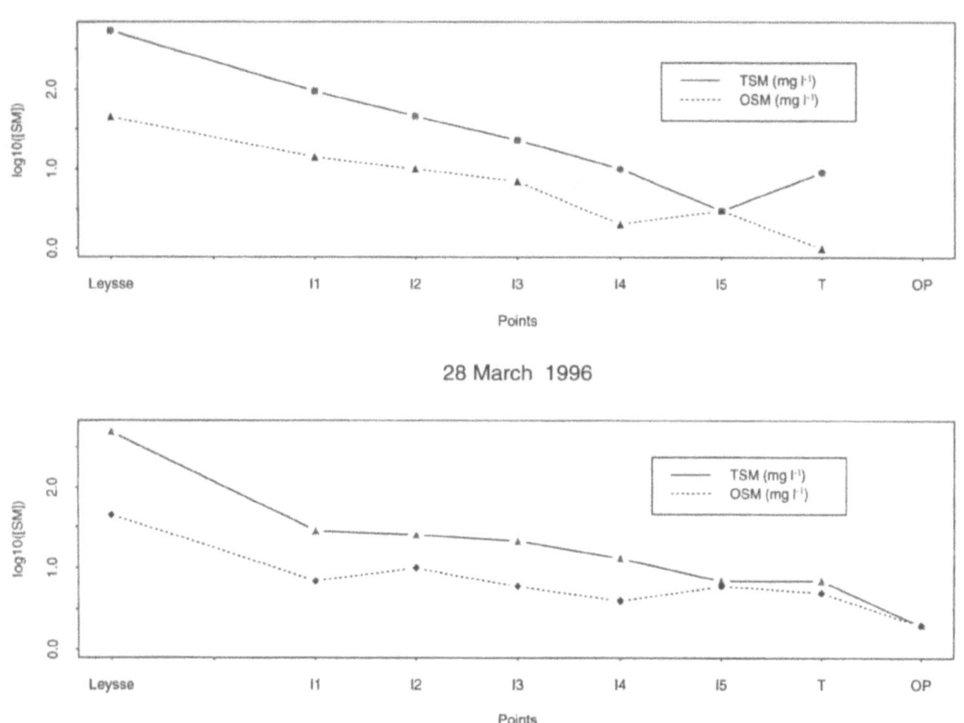

Figure 6. Decimal logarithm of total suspended matter (TSM) and organic suspended matter (OSM) in Lake Bourget on March 27th–28th 1996.

between a regular monitoring and a flood survey. Contrary to a regular monitoring based on large space and time scales and aimed at assessing a seasonal global evolution of the lake, a flood survey needs small space and time scale campaigns to understand the impact of a specific episode on lake seasonal evolution.

Moreover, despite well-defined campaigns, the flood meteorological conditions are generally so unfavourable that the survey may be incomplete. In order to obtain the most complete data set, we followed a methodological approach based on both numerical modelling and field surveys including continuous measurements and punctual sampling.

From a hydrological point of view, the river Leysse discharges were estimated by a hydrological conceptual reservoir model, *GR4j*, (De Oliveira Nascimento, 1995) whose four parameters were calibrated according to the 1980–88 measured discharges.

Concerning hydrodynamics of river entrance during the flood, a two-dimensional horizontal hydrodynamical model gives the direction, the depth and the extension of the river plume in the lake (Bournet, 1996).

Moreover, a one-dimensional vertical thermal model gives the temperature evolution, the onset of stratification and the occurrence of mixing in the water column. The time and space steps of the model are one day and one meter, respectively. This thermal model is coupled with a model of the phosphorus cycle which includes the processes involving the algal growth (Vinçon-Leite et al., 1995).

Continuous field data are useful to analyze the data collected during a flood event. Therefore, a thermistor chain and sediment traps were moored at point T (Figure 1). Figure 3 presents a global schema showing the interaction between the field point and continuous measurements and the numerical models.

The flood of March 1996

The Leysse flood peak was estimated by the hydrological model *GR4j* to approximately 50 $m^3 s^{-1}$ on March 27th. The flood lasted from 26th to 29th March. According to time series of air temperature and rainfall in the watershed, the flood was caused by rain and snow melting in the highlands associated with rain in the lower region of the watershed. This flow rate corre-

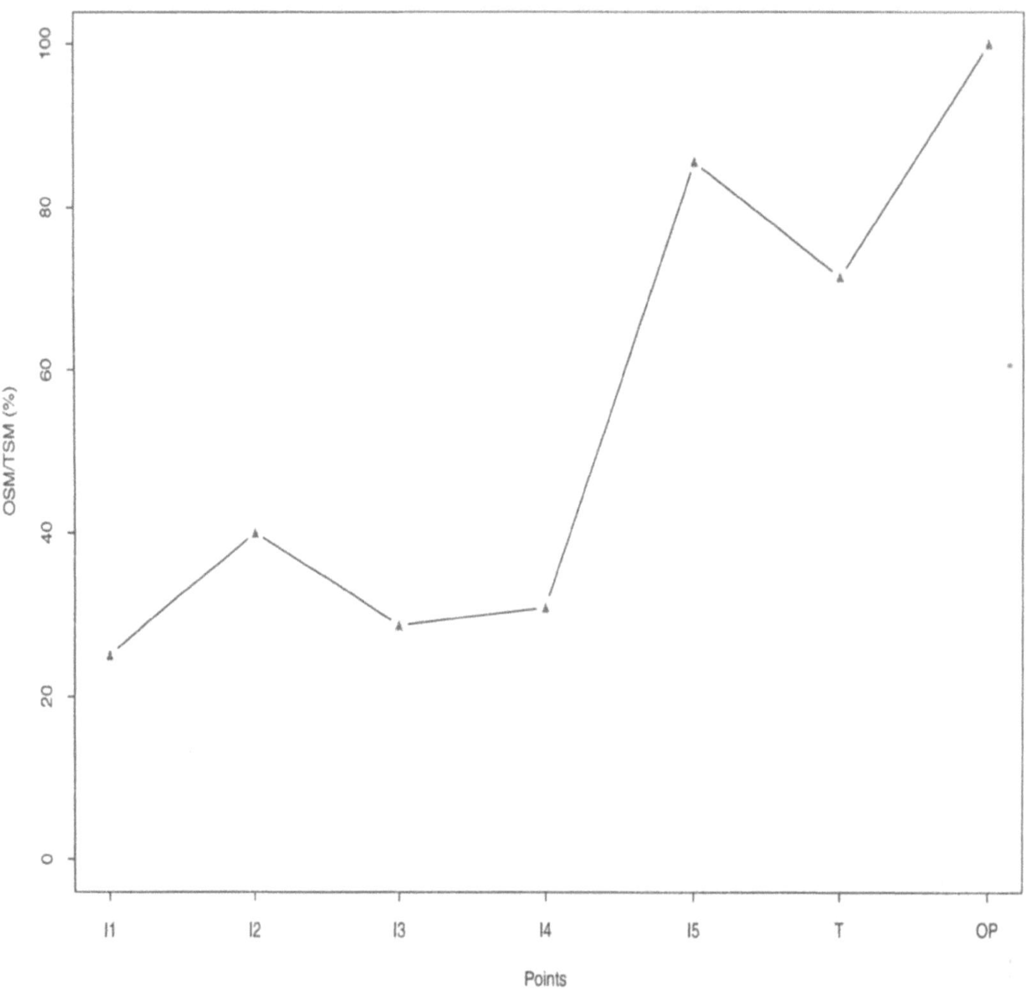

Figure 7. Evolution of the organic suspended matter fraction in the surface lake water on March 28th.

sponds approximately to a return period of six months. The impact of even a slight flood is potentially significant as it happens during a sensitive and biologically active period: the set-up of the thermal stratification and the beginning of algal production.

The meteorological data were provided by two Météo-France stations, one located in the highlands of the Leysse watershed (Les Déserts) et the other located at the southern end of the lake (Voglans).

Sampling points

During the March 1996 flood survey (Figure 1) a daily water sampling was done during two consecutive days (27th–28th) in the Leysse, at point T, and at six additional points (Figure 1).

The river Leysse temperature was continuously measured with a 15 minutes time step by a thermal probe located one km upstream of the river entrance. Water samples used for the determination of the physical and chemical data were gathered monthly at this same point.

Two reference points are located in the lake: point B, the deepest point (145 m) in the pelagial area and point T, in the eastern region of the southern basin (82 m). At point B, a monthly sampling was carried out and at point T, a thermistor chain (Anderaa) (10 thermistors from 10 m to 51 m depth) and three sediment traps (10, 30, 80 m) were moored.

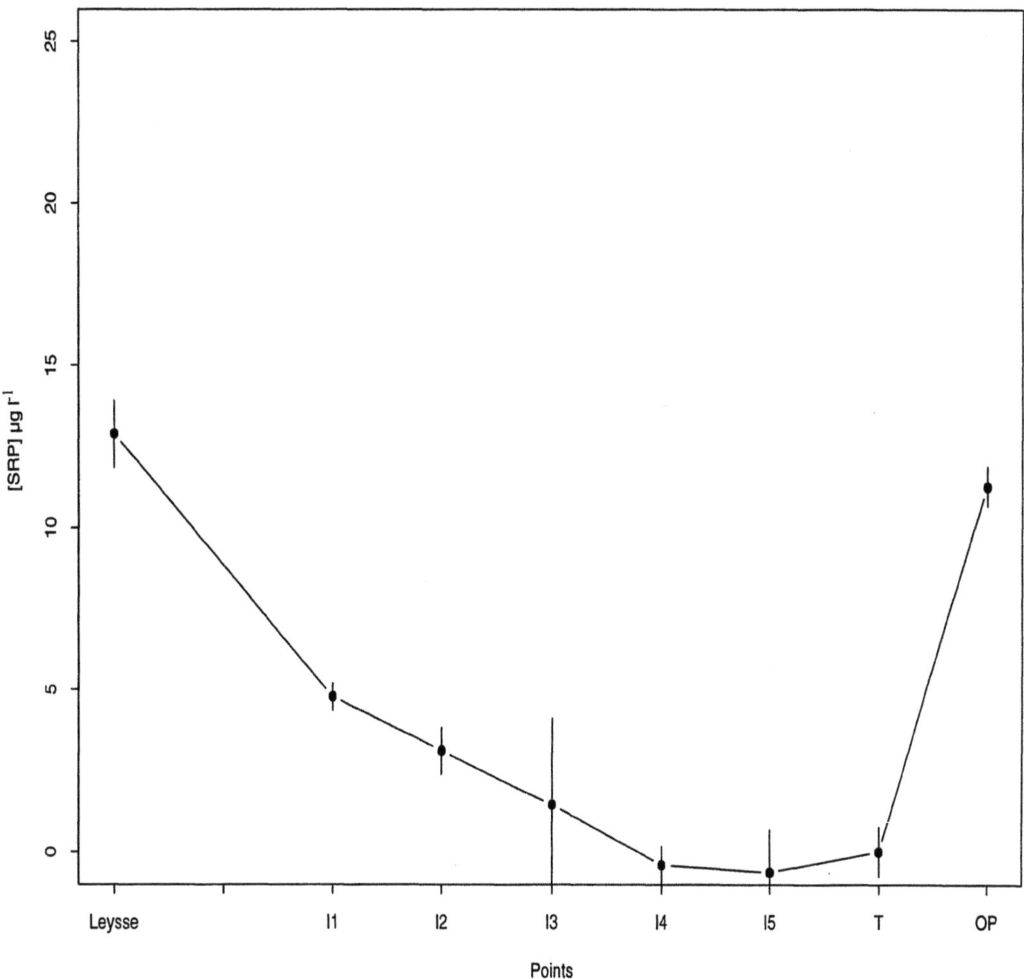

Figure 8. Evolution of the SRP concentration in the surface lake water on March 28th. Bars represent the standard deviation.

During the flood survey, five additional points (I_1, I_2, I_3, I_4, I_5) inside the river plume and one point (OP) located outside the plume were sampled. Sampling was done 50 cm under the water surface as, according to the Leysse and lake relative densities the river was entering the lake as a surface overflow. The location of point OP was determined near the western shoreline according to the hydrodynamical model results (Bournet, 1996) which show that the river plume is deflected towards the east and moves parallel to the shoreline.

Field data

The same physical, chemical and biological variables were measured in the river and in the lake.

Chemical data

The major ions (Cl^-, SO_4^-, NO_3^-, K^+, Ca^2, Na^+, Mg^2) were measured by capillary electrophoresis (Waters Quanta 4000). A 100 μm × 100 cm capillary (Chromate OFM-OH, 20 kV, 60 s) was used for anion, and a 75 μm × 60 cm (UV-CAT2, 20 kV, 10 s) for cation analysis.

The different forms of P: orthophosphate, total P in raw water and in filtered water (0.4 μm Cellulose Acetate Nuclepore filter), particulate P on 0.4 μm filter were analyzed. Orthophosphate determination was based on the ascorbic acid colorimetric method. The other forms of P were measured after transformation to orthophosphate by ammonium persulfate digestion (AFNOR, 1986).

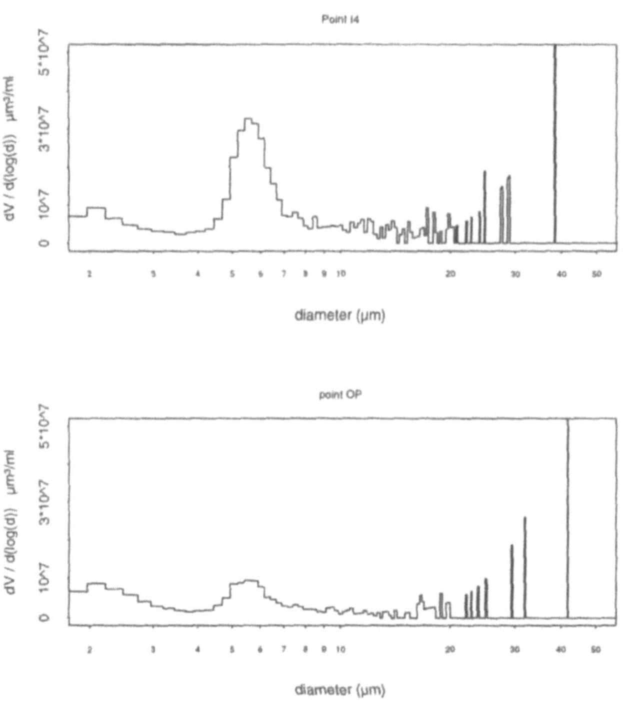

Figure 9. Particle size distribution in Lake Bourget: points I_4 and OP on March 28th.

Dissolved organic carbon (DOC) was measured with a Dohrman DC80 'Total Carbon Analyser' based on U.V.-promoted potassium persulfate oxidation after elimination of mineral carbon by orthophosphoric acid. Biodegradable (BDOC) and refractory (RDOC) fractions of the dissolved organic carbon were assessed using the methods of Servais et al. (1987) and Servais et al. (1989). BDOC is that part of the initial DOC concentration which is assimilated by autochthonous bacteria during a short-term incubation (Servais et al., 1987; Boissier & Fontvieille, 1995). A 0.2 μm cellulose acetate membrane (Millipore) was washed three times with 20 ml of organic-C free distilled water and then used to filter 130 ml of the water being investigated. 10 ml were kept for the initial DOC measurement. The remaining 120 ml were then transferred into a 250 ml pre-combusted (550 °C, 4 h) glass bottle with an aluminium closure. The bottles were inoculated with a suspension of autochthonous bacteria (removed from the cellulose acetate filter), and incubated in the dark, for 30 and 35 days, at 15 °C. The decrease of DOC concentration is rapid (around 4 or 5 days) and is followed by a stabilization at a constant value until the end of the incubation (Servais et al., 1987; Boissier & Fontvieille, 1993). The DOC concentration measured at the end of the incubation was considered as biologically refractory (RDOC) and the difference between initial and final values of DOC was considered as the biodegradable fraction (BDOC).

Characterization of particles

The following analysis were done on particulate matter collected in the river, in the lake and in the sediment traps.

- Total suspended matter (TSM) and organic suspended matter (OSM) were estimated by filtering water on glass fibre filters (Whatman GF/F) previously ignited at 525 °C. The filters were dried at 65 °C during 48 hours for TSM determination and combusted at 525 °C for one hour for OSM determination.
- Size distributions were determined by a Coulter Counter using a 100 μm orifice which measures particles from 2 to 50 μm.
- The morphology of the particles was studied with a scanning electronic microscope (ABT55). A semi-quantitative estimation of particle composition was obtained by a EDX micro-analyzer (Kevex).

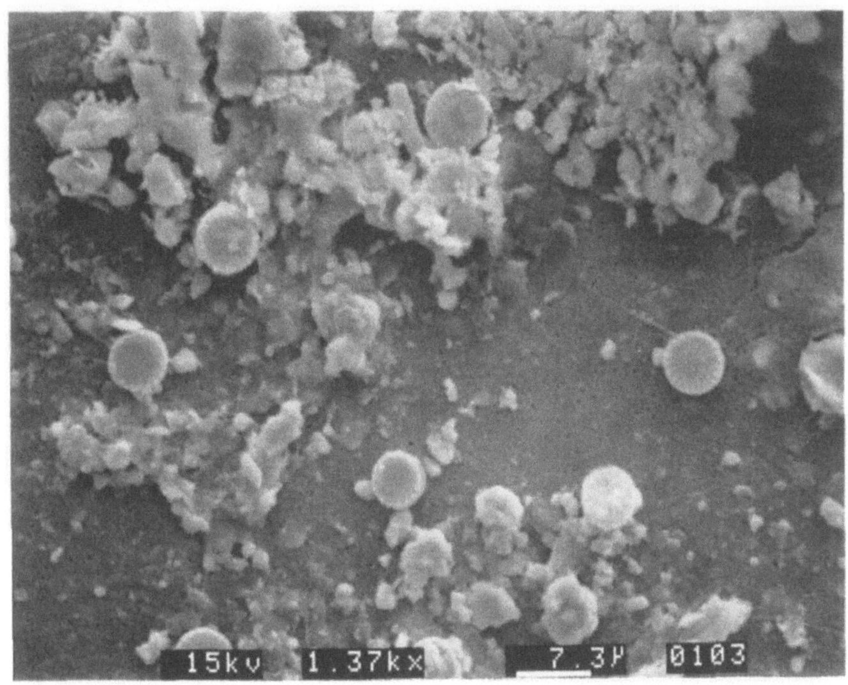

Figure 10. SEM photographs (a) Association of *Stephanodiscus* and allochthonous mineral particles at point I_1 on March 27th. (b) View of some *Stephanodiscus* cells attached on a large alumino-silicate particle.

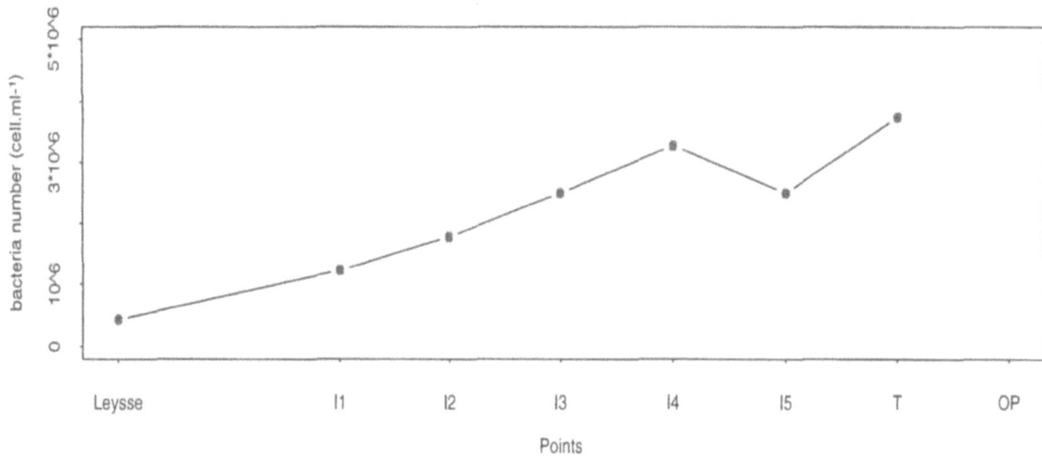

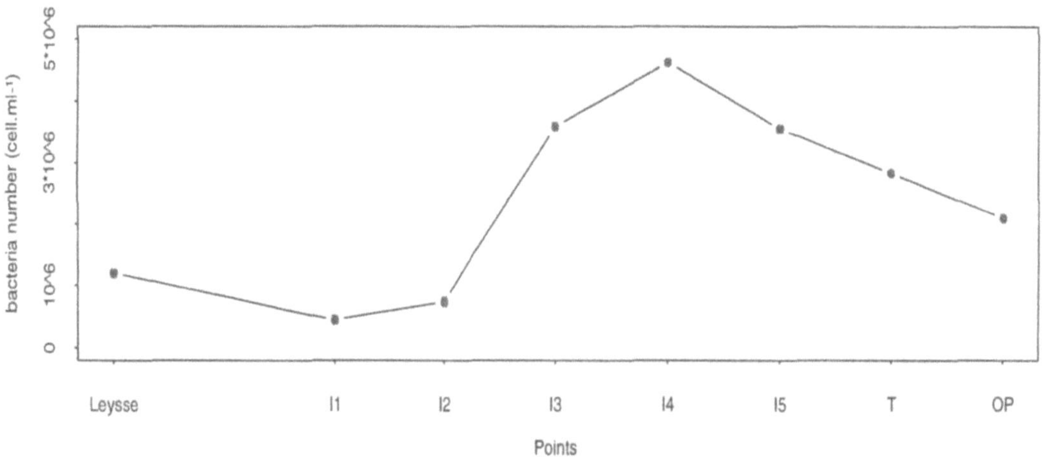

Figure 11. Bacterial biomass in Lake Bourget on March 27th–28th 1996.

Trap sediments

The particulate matter was collected in the sediment traps during approximately two week periods (Table 1). In addition to the characterization of the particles following the methods presented in Section 3.4.2, organic carbon, organic nitrogen, Ca, Mg, K, Na, Fe, Mn, P, Al, Si, were analyzed at the *Laboratoire d'analyse des sols (INRA*[1] *Arras)*, according to (AFNOR, 1996). Organic carbon and nitrogen were measured by elemental analysis (Carlo Erba CHN analyzer). Total Ca, Mg, K, Na, Fe, Mn, P, and Al were analyzed after HF/HClO4 digestion. P was determined by a colorimetric method. Ca, Fe, Mg and Mn were determined by flame atomic absorption spectrometry. K and Na were analyzed by flame atomic emission spectrometry. Al was analyzed by plasma emission spectroscopy and Si by X fluorescence. The chlorophyll *a* composition was determined by a spectrophotometric method after acetone pigment extraction (AFNOR, 1986).

Bacterial data

Bacterial counts were performed using epifluorescence microscopy. The total number of bacteria was estimated after DAPI staining (Porter & Feig, 1980; Fry, 1988). The incorporation of 3H-methyl thymi-

[1] INRA: French National Institute for Agronomical Research

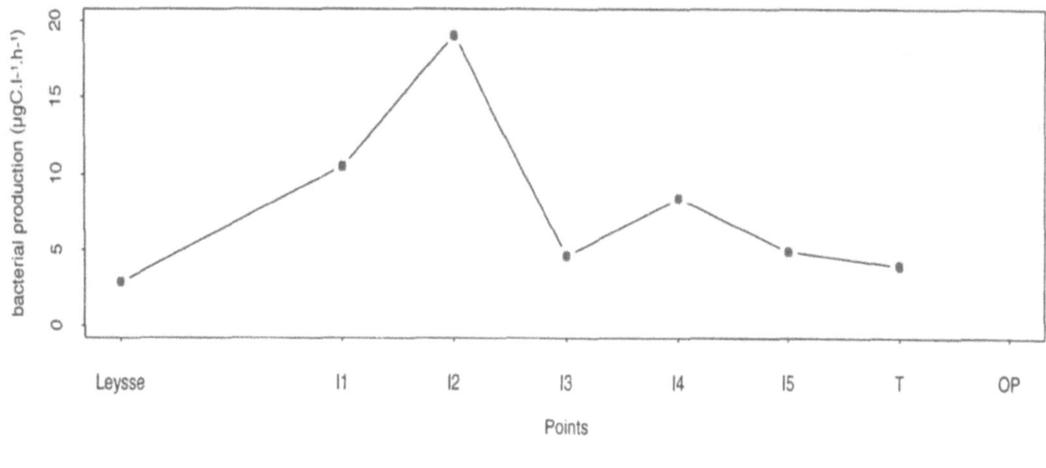

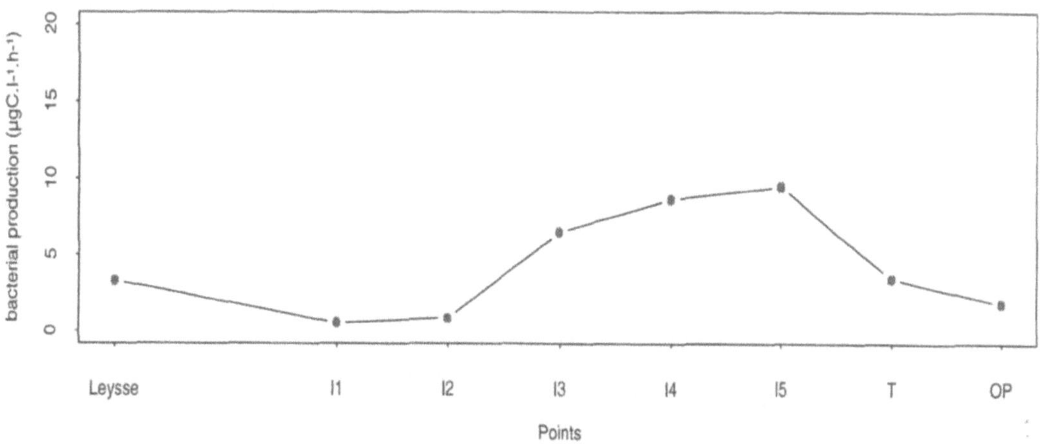

Figure 12. Bacterial production in Lake Bourget on March 27th–28th 1996.

dine into ice-cold trichloroacetic acid (TCA) insoluble material was used to estimate heterotrophic bacterioplankton production (Furhman et al., 1982).

Results

Pre-flood conditions: the limnological context

In order to characterize the context before the flood, physical, chemical and biological conditions in Leysse river and in lake Bourget are presented.

The Leysse river

The time series of the Leysse and the lake temperature for the days preceding the flood indicate that river temperature was close to lake surface temperature. These data and the hydrodynamical model results agree in indicating that on the peak flow day (March 27th), the river entered into the lake, spreading on the lake surface. According to the hydrodynamical model, the plume reached point I_5 ($\approx$ 1.5 km from the river mouth) after few hours and reached point T ($\approx$ 4 km) in one day. Leysse pre-flood concentrations are available on February 9th because of a sampling problem on March 5th. Chemical and bacterial data are presented in Table 2.

Lake Bourget

Data collected on March 5th 1996 at the pelagial point B (see Figure 1), as part of the regular monthly monitoring were used to estimate lake conditions before the flood. This campaign took place during a dry weather period when the Leysse discharge was estimated by the hydrological model to approximately 4 m^3 s^{-1}. The lake was considered horizontally homogeneous before the flood and point B conditions were extrapolated to the southern basin of the lake. At point B, on March 5th, the water column was vertically homogeneous with a mean temperature of 6 °C and a mean oxygen concentration of 10.2 mg l^{-1}. Lake chemical and bacterial data are shown in Table 2. The water transparency (secchi depth: 15 m) indicates that algal growth had not yet begun. Neither the SEM observations of suspended matter nor its granulometric distribution did show any specific particle population.

Information about algal growth was obtained from the suspended matter collected in the sediment traps. A tenuous 6 μm peak appears in the particles collected by the 10 m trap, suspended in the well mixed upper layer during 14–29 March, indicating that algal growth might have just begun during the second half of March. The chlorophyll flux estimated from the data of this sediment trap is very small in March ($\approx$ 2 mg m^{-2} day^{-1}). It increased to 17 mg m^{-2} day^{-1} in the following period (March 29th–April 9th).

Lake temperature data at 10m depth are available from the chain first thermistor. In order to estimate the lake surface temperature, a one-dimensional vertical thermal model (Vinçon-Leite et al., 1995) was used. The results show a slight temperature increase in lake surface water, approximately 1.4 oC, from 14th to 27th March. The thermistor data show also a light temperature increase of 0.2 °C at 10 m depth (Figure 4). Therefore, over the period between the 5th of March and the flood, there probably was only a slight surface thermal stratification. Finally, during the period preceding the flood, algal production had possibly begun but probably slowly.

Physical-chemical variables evolution during the flood

Dissolved minerals and organic matter

Among minerals, Cl^-, SO_4^- and Na^+ presented the highest changes along the river-lake ecotone. During dry weather, river and lake concentrations were close, reflecting the biogeochemical background of the watershed. During the flood, river concentrations decreased by about 40% of dry weather concentrations (Table 3) indicating a classic dilution effect (Meybeck, 1985). Flood lake concentrations remained close to mean 1996 concentrations.

RDOC displayed a near constant concentration both in time and space, although its concentration was higher in the river than in the lake (Figure 5). On the contrary, obvious BDOC concentration changes occurred on a daily time scale: consumption of 1 mg l^{-1} between the two days of the study in the plume region (Figure 5).

Particulate matter

Total suspended matter (TSM) and organic suspended matter (OSM) increased significantly during the flood (Figure 6). The fraction of organic matter in total suspended matter increased from the mouth of the river to the offshore points. Outside the river plume, on March 28th the suspended matter concentration was significatively lower than at the end of the plume region ([TSM] = 7 mg l^{-1} at point T and [TSM] = 2 mg l^{-1} at point OP). Moreover at point OP, nearly 100 % of the suspended matter was organic (Figure 7).

Particulate P (PP) increased during the flood, but total P (TP) increased only lightly. Total soluble P (TSP) and soluble reactive P (SRP) both decreased significantly. The decrease of SRP within the plume was clearly higher than at the outside-plume point OP (Figure 8).

Evolution of the particles during the flood

On the 27th and 28th, the number of particles strongly increased in the river and in the lake. The size distribution of the river particles was uniform but lake particle distribution was unimodal, with a clear 6 μm peak (Figure 9). This peak appeared at all points located inside the plume on March 27th. It was higher near the entrance of the river. SEM observations identified this population of particles as *Stephanodiscus*, a small centric diatom. On the 28th, the 6 μm particle population disappeared at points I_1 and I_2. It remained present at the offshore points but the number of particles decreased (Leroy, 1996). At point OP, a clearly smaller peak was observed (Figure 9). The SEM observations (Figure 10) also indicated an agglomeration of diatoms and coarse allochthonous mineral particles, which increased in settling rate and may explain the disappearance of the 6 μm particles near the river entrance on March 28th.

Bacterial dynamics

During both days of the survey, cell numbers were higher in the lake than in the river (Figure 11). On March 27th a gradient established in the mixing area between river and lake water. However, on March 28th, bacterial concentration displayed a maximum in this area together with the maximum of the bacterial production. At that time, bacterial production was half flat on March 27th (Figure 12).

Discussion

An overall decrease of nutrients (SRP and BDOC) was observed during this flood along the river-lake ecotone. This phenomenon seems to be due to biological, physical and meteorological factors.

Biological factors

The most significant influence of the flood on the lake seemed a stimulation of the biological activity in the region affected by the river plume. As seen in Paragraph 4.1.2, algal growth had only just begun when the flood occurred. The larger decrease of SRP concentration inside than outside the plume may be due to phosphate uptake by *Stephanodiscus*. According to Burkholder (1992), increasing suspended particle concentration enhances phosphate uptake by algae as a survival mechanism under episodic sediment loading. The granulometric data (Figure 9) and SEM observations corroborate this interpretation: the cell number of the dominant algae, *Stephanodiscus*, was much higher inside the plume than outside. Moreover, the P-cycle model was used to estimate the algal SRP uptake in the lake surface water during the 5–28 March period. No specific calibration of the model parameter set was performed for this simulation. The parameter values used were those obtained formerly according to the previous year field data (Vinçon-Leite et al., 1995). At 28th March, the model SRP concentration is 10 μgP l^{-1}, close to the concentration measured at point OP outside the river plume. This indicates that the model, which does not take into account the specific processes occurring during the flood, is not able to represent the SRP exhaustion by the stimulated algal activity observed in the plume region. In compensation, outside the plume, the model algal SRP uptake correctly explains the SRP decrease.

The microbial activity (pelagial bacteria) might also have contributed to the SRP concentration decrease in the plume where the highest values of bacterial production were found. Bacterial uptake of phosphate may be high enough to compete with algal consumption (Rhee, 1972; Vadstein et al., 1988).

The importance of BDOC consumption observed in the plume which was similar to that found in others fields (Servais et al. 1987; Boissier & Fontvieille, 1993; Gayte & Fontvieille, submitted) may also be explained by bacterial production if we apply bacterial growth yields obtained on natural DOC: 25% (Hannish et al., 1996).

Influence of allochthonous particles

Additional scavenging of phosphate by mineral clay particles could explain the large decrease of SRP in the plume. Indeed, mineral suspended matter concentration was higher inside than outside the plume and a fraction of phosphate could be adsorbed to mineral suspended matter. Phosphate partitioning between suspended sediment and phytoplankton is difficult to estimate. Nevertheless it can be assumed that the SRP pattern was mainly caused by biological uptake because mineral suspended matter is generally unable to adsorb SRP under 10 μg l^{-1} (Froelich, 1988).

Soluble unreactive P (SUP) did not increase, therefore phosphate adsorption by humic colloidal matter originated from the soil watershed leaching and transported into the lake by the river was probably negligible.

Another significant process was the aggregation of diatom cells to mineral particles (Figure 10b). This association increased the settling rate of the algae, which were rapidly eliminated by sedimentation wherever suspended sediment concentration was high.

Post flood conditions: Influence of meteorological perturbation

A flood episode is frequently accompanied by strong wind and air cooling. Meteorological conditions may cause a mixing of the water column which dilutes the biogeochemical impact of the flood. SEM observations of the particulate matter collected in the traps after the flood showed that *Stephanodiscus* cells rapidly reached the bottom of the water column. *Stephanodiscus* was abundant in the 10 m, 30 m and 80 m traps moored after the flood, during period 3 (March 29th–April 9th, see Table 1) and was absent during the previous period (March 14th–29th). Settling rate computed according to Stokes' equation is approximately 1 m day^{-1}, a value much lower than that necessary for

Stephanodiscus to reach the bottom of the lake within 2 weeks by sinking only. If we assume that algae are transported downwards in association with larger and denser mineral particles (Figure 10b), an apparent settling rate of 20 m day^{-1} can be estimated on the basis of size. This value may allow the algae to reach the 80 m traps. The hypothesis of vertical mixing of the lake after the flood can also explain the distribution of *Stephanodiscus* over the whole water column. The results of the unidimensional vertical thermal model corroborate this hypothesis. A complete mixing occurred at the beginning of April. Therefore, the algal population whose growth was stimulated by the flood and whose settling was increased by aggregation with mineral particles was transported still faster towards the lake bottom by water column mixing.

Conclusion

A flood event during the setup of thermal stratification (March 1996) was studied in mesotrophic Lake Bourget. The dominant process was the decrease of nutrient concentrations along the river-lake ecotone. This phenomenon seems due to physical and biological factors acting together. The flood stimulated both algal and bacterial growth. The algal growth production had probably begun some days before the flood. This would explain that the flood stimulation lasted a very short time (less than 24 hrs) leading to a fast consumption of SRP and of BDOC. But the role of the river plume was significant because biological activity was more intense in the plume than outside the panache. The decrease of SRP concentration to such a low level could hardly be explained by adsorption on clay particles or humic matter.

Another key process was the association of algal cells with allochthonous mineral particles, which resulted in an increased settling rate. After the flood, sedimentation was also accelerated by a mixing of the water column which distributed the particles over the whole water column.

The overall effect of this flood was to depress the available phosphate for the phytoplankton in the region crossed by the river plume. The high P uptake during the flood exhausted the epilimnetic phosphate. The large diatom population which developed was then rapidly eliminated by sedimentation. Thus, the P load brought into the lake by the river during this flood did probably not contribute to an increase of the open water P concentration.

This study demonstrated the importance of physical and biological factors acting together, in reducing nutrient loads coming from the watershed to lakes during flood events, even during a rather small flood. It indicates that management changes in the watershed which would alter the hydrological cycle and the suspended matter loading to the lake could have a significant impact on lake water quality. The study of these particular flood periods seems important to identify the processes which control lake dynamics and which may be different according to season, river discharge and lake physical and chemical conditions.

Acknowledgements

This work was conducted in collaboration between CERGRENE (ENPC-ENGREF) and GRETI (Université de Savoie). It was supported by the French Ministry of Agriculture, the Agence de l'Eau Rhône-Méditerranée-Corse and the lakeside intercommunal councils (SILB and SIAC). We thank C. Leroy and S. Halbout for their contribution in the laboratory and the field surveys. We are grateful to the *Cellule technique du lac du Bourget* and particularly Gérard Paolini for his essential contribution to the field surveys. We are grateful to the two referees whose valuable comments have largely improved this paper.

References

AFNOR, 1986. Eaux – Méthodes d'essai. AFNOR, Tour Europe, Paris.

AFNOR, 1996. Recueil de normes – Qualité des sols. AFNOR, Tour Europe, Paris, 3ème édition.

Balland, P., 1980. La lutte contre l'eutrophisation des lacs – Méthodologies de quantification des flux superficiels. Technical report, ENGREF, Département Formation Continue.

Boissier, J.-M. & D. Fontvieille, 1993. Biodegradable dissolved organic carbon in seepage waters from two forest soils. Soil Biol. Biochem. 25: 1257–1261.

Boissier, J.-M. & D. Fontvieille, 1995. Biological characteristics of forest soils and seepage waters during simulated rainfalls of high intensity. Soil Biol. Biochem. 27: 139–145.

Bournet, P.-E., 1996. Contribution à l'étude hydrodynamique et thermique du lac du Bourget – Courants de densité et ondes internes. PhD thesis, Ecole Nationale du Génie Rural, des Eaux et des Forêts.

Brock, T., 1966. Principles of microbial ecology. Prentice-Hall inc., Englewood New-Jersey, 306 pp.

Buffle, J. & G. Leppard, 1995. Characterization of aquatic colloids and macromolecules. 1. Structure and behavior of colloidal material. Envir. Sci. Techn. 29: 2169–2175.

Burkholder, J. M., 1992. Phytoplankton and episodic suspended sediment loading: phosphate partitioning and mechanisms for survival. Limnol. Oceanogr. 37: 974–988.

Carmack, E. C., C. H. Pharo & R. J. Daley, 1979. Importance of lake-river interaction on seasonal patterns in the general circulation of Kamloops Lake, British Columbia. Limnol. Oceanogr. 21: 634–644.

CEMAGREF, 1978. Le lac du Bourget – Bilan des observations menées en 1975–76–77. Technical report, Centre National du Machinisme Agricole, du Génie Rural, des Eaux et des Forêts, Lyon, France.

CEMAGREF, 1985. Le lac du Bourget – Etude 1983. Technical report, CEMAGREF – DQEPP.

De Oliveira Nascimento, N., 1995. Appréciation à l'aide d'un modèle empirique des effets d'actions anthropiques sur la relation Pluie-Débit à l'échelle d'un bassin versant. PhD thesis, Ecole Nationale des Ponts et Chaussées.

Dorioz, J.-M., E. Pilleboue & A. Ferhi, 1989. Dynamique du phosphore dans les bassins versants: Importance des phénomènes de rétention dans les sédiments. Wat. Res. 23: 147–158.

Elber, F. & F. Schanz, 1990. The influence of a flood event on phytoplankton succession. Aquat. Sci. 52/4: 330–344.

Fletcher, M., T. Gray & J. Jones, 1987. Ecology of microbial communities. Cambridge University Press, 440 pp.

Froelich, P. N., 1988. Kinetic control of dissolved phosphate in naturals rivers and estuaries: a primer on the phosphate buffer mechanism. Limnol. Oceanogr. 33: 649–668.

Fry, J., 1988. Methods in aquatic bacteriology. In B. Austin (ed.), Determination of biomass, John Wiley & Sons Ltd., New York, USA: 27–72.

Furhman, J., J. Ammerman & F. Azam, 1982. Thymidine incorporation as a measure of heterotrophic bacterioplankton production in marine surface waters: evaluation and field results. Mar. Biol. 66: 109–120.

Groleau, A., 1994. Contribution à l'étude de l'influence des crues sur le cycle du phosphore dans le lac du Bourget. DEA Sciences et Techniques de l'Environnement. Ecole Nationale des Ponts et Chaussées, Ecole Nationale du Génie Rural, des Eaux et des Forêts, Paris, France.

Hannish, K., B. Schweitzer & M. Simon, 1996. Use of dissolved carbohydrates by planktonic bacteria in a mesotrophic lake. Microb. Ecol. 31.

Imboden, D. M., 1990. Large Lakes: Ecological structure and function. In M. M. Tilzer & C. Serruya (eds), Mixing and Transport in Lakes: Mechanisms and Ecological Relevance. Springer-Verlag: 47–80.

INRA, 1972. Suivi du lac du Bourget. Technical report, Institut national de la Recherche Agronomique, Thonon-les-Bains, France.

Leroy, C., 1996. Caractérisation et évolution dans le temps des particules dans le lac du Bourget. DEA Sciences et Techniques de l'Environnement. Ecole Nationale des Ponts et Chaussées, Ecole Nationale du Génie Rural, des Eaux et des Forêts, Paris, France.

Meybeck, M., 1985. Variabilité dans le temps de la composition chimique des rivières et de leurs transports en solution et en suspension. Rev. Sci. de l'Eau 4: 93–121.

Porter, K. & Y. Feig, 1980. Use of DAPI for identifying and couting aquatic microflora. Limnol. Oceanogr. 25: 943–948.

Reynolds, C. S., 1984. The Ecology of Freshwater Phytoplankton. Cambridge University Press, Cambridge.

Rhee, G.-Y., 1972. Competition between an alga and an aquatic bacterium for phosphate. Limnol. Oceanogr. 17: 505–514.

Servais, P., A. Anzyl & C. Ventresque, 1989. A simple method for the determination of the biodegradable dissolved organic carbon in waters. Appl. Envir. Microbiol. 55: 2732–2734.

Servais, P., G. Billen & M.-C. Hascoët, 1987. Determination of the biodegradable fraction of dissolved organic matter in waters. Wat. Res. 21: 445–450.

Sigg, L., M. Sturm & D. Kistler, 1987. Vertical transport of heavy metals by settling particles in Lake Zürich. Limnol. Oceanogr. 32: 112–130.

Stabel, H. H. & M. Geiger, 1985. Phosphorus adsorption to riverine suspended matter, implications for the P-budget of lake Constance. Wat. Res. 19: 1347–1352.

Tilzer, M. M., 1988. Secchi disk – chlorophyll relationships in a lake with highly variable phytoplankton biomass. Hydrobiologia 162: 163–171.

Vadstein, O., A. Jensen & H. Reinertsen, 1988. Growth and phosphorus status of limnetic phytoplankton and bacteria. Limnol. Oceanogr. 31: 998–1009.

Vinçon-Leite, B., B. Tassin & J.-M. Jaquet, 1995. Contribution of mathematical modeling to lake ecosystem understanding: Lake Bourget (Savoy, France). Hydrobiologia 300/301: 433–442.

Zumstein, J. & J. Buffle, 1989. Circulation of pedogenic and aquagenic organic matter in an eutrophic lake. Wat. Res. 23: 229–239.

MIX
Papier aus verantwortungsvollen Quellen
Paper from responsible sources
FSC® C105338

If you have any concerns about our products,
you can contact us on
ProductSafety@springernature.com

In case Publisher is established outside the EU,
the EU authorized representative is:
Springer Nature Customer Service Center GmbH
Europaplatz 3, 69115 Heidelberg, Germany

Printed by Libri Plureos GmbH
in Hamburg, Germany